T5-DGK-319

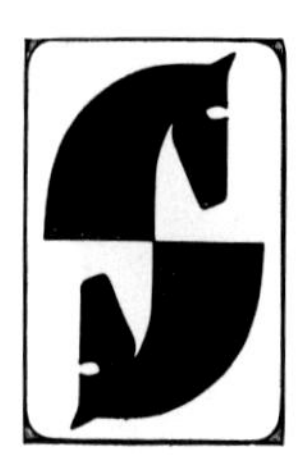

Sanford L. Palay · Victoria Chan-Palay

Cerebellar Cortex

Cytology and Organization

With 267 Figures including 203 Plates

Springer-Verlag
New York · Heidelberg · Berlin 1974

Prof. Dr. Sanford L. Palay and Dr. Victoria Chan-Palay
Havard Medical School, Departments of Anatomy and Neurobiology
25 Shattuck Street, Boston, Massachusetts 02115, U.S.A.

ISBN 0-387-06228-9 Springer-Verlag New York Heidelberg Berlin
ISBN 3-540-06228-9 Springer-Verlag Berlin Heidelberg New York

Library of Congress Catalog Card Number 73-77568. Printed in Germany.

Typesetting, printing, and binding: Universitätsdruckerei H. Stürtz AG, Würzburg
Type face: Monophoto-Times 10'/12', 8'/9', and Helvetica. Paper: Papierfabrik Scheufelen, Oberlenningen. Reproduction of the figures: Gustav Dreher, Württemb. Graphische Kunstanstalt GmbH, Stuttgart. Layout of the dust cover: W. Eisenschink, Heidelberg. Drawing for dust cover: Victoria Chan-Palay, Boston, U.S.A.

«... on ne peut, ni l'on ne pourra jamais parler du cervelet sans que Cajal ne vienne au devant, et quiconque l'ignorerait serait forcément obligé à coïncider avec lui sur beaucoup de points, soit dans l'interprétation, soit dans les faits qui constituent les fondaments de toutes les constructions scientifiques.»

CLEMENT ESTABLE 1923
Trab. Lab. Invest. Biol. (Madrid)
vol. XXI, p. 187.

«Malheureusement pour moi, d'autres, courant la même carrière, virent plusieurs des choses que j'avois vuës, & s'étant fait un plan moins étendu, m'enlevèrent, en publiant leurs observations, une espèce d'honneur que je croiois avoir également mérité.»

PIERRE LYONET 1762
from the preface to *Traité anatomique de la Chenille, qui ronge le Bois de Saule.*
Pierre Gosse, jr. et Daniel Pinet, La Haye.

Preface

The origins of this book go back to the first electron microscopic studies of the central nervous system. The cerebellar cortex was from the first an object of close study in the electron microscope, repeating in modern cytology and neuroanatomy the role it had in the hands of RAMÓN Y CAJAL at the end of the nineteenth century. The senior author vividly remembers a day early in 1953 when GEORGE PALADE, with whom he was then working, showed him an electron micrograph of a cerebellar glomerulus, saying "That is what the synapse should look like." It is true that the tissue was swollen and the mitochondria were exploded, but all of the essentials of synaptic structure were visible. At that time small fragments of tissue, fixed by immersion in osmium tetroxide and embedded in methacrylate, were laboriously sectioned with glass knives without any predetermined orientation and then examined in the electron microscope. After much searching, favorably preserved areas were studied at the cytological level in order to recognize the parts of neurons and characterize them. Such procedures, dependent upon random sections and uncontrollable selection by a highly erratic technique of preservation, precluded any systematic investigation of the organization of a particular nucleus or region of the central nervous system. It was difficult enough to distinguish neurons from the neuroglia. Even so, much was learned about the fine structure of the nerve cell, especially about the perikaryon, axons, and dendrites, and, most important for the purpose of this book, about the structure of the synapse.

During the past twenty years vast improvements in technique have made it possible to study all parts of the nervous system at the fine structural level. Now we are able to fix the tissue before fragmenting it, thus retaining the orientation of its components. That signal improvement has made it relatively easy to recognize types of cells and their processes with the minute clues found in thin sections, whereas before it was generally impossible. Improved fixing solutions, embedding and staining methods, as well as refinements in the electron microscope itself, have greatly increased the number and subtlety of the discriminations that can be made, so that many details formerly nonexistent in the electron micrographs were now useful in identifying structures. All of this technical advance, to which we have been privileged to contribute, has made it possible to attempt such a work as the present volume.

Concentrated labor on this book began in the autumn of 1969, when the senior author was joined in this endeavor by the junior author. With this collaboration, the pace of the investigation was greatly accelerated. Vast numbers of Golgi preparations and electron micrographs were made, permitting a fruitful interplay between traditional optical microscopy and electron microscopy.

Actually, the book was conceived a long time ago as a comprehensive demonstration presented to the annual meeting of the American Association of Anatomists in April 1964 at Denver (PALAY, 1964a). Some fifty electron micrographs of cells, fibers, and synapses in the cerebellar cortex were exhibited. During the display a well-known elder statesman of the traditional neuroanatomical school came by and, casting a scornful glance at the micrographs, came out with "Well, what have you learned that we didn't know before?" In vain to tell him about the synapses of parallel fibers on Purkinje cell thorns, or about a new understanding of the glomerulus, or a fresh view of the pericellular basket and the pinceau, or about a thousand other points. It had all been worked out by RAMÓN Y CAJAL almost a century before. Nor were anatomists the only ones who thought that such investigation was futile. In the late 1950's a well-known neurophysiologist chided the senior author for working on the structure of the cerebellum. "Why do anatomists," he asked, "always like to study the cerebellum? Nothing interesting goes on there!" In the intervening years a great many investigators, both morphologists and physiologists, have found new and interesting things in the cerebellar cortex, so much, in fact, that the literature on the subject has burgeoned beyond the ability of anyone to cope with it. Once again the notion is prevalent among neuroanatomists that the subject is exhausted. This book is testimony to the fact that although much has been learned, a great many questions still go unanswered.

In the present volume each of the cell types and afferent fibers in the cerebellar cortex is taken up in turn and described. Both optical and electron microscopy are used and illustrated. A careful study of the cerebellar cortex with these two methods indicates that considerable reliance may be placed on the Golgi technique for the general architecture and the three-dimensional form of the cells and fibers. All of the known synapses are characterized and their function is discussed from the anatomical point of view. It would be presumptuous on our part to undertake a review of the electrophysiology of this cortex, although we have drawn freely upon the results and insights derived from that discipline. All of the drawings are original India ink tracings made with the aid of a camera lucida at high magnifications. They were prepared by the junior author especially for this book or for recent journal articles.

Finally, a word may be in order to explain our use of the laboratory rat for this study instead of the cat, which has long been the favorite of neurophysiologists. Besides believing that the cat is a most peculiar animal, we could claim a dangerous sensitivity to feline dander. But, more pertinently, the cerebellum of the rat has all of the essential machinery of the mammalian cerebellum. It is much more reliably preserved than that of the cat and there is no cytological advantage in studying the larger animal. Furthermore, our work may encourage neurophysiologists to make use of this lowly beast in which a great deal of fundamental biology has been explored.

It remains to acknowledge the support of our laboratory by research grant NS 03659 and training grant NS 05591 from the National Institute of Neurological Diseases and Stroke, Bethesda, Maryland. The composition of the manuscript was made possible by a fellowship from the John Simon Guggenheim Memorial Foundation, granted to the senior author while he was on sabbatical leave from the Harvard Medical School. We wish to express our gratitude to CAROL WILUSZ for her patient assistance in making counts and measurements, tracing fibers in Golgi preparations, and preparing the graphs; to PHOEBE FRANKLIN for her meticulous preparation of the typescript for publication and for her bibliographic searching. Finally, we should like to thank VICTORIA LI MEI PALAY for the use of Fig. 10.

Boston, September 1, 1973 SANFORD L. PALAY and VICTORIA CHAN-PALAY

Table of Contents

Chapter I

Introduction

Probably no other part of the central nervous system has been so thoroughly investigated and is so well known as the cerebellar cortex. For nearly a century all of its cell-types have been recognized, and the course and terminations of their processes have been described countless times by numerous authors. Yet unanimity on many doubtful points has not been reached. Most of our knowledge of the cerebellar cortex derives from the early work of RAMÓN Y CAJAL. It was, in fact, during the period of his first successes with the Golgi method, when RAMÓN Y CAJAL was seized with what he described in his autobiography as a *fièvre de publicité*, that the basic plan for the organization of the cerebellar cortex was worked out. Subsequent research has confirmed many of his intuitions and added only details. For a long time these observations and the plan of the cerebellar cortex that he derived from them were far in advance of the physiological understanding of this organ. The information that this cortex is divisible into three layers, each with its own distinct populations of cells, and that they are interconnected in a few simple neuronal chains was already sufficient to baffle comprehension. No one had any idea of how the cerebellar cortex should work or what operations it should perform with the impulses coming into it from diverse sources. The Sherringtonian concept that the cerebellum was somehow related to the maintenance of muscle tone, muscle coordination, equilibration, and proprioception balanced securely on a knowledge of afferent and efferent pathways that was already too encumbered with detail. The interpretation of coarse experiments with ablations and evoked surface potentials had no need for intracortical circuits (see, for example, FULTON, 1949, and DOW and MORUZZI, 1958).

1. A New Morphology

It is only in the past decade that a more precise knowledge of the anatomy of the cerebellar cortex has been required. Today we still have only glimmerings of how the cerebellar cortex operates and no clear vision of what its function is, but the situation with respect to anatomicophysiological correlations has vastly changed. Physiologists are now probing the activities of individual neurons and eavesdropping on the coordinated interchanges between the members of neuronal assemblies. A much more detailed and precise knowledge of the morphology of the cerebellar cortex is required now than hitherto in order to guide these experiments and to inform their results, as well as to contain the speculations that they induce in physiologists, cyberneticists, and morphologists alike.

This new level of morphology can be achieved by the exercise of three technical procedures—one ancient, the others relatively recent—which provide complementary views of the cerebellar cortex. The old technique is the Golgi method, which figured so strongly in the early advances in our knowledge of the cerebellar cortex, and the centenary of which should be celebrated this year. Although this method is still poorly understood, it has undergone a revival of interest and confidence during the past decade such as it has not enjoyed since the end of the nineteenth century. Its survival through this long period of neglect and antipathy is owing to a handful of adepts, whose painstaking observations kept it alive until the modern era of neurocytology reasserted its value. Even now the method has few practitioners and unnecessarily appears esoteric to the uninitiated. Important data are still to be derived from the careful study of Golgi preparations, especially when they are coordinated with the results of the more modern methods. The second method of present usefulness is the method of experimental degeneration, particularly the Nauta technique and its recent variants, for locating the terminal arborizations of axons within the neuropil. This technique provides essential data about the distribution of nerve fibers connecting one region of the nervous system with another. While these data are necessary for the location of terminals belonging to specific systems, they do not identify the actual sites of synapses. This deficit is filled by electron microscopy.

When expertly prepared, electron micrographs of thin sections display all of the nerve terminals synapsing upon any particular neuron and its processes, as well as the neuroglia. The various profiles can be classified according to their disposition, shape, and cytological content, and usually can be readily attributed to one or another part of a neuron or neuroglial cell. The problem is to identify the terminals according to their source and to recognize the neuron on which they synapse. Tracing fibers or cell processes for long distances to the cell of origin is usually impossible in the thin sections used for electron microscopy. Reconstruction of three-dimensional models from serial sections has a very limited usefulness. It is here that the complementarity of the three methods comes into play. By correlating the profiles seen in electron micrographs with the form and distribution of the neurons and their processes as learned from the application of Golgi and experimental degeneration techniques, it is possible to build up the pattern of synapses on perikarya and in the neuropil and to classify the profiles according to their sources. Thus a complete map of intercellular connections can be constructed and used in coordination with neurophysiological data to learn how the cerebellar cortex operates. In this book we have endeavored to provide such a map for an ordinary laboratory animal, the normal adult white rat.

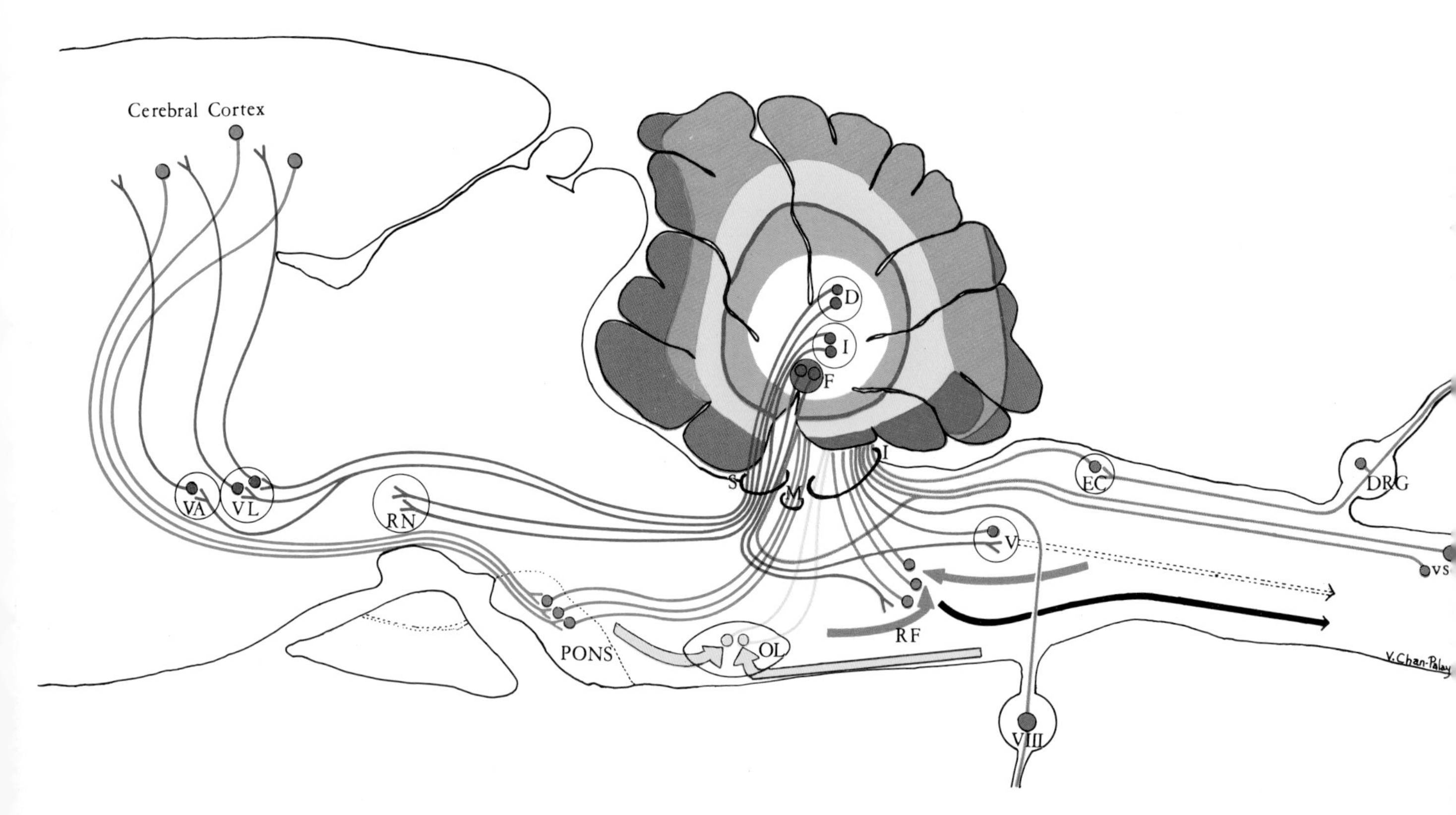

Fig. 1. Scheme of the afferent and efferent pathways of the cerebellum. This diagram summarizes the major cerebellar tracts. Although many details have been omitted for the sake of simplicity, the plan of the major cerebellar connections is presented. The afferent pathways to the cerebellum and the distribution of these fibers in the cortex of the vermis are indicated by their respective colors. Afferents from the spinal cord (brown) reach the cortex through (a) the dorsal spinocerebellar tract (*dsc*); (b) the cuneocerebellar tract; (c) the ventral spinocerebellar tract (*vsc*), a portion of which enters the cerebellum through the inferior cerebellar peduncle (*I*), and the rest through the superior cerebellar peduncle (*S*). A representative dorsal root ganglion cell (*DRG*) is shown projecting through the cervical spinal cord to the external cuneate nucleus (*EC*). Spinal afferents distribute in the cerebellar cortex in the anterior lobe and part of the posterior vermis (brown). Vestibular afferents (blue) include primary vestibular fibers from the VIIIth nerve ganglion (*VIII*) and secondary fibers from the vestibular nuclei (*V*). These fibers distribute to the fastigial nucleus (*F*) and flocculonodular lobe, nodulus, and uvula (blue). Afferents from the reticular formation (*RF*) enter the cerebellum and distribute throughout the cortex (orange). The reticular formation receives input from higher and lower centers (*orange arrows*). Fibers from the inferior olive (*OL*) (yellow) distribute to the entire cortex (yellow). Input to the olive comes from higher and lower brain centers (*yellow arrows*). All of the above inputs enter through the inferior cerebellar peduncle (*I*). Afferents to the cerebellum from the pons (green) enter through the middle cerebellar peduncle (*M*) to distribute to the entire cortex. The pons receives its input from the cerebral cortex (*green tracts*). The major efferents of the cerebellum (red) leave through the superior cerebellar peduncle (*S*). The fibers from the dentate nucleus (*D*) go to the ventrolateral nucleus (*VL*) and the ventroanterior (*VA*) nucleus of the thalamus. Fibers from the interpositus nuclei (*I*) go to the red nucleus (*RN*). Fibers from the fastigial nuclei reach the reticular formation (*RF*) and the vestibular nuclei (*V*). The reticulospinal tract (*black arrow*) and the vestibulospinal tract (*dotted arrow*) bring integrated information from the vestibular nuclei and reticular formation back down the spinal cord. (For further details, see text.)

As indicated above, the interpretation of electron micrographs requires the identification of nerve endings and other cell fragments in the random specimens provided by thin sections. Since any particular section contains only extremely small samples of thousands of fibers and cells, arranged in the most complex fashion and cut in various planes, how can they be recognized? What criteria have been developed to take advantage of the minimal clues in these sections? Of course, such discriminations proceed from established criteria for the recognition of the parts of neurons and neuroglial cells in electron micrographs. These have been reviewed in other publications (see PETERS, PALAY, and WEBSTER, 1970) and will not be repeated here. Once a profile has been recognized as a dendrite, axon, or other specific part of a cell, the next question concerns the identification of the cell type to which that part belongs. Although the chances for confusion at this point are great, the microscopist is very considerably assisted by the regular fabric of the cerebellar cortex. Our fundamental method consists of systematically cataloguing morphological features that consistently occur together until a constellation of traits has been discovered that is specific for each type of cell and its processes.

In preparations for the optical microscope, each cell type has a peculiar form, position, distribution, number, and internal structure. These distinctive characteristics must carry over to the level of electron microscopic analysis. Profiles in the thin sections must be consistent with the shapes of the cells or processes, their location, and course, as seen in Golgi preparations at the light microscope level. Thus, for example, the pattern of the Purkinje cell dendritic arborization is so distinctive in Golgi preparations that there is no difficulty whatever in recognizing its fragments in thin sections in the electron microscope. In addition, new characteristics emerge when the tissue is examined in the electron microscope—characteristics involving greater detail and differentiation of structures already known as well as new structures unsuspected because invisible or confused at the level of the light microscope. Thus the size, number, shape, and distribution of mitochondria can be important distinguishing characteristics of a particular dendrite or a perikaryon. The size, shape, and clustering pattern of synaptic vesicles, combined with the size, location, and interfacial characteristics of the synaptic junction, are critical features for the identification of the terminals of a particular type of fiber. The presence of neurofilaments or microtubules, the density or sparseness of a fibrillar cytoplasmic matrix, are other differential characteristics.

2. The Fiber Connections of the Cerebellar Cortex

Before beginning the fine structural analysis of the cerebellar cortex, it will be helpful to have an overview of its extrinsic and intrinsic connections. The fiber tracts leading to and from the cerebellar cortex connect it directly or indirectly with all of the major subdivisions of the central nervous system. These pathways are quite complicated, and many doubtful questions remain to be resolved. Even a superficial examination of these pathways could justifiably include a survey of the entire nervous system. Such a review is clearly beyond the scope of this volume, which is restricted to the fine structure of the cortex. For a detailed consideration of the pathways, the reader is referred to the chapter on the cerebellum in BRODAL'S (1969) comprehensive text on neurological anatomy and to Volume III of LARSELL and JANSEN'S (1972) monumental treatise on the comparative anatomy of the cerebellum. Citations of the voluminous literature and a critical evaluation of the subject are to be found in these two works. For the present purpose, however, a few general statements are required in order to indicate the wide variety of sources from which the cerebellar cortex draws its information and to contrast that with the apparent simplicity of its architecture.

The afferent and efferent fibers of the cerebellum are drawn together into three pairs of massive fiber bundles, known as the cerebellar peduncles and named according to their relative positions in the human brain. In general, fibers leading impulses into the cerebellum enter by way of the inferior and middle peduncles, while fibers leaving the cerebellum exit by way of the superior peduncles. There are, however, exceptions, for example, certain spinocerebellar fibers that enter through the superior peduncles and certain efferent fibers from the flocculonodular lobe and fastigial nuclei that exit by way of the inferior peduncle. In general, also, the fiber paths ascending from spinal, reticular, and vestibular sources are ipsilateral, whereas those coming from the inferior olive and the pontine nuclei are contralateral. Similarly, the efferent pathways that lead to higher centers, such as the red nucleus and the thalamus, cross to the other side, whereas those descending to the vestibular nuclei and reticular formation remain ipsilateral. Here, too, there are exceptions, however, such as the uncinate fasciculus, which on leaving the fastigial nucleus crosses to the other side and distributes to the contralateral vestibular nuclei and reticular formation. Table 1 summarizes the major pathways and classifies them according to the peduncle through which they pass. Finally, although the organization of the cerebellar cortex is everywhere alike, the

Table 1. Fiber pathways of the cerebellum

Origin	Type of fiber/tract	Destination
Inferior peduncle		
Afferents:		
vestibular afferents primary (vestibular ganglion) secondary (vestibular nuclei)	mossy fibers/juxtarestiform body	cortex: flocculonodular lobe, ventral uvula, fastigial nucleus
spinal afferents dorsal nucleus of Clarke, external cuneate nucleus, intermediate dorsal gray	mossy fibers/dorsal and ventral spinocerebellar tracts	cortex: anterior lobe, vermis; part of posterior lobe, pyramis, uvula
lateral reticular nucleus (cerebral cortex, spinal cord)	mossy fibers	entire cortex
reticular formation	mossy fibers	entire cortex
inferior olive (cerebral cortex, spinal cord, caudate nucleus, globus pallidus, periaqueductal gray)	climbing fibers (crossed)	entire cortex
Efferents:		
Purkinje cells of flocculonodular lobe, lateral vermis	juxtarestiform body	vestibular nuclei
fastigial nucleus	restiform body	vestibular nuclei, reticular formation
Middle peduncle		
Afferents:		
pontine nuclei (cerebral cortex)	mossy fibers/(crossed) pontocerebellar tract	entire cortex, except flocculonodular lobe
Superior peduncle		
Afferents:		
spinal afferents intermediate dorsal gray	mossy fibers/ventral spinocerebellar tract (1/3)	cortex: anterior lobe
Efferents:		
interpositus nucleus	brachium conjunctivum (crossed)	red nucleus, thalamic nuclei VL and VA, other thalamic and midbrain nuclei
dentate nucleus	dentatorubral tract (brachium conjunctivum, crossed)	thalamic nuclei VL and VA and other thalamic and midbrain nuclei
fastigial nucleus	hooked bundle (crossed)	vestibular nuclei, reticular formation

projections of the different afferent systems follow a longitudinal plan. Each system distributes into a characteristic pattern of parallel rostrocaudal zones that cross the transverse folds of the cortex (VOOGD, 1969).

The afferent supply to the cerebellar cortex can be subdivided into four principal groups (Fig. 1): (1) vestibular (blue), (2) ascending spinal (brown), (3) descending pontine (green), and (4) olivary (yellow) and reticular (orange). These groups end in particular, sometimes overlapping parts of the cerebellar cortex, as is indicated in the color-coded diagram in Fig. 1. No attempt will be made here to detail the precise topographical distribution of these several projections onto the cerebellar cortex.

(1) The cerebellar cortex receives direct afferents from the labyrinth by way of primary vestibular root fibers, which end in the flocculonodular lobe and the ventral part of the uvula. The same parts of the cerebellar cortex also receive secondary afferents from certain regions of the medial and descending (inferior) vestibular nuclei. Since these regions of the vestibular nuclei are supplied not with primary vestibular fibers but with ascending spinal inputs, this second contingent should be included in the next large group of cerebellar afferents, the spinocerebellar fibers. Like them, all of the inputs to the cortex except for group 1 are indirect, involving one or more synaptic interruptions in their ascent or descent into the cerebellum.

(2) Impulses originating from the activity of muscle spindles, Golgi tendon organs, pressure and tactile organs in the skin and deeper tissues ascend through the spinal

cord by various spinocerebellar pathways to end in the anterior lobe and the posterior vermis. The most direct routes are by way of the dorsal and ventral spinocerebellar tracts, conveying information from the hindlimbs, and the very similar cuneocerebellar and rostral spinocerebellar tracts, concerned with the forelimbs. There is also a corresponding pathway leading from the secondary trigeminal nuclei to the same parts of the cortex.

(3) A massive descending bundle of afferents comes from the contralateral pontine nuclei, bearing information from all lobes of the cerebral cortex. These fibers are distributed to all parts of the cerebellar cortex except for the lingula and the flocculonodular lobe.

(4) The last great group of afferents consists of projections from the contralateral inferior olive to all lobules of the cerebellar cortex, and conveys influences from both ascending and descending pathways. The olive receives most of its afferents from higher levels: the cerebral cortex, especially the motor cortex, the caudate nucleus, globus pallidus, red nucleus, and the periaqueductal gray. A smaller, but important, source of afferents reaches the olive from the spinal cord, another of the indirect spinocerebellar paths. It is important to note that these various contingents of olivary afferents are not diffusely dispersed through the olivary complex but terminate in specific parts with little overlap. Since the olive has a very precise topographical projection onto the cerebellar cortex, its fibers distribute onto this cortex a sharply localized pattern of the descending and ascending influences on it. Like the olive, the lateral reticular nucleus receives descending and ascending afferents, but they are much more diffusely localized in this cell group. It also projects widely to most parts of the cerebellar cortex.

In addition to these four major groups there are numerous small bundles originating from the tegmental reticular formation and the arcuate nuclei. There is also evidence for inputs from visual, auditory, and autonomic centers, although the anatomical pathways are obscure.

The efferent projections from the cerebellar cortex are considerably easier to summarize briefly. The axons of the Purkinje cells make up all of the efferents from the cerebellar cortex, and by far the greatest number terminate in the central nuclei. This projection is organized into orderly longitudinal (rostro-caudal) zones (JANSEN and BRODAL, 1942; VOOGD, 1964, 1969). A medial zone roughly equivalent to the vermis sends its axons to the fastigial nucleus, while a more lateral, intermediate zone is connected with the intermediate (interpositus) nuclei, and the most lateral zone, roughly coincident with the hemisphere, is connected with the lateral (dentate) nucleus. In addition, there is a thin band in the lateral portions of the vermis that projects directly to the vestibular nuclei, as does the cortex of the flocculonodular lobe. The distribution of these corticovestibular fibers follows, as might be expected, a specific pattern for each part. It is to be particularly noted that the fibers from the anterior vermis end mostly in the lateral vestibular nucleus, from which a massive tract descends into the spinal cord, whereas fibers from the flocculonodular lobe end mostly in the medial, inferior, and superior vestibular nuclei.

Similarly, the zonal distribution of the corticonuclear fibers takes on added significance when the pathways leading from the central nuclei are considered. The fastigial nuclei give rise to another projection to the vestibular nuclei by both crossed and uncrossed tracts, and also send fibers diffusely into the bulbar and pontine reticular formation and to some thalamic nuclei. The intermediate (interpositus) nuclei project their fibers forward to the contralateral red nucleus in which most of them end. The lateral (dentate) nucleus distributes its fibers principally in the thalamus, especially the nucleus ventralis lateralis (VL) and to some extent farther forward in the nucleus ventralis anterior (VA). These nuclei project to the precentral gyrus of the cerebral cortex, from which again fibers descend into the pons.

Thus the efferent fibers from the cerebellar cortex lead into numerous complicated loops that have the possibility of feeding information back into the cerebellum. The existence of these loops invites the building of model circuits involving the cerebellar cortex and the central nuclei as switching stations. The authenticity of such models, however, depends upon a precise knowledge of intercellular connections, both morphological and physiological, information that is at present almost entirely nonexistent. Each nuclear group must be examined and analyzed in detail by means of single unit recording and refined anatomical methods. Such studies have only just begun in a variety of these stations. Knowledge of the cerebellar cortex is much more advanced than that of its associated clusters of neurons.

3. The Design of the Cerebellar Cortex

In contrast to the rather complex pattern of the afferent and efferent pathways briefly reviewed above, the cerebellar cortex itself displays a surprisingly uniform and simple structure. The kaleidoscopic patterns of its fiber connections are hardly reflected, if at all, in its mono-

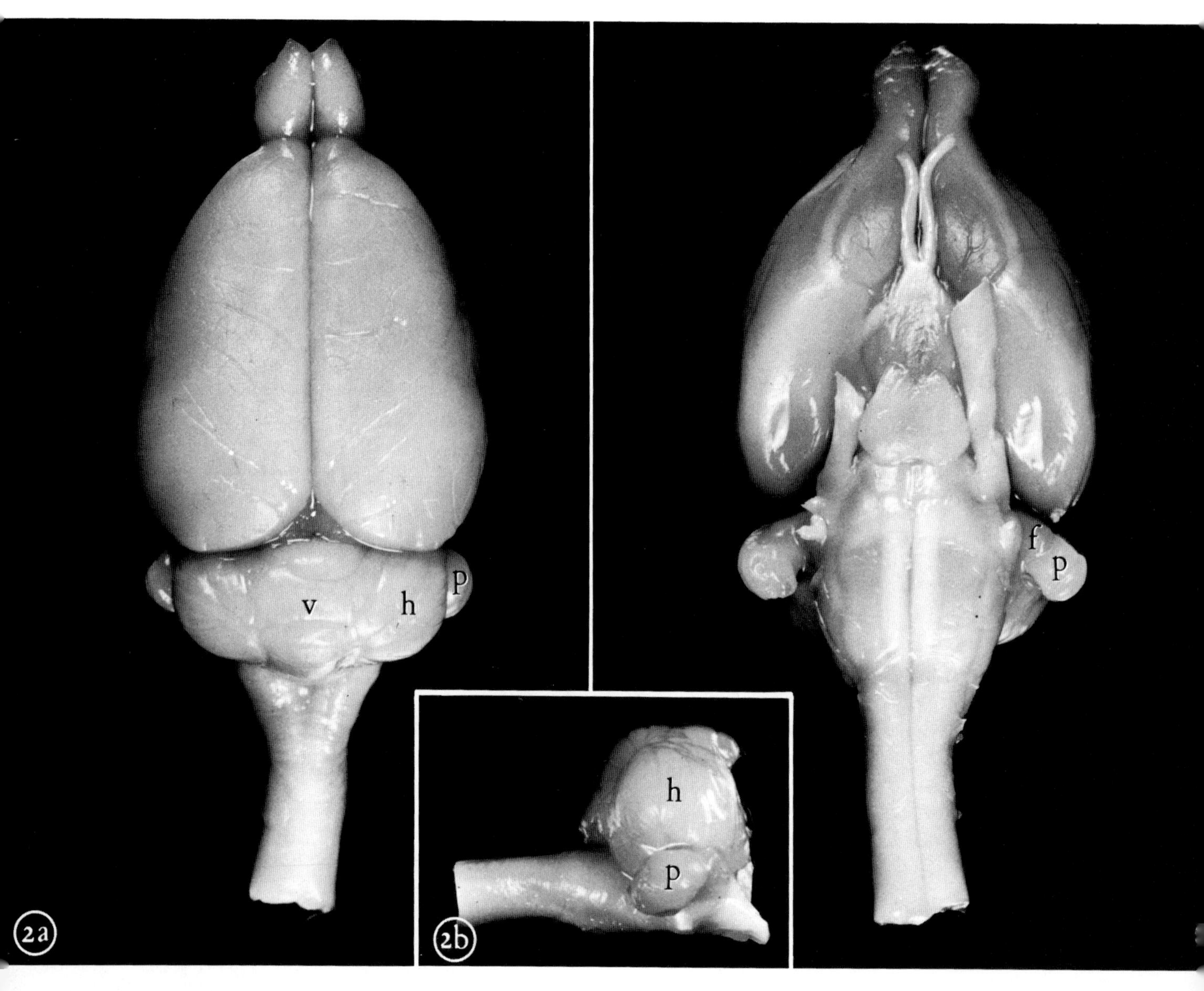

Fig. 2. a Superior aspect of the rat brain including the cerebellum. The vermis (*v*), cerebellar hemispheres (*h*), and paraflocculi (*p*) are well displayed. **b** Lateral aspect of the cerebellum and brain stem. The paraflocculus (*p*) and cerebellar hemisphere (*h*) are most prominent. **c** Inferior aspect of the rat brain. The paraflocculus (*p*) and flocculus (*f*) are well illustrated. Note that the folia of the cerebellar cortex run transversely across the cerebellum

tonous lamination. It consists simply of a highly folded sheet of cells and their processes, everywhere arranged into the same three layers (Fig. 3).[1] Furthermore, all of the folia extend in the transverse plane of the brain, from right to left, so that a parasagittal section cuts across them and a transverse section runs parallel with their longitudinal axes. The transverse folding of the cerebellar cortex is displayed in Fig. 2a–c, which shows the cerebellum of the adult rat in superior, inferior, and lateral view. At the center of each folium (Fig. 3) lies a thin lamella of white matter composed of the myelinated afferent and efferent fibers connecting the cortex with other centers. It should be noticed that the cortex, as a folded sheet covering these lamellae of white matter, has free edges and does not join up with itself to form a continuum (BRAITENBERG and ATWOOD, 1958). When unfolded, this sheet has been estimated to have an area of 50000 mm^2 in man (BRAITENBERG and ATWOOD, 1958), 2300 mm^2 in the cat (PALKOVITS *et al.*, 1971a), and only 270 mm^2 in the rat (ARMSTRONG and SCHILD, 1970).

1 There are, of course, certain exceptions to this generalization, which refers principally to mammals. In certain teleosts, e.g., mormyrids, regional variation in the form of the cerebellar cortex and its layers is carried to an extreme (KAISERMAN-ABRAMOF and PALAY, 1969; NIEUWENHUYS and NICHOLSON, 1969).

Fig. 3. Midsagittal section of the rat's cerebellum. ▶ This figure shows the three layers of the cerebellar cortex, as well as the various lobules of the cortex in the vermis. The rectangle indicates the region in lobules I and II (LARSELL, 1952) that have been enlarged in the next figure. 1 μm toluidine blue epoxy section. × 17

Fig. 4. Cortical layers in the cerebellum. Two lobules of the cerebellar cortex meet with their pial surfaces (*pia*) in apposition. The molecular layer (*mol*), Purkinje cell layer (*PC*), as well as the granular layer (*gr*) and white matter (*wm*) are indicated. An arrow points to the myelinated fibers of the anterior medullary lamina forming the roof of the IVth ventricle
▼

3

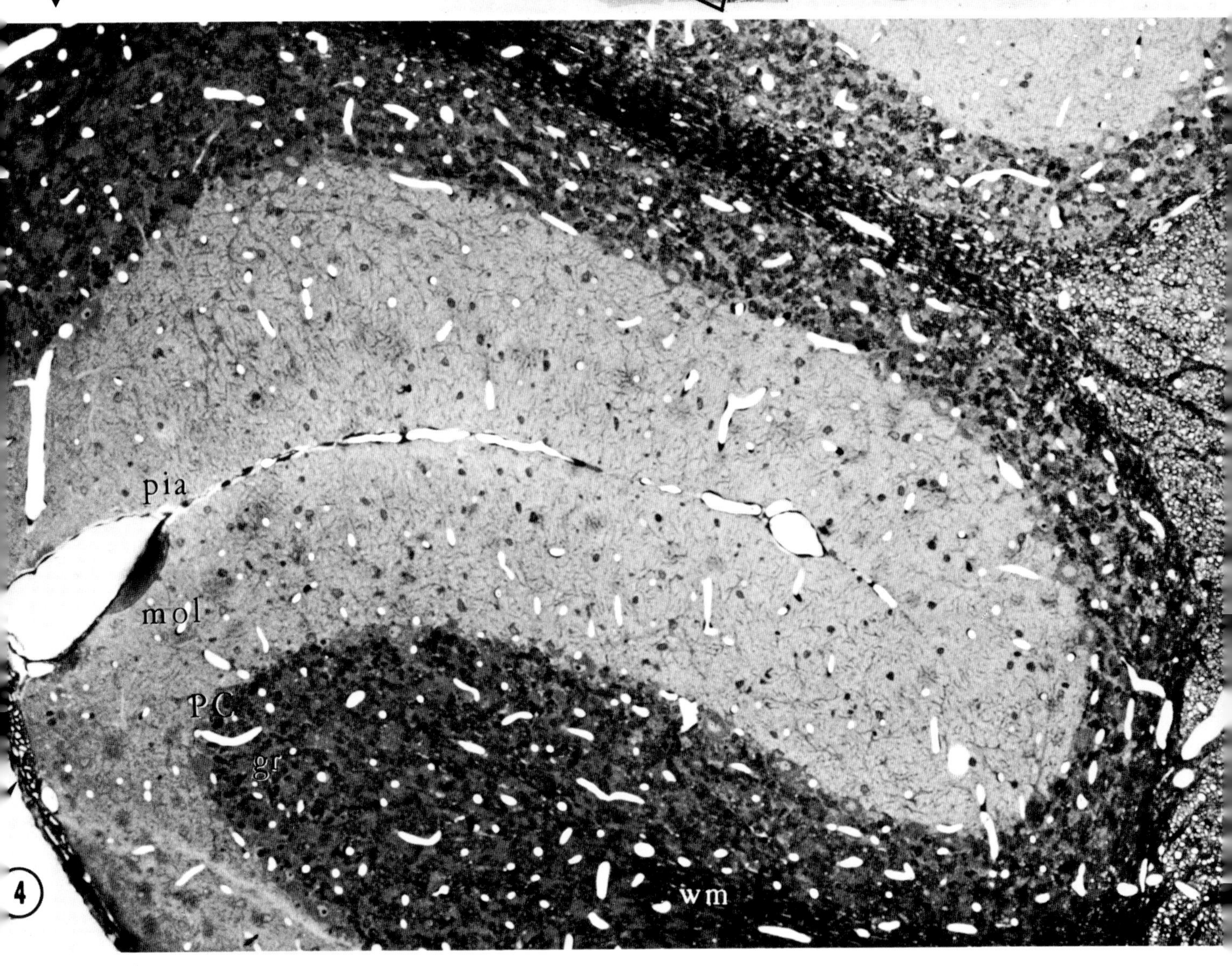

The cortical sheet, as indicated, is trilaminate (Figs. 3 and 4). Older authors considered the white matter in the center of the folium as one of the layers and divided the cortex itself into two layers, an outer plexiform and an inner granular layer. With the development of improved staining methods, the intervening layer of Purkinje cell perikarya became more conspicuous. RAMÓN Y CAJAL (1911) quite properly considered the Purkinje cells to be part of the plexiform layer, but conventional usage has not followed him in this regard, and usually they are thought of as constituting a separate layer, one cell thick. Also, the fibrous nature of the plexiform layer has been disregarded in favor of its fine, grainy appearance in the fresh or poorly stained parasagittal sections of an earlier period, so that it has retained its early name, the molecular layer, despite RAMON Y CAJAL's attempt at revision. The granular layer derives its name from the presence in it of numerous small nerve cells, which gave it a uniform granular texture in the early preparations.

Reference to the simplified diagram in Fig. 5 will show how the neurons in these three layers are arranged and interconnected. The small nerve cells in the granular layer, the *granule cells*, send their axons up into the molecular layer, where they divide like the letter *T* into two long branches that run parallel to the longitudinal axis of the folium. These parallel fibers make up the bulk of the molecular layer, and their cross sections give it its fine, grainy texture when viewed in parasagittal sections. The parallel fibers traverse the dendritic fields of all the other cells in the cerebellar cortex and synapse with them. The *Purkinje cell* bodies form a layer one cell thick lying on the upper margin of the granular layer. They extend into the molecular layer an elaborate dendritic arborization flattened in the parasagittal plane at right angles to the course of the parallel fibers. This intersection of perpendicular planes gives the cerebellar cortex a rectilinear, latticed plan that, as BRAITENBERG and ATWOOD (1958) point out, is unique for cortical structures. The axon of the Purkinje cell leaves the perikaryon from the pole opposite to the dendrites and descends through the granular layer to the white matter, whence it proceeds to its destination in one of the central or vestibular nuclei. The Purkinje cell is the only neuron in the cerebellar cortex whose axon projects beyond the confines of the cortical gray.

Just above the Purkinje cell bodies is a cell of intermediate size—the *basket cell*—which radiates dendrites through the molecular layer but projects its axon horizontally in the parasagittal plane above the Purkinje cell layer. As the axon passes successive Purkinje cell bodies (usually skipping the nearest one), it sends out collaterals that descend around the Purkinje cell perikaryon, forming with axons from other similar cells a pericellular basket. The fibers synapse on the cell body and end in a tangle of terminal branchlets surrounding the initial segment of the Purkinje cell axon. Slightly smaller cells with radiating dendrites—the *stellate cells*—are located higher up in the molecular layer. They extend their axons and collaterals to terminate on the shafts of Purkinje cell dendrites. Both of these cell types, of course, receive inputs from the parallel fibers.

In the granular layer two other fairly large cell types are found. One, the *Lugaro cell*, lies horizontally just beneath the Purkinje cell bodies. Its dendrites are very long, and are contacted mainly by recurrent collaterals from the Purkinje cell axon. Its axon, however, either enters the molecular layer or descends through the granular layer. Its terminals are still, unfortunately, unidentified. The other large neuron of the granular layer is the *Golgi cell*, which receives inputs from parallel fibers through its long, radiating dendrites, and which synapses with the dendrites of numerous granule cells through an extraordinary axonal plexus.

The afferent pathways to the cerebellar cortex resolve into two great classes of nerve fibers: climbing and mossy fibers (Fig. 5, red). Both kinds of fibers give off collaterals to the central nuclei, and this connection must be remembered in speculation on the mode of operation of the cerebellum. *Climbing fibers* originate largely, perhaps entirely, from the cells of the contralateral inferior olive. Although there is good reason to question the accuracy of this statement (see MLONYENI, 1973), other sources of climbing fibers have not yet been securely identified. Upon reaching the cortex they ramify meagerly in the granular layer, synapsing with both Golgi and granule cells, and ascend into the molecular layer. Here they burst into a highly divided terminal plexus that twines round the main trunks of the Purkinje cell dendritic tree. As a rule, there is only one climbing fiber for each Purkinje cell. Small branches also articulate with basket and stellate cells. All the other afferents to the cerebellar cortex—vestibular, spinal, medullary, reticular, pontine, etc.—apparently end in the cortex as *mossy fibers*. These ramify only in the granular layer and synapse by means of characteristic efflorescences, or rosettes, with granule and Golgi cells.

Both mossy and climbing fibers have been shown by recent physiological work (see ECCLES, ITO, and SZENTÁGOTHAI, 1967) to be excitatory. Of the intrinsic components in the cortex, only the granule cells and their parallel fibers are excitatory. All of the other cells produce inhibitory postsynaptic effects. Likewise, the Purkinje

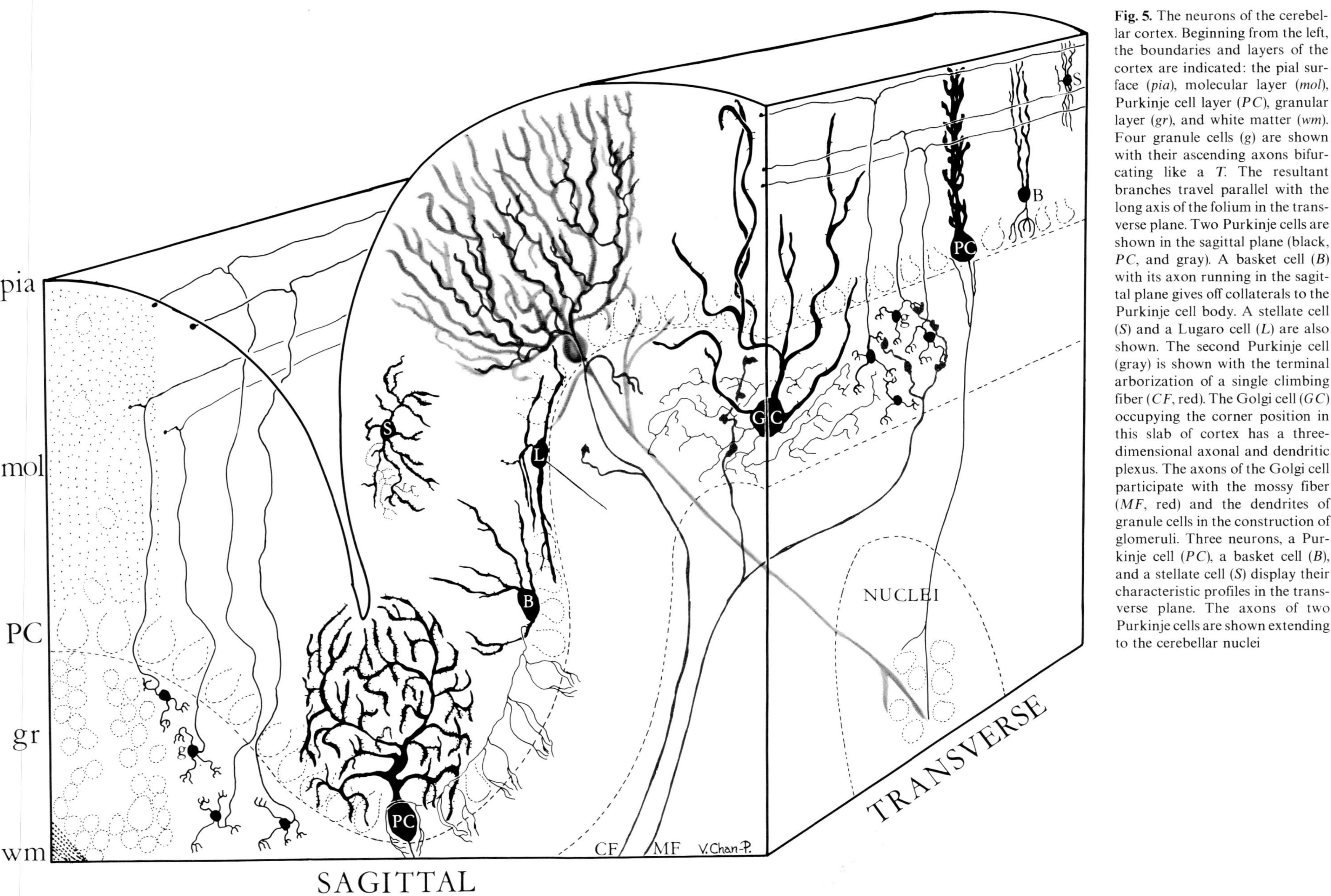

Fig. 5. The neurons of the cerebellar cortex. Beginning from the left, the boundaries and layers of the cortex are indicated: the pial surface (*pia*), molecular layer (*mol*), Purkinje cell layer (*PC*), granular layer (*gr*), and white matter (*wm*). Four granule cells (*g*) are shown with their ascending axons bifurcating like a *T*. The resultant branches travel parallel with the long axis of the folium in the transverse plane. Two Purkinje cells are shown in the sagittal plane (black, *PC*, and gray). A basket cell (*B*) with its axon running in the sagittal plane gives off collaterals to the Purkinje cell body. A stellate cell (*S*) and a Lugaro cell (*L*) are also shown. The second Purkinje cell (gray) is shown with the terminal arborization of a single climbing fiber (*CF*, red). The Golgi cell (*GC*) occupying the corner position in this slab of cortex has a three-dimensional axonal and dendritic plexus. The axons of the Golgi cell participate with the mossy fiber (*MF*, red) and the dendrites of granule cells in the construction of glomeruli. Three neurons, a Purkinje cell (*PC*), a basket cell (*B*), and a stellate cell (*S*) display their characteristic profiles in the transverse plane. The axons of two Purkinje cells are shown extending to the cerebellar nuclei

cells have an inhibitory influence on the cells of the central nuclei and the vestibular nuclei.

Further study of Fig. 5 will reveal some of the architectonic principles expressed in the design of the cerebellar cortex. First, it should be noted that although the afferent fibers transmit their impulses to a variety of cells, the efferent fibers of the entire cortex originate in only one type of cell, the Purkinje cell. Therefore, the Purkinje cell is the focal point of all the neuronal circuits in the cerebellar cortex. Like the motor neuron in the spinal cord, it is a final common pathway. It performs and transmits to other centers the final computation carried out in the cerebellar cortex. Second, both of the afferent pathways project upon the same cells but by different routes. The climbing fiber has direct access to the Purkinje cell by way of its extensive terminal arborization. The mossy fiber can affect the Purkinje cell only through its activation of granule cells and their parallel fibers. Both of these afferent fibers result in excitation of the Purkinje cells. However, other routes can be traced that end in inhibition or suppression of Purkinje cell activity, for example, a circuit leading from mossy fiber to granule cells to basket or stellate cells to the Purkinje cell. Or another circuit can be constructed going from the mossy fiber to granule cells to Golgi cell to granule cells again, or directly from mossy fiber to Golgi cell to granule cells, which in shutting down the granule cells can secondarily silence the Purkinje cell by a process called disfacilitation. Similar inhibitory circuits can be traced for the climbing fiber.

Third, both divergence and convergence are exemplified in the circuits of the cerebellar cortex. The synapses of the inputs to the granular layer are disposed in such a way as to diverge the incoming information to a large number of cells. Mossy fibers display this feature more extensively than climbing fibers. The small granule cell in a modest way illustrates both convergence and divergence, as each of its dendrites articulates with a different mossy fiber or possibly with a climbing fiber, and its axon synapses with a large number of Purkinje cells. The Purkinje cell dramatically illustrates the principle of convergence, as it receives thousands of inputs from the axons of granule cells (parallel fibers), basket and stellate cells, as well as from the climbing fiber. Fourth, the various afferents impinging on the Purkinje cell are deployed in quite specific patterns over its surface. Each type of afferent has its exclusive territory—parallel fibers on the thorns of spiny branchlets, stellate cell axons on the shafts of the smaller dendrites, basket cell axons on the shafts of larger dendrites, soma, and axon initial segment; climbing fibers on the thorns of the larger dendrites. Fifth, the molecular layer is organized in a spectacular rectilinear lattice, the physiological value of which has yet to be elucidated. The fact that this fundamental order has been preserved throughout the vertebrate phylogenetic series suggests that it is essential to the operations performed by the cerebellum. The order of the granular layer is nowhere near as geometric, and is therefore much more difficult to discern.

Chapter II

The Purkinje Cell

The apparent simplicity of the cerebellar cortex reflects in part the small number of neuronal types comprised in it. Most descriptions credit it with only five types of nerve cells: Purkinje cells, granule cells, basket cells, stellate cells, and Golgi cells. The first three of these classes are remarkably homogeneous, but the last two probably include several subtypes each, as has been indicated by RAMÓN Y CAJAL (1911). Nevertheless, stellate and Golgi cells have not received as thorough study as the others. Probably a sixth cell type, the Lugaro cell, should be added to this list. Despite the small number of cell types, the cerebellar cortex includes examples of the largest and the smallest, the most complex and the simplest kinds of neurons in the central nervous system, as well as cells from the middle range of the spectrum. In this and the following chapters, each cell type will be considered in turn and will be described as completely as possible with the information available at present.

The Purkinje cell is certainly one of the most spectacular nerve cells known (Figs. 6 and 7). Because of its focal position in the architecture of the cerebellar cortex and its beauty at all levels of morphological study, it has been the object of countless investigations. In the present work, it will be given a more detailed treatment than the other cell types because it displays the whole range of cytological features that are found in neurons. The description of similar structures in other cell types can then be referred to the Purkinje cell as a standard.

1. A Little History

PURKINJE was among the first to distinguish nerve cells in the central nervous system. In an epochal paper delivered before an assembly of German scientists and physicians held at Prague in September 1837, he described cellular elements in various parts of the brain, including the large neurons of the cerebellar cortex that have been known by his name ever since. Although Purkinje's observations were briefly recorded in 1837 in the Proceedings of the Prague meeting, they were already well known before this time; a year earlier JOHANNES MÜLLER had reported Purkinje's discovery in almost identical words, in his annual review of anatomy and physiology for the year 1836 (MÜLLER, 1837). What PURKINJE had seen were numerous fig-shaped corpuscles arranged in a row, with their rounded ends directed towards the white center of the folium and their tail-like apical processes directed towards the surface and disappearing in the gray matter. He was able to make this discovery because he succeeded in cutting freehand thin sections of fresh brain, which he then placed under the microscope without losing the original orientation of the tissue or the interrelations of its components (MÜLLER, 1838). The sections were thin enough to reveal cells, nuclei, nucleoli, and pigment granules, without staining, which had not yet been invented. Although LEEUWENHOEK had also used this technique well over a century and a half earlier, most of Purkinje's contemporaries were still compressing their fresh samples of tissue between glass plates, a practice that contributed to the slow progress of neurocytology during the first half of the nineteenth century. Purkinje's successors saw little more of this cell than he did until the last quarter of the nineteenth century, when staining with basic dyes, silver impregnation methods, and the Golgi methods came into use.

2. The Soma of the Purkinje Cell

Purkinje cells are among the largest neurons in the central nervous system. According to RAMÓN Y CAJAL (1911), their cell bodies are between 35 and 65 μm in man. JAKOB (1928) gives their dimensions as 30–35 μm in diameter and 50–70 μm in length. In the cat they are 29 μm in diameter (PALKOVITS *et al.*, 1971a). In the rat the cell bodies are 21 μm in diameter and 25 μm long on

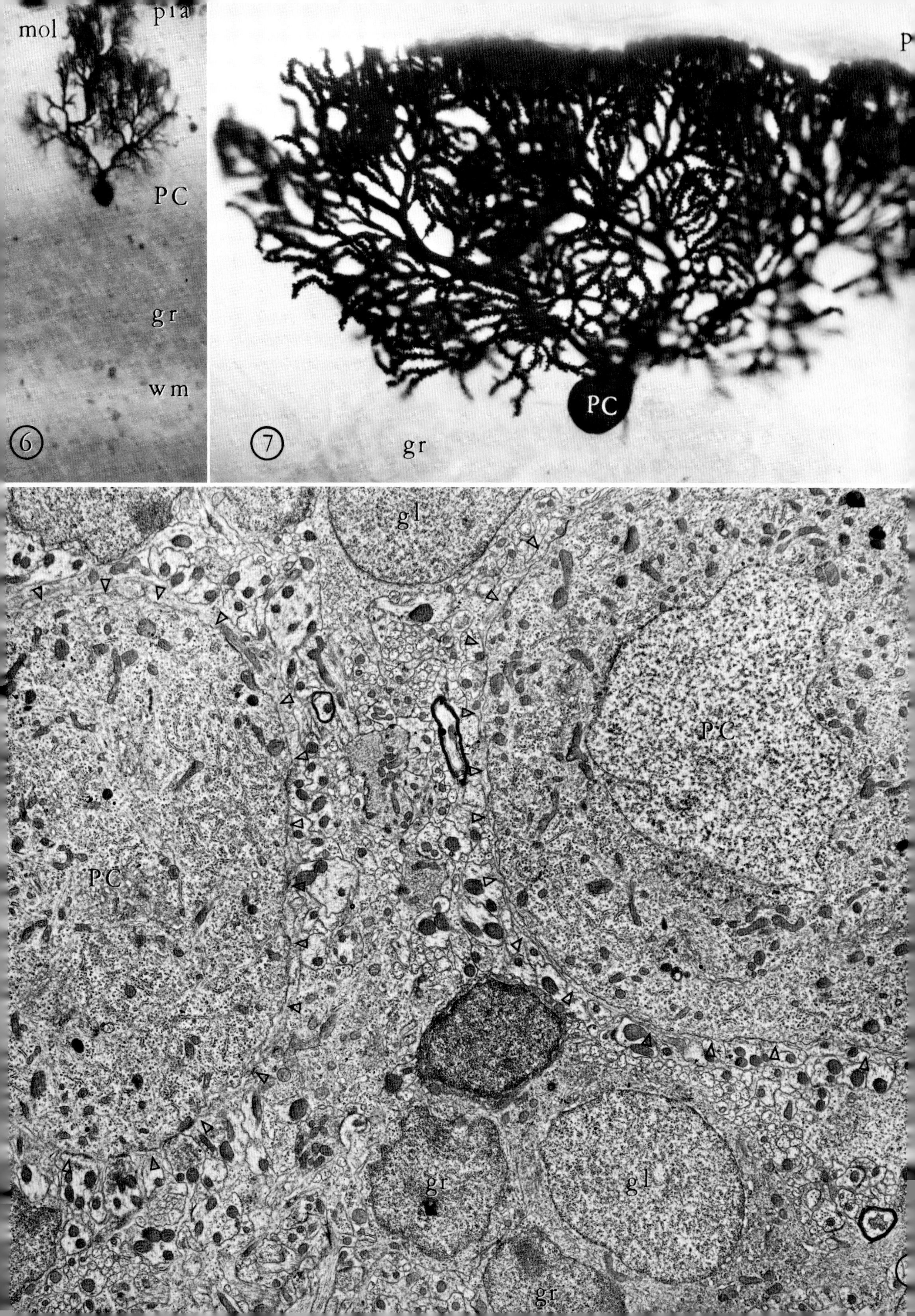
mol
pia
PC
gr
wm
6
PC
gr
7
gl
PC
PC
gr
gl
gr

the average.[2] BRAITENBERG and ATWOOD (1958) give the number of Purkinje cells in man as 15×10^6. There are about 5×10^5 in the cat (BELL and DOW, 1967; 1.2–1.3×10^6, according to PALKOVITS *et al.*, 1971a), and 3.5×10^5 in the rat (ARMSTRONG and SCHILD, 1970; 4.5×10^5, according to SMOLYANINOV, 1971). The cell bodies are arranged in a sheet one cell thick at the interface between the molecular and the granular layers without any obvious pattern or clustering (BRAITENBERG and ATWOOD, 1958; ARMSTRONG and SCHILD, 1970; SMOLYANINOV, 1971). There are, however, local variations in the density of Purkinje cells according to their position on the folium. BRAITENBERG and ATWOOD (1958) showed in man, and ARMSTRONG and SCHILD (1970) made similar findings in the rat, that the cells are farther apart in the depths of the cerebellar fissures than on the sides and crests of the folia. In the rat ARMSTRONG and SCHILD obtained a mean density of 1200 Purkinje cells per mm^2 of Purkinje cell sheet by one method of counting, and by another method a mean density of 1080 cells per mm^2 in the anterior lobe and 930 cells per mm^2 in the posterior lobe.[3] These values are to be compared with the much lower densities found in primates, 510 cells per mm^2 in monkey (FOX and BARNARD, 1957) and 300 cells per mm^2 in man (BRAITENBERG and ATWOOD, 1958; SMOLYANINOV, 1971). As in the cerebral cortex, where a similar comparison between species can be made (TOWER, 1954), the wider spacing of cells in primates is a consequence of the increased volume occupied by preterminal axons and dendrites, synaptic terminals, and neuroglial cells. In the cat PALKOVITS *et al.* (1971a) obtained a mean Purkinje cell density for the whole cerebellum of 567 per mm^2, which, when corrected for shrinkage in the histological preparation, became only 330 per mm^2.

Despite the admittedly irregular distribution of the Purkinje cell bodies in their layer (BRAITENBERG and ATWOOD, 1958; ARMSTRONG and SCHILD, 1970), various authors have tried to rationalize their arrangement into some idealized pattern. The impetus for these attempts derives from the alluringly rectilinear pattern of the molecular layer and from the obvious need to provide physiologists and cyberneticists with a standard model for their calculations. In the parasagittal plane the Purkinje cell bodies are usually much closer together than the breadth of their dendritic trees should allow if they are not to overlap or interpenetrate (Fig. 8). In fact, the cell bodies often appear to be nearly touching (Fig. 9), and in electron micrographs it can be seen that they are separated only by their thin neuroglial envelopes and a few ascending or descending nerve fibers. In view of these proximities and the frequent irregularities in their array, the ranks of cell bodies might be expected to be staggered in order to provide minimal overlap of the dendritic trees in the parasagittal plane. Nevertheless, SMOLYANINOV (1971) assumed a strict hexagonal grid-like arrangement of the cell bodies, and he predicted that the resultant crowding of the major dendrites would produce a network of Purkinje cells bound together by dendrodendritic junctions. Such junctions have never been seen in electron micrographs.

PALKOVITS *et al.* (1971a) proposed an ingenious model for the cat's cerebellum, also involving hexagonal packing of the cell bodies. With values taken from their actual measurements and calculations, they set 8.9 μm as the thickness of the plate filled by the dendritic tree, 292.5 μm as the width of this plate in the parasagittal plane, and 51 μm as the center-to-center distance between perikarya. They then projected upon the plane surface of the cortex tangential sections of these plates represented as oblongs, 292.5 μm × 8.9 μm, each centered on the circular profile of the perikaryon. Since they had counted 384 perikarya in 1 mm^2 of flat folial surface, they packed 384 oblongs into an area 1 mm square, maintaining an average distance of 51 μm between perikarya. They obtained a regular array with the Purkinje cells staggered in rows that deviated 11° from the transverse axis of the folium (the parasagittal plane). The neighboring cell bodies were arranged at the corners of a slightly anisotropic rhomboidal unit figure. Although this order is much more rigid

2 Profiles of 21 perikarya were measured in Epon-embedded 1 μm sections stained with toluidine blue.

3 The small area of the Purkinje cell sheet in the rat was calculated by ARMSTRONG and SCHILD as 270 mm^2; SMOLYANINOV (1971) obtained an area of 322 mm^2 for the rat, and 5500 mm^2 for the cat, but PALKOVITS *et al.* (1971a) accepted 2285 mm^2 as the area of the Purkinje cell layer in the cat. According to BRAITENBERG and ATWOOD (1958), it is 50000 mm^2 in man, but SMOLYANINOV (1971) gives it as 72000 mm^2.

◄

Fig. 6. A Purkinje cell in the cerebellar cortex. A single Purkinje cell is shown with its dendritic arborization extending throught the molecular layer (*mol*). The pial surface (*pia*), Purkinje cell layer (*PC*), granular layer (*gr*), and white matter (*wm*) are indicated. Rapid Golgi, sagittal, 90 μm, lobule VII. × 240

Fig. 7. A Purkinje cell in the cerebellar cortex. The entire dendritic arborization of this Purkinje cell (*PC*) has been impregnated. Myriads of thorns on spiny branchlets are visible. The dendrites reach from the Purkinje cell layer to the pial surface (*pia*). The granular layer (*gr*) is indicated. Rapid Golgi, sagittal. 120 μm, lobule V. × 540

Fig. 8. Two Purkinje cells. The two Purkinje cells (*PC*, encircled with arrowheads) are placed among neuroglial cells (*gl*), granule cells (*gr*), and dendrites, axons, and neuroglial processes of the neuropil. × 3500

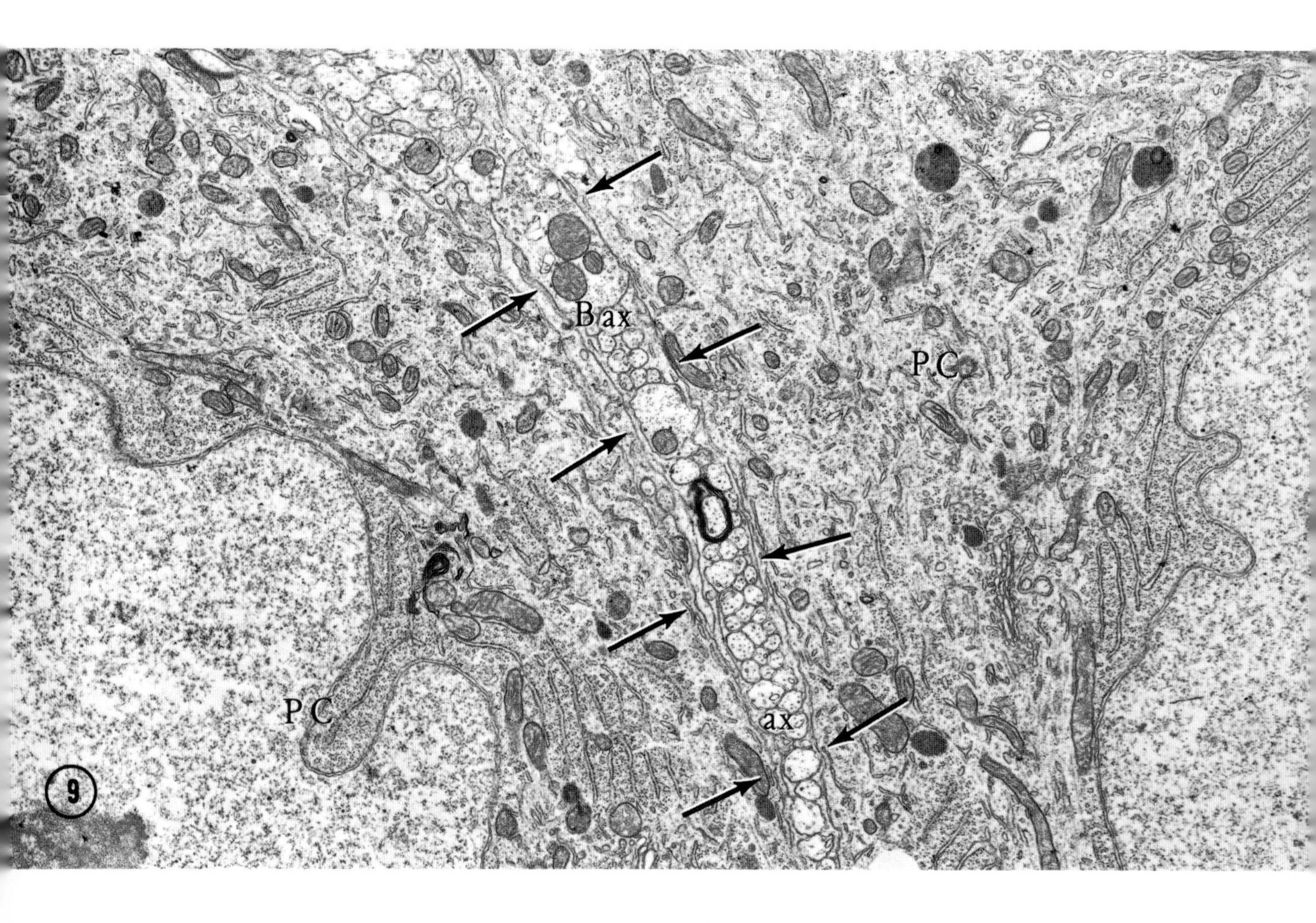

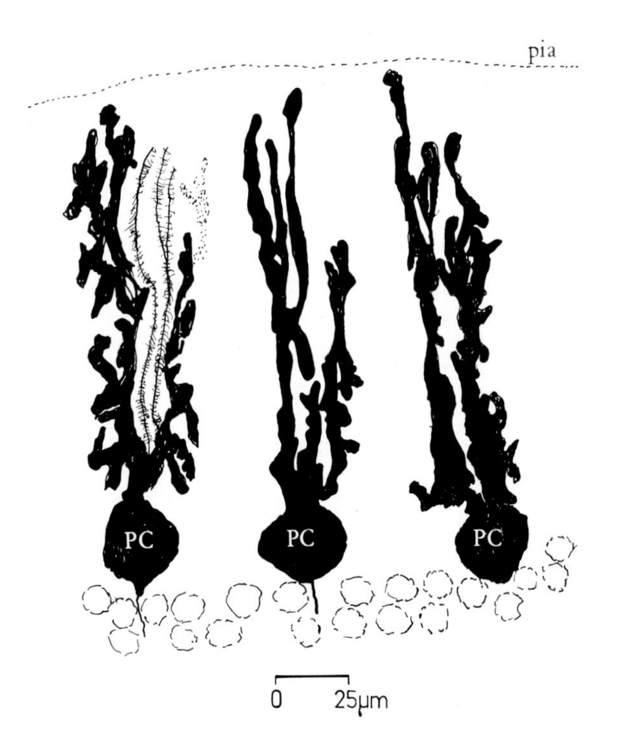

▲

Fig. 9. Between two Purkinje cells. In many instances the space between the somata of adjacent Purkinje cells (*PC*, between arrows) is occupied by slips of neuroglia, ascending axons of granule cells (*ax*), a rare myelinated fiber, and a terminal of a basket cell axon (*Bax*). Lobule X, nodulus. × 1 200

◀ **Fig. 10.** The Purkinje cell dendritic tree in the transverse plane. The dendritic trees of three Purkinje (*PC*) cells are shown. The pial surface (*pia*) as well as the granular layer (dotted outlines of granule cells) is indicated. Rapid Golgi, transverse plane, 90 μm, lobule IV, adult rat. Camera lucida drawing

than the observed arrangement of the cell bodies, it does allow for the observed minimal intermingling of dendritic trees at the same time as it fills nearly all of the molecular layer with a regular lattice of Purkinje cell dendrites. Distortions in the regular arrangement of the cell bodies could be compensated for by bending, curving, or elongating the main stem dendrites. The model also fits with FRIEDE'S (1955) assertion that the disposition of the Purkinje cell bodies is determined by the spatial relations of the dendrites in the molecular layer. It seems, however, that 8.9 μm is an exceedingly small value for the thickness of the whole dendritic plate of a Purkinje cell. It might represent the thickness of only a single frond ramifying from a large dendrite. Our own observations give a quite different value. In Golgi-Kopsch material we measured the thickness of the dendritic trees in a row of 20 Purkinje cells found in a section passing parallel with the long axis of the folium (see Fig. 10). The three trees oriented most precisely perpendicular to the section plane measured 15, 19, and 21 μm in thickness respectively; in other words, only slightly thinner than the diameter of the cell body. Such measurements make it difficult to accept the model suggested by PALKOVITS *et al.* without further refinement.

In Golgi preparations the cell body of the Purkinje cell appears globular or ovoid. It has been variously described as fig-shaped, pear-shaped, pyramidal, or beet-shaped. The usually depicted pear shape originates from the presence of a single dendritic trunk issuing from the apical pole of the soma. Generally, however, the number of dendrites is greater than one, and with its single axon dangling from its lower pole, the Purkinje cell body in Golgi preparations resembles a kohlrabi much more than any pear.

3. The Nucleus

In Nissl preparations the cell body is characterized by its large, pale nucleus, an intensely basophilic nucleolus, and scattered, rather small, polygonal Nissl bodies. The nucleus is outlined by a distinct membrane, usually shown as smooth and round in drawings and photomicrographs of stained sections. In thin plastic sections stained with toluidine blue it is evident, however, that the nuclear membrane is not smooth all the way around the nucleus but is wrinkled or puckered on the side facing the origin of the dendritic tree. The wrinkles vary from a shallow dimple in the surface of the nucleus to a deep and complicated crease with many tributary folds. The irregular depression in the nucleus is stuffed with Nissl substance, forming a prominent basophilic nuclear cap, which both RAMÓN Y CAJAL (1911) and JAKOB (1928) especially noted. The association between the position of the nuclear cap and the dendritic pole of the cell is quite consistent in all Purkinje cells. When the axis of the cell is tilted away from the vertical so that the dendrites issue from the lateral aspect instead of the apical aspect of the cell body, the nuclear depression and the nuclear cap are also rotated into the corresponding position. This consistent relationship may be taken as evidence that the wrinkling of the membrane is not due to shrinkage of the nucleus or some other technical artifact. Furthermore, since Nissl bodies—although less extensive ones—are also attached to the smooth parts of the nuclear surface, the wrinkling of the nuclear membrane cannot be necessary for the attachment of the Nissl substance.

Electron microscopy demonstrates that, as expected from the optical microscopic observations just mentioned, the dendritic pole of the nucleus is deeply invaginated, with many secondary folds, and that the cytoplasm filling the invagination consists almost exclusively of granular endoplasmic reticulum, sometimes of clustered ribosomes alone. Fig. 11, an electron micrograph of the dendritic pole of a Purkinje cell, shows the wrinkled nuclear envelope and the associated Nissl substance composing the nuclear cap. As in other cells, the nuclear envelope consists of a single cisterna of the endoplasmic reticulum encircling the karyoplasm (see PETERS, PALAY, and WEBSTER, 1970). The outer, cytoplasmic, wall of this cisterna undulates irregularly, and occasionally gives off tubular projections, which join with the nearest cisterna of the endoplasmic reticulum in the Nissl substance. Since the inner, karyoplasmic, wall of the envelope is smooth, the lumen of the cisterna making the nuclear envelope varies considerably in depth from place to place. Microscopists at the beginning of this century (e.g., RAMÓN Y CAJAL, 1909; MARINESCO, 1909) described and depicted the nuclear membrane of many neurons as having a double contour. Their drawings show the

Fig. 11. The Purkinje cell nucleus. The nuclear chromatin is thinly and uniformly distributed throughout, except for two sites (*arrows*) where it is collected into small masses. The wrinkle at the dendritic pole of this cell is capped by a small Nissl body (*NB*). Lobule X, nodulus. ×14000

Fig. 12. The nuclear cap region of the Purkinje cell. Scores of ribosomes (*r*) and granular endoplasmic reticulum (*ER*) collect against the wrinkled nucleus (*PC Nuc*) at the dendritic pole of this cell, forming the nuclear cap. A lysosome (*Ly*) is also indicated. Crus I, lobule VII. ×18000

Fig. 13. The Purkinje cell perikaryon. The section passes through a fold in the nucleus of this Purkinje cell (*PC Nuc*) and includes the nuclear cap. The nucleus contains a prominent mass of chromatin (*arrow*) that is normally associated with the nucleolus. Two large Nissl bodies (*NB*) flank the nucleus on either side. Lobule X. ×13000

Fig. 14. Ribosomal arrays and nuclear pores in the Purkinje cell perikaryon. This section grazes the Purkinje cell nucleus (*PC Nuc*), exposing many nuclear pores (*np*). Polysomal arrays are attached to the surface of the endoplasmic reticulum and free ribosomes (*r*) are clustered in the intervening matrix. Lobule X, vermis. ×33000

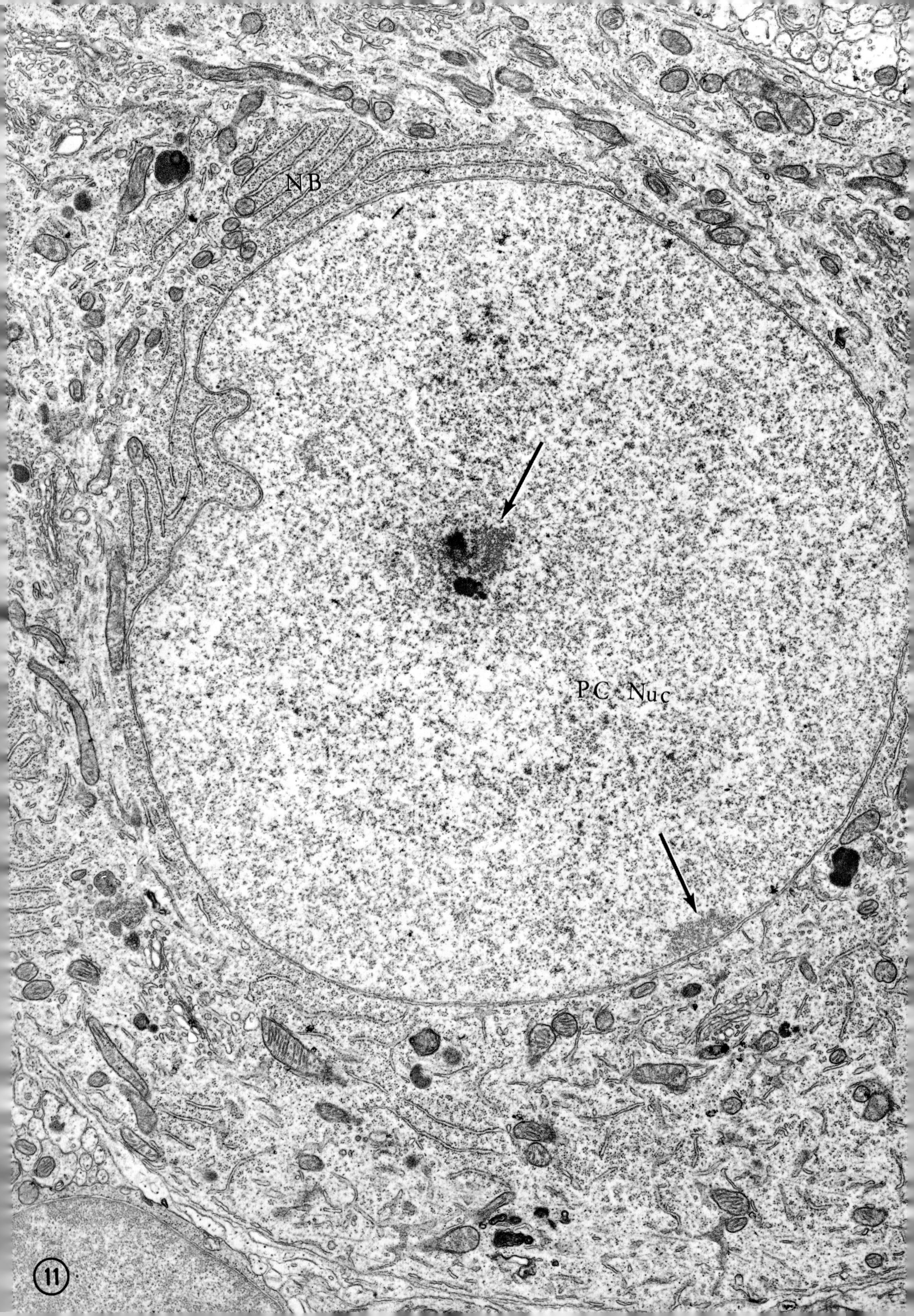
NB
PC Nuc
11

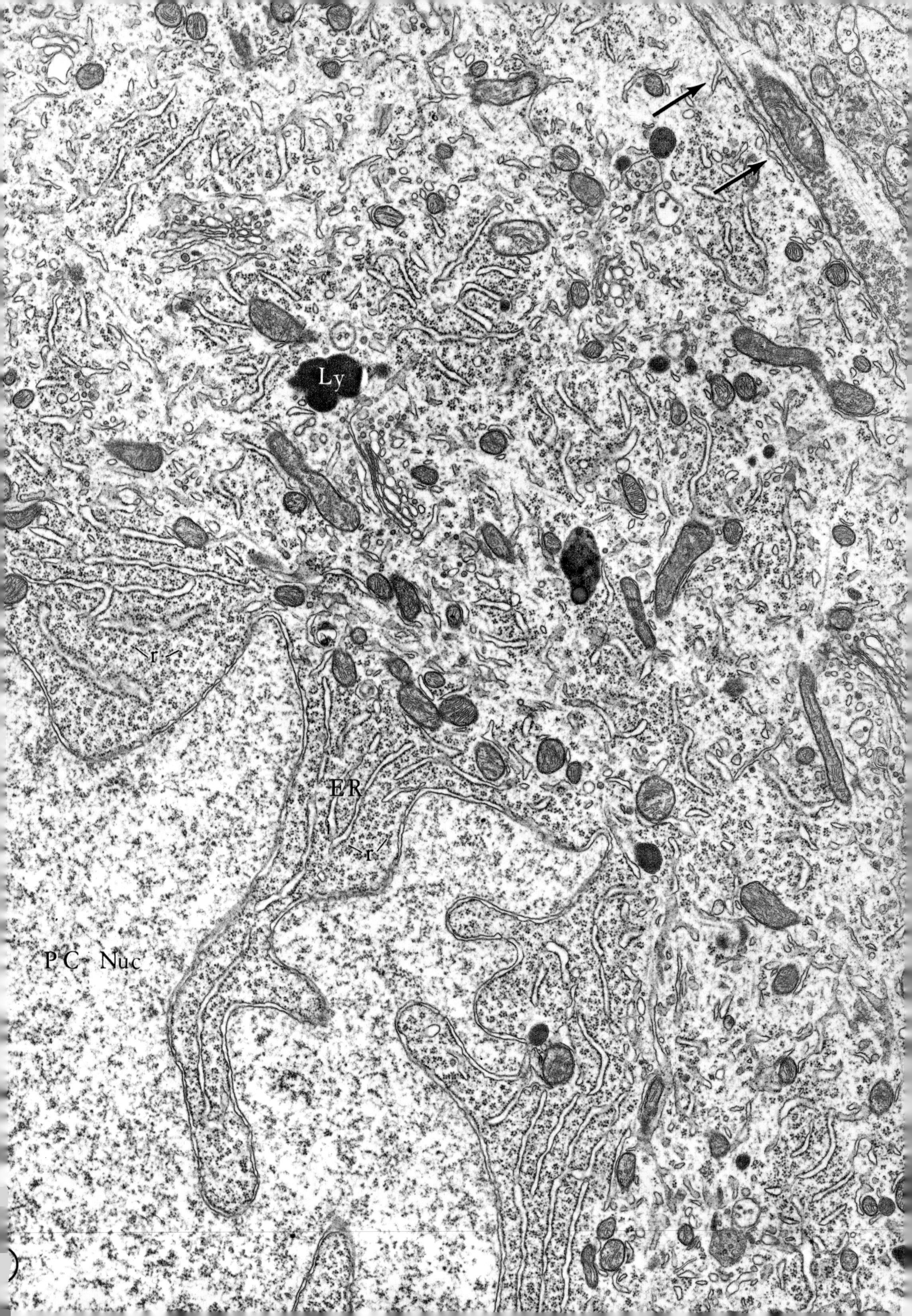

Ly
r
ER
r
PC Nuc

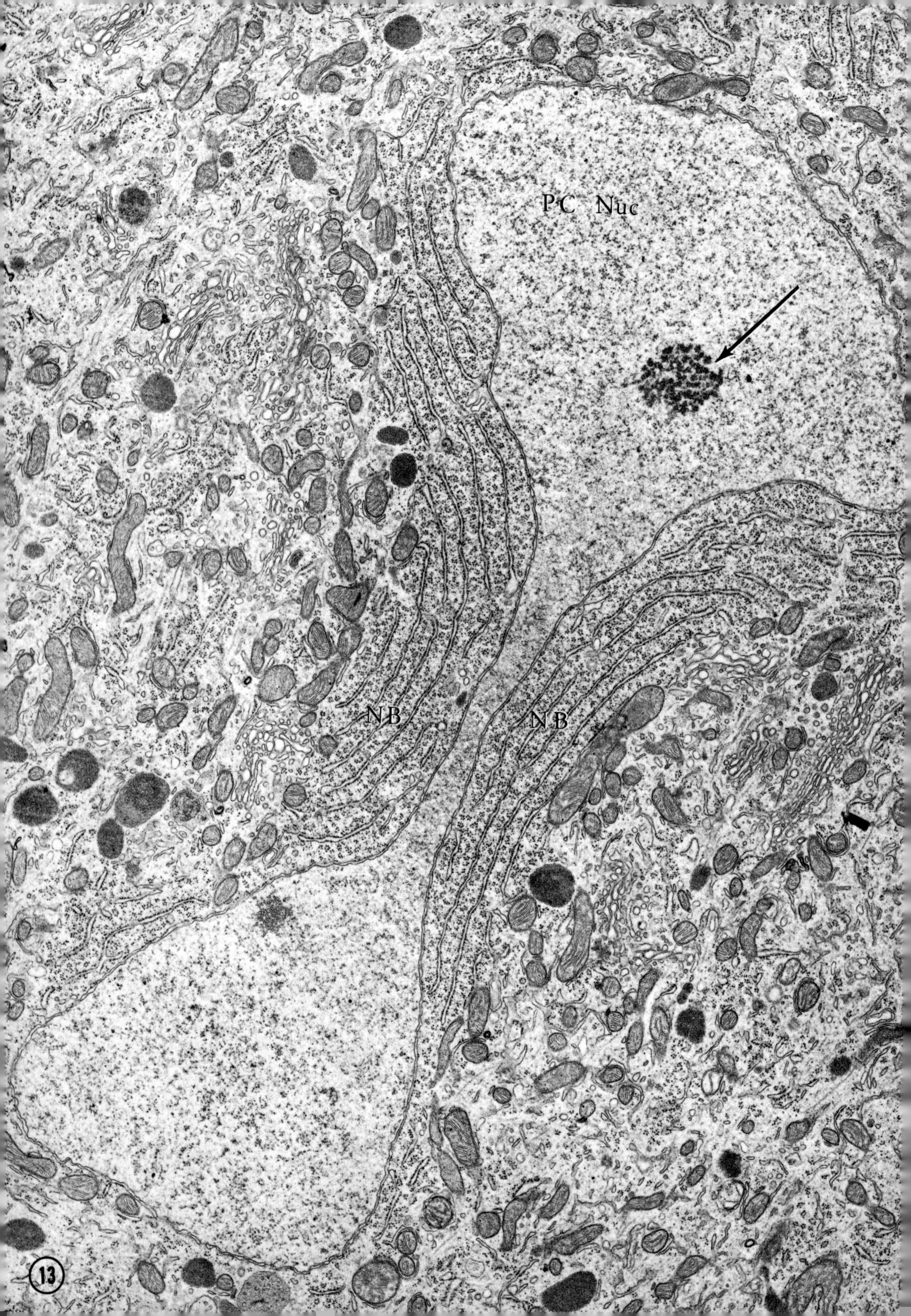
PC Nuc
NB
NB
13

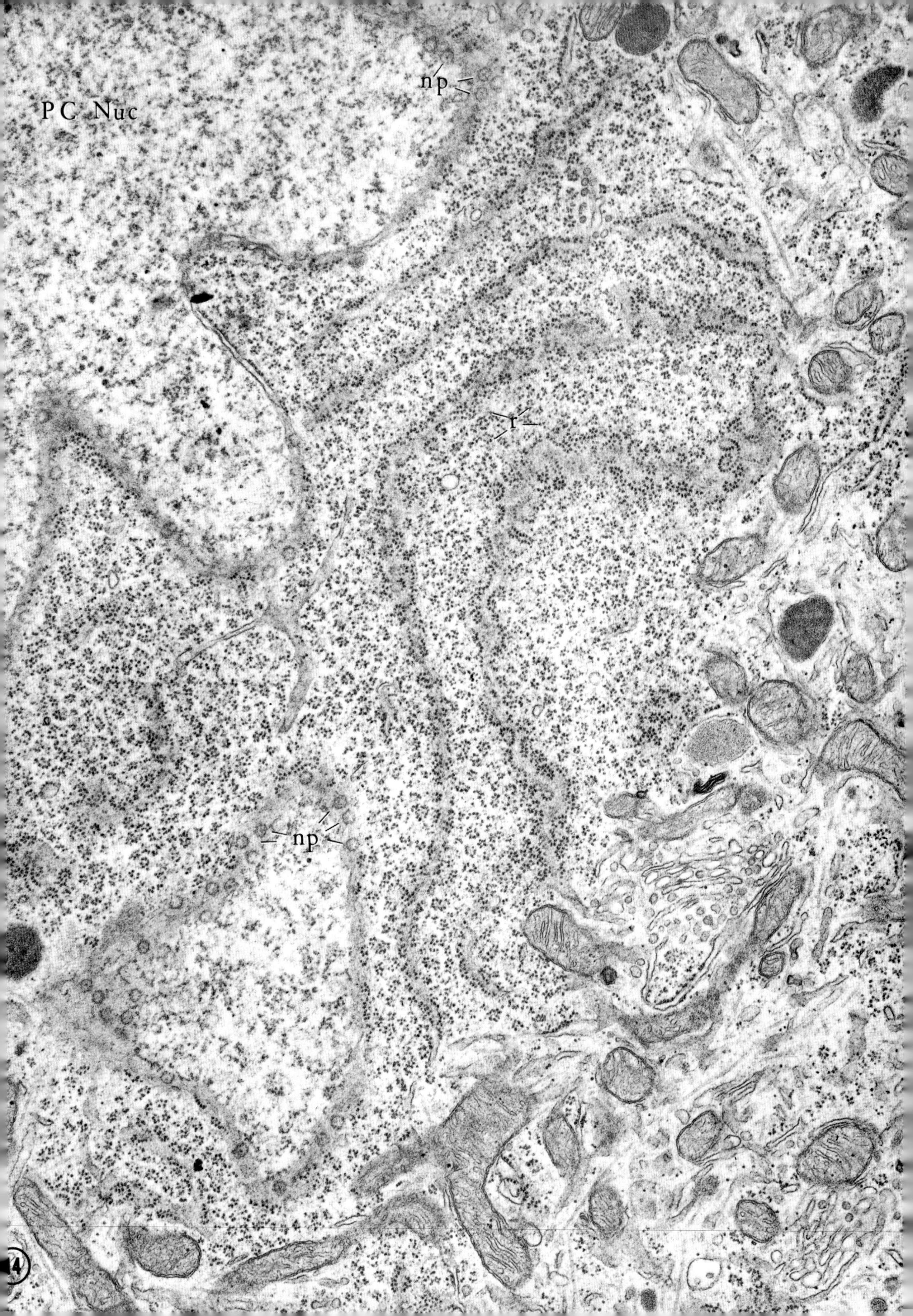
PC Nuc
np
r
np
14

nucleus enclosed by two parallel lines. It is difficult to know precisely what produced this appearance in their preparations. Although it is possible that they merely recorded a diffraction image forming in their thick sections, it is also possible that they observed the separation of the cytoplasmic and karyoplasmic walls of the perinuclear cisterna, especially if it was enlarged by the methods of fixation that they used.

In the nuclear cap region (Figs. 11–15), the outer surface of the nuclear envelope regularly bears short rows and spiral arrays of ribosomes, while the rest of the nuclear surface is free of ribosomes. The nuclear pores also appear much more frequently in the nuclear cap region than in the smooth part of the nucleus. These differences may, however, be spurious, since the higher frequency of tangential sections (e.g., in Fig. 14) given by the folded nuclear envelope in the cap region would greatly increase the opportunities for nuclear pores and polysomes to be included in sections of that region. Such tangential sections, which graze the surface of the nucleus, provide an opportunity to study the arrangement of the pores in the nuclear envelope. The pores are about 650 Å in diameter and are set out in ranks regularly spaced at intervals of about 1500 Å from center to center. The distance between ranks is also about 1500 Å. As the pores in adjacent ranks are staggered, alternating ranks fall into register with one another. Thus the overall pattern is a hexagonal arrangement, similar to that found in the rat's kidney by MAUL, PRICE, and LIEBERMAN (1971). The nuclear pores of neurons have already been described (PETERS, PALAY, and WEBSTER, 1970).

The inner wall of the perinuclear cisterna is lined with an extremely tenuous fibrous lamina (FAWCETT, 1966a and b), which is perceptible only at relatively high magnification (Figs. 15–17).

a) The Chromatin

The nuclear chromatin is homogeneously and thinly dispersed throughout the karyoplasm. In preparations stained for the optical microscope it takes up hardly any color at all, so that the nucleus appears to be clear and vesicular. Some small pale masses, more acidophilic than basophilic, can be seen here and there, especially near the nucleolus, which stands out because of its size and intense basophilia. In male animals a conspicuous small basophilic particle is sometimes associated with the nucleolus, and in female animals there are usually two particles of nucleolus-associated chromatin.

Since this apparent correlation between the sex of the animal and the number of heterochromatic bodies associated with the nucleolus was based on casual observations, a more systematic investigation was undertaken by one of us with light microscopic techniques in order to see whether it was true (CHAN-PALAY, 1968, previously unpublished results). Measurements were made on the incidence, position, number, and size of the nucleolus-associated chromatin in rats, mice, cats, and monkey. A variety of large cells were compared in each animal: Purkinje cells, large pyramidal cells in the motor cortex, anterior horn cells, Deiters giant cells in the lateral vestibular nucleus. Only the results on the Purkinje cell will be presented here. The Feulgen reaction for DNA and methyl-green pyronin staining (distinguishes between DNA- and RNA-containing structures) as well as cresyl violet were used. The masses of nucleolus-associated chromatin were Feulgen-positive (Figs. 19 and 20) and methyl green-positive. These reactions indicate the presence of DNA. Digestion with DNA'ase eliminated the stainability by these methods, thus proving the presence of DNA. In each preparation, 500 cells were counted. Each cell was chosen for having a well-defined nucleus and nucleolus. The position of any heterochromatic mass within the nucleus was recorded for each cell. Five categories of positions were used to classify these cells: (1) two heterochromatic masses, one on either side of a single large nucleolus; (2) two heterochromatic masses, each attached to a separate nucleolus, two small nucleoli being present; (3) a single heterochromatic mass on a single nucleolus; (4) two heterochromatic clumps, one on a single nucleolus and a second either at the nuclear envelope or free in the nucleoplasm; (5) a single heterochromatic mass either floating free in the nucleoplasm or attached to the nuclear envelope. The results obtained for Purkinje cells are summarized in Tables 2 and 3 and Figs. 19a–d (mouse) and 20a, b (rat).

Female rats generally have one or two bodies of nucleolus-associated chromatin. When two are present, each averages 1 μm in diameter. When only a single mass is present, it tends to be larger, 1.3 μm. In male rats, single heterochromatic bodies were found in only 19–30% of the Purkinje cells examined. Double heterochromatic bodies are rare; however, an example is illustrated in Fig. 20b. The majority of the cells contained no obvious nucleolus-associated chromatin. The amount of DNA in the nuclei of Purkinje cells from both male and female rats was proved to be tetraploid by measurements of Feulgen preparations (CHAN-PALAY, 1970, previously unpublished data) with a Barr and Stroud integrating microdensitometer (type GN 2 No. 3). Thus the sex differences in nucleolus-associated chromatin are not correlated with the amount of DNA present.

Table 2. A comparison of the incidence and position of nucleolus-associated chromatin bodies in Purkinje cells of various male and female animals[a]

Animal	Sex	Reagent/stain	% Incidence and Position of Nucleolus-associated Chromatin Bodies					% Total
Rat	♀	Feulgen	12.5		43.5	1	22	79
	♀	Cresyl violet	11.5		46		1	58.5
	♂	Feulgen	2.3	0.3	21.7		6	30.3
	♂	Cresyl violet			19			19
Mouse	♀	Feulgen	57.3	1.5	39.3	0.3		98.4
	♀	Methyl green-pyronin	33	1.3	30.7	1.5		66.5
	♀	Cresyl violet	27.7	1.3	47			76
	♂	Feulgen	48.5	2.8	33.3			84.6
	♂	Methyl green-pyronin	45	9	28.9	5.7		88
	♂	Cresyl violet	22.4	1	37.2		0.2	60.8
Cat	♀	Feulgen			39			39
	♀	Cresyl violet			42.7			42.7
	♂	Feulgen			0.7			0.7
	♂	Methyl green-pyronin						0
	♂	Cresyl violet			5			5
Monkey	♂	Feulgen						0
	♂	Cresyl violet						0

[a] A total of 500 cells were counted after reactions with each of the following: (1) Feulgen reagent, (2) methyl green-pyronin, and (3) cresyl violet. Every cell counted contained a distinct nucleolus. Five positions of heterochromatin were recorded, and are presented in the table above as follows: (1) two DNA masses, one on either side of a single nucleolus, (2) two DNA masses, each attached to a separate nucleolus in a cell containing two nucleoli, (3) a single DNA mass on a single nucleolus, (4) a single DNA mass on the nucleolus, a second at the nuclear envelope, (5) a single DNA mass, either free in the nucleus or lying against the nuclear envelope.

In the mouse a majority of the Purkinje cells in both males and females presented heterochromatin particles in association with the nucleolus. Among these, single and paired bodies appeared with approximately equal frequency. The size of these bodies in the mouse is impressive (see Table 2 and Fig. 19): in pairs each mass averaged 1.3 µm in diameter; single masses could be up to 1.9 µm in diameter. Figs. 19 and 20 are photomicrographs of Purkinje cells in mouse and rat displaying Feulgen-positive clumps of heterochromatin.

Only about 40% of the Purkinje cells in female cats have heterochromatic masses associated with the nucleolus, and then only as a single clump. Nucleolus-associated bodies have only rarely been observed in the Purkinje cells of male cats. When present, they are only very small particles. No nucleolus-associated bodies have been found in the Purkinje cells of the male monkey.

The data presented here for Purkinje cells agree with similar quantitative and distributional observations from other large cells in each of the species studied. The pattern of the nucleolus-associated chromatin is dependent on the species, and does not reflect the overall DNA content or the sex of the animal in all species. The cat appears to have a very strong tendency for large single heterochromatin bodies to occur in the female (Barr, Bertram, and Lindsay, 1950), but almost 60% of the cells show no sex difference. In the rat nearly 80% of the Purkinje cells have some kind of nucleolus-associated chromatin in the female, whereas in the male only about 30% of the cells display this feature. Such percentages suggest that the differences are less related to sex than to some metabolic cycle of the Purkinje cell and its nucleus.

In the usual electron micrographs of thin sections very little can be learned about the structure and organization of the chromatin. The nucleus appears replete with extremely fine, filamentous material, more or less uniformly dispersed. Fine grains are strung on these filaments, and among them are clumps of larger and denser rounded particles, 150–200 Å in diameter, and often regularly spaced in contorted rows (Figs. 11–13, 15–17). The heterochromatin consists of extremely dense helical aggregates of particles of approximately the same size. In addition, there are scattered individual perichromatinic particles 300–350 Å in diameter (see Peters, Palay, and Webster, 1970).

One or two rounded masses of extremely dense hetero-

Table 3. A comparison of the diameters of nuclei and of single and double nucleolus-associated chromatin bodies in the Purkinje cells of various male and female animals

Animal	Sex	Reagent	Nuclear Diameter (µm)	Diameter of Nucleolus-associated Chromatin (µm)	
Rat	♀	Feulgen	12.2	1	1.3
Rat	♂	Feulgen	11.6	—	1.5
Mouse	♀	Feulgen	10.3	1.3	1.9
Mouse	♂	Feulgen	10.3	1.3	1.5
Cat	♀	Feulgen	11.5	—	1.3
Cat	♂	Feulgen	11.3	—	0.7
Monkey	♂	Feulgen	13.5	—	very few

chromatin are usually associated with the nucleolus. At the light microscope level, these have already been described in detail above. They are frequently encountered in electron micrographs of sections that include the nucleolus (Figs. 15, 16, 34). Sometimes they appear in the karyoplasm detached from the nucleolus, but when they are found alone in a random section, as in Fig. 13, it is impossible to tell whether the nucleolus might not have been in a neighboring section. The fine structure of these bodies is unlike that of the nucleolus. They consist of a loosely twisted strand, about 600 Å thick and very dense. Because of its contortions, only short segments of it are captured in a single plane of section.[4] The strand itself is composed of dense granules, 150–200 Å in diameter, strung close together and spirally wound around a less dense core, about 200 Å across. This helix is immersed in a fluffy matrix of fine filaments resembling those found dispersed all over the karyoplasm but somewhat more condensed in this location.

The pallor of the Purkinje cell nucleus contrasts with the density and basophilia of the nucleus in the much smaller granule cell. These two cells stand at opposite extremes in the spectrum of neurons in the central nervous system. The volume of the Purkinje cell nucleus in the rat is about 864 μm^3, whereas the volume of the granule cell nucleus is about 100 μm^3, or roughly one eighth. PALKOVITS *et al.* (1971a and b) give the corresponding measurements corrected for the living cat as 2202 μm^3 for the Purkinje cell and 100 μm^3 for the granule cell, or a ratio of 1:22. Microspectrophotometric analyses have demonstrated that the Purkinje cell nucleus in all species examined (including the rat) contains the tetraploid amount of deoxyribonucleic acid; in other words, twice as much as the granule cell nucleus contains (SANDRITTER *et al.*, 1967; LAPHAM, 1968; BILLINGS and SWARTZ, 1969; CHAN-PALAY, 1970, unpublished; LENTZ and LAPHAM, 1970). Since the Purkinje cell nucleus contains only twice as much DNA in about eight times as much space as the granule cell nucleus, the difference in staining and density between the two nuclei may be partially explained by dilution. But that is not the only consideration. Both optical and electron microscopy show that the chromatinic material is differently dispersed in the two nuclei. In the Purkinje cell nucleus the chromatin is widely and almost uniformly distributed. Blocks of heterochromatin are few and rather loosely aggregated (Figs. 11–13, 15–16). In the granule cell nucleus (Fig. 63), the heterochromatin is characteristically compact and arranged in checkerboard fashion. Cytochemical studies of these two neighboring nuclei can be expected to yield interesting and helpful insights into neuronal metabolism. It is not unreasonable to conjecture that a large proportion of the DNA in the granule cell nucleus is condensed in the heterochromatin and is mostly inactive, while in the Purkinje cell a large proportion of the DNA is in the extended form and is probably active. Is this difference related to the tetraploidy of the Purkinje cell nucleus, and is it involved in some specific processes that the granule cell cannot participate in?

b) The Nucleolus

The nucleolus of the Purkinje cell is an impressive, approximately spherical body usually lying near the center of the nucleus, but occasionally near the nuclear envelope. In the latter position it is separated from the nuclear envelope by heterochromatin. In a few cells there are two nucleoli. By measuring the nucleoli in 50 Purkinje cells in the optical microscope, ARMSTRONG and SCHILD (1970) obtained a mean diameter of 2.3 µm, a value that we have confirmed in measurements on

4 NOSAL and RADOUCO-THOMAS (1971) have found an apparently similar body in young Purkinje cells, which they call "the spotted body" and which they claim to have discovered. They have evidently missed seeing published pictures of this structure in adult neurons (PETERS, PALAY, and WEBSTER, 1970). They also fail to recognize this structure as heterochromatin, and prefer to regard it as a nucleolar product.

Fig. 15. The Purkinje cell nucleolus. The nucleolus, lying near the nuclear envelope, consists of the characteristic pars fibrosa (*pF*) and pars granulosa (*pG*). In the perikaryon a small Nissl body (*NB*) is indicated. Lobule IX, nodulus. ×27000

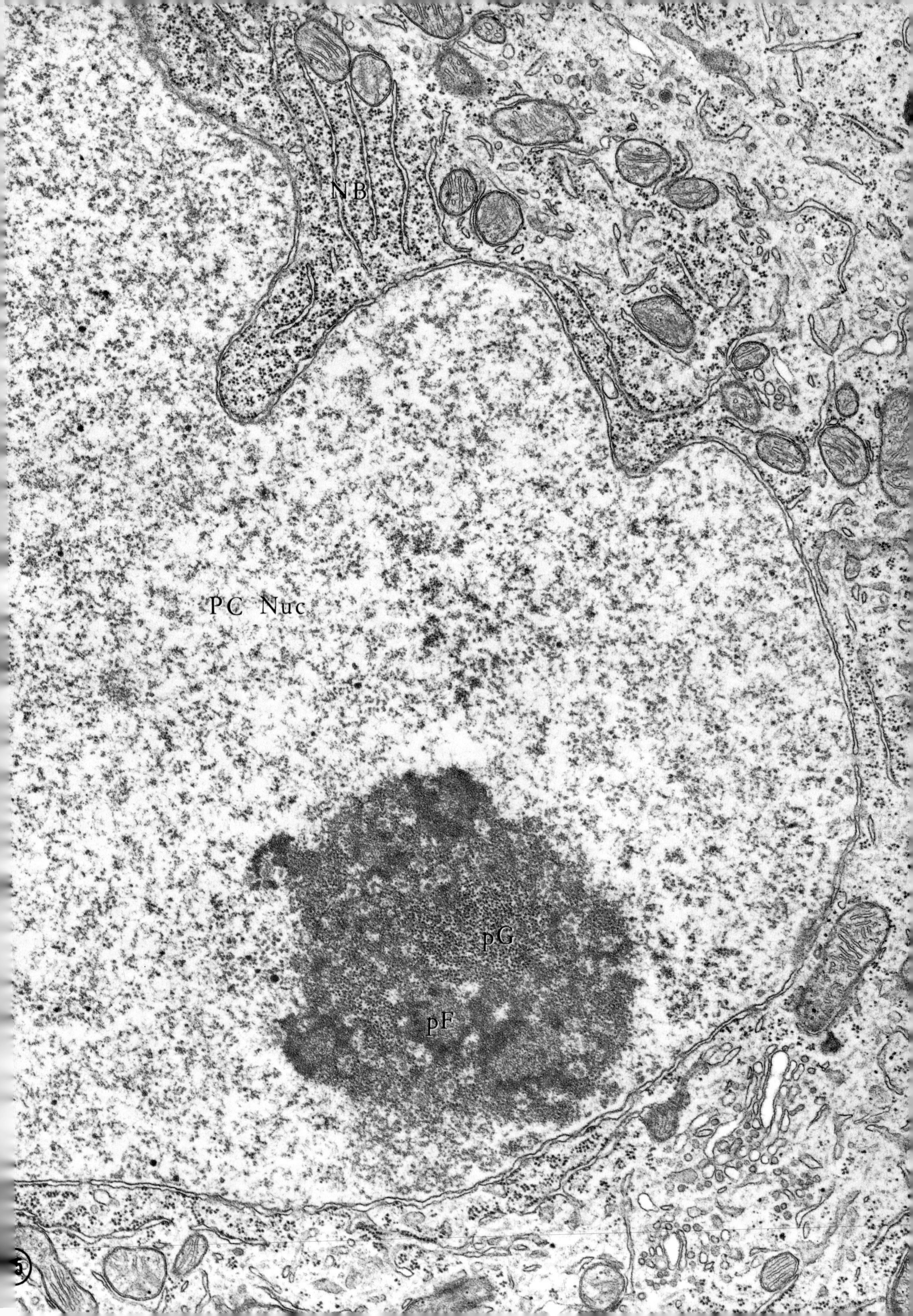
NB
PC Nuc
pG
pF
5

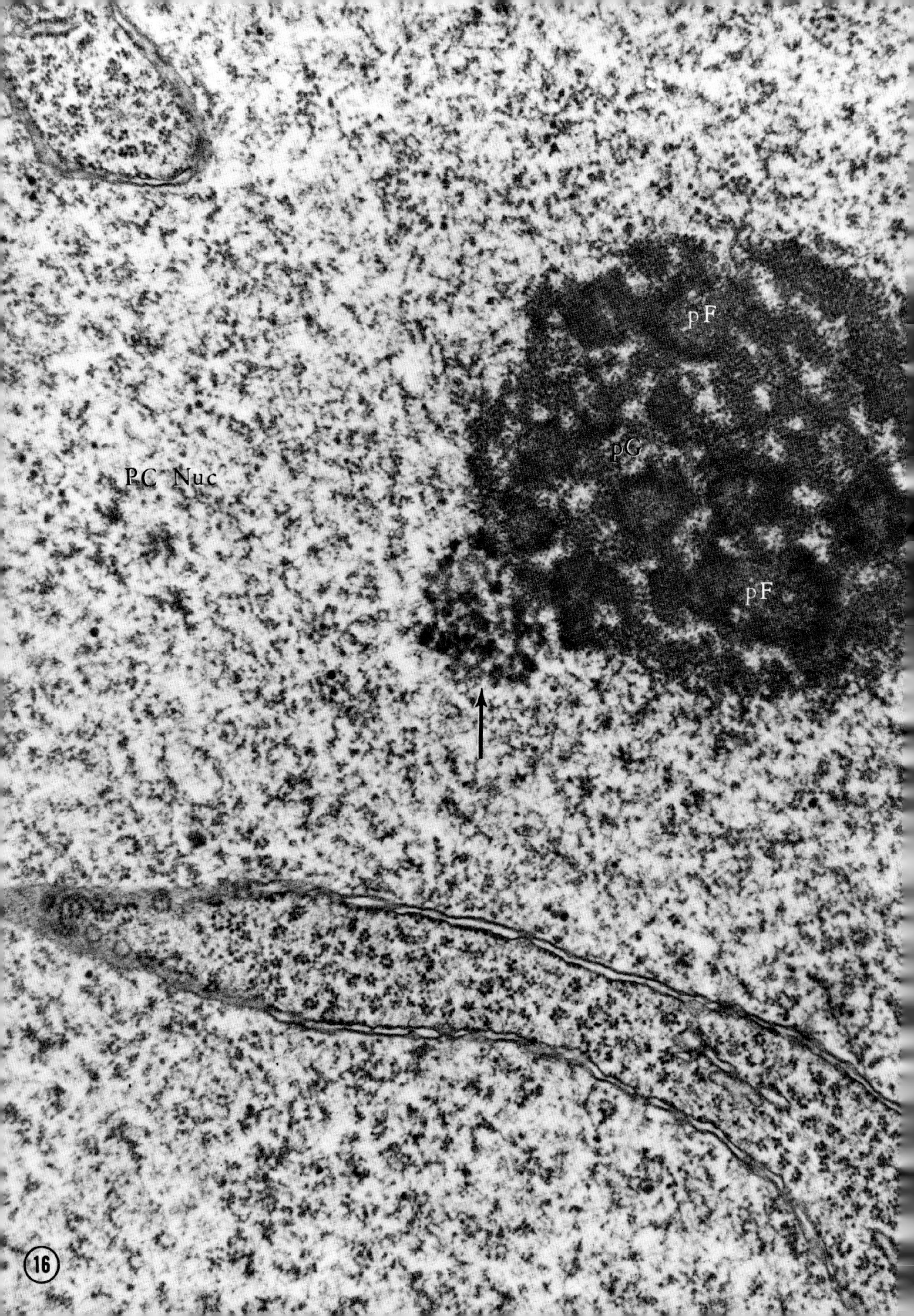
pF
pG
PC Nuc
pF
16

electron micrographs (range 2.1 to 3.1 µm). PALKOVITS *et al.* (1971a) obtained a mean nucleolar diameter of 3.6 µm in the cat.

As in all other cell types (HAY, 1968), the nucleolus consists of a dense conglomerate of granules and filaments. The composite nature of the Purkinje cell nucleolus is shown in Figs. 15 and 16. The pars fibrosa, composed of densely matted filaments with a sprinkling of dark granules, takes the form of several irregularly rounded lumps, about 0.4 µm in diameter, that are distributed throughout the nucleolus, even at its periphery. Each of these lumps has a compact dark shell, about 0.05 µm thick, and a much less dense center. Occasionally two or three such masses fuse to form one large lump with a single continuous shell. The pars fibrosa merges imperceptibly into the pars granulosa, which consists of myriads of small, extremely dense granules, 150–200 Å in diameter, suspended in a matting of fine filaments. Both the granules and the filaments are the same as those in the pars fibrosa; only the proportions are reversed. The pars granulosa is arranged as a tangle of coarse strands (about 0.1–0.15 µm thick) plaited loosely into a ball. The openings between the strands vary greatly in width, being 0.1 µm or more wide. They are often hexagonal or cylindrical spaces, occupied by filaments and particles resembling those in the karyoplasmic matrix. These inclusions may therefore be termed the third component of the nucleolus, the pars chromosoma.

Like the chromatin pattern, it is profitable to compare the nucleolar pattern of the Purkinje cell with its counterpart in the granule cell. In the latter the nucleolus is not recognizable in stained preparations for the optical microscope. It is made out with less difficulty in electron micrographs of thin sections (Fig. 63). The pars fibrosa appears as small, rounded lumps similar in size and composition to those in the Purkinje cell nucleolus, but they are individually dispersed in the granule cell nucleus and are not necessarily associated with the pars granulosa. This part is encountered seldom because it is evidently widely dispersed among the chromatin masses and is difficult to distinguish. The nuclear cytochemistry of these two cell types would reward further study.

◄

Fig. 16. The Purkinje cell nucleolus and nucleolus-associated chromatin. This prominent nucleolus is almost spherical in shape, and is composed of several masses of pars fibrosa (*pF*) interwoven with the pars granulosa (*pG*). A chromatin mass (*arrows*) is attached to the nucleolus. Such masses are frequently found in pairs in female rats. A fold of cytoplasm is invaginated into the nucleus (*PC Nuc*) but is bounded by the fenestrated nuclear envelope. Lobule V. ×49000

4. The Perikaryon of the Purkinje Cell

The cytoplasm of the Purkinje cell is characterized by a whorl-like arrangement of the organelles centered on the nucleus. This pattern is evident in preparations for the light microscope, but is most obvious in low power electron micrographs (Figs. 8 and 13). As the Purkinje cell has no axon hillock and gives off one or several thick dendritic trunks, the cell body bears almost no sign of the structural polarity that is usually shown by large neurons. Almost all of the organelles encircle the nucleus. In preparations stained with toluidine blue for optical microscopy, the Nissl bodies appear to be concentrated in the perinuclear cytoplasm, leaving the peripheral cytoplasm almost clear. The largest Nissl bodies lie close to the nucleus, especially against its dendritic pole where the cytoplasm invaginates the nuclear envelope (Figs. 11–15). At successively greater distances from the center, the Nissl bodies diminish in size and become quite inconspicuous in the peripheral cytoplasm. The long axes of the larger bodies tend to lie parallel to the nuclear envelope. The alternation of these flattened or ellipsoidal basophilic masses with apparently clear intervening spaces—the *Plasmastraßen* or "roads" (ANDRES, 1961)—impart a striated appearance to the cytoplasm. The mitochondria, confined largely to these *Plasmastraßen*, conform to the general pattern by aligning themselves more or less parallel to the nuclear membrane, and the Golgi apparatus, as is well known, forms a reticulate shell about the nucleus. Electron micrographs show that the microtubules and neurofilaments also follow circumferential trajectories in the *Plasmastraßen*. The whorl-like appearance of the cytoplasm suggests that the orientation of the organelles is related to the rotation of the nucleus, which is shown so dramatically in time-lapse cinematographic records of neurons in tissue culture (POMERAT *et al.*, 1967).

a) The Nissl Substance

Although the Purkinje cell is often cited as typical of large neurons, which usually have large block-like Nissl bodies, its Nissl bodies are generally small and are widely dispersed throughout the perikaryon. In sections stained with basic dyes, some Purkinje cells display a diffuse basophilia with only small, more intensely basophilic areas representing the Nissl bodies. Other Purkinje cells contain a few large Nissl bodies and a multitude of small ones, distinctly separated from one another by clear spaces. It is also readily apparent in electron micrographs

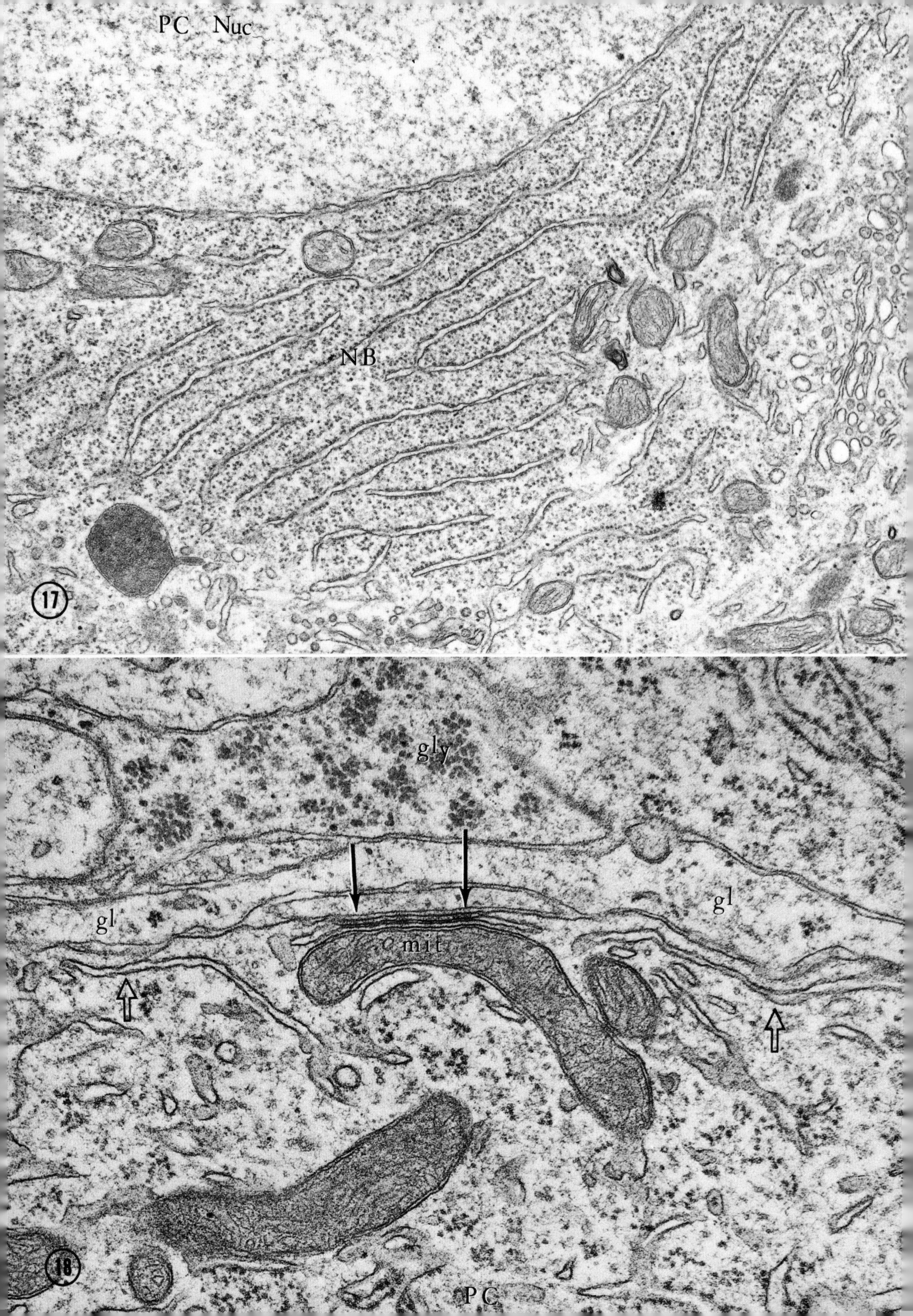
PC
Nuc
NB
17
gly
gl
mit
gl
18
PC

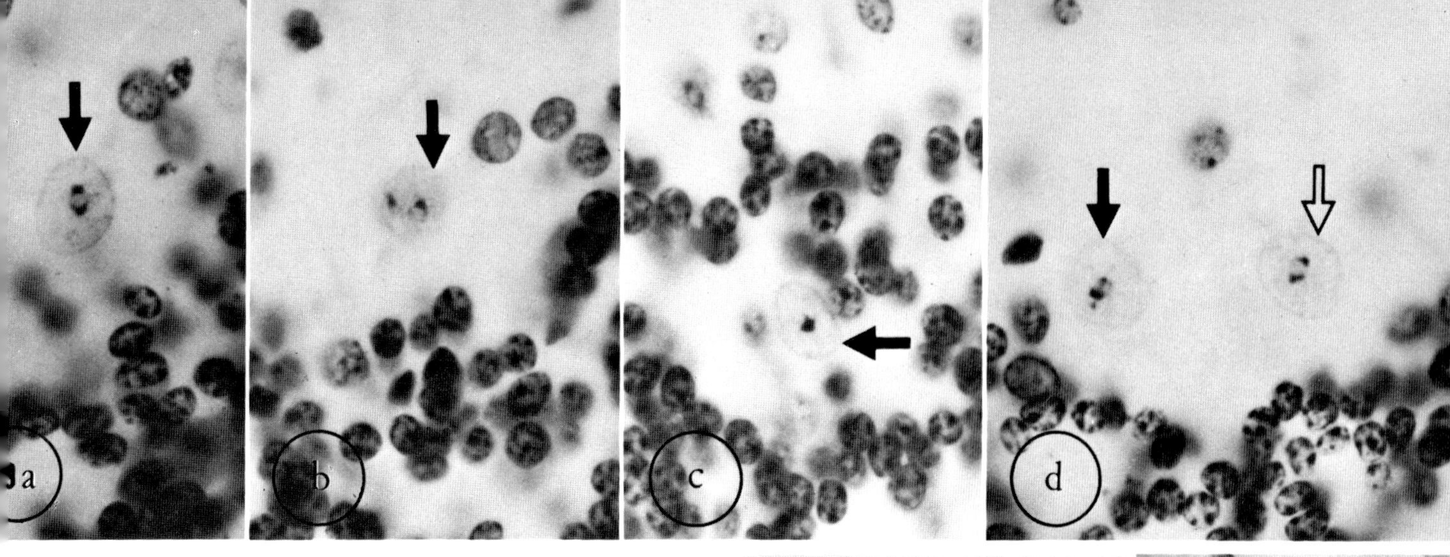

▲

Fig. 19a–d. Nucleolus-associated heterochromatin in the Purkinje cells of a male mouse. These heterochromatic bodies contain DNA and are found as: **a** a double body present on either side of the nucleolus; **b** a single body on each of two nucleoli; **c** a single, very large body on the nucleolus; **d** two asymmetrical bodies, one on each side of the nucleolus. (See text.) Feulgen reaction, male mouse, cerebellar cortex, 10 μm, paraffin section

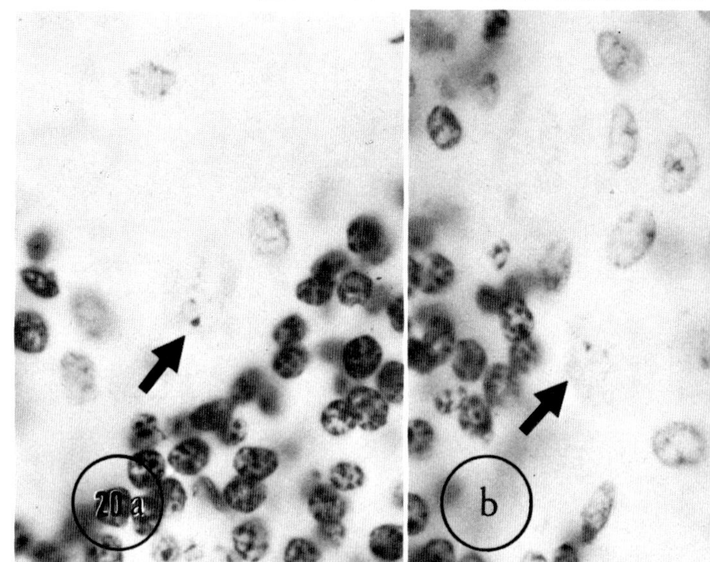

Fig. 20a and b. Two nucleolus-associated heterochromatin bodies in the Purkinje cells of (**a**) a female rat, (**b**) a male rat. (See text.) Feulgen reaction, rat cerebellar cortex, 10 μm, paraffin section ▶

of thin sections that the Nissl substance in all of these cells is highly diffuse, forming what amounts to either a close- or loose-meshed network throughout the cytoplasm.

As is well known, the Nissl substance in all nerve cells consists of the granular endoplasmic reticulum together with clouds of ribosomes that populate the surrounding matrix (PALAY and PALADE, 1955; PETERS, PALAY, and WEBSTER, 1970). The most variable component of this complex is, of course, the endoplasmic reticulum. The diverse arrangements of this organelle in the Nissl substance of an individual Purkinje cell often encompass the entire gamut of variations that can be encountered in the nervous system. The most highly ordered arrays are usually found in large masses located near the nucleus and making up the nuclear cap (Figs. 11–15), but they are not restricted to this region. They consist of from one to twelve, or even more, imbricated, broad, flat cisternae all parallel one with another and spaced about 0.2 μm apart. In those regular arrays situated in the perinuclear zone, the cisternae lie parallel to the nuclear envelope. Farther away they can be arranged in other planes, but there is a strong tendency even peripherally to conform to the general perinuclear parallelism. Anastomoses between overlapping cisternae are frequent, especially at the edges of the array where two or three cisternae can empty into a common pool or sinus. In the nuclear cap the innermost cisternae bend into the

◀

Fig. 17. The Nissl substance in the Purkinje cell. The Nissl body (*NB*) consists of imbricated branching and fenestrated cisternae of the endoplasmic reticulum associated with ribosomes, some attached to the membranes and some free in the cytoplasmic matrix. The Nissl body abuts against the nuclear envelope of the Purkinje cell nucleus (*PC Nuc*). Lobule X. × 34000

Fig. 18. Hypolemmal cisterna and a subsurface cisterna of the Purkinje cell. The Purkinje cell (*PC*) surface is surrounded by neuroglial cytoplasm (*gl*), part of which is well supplied with glycogen particles (*gly*). Deep to the plasmalemma, two elements of the hypolemmal cisterna (*large arrows*) are indicated. A third cisterna (*small arrows*) approaches very close to the plasmalemma and is also partially collapsed. The mitochondrion (*mit*) apposed to this subsurface cisterna completes an association that is characteristic of the Purkinje cell. Lobule V. × 64500

valleys formed by the wrinkled nuclear envelope, maintaining the usual regular distance of 0.2 μm between them (Fig. 17). The outer wall of the nuclear envelope throws out occasional tubular projections which join with the nearest cisterna of the endoplasmic reticulum.

The granular endoplasmic reticulum of the smaller Nissl bodies is much more pleomorphic and less regularly arrayed (Figs. 12, 13, and 18). One or two branching cisternae, contorted and ragged, suffice to form the skeleton of these bodies. The cisternae are sometimes rather strung out, joining at their ends with more orderly arrays or with agranular elements of the endoplasmic reticulum.

The cisternae are remarkably constant in depth, varying from about 200 to 400 Å. Greater depths occur at sites of confluence or branching. Cisternae throughout the granular endoplasmic reticulum are irregularly fenestrated by small circular interruptions, 400 Å or more in diameter. In the more orderly arrays these fenestrations are sometimes seen to be in register from one cisterna to another.

The outer surfaces of the cisternae in the Nissl bodies are studded with ribosomes, disposed in rows, double rows, coils, and rosettes. The fenestrations of the cisternae are typically marked by a chain of ribosomes laid out in a circle around the rim of the openings. The number of ribosomes in these linear arrays ranges from two or three up to twenty-six. Although it is impossible to determine the maximal length of these rows from micrographs of thin sections, the most common sequence is between six and twelve ribosomes long. Between the arrays, the surface of the cisterna is smooth. Regions of anastomosis or branching are also generally smooth.

The cisternae of granulated endoplasmic reticulum are immersed in a matrix containing multitudes of free ribosomes. In contrast to the ranked curlicues of the attached ribosomes, these polysomes are nearly all small clusters of four to six particles, often with one particle in the center of the group. The spaces between cisternae and surrounding their arrays are crowded with these small polysomes (Fig. 14). They occupy most of the cytoplasm that fills the invaginations at the dendritic pole of the nucleus. They pervade the cytoplasm, from the nuclear envelope almost to the very surface of the cell, always associated with some fragmentary elements of the granular endoplasmic reticulum, but extending far beyond the packed arrays of this membranous organelle. It is these free ribosomes that are responsible for the diffuse basophilia of the Purkinje cell perikaryon.

b) The Agranular Reticulum

The Purkinje cell possesses a highly developed agranular endoplasmic reticulum. Like the granular endoplasmic reticulum in the Nissl bodies with which it is confluent, the agranular endoplasmic reticulum forms a loose-meshed network throughout the perikaryon (Figs. 12, 13, and 18). In contrast to its granular counterpart, however, it extends into all parts of the Purkinje cell, even into the farthest reaches of its processes.

As was mentioned above, the limiting membrane of the endoplasmic reticulum in the Nissl bodies is not everywhere covered with ribosomes. Stretches of the membrane are free of these granules, especially the portions where the cisternae branch or anastomose. From these smooth or agranular regions, irregular cisternae and tubules arise that extend into the surrounding matrix and detach themselves from the compact masses of Nissl substance. This agranular endoplasmic reticulum is much less orderly than the granular variety. Its elements branch and anastomose more frequently, have a more irregular caliber, and pursue a course through the cytoplasm that is generally independent of the ribosomal population. In most Purkinje cells they are distributed singly or in small ragged groups throughout the perikaryon. They are more numerous, however, in the peripheral cytoplasm, where the Nissl bodies are likely to be small and the intervening spaces larger than in the immediately perinuclear zones. They appear in the electron micrographs as simple circular or elongated profiles, representing cisternae and long, stringy tubules of widely varying diameters that join up with one another, with the granular endoplasmic reticulum, the perinuclear cisterna, and the Golgi apparatus.

c) The Hypolemmal Cisterna

At the periphery of the cell the agranular reticulum spreads out into a broad, discontinuous (or irregularly fenestrated) cisterna that lies about 600 Å beneath the plasmalemma and parallel with it (Figs. 12 and 18). This hypolemmal cisterna is so well developed in the Purkinje cell that it may be considered a specific characteristic and can be used to identify fragments of this cell in thin sections, since it extends under the plasmalemma of the entire cell body, the dendritic tree, and the axon. Although it is represented in other large nerve cells by a few isolated cisternae, in no other cell is it so extensive as in the Purkinje cell. The hypolemmal cisterna was first recognized in 1969 in the Purkinje cells of the mormyrid fish, *Gnathonemus petersi*, where it is especially well developed

(KAISERMAN-ABRAMOF and PALAY, 1969). In the rat the cisterna is highly interrupted. Its components vary from short, almost spherical vesicles to long, flat profiles separated by greatly varying distances and all ranged in a single plane at a fairly constant distance from the cell surface. The cisterna receives frequent tributaries from the depths of the cytoplasm and from neighboring Nissl bodies. Occasionally it bears a row of ribosomes on its deep surface.

The hypolemmal cisterna constitutes a secondary membranous boundary within the plasmalemma. No other organelle intervenes between it and the surface membrane. At frequent intervals around the perimeter of the cell the hypolemmal cisterna approaches the inner face of the plasmalemma even more closely. The outer surface of the cisterna and the inner surface of the plasmalemma can be only 100–200 Å apart, the interstice being occupied by thin filaments extending across it. In these places the cisterna is often collapsed so that its lumen is almost entirely obliterated. Such cisternae occur in all nerve cells. They were first discovered in acoustic ganglion cells by ROSENBLUTH, who described a number of variations in their form (ROSENBLUTH, 1962). In nerve cells other than the Purkinje cell these "subsurface cisterns", as ROSENBLUTH named them, are usually the only representatives of the hypolemmal cisterna.

In the Purkinje cell a curved mitochondrion is frequently applied against the inner face of the subsurface cisterna (Fig. 18), an association first reported by HERNDON (1963). The interval between the two organelles is about 100–200 Å deep. Sections passing normal to the plasmalemma in these locations show a striking complex of five parallel membranes: the inner and outer mitochondrial membranes, the walls of the subsurface cisterna, and the plasmalemma, all about equal distances apart or with the collapsed membranes of the cisterna forming a dense line in the center of the complex. Sometimes the plasmalemma of the neuroglial cell covering the nerve cell adds a sixth parallel membrane. Because of the parallelism of all of these membranes, oblique sections through this complex appear as conspicuous dark patches. As HERNDON noted, there is no consistent association of these subsurface complexes with any particular structure on the surface of the Purkinje cell. Although nerve terminals can cover these structures, synaptic junctions do not occur immediately over them. It is worth noting that while the hypolemmal cisterna extends throughout the dendritic and axonal arborization of the Purkinje cell, including the myelinated axon, the association of mitochondria with subsurface cisternae does not occur beyond the proximal segments of the major dendritic trunks.

d) The Golgi Apparatus

In specimens specially impregnated with heavy metals the Golgi apparatus of the Purkinje cell appears in the optical microscope as a complicated network enclosing the nucleus as if in a cage. It is the archetype of the Golgi apparatus, having been described and illustrated by GOLGI in the primary publications of 1898 (a and b). This structure is resolved in electron micrographs of thin sections as a set of compact aggregates of parallel, flattened cisternae and assorted vesicles scattered over a ring-shaped field midway between the nucleus and the plasmalemma (Figs. 11–13). The larger of these assemblages correspond to nodal points and heavy trabeculae in the Golgi apparatus of light microscopy, while the smaller ones correspond to the tenuous connecting links. Each complex is organized according to the same plan: a compact stack of broad, flat cisternae surrounded by variable numbers of small vesicles. Grazing sections show that some cisternae are pierced with regularly spaced holes (Figs. 21 and 22). As a result, they appear in normal sections as a row of vesicular profiles like a string of pearls. In tangential sections it can be seen that the fenestrations are generally circular in outline, but some are elliptical or even rectangular. Their diameters range from 400 to 700 Å, and a few can measure up to 950 Å across. They are distributed at intervals of 1500 to 2000 Å in fairly regular arrays. These perforate cisternae are mingled with broad, continuous cisternae to form curved or circular enclosures. Small vesicles, vacuoles, multivesicular bodies, and lysosomes are dispersed within these fields and in the surrounding matrix. In some somatic cells the substructures in the Golgi apparatus are polarized so that the flat cisternae face the nucleus and the vesicular elements are restricted to the outer face of the aggregate. In the Purkinje cell there is no consistent arrangement, and it is very difficult to decide which face of the complex should be designated as external, since vesicles occur on both sides.

The margins of the Golgi complexes are also not very sharp. The wide dispersal of the Nissl substance as well as the ordinary agranular reticulum and the Golgi complexes brings all of these reticular organelles into close proximity. Confluence of their cisternal and tubular components can be demonstrated at the borders of the Golgi complexes.

Although the cisternae in the Golgi complexes rarely anastomose with one another or branch, they frequently display rounded protuberances that suggest budding or confluence of vesicles. However, such forms are often presented by cisternae in the middle of a stack, where no

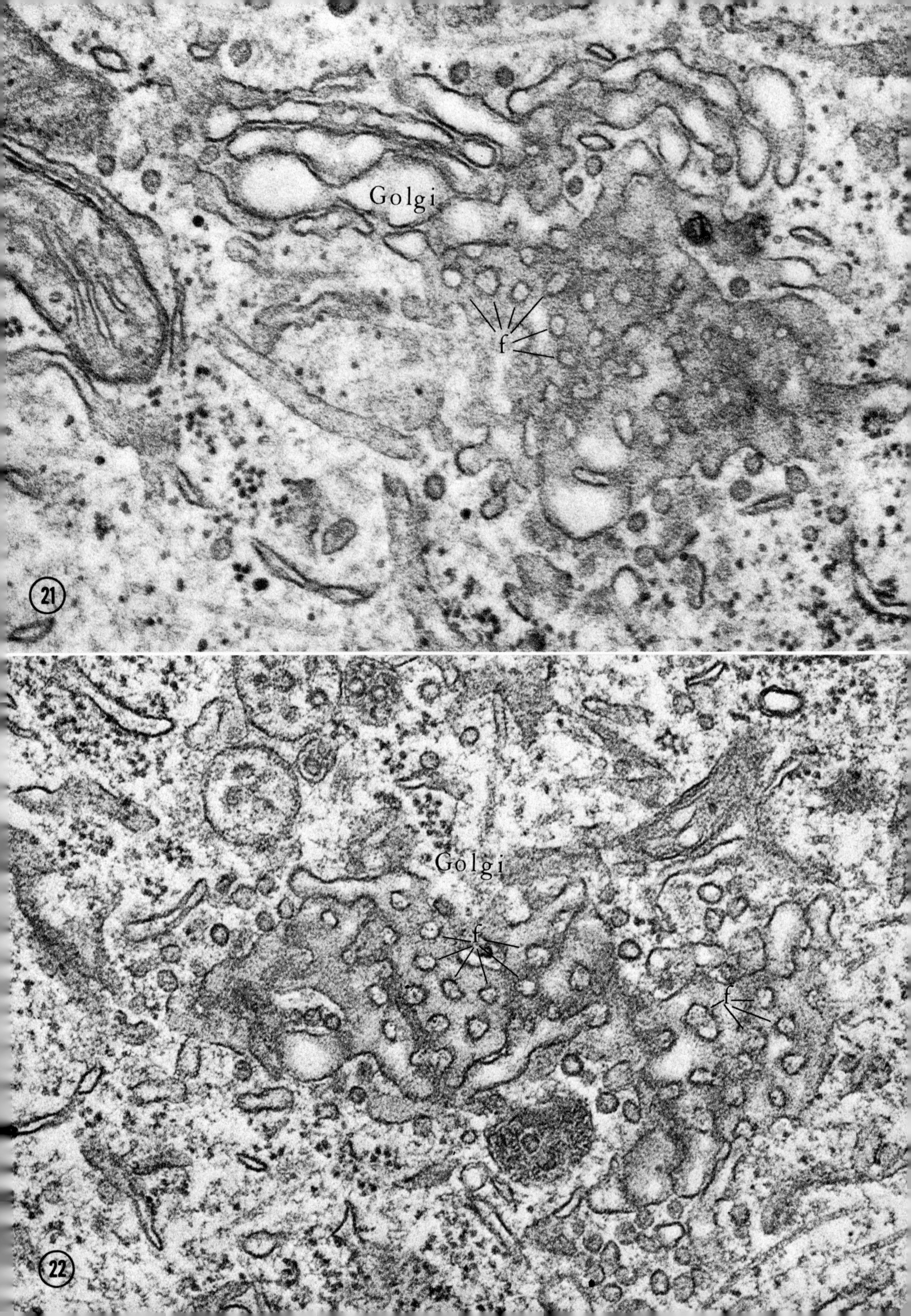
Golgi
f
21
Golgi
f
f
22

free vesicles are to be seen. The vesicular components of the Golgi apparatus are of three types: small, rounded vesicles with clear centers, small vesicles with dense centers, and alveolate vesicles. The small, round ones are about 200–400 Å in diameter, and their number varies a great deal from one aggregate to another in the same cell. They usually cluster about the periphery of the aggregate. Vesicles with dense centers are not common in Purkinje cells, and they occur not only in the Golgi apparatus but also in the Nissl substance. They measure about 400–600 Å in diameter, and the dense core is usually separated from the wall of the vesicle by a clear halo. Alveolate vesicles are very common in Purkinje cells (PALAY, 1963). They occur singly or in clusters, and they are often encountered attached to tubules or broad cisternae of the Golgi complex, where they appear to be either fusing with these structures or detaching from them. Alveolate vesicles are also commonly found at the surface of the Purkinje cell body and its dendrites. Recently it has been suggested that they are involved in the assembly of somatic and dendritic attachment sites for synaptic junctions (ALTMAN, 1971). All of these vesicular structures have been implicated in the process of forming lysosomes (FRIEND and FARQUHAR, 1967; HOLTZMAN *et al.*, 1967; NOVIKOFF, 1967a and b; HOLTZMAN, 1971).

e) Lysosomes

Purkinje cells are particularly rich in lysosomes (Figs. 11 to 13). In young adult animals, which have been extensively studied for this book, these inclusions are scattered all over the perikaryon. They vary from small, dark vesicles, 0.1 µm in diameter, to large, irregular masses, several microns in diameter. The larger ones are visible in the optical microscope and have been mistaken for neurosecretory granules because of their capacity for staining with chrome alum haematoxylin (SHANKLIN *et al.*, 1957). Lysosomes all contain a dense, finely granular material that obscures the limiting membrane unless it is sectioned normally. Subsidiary inclusions in the form of spherical lipid droplets, stacks of dense parallel membranes, and crystalline arrays of fine granules are common. With advancing age of the animal the lysosomes become larger and more complex, as lipofuscin pigment accumulates in them (SAMORAJSKI *et al.*, 1965). As in other cells, the lysosomes of Purkinje cells contain acid hydrolases.

f) Mitochondria

The mitochondria of the Purkinje cell are among the most pleomorphic in neurons. They vary in diameter from 0.1 to 0.6 µm, and most of them fall into the lower half of this range, but occasionally a giant, one micron across, is encountered. Their length is also varied, but this is more difficult to determine from sectioned material. Most of them are rather short, about 2 µm long or less, but again elongated mitochondria, more than 4 µm long, are not infrequent. The shapes of the profiles of the larger mitochondria are highly diversified. Many are contorted, bent, or branched. In some cells small, round profiles about 0.2 µm in diameter seem to predominate. As may be seen by study of Figs. 12 and 13, the mitochondrial profiles tend to occur in clusters of three to six members. Some clusters consist mainly of rounded profiles all about the same size. Others are more varied, and clusters consisting of a larger, bent or branched profile and four or five small, round ones are not uncommon. The significance of these patterns is revealed by a study of serial sections. Even two or three neighboring sections are sufficient to show that each cluster of profiles is a section through a single, highly ramified, or tangled mitochondrion. The various shapes in sections are simply the result of random orientation of the mitochondria with reference to the plane of section. Evidently a common shape is that of a hand with the digits extended in parallel fashion. This shape is characteristic of the perikaryon and is hardly ever found in the processes of the Purkinje cell. As might be expected from such complex profiles, the orientation of the cristae inside the mitochondria is also highly irregular. In the small, round profiles and many of the elongated ones, the cristae are transversely oriented, whereas in the larger profiles longitudinally oriented cristae are common. Profiles having a mixture of the two orientations, as well as profiles with fenestrated or tubular cristae, are frequently seen.

Although the mitochondria are generally excluded from the Nissl bodies and the Golgi complex, they nevertheless have close topographic relations with the agranular endoplasmic reticulum and even with the granular reticulum. It could hardly be otherwise in the crowded cytoplasm of the Purkinje cell. It is common to see a long mito-

◄

Figs. 21 and 22. The fenestrated cisternae of the Golgi apparatus of the Purkinje cell. Grazing sections tangential to the surface of the broad, flat cisternae of the Golgi apparatus show the fenestrations (*f*) that are characteristic of these membranes. They appear as circular, elliptical, or somewhat rectangular profiles with diameters ranging between 400–700 Å. They are distributed in a regular array about 1500–2000 Å apart. Lobule IX. ×71000

chondrion accompanied by a cisterna of the agranular reticulum, sometimes almost entirely enclosed by it. A special example of this configuration appears at the cell surface and has already been noted above. Mitochondria hooked around tubular forms of the endoplasmic reticulum are frequently encountered, as if during their migrations they had been caught on a snag in the stream.

g) Microtubules and Neurofilaments

Arranged in a roughly perinuclear fashion, the microtubules and neurofilaments fill up the interstices between the other organelles. In most cells they are relatively inconspicuous, whereas in some, with small amounts of Nissl substance concentrated around the nucleus, they are much more prominent. Of the two filamentous organelles the microtubules are the more noticeable in the perikaryon, to some extent because of their larger caliber but also because they are more numerous (Figs. 12 and 13). There is also a diffuse cottony material dispersed throughout the matrix of the cytoplasm and more salient in the *Plasmastraßen*, where the more organized organelles are absent. This fuzzy material adheres to the microtubules and the neurofilaments, and may be chemically related to both of these components.

5. The Dendrites of the Purkinje Cell

The dendritic tree of the Purkinje cell arises from one to four trunks that issue out of the apical pole of the cell body. The trunks extend directly outward or at an angle toward the surface of the folium, depending upon the location of the cell in the folium. Cells at the crests of the folia or in the depths of the sulci tend to have more oblique and widespreading dendritic trunks than those on the sides of the folia, where in any case the more typical examples are always situated. The trunks before the first major division are termed primary dendrites. If there is only one primary dendrite, it branches into two or more secondary dendrites that tend to run horizontally, parallel with the folial surface. Tertiary and successive branches given off from these rise toward the surface or curl downwards toward the granular layer, forming an extraordinarily rich arborization that extends through the whole thickness of the molecular layer. The pattern of branching of the Purkinje cell dendrites is usually described as dichotomous. It is evident, however, from a glance at Figs. 6 and 7 that this term can be applied only to the major divisions. Both the trunk and the secondary and tertiary dendrites give off three, four, or five branches of different calibers at a single branch point. The smaller or narrower branches are usually terminal and subdivide in the immediate vicinity of their origin. The heavier branches continue to ascend and divide again in the same manner. The terminal branchlets are 1–2 μm in diameter, and are fitted with numerous spines or thorns, which project from all sides, like the bristles on a bottle brush. For this reason they were called spiny branchlets by Fox and Barnard (1957). The general form of all branches, both large and small, is gently curving, away from the parent branch. Terminal dendrites in the lower third of the tree tend to curve downwards, the lowest ones filling up the intervals between Purkinje cell bodies, while those in the upper two thirds tend to curve upwards toward the pial surface. Many of the terminal branchlets close to the surface turn downward again. It should be noted that terminal branchlets are given off by secondary and tertiary dendrites directly as well as by their final subdivisions.

a) The Form of the Dendritic Arborization

The most remarkable characteristic of the Purkinje cell dendritic tree is its three-dimensional form. It is spread out in a vertical plane at right angles to the longitudinal axis of the folium, and is therefore displayed best in parasagittal sections (Stieda, 1864). In this plane the tree extends for 300 to 400 μm, while in the longitudinal axis of the folium it is only 15 to 20 μm wide (see Fig. 10). It therefore resembles an espaliered fruit tree, as Ramón y Cajal (1911) remarked, or perhaps more aptly, it resembles the clipped plane trees that line the country roads of southern France. The arborization is so rich and the terminal branchlets appear so crowded in Golgi preparations that there seems to be little room for the other components of the molecular layer that lie in the same region. But on careful examination with continuous manipulation of the fine focus, it becomes obvious that there are many open spaces in the tree, some large enough to accommodate blood vessels; the cell bodies of basket, stellate, or neuroglial cells; and parallel fibers on their way through the tree. Drawings and photographs of the Purkinje cell dendritic tree tend to exaggerate the flatness of the arborization, because they project it upon the plane of the paper. It is necessary to study the original preparations to acquire a true appreciation of its three-dimensional configuration.

Fig. 23. Thorns on dendrites of a Purkinje cell. The dendrites in this illustration are studded with many thorns all over the surface of each spiny branchlet (*black arrow*). The larger trunks of these dendrites also bear thorns of a stubbier variety (*open arrows*). Climbing fibers synapse upon thorns of this second type, whereas parallel fibers synapse with the thorns of the spiny branchlets. Rapid Golgi, 90 µm. × 1 500

b) Dendritic Thorns

The second remarkable characteristic of the Purkinje cell dendritic tree is its rich complement of thorns (Figs. 7 and 23). These delicate appendages projecting from all of the smaller dendrites were discovered by RAMÓN Y CAJAL (1891) in the cerebral cortex as part of the rich harvest from his first investigations of the brain. He later wrote (1911, p. 10) that the dendritic thorns of Purkinje cells form a distinguishing characteristic of these cells, "for in no other cells are they so short, so stout, or so abundant." FOX and BARNARD (1957) estimated that each Purkinje cell in the monkey had on the average 60 000 thorns (later the estimate was doubled, FOX, SIEGESMUND, and DUTTA, 1964). They found two to three thorns per linear micron of spiny branchlet. PALKOVITS *et al.* (1971 c) calculated that the average Purkinje cell in the cat should have about 80 000 thorns, and they found 4.5 thorns per linear micron of dendrite. SMOLYANINOV (1971), however, recorded much lower values: in man, 60 000; monkey, 34 000; cat, 18 000; rat, 7 500; and mouse, 4 000. His calculations indicated a density of only 1 to 1.5 thorns per linear micron of dendrite in all animals. Apparently he failed to take into account the thorns projecting from the front and back of the dendrite, and he did not distinguish between the smooth and spiny parts of the dendritic tree.

In order to make comparisons with published figures on other animals, we obtained direct counts of spines in Golgi preparations of the white rat. We made accurate camera lucida drawings, at a magnification of 1 800 ×, of two Purkinje cells selected from neighboring folia for the apparent completeness of the impregnation and their orientation in the plane of the section. Similar measurements were made on both cells. All the branches of the dendritic tree were traced on drawing paper and every thorn indicated. The lengths of the smooth and spiny portions of the tree were measured with a cartographer's wheel and adjusted for the magnification. The number of thorns per unit length of dendrite was obtained by averaging the number of thorns noted on ten 10 µm stretches of spiny branchlets selected for displaying their appendages clearly. The total number of thorns was calculated by multiplying the total length of spiny dendrite by the number of spines per unit length. In addition, camera lucida drawings were made at 1 800 × of segments of spiny branchlets in order to estimate the distance between thorns. One hundred measurements were made on one cell, and 80 on the other, of the distance from the center of a thorn to its adjacent neighbor, and the measurements were averaged. The results are given in Table 4. The similarity between these two cells is impressive. Even the total lengths of the dendritic trees are almost identical (10 622 and 10 978 μm, respectively). It is also worth noting that the proportion of the dendritic tree occupied by spiny branchlets is 95.4% in one cell and 88.8% in the other. The number of thorns per cell recorded in our study is more than twice the number obtained by SMOLYANINOV (1971) in the same species, but is only about one fourth the number reported by PALKOVITS *et al.* (1971 c) in the cat. Both the density of thorns (number per unit length of dendrite) and the total length of the spiny dendrites are greater in the cat.

It should be noted, however, that counts of thorns made on Golgi preparations are fraught with difficulties. The best method is to trace the outline of the dendrite, as we have done, with the aid of a camera lucida at high magnification, and then to do the counts on the drawing. Even so, it is not easy to count every thorn. Those that project from the near and far sides of the dendrite are necessarily obscured by the shaft and may be omitted. Local differences in the impregnation can obscure neighboring thorns (CHAN-PALAY and PALAY, 1972 b), and bent or twisted ones can appear as mere nubbins that escape recognition. All of these difficulties tend to lower the counts from the true values. But calculating the number from sample counts tends to an overestimate because the thorns are not distributed uniformly. Some of these factors can be corrected by making counts on electron micrographs, but here the samples are so small that only corroborative estimates could be obtained. Determining the number of thorns on a Purkinje cell is not merely an arid exercise in quantitation. Since each thorn is the site of a synaptic junction, an estimate of the number of thorns gives at least a minimal estimate of the number of synapses impinging upon the dendritic tree. In this particular case, each synapse is probably made by a different parallel fiber. Consequently, the number of thorns indicates the wealth of information bombarding a single Purkinje cell and suggests the immensity of the task assumed by any investigator bent on understanding the physiology of a Purkinje cell.

Table 4. Sample measurements on dendritic trees of two Purkinje cells

	Cell A	Cell B
Total length of smooth branches	489 μm	1 228 μm
Total length of spiny branches	10 133 μm	9 750 μm
Average number of thorns per 10 μm unit of spiny branch	17.8	17.1
Total number of thorns per cell	18 037	16 672
Average distance between centers of thorns	1.33 μm (0.55–3.0 μm)	1.10 μm (0.55–1.94 μm)

A careful study of Golgi preparations reveals that the Purkinje cell dendritic tree possesses two kinds of thorns. The first type is the more common, and is the only one that has been considered in the estimates given above. This type is composed of the very numerous projections from the spiny or terminal branchlets, and is the one that articulates with the parallel fibers. It has a slender, thread-like stem and terminates in a rounded knob. In photographs, such as Fig. 23, it is difficult to record the stem unless the thorn appears in profile. The thorns seldom project at right angles to the dendritic stem, but are tilted or bent at various angles. RAMÓN Y CAJAL (1911, p. 10) records that in the mouse the thorns projecting from neighboring spiny branchlets come so close together as to be nearly touching, and he declares that the little spaces between them could hardly contain more than two or three parallel fibers. In man, he says, the intervening spaces are larger and could probably allow four or six fibers to pass through. RAMÓN Y CAJAL (1911, p. 10) also states that the thorns "almost all originate from the sides of the dendrites." A few project from the faces of the dendrites and are therefore visible in sections parallel to the longitudinal plane of the folium. This description cannot be borne out by a study of our Golgi material with high magnification oil immersion lenses. Furthermore, electron microscopy of thin sections shows that thorns originate all over the surface of the terminal branchlets in a roughly spiral pattern. Although similar thorns are given off by the larger dendrites here and there, in general this type of thorn is restricted to the spiny branchlet.

The second type of thorn is also visible in Fig. 23. It has a shorter stem, sometimes none at all, and a somewhat larger head than the thorns just described. Consequently, it appears as a stubby protuberance on the surface of the dendrite. Thorns of this second type are restricted to the larger dendrites, the primary and secondary branches of the main trunks, and they are the exclusive sites for the synapses of climbing fibers (see Chap. X).

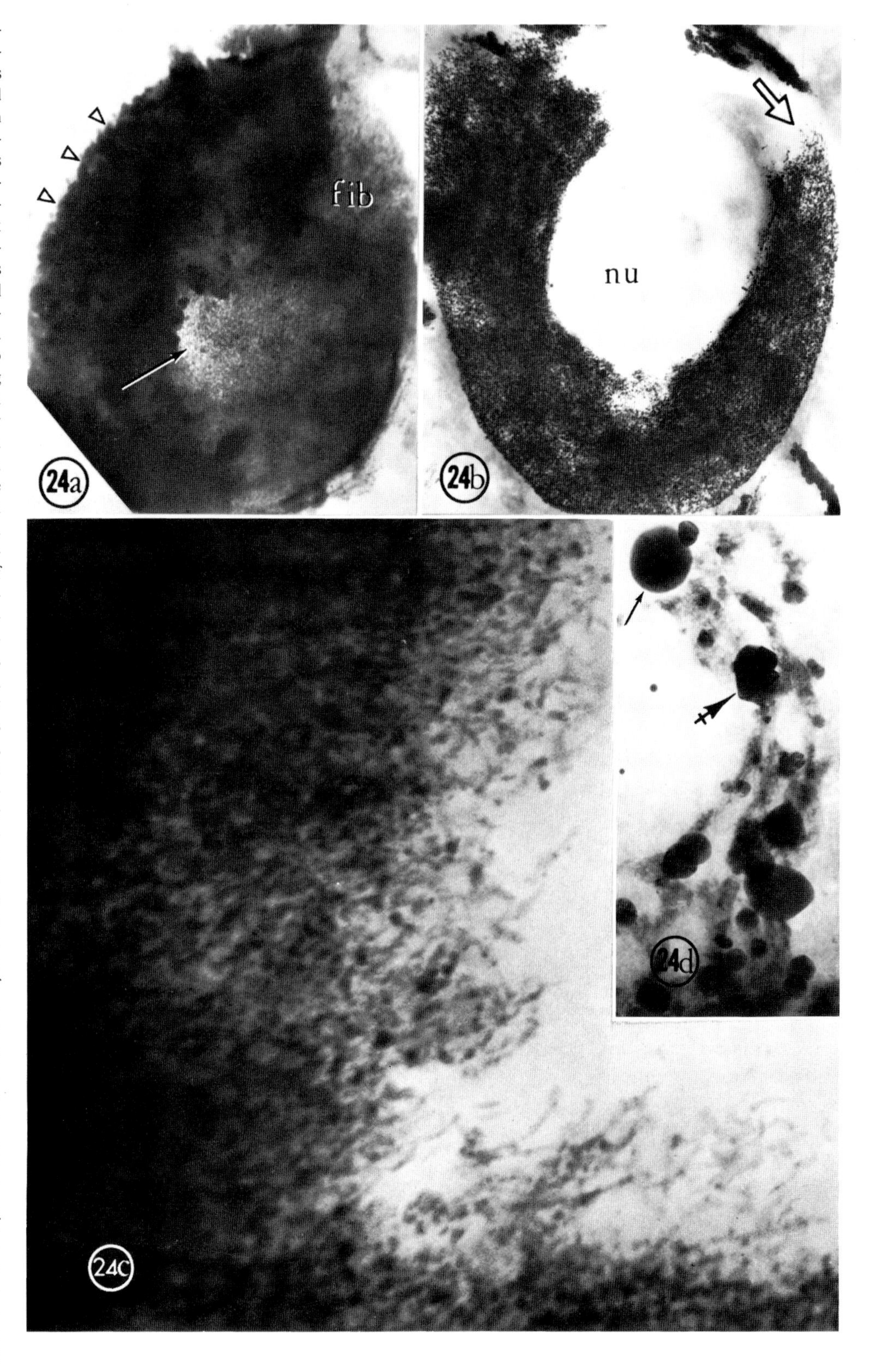

Fig. 24 a–d. Rapid Golgi impregnations in Purkinje cell by high voltage electron microscopy. **a** shows the surface of an impregnated Purkinje cell body that has been grazed by the knife during sectioning. The left half of this cell is heavily incrusted with conglomerates of globular crystalline particles (Δ). This crust is almost black under the light microscope, whereas the right half of the cell appears deep red. This part of the cell shows the typical fibrillar meshwork that fills the cell matrix (*fib*). A light area between the two halves shows through, indicating the location of the unimpregnated nucleus (*arrow*). Rhesus monkey cerebellar cortex, 5 μm. ×3700. **b** shows a thinner section through another Purkinje cell body whose matrix has been well impregnated with only the fibrillar meshwork. The filaments are bounded externally by the plasma membrane of the cell and internally by the membrane of the unimpregnated nucleus (*nu*). The details of the crystals are best examined at the edges of the fibrillar meshwork. The area at the tip of the arrow is greatly magnified in Fig. 24d. Rhesus monkey cerebellar cortex, 1 μm, lobule X. ×2625. **c and d** present at a higher magnification the fibrillar meshwork in the cellular matrix of a Purkinje cell impregnated red with the rapid Golgi method (high voltage electron micrographs). **c** shows that the meshwork is a complex reticulum of dense, branched, filamentous rodlets 410–440 Å in diameter. It is peppered with an assortment of small (250–600 Å), very dense globules. Rhesus monkey cerebellar cortex, 5 μm. ×15000. **d** shows a portion of this fibrillar material spread out in the section. Each fibril or rodlet has a patchy distribution of density along its extent, and the patches have a filamentous substructure. Each fibril is approximately 0.5 μm long; however, there is a great deal of overlap of one fibril with another. The crystals that pepper the fibrils are usually globular (*arrow*) and occasionally polyhedral (*crossed arrow*). Rhesus monkey cerebellar cortex, 1 μm, lobule X. ×31000

These thorns are clustered together in elongated groups of five or six, leaving most of the dendritic surface smooth. Since these thorns are a relatively new discovery, they were not noted in the classical literature, which always described the large dendrites as being smooth, in contrast to the spiny branchlets. Larramendi and Victor (1967) were the first to call attention to them, in the course of an electron microscopic study of climbing fibers in the cerebellum of the mouse. The first photomicrographs of them in Golgi preparations were published even more recently (SOTELO, 1969, in frog; MUGNAINI, 1970 and 1972, in cat). Their existence has now been confirmed in a variety of species, including the rat.

High Voltage Electron Microscopy of Dendritic Thorns. Examination of rapid Golgi preparations in the high voltage electron microscope sheds an interesting sidelight on the form and number of the dendritic thorns. In two recent publications (CHAN-PALAY and PALAY, 1972b and c), we have shown that the impregnation of cells in rapid Golgi preparations takes the form of fine fibrils and small particles deposited in the cytoplasmic matrix of certain cells and their processes. Continued accretion of fibrillar material results in a dense mattress of filaments, which gives the cells and their processes a deep red color in the light microscope (see Fig. 24a–d). The nuclei, mitochondria, and probably synaptic vesicles are not impregnated. Further development of the impregnation results in the formation of small crystalline particles, most of them globules and some polyhedral, on the surface of the structure. These grow into larger individual globules, which then agglomerate to form a crust around the impregnated structure. A heavy black impregnation results when this second stage is advanced.

The accumulation of globules on the surface of small structures like thorns can cause them to appear larger in the light microscope than they really are, particularly because the individual globules are not resolved by optical microscopy even with the best oil immersion lenses. For example, in Fig. 25a, a large globule located near the top of the stem of thorn t_1 would make the stem appear shorter than it really is (Figs. 25b and c). A short thorn might be obscured altogether in the incrustation of the dendritic stalk. This could explain why the thorns that articulate with climbing fibers were not seen earlier. At magnifications comparable to those of light microscopy, high voltage electron micrographs of red Golgi impregnations show Purkinje cell thorns as delicate bulbs tethered to the stalk of the dendrite by slender stems. They are tilted and bent, leaning toward one another or diverging, rather irregularly, like bristles on a bottle brush. Stereoscopic pairs of high voltage electron micrographs give an enhanced impression of this brush-like formation. The thorns appear similarly in the optical microscope (Figs. 23 and 25f). Black impregnations, however, result in stubby thorns with thick bases (Fig. 25g). Such images would account for RAMÓN Y CAJAL's now surprising characterization of Purkinje cell thorns as short and stout that was cited above (RAMÓN Y CAJAL, 1911). Another situation that can affect the apparent shape and number of thorns in Golgi preparations is shown in Fig. 26b. Here the spiny branchlet is enshrouded with tattered velamentous neuroglial processes. At the level of optical microscopy the thorns may not be discernible under this covering.

c) The Fine Structure of Dendrites and Thorns

The origin of a major dendrite of a Purkinje cell may be studied in Fig. 27. At the point where the trunk issues from the apical pole of the perikaryon there is a more or less abrupt shift in the distribution and arrangement of the cytoplasmic organelles. The Nissl substance detaches itself from the diffuse continuum of the perikaryon and appears in progressively smaller separate fragments. Although it can still be found at considerable distances from the cell body, and it appears as triangular masses lodged at the bifurcation points of the larger dendrites (Fig. 28; and RAMÓN Y CAJAL, 1909, p. 167), the Nissl substance rather quickly declines to an unimportant station in the list of dentritic components. In Nissl preparations the dendrites of the Purkinje cell are almost colorless and can not be followed beyond the first few subdivisions. The disappearance of the Nissl substance produces a sensible clearing of the cytoplasm that is evident in electron micrographs as well as in Nissl preparations. In this clearing, which begins at the very onset of the dendrite, it can be seen that the predominantly circular or perinuclear orientation of the organelles in the cell body gives way to an alignment parallel with the major axis of the dendrite. This longitudinal alignment is especially well shown by the mitochondria, the neurofilaments, and microtubules, but it is also displayed by the tubules of the agranular endoplasmic reticulum and even by the portions of the Golgi apparatus that enter into the dendrite (Fig. 29). The microtubules and neurofilaments run into the tapering dendrite from all directions, as if drawn into a stream pouring through a funnel. All other structures are trapped in the same stream and soon conform to the prevailing longitudinal disposition.

The agranular endoplasmic reticulum forms a loose meshwork with predominantly longitudinal tubules and narrow cisternae running the whole length of the dendrites and entering into their most delicate ramifications (Fig. 28). Irregular, branching tubules and cisternae run transversely at various angles to connect these longitudinal members with one another and with the hypolemmal cisterna. The endoplasmic reticulum of the Purkinje dendrites displays two curious patterns in relation to the microtubules: (1) a broad cisterna (Fig. 30) regularly fenestrated with rounded perforations, 500–600 Å across, through the

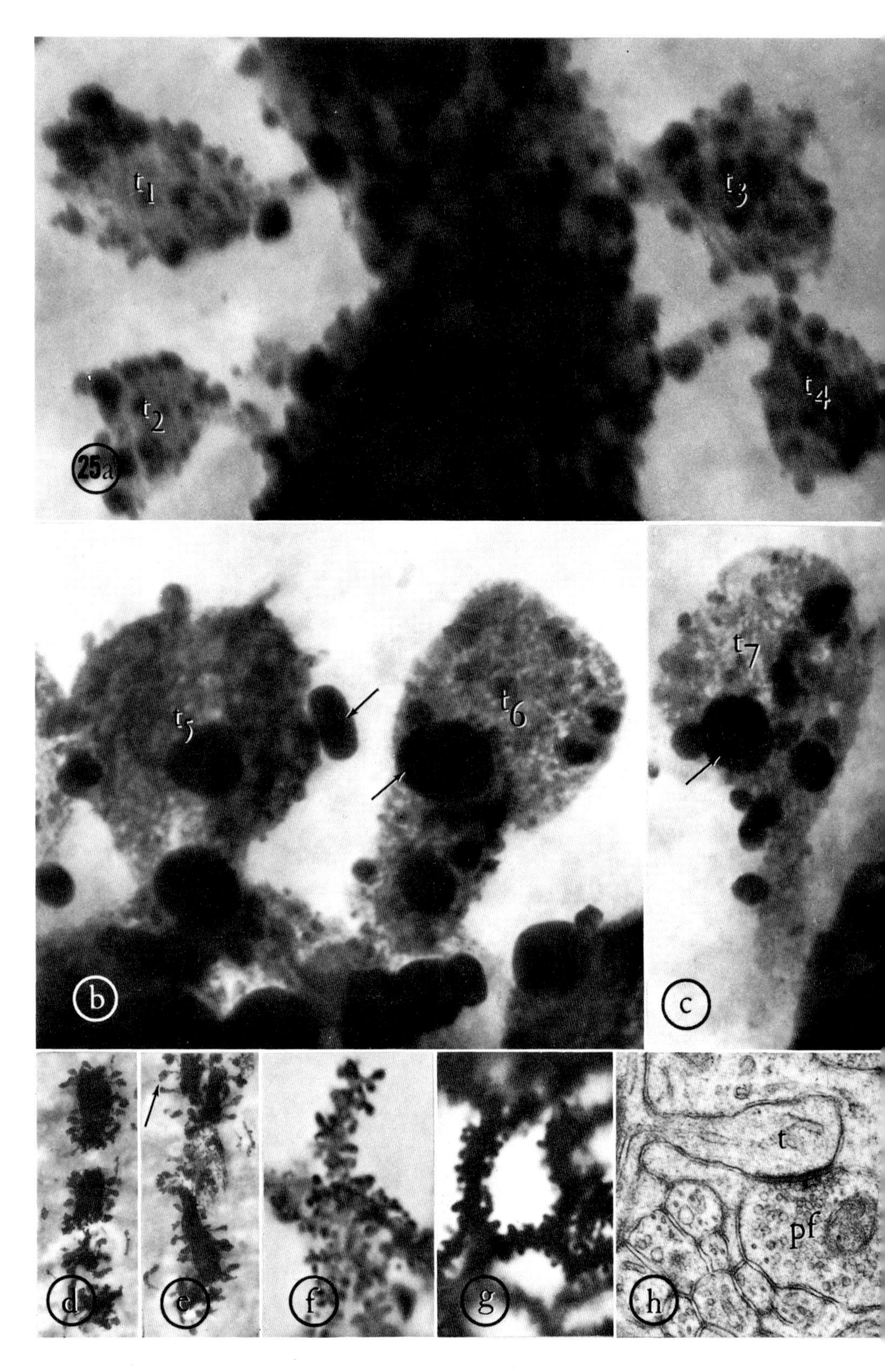

Fig. 25a–h. High voltage electron micrographs of spiny branchlets and dendritic thorns of Purkinje cells after rapid Golgi impregnation. The thorns (t_1–t_7) and their thin necks are clearly defined by the fibrillar background meshwork. Upon this meshwork globular particles are superimposed. The particles are sometimes rather large (*arrows*) and can be indiscriminately placed over the surface of the thorn. Because these globules are generally less than 3000 Å in diameter and very close together, they are not resolved as individual particles in the light microscope. In **a** one thorn (t_1) has a large globule at the base of its head, suggesting that in the light microscope the neck of this thorn might appear to be shorter and thicker than it actually is. Rhesus monkey cerebellar cortex, 5 µm. ×34000. **b and c** show thorns with globules incrusting their necks. These thorns might appear to be stouter with a thicker neck in the light microscope than in these high voltage electron micrographs, where the background meshwork can be readily discerned. Rhesus monkey cerebellar cortex, 1 µm, lobule X. ×52000. **d and e** Low magnification high voltage electron micrographs of Purkinje cell spiny branchlets after rapid Golgi impregnation. These dendrites were reddish brown in the light microscope. Each thorn is resolved clearly with head and thin neck portions. Even the surface globules that occasionally stud the thorn surface can be resolved at this magnification (*arrow*). Rhesus monkey cerebellar cortex, 1 µm, lobule X. ×2100. **f** Light micrograph of Purkinje cell dendritic thorns (spiny branchlets) after a red rapid Golgi impregnation. The thorns studding these dendrites show delicate, slender necks, although naturally without the resolution achieved in Figs. (d) and (e). Rat cerebellar cortex, 90 µm, nitrocellulose embedding, lobule X. ×2000. **g** Light micrograph of Purkinje cell dendritic thorns (spiny branchlets) after a black rapid Golgi impregnation. The thorns on these dendrites are stumpy, and few or no necks are apparent; the processes are coarser and heavier than those in Figs. (d) and (e). (Compare these with the drawings of Ramón y Cajal, 1911, Vol. II, Figs. 9, 10.) Rat cerebellar cortex, 90 µm, nitrocellulose embedding, lobule X. ×2000. **h** A dendritic thorn on a Purkinje cell spiny branchlet from thin section electron microscopy. This thorn, illustrated at a magnification less than that of the thorns in Fig. 25a, shows the fine structural morphology of the thorn and its surrounding neuropil. The thorn (*t*) synapses with the surface of a parallel fiber varicosity (*pf*) through a typical Gray's type 1 junctional complex (Gray, 1959). Rat cerebellar cortex, lobule X. ×31000

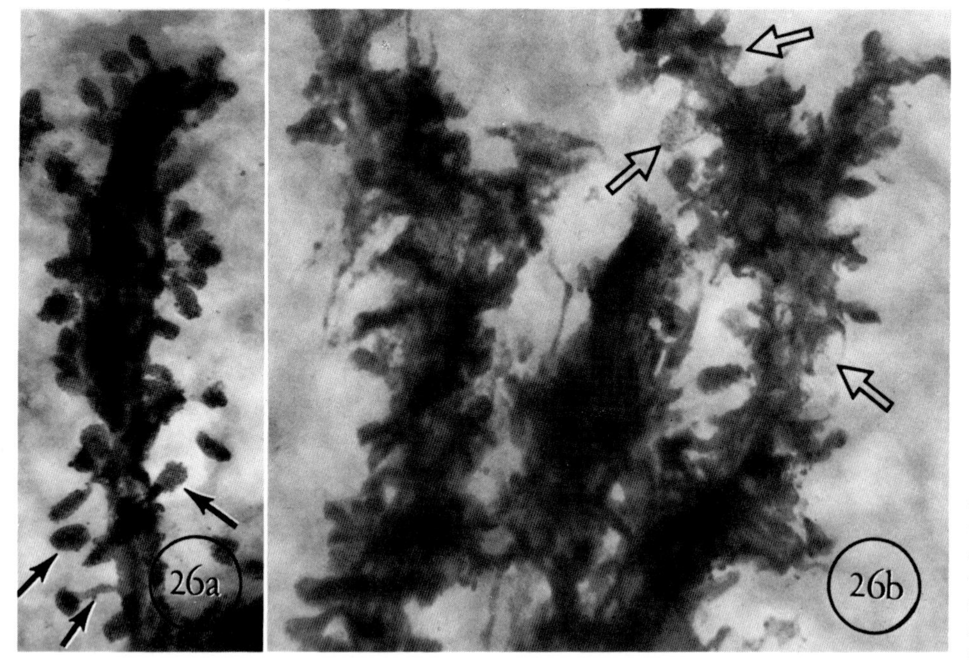

Fig. 26. a Spiny branchlet and dendritic thorns of a Purkinje cell dendrite. The dendrite and thorns (*arrows*) borne on its surface have been impregnated by the rapid Golgi method. The neuroglial sheath formed by processes of Golgi epithelial cells, however, have not been impregnated. Compare this with Fig. 26b. High voltage electron micrograph, rhesus monkey cerebellar cortex, 3 µm, lobule X. ×7250. **b** Spiny branchlet, dendritic thorns, and neuroglial sheath of a Purkinje cell dendrite. The laminar processes of Golgi epithelial cells that ensheathe the dendrites of Purkinje cells are seen overlying the impregnated dendrite and thorns (*arrows*). In the light microscope this ragged appearance resulting from the combination of dendrites and neuroglia could be passed off as an inadequate impregnation; however, high voltage electron microscopy shows the true composition of such areas. Rhesus monkey cerebellar cortex, 3 µm, lobule X. ×7250

centers of which the microtubules pass in their inexorably longitudinal trajectory; (2) a set of three to six cisternae (Fig. 32) arranged like the petals of an aster, around a single central microtubule. These patterns are fairly common and may be characteristic of the Purkinje cell dendrite. At least, we have not seen the second configuration, the floret, in the dendrites of any other cell type. The hypolemmal cisterna is a continuation of the same cisterna in the cell body (Figs. 29 and 31). Highly perforated, it forms a tattered sleeve beneath the surface membrane of the dendrite, extending into all of its branches, even into the thorns (Fig. 33). This distinctive structure is an aid to the identification of Purkinje cell dendrites in either longitudinal or transverse sections.

In the large Purkinje dendrites the mitochondria are preferentially, but not exclusively, arranged in a peripheral cylinder of cytoplasm, their long axes parallel with the major axis of the dendrite. Usually they are long, slender, bacillary organelles, 1–5 µm long and 0.2 µm across, with longitudinally oriented cristae and a dense matrix. However, as in the perikaryon, larger plump forms are sometimes seen, and mitochondria with transverse or heterogeneous cristae are not uncommon. Clustered mitochondria are also fairly frequent, appearing as several long profiles packed closely together like sausages. Occasionally a mitochondrion associated with a subsurface cisterna is seen in the large dendrites as in the cell body, but these rarely occur beyond the first bifurcation.

The mitochondria in Purkinje cell dendrites are really very numerous. In the larger dendrites they are so diluted in the large volume of the processes that they do not seem impressive. But as the dendritic trunks arborize, the complement of mitochondria appears to increase instead of being partitioned among the successive subdivisions (Fig. 30). As a result the terminal branchlets are often choked with longitudinal mitochondria and have barely space for the most slender tubules of the endoplasmic reticulum and a few microtubules (Fig. 33). This appearance is what one would expect if there is a net outward protoplasmic flow from the cell body toward the terminals, carrying the mitochondria distally, like logs on a flume. This concentration of mitochondria produces such a distinctive, dense appearance in one-micron plastic sections stained with toluidine blue, that the finer dendrites of the Purkinje cell can be recognized with the light microscope in either longitudinal or transverse section. Despite this crowding in the terminal branchlets, the mitochondria do not enter the thorns (Fig. 33).

As in dendrites of all other neurons, the most characteristic components of the Purkinje cell dendrites are the microtubules. These organelles, seemingly endless, extend out of the perikaryon into all of the dendritic

Fig. 27. The origin of two primary dendrites of a Purkinje cell. Two dendrites (*Pc* d_1 and *Pc* d_2) are shown emerging from a Purkinje cell soma (*PC*). The dendrite *Pc* d_1 bears two short-necked thorns (*arrows*). One of these is shown in synapse with a climbing fiber profile (*cf*). The boundary of the Purkinje cell is outlined in this illustration with arrowheads. Lobule VI. ×11500

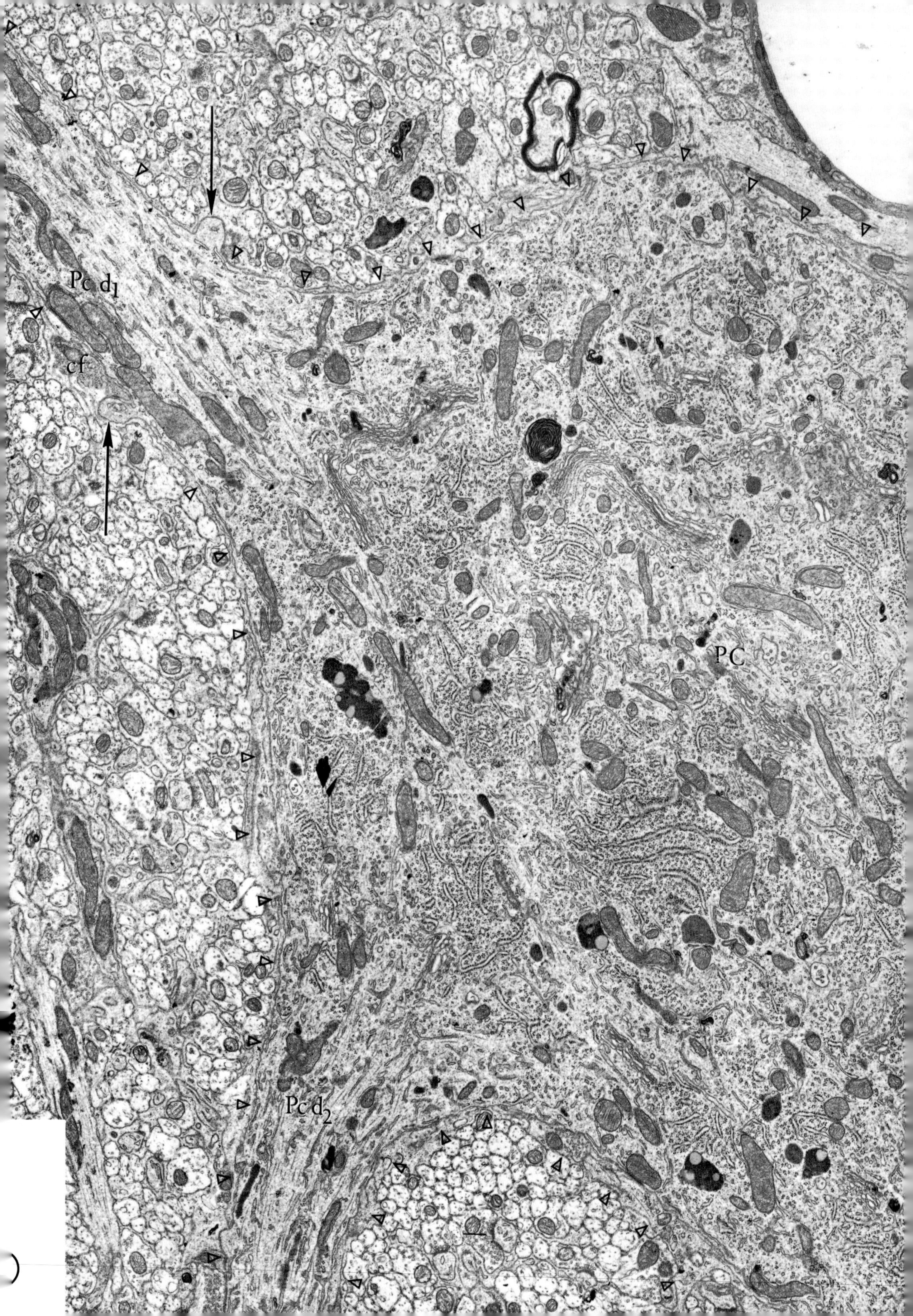
Pc d1
cf
PC
Pc d2

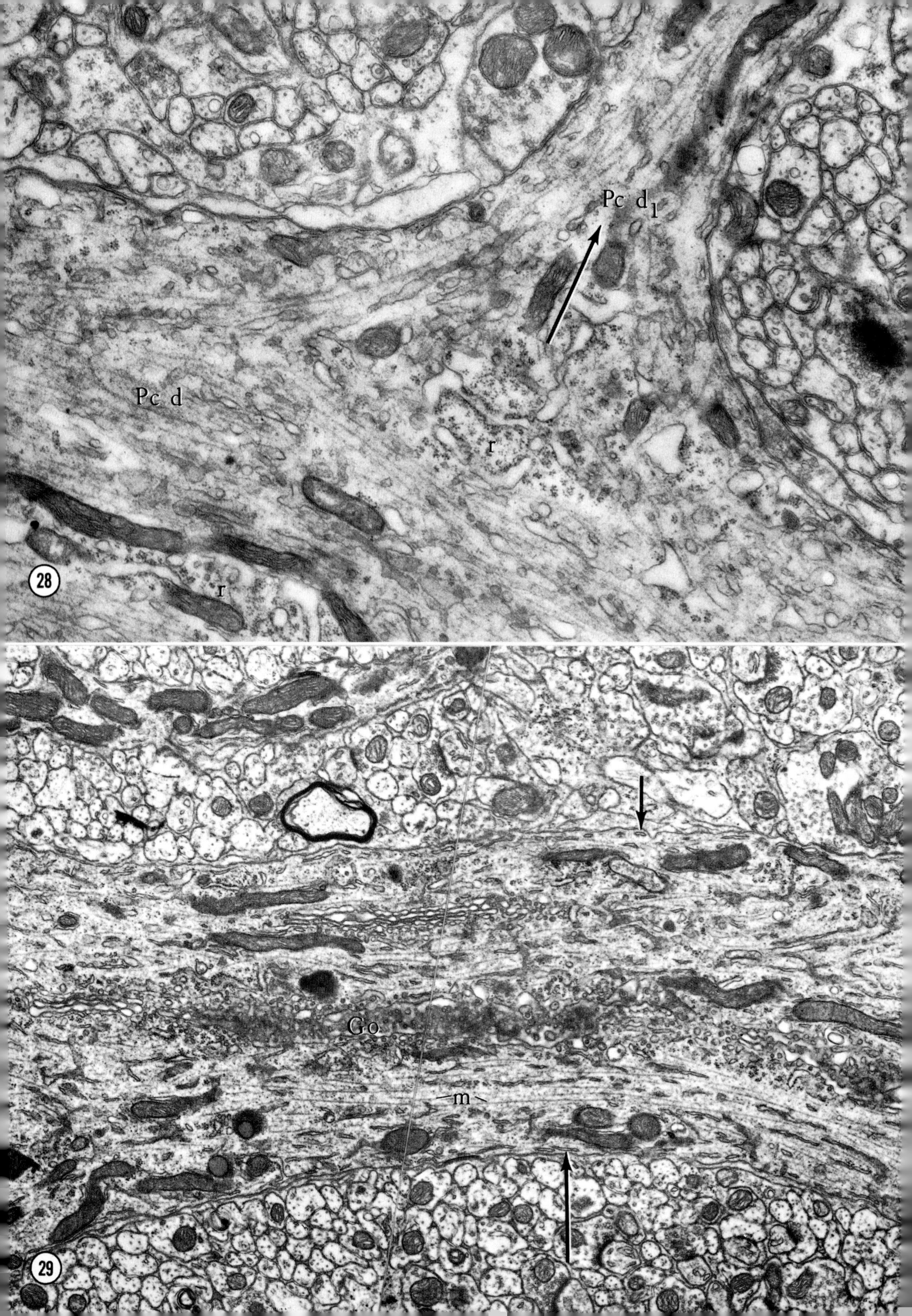

Pc d_1
Pc d
r
r
28
Go
m
29

branches, except the thorns. They are remarkably straight and are arrayed parallel to the major axis of the dendrite in hexagonal array. Each microtubule is about 1000 Å distant from its neighbors, and the order deviates only for the occasional mitochondrion in their path (Fig. 32). Most of the microtubules lie in the core of the dendrite. Their semicrystalline array is less rigid in the peripheral zone of the process where mitochondria are numerous. At branching points the complement of microtubules in the parent trunk is partitioned among the daughter branches according to their cross-sectional area. The distance between microtubules remains approximately the same. The smaller branches, the spiny branchlets, full of mitochondria, contain only a few microtubules. Everywhere the microtubules are surrounded by a cloud of cottony material that stretches between neighbors, a feature found in all neurons (see WUERKER and PALAY, 1970).

The neurofilaments are sparse and hence inconspicuous. They run parallel to the microtubules and do not enter the dendritic thorns.

As can be seen in Figs. 31 and 33, electron micrographs prove conclusively that the dendritic thorns are integral parts of the dendrite and not an artifact of the metallic impregnation methods (GRAY, 1959, 1961; see RAMÓN Y CAJAL, 1909, pp. 67–70 for a penetrating discussion of that problem at the level of the optical microscope). All of the thorns have the same internal fine structure; only their shapes and sizes are varied. The thorn consists of a slender pedicel with a rounded, subspherical or flattened enlargement at its free extremity, the whole protrusion being surrounded by the plasmalemma of the dendrite. The thorns on the spiny branchlets are from 1.5 to 2.0 μm long, and their heads are 0.4 to 0.5 μm across. Those on the larger dendrites are shorter, about 1 μm long, and are more clavate in form, having a shorter, more sharply tapering stalk. Their heads are about 0.6 μm across. The stalks of both forms are 0.2 to 0.3 μm in diameter. Most of the dendritic organelles do not enter the thorns, as has already been mentioned. Within the thorn are a few cisternae and tubules of the agranular endoplasmic reticulum immersed in a finely filamentous matrix. The elements of the reticulum can frequently be seen penetrating through the stalk of the thorn and branching or expanding within the head (Fig. 33). Ordinarily only one to three small profiles of tubules or flat cisternae are encountered in the head. They are never closely appressed as in the spine apparatus of pyramidal cells in the cerebral cortex. Since the discovery of the spine apparatus by GRAY in 1959, it was long thought that this organelle is restricted to the cerebral cortex (GRAY and GUILLERY, 1963), but more recent investigations (PETERS and KAISERMAN-ABRAMOF, 1970) make it evident that the spine apparatus is only a special configuration of the endoplasmic reticulum that exists in thorns of cells in other parts of the nervous system.

The synaptic specializations in the thorns will be deferred for discussion to the next chapter.

6. The Purkinje Cell Axon

The axon of the Purkinje cell arises from a barely perceptible conical projection on the basal pole of the soma. As RAMÓN Y CAJAL (1911) remarked, there is no clear line of demarcation between the cell body and the axon, as

Fig. 28. The bifurcation of a dendrite of a Purkinje cell. A primary Purkinje cell dendrite (*Pc d*) gives rise (*arrow*) to a secondary dendrite (*Pc* d_1). At the point of bifurcation, a small Nissl body is represented by clusters of ribosomes (*r*) and cisternae of granular endoplasmic reticulum. Osmium tetroxide fixation. ×10000

Fig. 29. Primary dendrite of a Purkinje cell. The trunk of this primary dendrite shows the characteristic features of a Purkinje cell, particularly the well-developed Golgi apparatus (*Go*) (see Figs. 21, 22), microtubules (*m*), and prominent elements of the hypolemmal cisterna (*arrows*). Lobule V. ×16000

Fig. 30. Tertiary dendrites and spiny branchlets of a Purkinje cell. Both the tertiary dendrite (*Pc d*) and the several spiny branchlets (*sb*) arising from it are covered by a veil of neuroglia (*gl*). The Purkinje cell cytoplasm contains its characteristic complement of mitochondria, microtubules, hypolemmal cisterna, and smooth endoplasmic reticulum (*SR*). Note the increased concentration of mitochondria in the more distal dendritic profiles. Lobule III. ×14500

Fig. 31. Synapses on the Purkinje cell dendrite. A single dendrite of a Purkinje cell is seen in cross section on the right as well as in longitudinal section on the left. Tubules of the smooth endoplasmic reticulum (*SR*) as well as elements of the hypolemmal cisterna (*arrows*) are shown. Two thorns (*t*) emerge from the trunk of the dendrite to synapse on a climbing fiber terminal (*cf*). Terminals of basket axons (Bax_1, Bax_2, Bax_3) synapse directly on the shaft of the Purkinje dendrite. Lobule X. ×22000

Fig. 32. The arrangement of microtubules and smooth endoplasmic reticulum in a Purkinje dendrite. This cross section of a Purkinje cell dendrite shows an array of microtubules (*m*) interspersed with mitochondria (*mit*) and tubules or cisternae of the endoplasmic reticulum (*SER*). The rectangle indicates a floret arrangement of six profiles of the smooth endoplasmic reticulum surrounding a microtubule. This seems to be peculiar to Purkinje cell cytoplasm, particularly after osmium tetroxide treatment. ×100000

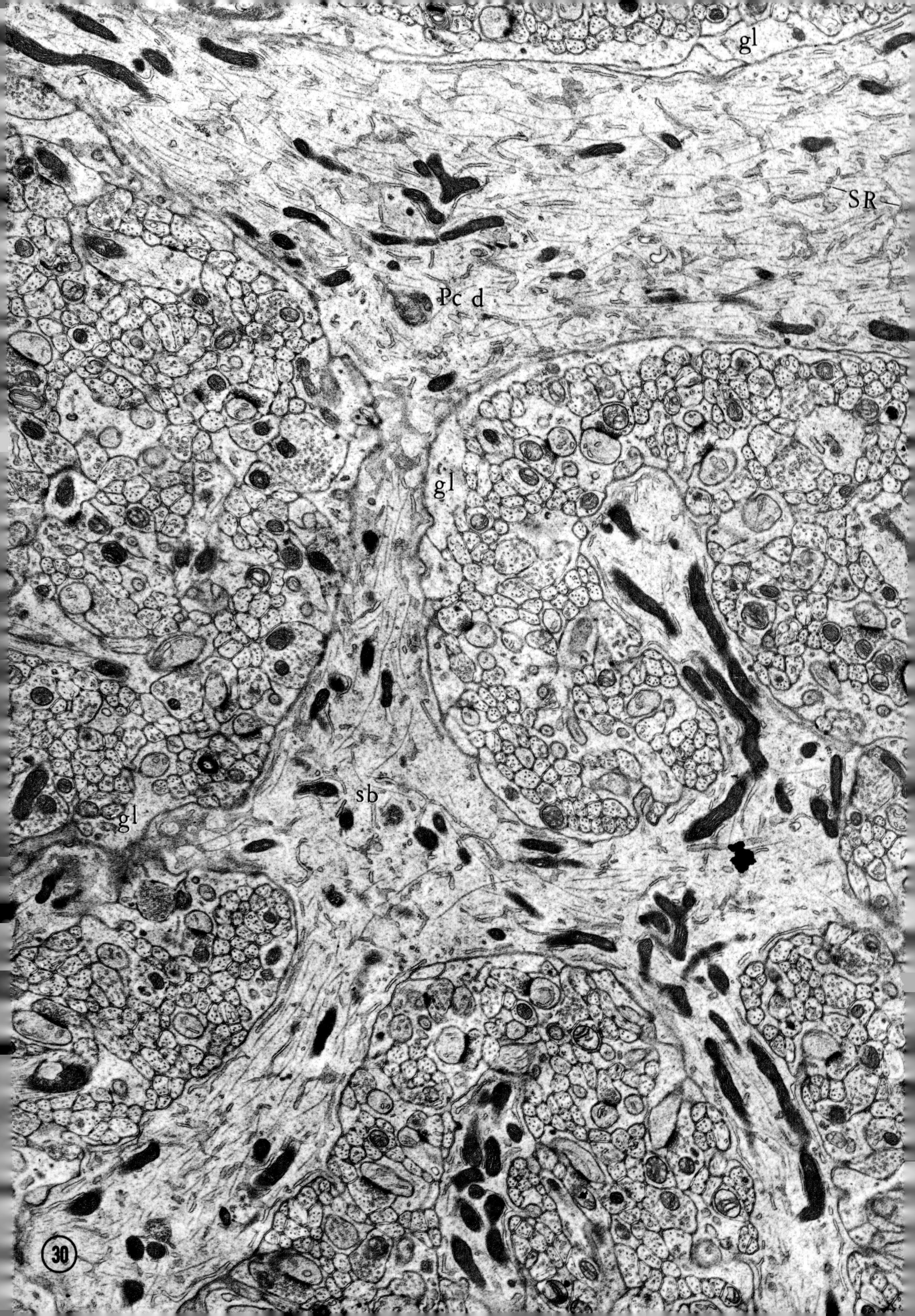
gl
SR
Pc d
gl
sb
gl
30

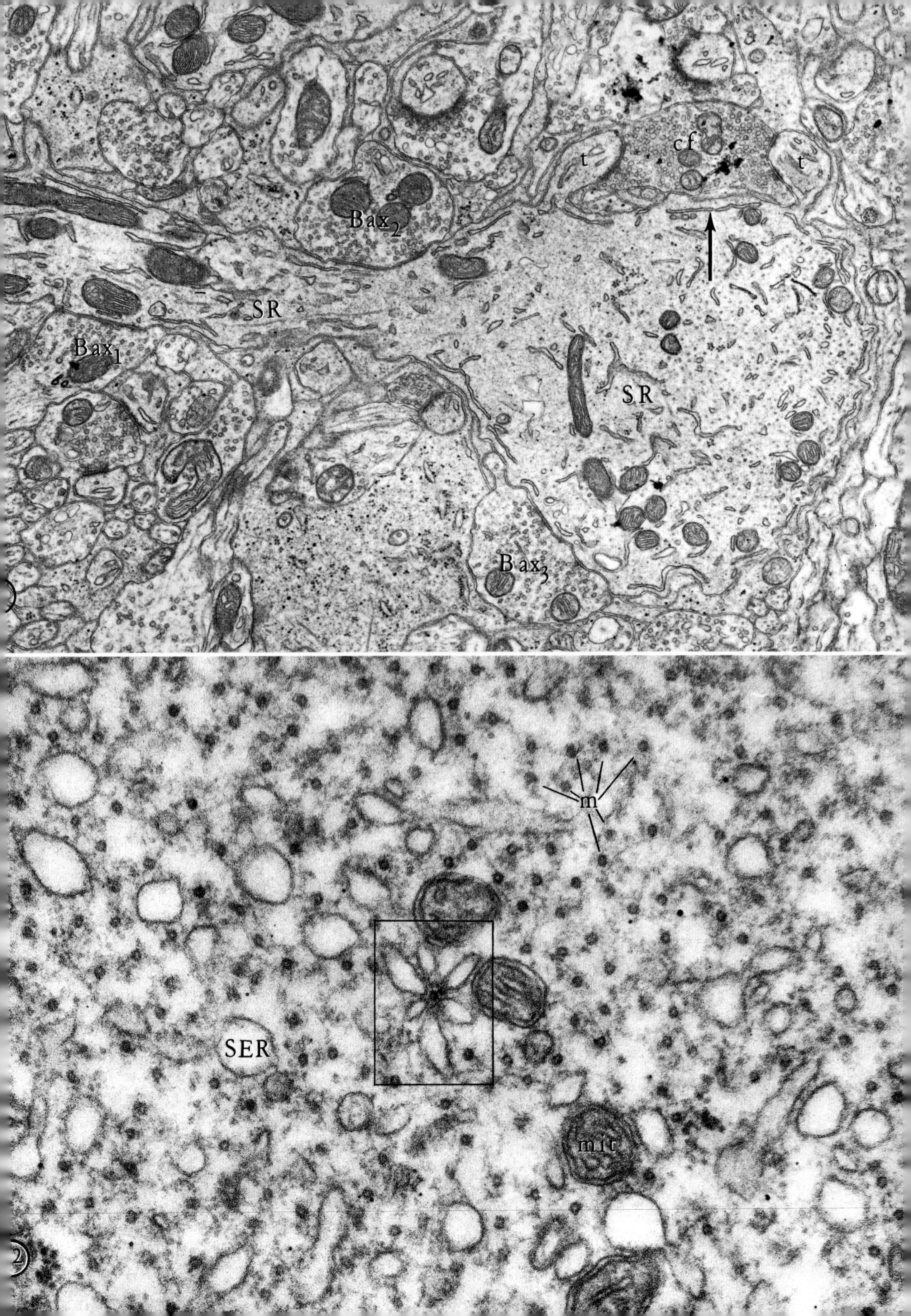

t
cf
t
Bax2
SR
Bax1
SR
Bax3
m
SER
mit

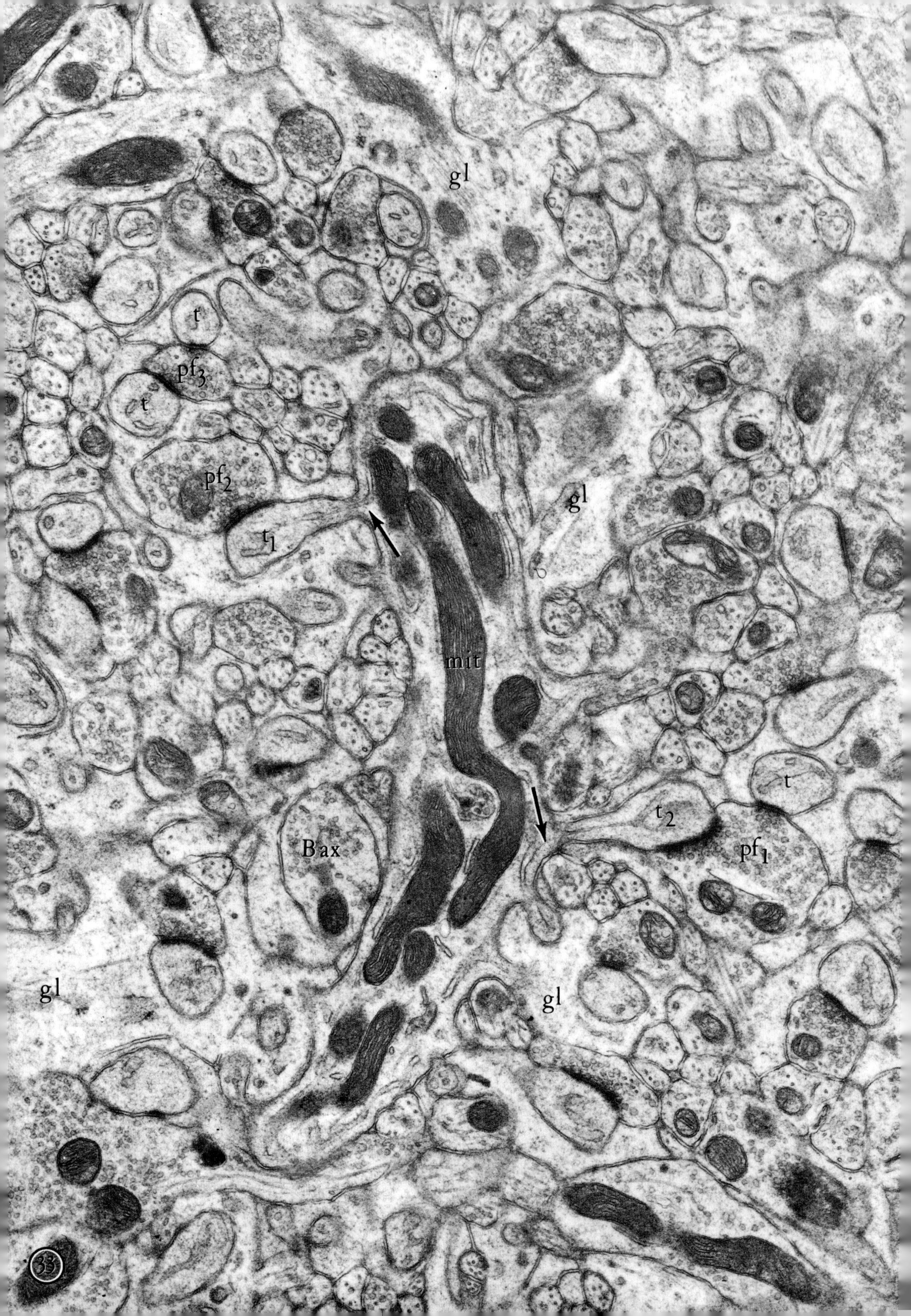
gl
t
pf3
t
pf2
gl
t1
mit
t
t2
Bax
pf1
gl
gl
33

small granules of Nissl substance enter into the first part of the latter. In electron micrographs the origin of the axon is signaled by a slight clearing in the most basal cytoplasm into which the microtubules veer from their usual perinuclear course (Fig. 34). This appearance must correspond to the description of the initial segment by RAMÓN Y CAJAL (1911, p. 12): "... the axon is the result of the confluence and anastomosis of a considerable number of fine fibrils from the perikaryal network." The axon issues forth from the perikaryon in a gentle arc or a nearly straight line, plunging through the pinceau of basket fiber terminals to reach its myelin sheath, about 40 or 50 microns beneath the cell body. The initial, unmyelinated segment of the axon is about one micron in diameter at its onset and gradually becomes narrower as it proceeds towards the myelin sheath. At the point just before it enters into the sheath it attains its smallest diameter, about 0.5 μm, and then it expands again.

a) The Initial Segment

The initial segment of the axon of the Purkinje cell does not differ in its fine structure from those of other multipolar nerve cells (PETERS, PALAY, and WEBSTER, 1970). In fact, the initial segment of the Purkinje cell was the first one to be recognized in electron micrographs (PALAY, 1964; KOHNO, 1964; PALAY, SOTELO, PETERS, and ORKAND, 1968). The plasmalemma is lined with a thin, dense undercoating that begins at the apex of the small axon hillock and ends at the onset of the myelin sheath. Although this undercoating was originally described as consisting of finely granular material, a more detailed study (CHAN-PALAY, 1972a) has recently shown that it has a tripartite structure (Figs. 34–40). Immediately beneath the axolemma is a layer of granules, 75 Å in diameter, and regularly spaced 75 Å apart in the longitudinal direction and 95 Å apart in the circumferential direction. The granules are seated upon a thin, dense lamina, 75 Å thick, which follows the contours of the axolemma. Coiling beneath this layer is a tight spiral, the gyres of which have a triangular cross section, 550 Å wide at the base, and are spaced 600–700 Å apart at their apices. The spiral consists of a fine fabric made up of filaments 50 Å thick. These constituents of the undercoat are illustrated in Figs. 35–39, and a drawing synthesizing these observations is shown in Fig. 40.

Fasciculated microtubules course longitudinally through the entire length of the initial segment. The number of microtubules in a fascicle varies from 2 to 16 or more, and they are usually arranged as a branching row or a compact bundle. Crossbars join the microtubules

Fig. 33. The Purkinje cell spiny branchlet. This spiny branchlet, like all Purkinje cell dendrites, is covered by a layer of neuroglia (*gl*). A basket cell axon (*Bax*) synapses directly upon the spiny branchlet. Two thorns (t_1 and t_2) emerge from the spiny branchlet to synapse upon parallel fiber varicosities (pf_1, pf_2). The hypolemmal cisterna extends from the dendrite into the neck and body of the thorns (*arrows*). Mitochondria (*mit*), however, are not found in thorns, but are confined to the shaft. Two parallel fiber profiles (pf_1 and pf_3) each form synapses with two Purkinje cell thorns. Lobule III. × 19000

Fig. 34. The axon hillock and initial segment of the Purkinje cell. The initial segment (*IS*) arises from a barely perceptible axon hillock (*H*) and extends into the pinceau made up of axons of basket cells (*B ax*) and neuroglia (*gl*). The nucleus of the Purkinje cell (*PC Nuc*) contains a prominent nucleolus (*ncl*) and a mass of nucleolus-associated chromatin (*arrow*). Lobule I, vermis. × 19000

Fig. 35. Longitudinal section through initial segment of Purkinje cell axon. This illustration displays various longitudinal aspects of the tripartite nature of the undercoat in relatively low magnification. Firstly, like most initial segments, this one is gently curved as well as slightly contorted, thus passing through the section plane at various angles and showing different aspects of the undercoat. The segments indicated between arrows show parts of the undercoat where the tripartite nature can be discerned in true longitudinal section. The layers are (*1*) the layer of granules directly beneath the plasmalemma, (*2*) a dense lamina deep to it (*3*) followed by the triangular tufts made up of filaments (see Fig. 36 and inset). The segments of undercoat indicated between asterisks show glancing sections through the layer of tufts. The periodic repeating nature of these tufts can be distinguished. Often the layer of tufts is approached by microtubules (*m*) that lie close to it. Lobule II. × 37000

Fig. 36 and Inset. High magnification of the tripartite undercoat in longitudinal section. A portion of the undercoat has been shown here, together with an inset drawing made from a tracing of this illustration, labelled to match the original micrograph. The double-layered structure of the plasmalemma (*P*) is followed immediately by the layer of granules (*1*). Underlying this is the dense lamina (*2*) followed by the layer of triangular tufts made up of interwoven filaments about 50 Å in diameter (*3*). Lobule II. × 752000

Fig. 37. Cross section of the initial segment of a Purkinje cell axon. This illustration shows the two features of initial segments in cross section—firstly, the undercoat, and next the fasicles of microtubules (*m*). The first and second layers of the tripartite undercoat are seen well in the section of the axon between arrows. The third component of the undercoat, the spiral of filaments lying on the second layer, is shown in the portion of the axon between asterisks. Lobule VIII. × 75000

Fig. 38. The dense lamina or second layer of the undercoat. This portion of an initial segment seen in a section nearly normal to the longitudinal axis of the axon shows the dense lamina or the second layer of the tripartite undercoat (between arrows). The space occupied by the first lamina (the granular layer) is clear, perhaps because the section largely passes through the interval between ranks of granules. Lobule X. × 268000

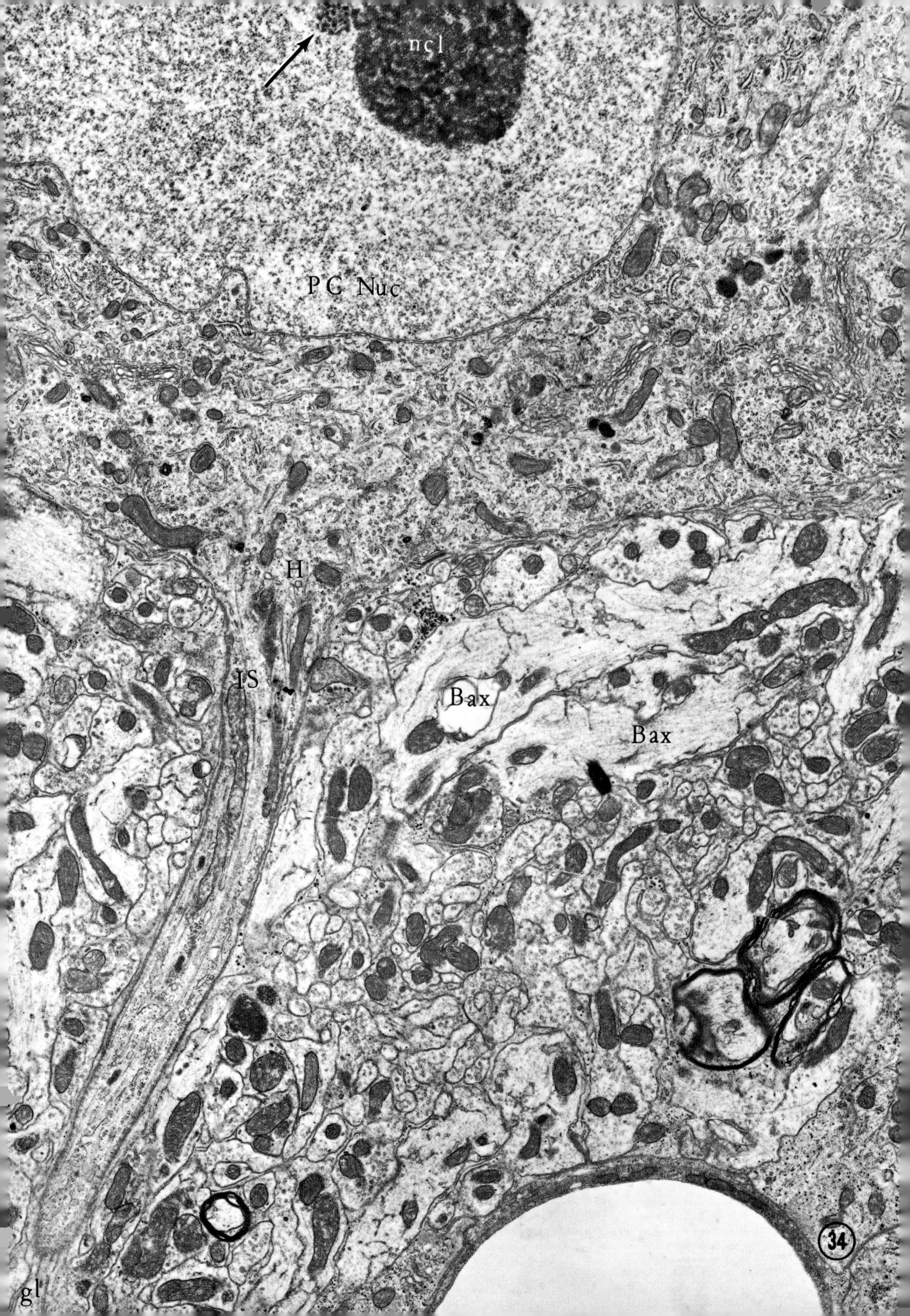
ncl
PC Nuc
H
IS
Bax
Bax
gl
34

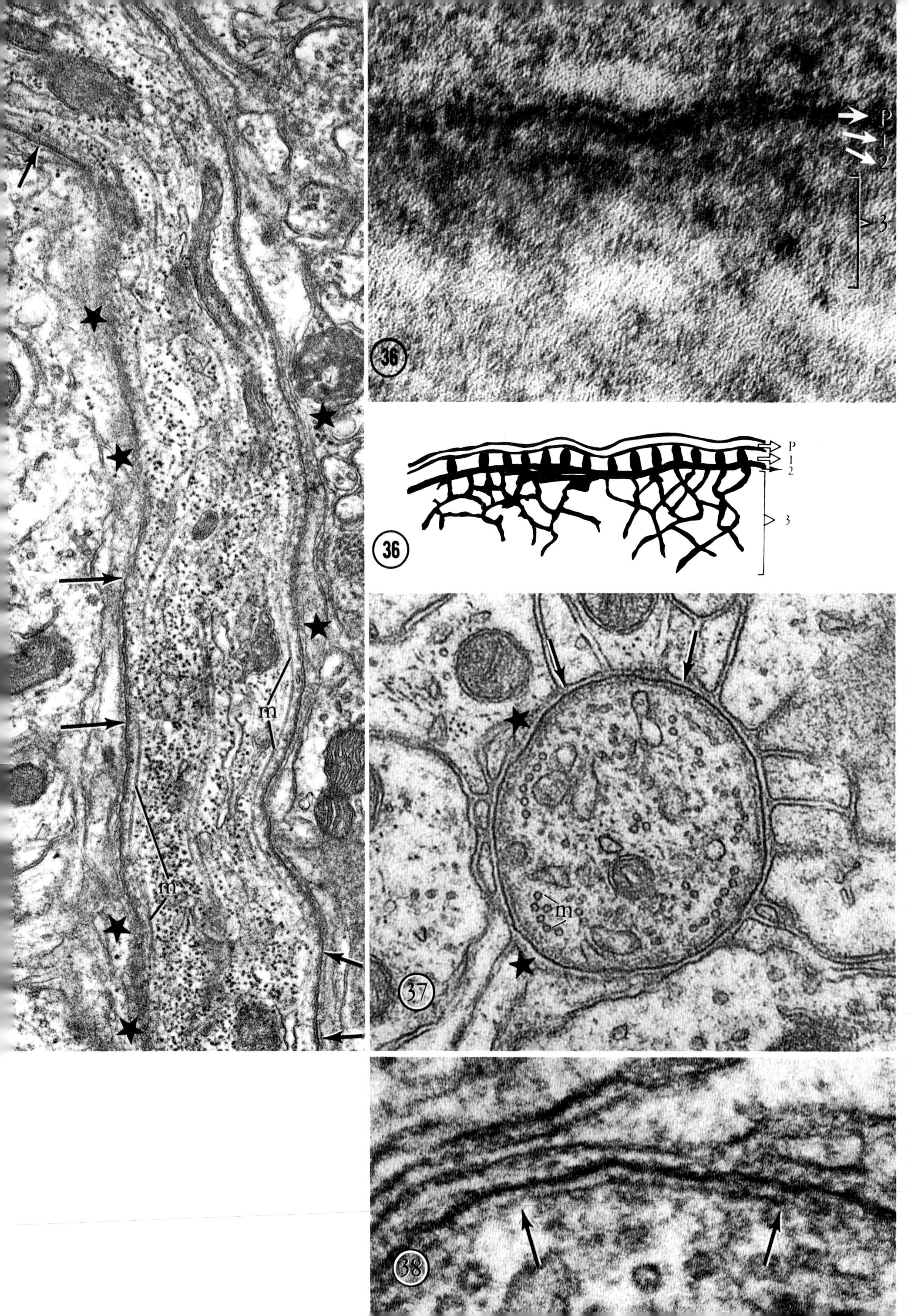

m
m
m
P
1
2
3
36
P
1
2
3
36
m
37
38

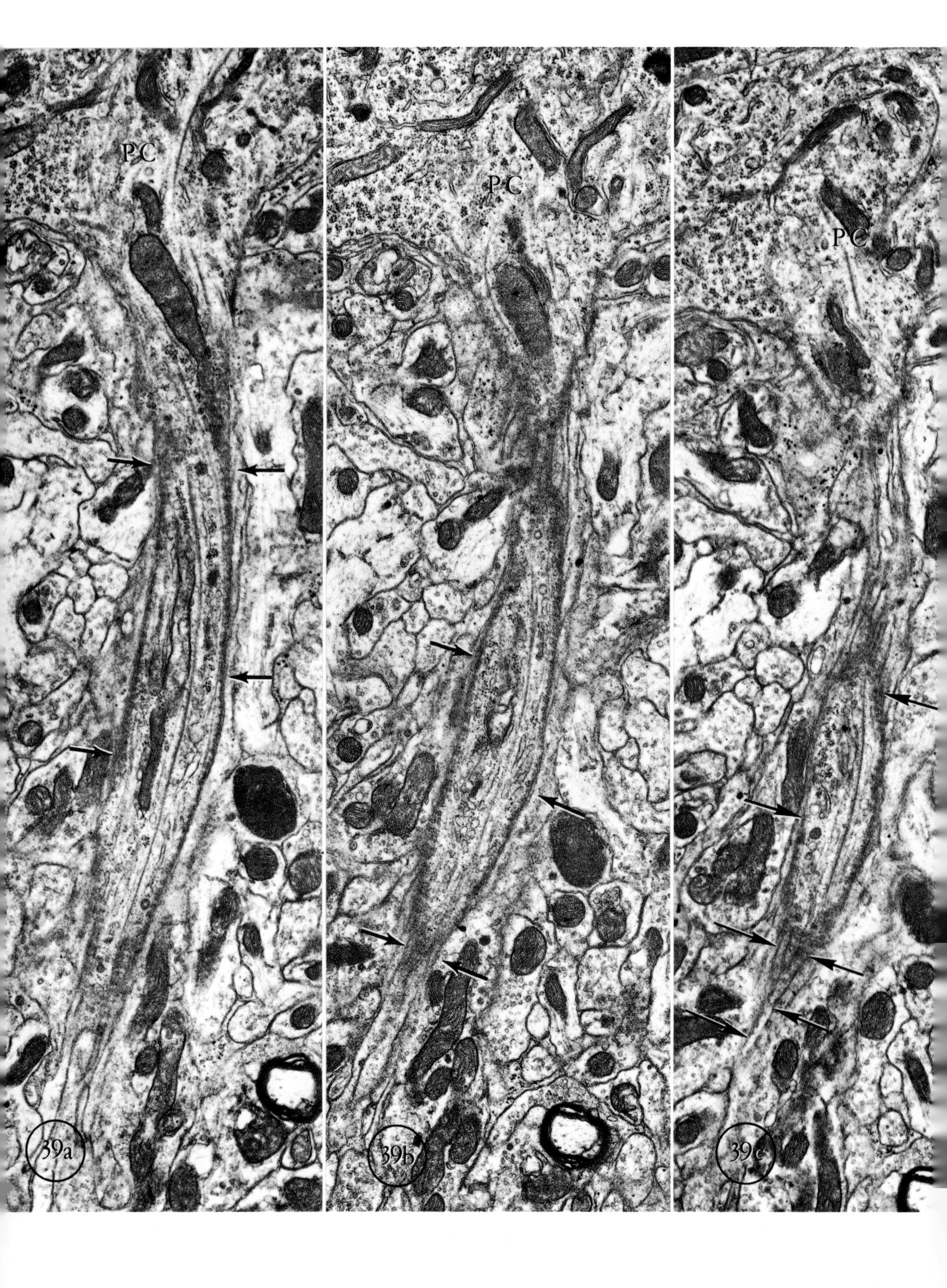

PC
PC
PC
39a
39b
39c

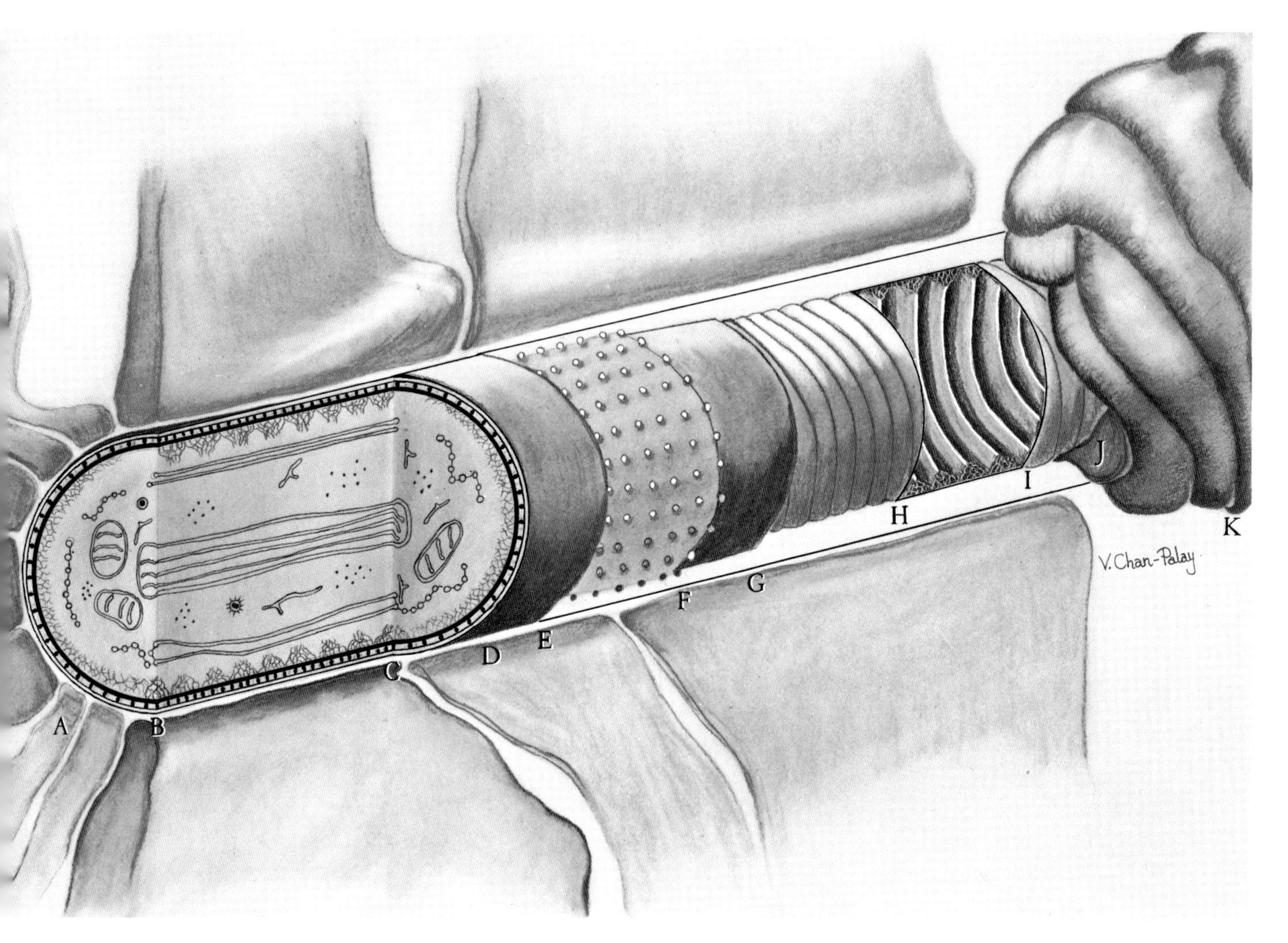

Fig. 40. The tripartite undercoat of initial segments in Purkinje cell axons. This initial segment begins on the left where it is seen cut first in cross section (*A–B*), then in longitudinal section (*B–C*), and then again in cross section (*C–D*). Here the characteristic features of the initial segment such as fascicles of microtubules, and especially the tripartite undercoat, are diagrammed in cross section and longitudinal section. Beginning with the plasmalemma (*D–E*), each lamina is peeled off to reveal the next one lying beneath it, thus exposing the three components of the undercoat. For purposes of clarity, the drawing was not made to scale

◀

Fig. 39 a–c. Longitudinal sections through an initial segment of a Purkinje cell (*PC*) axon. These three electron micrographs are non-adjacent serial thin sections. The plane of section catches the axon as it twists gently along its course. Between the pairs of arrows are regions where the section just grazes the third layer of the undercoat—the layer of triangular tufts. Here it is seen that a series of periodic alternating dense and clear bands encircles the initial segment. This signifies that the third layer may be a ridge made up of a web of 50 Å filaments spiralling internal to the dense lamina of the initial segment. Lobule V. × 26000

one with another. The mitochondria in the initial segment tend to be few and very long. The endoplasmic reticulum is agranular and consists of branching tubules extending longitudinally. The hypolemmal cisterna is rather fragmented and is poorly represented in the initial segment. Clusters of ribosomes appear here and there. They are more numerous in the more proximal portion of the initial segment and become less common with progressive distance from the axon hillock. Groups of randomly arranged ribosomes also appear in the more distal portions of the initial segment. This is the only part of a nerve cell where the ribosomes may not be encountered in polysomal array. In the rat, ribosomes do not occur very far beyond the beginning of the myelin sheath. Evidently in the frog they persist farther along, into the first myelinated segment (SOTELO, 1969). Neurofilaments are not conspicuous in the initial segment of the Purkinje cell axon, but they do occur individually. Their number increases, as the number of microtubules decreases, in the early myelinated segments of the axon. Finally, although the Golgi apparatus does not extend into the

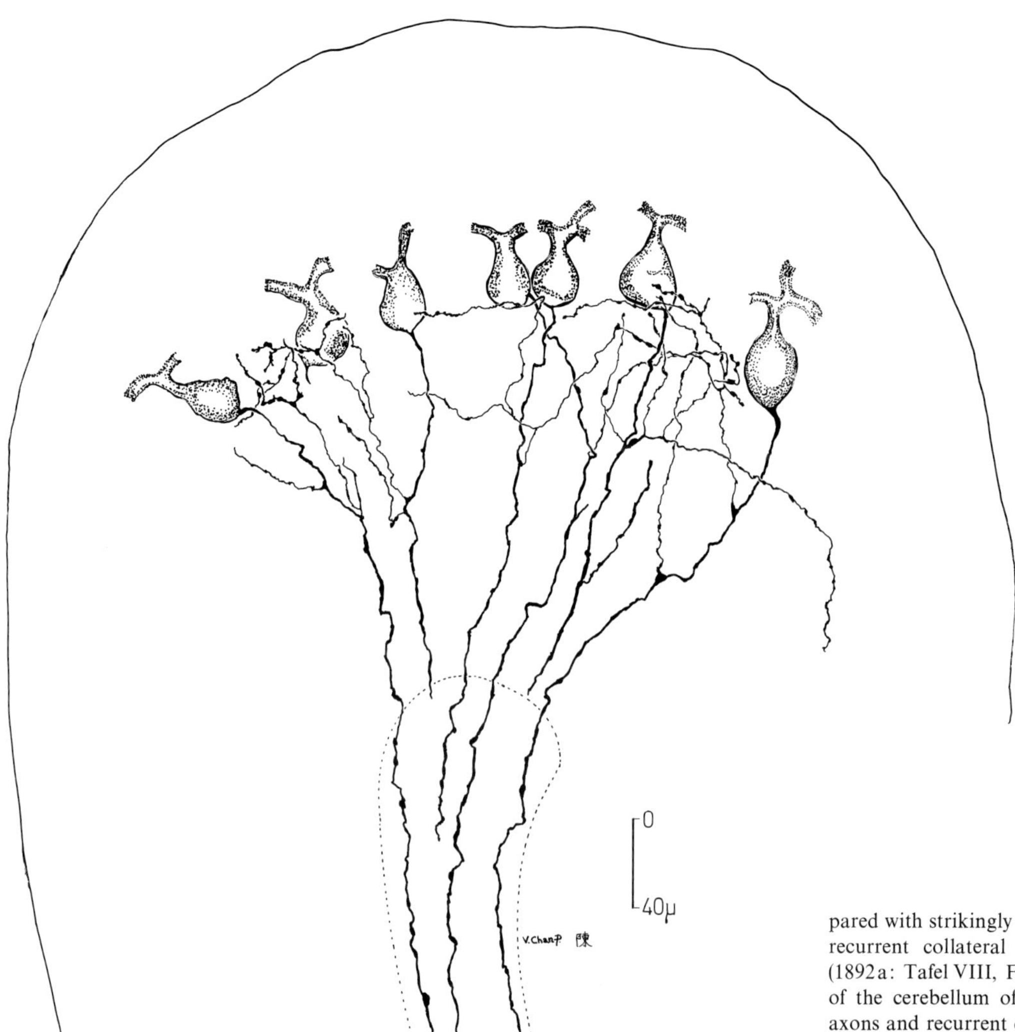

Fig. 41. The recurrent collateral plexuses of Purkinje cell axons. Camera lucida drawing of a Golgi preparation made from a single section of lobule IX in the ventral paraflocculus. Each axon emerges from the inferior pole of the Purkinje cell and runs through the granular layer toward the white matter. At various points along its length collaterals emerge which fan out to return towards the molecular layer. Each axon together with its recurrent collaterals occupies a roughly triangular space in the folium with the apex pointed towards the white matter and the base against the Purkinje cell layer. Three fiber plexuses are formed by branches of these recurrent collaterals, one in the granular layer, a second profuse plexus in the infraganglionic region, and a sparse supraganglionic plexus in the molecular layer. The recurrent collateral plexuses in this illustration should be compared with strikingly similar though more exuberant recurrent collateral networks drawn by RETZIUS (1892a: Tafel VIII, Fig. 1) from Golgi preparations of the cerebellum of an infant (7 day) rabbit. The axons and recurrent collaterals of the first and third Purkinje cells on the left are reproduced in larger detail in the next figure. Parasagittal section, 120 μm, $4\frac{1}{2}$ week old rat, 90 g in weight

initial segment, alveolate vesicles and lysosomes are found there. The narrowest portion of the initial segment, just before the onset of the myelin sheath, usually contains one or two large lysosomes. This constricted region is probably a functional obstruction, very similar to the experimental ligatures that have been placed on peripheral nerves (WEISS and HISCOE, 1948) but more gentle and more selective.

The elaborate axo-axonal synapse formed by the basket cell axons ending on and around the initial segment will be described in Chap. VII on the basket cell.

b) The Recurrent Collaterals

Beyond the initial segment the myelinated portion of the Purkinje cell axon proceeds directly through the granular layer into the white matter. The fibers are usually easily identified because they are somewhat larger than any other myelinated fibers. In Golgi preparations the axon of the Purkinje cell begins as a smooth, tapering thread but soon becomes twisted and varicose as it crooks its way among the granule cells (Figs. 41 and 42). Beginning with the third or fourth node of Ranvier (a fact noted by RAMÓN Y CAJAL in 1896), the axon gives off collateral branches from triangular, gnarled distensions in its course. These branches make an acute angle with the axon and return upwards through the granular layer, hence their name, recurrent collaterals. Each collateral consists of an exceedingly tenuous thread bearing small, rounded or ellipsoidal beads, about 0.4–0.5 μm in diameter. Despite their tenuity, these fibers are also myelinated. As Figs. 41 and 42 show, each collateral radiates into the granular layer, sometimes bifurcating on the way, and enters into plexuses in the granular and molecular layers, where it

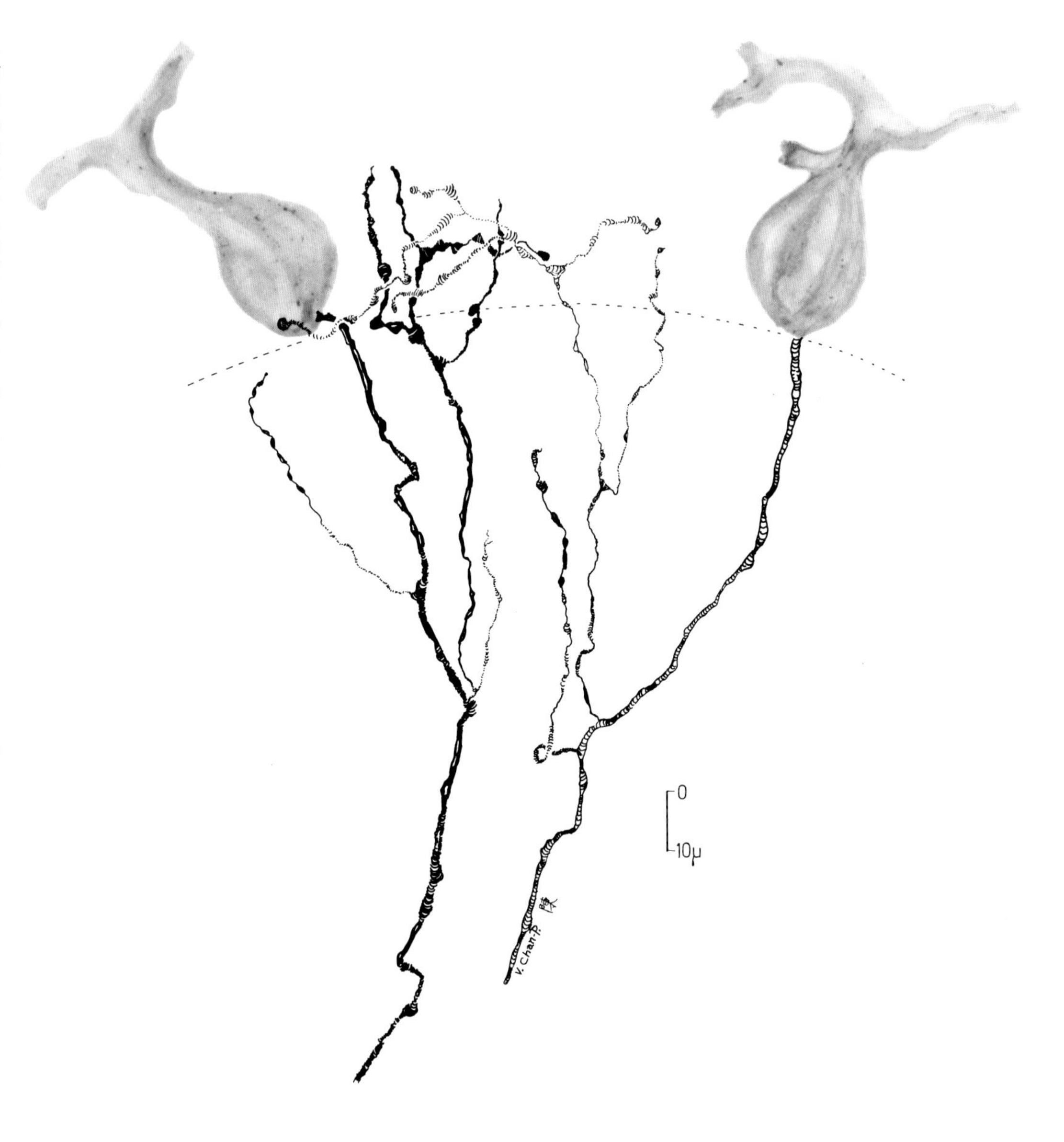

Fig. 42. Details of the recurrent collateral plexuses of two Purkinje cell axons in the rat cerebellar cortex. The main axon of the Purkinje cell is smooth initially but soon develops slight distensions along its length which are more obvious at twists and at the bifurcations of collaterals. Here the axon enlarges to form a triangular, gnarled distension, one arm giving rise to a thin, very finely beaded collateral. These collaterals ramify and terminate in the granular layer, or ascend to terminate in the vicinity of the Purkinje cell. The infraganglionic plexus is the most profuse plexus, constructed mainly of branches with large varicosities that ramify and intertwine in the region of the Purkinje cell somata. Some branches do not end in the infraganglionic region but bend away at right angles and ascend vertically to the apical pole of the Purkinje cell, where a very sparse supraganglionic plexus is generated. Parasagittal section, 120 μm, rat $4\frac{1}{2}$ weeks old, 90 g

terminates in a burst of varicose subdivisions. The network of recurrent collaterals belonging to the Purkinje cells, like their dendritic arborization, is best displayed in the parasagittal plane, that is, perpendicular to the axis of the folium. The arrangement of each axon with its family of recurrent collaterals is reminiscent of the flowerhead of Queen Anne's lace, compressed between the pages of a book. Each axon with its recurrents occupies a roughly triangular field with its apex directed toward the white matter and its base expanding against the Purkinje cell layer. Although its main extent is in the parasagittal plane, some of the collaterals can be traced for a short distance beyond the confines of this plane. If the entire plexus were viewed in a plane tangential to the surface of the folium, it would appear elliptical in outline.

It is noteworthy that the dendritic and axonal patterns of the Purkinje cell are both characteristic of the relative position of the cell in the folium (Fig. 43). At the summit of the folium the dendritic tree is fully expanded in the parasagittal plane with branches more or less symmetrically arranged on either side of an imaginary axis running vertically through the cell body. On the slopes of the folium the dendritic arborization is widest near the summit but trimmed and slanted as the sulcus approaches. In the depths of the sulcus the dendritic field is flattened out on either side of the sulcus and may be quite asymmetrical. Here the dendritic expansion may be either very large or constricted in the parasagittal plane. Similarly, the trajectory of the main axon of the Purkinje cell and the arrangement of its collaterals depend on the location of the cell in the folium. The axon of a cell at the summit of a folium takes the most direct route to the white matter, usually a gentle arc through the granular layer. Its recurrent collaterals fan out like an equilateral triangle with

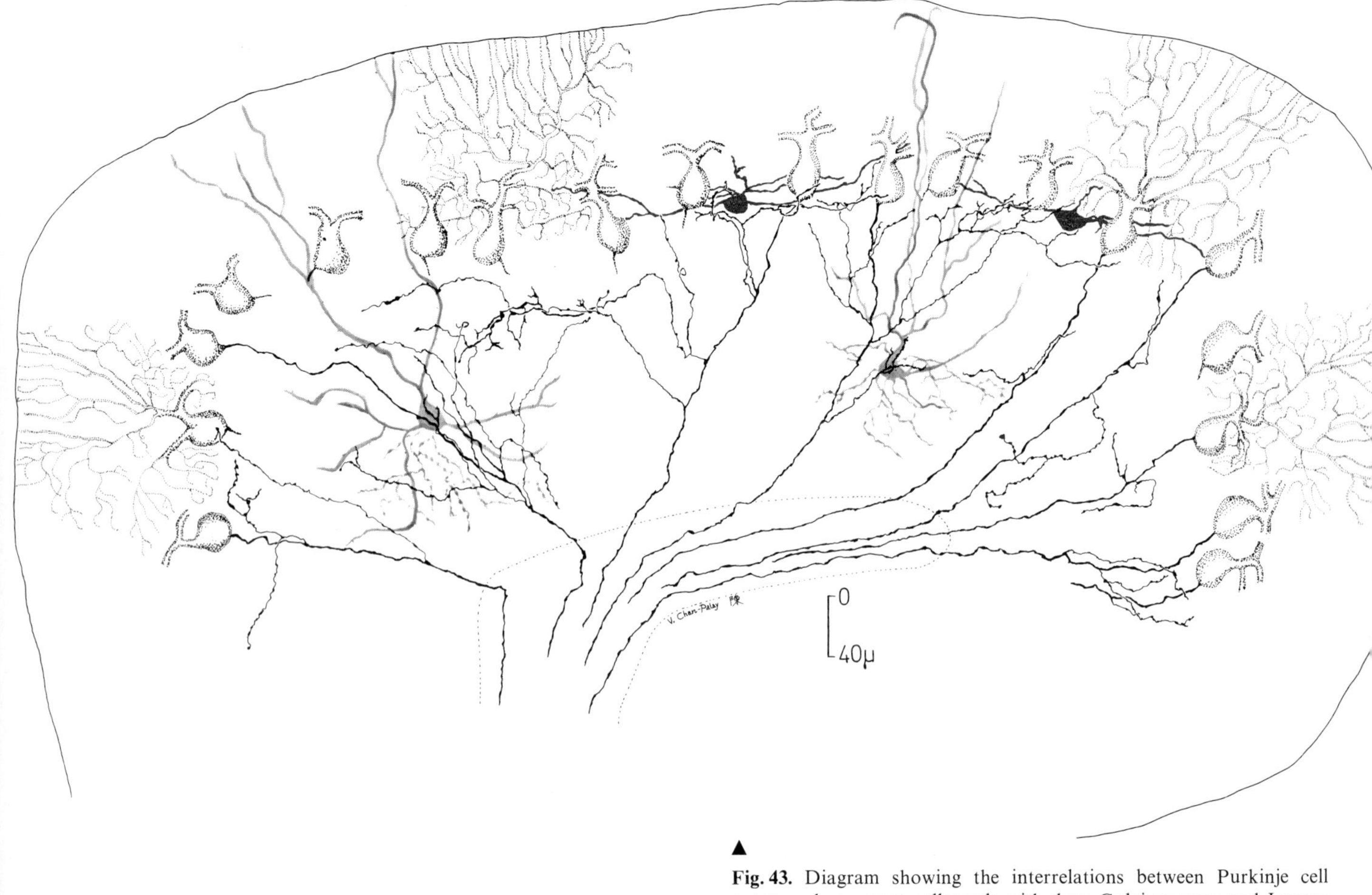

▲ **Fig. 43.** Diagram showing the interrelations between Purkinje cell axons and recurrent collaterals with deep Golgi neurons and Lugaro cells. Camera lucida drawings of recurrent collateral arborizations of several Purkinje cells in a folium (lobule IX ventral paraflocculus; stipple and black). Two deep Golgi cells (gray), and two Lugaro cells (red) were superimposed to produce this diagram. The interrelations of the processes of these cells have been observed in the electron micrographs, and are reproduced in this figure in order to provide the light microscope equivalent. Recurrent collaterals that originate deep in the granular layer generally divide several times, giving rise to varicose terminal branches that form a plexus in this layer. The varicosities or enlargements of these terminal branches synapse with dendrites and somata of deep Golgi neurons. Recurrent collaterals that emerge from the main axon in the middle or upper thirds of the granular layer generally ascend to the zone of the Purkinje cells. Here their subsequent branchings contribute to the formation of two plexuses, the infra- and the supraganglionic plexuses, situated below and above the level of the Purkinje cell body respectively. The infraganglionic plexus consists of varicose branches of collaterals, some of which bend to travel a tortuous horizontal course in the parasagittal plane of the folium. The varicosities end in relation with the somata and dendrites of Purkinje cells and Lugaro cells and also in complex synaptic arrangements in the surrounding neuropil. The supraganglionic plexus is very sparse and consists of occasional varicose branches that ascend into the low molecular layer to end on dendrites of Purkinje cells and rarely on basket cells. Parasagittal section, 150 μm thick, $4\frac{1}{2}$ week old rat

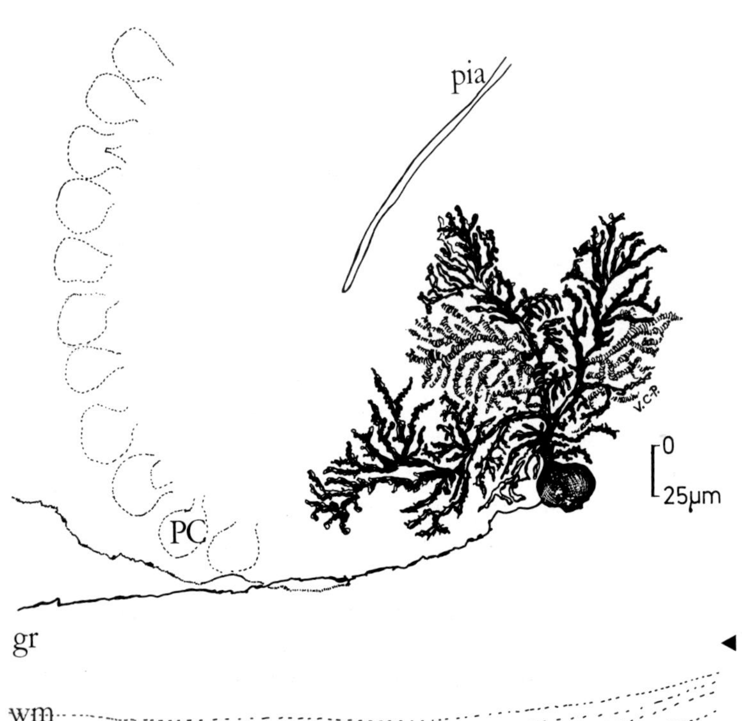

◀ **Fig. 44.** Purkinje cell with axon at the base of a folium. The axon of this Purkinje cell arises from a side of the cell body, travels obliquely through the Purkinje cell layer (*PC*) and the granular layer (*gr*) to the white matter (*wm*). A collateral is emitted along this course. Rapid Golgi, 90 μm, parasagittal plane, adult rat. Camera lucida drawing

its apex pointed towards the medullary center and its base against the Purkinje cell layer. The axon of a cell on the shoulders of a folium takes a curving diagonal traverse through the granular layer. Its recurrent collaterals form a tall triangle with a narrow base. Finally, the axon of a cell in the depths of a sulcus (Fig. 44) enters the white matter only after making a right-angled bend, and the recurrent collaterals form a narrow, constricted field (CHAN-PALAY, 1971).

The recurrent collaterals of the Purkinje cell were discovered by GOLGI (1874) in one of his first studies with his new method for staining nervous tissue. RAMÓN Y CAJAL (1888a, 1889a, 1890a and b, 1896, 1911, 1912; RAMÓN Y CAJAL and ILLERA, 1907) confirmed GOLGI'S discovery in a variety of birds and mammals, and extended his description with a more thorough analysis of the horizontal plexuses formed by the recurrent collaterals above and below the layer of Purkinje cell bodies. He was not able, however, to provide a definitive description of the mode of termination of these fibers. In preparations stained by the neurofibrillary methods, certain fine fibers of uncertain origin show up, which end in small boutons or ring formations around the cell bodies and major dendrites of the Purkinje cells. These end formations RAMÓN Y CAJAL proposed as the terminals of the recurrent collateral plexuses, although they could not be traced to recognizable recurrents (1911). His identification was unconvincing, and the question remained open until recently, although the recurrent collaterals have been studied in various species for many years. RETZIUS (1892a) presented detailed drawings of the collaterals in the dog, as did HELD (1897) in the rabbit. BIELSCHOWSKY (1904), SCHAFFER (1913), WINKLER (1927), and JAKOB (1928) also studied this question with neurofibrillary methods. More recently, they have been studied with the Golgi method in the monkey (FOX, HILLMAN, SIEGESMUND, and DUTTA, 1967), several mammalian and avian species (O'LEARY *et al.*, 1968), and the mormyrid fish (NIEUWENHUYS and NICHOLSON, 1969). Only in the past few years has the question been settled with the identification of the terminals in electron micrographs of mouse cerebellar cortex (LARRAMENDI and VICTOR, 1967; LEMKEY-JOHNSTON and LARRAMENDI, 1968; LARRAMENDI and LEMKEY-JOHNSTON, 1970). A thorough description of these fibers in the rat has been published by CHAN-PALAY (1971), who studied them in both Golgi and electron microscopic preparations. The present account is derived from that description.

Recurrent collaterals that originate deep in the granular layer, more distally along the Purkinje cell axon, generally divide several times in this layer, thus giving rise to a deep plexus that is distinct from the supra- and infraganglionic plexuses found above and below the layer of Purkinje cell bodies. The secondary and tertiary ramifications that make up this deep plexus in the granular layer are fine threads furnished with a succession of numerous globular or irregular ellipsoidal varicosities, about 0.8 to 1 μm in diameter. The terminal branches become fleshier as they approach their terminations, which are large clavate bulbs, about 1.5 μm in diameter, almost twice the size of the small beads found along their more proximal parts. The electron micrographs indicate that these varicosities, both terminal bulbs and distensions in the course of the fibers, are the sites of synapses with the dendrites (see Fig. 48) and somata of Golgi cells and possibly with the dendrites of granule cells.

Collaterals that emerge from Purkinje cell axons in the middle or upper third of the granular layer generally ascend to the layer of Purkinje cell bodies. Here their ramifications contribute to the formation of two plexuses of myelinated fibers, the supra- and infraganglionic plexuses above and below the Purkinje cell bodies respectively. The collaterals from the main axon ascend to a level of about 8–10 μm below the inferior poles of the Purkinje cells, where they bend horizontally to travel a somewhat tortuous course in the parasagittal plane. They may terminate shortly, but more often they continue for about 150–200 μm across the territories of several successive Purkinje cells. Generally they bifurcate several times, giving rise to shorter, stouter, and varicose secondary and tertiary branches that weave in and out among the neurons of the area to end in large boutons or small, rounded nodules in the infraganglionic region. The plexus produced by all these branches is dense and rich in terminals. Electron micrographs show that the recurrent collaterals in the infraganglionic plexus terminate in synaptic relation with the somata and initial segments of dendrites belonging to neighboring Purkinje cells (Figs. 46 and 47) and Lugaro cells (Fig. 12).

The supraganglionic plexus originates from vertical tertiary branches of the recurrent collaterals that continue through the infraganglionic region to the level of the apical poles of the Purkinje cells. A few short preterminal segments are emitted from these vertical branches and end in either large, clavate bulbs or smaller, round boutons. The supraganglionic plexus is extremely sparse and poorly developed in the adult rat, as it is also, according to RAMÓN Y CAJAL (1911), in the mouse and the rabbit. In electron micrographs terminals belonging to this plexus are found in synaptic relation with primary dendrites, some smaller dendrites, and, occasionally, thorns of Purkinje cells. Rarely they can be recognized synapsing on dendrites and somata of basket cells. Careful searching

(CHAN-PALAY, 1971) has not disclosed fibers in this plexus coursing longitudinally, parallel with the major axis of the folium for any considerable distance. This result is in apparent disagreement with RAMÓN Y CAJAL, who described fibers in the supraganglionic plexus that "become longitudinal after a transverse course of some extent" (1911, p. 14). Possibly the discrepancy reflects merely differences in the species examined, corresponding to the sparseness of the supraganglionic plexus in the adult rat in comparison with the extent of the plexus in the infant animals examined by RAMÓN Y CAJAL. Similarly, JAKOB (1928, p. 781) describes the field of the recurrent collaterals as three-dimensional, with perhaps some predominance in the frontal plane (parallel to the longitudinal axis of the folia). Apparently JAKOB was referring to the human cerebellum here, another example of differences in detail among different species. Because of the lack of fiber components running parallel with the folium, the recurrent collateral network in the rat is not three-dimensional, but rather is confined to a narrow parasagittal plane. Thus the axonal arborization, like the dendritic, is arranged like a fan opened out in a plane transverse to the longitudinal axis of the folium.

c) The Terminal Formations of the Collaterals

The terminal ramifications of the recurrent collaterals in the infra- and supraganglionic plexuses are each composed of a series of large, globular varicosities (about 0.8–1 µm in diameter) connected by a slender thread. As was mentioned above, rapid Golgi preparations indicate that these terminals are highly concentrated in the upper granular layer just beneath the Purkinje cells, and in electron micrographs it is just this region, the infraganglionic plexus, that is crowded with distinctive synaptic terminals identifiable as those of the recurrent collaterals. The terminal branchlets are most readily recognized in electron micrographs when they are found in continuity with myelinated segments of the collaterals coursing in the plexuses (Fig. 48). These branchlets—the tenuous threads of the Golgi preparations, 0.7–1 µm in diameter—slip out of the myelin sheath and soon become varicose. They contain a mixture of neurofilaments, microtubules, and slender mitochondria, all oriented longitudinally. The knot-like or clavate varicosities appear in the electron micrographs as rounded or elongate distensions, from 0.8 to 2 µm across, containing a dark, filamentous matrix in which mitochondria and a loose array of synaptic vesicles are suspended. Often the varicosity is hardly larger than the axonal branchlet and the vesicles are dispersed over a long distance, forming a kind of cloud around a core of neurofilaments, microtubules, and mitochondria. The vesicles are characteristically pleomorphic, ranging in shape from flat and rectangular to elliptical or round (Fig. 45). These characteristics were first recognized by LEMKEY-JOHNSTON and LARRAMENDI in 1968 (see also LARRAMENDI, 1969a, and LARRAMENDI and LEMKEY-JOHNSTON, 1970).

The synaptic interface formed by the recurrent collaterals is also distinctive. The synaptic complexes tend to be small, circular spots, 0.05–0.1 µm in diameter. Where the junction between the varicosity and the postsynaptic surface is broad, as on the somata of Purkinje cells (Figs. 46 and 47) or Lugaro cells (Fig. 121), there are several synaptic complexes. In these spots the interstitial cleft is slightly widened from the usual 125 Å to about 230 Å, and it contains an accumulation of dense, fibrillar material stretching between the synaptic membranes. Thin, symmetrical densities are attached to the pre- and postsynaptic surfaces bounding these spots. The synaptic interface is intermediate between Gray's types 1 and 2, having some features reminiscent of each. A second type of synaptic junction has also been observed between these terminals and large dendrites probably belonging to Golgi cells, in the granular layer. Here the synaptic junction is large and occupies nearly the whole of the oppositional surfaces (Fig. 49). The junctional complex itself resembles that seen in the macular form of synapse described above. In summary, the terminal structure of the Purkinje cell recurrent collateral is a small, irregular varicosity containing a dense, filamentous matrix and pleomorphic, mostly flattened vesicles; it forms macular

Fig. 45. Terminal branches of recurrent collaterals in the infraganglionic region. A long, distended terminal (*Prc*$_1$) is identified by the dark appearance of its axoplasmic matrix and the polymorphic population of flat, elliptical, and round synaptic vesicles. Embedded in this matrix are neurofilaments, microtubules, occasional dense core vesicles, and slender mitochondria. Two other axons belonging to the recurrent collateral plexus (*Prc*) can also be seen in this field among the long profiles of basket axons (*Ba*, *B ax*). Lobule IV, vermis. × 17000

Fig. 46. Synapses of recurrent collaterals on the soma of a Purkinje cell. Three profiles of boutons belonging to the infraganglionic plexus of recurrent collaterals (*Prc*$_1$, *Prc*$_2$, *Prc*$_3$) synapse on the soma of a Purkinje cell (*PC*). Two of the three terminals (*Prc*$_1$, *Prc*$_3$) each have three macular synaptic complexes (←) on the surface apposed to the Purkinje cell. In these spots the interstitial cleft is widened and contains dense fibrillar material. On the pre- and postsynaptic surfaces very shallow symmetrical densities are present. Just deep to this the hypolemmal cisterna of the Purkinje cell is evident. The free surface of each terminal is surrounded by a neuroglial sheath. Many thick basket cell axons (*B ax*) sweep around the terminals to end in the *pinceau* at the upper left. Lobule IV, vermis. × 17000

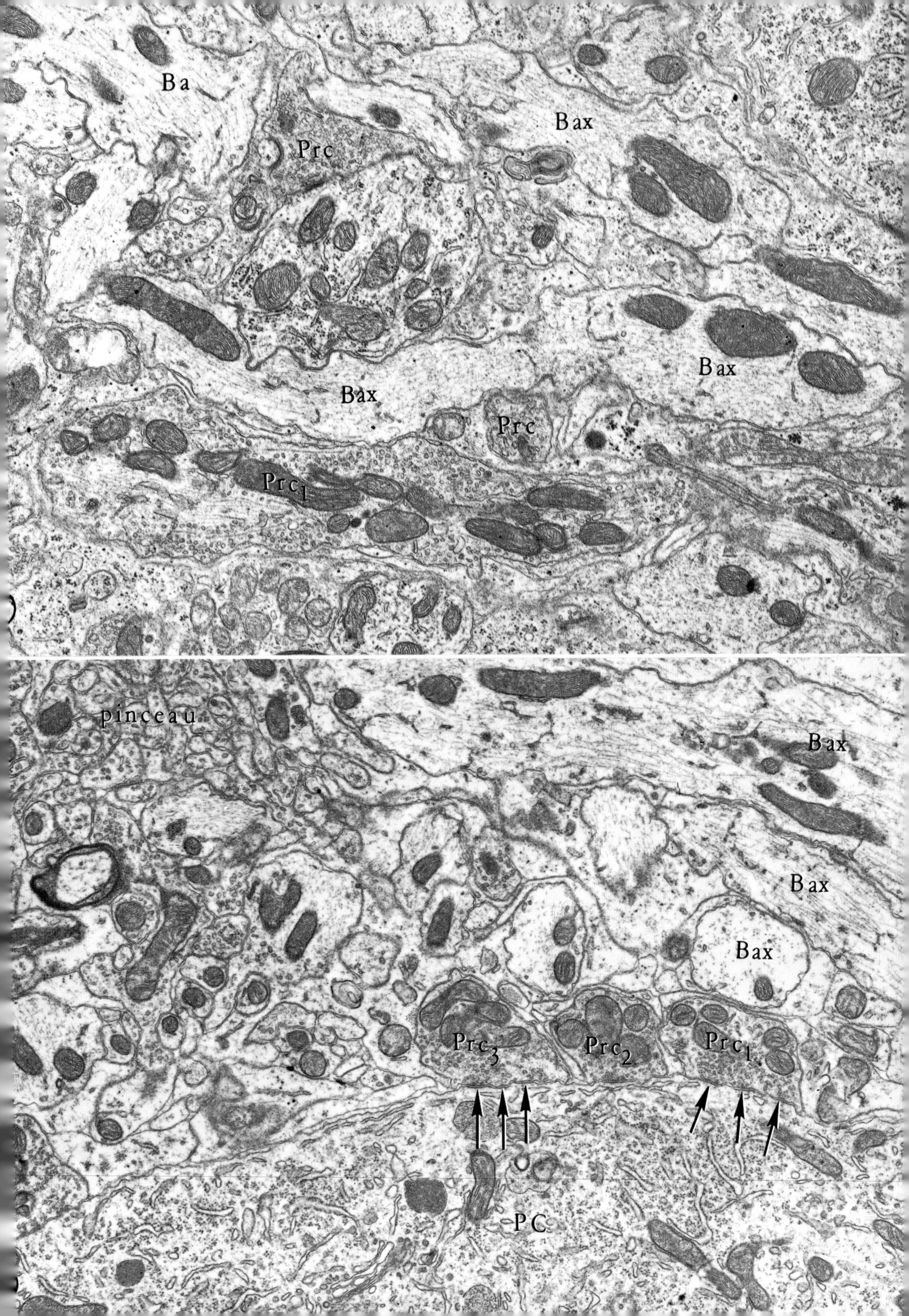
Ba
Prc
Bax
Bax
Bax
Prc
Prc_1
pinceau
Bax
Bax
Bax
Prc_3
Prc_2
Prc_1
PC

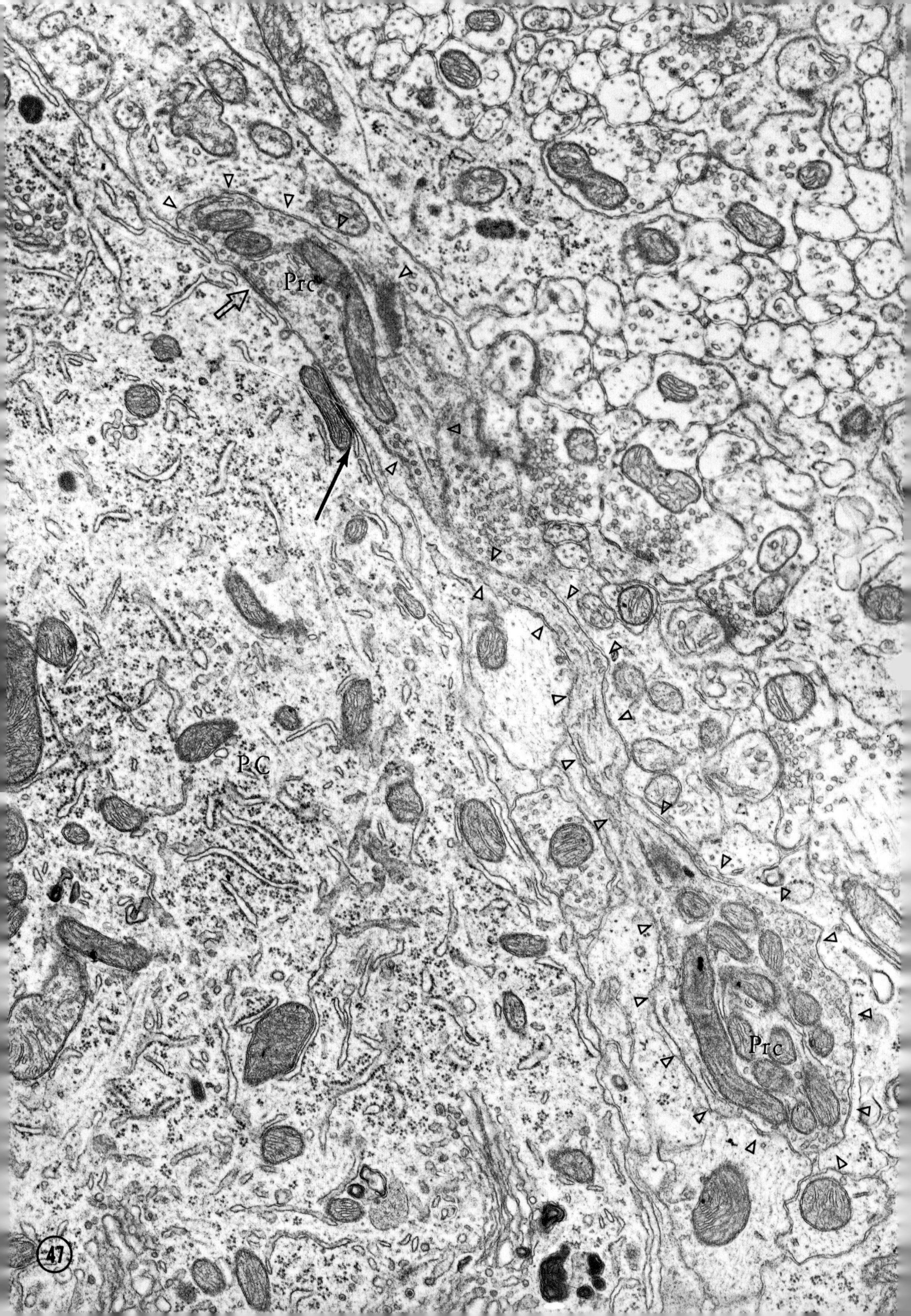

Prc
PC
Prc
47

synaptic complexes with some postsynaptic elements, e.g., Purkinje cells, Lugaro cells; it forms larger synaptic complexes with other cells, e.g., Golgi cells.

Several authors have reported that recurrent collaterals of Purkinje cells contain stacks of tubular membranes similar to those that were described by DUNCAN and WILLIAMS (1962), ANDRES (1965), and others in myelinated nerve fibers. These authors proposed that these formations should be used as criteria for the identification of recurrent collaterals in electron micrographs (ECCLES, ITO, and SZENTÁGOTHAI, 1967, p. 182; HÁMORI and SZENTÁGOTHAI, 1968; O'LEARY *et al.*, 1968). LARRAMENDI (1969a) reported that in the cerebella of the mice that he studied, these tubular systems were extremely infrequent and were therefore not useful criteria for recognizing the collaterals of the Purkinje cell axon. In our own studies on the rat's cerebellar cortex, we have encountered such tubules in only one animal, which displayed a few myelinated axons containing them in lobule X. It should be pointed out here that the tubules discovered by DUNCAN and WILLIAMS have been seen in various sites in the central nervous system. An intensive study of them in the lateral vestibular nucleus of the rat (SOTELO and PALAY, 1971) showed that they are usually found in the company of several degenerative phenomena, such as hyperplastic endoplasmic reticulum, whorled membranes, and gigantic and distorted mitochondria. It is therefore questionable whether they should be considered typical or normal constituents of Purkinje cell axons.

There are, however, more reliable criteria for the identification of the myelinated segments of the Purkinje cell axons in electron micrographs. These are the presence of the hypolemmal cisterna and the distinctive patterns of smooth endoplasmic reticulum that are so highly developed in both the perikaryon and dendrites of the Purkinje cell. They are also prominent in the axons. Furthermore, as in their terminals, the myelinated Purkinje axons have a relatively dense axoplasm containing longitudinally oriented neurofilaments and microtubules (Figs. 50 and 157).

Synaptic Relations of the Recurrent Collaterals. In the electron micrographs of the molecular layer, terminals of the Purkinje cell recurrent collaterals must be distinguished from axons belonging to basket, stellate, and Lugaro cells, parallel fibers, and climbing fibers. All of these axons, with the exception of those from Lugaro cells (which have yet to be recognized in electron micrographs), have cytological characteristics and synaptic patterns that set them apart from one another and from the recurrent collaterals. Most of them can be seen together in Fig. 207. As the other fibers will be described in detail in later chapters, their distinguishing characteristics will be only briefly summarized here for the purpose of comparison. The varicosities of parallel fibers are small and globular, contain a loose collection of round synaptic vesicles, and articulate with the thorns of Purkinje cell dendrites. Climbing fiber varicosities are large, distended sacs jammed with round synaptic vesicles and microtubules in a dense matrix. These, too, synapse with dendritic thorns of Purkinje cells, but they are thorns projecting from the larger dendritic trunks. Basket fibers are recognized by their fleshy proportions, abundance of neurofilaments, and loose aggregates of elliptical and round synaptic vesicles in a light matrix. The majority of these synapse on the shafts of Purkinje cell dendrites in the lower molecular layer. The axons of stellate cells are slender, crooked fibers, having closely spaced varicosities that contain round and elliptical synaptic vesicles. The orientation of the parallel fibers in the longitudinal axis of the folium serves to eliminate the most numerous type of axonal profile from any ambiguity.

In the supraganglionic plexus in the rat, recurrent collaterals of the Purkinje cell axon are relatively sparse. A few profiles of myelinated secondary, tertiary, and quartenary branches of collaterals can be found vertically oriented between the apices of the Purkinje cell somata. At a level of about 10–20 μm above the layer of Purkinje cells, these myelinated axons bend away from the vertical and end in fleshy varicose terminals. Occasionally they can be found synapsing on the shafts of medium-sized and smaller Purkinje cell dendrites and, more rarely, on dendritic thorns (see Fig. 6 in CHAN-PALAY, 1971). Synapses of recurrent collaterals on the somata of basket cells are equally rare (see Fig. 7 in CHAN-PALAY, 1971). However, when they are found, they usually occur in clusters of two or three club-shaped boutons.

The infraganglionic plexus lies in a transition zone between the upper granular layer and the layer of Purkinje cell bodies. Through this zone all axons and dendrites entering or leaving the molecular layer must pass. The fusiform intermediate cells of Lugaro are nestled here, spreading their dendrites horizontally, largely within it. By far the most distinctive structure in the infraganglionic zone is the basket that cradles each Purkinje cell and ends

◄

Fig. 47. Terminals of a Purkinje axon recurrent collateral. A fine axonal thread (ΔΔ) connects two varicosities (*PrC*) in the course of a terminal collateral. The terminal synapses upon the soma of a Purkinje cell (*PC*) by means of a macular synaptic junction (open arrow). The large arrow indicates a subsurface cisterna. Lobule V. ×40000

in the pinceau around the initial segment of the emerging axon. But the recurrent collaterals of Purkinje cell axons also ramify in this zone, forming a large part of the neuropil. Such collaterals are more abundant in the infraganglionic region than anywhere else in the rat's cerebellar cortex. Unlike basket fibers, in which the axoplasm can be filled with neurofilaments and in which synaptic vesicles occur only near synaptic interfaces, the fleshy terminal branches of recurrent collaterals have synaptic vesicles dispersed over considerable lengths (Fig. 45). Synaptic boutons originating from such collaterals penetrate the pericellular neuroglial sheath to form axo-somatic synapses on some Purkinje cell bodies (Figs. 46 and 47). These usually occur in clusters on the lower surface of the cell. The boutons are impressed lightly into the convexity of the cell surface, and are covered over by slips from the neuroglial sheath. Each bouton, about 0.8–1 μm in diameter, contains at least several profiles of slender mitochondria and a collection of polymorphic synaptic vesicles in a relatively dense matrix. At the synaptic interface each profile of a bouton may display from one to three small, macular synaptic complexes, with widened synaptic clefts as described above. The Lugaro cells and their dendrites receive considerable numbers of synapses from the recurrent collaterals. The terminals of these collaterals are usually clustered on the cell surface, each having a series of the characteristic small, macular synaptic complexes on the junctional interface.

Besides these synapses with the somata and initial portions of dendrites belonging to Purkinje and Lugaro cells, terminals of the recurrent collaterals articulate in a complex way with a number of other dendrites in the infraganglionic region. These dendrites presumably belong to Golgi cells and granule cells in the vicinity. There are two basic patterns of synaptic relationship with these dendrites. Sometimes a recurrent collateral articulates with two or more dendrites, or more commonly a single dendrite receives several boutons only one of which belongs to a recurrent collateral, the others being granule cell axons or even mossy fibers. Although each of these patterns can be found separately, it is more common to find clustered arrangements in which the recurrent collateral boutons are involved in pseudo-glomerular formations combining a number of fiber types, including mossy and climbing fibers (see Figs. 14 and 15 in CHAN-PALAY, 1971). The sets of junctions, grouped in axodendritic pairs, are partially or completely enclosed in thin lamellae of neuroglial cytoplasm.

Thus the electron micrographs display an extremely dense fiber plexus beneath the layer of Purkinje cell bodies, the complexity of which is hardly adumbrated by the tangle of varicose nerve terminals seen in Golgi preparations with the light microscope. Here, intermingled with the descending basket fibers and ascending granule cell axons and in between the ranks of pinceaux, the ramifying recurrent collaterals synapse with the somata of Purkinje cells and Lugaro cells, as well as with the dendrites of Lugaro, Golgi, and granule cells. Climbing fibers, mossy fibers, and axons of Golgi cells also articulate with some of the same dendrites and somata in this region. This complex region is dominated by the synapses between recurrent collaterals and the Lugaro cell. We believe that an understanding of the physiology of the infraganglionic plexus would tell a great deal about the role of the Purkinje cell recurrents and the Lugaro cell.

As described earlier, a plexus of recurrent collaterals of Purkinje cells also ramifies in the middle and deeper granular layer. In electron micrographs of this layer it can be seen that the fibers of this plexus terminate on the somata and dendrites of deep Golgi cells with clustered boutons. The contact zone is characterized by discrete, macular synaptic complexes with a somewhat widened synaptic cleft and symmetrical, thin pre- and postsynaptic densities. Recurrent collaterals also form extensive *marron* synapses with some deep Golgi cells (CHAN-PALAY, 1971; CHAN-PALAY and PALAY, 1971a and b). This type of synapse will be described in the chapter on the Golgi cell, and will not be discussed further here, except to say that unlike mossy fiber and climbing fiber synapses *en marron*, the free surface of a recurrent collateral in a synapse *en marron* does not articulate with other dendrites as in a glomerulus.

While considering these terminals of recurrent collaterals on deep Golgi cells, it is well to recall certain observations of RAMÓN Y CAJAL (1912). In a reinvestigation of the granular layer with neurofibrillary stains, he discovered extensive pericellular plexuses around a particular kind of short-axon Golgi cell in young dogs, cats, and man, and in animals with experimental lesions. HÁMORI and SZENTÁGOTHAI (1968) and LLINÁS and PRECHT (1969) both call attention to this paper, with some justice, as it has been ignored. There is no question, however, as to RAMÓN Y CAJAL's opinion on the identity of these Golgi cells. He states clearly that the neurons around which the plexuses are found are *not* the larger, deep Golgi cells of the granular layer, but the smaller, fusiform neurons found beneath the Purkinje cells, whose axons traverse the granular layer and enter the white matter. We presume on the basis of the location and shape of the cell bodies that he meant the Lugaro cells, even though his description of their axons differed from

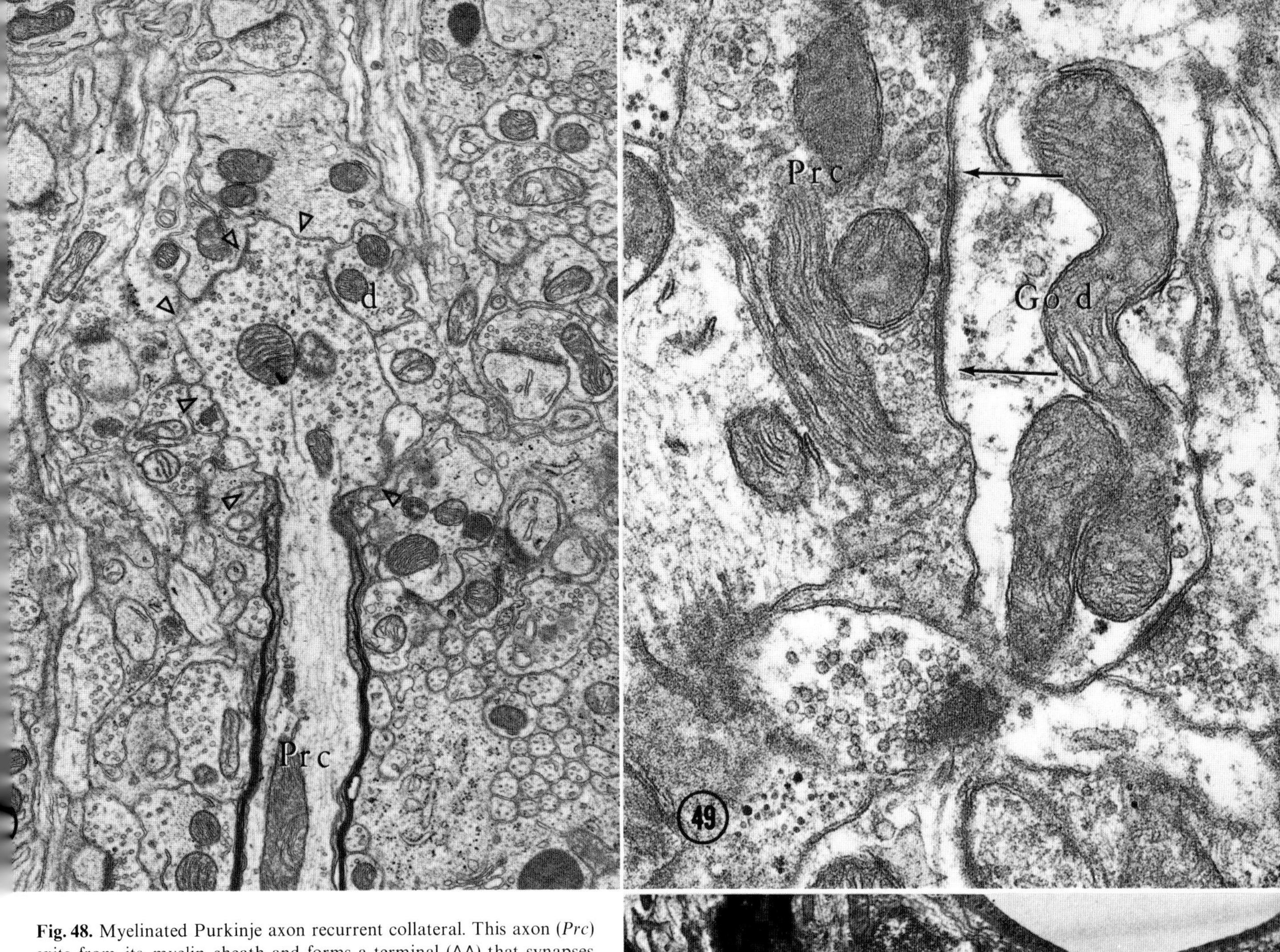

Fig. 48. Myelinated Purkinje axon recurrent collateral. This axon (*Prc*) exits from its myelin sheath and forms a terminal (ΔΔ) that synapses with a dendrite (*d*) in the molecular layer. Lobule X. ×16000

Fig. 49. Synapse of a Purkinje axon recurrent collateral upon a Golgi dendrite in the granular layer. The dendrite of a Golgi cell (*Go d*) is shown in a synapse with a terminal of a Purkinje axon (*Prc*). The synaptic junction (between arrows) is longer than the usual macular synapses between Purkinje axon terminals and Purkinje cells or Lugaro cells. Lobules II and III. ×44000

Fig. 50. The myelinated Purkinje cell axon. The axoplasm of the Purkinje cell (*PC ax*) is distinctive. The matrix of the axoplasm is relatively dark, with many longitudinally oriented neurofilaments, a few microtubules and mitochondria. The hallmark of these axons is the presence of the system of tubules of the smooth endoplasmic reticulum throughout the axoplasm and particularly the well-developed hypolemmal cisterna. Lobule V. ×10000

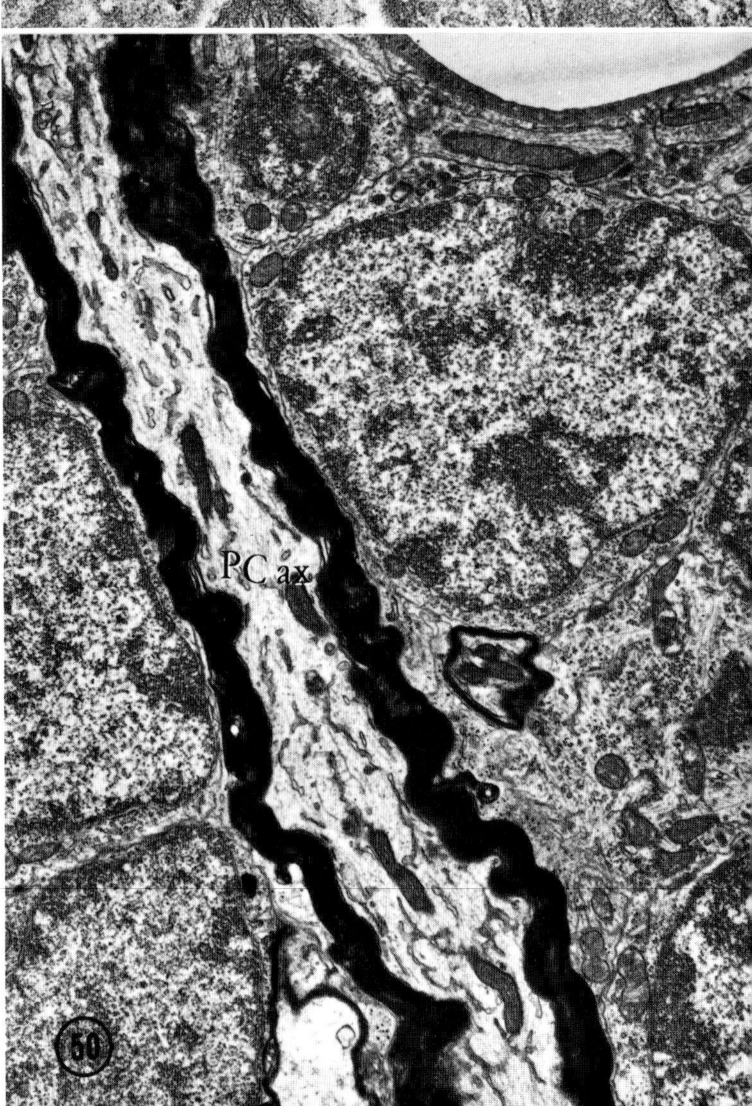

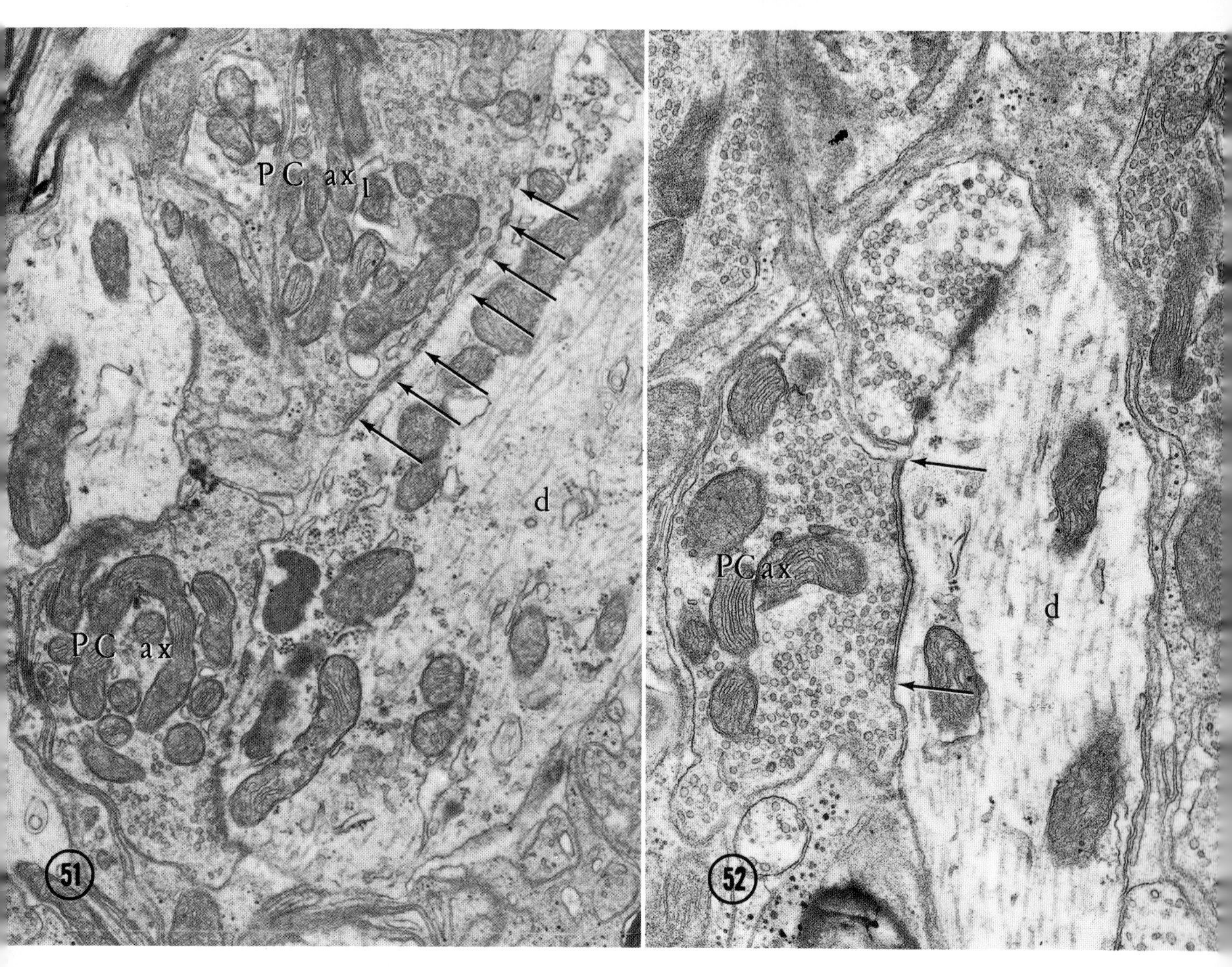

Fig. 51. Purkinje axon terminals in the nucleus interpositus. The axon terminals of the Purkinje cell resemble terminals of the recurrent collaterals of Purkinje axons in the cortex. There is a dense axoplasmic matrix and a polymorphic array of synaptic vesicles. The junction of one terminal (*PC* ax_1) in contact with a dendrite of a large neuron (*d*) displays six attachment plaques and one synaptic complex. Compare this with Figs. 46 and 47. × 30000

Fig. 52. The termination of a Purkinje axon in the nucleus interpositus. The axon (*PC ax*) terminates on a dendrite (*d*) and forms a large synaptic junction (*between arrows*). Compare this with Fig. 49. × 30000

that of LUGARO (1894) and FOX (1959). The Lugaro cell is discussed in Chap. V of the present work. LEMKEY-JOHNSTON and LARRAMENDI (1968) and LARRAMENDI and LEMKEY-JOHNSTON (1970) have provided confirmation of this interpretation by showing that a category of "deep basket cells" receive a large proportion of the recurrent collateral synapses. Their description of the "deep basket cells" conforms with the descriptions of LUGARO, RAMÓN Y CAJAL, and FOX for the cell-type that must be the Lugaro cell. At the same time, they found almost no recurrent collateral synapses on true Golgi cells, that is, large stellate cells with highly ramifying axons in the granular layer. It is difficult to reconcile these observations and those of RAMÓN Y CAJAL with our own observations showing a dense plexus of recurrent collaterals around the true Golgi cells (Figs. 43 and 96). Both electron microscopy and Golgi preparations show that the Golgi cell is well supplied with endings of recurrent collaterals. They contact the Golgi cell by one or all of these three modes: axo-somatic boutons terminaux, axo-somatic synapses *en marron*, and axo-dendritic synapses. The Lugaro cell in our preparations is also well covered by endings of recurrent collaterals, both axo-somatic and axo-dendritic.

A great deal of confusion has been introduced by RAMÓN Y CAJAL'S failure to recognize the Lugaro cell as a separate cell type and by his unfortunate failure to confirm LUGARO'S (1894) description of the axon of this cell. FOX (1959) clarified the situation somewhat, but many writers on the cerebellar cortex have merely continued to classify all large neurons in the granular layer as Golgi cells. As a result it is difficult, often impossible, to be sure which cells they have in mind when they discuss the endings of Purkinje cell recurrents. It is particularly difficult to learn which cell the neurophysiologist is dealing with, as a simple statement of depth of the recording electrode is not sufficient to characterize the cell recorded from.

Purkinje Cell Axon Terminals in the Central Nuclei. All of the structural characteristics of the recurrent collaterals and their terminals can be found again in the central nuclei of the cerebellum where the axons of the Purkinje cells converge on cell bodies and dendrites alike (CHAN-PALAY, 1973a). The myelinated fibers belonging to Purkinje cells contain the same dense matrix and conspicuous hypolemmal cisterna that they do in the granular layer. Upon entering the nuclei they soon give rise to elongated terminals that bear either interrupted macular synapses or larger synaptic junctions on their junctional faces (Figs. 51 and 52). The terminals contain a relatively dense matrix in which pleomorphic synaptic vesicles are dispersed. These endings correspond precisely to their congeners, the recurrent collaterals, in the cortex.

7. The Neuroglial Sheath

One of the most peculiar characteristics of the Purkinje cell is that the perikaryon and all of its processes are almost completely enshrouded in neuroglial processes. This sheath is generated principally by the specialized neuroglial cells lying in the Purkinje cell layer, the Golgi epithelial cells, and in the deeper molecular layer, the Fañanas cells. Both of these are varieties of protoplasmic astrocytes (CHAN-PALAY and PALAY, 1972c). They send out velamentous sheets that wrap round the cell body, the dendrites, and the initial segment (see Fig. 40), isolating the Purkinje cell from the sea of axons in which it is immersed. The dendritic thorns poke their heads through this sheath in order to contact the parallel fibers and climbing fibers (Fig. 33), and the basket fibers must also penetrate it in order to synapse with the larger dendrites, perikaryon, and initial segment. It should, of course, be recognized that the great majority of the axon and its collaterals are covered with myelin, which is also a neuroglial sheath. A more detailed examination of the neuroglial sheath will be deferred until the other cells and nerve fibers have been considered (Chap. XI).

8. Some Physiological Considerations

Since it is well established that the action of Purkinje cell axon terminals upon neurons outside of the cerebellar cortex is inhibitory in nature (ITO and YOSHIDA, 1964, 1966; ITO, YOSHIDA, and OBATA, 1964), it is reasonable to assume that the recurrent collaterals exert a similar inhibitory action on their target cells within the cerebellar cortex. Data from physiological experiments in cats show that following juxtafastigial stimulation, a weak inhibitory response is evoked in Purkinje cells, presumably through the antidromic excitation of their recurrent collaterals (ECCLES, LLINÁS, and SASAKI, 1966e; ECCLES, ITO, and SZENTÁGOTHAI, 1967, p. 185). This feeble effect may be related to the relatively small number of terminals synapsing with Purkinje cell somata and dendrites. It is necessary to point out here that some Purkinje cells have been observed to have more terminals from recurrent collaterals of Purkinje cell axons than others. There is so far no evidence from physiological work that the recurrent collaterals from a particular axon synapse with the soma or dendrites of the cell of origin. The distribution of the supra- and infraganglionic plexuses in the rat, and probably in other mammals, indicates that the inhibitory effects should be felt principally in the neighboring Purkinje cells in the same row on either side of the cell of origin. Since the axonal field is an ellipse (CHAN-PALAY, 1971) the major axis of which lies in the parasagittal plane, the extent of inhibition in the longitudinal axis of the folium should be limited to only a few cells in that direction. This is in apparent disagreement with RAMÓN Y CAJAL (1911), who declared that the fibers of the supraganglionic plexus extended for considerable distances in the longitudinal axis of the folium. In addition to the possible existence of serious species differences, however, it is probable that the thicker parallel fibers in the deep molecular layer were confused with the recurrent collaterals (MUGNAINI, 1972).

Similarly, stimulation of the white matter in the depths of the cerebellum results in strong inhibition of basket cells and other interneurons in the molecular layer (LLINÁS and AYALA, 1967; LLINÁS and PRECHT, 1969). ECCLES, LLINÁS, and SASAKI (1966a; and ECCLES, ITO, and SZENTÁGOTHAI, 1967) also observed a highly effective inhibitory action of antidromic stimulation on interneurons presumed to be Golgi cells. Since the action of these interneurons is inhibitory, their inhibition by the collaterals of Purkinje cell axons would result in the release of granule cells and Purkinje cells, by a process called disinhibition. Specifically, the inhibition of Golgi cells would depress the inhibitory influence of their axons in the glomeruli of the granular layer, allowing incoming excitation from mossy and climbing fibers to activate granule cells. The inhibition of basket cells by recurrent collaterals would also mitigate the control of basket cells over the Purkinje cell itself. This relaxation would permit the restoration of the Purkinje cell to its previously responsive state and allow afferent inputs to affect the Purkinje cell again. Thus, the Purkinje cell would contribute a tempering influence, tending to cut off or terminate the inhibitory side-effects of discharges in the climbing and mossy fiber pathway to its own dendritic tree.

The role of the Lugaro cell in the recurrent collateral circuits remains an enigma, partly because physiological data on this cell are lacking. We may speculate on the basis of their anatomy that these cells also function as coordinating neurons in the cerebellar circuits. Their location between and beneath the Purkinje cells and their dendrites extending in the parasagittal plane are suited to the task of gathering information from specific strips aligned at the junction of the granular and molecular layers. By way of their axons in the molecular layer their influence is transmitted to unidentified neurons in the lower molecular layer. As these can only be basket or Purkinje cells or both, we can suppose that the Lugaro cell fits into a positive feed-forward circuit focussed on the Purkinje cell itself.

9. Summary of Intracortical Synaptic Connections of Purkinje Cells

Afferents	granule cell	———	Purkinje cell
	basket cell	———	Purkinje cell
	stellate cell	———	Purkinje cell
	Purkinje cell axon	———	Purkinje cell
	climbing fiber	———	Purkinje cell
Efferents	Purkinje cell	———	Lugaro cell
	Purkinje cell	———	Purkinje cell
	Purkinje cell	———	Golgi cell
	Purkinje cell	———	granule cell

Chapter III

Granule Cells

The granule cells of the cerebellar cortex are among the smallest nerve cells in the body. In the middle of the nineteenth century they were still considered to be connective tissue cells, and as late as 1900 BETHE doubted their neural nature. Our present knowledge of them is due first to GOLGI (1874 and 1883), who demonstrated their dendritic and axonal processes, and second to RAMÓN Y CAJAL (1888a and b, 1911), whose precise description of their form and synaptic relations has hardly been modified by subsequent work.

The number of granule cells is enormous, and they are densely packed in the cerebellar cortex of all vertebrates. In the human cerebellum BRAITENBERG and ATWOOD (1958) estimated a density of 3 to 7×10^6 granule cells per mm^3 of granule cell layer, or with an average thickness of 0.3 mm, from 3000 to 9000 granule cells for every Purkinje cell. They calculated that in the entire cerebellar cortex of man there should be 10^{10}–10^{11} granule cells. SMOLYANINOV (1971) found considerably smaller numbers. He gives the density of granule cells in man as 2.1×10^6 per mm^3, and the ratio of granule cells to Purkinje cells as 1600. In the monkey FOX and BARNARD (1957) found an average of 2.4×10^6 granule cells per mm^3 of granular layer or, with an average thickness of 0.2 mm, 960 granule cells for every Purkinje cell. In the cat PALKOVITS *et al.* (1971b) counted 6.2×10^6 granule cells per mm^3, which, when corrected for shrinkage in the preparations, gave 2.8×10^6 per mm^3 in the living animal. They obtained a total number of 2.2×10^9 granule cells in the cerebellum of the cat. Although they agreed with SMOLYANINOV (1971) on the density of granule cells, they recorded a ratio of 1769 granule cells for every Purkinje cell, nearly three times the value reported by SMOLYANINOV for the same species. In the rat we have only the numbers given by SMOLYANINOV, who found a density of 3.2×10^6 granule cells per mm^3 and a ratio of 250 granule cells for every Purkinje cell. The lower ratio in the rat as compared with the cat might prove advantageous in physiological studies.

1. The Granule Cell in the Optical Microscope

These nerve cells are unusual in many respects. In Nissl preparations their cell bodies, from 5 to 8 µm in diameter (in the rat, most are about 5–6 µm), seem to consist of nothing more than their round nuclei, and no Nissl substance can be seen (FOX and BARNARD, 1957). They are pressed so close together into rounded clusters or elongated columns that there would appear to be no space between them for their processes. It is no wonder that they were once thought to resemble lymphocytes. Furthermore, their nuclei, contrasting with those of nearly all other neurons, are darkly basophilic; the chromatin is condensed into massive blocks, as in small lymphocytes or plasma cells, that are attached to the margin of the nucleus. Sometimes the center of the nucleus is occupied by a block of chromatin, which RAMÓN Y CAJAL (1911) suggested might represent the nucleolus. Certainly in ordinary sections stained for the light microscope the nucleolus is usually impossible to recognize. An intranuclear rodlet is visible in many granule cells (see RAMÓN Y CAJAL, 1911, Fig. 2) as a bent or twisted line across the nucleus.

Rapid Golgi preparations show that the granule cell has an unmistakably characteristic shape, a globular cell body with three or four short, radiating dendrites. These processes are typically sinuous, branching only at their ends, where they produce a gnarled, claw-like, sometimes varicose inflorescence. An astounding variety of forms is generated from the disposition of these few processes, as may be seen in Figs. 5 and 53, in which the granule cells, hanging by their axons from the parallel fibers in the molecular layer, appear like marionettes on strings, gesticulating absurdly. The terminal digitiform branches of the dendrites tend to curve around and enclose a small space, which is occupied by the expanded terminal of a mossy fiber. The dendrites from several granule cells, perhaps as many as six, converge upon the mossy fiber terminal. This articulation, a ball-like tangle of dendrites and axons, was given the name *glomerulus cerebellosi* by HELD (1897). It will be described in a later section (see Chap. VI).

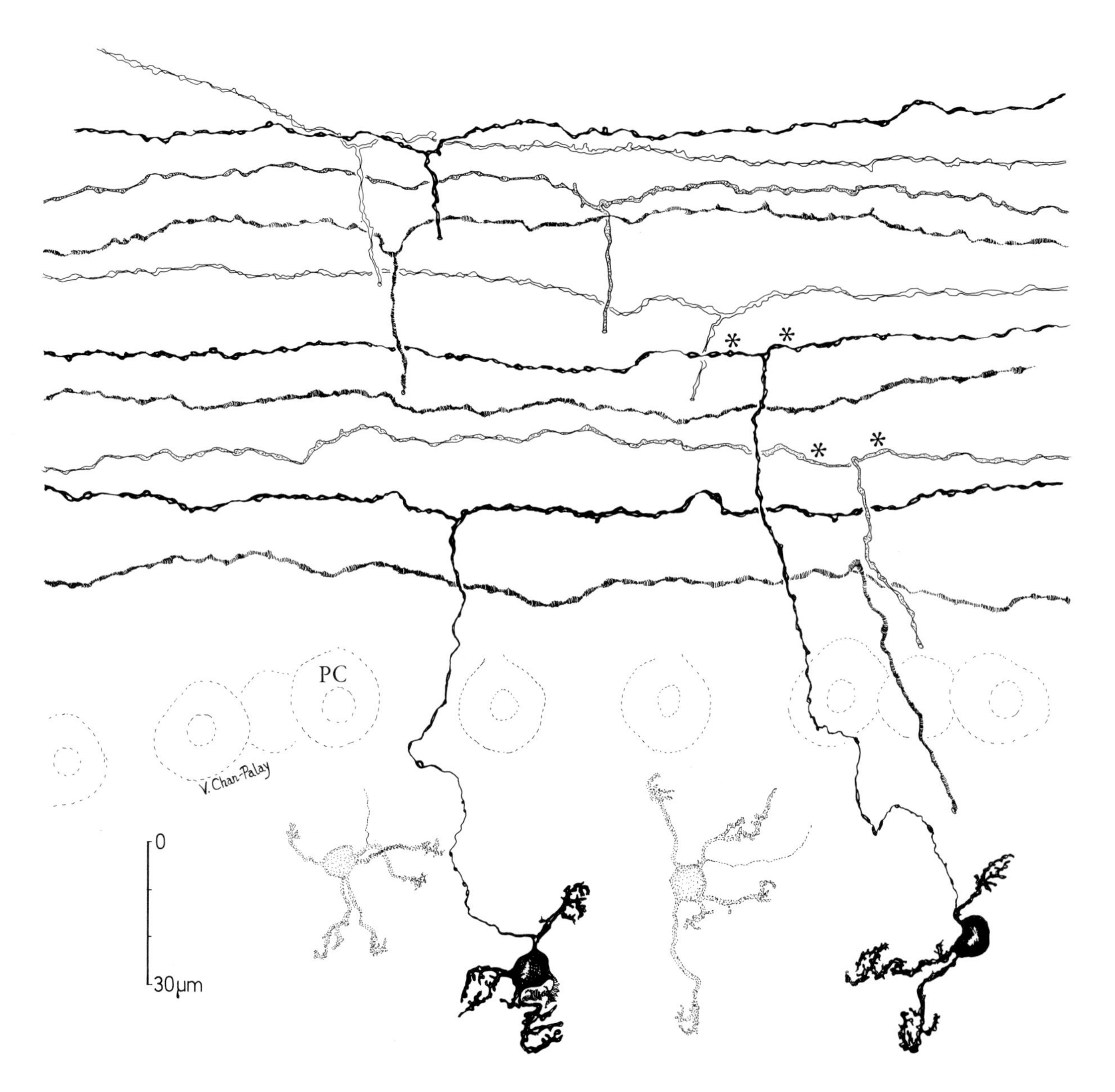

Fig. 53. Granule cells and their axons; parallel fibers. The axon of the granule cell usually issues from a dendrite or the soma. It rises in a tortuous course to the molecular layer where it bifurcates like a T (*asterisks*) or a Y to produce a parallel fiber. The layer of Purkinje cell bodies (*PC*) is indicated. Rapid Golgi, 90 µm, frontal, adult rat. Camera lucida drawing

The axon of the granule cell originates from the cell body, or frequently from the thicker stem of a dendrite, and snakes its way up through the granular layer. It sometimes undertakes wide divagations, descending or looping around, before settling into its typical vertical course (Figs. 53 and 54). In the upper third of the granular layer, axons from neighboring granule cells come together to form thin bundles, which penetrate between the Purkinje cell bodies and ascend into the molecular layer. Each bundle is accompanied by one or more neuroglial cells oriented parallel with it (Bergmann fibers, Golgi epithelial cells, Fañanas cells, see below). The ascending portion of the granule cell axon is 0.1–0.3 µm in diameter, and it is initially smooth. While still in the granular layer it develops kinks and twists marked by warty excrescences, a characteristic that has also been recently noted by MUGNAINI (1972). After it enters the molecular layer, the axon becomes less gnarled, perhaps guided by its neuroglial sheath (MUGNAINI and FORSTRØNEN, 1966 and 1967; RAKIC, 1971); elongated varicosities, particularly near its division, are common. Electron microscopy shows that all of these irregularities are the sites of synaptic articulations with the dendrites of Golgi cells in the granular layer and with dendritic thorns of Purkinje cells after it has entered the molecular layer. In this layer each granule cell axon bifurcates like a *T*, giving rise to

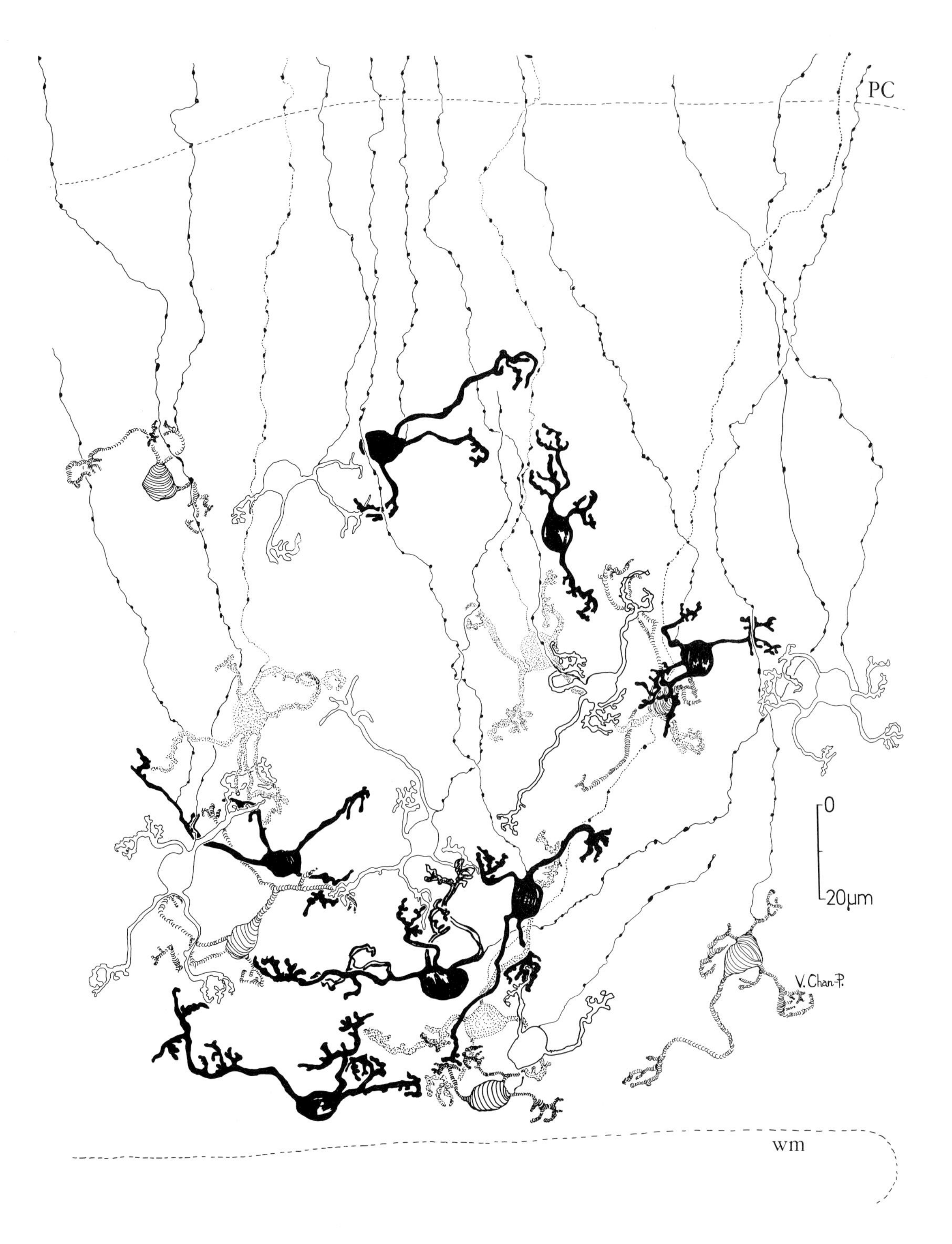

Fig. 54. The granule cells. The dendrites of many granule cells are shown with their claws bent in a contorted fashion round clear foci in the neuropil. These foci are unimpregnated glomeruli. The axons of these cells ascend in a somewhat wavy course through the granular layer. Along this course, small varicosities occur, probably corresponding to synaptic sites (see Fig. 68). The white matter (*wm*) and the layer of Purkinje cells (*PC*) are indicated. Rapid Golgi, sagittal, 90 µm, adult rat. Camera lucida drawing

a pair of long, thin fibers, 0.1–0.2 µm in diameter, running in opposite directions parallel to the longitudinal axis of the folium (Figs. 5 and 53). For this reason they were termed parallel fibers by RAMÓN Y CAJAL (1888 b). The entire thickness of the molecular layer is filled with these fine parallel fibers coursing at right angles to the plane of the Purkinje cell dendritic expansion. As each parallel fiber passes through the Purkinje cell dendritic tree, it makes one or more synapses on the thorns projecting from the spiny branchlets. These articulations are effected by means of small varicosities, nubbins, and excrescences which protrude from the parallel fiber (ESTABLE, 1923; FOX and BARNARD, 1957). Synapses are also made with the dendrites of stellate, basket, and Golgi cells by means of these appendages.

In general the parallel fibers in the lower third of the molecular layer are thicker than those higher up (FOX, UBEDA-PURKISS, and MASSOPUST, 1950; FOX and BARNARD, 1957). This difference is more striking in the cerebella of large animals such as cat, monkey, and man than it is in the rat or mouse. Nevertheless, it is evident in both Golgi preparations and electron micrographs when measurements are made. In the cat, PALKOVITS *et al.* (1971c) found that the average thickness of parallel fibers, measured in electron micrographs of cross sections, was 0.1917 µm in the outer third of the molecular layer, 0.1921 µm in the middle third, and 0.2067 µm in the inner third. Only the last value was thought to be significantly different from the others, but for some unstated reason they considered that this difference in thickness had no "biological relevance". It is unlikely that they recognized and measured the myelinated parallel fibers in the lower molecular layer (MUGNAINI, 1972, see below). Perhaps by including these myelinated fibers, SMOLYANINOV (1971) measured parallel fibers up to 1.0 µm in diameter in the cat.

The differential distribution of fiber calibers corresponds to the distribution of the cell bodies from which the parallel fibers originate. Granule cells located deeper in the granular layer give rise to the parallel fibers running in the deeper molecular layer; granule cells in the upper granular layer provide the parallel fibers for the upper molecular layer. While this distribution is true in general, as can be verified by a study of well-impregnated rapid Golgi preparations, there are numerous exceptions, in which axons originating from deep granule cells give off their parallel fibers in the more superficial molecular layer, and axons coming from the upper granular layer already divide in the lower molecular layer. Nevertheless, the prevalent rule means that the ascending axons of the deeper granule cells are a little shorter than those arising from upper granule cells, because the granular layer is on the average somewhat thinner than the molecular layer. Since the ascending axons of the granule cells are gathered together into small fascicles, the bifurcations of the fibers in these bundles are also arranged in a vertical sequence, with the lowest bifurcations being those of the deepest granule cells. It will be noticed here that the granular layer is often greatly attenuated in the troughs of the sulci, whereas the molecular layer is not much changed in these regions. At the depths of the sulci the ascending granule cell axons approach the molecular layer from the sides at a great angle, and their trajectory is at least as long as it is in the more regular parts of the cortex, on the walls and crests of the folia.

Not only are the stems and parallel fibers of the deeper granule cells thicker than the others, they are also frequently myelinated. The number of myelinated parallel fibers varies a great deal among the mammalian species. In the cat, myelinated parallel fibers constitute an important fraction of the supraganglionic plexus (MUGNAINI, 1972). In the rat, myelinated stems and parallel fibers are often encountered in the deeper molecular layer, but they are not so numerous as in the cat.

Certain additional details on the form of the parallel fibers deserve mention. They are by no means so regular and rigidly linear as circuit diagrams lead one to expect. Although the dichotomy of the ascending granule cell axon is usually likened to a *T*, it is much more often a Y-shaped bifurcation (Fig. 53). It is also frequently tilted from the frontal plane, and the two branches can be given off asymmetrically so that they attain different levels in the molecular layer before they turn to run longitudinally. The two branches can be of asymmetrical lengths, especially if they originate in the more lateral parts of a folium. Finally, the fibers deviate frequently from a straight line, especially in the horizontal plane, by flexuosities, kinks, and twists, introduced on account of the necessity of avoiding dendrites, cell bodies, and blood vessels in their paths.

a) Some Numerical Considerations

It is difficult to obtain true measurements of the lengths of parallel fibers. Judging by the number of granule cells and their packing density in a typical folium, BRAITENBERG and ATWOOD (1958) concluded that the length of parallel fibers in the human brain must be between 1 and 10 mm. FOX and BARNARD (1957) measured fibers in Golgi preparations of the monkey and concluded that the maximal length is 3 mm, but were unable to establish a minimal length. They thought that the average length in the monkey should be close to 1.5 mm. Both degeneration experiments (SZENTÁGOTHAI, 1965) and electrophysiological studies (DOW, 1949; ECCLES, ITO, and SZENTÁGOTHAI, 1967) indicate that in the cat the maximal length could be no more than 3–5 µm. In more recent calculations based on statistical data, SMOLYANINOV (1971) obtained the following average lengths of parallel fibers for a variety of species: man, 2.6 mm; monkey, 2.2 mm;

cat, 1.6 mm; rat, 0.9 mm; and mouse, 0.6 mm. According to SMOLYANINOV, the average length of the parallel fibers increases steadily with increasing depth in the molecular layer, but he provided no evidence for this statement, and it is difficult to see how it could have been obtained with currently available techniques. Nevertheless, his figure for the cat agrees fairly well with values for the living cat reported by PALKOVITS *et al.* (1971c), who obtained lengths close to 2.0 µm by two different methods of calculation.

It had always been thought that since the parallel fibers run longitudinally in the folia, each one should synapse with each of the successive Purkinje cells that it encounters in its traverse. FOX and BARNARD (1957) suggested that the spiny branchlets are set in a staggered arrangement in order to contact as many parallel fibers as possible as they pass through the tree. Their counts, however, revealed that there were 200000–300000 parallel fibers coursing through the dendritic tree of a typical Purkinje cell in the monkey and only 60000 dendritic thorns. If it is assumed that every thorn on the spiny branchlet synapsed with a different parallel fiber and that each parallel fiber synapsed only once in its passage through the tree, then the total number of synapses made must equal the number of thorns, or only one in every three to five parallel fibers can synapse in any one Purkinje cell dendritic tree. Thus, although a single parallel fiber could pass through 230–460 Purkinje cell arborizations, depending on its length, it would synapse with only one in three to five, or 46–92 cells. PALKOVITS *et al.* (1971c) arrived at a very similar number for the cat, by taking into consideration the ratio of profiles of synaptic thickenings to those of nonsynaptic portions of parallel fibers in electron micrographs of transverse sections. They estimated that of the 400000-odd parallel fibers traversing the dendritic arborization of a single Purkinje cell in the cat, only 20% (80000) synapsed with it. Thus an average parallel fiber, 2 mm long, would pass through the dendritic trees of 225 Purkinje cells but would synapse with only every fifth one, or 45 cells altogether, a number close to the minimal number estimated from the data of FOX and BARNARD in the monkey.

These calculations are interesting in view of the curious stability discovered by SMOLYANINOV (1971) in his statistical study. He found that while the average length of the parallel fiber varied by a factor of 4 in the series of animals he considered (see above), the number of Purkinje cells contacted synaptically remained nearly constant: from 29 in mouse to 37 in man. This constancy may only be apparent, as it would require that the thickness of the dendritic plate or the distance separating successive plates, or both, increase with the increase in length of the parallel fiber in different species. There seems to be no evidence that this requirement is fulfilled by the data.

These various estimates are based partly on empirical data and partly on certain simplifying assumptions about the parallel fiber-Purkinje dendrite synapse. It is assumed, for example, that a parallel fiber synapses only once as it crosses a dendritic tree, that each synaptic swelling articulates with only one thorn, and that each thorn synapses with only one fiber. It is furthermore assumed that the synaptic varicosities are spaced regularly at about 10 µm intervals along the parallel fiber, and that only a very small proportion of the varicosities articulate with stellate, basket, or Golgi dendrites.[5]

We considered that direct measurements on parallel fibers in rapid Golgi preparations of the rat might provide some data that could modify these assumptions. Even though Golgi preparations cannot tell us whether a particular varicosity synapses with a Purkinje cell thorn or a basket cell dendrite, it does rather accurately localize the synaptic region of the fiber, as correlation with electron microscopy demonstrates. The question to be treated then was the number of varicosities per unit length of parallel fiber. This question was attacked at two levels of observation, first in the light microscope and then in the high voltage electron microscope.

With the aid of a camera lucida, careful tracings were made, at a magnification of 1800×, of 95 parallel fibers in three longitudinal sections taken from three animals. Fields with rather dense impregnation of the whole molecular layer were selected, and all of the fibers were drawn. The molecular layer was subdivided into equal thirds in order to compare parallel fibers at different depths. The total length of each fiber was measured with a cartographer's wheel. In all, 2222 µm of parallel fibers were measured. Maximal and minimal dimensions of each varicosity were measured, as well as the interval between varicosities, i.e., the length of the fiber from the end of one varicosity to the beginning of the next. The area of each varicosity in the tracing was calculated as the simple product of the maximal and minimal dimensions, in order to give an index of size. These areas are therefore larger than the actual areas, but are valid for a comparison of the sizes of varicosities in different levels of the molecular layer. Nine hundred varicosities and 877 intervals

5 PALKOVITS *et al.* (1971c) tried to correct for the proportion of varicosities that articulate with cells other than Purkinje cells, but they counted on the axons terminating only on dendritic spines of these cells and ignored the larger proportion that synapses on the shafts of the dendrites.

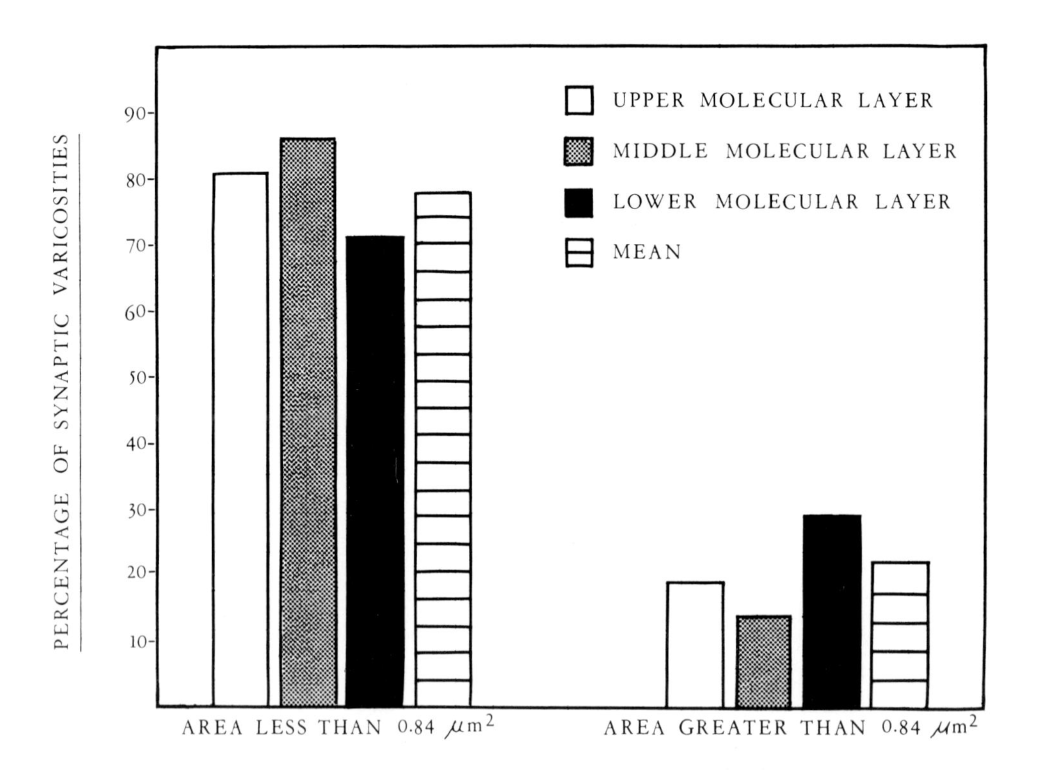

Fig. 55. Histogram showing the distribution of large and small varicosities of parallel fibers according to their position in the molecular layer. The areas of synaptic varicosities in the three portions of the molecular layer, upper (white bar), middle (stippled), and lower (black) were measured. Most of the synaptic varicosities (78%) had areas less than 0.84 µm^2. The remainder had areas greater than 0.84 µm^2. The lower third of the molecular layer tended to have a larger number of large synaptic varicosities (black bar)

were measured, approximately 300 of each in each of the three subdivisions. After inspection of the results, it appeared that most of the varicosities were less than 0.84 µm^2 in area and only a few were larger (Fig. 55). The areas were therefore tabulated according to this difference. The results are shown in Table 5.

According to these measurements in the rat, the synaptic varicosities in parallel fibers are much closer together than studies in other animals have suggested. Nearly 40% of the length of the average parallel fiber is occupied by synaptic varicosities, separated from one another by a mean distance of 1.7 µm. The histogram in Fig. 56 displays the distribution of the measured intervals between varicosities. It shows that slightly more than half of the varicosities are separated from one another by linear distances of only 1 µm or less. On the average one would expect to encounter one varicosity in every 2.5 µm along the fiber. If the plate occupied by the average Purkinje cell dendritic tree is taken as 15–20 µm thick, then an average parallel fiber could be expected to develop 6–8 varicosities as it traverses the tree. Any tendency to cluster the varicosities would increase the number of synapses made in any particular tree. Our figures suggest that the number of contacts differs by an order of magnitude from that calculated by SMOLYANINOV (1971) and PALKOVITS *et al.* (1971c). Of course, not all of the varicosities signify synaptic contacts with the thorns of Purkinje cells, and our estimates need to be reduced by some unknown amount. If we apply the correction used by PALKOVITS *et al.* (1971c), i.e., 6% of varicosities synapse with other than Purkinje cells, there would still be an average of 5.6–7.5 synapses with the Purkinje cell tree for each parallel fiber.

Table 5. Comparison of parallel fiber varicosities at different levels of the molecular layer in Golgi preparations: optical microscopy

Level	Upper 1/3	Middle 1/3	Lower 1/3	Whole Molecular Layer
Total fiber length measured (µm)	742[a]	675[a]	805[a]	2222[a]
Total non-varicose length (µm)	530[a]	392[a]	556[a]	1478[a]
% fiber length occupied by varicosities	28.4	41.8	30.9	37.9
Number of varicosities	300	300	300	900
Number of varicosities per 10 µm length of fiber	4.0	4.4	3.7	4.0
Mean fiber length per varicosity (µm)	2.5	2.2	2.7	2.5
Mean interval between varicosities (µm)	1.7	1.4	1.9	1.7
Median area of varicosities (µm^2)	0.48	0.48	0.56	0.48
% of total varicosities with area smaller than 0.84 µm^2	81	86	71	78
% of total varicosities with area larger than 0.84 µm^2	19	14	29	22

[a] Values have been rounded off to the nearest whole number.

Examination of the measurements from the different levels of the molecular layer did not reveal any salient differences between the populations of parallel fibers, except for the somewhat higher percentage of large varicosities in the lower third of the molecular layer. A histogram comparing the distribution of large and small varicosities is given in Fig. 55. This difference may be related to the tendency for the thicker fibers to be more numerous in the deeper third than in the outer two thirds. It may also signify that the larger varicosities articulate with basket cells and their larger dendrites, which are more common in the lower third than in the others.

An opportunity to amplify these results is afforded by high voltage electron microscopy of Golgi preparations. By this method synaptic sites can be identified with greater reliability than in the optical microscope, since they are indicated by small, circular clearings in the impregnation (see below; CHAN-PALAY and PALAY, 1972b). Unfortunately, this improvement is balanced by a considerably smaller sample for measurement. Electron micrographs similar to Fig. 57 were printed at a final magnification of 8000×, and the following measurements were made: total length of parallel fibers in sample, number of varicosities per fiber, number of synaptic sites per fiber, and the number of varicosities displaying synaptic sites (Table 6).

Table 6. Varicosities and synaptic sites in parallel fibers in Golgi preparations: High voltage electron microscopy

Fiber No.	Total length of fiber (μm)	Varicosities along fiber	Varicosities per 10 μm	Varicosities with visible synaptic sites	Synaptic sites along fiber	Synaptic sites per varicosity
1	20.00	7	3.5	5	8	1.6
2	18.75	4	2.1	2	3	1.5
3	25.00	9	3.6	6	8	1.33
4	31.25	7	2.2	5	5	1.0
5	31.25	9	2.8	3	5	1.66
6	35.00	9	2.5	5	9	1.8
7	33.75	11	3.2	6	9	1.5
8	20.00	5	2.5	5	8	1.6
9	28.25	7	2.4	4	8	2.0
Total	243.25	68	2.8	41	63	1.5

Although the sample is very small and originates from only a small area in the middle of the molecular layer, the incidence of varicosities per unit length of fiber (2.8/10 μm) is not very far from the value obtained from the more extensive measurements in the optical microscope. According to these measurements on the high voltage electron micrographs, one could expect to encounter a varicosity in every 3.6 μm along the average parallel fiber, whereas in the light microscope measurements the period was 2.5 μm. If each varicosity effected one synapse with the Purkinje cell dendritic tree, then in passing through an arborization 15–20 μm thick the average parallel fiber would make 4–5 synaptic contacts instead of 6–8. In either case, the number is higher by an order of magnitude than recent estimates in the cat and other animals.

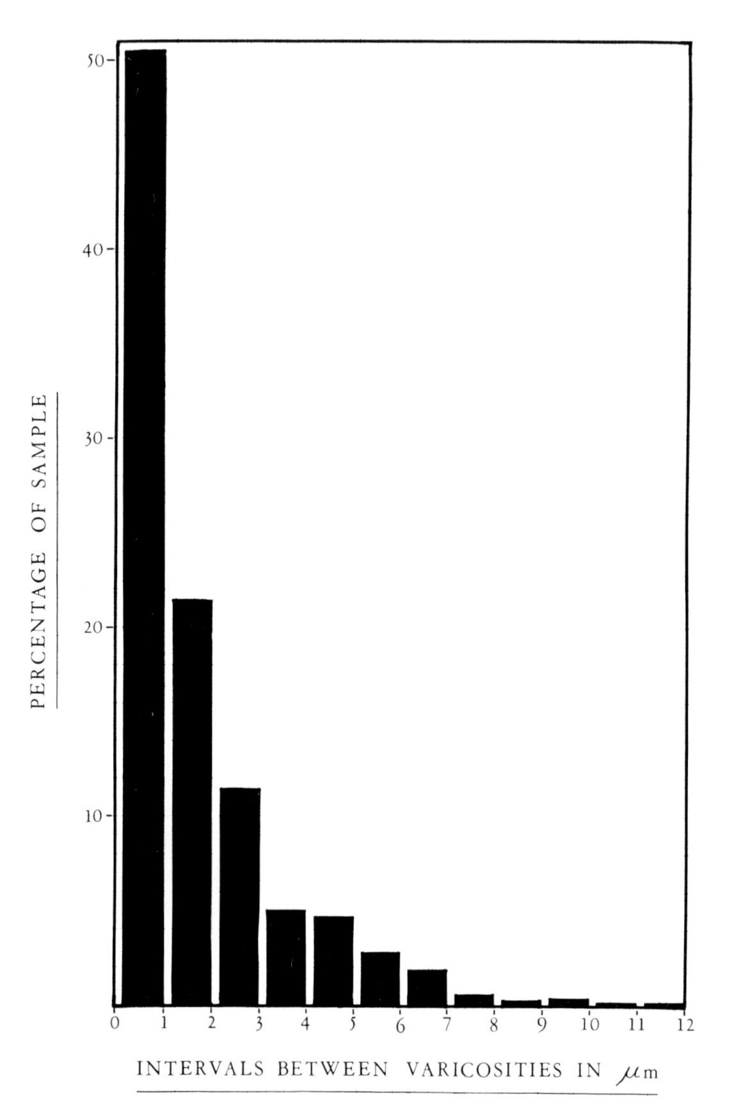

Fig. 56. Histogram showing the distribution of the measured intervals between synaptic varicosities of parallel fibers. More than 50% of the sample of synaptic varicosities considered were separated from one another by linear distances of 1 μm or less. More than 20% were separated by 2 μm and more than 12% were separated by 3 μm

It is furthermore interesting and quite revealing to take a closer look at the varicosities in the high voltage electron micrographs of the Golgi preparations (Figs. 57 and 58), for they show that a high proportion of the parallel fiber

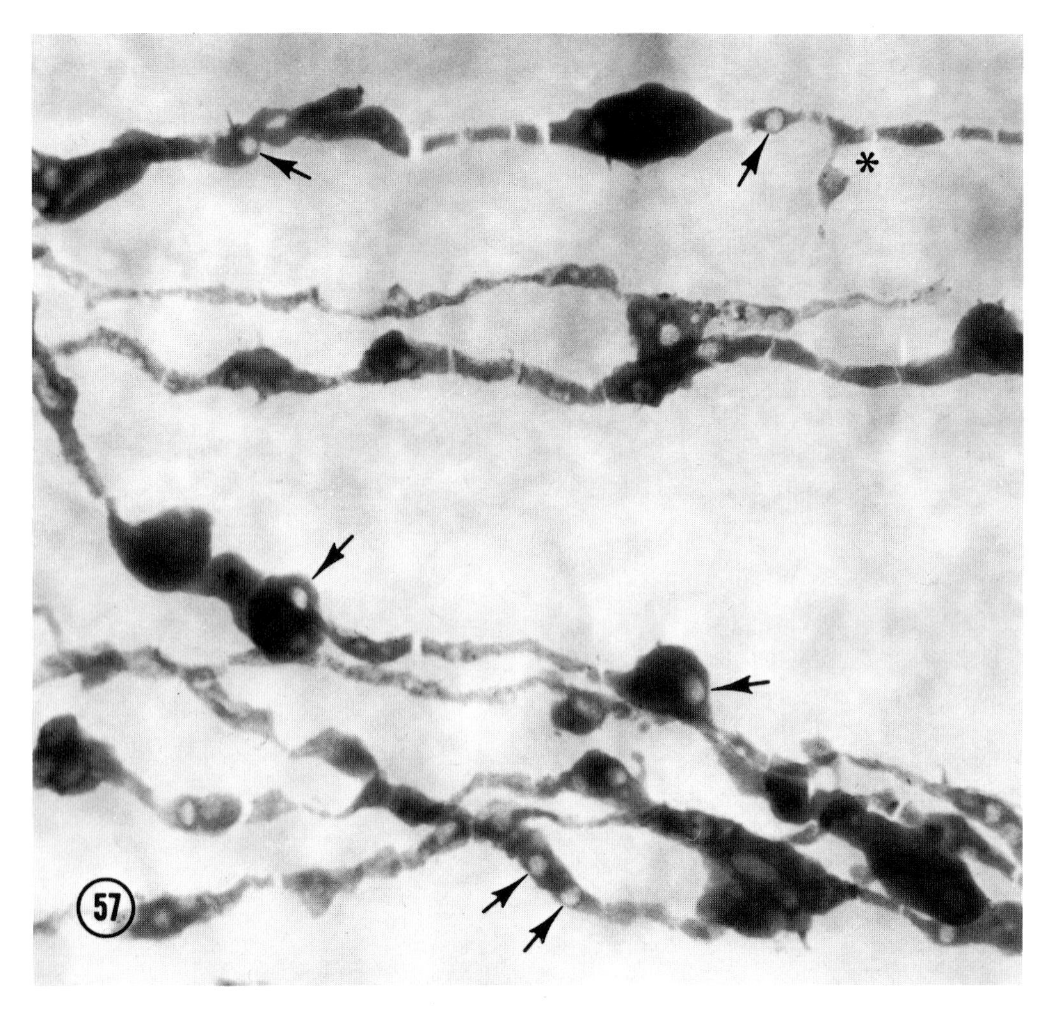

Fig. 57. High voltage electron micrograph of parallel fibers after rapid Golgi impregnation. These parallel fibers are well filled with the red filamentous matrix. No black globular particles are present. Along each axon the sites of synapses are defined by circular clear spots (*arrows*). The number of synapses along a particular stretch of each parallel fiber can be estimated by counting these spots. The asterisk (*) indicates the ascending stem of the granule cell axon, which divides to give rise to the parallel fiber. 5 μm, lobule X. ×8000

varicosities offer more than one synaptic site. Conventional electron microscopy of thin sections proves that many varicosities synapse with two or three dendritic thorns (Figs. 58d and e, 83, and 85). This finding undermines the assumption that each varicosity represents a single synapse. In the high voltage electron micrographs (Figs. 58a–c) several can be seen with two and even three synaptic sites. Measurements in our small sample (Fig. 57) indicate that nearly half of the varicosities have two synaptic sites and that a few have three (average 1.5 sites per varicosity). The round outline and fairly uniform size of these sites suggest that most of them articulate with Purkinje cell thorns. At present we have no method for distinguishing the synaptic sites for other varieties of postsynaptic elements. It should be emphasized, however, that high voltage electron microscopy of Golgi preparations is the only technique available that enables one to quantitate synapses by direct visualization. It could be used with profit by those who wish to quantitate the architectonics of the cerebellar cortex.

2. The Granule Cell in the Electron Microscope

Electron microscopy of thin sections confirms the general impressions of optical microscopy and adds valuable details about the organization of the granular layer and about the internal structure of granule cells and their processes. The cell bodies are generally packed so close together in their clusters that they abut one against the other with only a thin interstitial space between them (Figs. 59–61). Granule cells usually lack a complete neuroglial sheath. Rarely, a granule cell has been found surrounded by a myelin sheath (Fig. 138). Similar cells have been reported by ROSENBLUTH (1966) in the toad and MUGNAINI (1972) in the cat. The dendrites of granule and other nerve cells, as well as the processes of neuroglial cells, pass through the prismatic crevices left between the spherical cell bodies. Despite this close packing, the electron micrographs disclose no sign of synaptic specializations involving neighboring somata. Thin, dense fila-

Figs. 58 a–e. High voltage electron micrographs of the synaptic sites on parallel fibers. **a** shows parts of three parallel fibers impregnated by the rapid Golgi technique. The two areas marked by arrows have been reproduced at higher magnifications in **b and c** respectively. Rat cerebellar cortex, 1 μm. × 11 500. **b and c** show circular unstained areas in the varicosities of a parallel fiber. These areas correspond to the synaptic portion of the axon, containing many synaptic vesicles and mitochondria, which are not impregnated with the Golgi reaction. A varicosity may have more than one synaptic site on its surface. (Compare with Figs. 58 d and e.) 1 μm, lobule X. × 30 000. **d and e** Parallel fiber varicosities synapsing with thorns of Purkinje cell dendrites. These electron micrographs show two parallel fiber enlargements (pf_1, pf_2) synapsing with dendritic thorns (t). Both varicosities entertain two thorns each upon their surfaces. Lobule X. × 31 000

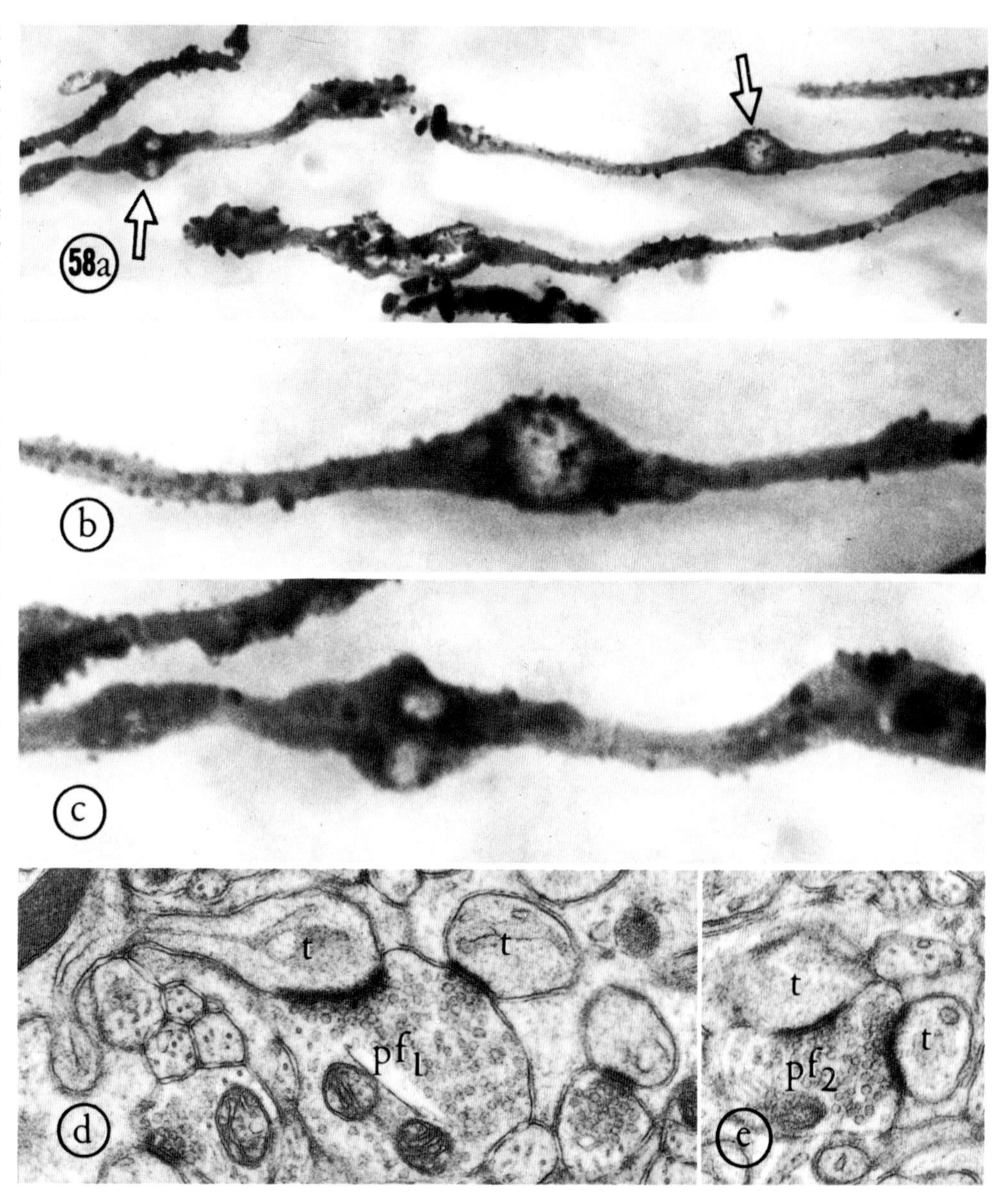

ments may be present, however, in the interstitial gap between the opposing granule cell membranes (Figs. 61 and 62). There are, in addition, frequent *puncta adhaerentia*, tiny spots of filamentous densities attached symmetrically to apposed membranes, marking the surfaces of adjoining granule cell bodies. Cells at the periphery of the clusters are necessarily contiguous to elements in the intervening neuropil, for example, mossy fiber terminals (Figs. 128 and 129) or Golgi axons (Fig. 108) in the glomeruli, basket fibers in the pinceau surrounding the initial segment of the Purkinje cell axon. In no case have synaptic specializations been seen along these contact zones, but patches of dense filaments are frequently found crossing the interstitial gap between mossy fiber terminals and granule cell bodies (Figs. 128 and 129).

These observations call to mind the hypothetical structure of the glomeruli proposed by the Scheibels on the basis of their findings in Golgi preparations. According to this analysis, which was reported in JANSEN and BRODAL's review of cerebellar structure (1958) but never published independently, the somata of granule cells form an integral part of the glomerulus and are the sites of massive axo-somatic synapses with the mossy fiber terminals. In the electron micrographs, however, typical synaptic contacts between mossy fibers and granule cells have not been seen (GRAY, 1961; PALAY, 1961; ECCLES, ITO, and SZENTÁGOTHAI, 1967; MUGNAINI, 1972), and, although contiguity between these elements is common enough, the interpretation of the focal intercellular densities mentioned here remains debatable.

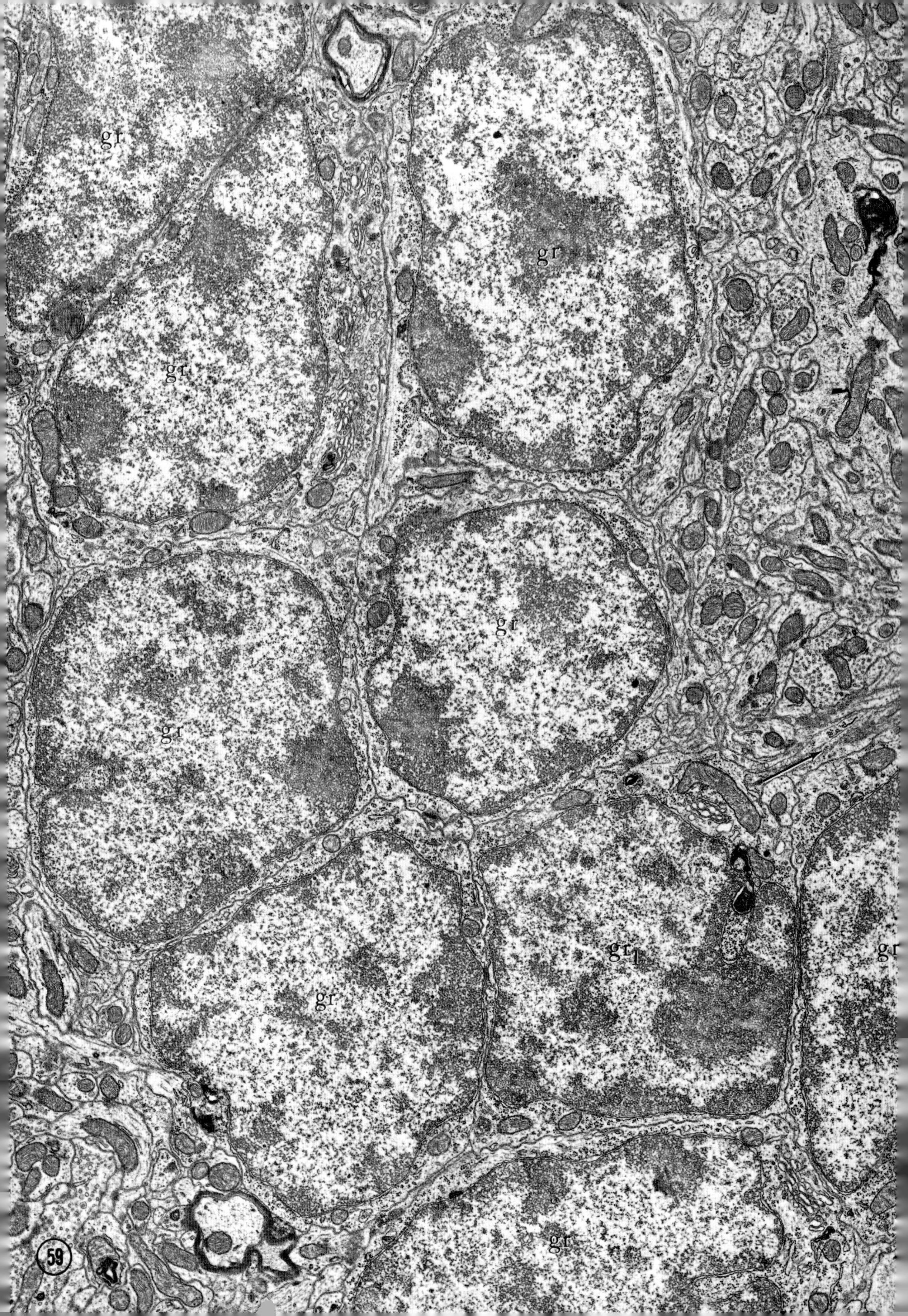
gr
gr
gr
gr
gr
gr
gr1
gr
gr
59

a) The Nucleus

Almost the entire cell body of a granule cell is occupied by the nucleus, and the perikaryal cytoplasm is confined to a thin shell often no more than 0.1 to 0.3 μm in thickness (Figs. 59, 60, 69, and 70). Frequently the nucleus is eccentric, abutting directly against the plasmalemma, and the usual cytoplasmic organelles are assembled in a pool beside the nucleus. Furthermore, the nucleus, seen as spherical and smooth-surfaced in the light microscope, often proves to be flattened, dimpled, or creased on one side, producing a pocket filled with Golgi apparatus, mitochondria, and Nissl substance (Figs. 59 and 60), all of which are hidden by the deeply staining nucleus in preparations for the optical microscope. Deep creases and tubular invaginations, however, are filled with polysome-rich cytoplasm, which might be the basis for the intranuclear rodlets seen in the light microscope (RAMÓN Y CAJAL, 1911; COLONNIER, 1965).

The nucleus of the granule cell ordinarily contains large blocks of condensed chromatin, often distributed in a cart-wheel or clock-face pattern inside the nuclear envelope. Although the granule cell nucleus would be a good model for the structure of heterochromatin in an interkinetic nucleus, no one to our knowledge has given it serious attention. The chromatin blocks consist of linear chains of small globular units about 200 Å in diameter placed close together without any visible connecting thread. The units are not spherical but have a quadrangular profile with a lighter center. There are also chains and coils of smaller granules, 100–150 Å, dispersed throughout the looser parts of the nucleus, and widely scattered larger interchromatic granules, 300 to 400 Å in diameter and very dense (Fig. 60).

The nucleolus is usually represented by a small mass of almost pure pars fibrosa, amalgamated with a more centrally placed block of chromatin or attached to the inner face of one peripherally placed (Figs. 60 and 63). It is rather unusual to discern the pars granulosa of the nucleolus in these cells. One cannot help wondering whether this is a manifestation of low ribosomal and protein production consistent with the presumably inactive condensed chromatin of the nucleus and the relatively poorly developed Nissl substance of the cytoplasm.

The nuclear envelope has the usual fenestrations and is furthermore characterized by rather frequent streamers extending into the cytoplasm, and occasionally across to the plasmalemma, connecting the perinuclear cisterna with the cisternae of the endoplasmic reticulum and the interstitial space. Ribosomes in rows and spirals stud the outer surface of the nuclear envelope, much more commonly than in most other neuronal types (PALAY and PALADE, 1955; PETERS, PALAY, and WEBSTER, 1970).

b) The Perikaryon

The meager cytoplasm contains a few mitochondria varying in shape from small spherules, 0.1 to 0.3 μm across, to slender rodlets, from 2 to 4 μm long. The longer mitochondria tend to curve parallel to the nucleus, or, if lodged in the more ample cytoplasmic space provided by an indentation of the nucleus (Fig. 60), they can extend straight across the perikaryon well into the core of a dendrite (Fig. 69). The granular endoplasmic reticulum, as would be expected from optical microscopy, is scanty, consisting of a few tubules and short, branching cisternae bearing ribosomes. It is often confluent with the nuclear envelope on the one hand or the surface membrane on the other. Free clusters, rosettes, and disordered groups of ribosomes are common. The smooth endoplasmic reticulum is reduced to a few apparently discontinuous tubular profiles, often clearly continuous

◄

Fig. 59. A cluster of granule cells. Nine granule cells (*gr*) are closely packed in a cluster with their somatic surfaces in juxtaposition. The nucleus of one of the cells (gr_1) displays a dimple filled with Golgi apparatus, a mitochondrion, and ribosomes. A dendrite issues (*arrow*) from the apex of this collection of cytoplasmic organelles. Lobule IX, nodulus. ×14000

►

Fig. 60. Cluster of granule cells. The nuclei of these cells are conspicuous because of their clumps of condensed chromatin (*chr*). One of the granule cells (gr_1) shows a dimple in the nucleus with the Golgi apparatus (*Go*) tucked into it. A second cell (gr_2) is sectioned through the dimple in the nucleus, and ribosomes (*r*) and a mitochondrion (*mit*) are found in the perinuclear cytoplasm. Nuclear pores (*np*) are evident in this cell, as well as a centriole (*arrow*). Lobule X, nodulus. ×20000

Fig. 61. The apposition of two granule cell somata. The somata of two granule cells (gr_1 and gr_2) are closely apposed, and a fibrillar material (*arrows*) crosses the gap between the apposed surface membranes. (See next figure.) Lobule X. ×60000

Fig. 62. The granule cell apposition. At higher magnifications, the interstitial space at the sites of apposition between two granule cells (gr_1 and gr_2) displays small amounts of a fibrillar material (*arrows*). Lobule V. ×210000

Fig. 63. The nucleolus of a granule cell. The granule cell nucleolus consists almost purely of pars fibrosa (*PF*) attached to a clump of chromatin (*chr*). Lobule V. ×82000

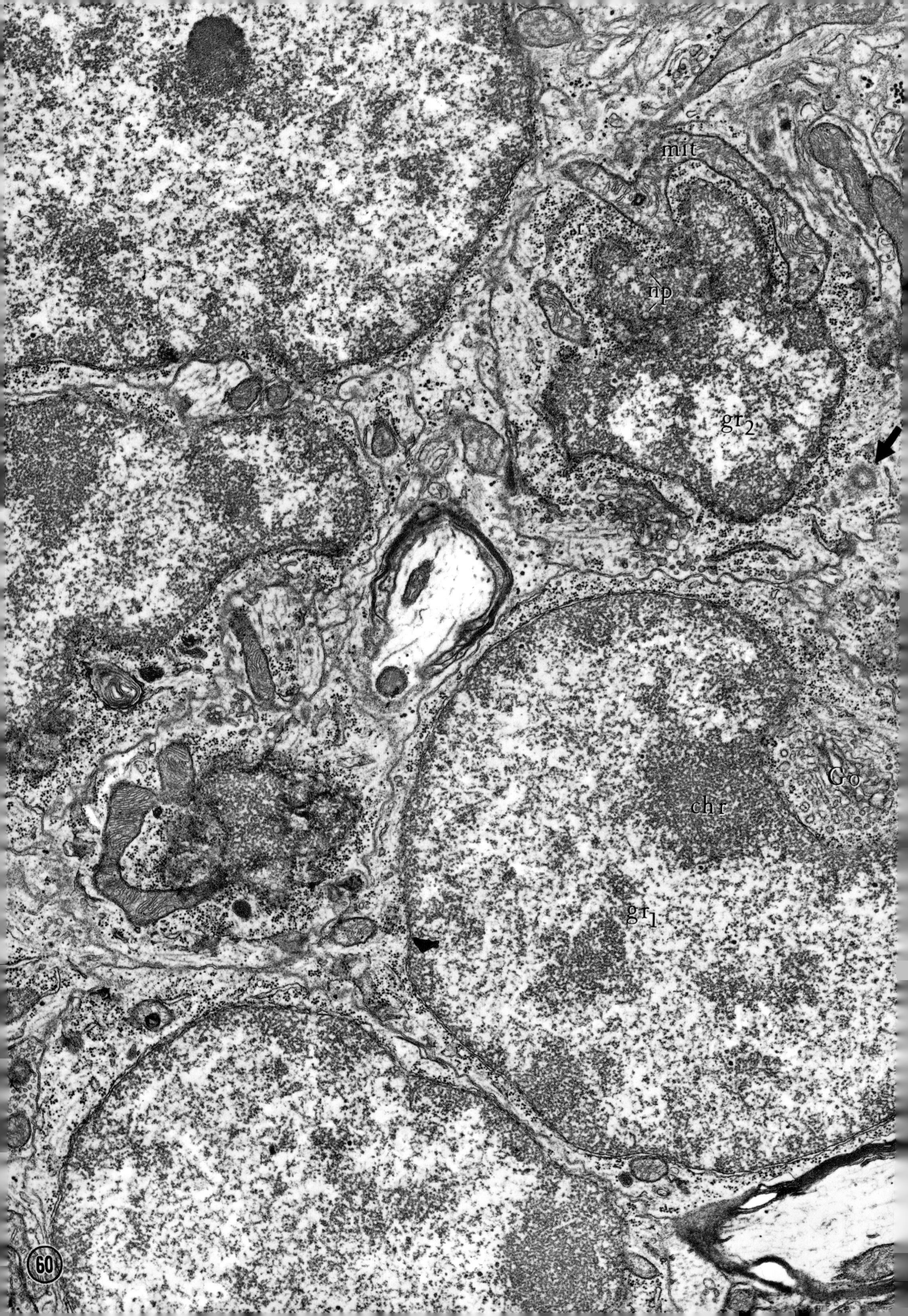
mit
r
np
gr_2
Go
chr
gr_1
60

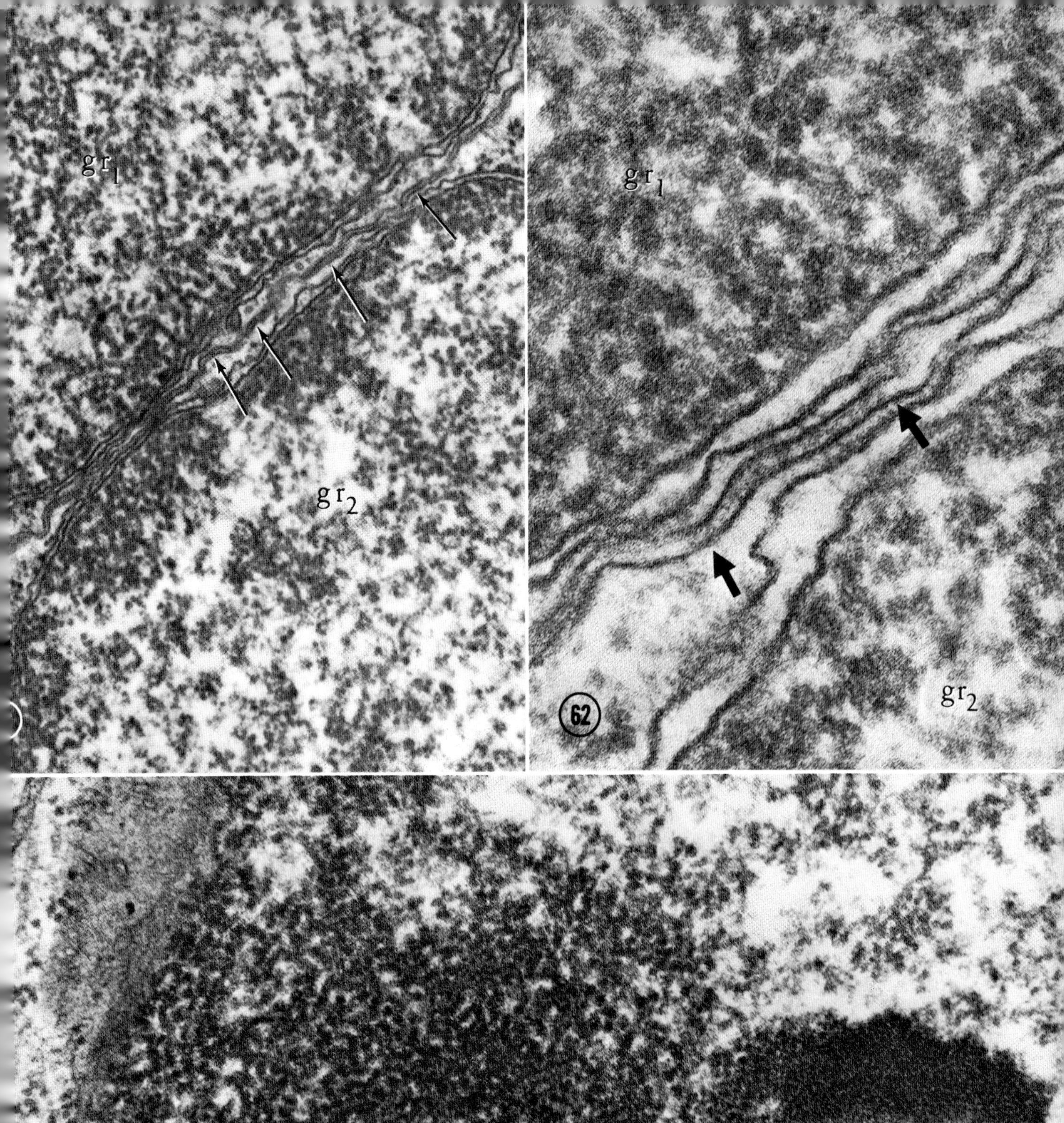
gr1
gr2
gr1
gr2
62

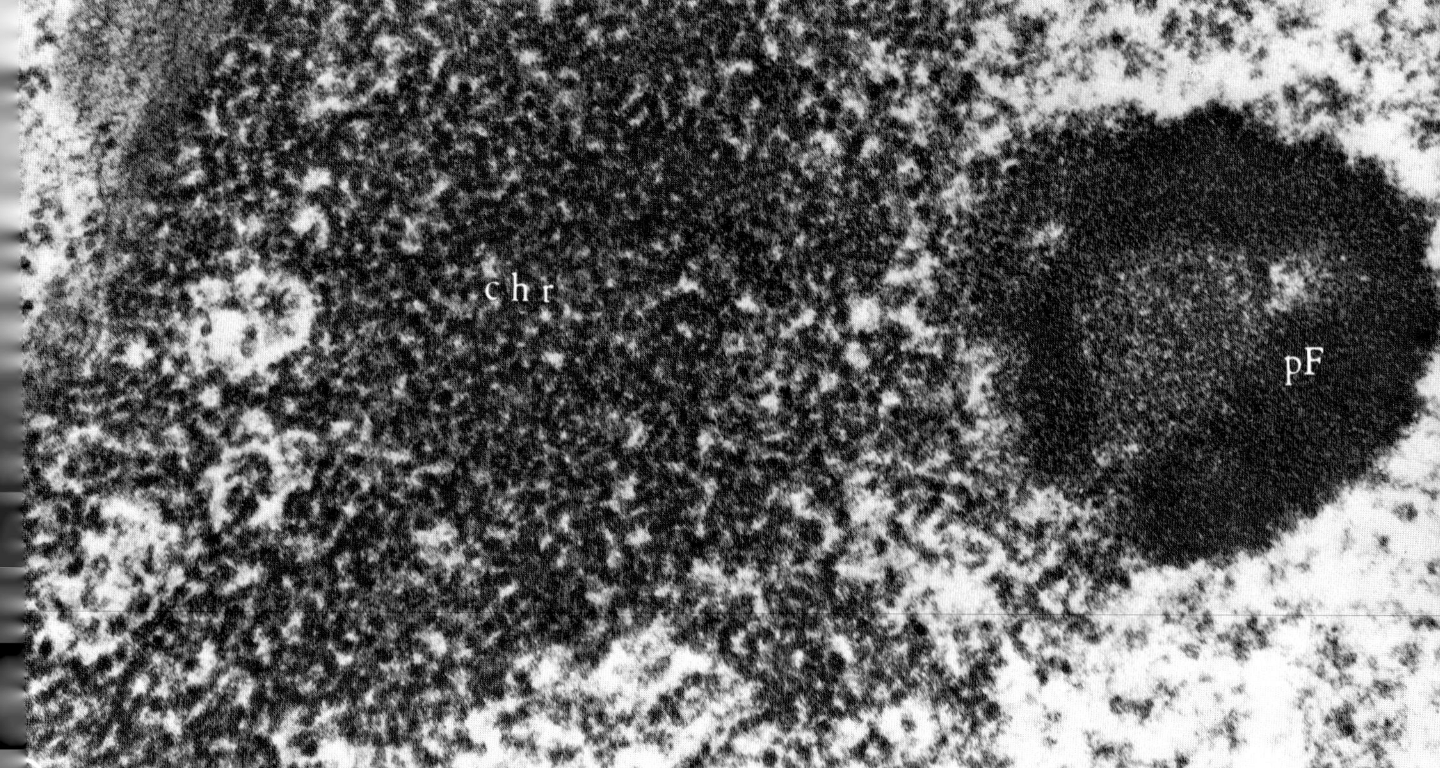
chr
pF

with the granular reticulum. The hypolemmal cisterna is represented principally by subsurface cisternae, which are sometimes associated with an adjacent agranular cisterna lying just beneath it, 200–400 Å away. The density resulting from the closely spaced membranes of the plasmalemma, the shallow or collapsed subsurface cisterna, and the associated cisterna of the endoplasmic reticulum makes these organelles stand out in low power electron micrographs of the granule cells. A characteristic arrangement appears in the clusters of granule cells where the cells are pressed closely together. It frequently happens that subsurface cisternae in adjacent cells lie in apposition (Fig. 64). A symmetrical figure results in which the axis is formed by the apposed parallel surface membranes, with a parallel collapsed subsurface cisterna on either side. Sometimes each cisterna has a dilated end opposite that of its symmetrical counterpart.

The Golgi apparatus is generally restricted to a single stack of four to six curved, flat cisternae and some associated vesicles. This complex is wedged into one of the thicker parts of the perikaryon that give rise to dendrites (Figs. 59 and 69), or it is the principal component of the "Hof" created by an invagination or simple flattening of the nucleus (Fig. 60). Its disposition corresponds quite well to the appearance of the Golgi apparatus in preparations for the light microscope as reported by SÁNCHEZ (1916). Accompanying the Golgi complex are multivesicular bodies and an assortment of vesicles and small lysosomes. The most conspicuous organelle in this "Hof" is, however, likely to be a pair of centrioles (Fig. 60). These structures, usually thought to be absent in neurons, appear quite commonly in thin sections of granule cells. In some fields of the granular layer, nearly every granule cell can display at least one of its two centrioles. The reason for this high frequency is surely the small size of these cells. The chance of encountering the centrioles in random sections through small cells is much greater than it is in sections through the large cells that have been the favorite objects of electron microscopy. It is evident that the assembly of centrioles and surrounding cisternae of the Golgi complex, together with lysosomes, multivesicular bodies, and assorted vesicles, constitute a centrosphere such as is usual in glandular cells and developing gametes (WILSON, 1927; FAWCETT, 1966 b).

The demonstration of centrioles in neurons still occasions some surprise, although they were first described in nerve cells as long ago as 1895 by LENHOSSÉK and since then by numerous other authors. An extensive review of this topic is given by DEL RÍO HORTEGA (1916) in a paper in which he demonstrated the existence of centrioles in all cell types of the cerebellar cortex. His figures show a pair of centrioles in each granule cell and even in each Purkinje cell, both in adult and infant man. We are inclined to agree with DEL RÍO HORTEGA that the persistence of centrioles and centrosphere in nerve cells long after the possibility of mitosis has passed indicates that these structures are concerned with some other more permanent and probably important function. What this function might be, we have hardly more inkling than he had. The topographical relation of microtubules to the centrioles and centrosphere might, however, suggest that this part of the cell is a staging ground for the assembly of microtubules.

Microtubules are rather infrequent in the perikaryon of the granule cell. There simply is not much room for them. As in the Purkinje cell, they are oriented in arcs encircling the nucleus and they can be traced from such curving paths into the processes (Fig. 69).

c) The Dendrites of Granule Cells

The dendrites of the granule cell are relatively fleshy processes, considering the small size of the parent cell. They arise abruptly from the spherical cell body and slip between the neighboring cells to reach the nearest glomerulus (Figs. 54, 65–67, 69). As the granule cells are

Fig. 64. Adjacent subsurface cisternae in granule cells. Two adjacent subsurface cisternae (*arrows*) occur in parallel in these two granule cells. Lobule X, nodulus. ×30000

Fig. 65. High voltage electron micrograph of a granule cell and one of its dendrites. This cell (*gr c*) shows the fibrillar precipitate on the left half of the cell body; on the right, the cell is covered with a crust of the globular particles. From the top right a dendrite (*gr d*) emerges which gives rise to a set of varicose branches or claws. These claws are entangled in a complex manner around the unimpregnated terminal of either a mossy or a climbing fiber in a glomerulus. This set of dendrites is best appreciated by studying it with its paired stereoscopic image under a viewer. Rat cerebellar cortex, 5 μm, lobule X. ×6500

Fig. 66. The granule cell dendrite and axons of Golgi cells. This high voltage electron micrograph of a rapid Golgi preparation shows a granule cell dendrite ascending from the lower margin of the picture and becoming entangled with axons of Golgi cells in a glomerulus. Rat cerebellar cortex, lobule X. ×30000

Fig. 67. The dendrite of a granule cell. The dendrite (ΔΔ) emerges from this granule cell (*gr*) and runs alongside the terminal of a mossy fiber (*MF*). Lobule IV. ×20000

Fig. 68. Synapse of an ascending axon of a granule cell in the granular layer. A dendrite probably belonging to a Golgi cell (*Go d*) synapses with a varicosity of an ascending granule cell axon (*gr ax*). Lobule IV. ×22000

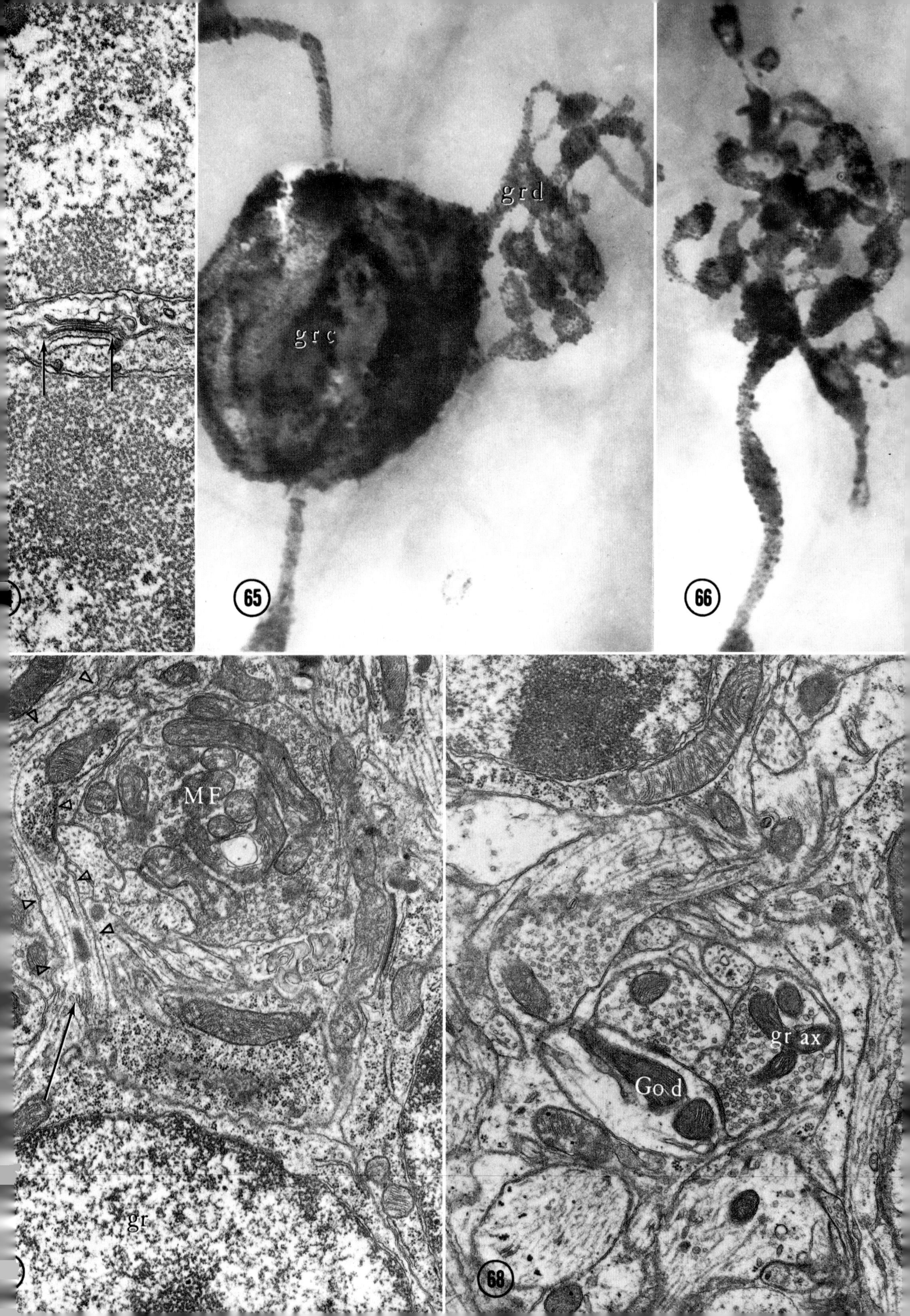
grd
grc
65
66
MF
gr
gr ax
Go d
68

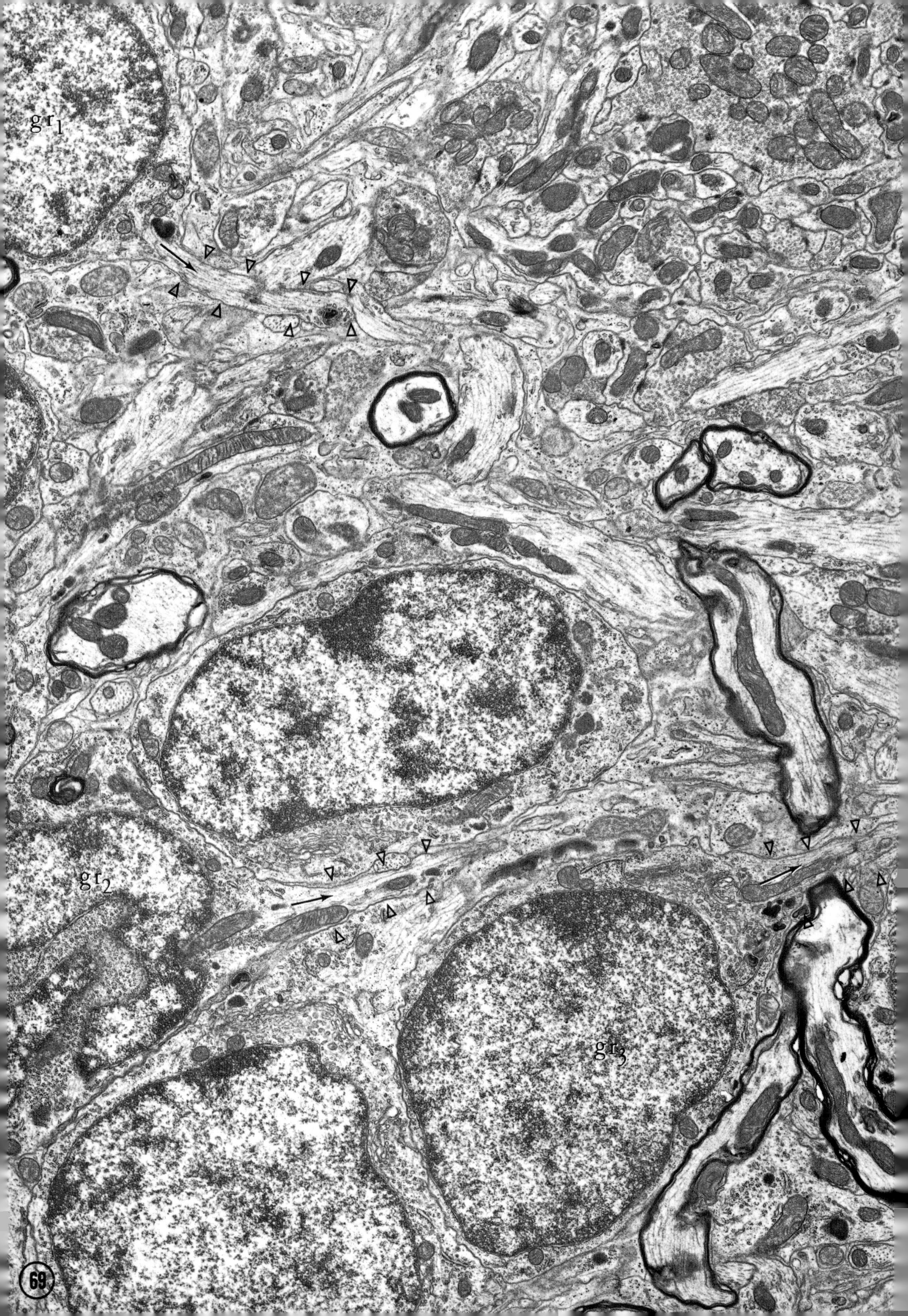

gr1
gr2
gr3
69

not completely encapsulated by neuroglial sheets, the dendrites come into direct contact with other granule cells and their dendrites. Puncta adhaerentia occur along such areas of contact. Unbranched until their terminal divisions, the dendrites can often be traced for surprisingly long distances in electron micrographs of thin sections. Like the cell bodies of the granule cells, they are not enveloped by neuroglial processes, although thin lamellae of astrocytic cytoplasm often accompany them. They have no synapses on their surfaces until their terminal branches.

They are characterized by smooth contours and the usual dendritic complement of longitudinal microtubules, tubular smooth endoplasmic reticulum, and long mitochondria. Here and there a cluster of ribosomes appears without any relation to the endoplasmic reticulum. All of these components pass into the digitiform branches, which enclose the mossy fiber terminal in a glomerulus (Figs. 141 and 142). These branches are only slightly more slender than the dendrite from which they arise. Each one contains a single long mitochondrion, around which are located one or two tubules of the endoplasmic reticulum, and several microtubules. Electron micrographs (Figs. 67, 133, and 134) of thin sections through the glomeruli usually display the terminal arbor of the granule cell dendrites as a peripheral collection of circular or polygonal transverse profiles. These fit neatly into the coves and inlets in the surface of the axon, each profile with its mitochondrial stuffing as like another as slices of a sausage. Contiguous branches are united by puncta adhaerentia, and on the face contacting the axonal terminal is a thick subsynaptic mat. A detailed treatment of this synapse will be deferred until the mossy fiber has been described.

d) The Ascending Axons of Granule Cells

The axon of the granule cell originates, with a slight axon hillock, from either the cell body or a dendrite (Fig. 70). The initial segment is identifiable in electron micrographs only by careful comparison with the dendrites. The initial segment is usually 0.1 to 0.2 μm in diameter, i.e., more slender than the dendrites, and typically it contains one fascicle of 4 to 6 cross-linked microtubules together with longitudinally directed tubules of the endoplasmic reticulum and a thread-like mitochondrion. The plasmalemma is undercoated with a thin tripartite layer of granular, laminar, and filamentous material, which extends for a variable distance into the Purkinje cell and molecular layers. This undercoat usually begins only 4–5 μm after the axon has left its point of origin. The initial segments of deeper granule cells lose their distinctive internal features in the upper third of the granular layer.

Beyond the end of the initial segment the axon becomes irregularly varicose as it enters into synaptic contact with dendrites of Golgi cells in the granular layer (Fig. 68) and with the thorns of the lower Purkinje cell dendrites in the deep molecular layer. The varicosities contain an aggregate of round synaptic vesicles, ranging in size from 240–440 Å, which collect against the axolemma along the synaptic junction. Tufts of dense, fibrous material extend inward from the axolemma among the vesicles, producing a hexagonal pattern of densities, a sort of grid in which the vesicles appear to be trapped (Gray, 1963, 1969; Pfenninger *et al.*, 1969; Peters, Palay, and Webster, 1970). One or two mitochondria are also lodged in the varicosity, usually in the core of the axon among the microtubules. The junctional surface of the varicosity is slightly concave so that the synaptic cleft between the axon and the dendritic thorn is somewhat widened, about 300 Å deep, compared with the usual 150 Å interstitial space. A set of fine filaments crosses the cleft, sometimes filling it with an irregular web, but most often stretching more or less vertically between the apposed plasmalemmas. Sometimes a thin, dense plate can also be seen in this cleft running midway between the pre- and postsynaptic surfaces. On the dendritic side of the cleft a thick matting of dense filamentous material is attached to the cytoplasmic surface of the plasmalemma. The difference in the thickness and continuity of the densities on the two sides of the synaptic cleft produces

Fig. 69. Dendrites of granule cells in the neuropil. Three granule cells in this cluster (*gr*₁, *gr*₂, and *gr*₃) give rise (*arrows*) to dendrites (ΔΔ) within this section. The third cell (*gr*₃) shows the nuclear pocket region with a mitochondrion issuing into the dendrite. Lobule V. × 11000

Fig. 70. The initial segment of a granule cell axon. The initial segment (ΔΔ) emerges (*arrow*) from the cell body of the granule cell (gr_1) and runs for about 10 μm before being lost from the section. A second granule cell (gr_2) shows a slight dent in the nucleus with the Golgi apparatus (*Go*) in the pocket. Lobule V. × 15000

Fig. 71. Parallel fiber varicosities in the molecular layer. This illustration shows a typical section through the molecular layer. The many profiles of parallel fiber synaptic varicosities are indicated (*pf*). Most often, they are in synaptic contact with thorns (*t*) of Purkinje cell dendrites (*Pc d*). Lobule X. × 19000

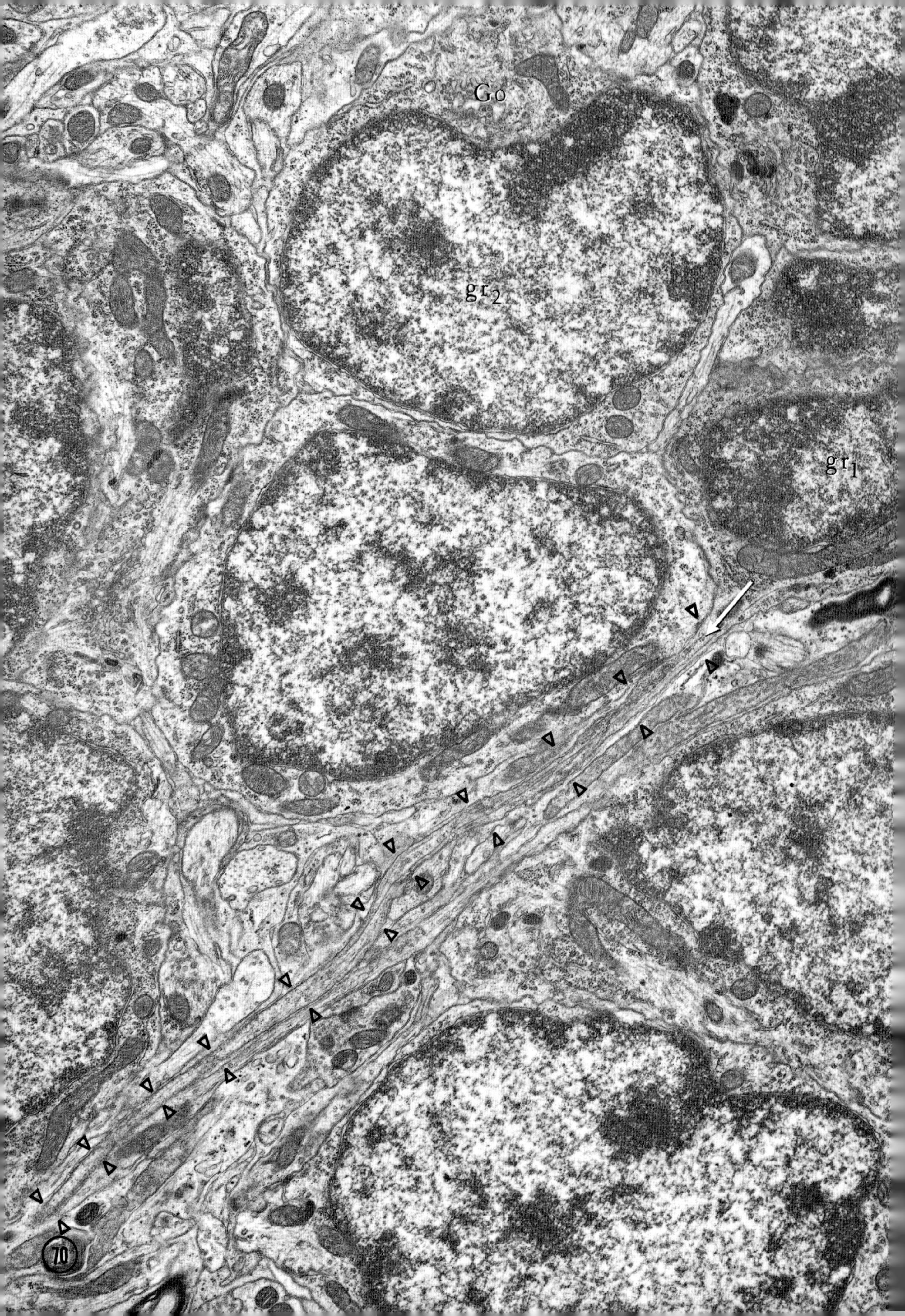

Go
gr2
gr1
70

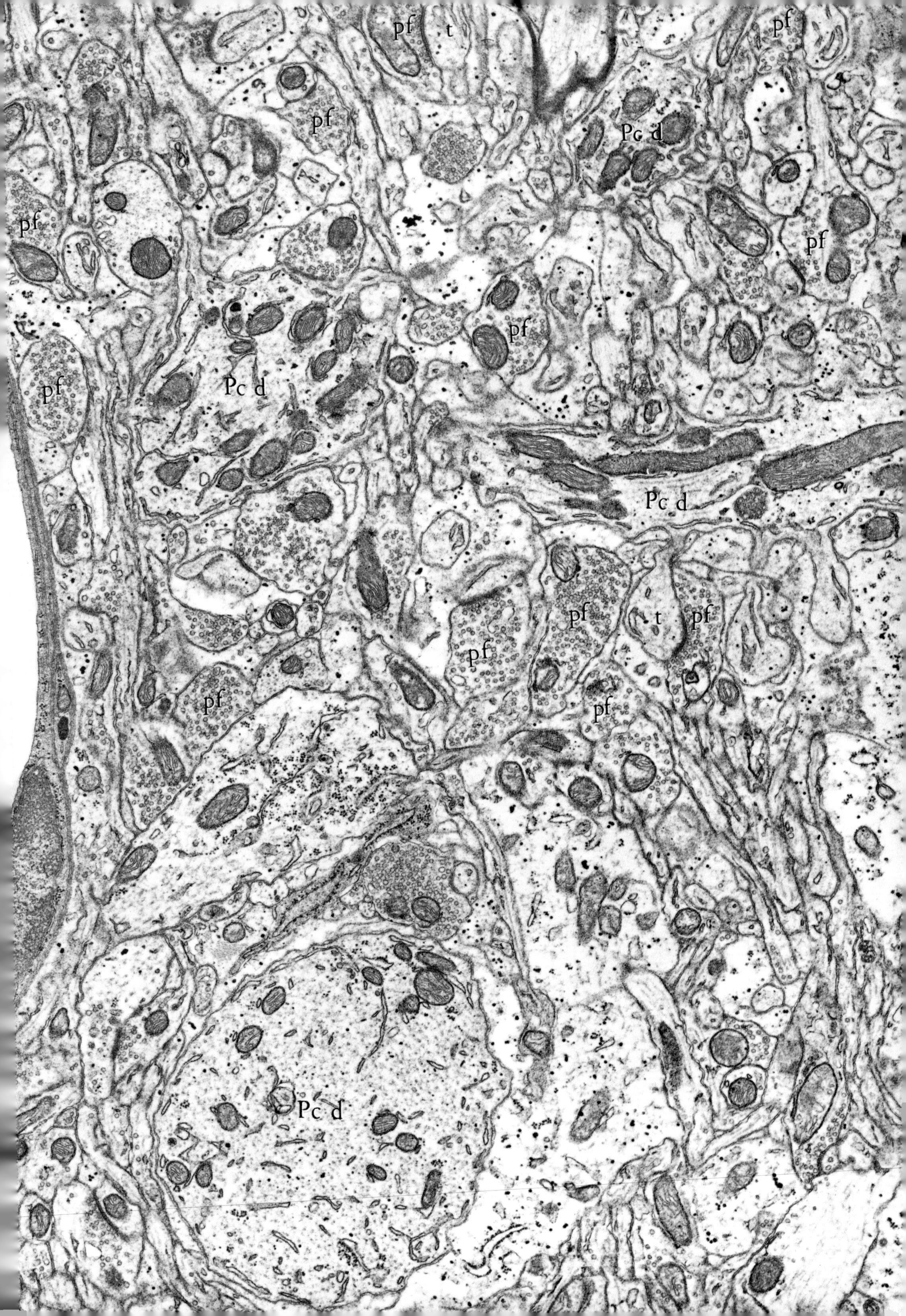
pf
t
pf
pf
Pc d
pf
pf
pf
Pc d
pf
Pc d
pf
t
pf
pf
pf
pf
Pc d

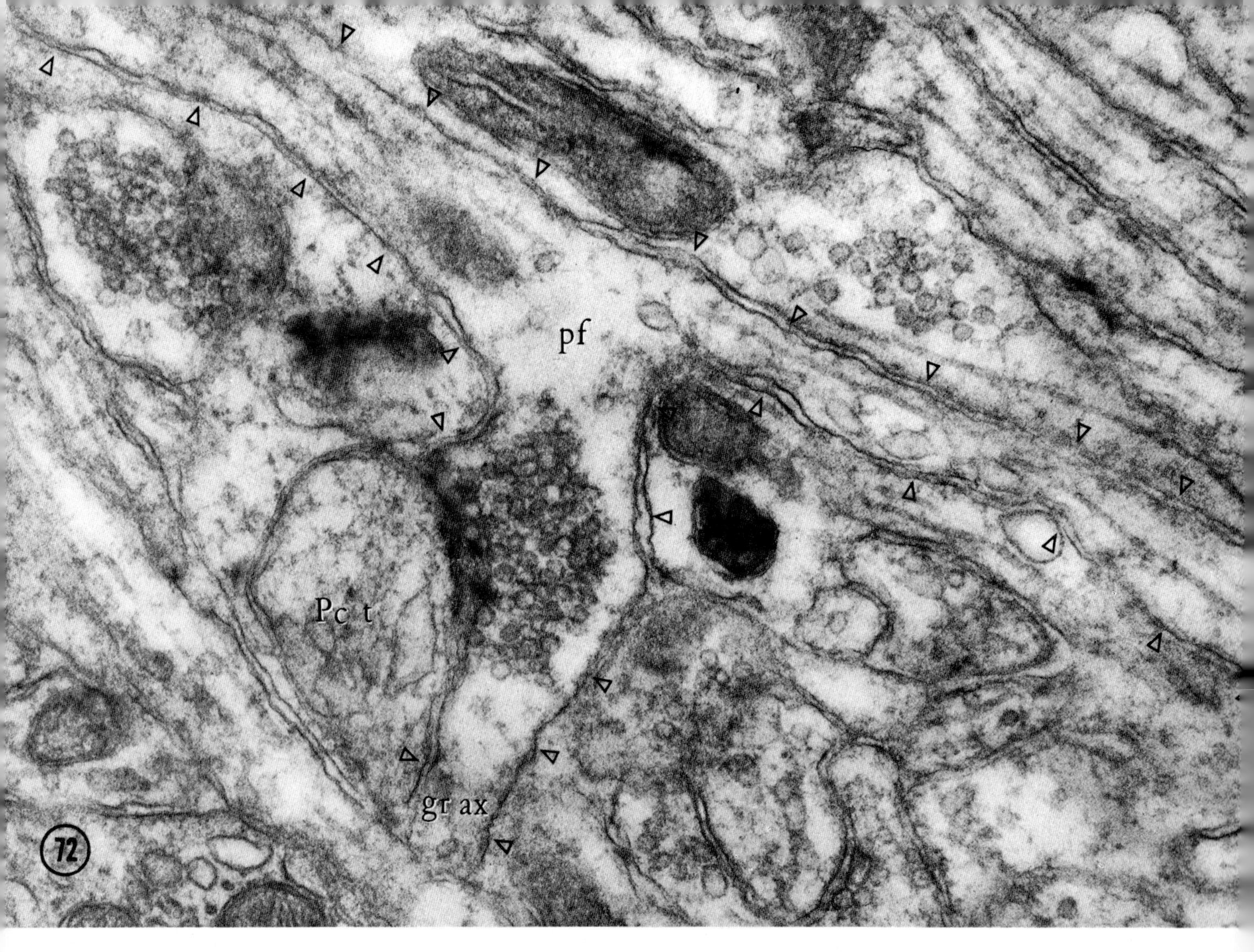

Fig. 72. Synapse at the bifurcation of a granule cell axon. The division of the ascending granule cell axon (*gr ax*) results in the formation of a parallel fiber (*pf*). Both parts of the axon are outlined by arrowheads. Just before the bifurcation the ascending portion of the axon swells into a varicosity where a synapse occurs with a Purkinje cell dendritic thorn (*Pc t*). Osmium tetroxide fixation. × 64000

an asymmetrical junction which, together with the widened cleft, fulfills the criteria for Gray's type 1 synapse. The varicosity encloses or caps the bulbous end of a thorn, which pokes into the distension in the axon like a finger into an inflated balloon (Figs. 71 and 33). Although this arrangement provides a broad interface for a synaptic junction, the synaptic specializations are usually developed only in a restricted macular region on the side of the bulbous end and not on its very tip. A similar synaptic varicosity regularly occurs in the granule cell axon just before its bifurcation to form a parallel fiber (Fig. 72).

It was mentioned earlier that the stems of the granule cell axons are gathered together into slender bundles as they pass between the Purkinje cell bodies into the molecular layer. In their ascending course they are accompanied by the vertical processes of the Golgi epithelial cells—the Bergmann fibers—which rise to the surface of the folium (Figs. 73a and b). This is not the only part of the granule cell that is closely associated with neuroglial cells, but it is the part that is most consistently and intimately related to them. The neuroglial processes do not, however, ensheathe the axons individually; they merely enclose a group or run alongside them. Since the Purkinje cell dendrites are almost completely enveloped in neuroglia, the thorns must perforate this sheath in order to articulate with the granule cell axons (Fig. 33). Mugnaini and Forstrønen (1966 and 1967) and Rakic (1971) have suggested that the vertically directed Bergmann fibers guide the developing granule cells in their descent through the molecular layer to their mature location in the granular layer. If this is true, the vertical fascicles of granule cell axons and their accompanying neuroglial processes remain as traces of an early migration.

Fig. 73a and b. Bundle of ascending granule cell axons. The granule cell axons (*gr ax*) ascend through the molecular layer wrapped into a bundle by the processes of neuroglial cells. Along their course, these axons display swellings which synapse with Purkinje cell thorns (*t*), a × 9000; b × 21000

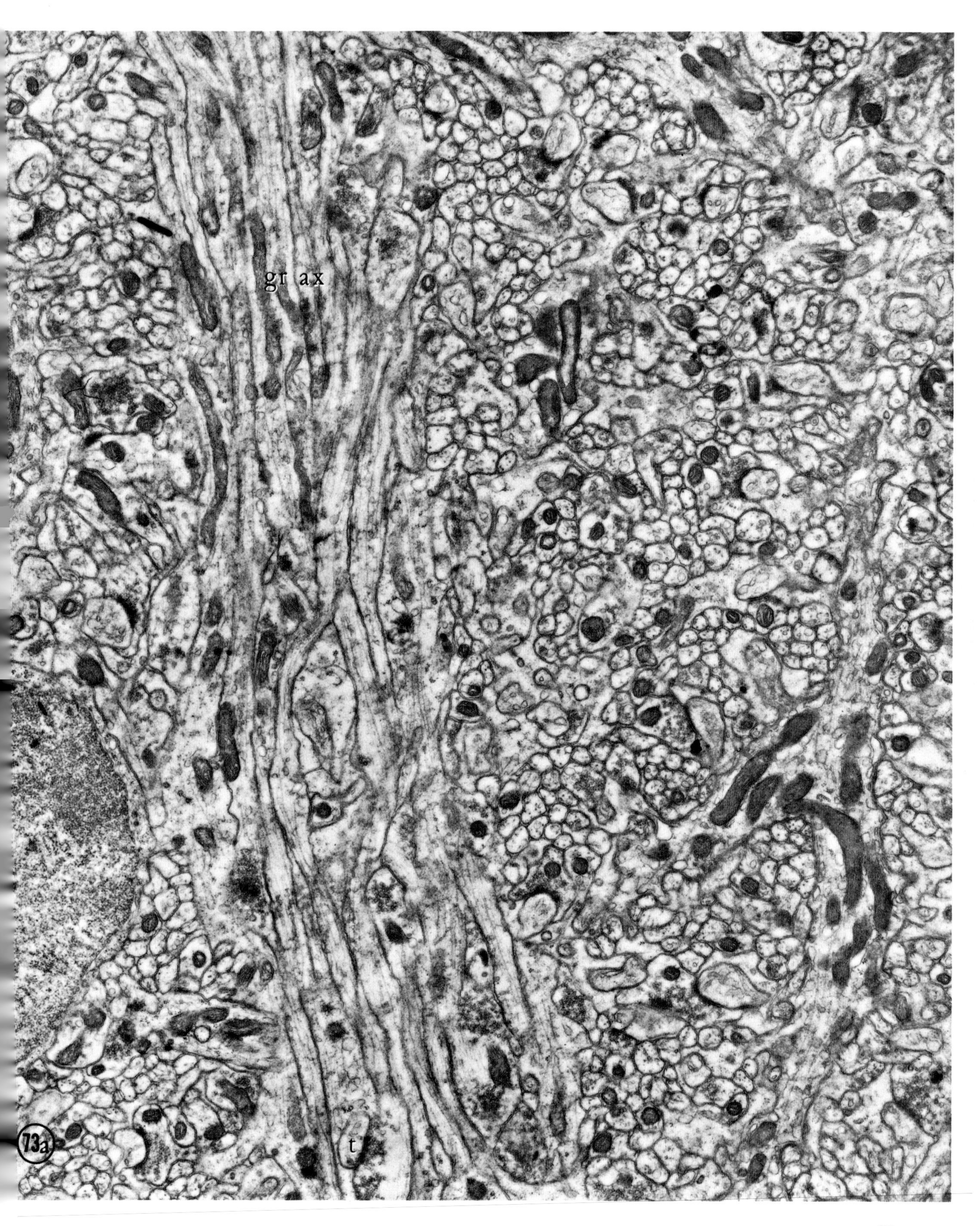
gr ax
t
73a

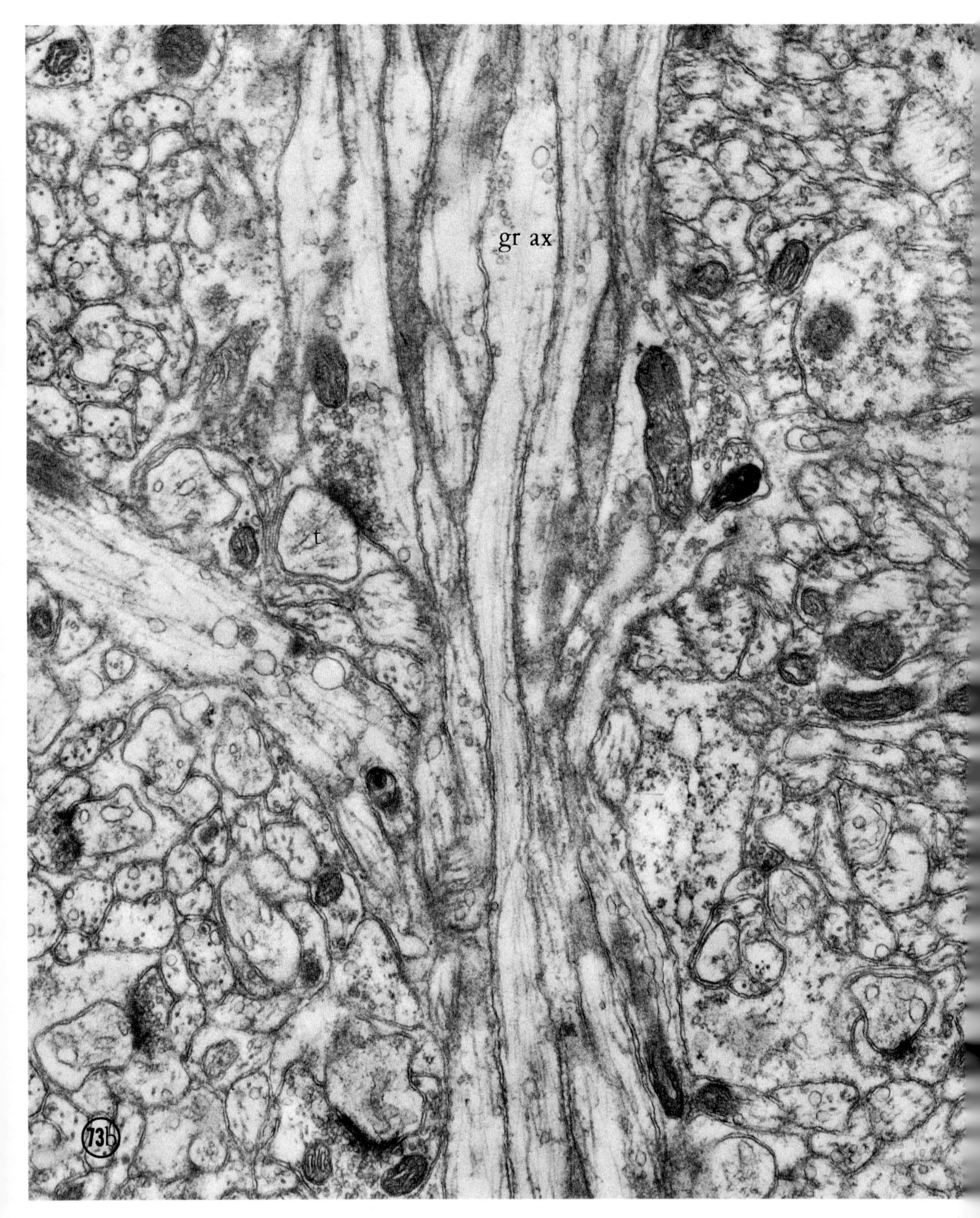
gr ax
t
73b

It should be recalled in this connection that the ascending segments of the granule cell axons often meander and loop about in an apparently casual manner through the granular layer before they reach their places in the radial fascicles of the molecular layer (Fig. 54). These divagations should not be viewed as a consequence of aimless wandering, as if spun out behind a cell that had been released from the restraint of the neuroglia and was migrating through an indifferent matrix. The granules that have descended below the level of the Purkinje cell bodies are tethered by their axons, which are held like a leash by the Golgi epithelial cells and Bergmann fibers, at the lower margin of the molecular layer. As further granule cells descend, the earlier arrivals are displaced farther away from the Purkinje cell bodies. Therefore, not only do the Purkinje cell axons have to lengthen in order for the perikarya to maintain their relative position, but also the axons of the granule cells below the Purkinje cells must grow and lengthen as the granular layer thickens. The divagations of the axons in this layer reflect the jostling and shifting that must go on during the developmental period as the cell bodies settle into their definitive positions. It is, of course, only an assumption without evidence that they have definitive positions.

e) Ectopic Granule Cells

An interesting irregularity bearing on this point is shown in Fig. 74. This low power photomicrograph of the cerebellar cortex from an adult rat shows a focus of granule cells that have failed to descend into the internal granular layer and have remained at the pial surface. Such foci appear consistently in certain locations, for example, in lobule V (vermis) near the midsagittal plane and in lobule IX (vermis). These clusters of cells are readily spotted even at low magnifications in the light microscope (Fig. 74), and they have been analyzed in a recent publication by CHAN-PALAY (1972b).

The usual focus is a conical mound of small cells with its base flat against the pial surface and its apex formed by a few fusiform cells pointing into the molecular layer. Where two pial surface meet, as in a sulcus, the focus of cells is confined to only one of the folia. Each focus measures about 250 μm in length and contains approximately 500 cells. The dense heterochromatin of the nuclei makes the focus stand out intensely in the lightly stained molecular layer. Careful examination with oil immersion lenses shows that this group of cells is not a pial plaque or some inflammatory focus. The cells are arranged in clusters, and between these clusters is a neuropil that stains more intensely than the surrounding molecular layer. In fortunate sections a small bundle of myelinated fibers may be seen extending from the internal granular layer through the molecular layer and reaching the subpial focus (Fig. 75).

In the electron microscope at low magnifications, a section through such a focus of cells (Fig. 76) shows two populations of neurons. The majority of cells are small, about 6 μm in diameter, with large, round, prominent nuclei and meager cytoplasm. The nuclei contain large blocks of condensed chromatin distributed on the inner side of the nuclear envelope. The cytoplasm forming a rim around the nucleus contains a few mitochondria, free ribosomal clusters, and some short tubules of the granular endoplasmic reticulum. No synapses are evident on the cell bodies (see Figs. 76 and 77). Morphologically these cells clearly resemble the granule cells of the internal granular layer.

Occasionally a few other neurons may be scattered among the granule cells. The neurons of this second class are larger (10–12 μm), with more ample cytoplasm, and they contain lobulated nuclei with less dense chromatin than the granule cells (Fig. 76). The somata bear synaptic junctions with parallel fibers. The identity of these cells is uncertain; they could be stellate or basket cells that also failed to descend.

Fig. 74. A focus of undescended granule cells at the pial surface of the cerebellar cortex. This mound of arrested granule cells (*arrow*) lies at the top of the molecular layer in lobule VII in the vermis. Such foci are easily detected, even at low magnifications, because of the basophilia of the clustered cells. In size and staining properties they resemble the granule cells of the internal granular layer. The pial surface (*pia*) and the layers of the cortex, molecular (*mol*), Purkinje cell (*PC*), and granular (*gr*) layers, are indicated. 1 μm epoxy section, toluidine blue. ×137

Fig. 75. Another focus of undescended granule cells at higher magnification. This cluster of cells is about 120 μm long and contains about 100 cells in this section alone. The cells (*gr c*) resemble granule cells of the internal granular layer. Between clusters of cell bodies lies a neuropil (*neu*) that even in photomicrographs appears different from the surrounding molecular layer. The mound of cells lies against the pial surface, and a bundle of myelinated mossy fibers (*arrow*) traverses the molecular layer to reach it. A few straggling cells can be separated from the main cluster (*rectangle*). 1 μm epoxy section, toluidine blue. Lobule II. ×2025

Fig. 76. Granule cells, larger neurons, and glomeruli in a focus of cells at the pial surface. The undescended granule cells (*gr*) are clustered close together, as in the granular layer. Scattered among them are larger neurons (*N*), probably stellate or basket cells isolated in the focus. The somata of these cells bear synapses with parallel fibers. Glomeruli are interspersed among the cell bodies (*arrowheads*). Thus each focus is an ectopic mass of granular layer displaced into the molecular layer. Lobule IV. ×13000

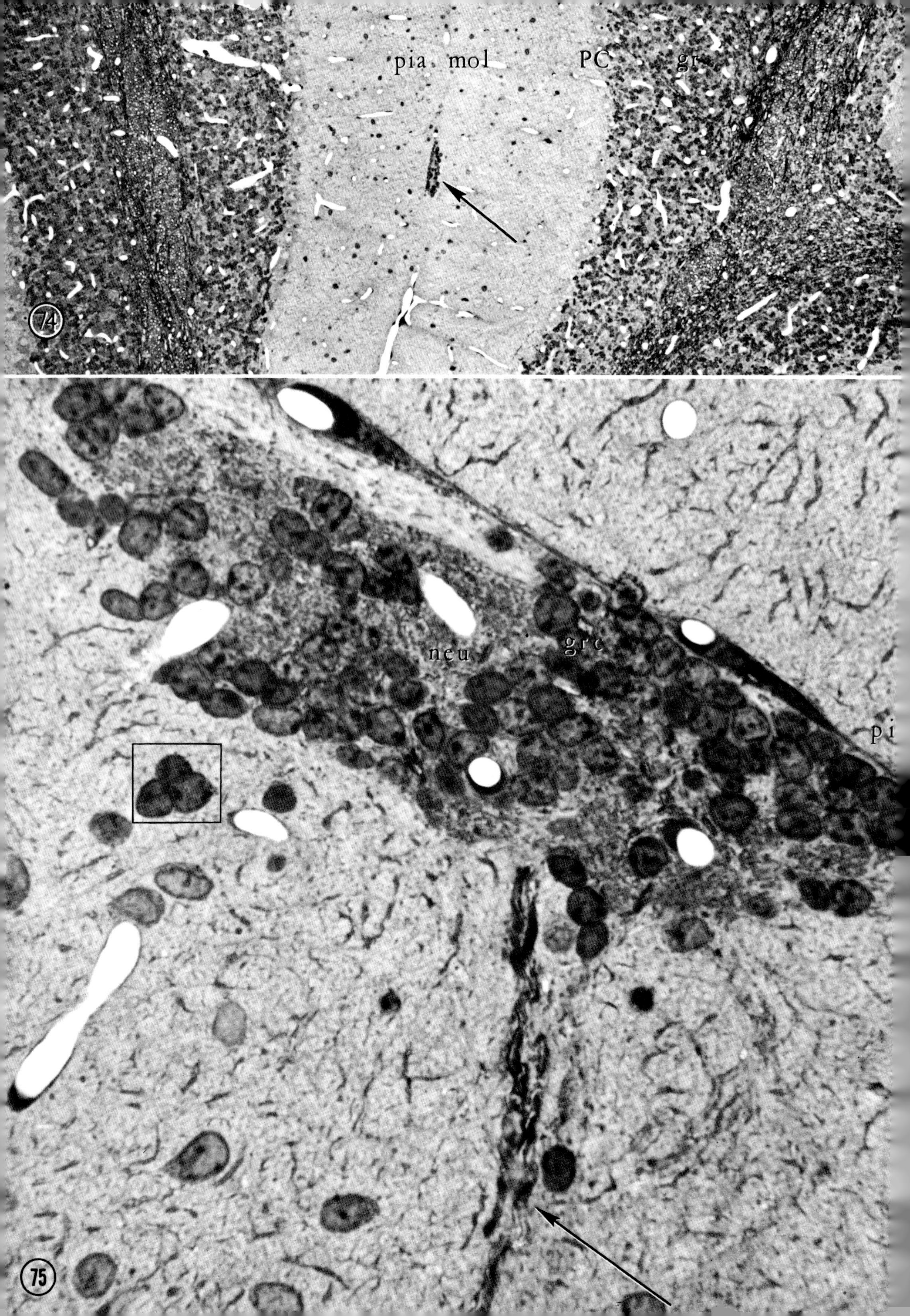
pia
mol
PC
gr
74
neu
grc
pi
75

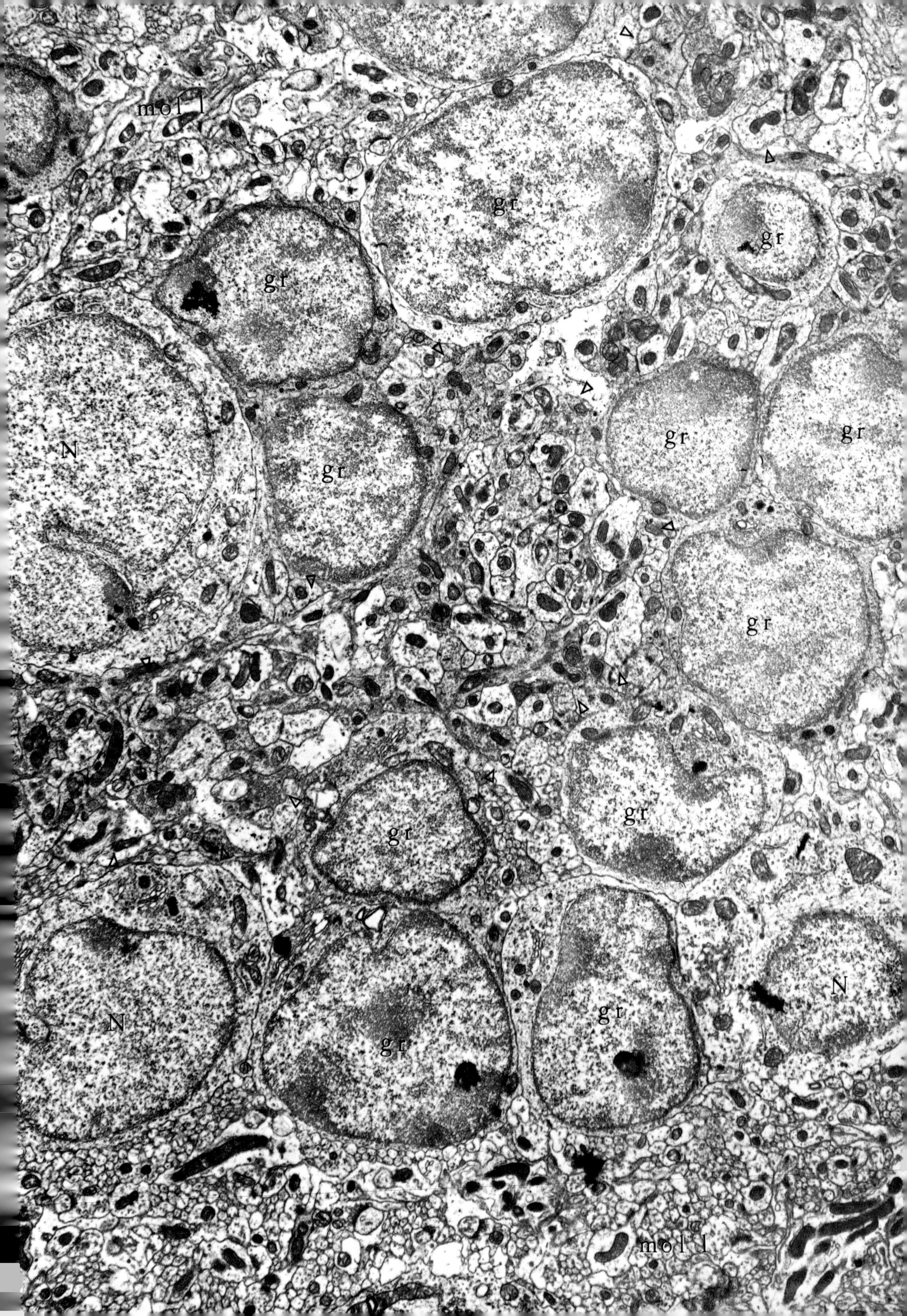
mol l
gr
gr
gr
gr
N
gr
gr
gr
gr
gr
N
gr
gr
N
mol l

Between the clusters of granules is a neuropil that is distinctly different from the surrounding molecular layer. It closely resembles the neuropil of the internal granular layer with its glomeruli. The myelinated fibers ascending through the molecular layer and entering a focus terminate in this neuropil as the central components of glomerular formations between the granule cells. The terminals contain large clusters of round synaptic vesicles, 300–460 Å in diameter (Fig. 78), similar to those found in mossy fiber rosettes (e.g., Figs. 133, 134, and 136). Numerous profiles of small dendrites are ranged around the terminals and synapse with them by means of asymmetrical, Gray's type 1 junctions. The dendrites apparently belong to the local, undescended granule cells. The resemblance between these glomeruli and those in the internal granular layer is confirmed by the presence of a set of smaller axonal profiles, containing flattened or elliptical vesicles, which synapse with the small dendrites through symmetrical Gray's type 2 junctions. These axons probably issue from the second type of nerve cell in these foci tentatively identified as either stellate or basket cells. Slips of neuroglial cytoplasm are insinuated around and between glomeruli (Fig. 79).

A few granule cells are also found straggling in the molecular layer between the main focus of cells and the definitive granular layer. These cells are round or fusiform, elongated in the vertical direction. Their dendrites also participate in small, poorly defined glomerular formations with collaterals from the ascending bundle of mossy fibers. The stragglers suggest that they have been arrested after they had begun their descent from the external granular layer.

The location of these ectopic masses of granular layer at the pial surface indicates that these cells are among the last to mature in the external granular layer. Perhaps they were also late in proliferating. Their failure to descend into their proper places deep in the cortex suggests that the migratory phase of development is independent of the maturation of the nerve cell, since their abnormal location has not interfered with their ability to form synapses with appropriate partners. We are, however, unable to describe the axons of these undescended granule cells, because they have not yet been impregnated in Golgi preparations.

However, a number of other conclusions may be proposed on the basis of these observations. The fact that these groups of undescended granule cells occur regularly in a certain few of the lobules of the rat's cerebellar cortex suggests that the development of such foci is under genetic control. Similar foci have also been observed in the mouse cerebellum (V. Friedrich, personal communication). Lemkey-Johnston and Larramendi (1968a) claim to have seen scattered granule cells in the molecular layer of the mouse. One of these cells was said to be in contact with a displaced mossy fiber terminal. Mugnaini (1972) mentions that granule cells have on rare occasions been observed in the cat in the low molecular layer, and that these cells may contact mossy fibers. Large collections of cells and glomeruli like those described by Chan-Palay have not been seen in the cat.

The fact that mossy fibers are able to grow through the molecular layer and synapse with a superficial group of granule cells provides strong evidence for the specificity of interneuronal connections, which is preserved despite the failure of one of the components to migrate to its proper place. As the granule cells have been arrested, the mossy fibers have continued their search for their appropriate partners by growing toward them. There must be some kind of specific signal that makes this interaction possible. Lastly, it is conceivable that the migration of cells is discontinued generally at some preset moment, and that cells delayed for one reason or another are stopped in their tracks wherever and in whatever stage they happen to be. It is important to notice that a defect in the Golgi epithelial cells and Bergmann fibers cannot be responsible for the failure of the granule cells to descend, for the mossy fibers have followed just that path in their vertical ascent through the molecular layer.

f) Parallel Fibers

The parallel fibers that result from the bifurcation of the ascending granule cell axons run immediately next to one another without any intervening neuroglia. As they pass through the dendritic trees of the successive Purkinje cells, stellate, basket, and Golgi cells in their path, they diverge to circumvent the dendritic branches or to synapse with them (Fig. 80). Between dendritic trees the molecular layer appears to be a solid array of parallel fibers packed close together, like a crystal (Figs. 81 and 82). The smooth shafts of the axons vary from 0.1 to 0.2 μm in diameter, with the thicker fibers generally in the deep molecular layer. A few of these deep fibers are covered with thin myelin sheaths, three or four lamellae thick.

Fig. 77. Four undescended granule cells and glomeruli in the molecular layer. The four granule cells (*gr c*), with their characteristic, round, heterochromatic nuclei and meager cytoplasm, bear no axosomatic synapses, but their dendrites participate with mossy fibers in the formation of small, loosely organized glomeruli (Δ). The molecular layer is indicated (*mol l*). Lobule IV. × 10000

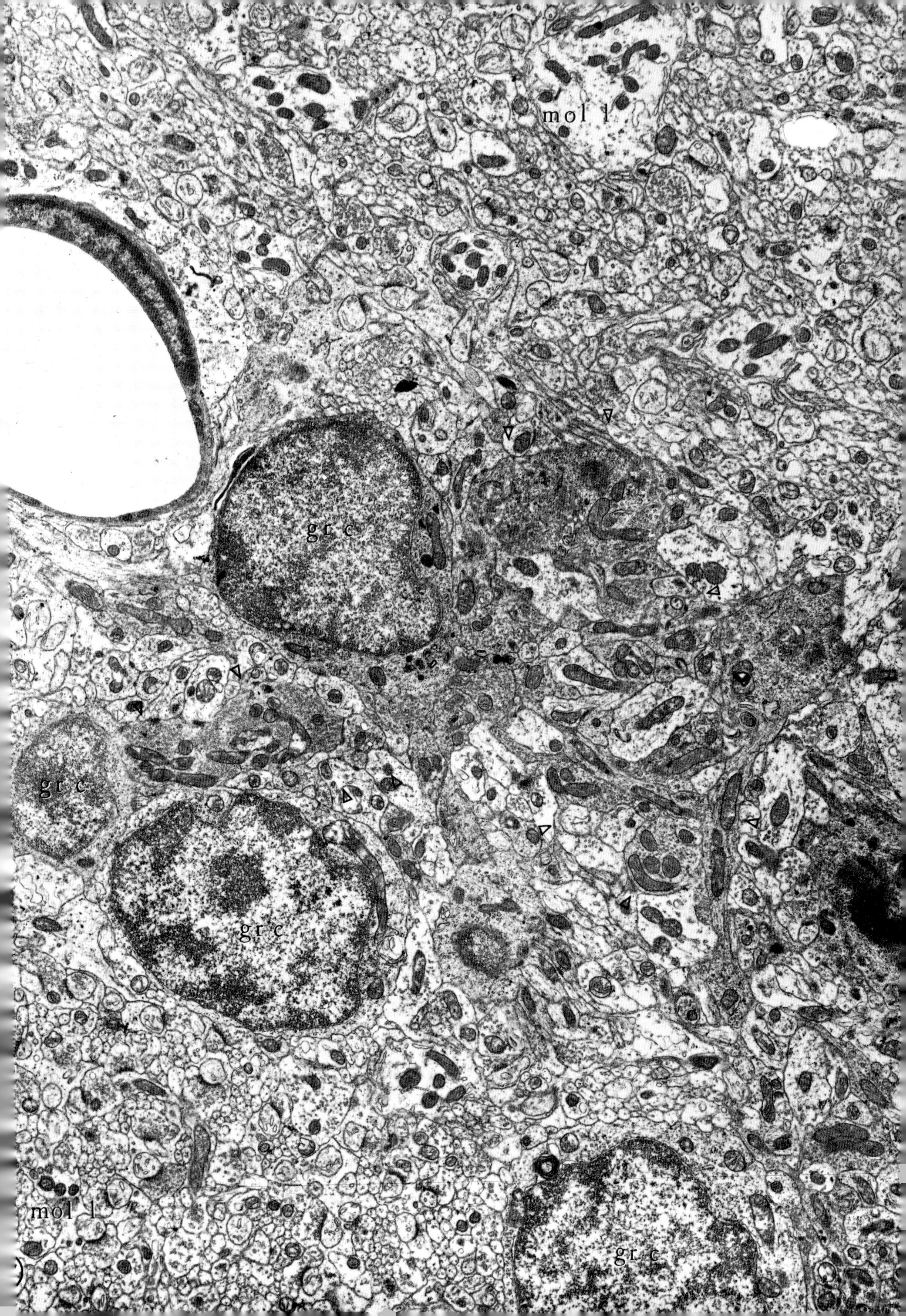
mol l
gr c
gr c
gr c
mol l
gr c

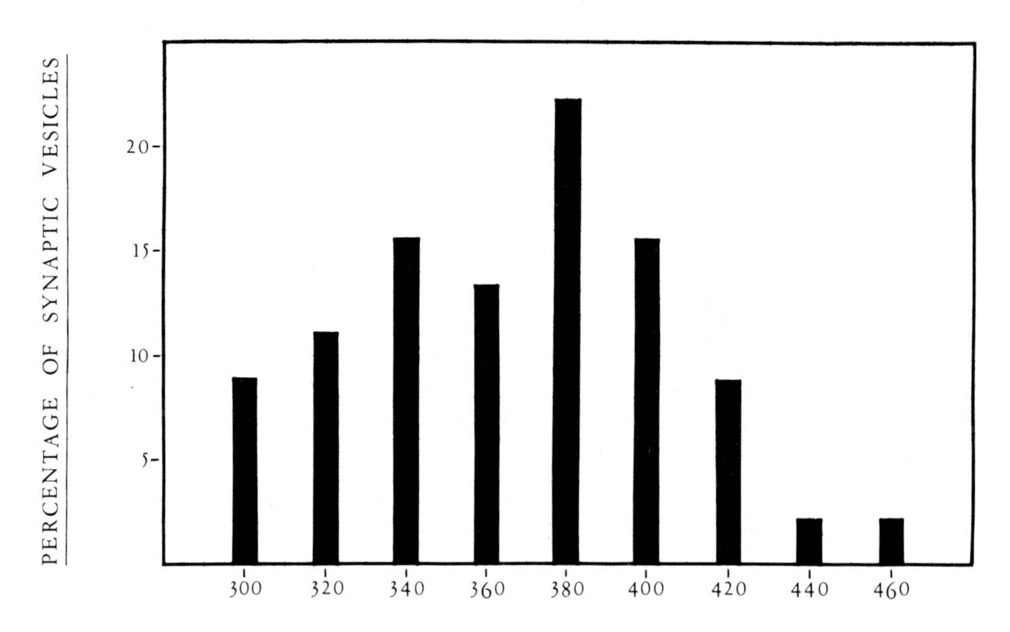

Fig. 78. Histogram showing the size distribution of synaptic vesicles in the mossy fiber terminals within a focus of undescended granule cells. The diameters of profiles of 50 synaptic vesicles in an axon terminal were measured. The vesicles ranged in size between 300 and 440 Å, with a peak at about 380 Å. The actual diameter of the synaptic vesicles is probably about 400 Å

Freeze-fractured preparations of the molecular layer convey an almost three-dimensional impression of the parallel fibers. In fractures passing through the longitudinal axis of the folium (Fig. 82A) the closely packed fibers appear in relief, displaying their wavering course and their frequent varicosities. The diameters of the fibers vary continuously along their course, and many of the enlargements can be shown to be presynaptic varicosities related to Purkinje cell dendritic thorns (Fig. 85A and B).

The axoplasm contains two or three straight microtubules, a meandering tubule or two of the smooth endoplasmic reticulum, and single, long mitochondria spaced out at intervals. As the fiber is barely large enough to contain a mitochondrion, it is usually somewhat distended in the regions where these organelles are located. Some of the irregularities seen in the Golgi preparations may correspond to the positions of mitochondria. But the most obvious irregularities are caused by the varicosities and several kinds of sessile or pedunculated appendages that mark the positions of synaptic junctions (Fig. 83). In electron micrographs these appendages are clearly expansions of the axon filled with round synaptic vesicles and one or two mitochondria. Often the vesicles are collected at one side of the fiber (facing a thorn or dendrite) and the mitochondrion is lodged in the axis of the fiber with the microtubules passing by around it. The number of vesicles varies considerably from just a few packed against the axolemma to several dozen loosely aggregated. The diameters of their profiles range from 260 to 440 Å (Fig. 84). The diameter of the vesicles is probably about 380–400 Å. The junction between the parallel fiber and Purkinje cell thorns is identical to that already described above between the stem of the granule cell axon and the thorns. The interstitial cleft is widened by the concavity of the axonal membrane, and short, tufted densities on the axonal side of the membrane asymmetrically offset a thick, matted postsynaptic web (Figs. 34, 35, and 83). Fine filaments can be seen crossing the synaptic cleft, and in favorably oriented sections the thin profile of a dense plate can be seen midway between the two membranes. The axonal varicosity or appendage caps or hooks around the expanded end of the thorn, but only a small part of the apposed surfaces displays the morphological characteristics of the synapse. The synaptic interface is apparently a single macular specialization, about 0.25 to 0.5 μm in diameter, and it is usually located on the side of the head of the thorn but not at its apex.

Not rarely, thorns are encountered that synapse with two parallel fibers. Fig. 85 shows an example of such a double synapse. The head of the thorn is somewhat flattened between the two parallel fibers, each of which expands opposite a separate postsynaptic web. The

Fig. 79. A mossy fiber glomerulus in the molecular layer. This glomerulus (ΔΔ) is tucked away into the molecular layer (*mol l*), and is therefore surrounded by parallel fibers (*pf*), thorns, and dendrites of Purkinje cells and even a climbing fiber profile (*CF*). The central mossy fiber terminal (*MF*) synapses with many dendrites of granule cells (*g*) by means of Gray's type 1 junctions. Lobule IV. ×18000

Fig. 80. Interrelations of parallel fibers with Purkinje cell dendritic thorns. This high voltage electron micrograph shows parallel fibers (*pf*) that course across the molecular layer to synapse with thorns (*t*) projecting from the dendrites of Purkinje cells (*Pc d*). A film of neuroglia (*gl*) coats the surface of the Purkinje cell dendrites. Lobule X. ×7500

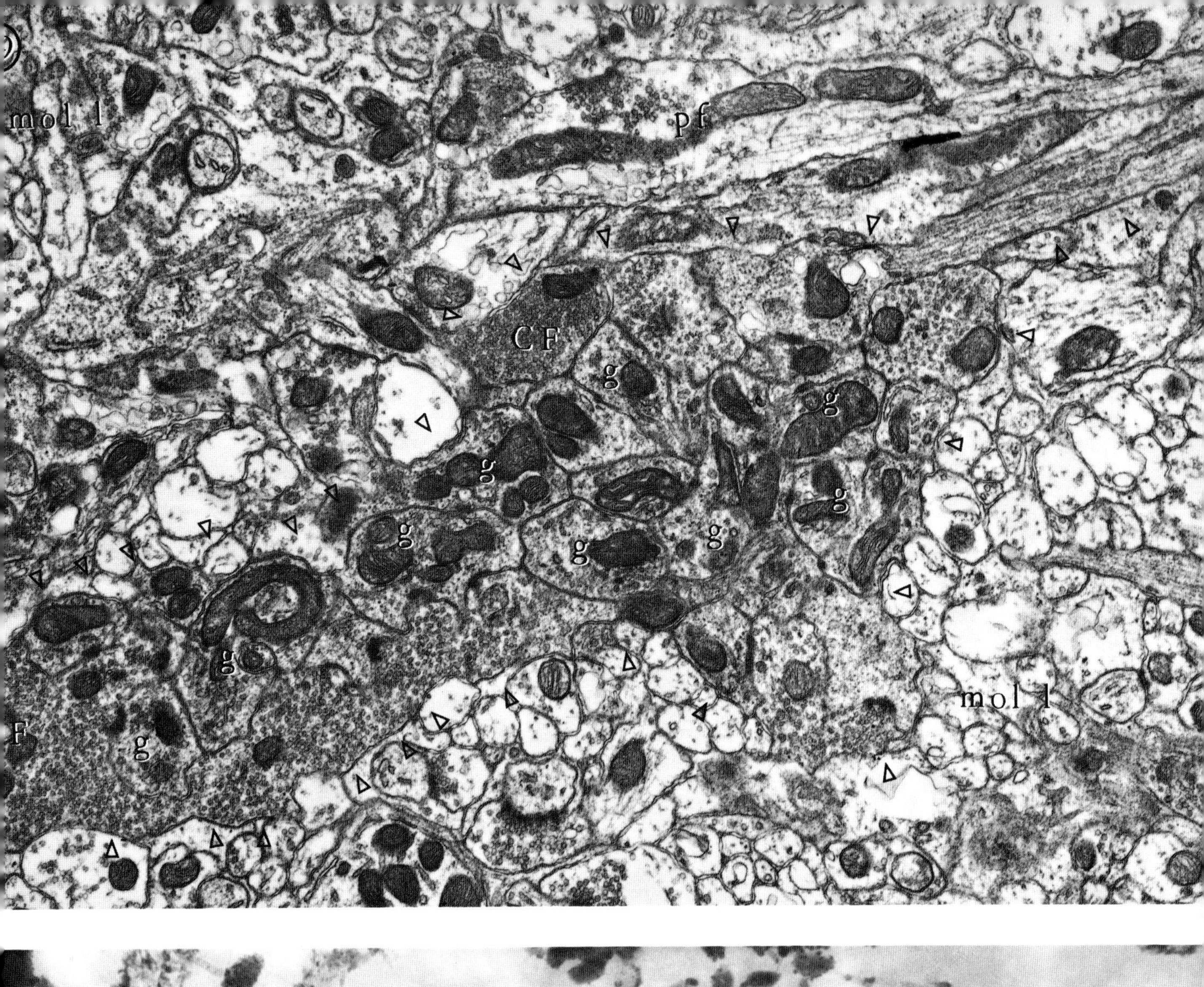
mol l
pf
CF
g
g
g
g
g
g
g
g
g
mol l
F

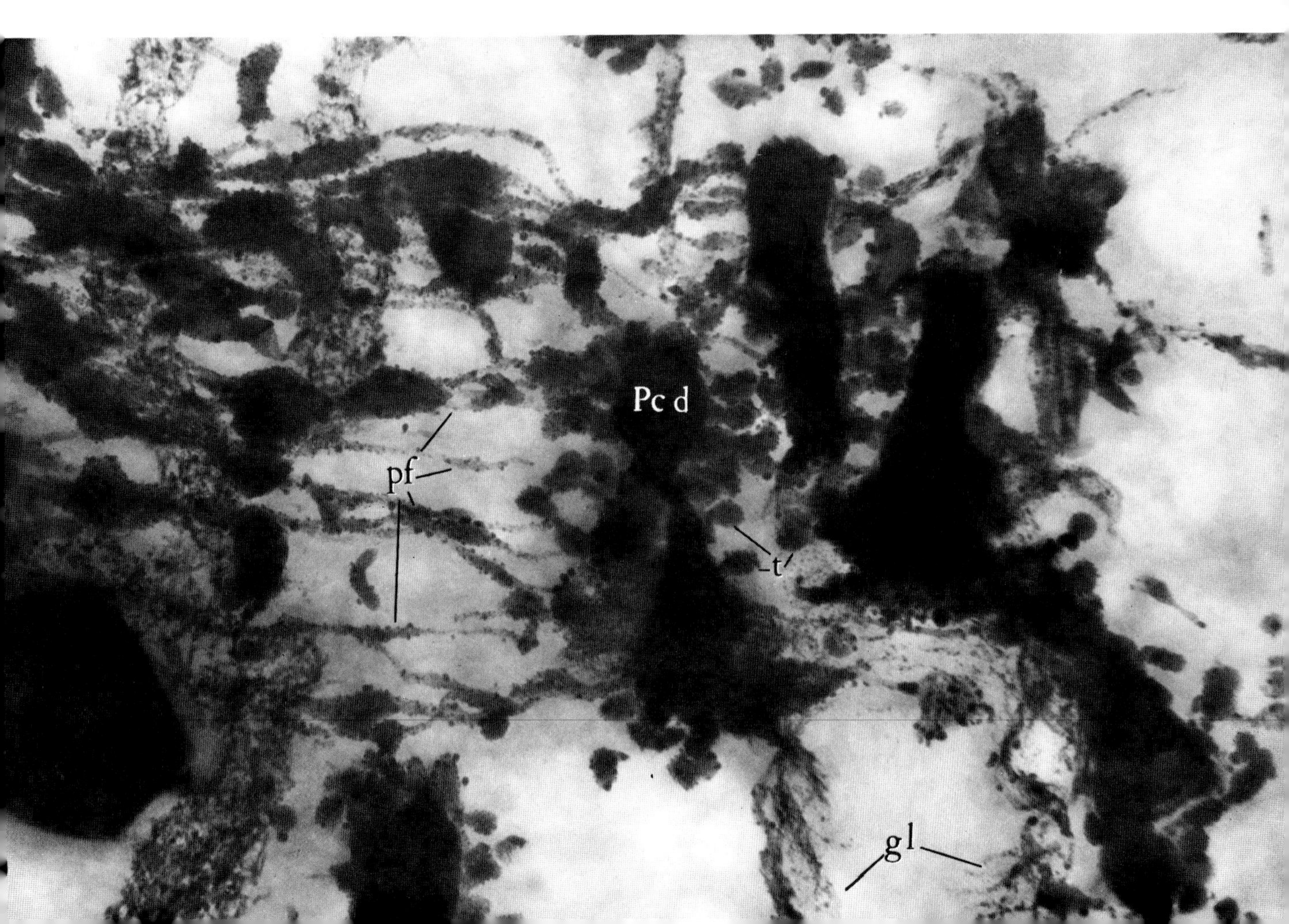
Pc d
pf
t
gl

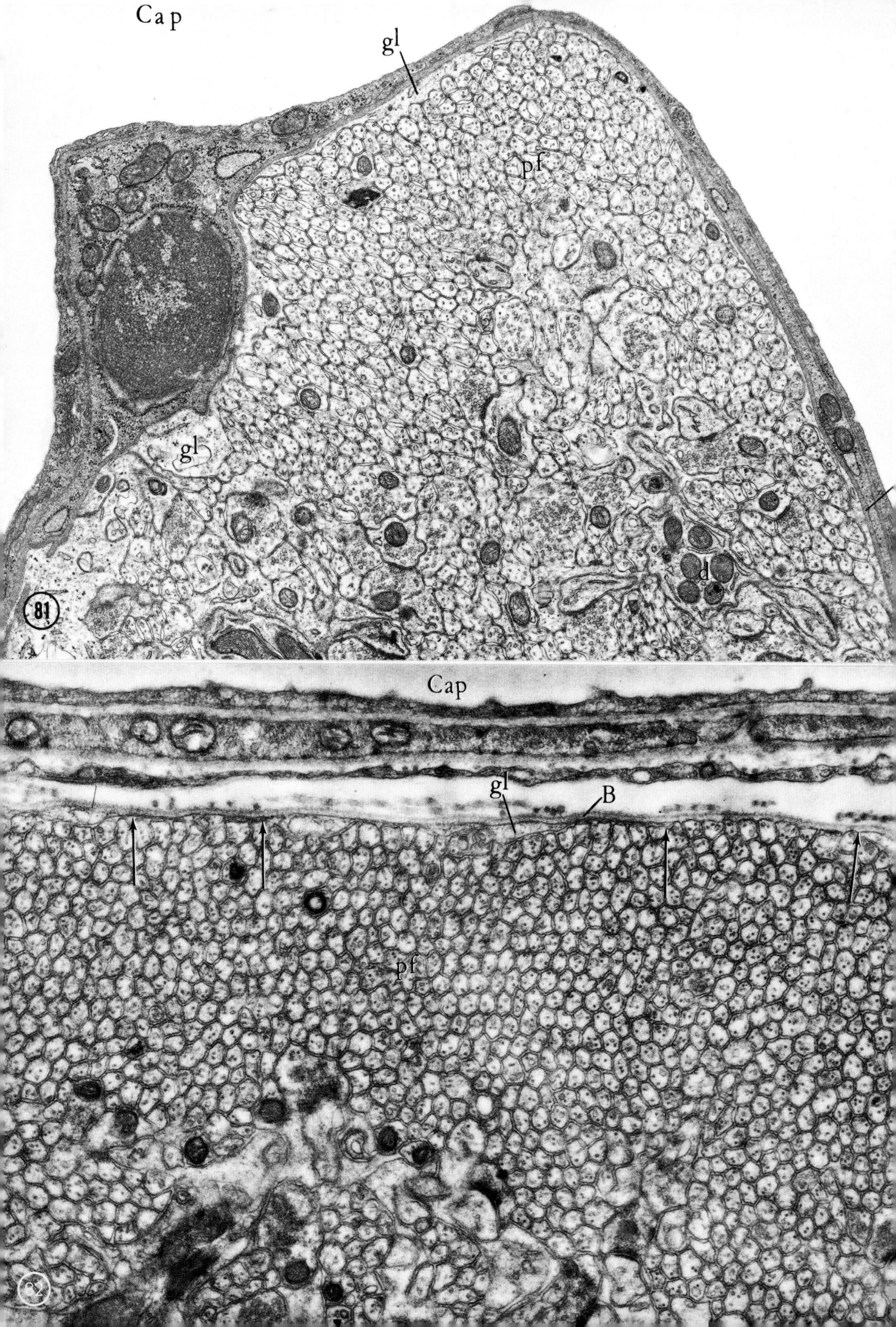

Cap
gl
pf
gl
d
81
Cap
gl
B
pf
82

reciprocal condition is much more frequent. In the same figure, a single varicosity of a granule cell axon can be seen in synapse with two thorns, each with its own synaptic interface. Synapses do not occur on the stems of Purkinje cell thorns. The junctions between parallel fibers and basket, stellate, and Golgi cells will be considered in the sections dealing with those cells.

Further information about the articulation between parallel fibers and Purkinje cell dendritic thorns has been obtained from recent studies of freeze-fractured preparations (PALAY, RAVIOLA, and CHAN-PALAY, 1973). In tissues prepared according to this method, the surface membrane of a cell splits along a cleavage plane that runs parallel with the cell surface (BRANTON, 1966; PINTO DA SILVA and BRANTON, 1970). Thus the fracture exposes one of two possible faces as it passes through a membrane: (1) the A face, representing the external aspect of the inner cytoplasmic leaflet of the plasmalemma, or (2), the B face, representing the internal aspect of the outer cytoplasmic leaflet. The cleavage plane apparently does not run along the interstices between cells or along the cytoplasmic surfaces, but cuts across the interstitial space or the cytoplasm within the cell. Therefore the true outer and inner aspects of the plasmalemma are not exposed, but they can be revealed to some extent by subliming the water from the cytoplasmic matrix or the intercellular space by a procedure called freeze etching. The A face of any plasmalemma is generally characterized by numerous randomly dispersed particles, 60–100 Å in diameter, whereas the B face is usually smooth, decorated with only an occasional particle. On either face, the arrangement of the particles into patterns is usually taken to indicate the presence of some specific membrane substructure such as might occur at an intercellular junction.

The plasmalemmal substructure at synapses in the subcommissural organ and in the spinal cord has recently been studied by AKERT and his coworkers and reported in a series of papers (AKERT and SANDRI, 1970; AKERT *et al.*, 1972; PFENNINGER *et al.*, 1972; SANDRI *et al.*, 1972; STREIT *et al.*, 1972). According to these authors, the A face of the presynaptic membrane is recognizable as a slight saucer-shaped depression in which small pits are disposed in a triangular array. The usual A face particles are nearly absent in this area and the fracture surface is barren except for a few particles clustered about the rims of some of the pits. In contrast, the complementary B face is slightly elevated and displays small protuberances arrayed in a triangular or hexagonal array. On the summit of many of these protuberances there is a ring of particles. The postsynaptic membrane is similarly identifiable: the A face exhibits a poorly defined assemblage of slightly larger particles, while the B face possesses a dense aggregation of particles in a clearly circumscribed patch.

With these reports in mind, we searched for synaptic junctions in freeze-fractured preparations of the rat's cerebellar cortex. Our prior analysis of the synaptic connections in conventional thin sections and in Golgi preparations provided criteria not merely for recognizing synaptic junctions but also for identifying the source of the cellular elements participating in the synapse. Thus it was possible to recognize Purkinje cell thorns and parallel fiber varicosities, stellate axons, Golgi dendrites, mossy fibers and so on.

A quite instructive field appears in Fig. 85A. In this micrograph a spiny branchlet (*PC d*) with its numerous projecting thorns (*t*) and several passing parallel fibers (*pf*) all are easily recognizable. The preparation offers a three-dimensional image that confirms the reconstructions

◄

Fig. 81. Parallel fibers around a blood vessel: cross section. The parallel fibers are packed in an almost crystalline array around dendrites (*d*) and synaptic varicosities of the molecular layer. A thin film of neuroglial cytoplasm (*gl*) intervenes between parallel fibers and the basal lamina (*B*). The cytoplasm of the endothelial cell lines the blood capillary (*Cap*) and the perikaryon of a pericapillary mesenchymal cell appears at the left. Lobule VI. × 18000

Fig. 82. Array of parallel fibers: cross section. An almost crystalline array of parallel fibers fills the molecular layer. In certain places the neuroglial layer is incomplete and the parallel fibers (*between arrows*) lie directly against the basal lamina (*B*). Elsewhere a neuroglial sheath intervenes (*gl*). The endothelium of a blood capillary is indicated (*Cap*). Osmium tetroxide fixation. × 20000

►

Fig. 82A. Parallel fibers in mid-molecular layer. This freeze-fractured preparation shows parallel fibers (*pf*) coursing diagonally across the field parallel with the longitudinal axis of a folium. Both A and B fracture faces are seen. The A faces display small particles irregularly strewn over the surface, but apparently concentrated along linear tracks corresponding to free edges of the close-packed fibers. The B faces are almost completely smooth. A Purkinje cell dendrite (*PC d*) appears in the right upper corner of the field and gives off four thorns, the stems of which are broken off. The cytoplasm of a neuroglial process (*ng*) is exposed in the middle of the field and another is seen at the right. The large arrow indicates the direction of shadowing. Replica, freeze fracture. × 67000

Fig. 83. A horizontal section through the molecular layer: parallel fibers. This illustration shows a portion of the molecular layer. Parallel fibers course across the neuropil in several bundles, only one of which is indicated (*pf, arrows*). Between these bundles, dendrites of Golgi cells (*Go d*) and Purkinje cells (*Pc d*), and neuroglial cytoplasm intervene (*gl*). Varicosities of parallel fibers (*s*) synapse upon thorns (*t*) of Purkinje dendrites. Lobule X. × 16000

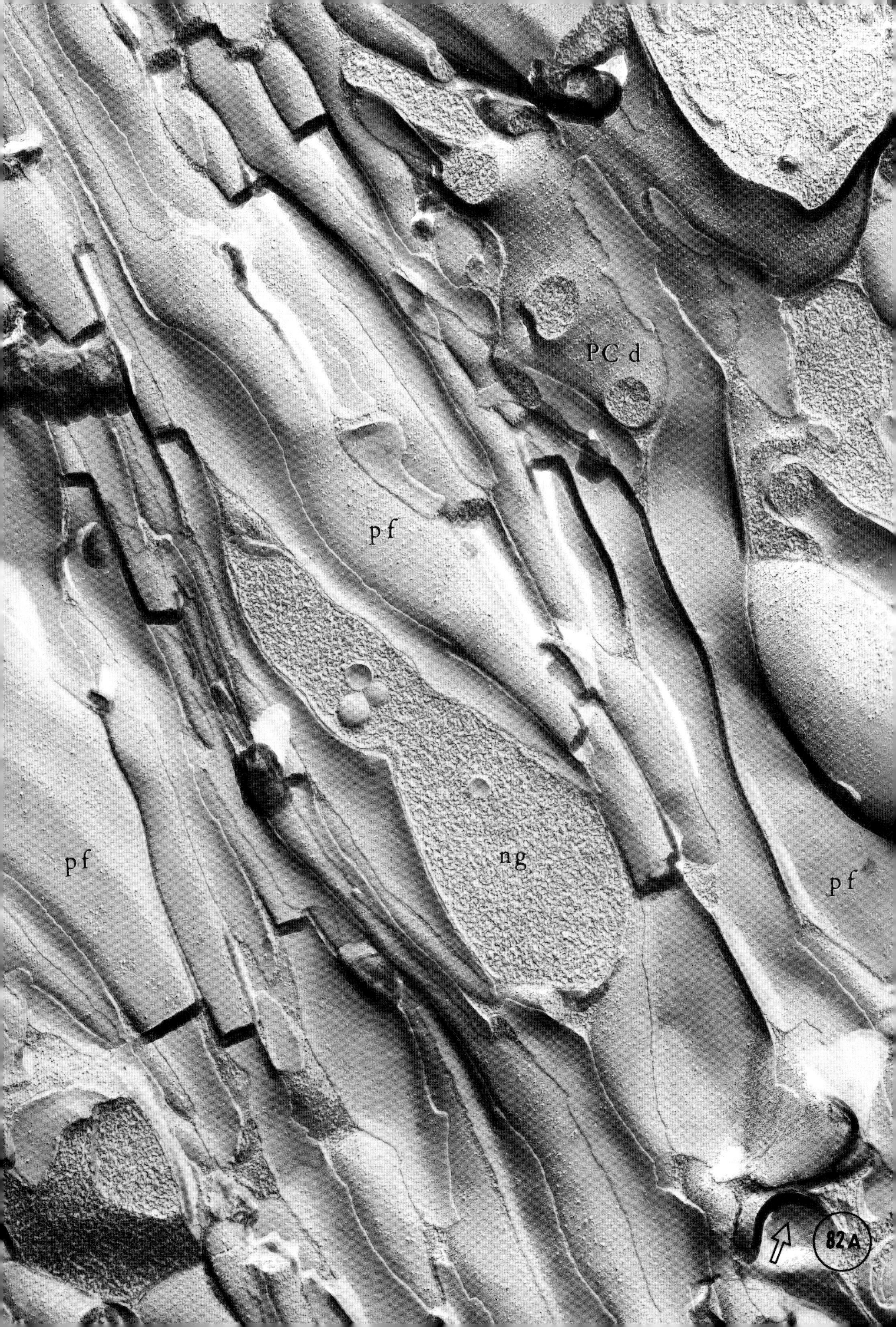
PC d
pf
pf
ng
pf
82A

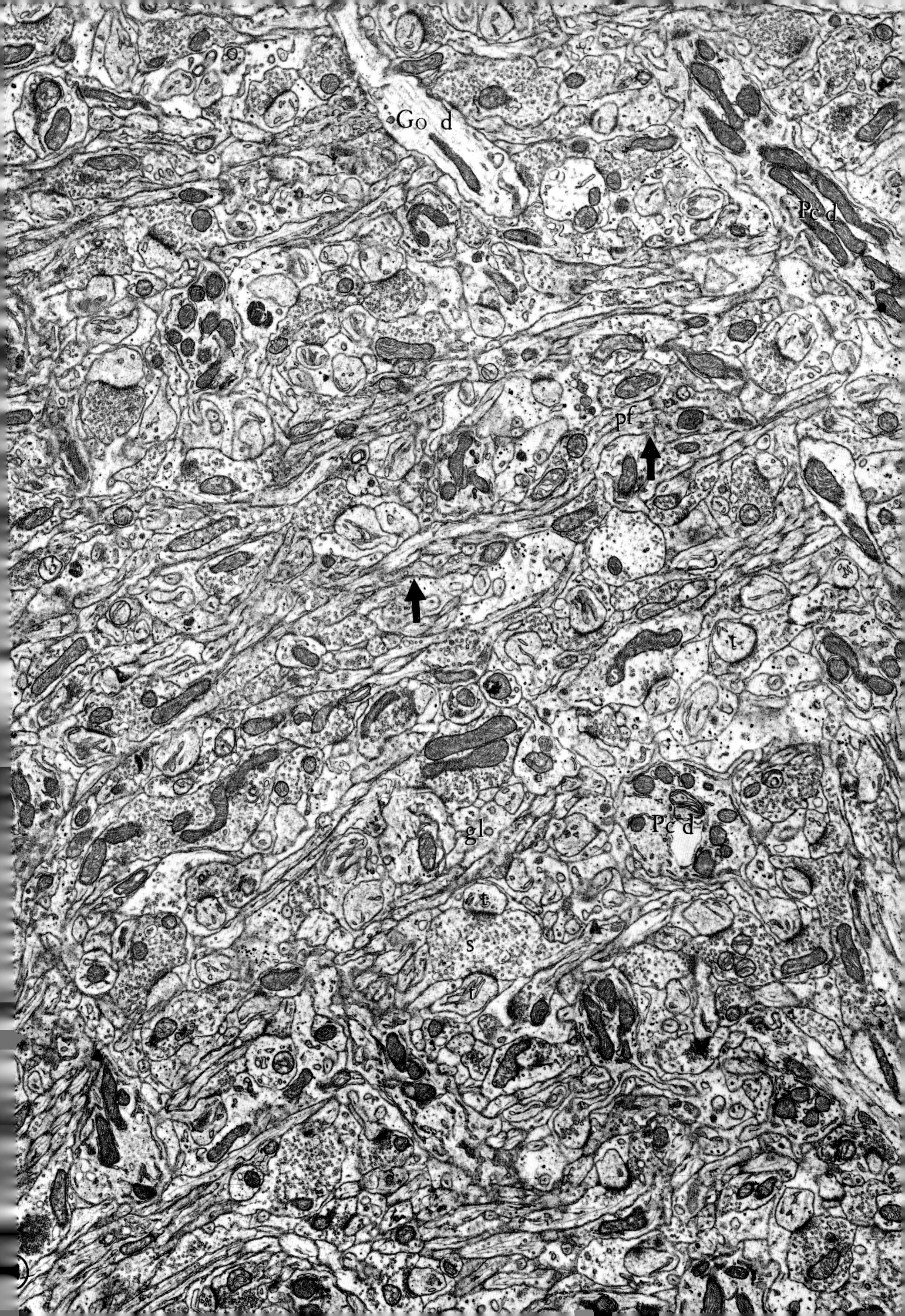
Go d
Pc d
pf
t
s
gl
Pc d
t
s
t

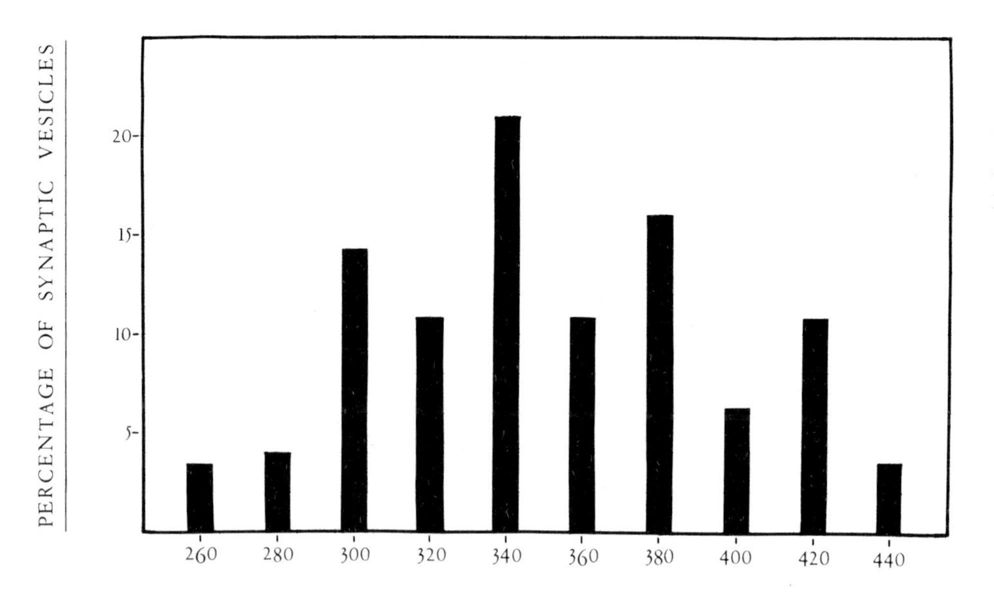

Fig. 84. Histogram showing the size of synaptic vesicle profiles in parallel varicosities. The vesicular profiles in parallel fibers are all round, and they range in size between 260–440 Å. The diameter of these synaptic vesicles is probably about 380–400 Å

obtained from Golgi preparations, high voltage electron microscopy, and conventional thin-section electron microscopy. The articulations between Purkinje cell thorns and parallel fiber varicosities are best studied by inspecting thorns t_2 and t_3 in this figure. In thorn t_2, the cleavage plane exposes first the smooth B face of the stem, and then cuts across it to expose the A face of the bulging head coated with fine particles. Near the tip of the thorn is a flattened oval zone that is relatively clear of A face particles. Thorns t_1 and t_4 similarly show the A face covered with particles but the synaptic junction is not exposed. The complementary aspect appears in thorn t_3 where the B face of the entire thorn is exposed. Whereas most of the B face is smooth, it is decorated with a dense collection of particles at the apex and side of the thorn. Similarly thorn t_5 shows an aggregation of B face particles at the synaptic site.

The opposite side of the synaptic junction can be studied by following the parallel fibers. Fiber pf_2, for example, which is represented by a rather fragmentary B face, extends over the summit of thorn t_2. The synaptic junction is indicated by the remains of a sharp elevation in the axonal B face opposite the smooth, particle-free zone on the tip of the thorn. The intercellular space is distinctly enlarged as would be expected from the thin section electron microscopy, which shows the parallel fiber-Purkinje cell dendrite thorn synapse to be a Gray's type 1 junction. A more complex version of the B face of the parallel fiber varicosity is shown in Fig. 85B in which the junction is seen head-on. The thorn has been cracked off near its stem and only the head remains in the field. Its B face shows a smooth zone near its tip which is partially covered by the B face of an expanding parallel fiber varicosity. The slightly enlarged synaptic cleft is discernible. The B face of the varicosity exhibits a slight elevation corresponding to the enlargement of the synaptic cleft and a nodular or cobbly surface similar to that described by PFENNINGER *et al.* (1972) in spinal cord presynaptic terminals. Returning to Fig. 85A, parallel fiber pf_3 shows the complementary A face of the presynaptic fiber. Here the A face, viewed from an extracellular position, appears somewhat depressed, corresponding to the widened synaptic cleft, but the specific configuration of the presynaptic locus is obscured by the overlying membrane of the dendritic thorn. Similarly, the junction between thorn t_5 and fiber pf_5 is obscured by an unfavorable angle of shadowing. However, the presynaptic A face is exposed in fiber pf_6, where the overlying thorn has been almost completely stripped away. The A face in this location possesses small shallow depressions associated with irregular clusters of particles that seem to correspond to the cobbly protuberances on the complementary presynaptic B face. Finally it should be noted that the thorns of the Purkinje cell dendrite all project through a sheath of neuroglia (*ng*).

Similar features are to be found in the synaptic junctions effected by the parallel fibers on the dendrites of basket and Golgi cells. Although these junctions are located on the shafts of the dendrites, their intramembranous structure is the same as that of thorn synaptic junctions. In all of these cases, as in the thorn synapses, the intramembranous structure displays a focal differentiation corresponding in extent roughly to the part of the junctional interface that is termed "the synaptic complex," that is, the region occupied on the postsynaptic side by densities adherent to the postsynaptic membrane and

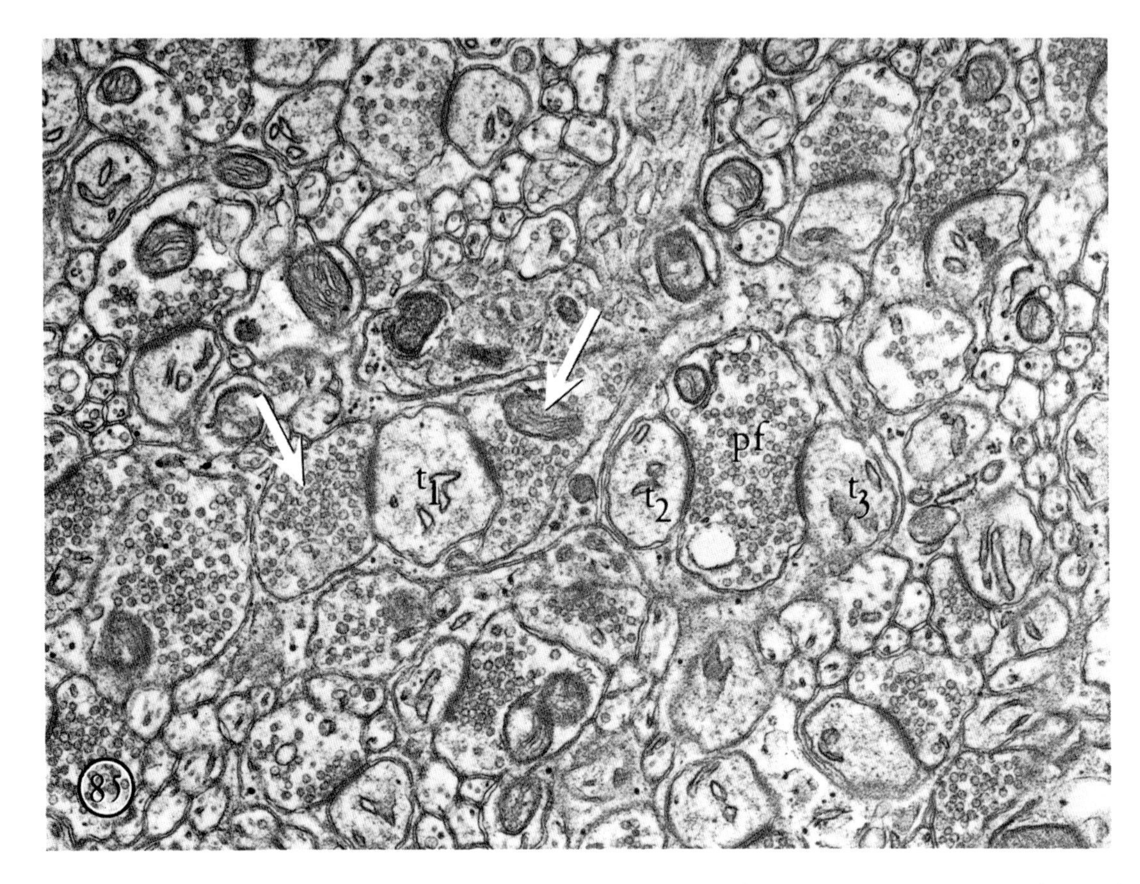

Fig. 85. Synapses of parallel fibers. This illustration shows common synaptic arrangements of parallel fibers. First, a single Purkinje cell thorn synapses with a single parallel fiber. Second, a single Purkinje cell thorn (t_1) synapses with two parallel fiber varicosities (*arrows*). Third, two Purkinje cell thorns (t_2 and t_3) can synapse upon a single parallel fiber varicosity. The first is the most common relationship and the third is more common than the second. Lobule IX. ×24500

on the presynaptic side by synaptic vesicles aggregated close to the presynaptic membrane. The existence of focal differentiations in the intramembranous structure at these sites lends some support to the original idea of the synaptic complex (PALAY, 1956, 1958) as the "active zone" (COUTEAUX, 1961) of the synaptic junction, the point where chemical transmission actually occurs.

At this point it is worth while to return to the high voltage electron micrographs of Golgi preparations that were mentioned earlier in this chapter (Figs. 57 and 58). These pictures show the varicosities in the course of the parallel fibers as fusiform or globular swellings, as in the light microscope. Occasionally a pedunculated appendage is also encountered. The higher magnification and resolution of the high voltage microscope, however, yield more detailed information in these preparations, which is not available by means of light microscopy and which could be achieved only at the cost of much labor from conventional electron microscopy. Almost every varicosity presents one or more discrete lucent areas resembling holes. Generally these light patches are rounded, about 0.5–1.0 μm in diameter; some are elongated, with rounded ends. Occasionally the negative image of a mitochondrion can be clearly recognized. A comparison of these images with conventional electron micrographs of thin sections leads to the inescapable conclusion that the rounded lucent areas represent the sites of synapses with dendritic thorns (CHAN-PALAY and PALAY, 1972b). The Golgi impregnation fills the cytoplasmic matrix with a fine, fibrillar deposit. Apparently the precipitate does not pass through membrane-limited compartments within the cell. Consequently the nucleus, mitochondria, and other membrane-limited organelles remain unimpregnated, at least during the initial stages of the reaction that produce a predominantly red coloration. The synaptic varicosities of parallel fibers contain both mitochondria and synaptic vesicles, the latter in large numbers clustered against the dense filamentous plaque attached to the presynaptic membrane. None of these structures becomes opacified by the Golgi reaction, while the axoplasmic matrix is impregnated. Thus the clearings correspond to synaptic sites. The circular outlines of the lucent areas match the junctional interface between the dendritic thorns and the parallel fibers (Figs. 58, 71, 83, and 85). Since a high proportion of the varicosities offer more than

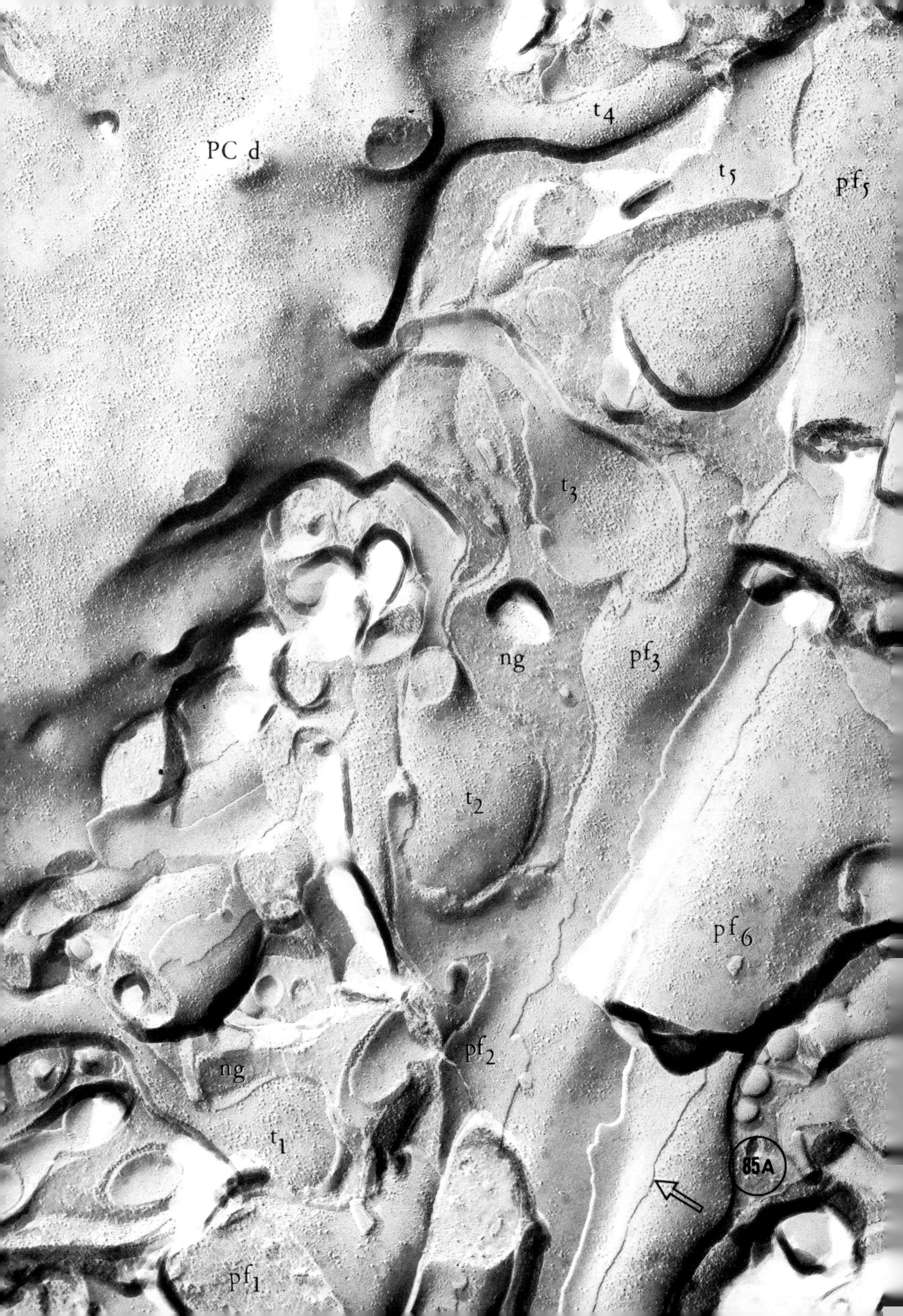

PC d
t4
t5
pf5
t3
ng
pf3
t2
pf6
pf2
ng
t1
85A
pf1

Fig. 85 B. Junction between a parallel fiber and a Purkinje cell thorn. The parallel fiber (*pf*) is broken at the left upper corner, and the B face of its plasmalemma expands round a varicosity. A circular elevated area, marked by a cobbly texture, indicates the presynaptic membrane. This face is partially broken away, exposing the A face of a thorn (t) underneath. The thorn is covered with particles except in the center of the postsynaptic area, where it is smooth. The synaptic cleft is indicated by an arrow. The large arrow indicates the direction of shadowing. Replica, freeze fracture. × 78 000

◄

Fig. 85 A. Spiny branchlet and parallel fiber-thorn synapses. A Purkinje cell spiny branchlet (*PC d*) crossing the left upper part of this field gives off four thorns (t_{1-4}), which extend toward the center of the picture. The connection of another thorn (t_5) with the spiny branchlet is not included in this preparation. The spiny branchlet and its thorns are coated by neuroglial processes (*ng*). A group of parallel fibers (pf_{1-6}) runs across the right side of the field. Thorns t_1 and t_4 are intact, displaying their A faces along their entire length. Thorn t_2, however, is broken; its stem is represented only by its B face, whereas the head presents an intact A face. Thorns t_3 and t_5 are represented by their B faces only. Near the top of thorn t_2 is a flattened, rounded area almost clear of A face particles. This thorn is partially covered by the torn B face of parallel fiber pf_2. The widened synaptic cleft of this junction is visible. The B faces of thorns t_3 and t_5 both show aggregates of particles at the postsynaptic position opposite varicosities in parallel fibers pf_3 and pf_5. The widened synaptic cleft is clearly shown in the junction of thorn t_5 with parallel fiber pf_5. The large arrow indicates the direction of shadowing. Replica, freeze fracture. × 82 000

a single synaptic site, the parallel fiber must make several synapses as it traverses the dendritic arborization of a single Purkinje cell. Some of these synapses will be junctions with basket, stellate, and Golgi cell dendrites, but probably most of them are axo-dendritic synapses with Purkinje cell thorns.

3. Summary of Synaptic Connections of Granule Cells

Afferents	mossy fibers	————	granule cells
	climbing fibers	————	granule cells
	Golgi cells	————	granule cells
	Purkinje cells	————	granule cells
Efferents	granule cell	————	Purkinje cell
	granule cell	————	basket cell
	granule cell	————	stellate cell
	granule cell	————	Lugaro cell
	granule cell	————	Golgi cell

Chapter IV

The Golgi Cells

1. A Little History

In 1874 GOLGI described two kinds of distinctive large nerve cells in the granular layer of the human cerebellar cortex. Although their cell bodies had been noted by others (e.g., by DENISSENKO, 1877), their processes had never been seen before, and their discovery was one of the many first fruits of his impregnation method that he was able to harvest long before other histologists became interested in it. The first kind of nerve cell had a long, fusiform perikaryon that lay directly beneath the layer of Purkinje cell bodies and extended transversely across the folium. The major dendrites issued from opposite poles of the perikaryon, and, continuing the longitudinal axis of the cell body, they ran for long distances before bifurcating and turning upward into the molecular layer or downward into the granular layer. GOLGI was unable to observe their terminations. The second kind of cell that he described was irregularly rounded or polygonal, almost as large as the Purkinje cell, and furnished with numerous dendrites that, in contrast to those of the fusiform cells, tended predominantly to run into the molecular layer. GOLGI pointed out that the dendrites of the polygonal cells reached the outer zone of the molecular layer and that many of them ended at the pial surface. The axons of both kinds of cells emerged from the deeper side of the cell body or one of the large dendrites, and after a fairly short course broke up into a multitude of fine branchlets, forming an elaborate and dense plexus that extended through the entire thickness of the granular layer.

Adding a few details to this description and several magnificent drawings made with the camera lucida, GOLGI repeated his first account of the large cells in the granular layer as a small part of an extensive paper in 1883 on the fine structure of the central nervous system. A few years later his discovery was confirmed by VAN GEHUCHTEN (1891), KÖLLIKER (1890), RAMÓN Y CAJAL (1888a, 1889a, 1890a), and RETZIUS (1892a), all of whom found many examples of the second kind of large cell, the polygonal cell, and perhaps a few of his first kind, the fusiform cell. KÖLLIKER called them large granule cells, while RAMÓN Y CAJAL, without referring to Golgi's descriptions of five and sixteen years earlier, gave them the name "large stellate cells of the granular layer". RETZIUS proposed that they should be called "the Golgi cells of the cerebellum", and this is the name that they have retained to the present day. It should be noted, however, that RETZIUS did not describe or figure the large fusiform cells that GOLGI designated as belonging to his first group, and RAMÓN Y CAJAL, even in his comprehensive descriptions of 1911, writes grudgingly about them as if he had seen very few. LUGARO in 1894 gave a more complete description of these fusiform cells, since he was successful in tracing their axons into the molecular layer. Because of the ease with which the form of the large stellate cells was confirmed, the term *Golgi cells* has been restricted to them in most accounts. In a recent book they are called "typical Golgi cells" (ECCLES, ITO, and SZENTÁGOTHAI, 1967), but in most descriptions the other large neurons of the granular layer are either ignored or are given other names, for example, "Lugaro cells," or "fusiform horizontal cells." In the present account we shall follow the traditional terminology and restrict the term *Golgi cells* to the large stellate or polygonal cells that Golgi listed as his second group.

The question of terminology is further confused by GOLGI's (1882) classification of all nerve cells in the central nervous system into two broad types according to their axonal patterns. Type I cells were those nerve cells whose axons give off few collaterals close to the perikaryon and pass directly into myelinated nerve fibers. Type II cells were those whose axons shortly after emerging from the cell body arborize into an intricate plexus pervading the gray matter. The division of nerve cells into these two morphological groups arose out of Golgi's concern to find a morphological representation of his great functional division of the nervous system into motor and sensorial-psychic aspects. This classification appeared again in 1883 in the same paper in which Golgi repeated his description of the large cells of the granular layer. He referred to his illustration of a large, rounded or polygonal cell in the cerebellar cortex, with its many dendrites and its remarkable axonal plexus, as one of the most striking examples of his second type of nerve cell. Perhaps for this reason it has become known as *the* Golgi Type II cell of the cerebellar cortex. This usage unjustifiably and unnecessarily restricts the significance of the term, for many other cells, such as the basket and stellate cells of the molecular layer, are also Golgi Type II cells according to the original definition. In fact, it is quite clear that both of the large nerve cells found in the granular layer and originally described by Golgi are Type II cells. Therefore it is a mistake, which has recently crept into common usage and appears in some textbooks, to designate these cells by the name "Golgi Type II cells" (although they belong to that class), since it is not distinctive enough. The mistake reflects a confusion of an instance with its class, and is compounded by the similarity of names.

On the basis of his own observations RAMÓN Y CAJAL (1911) described four kinds of large stellate cells (*grandes cellules étoilées*) in the granular layer: typical stellate cells (Golgi cells), horizontal fusiform cells, displaced stellate cells, and fusiform or stellate cells with long axons. The first two kinds of cell are the same as those Golgi described. The displaced stellate cells are a group that RAMÓN Y CAJAL found only in the rabbit; their cell bodies and dendrites lie in the molecular layer while their axons ramify in the granular layer like those of ordinary Golgi cells. RAMÓN Y CAJAL interpreted them as being Golgi cells displaced into the molecular layer during development. The fusiform or stellate cells with long axons constitute a heterogeneous group of cells encountered at various depths in the granular layer and even in the subjacent white matter (where GOLGI, 1883, and RETZIUS, 1892b had also noted similar cells). All of these cells have axons that pass into the white matter, and the perikarya of some of them are surrounded by nests of basket fibers. These cells are not common, but tend to be seen more often in the cerebella of young animals than in adults, especially in certain species. RAMÓN Y CAJAL proposed that all of them are displaced cells from the roof nuclei, but some, specifically those surrounded by pericellular baskets, are clearly displaced Purkinje cells (ESTABLE, 1923; JAKOB, 1928; ECCLES, ITO, and SZENTÁGOTHAI, 1967), although FOX (1959) considers the possibility that some may be the deeper horizontal fusiform cells. The cells of RAMÓN Y CAJAL's third and fourth kinds are certainly atypical and may well be manifestations of incomplete maturation in the specimens used for Golgi impregnations. They are very rare in the brains of the young adult rats that we have studied.

Still another group of stellate cells was distinguished by PENSA (1931). These were large multipolar cells, especially common in young cats, which alternated with the Purkinje cells at the margin of the molecular and granular layers, and which spread out their dendrites in the transverse plane of the folium while their axons arborized in the granular layer. These "intercalated cells" have remained somewhat mysterious, as other observers have not agreed on either their existence or their identity (JANSEN and BRODAL, 1958; ECCLES, ITO, and SZENTÁGOTHAI, 1967). GOLGI had figured such cells in 1883, and expressly states in the figure legend that they belong to the same type as the large polygonal cells, now designated by his name. ECCLES, ITO, and SZENTÁGOTHAI (1967) suggest that they may be displaced basket cells or Golgi cells, or alternatively that they may be misidentified Golgi epithelial cells (neuroglia), a large number of which lie in just this position. In the rat we have not seen any neurons in this position that answer to the descriptions and drawings of PENSA. BAFFONI (1956) proposed that during the life of the animal they were transformed into Purkinje cells, and it seems likely that they are immature states of Purkinje cells, as Pensa's specimens were taken from animals one month old or younger.

CAJAL (1911) further subdivided his first class of large stellate cells, the Golgi cells, into four types according to the extent and location of their axonal plexuses. The first type consists of the typical Golgi cells, large neurons whose axonal arborization fills the entire thickness of the granular layer and extends laterally across the folium for a considerable distance. The second type includes smaller cells whose axonal plexus is restricted to only a fraction of the thickness of the granular layer. The third type is a peculiar large cell lying near the margin of the white matter. The axon of this cell either generates two distinct plexuses separated by a short distance or it arborizes first close to the cell body and then crosses the white matter in the center of the folium to supply a second plexus to the granular layer of the other side. This type is probably related to the synarmotic cells of LANDAU (1928, 1932, 1933), described as interconnecting the granular layer on the two sides of a folium by bridging across the white matter. RAMÓN Y CAJAL's fourth type is a smaller cell whose axon produces a thin, elongated plexus among the granule cells near the white matter.

It must be clear from the foregoing historical review and terminological discussion (see also MUGNAINI, 1972) that the large cells of the granular layer constitute a heterogeneous group. Many of the cells in this group are not true Golgi cells at all, but represent other types of cells discovered in unusual locations. The tendency of neurohistologists to multiply cell types has been exacerbated by the success of Golgi impregnations in the brains of immature animals in which both the form and the position of many nerve cells may not have attained their definitive condition. In view of the complex migration of neuroblasts during the development of the cerebellar cortex, it is not surprising that atypical forms should be encountered in unaccustomed places during the early postnatal period. In the present account we shall accept as Golgi cells only those large cells in the granular layer having a finely divided axonal plexus in that layer. Cells whose axons pass out of the cortex into the white matter and cells that do not generate a highly ramified axonal plexus in the granular layer will not be considered in this section.

We have, however, measured the sizes of their cell bodies in both Golgi preparations and electron micrographs. Maximal and minimal diameters of 37 Golgi cells were measured in careful camera lucida drawings of Golgi preparations at a magnification of 1300×. Only cells with characteristic dendritic and axonal patterns were included. Since the perikarya of both cell types are nearly globular and the origins of their processes are distinct, it was possible to measure the circumference of the perikaryal silhouettes with a cartographer's wheel. These measurements were made twice with two different wheels, and the results averaged. These values gave a reliable index of the size of the cells directly without recourse to

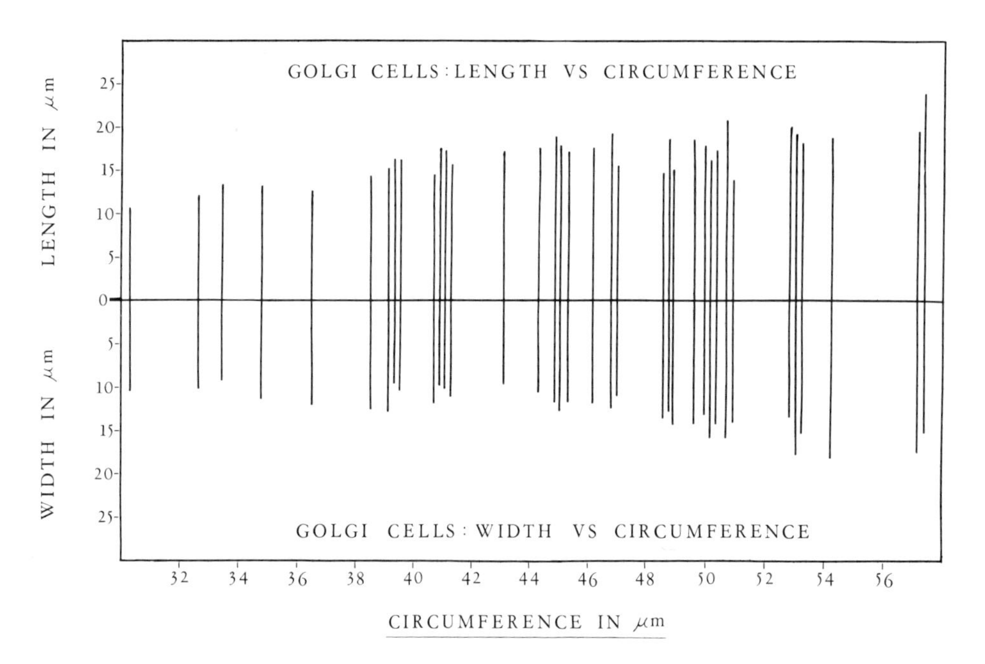

Fig. 86. Histogram showing the correlation of length and width with the circumference of individual Golgi cells. These parameters were measured in camera lucida drawings of Golgi cells in rapid Golgi preparations. Each line on the graph signifies the measurements of length and width of an individual cell, distributed along the horizontal axis according to its circumference. Only one or two cells in the lower range fit the description of small Golgi cells. The remaining cells are all large. These numbers do not reflect actual representative counts of all Golgi cells in the cerebellar cortex but only one population that has been selectively stained by the vagaries of the Golgi method

2. The Large Golgi Cell

In the adult rat two principal kinds of Golgi cells can be distinguished: large cells lying in the upper half of the granular layer and small cells in the deeper half. Neither cell, however, is absolutely restricted to these locations. In their counts of (large) Golgi cells in the cat, PALKOVITS *et al.* (1971b) found that 89% were located in the upper half of the granular layer. They counted about 1000 large Golgi cells per mm^3 of granular layer or 1 for every 3 Purkinje cells and 1 for every 5701 granule cells. Since they included the small Golgi cells among their counts of astrocytes, the actual number of Golgi cells in the cat may be larger than these counts indicate.

Comparable figures are not available for the rat, and we have not tried to count the Golgi cells in our animals.

calculations of area or volume that are fraught with assumptions. These data are displayed in Fig. 86.

The cell bodies that were measured varied from 10.7 μm to 23.8 μm in maximal dimension and from 9.4 μm to 17.9 μm in minmal dimension. The figure shows that the sample included only a few small Golgi cells, reflecting the fact that only a few were impregnated in the specimens used for the measurements. The lower values, however, correspond to the upper limits for the small cells seen in electron micrographs, in which the distinction between Golgi cells was first made, on the basis of cytological pattern and fine structure. The correlation with a difference in size was only subsequently discovered.

In addition to size, there are other differences discernible in the Golgi preparations (see Figs. 87–90). Golgi cells can display one of two modes of dendritic arboriza-

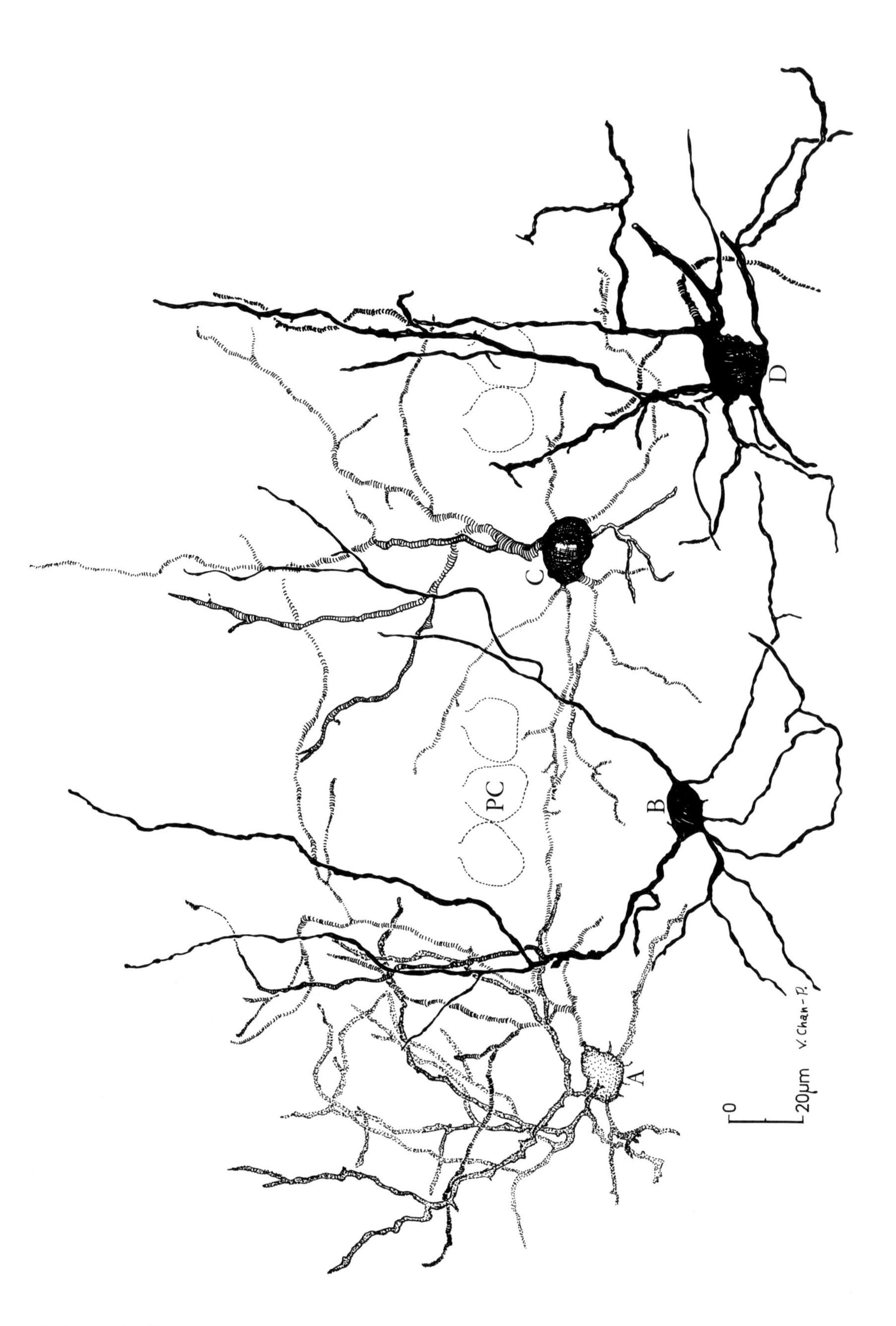

Fig. 87. The dendritic arborizations of small and large Golgi cells. Two patterns of dendrites exist: first, radial branching of more or less equal dendrites from the cell body; second, emission of one or two main trunks from which subsidiary branches arise. Two small Golgi cells (*A* and *B*) show the radial branching pattern. One large Golgi cell (*C*) shows the main dendrite pattern, whereas another large Golgi cell (*D*) shows the radial dendritic pattern. Rapid Golgi, 90 µm, sagittal, adult rat. Composite camera lucida drawing

tion. In the first, the dendrites radiate out more or less equally from the cell body, while in the second there are one or two main trunks which give off a series of subsidiary branches on their way to the molecular layer. The large cells can have either of these patterns, whereas the small cells usually exhibit only the first pattern. The small cells also tend to have a more restricted dendritic field.

a) The Form of the Large Golgi Cell

The large Golgi cells are situated at a little distance below the layer of Purkinje cells, not immediately beneath them; that level is occupied by the Lugaro cells and the infraganglionic plexus. Their cell bodies are almost as large as those of the Purkinje cells and could be mistaken for them on account of their rounded shape. But the configuration of the processes emerging from them easily betrays their true identity. Each cell has two to four thick dendrites that radiate from the cell body, usually from its upper surface. A short distance from their origin the dendrites bifurcate, and the resultant branches generally turn upward toward or into the molecular layer. It is characteristic of the Golgi cell dendrites that the branches first adopt a nearly horizontal course as they leave the bifurcations and then turn toward the surface of the folium. In some preparations the bifurcation points are marked by little triangular expansions of the stem dendrites. These characteristics can be seen in Figs. 87 and 89.

As the main dendrites reach the molecular layer they give off relatively few secondary and tertiary branches, subdividing always at an acute angle. Owing to the divergence of the primary dendrites and their branching pattern, the dendritic tree is rather thin and widespread. Unlike the dendritic trees of the other cells expanding in the molecular layer, the stellate, basket, and Purkinje cells, that of the Golgi cell is not confined to a single plane, but opens out into a three-dimensional, ungulate field (Fig. 5). This field is invaded by the vertical ranks of stellate, basket, and Purkinje cell dendrites, spreading like open fans through the sparse forest of radiating Golgi cell dendrites, and it is crossed by the multitudes of parallel fibers, which articulate with all of them by means of varicosities, bulbous projections, and other contrivances. The dendrites of Golgi cells are generally stout, and although they are not without small kinks and twists, loops and hooks, they pursue a relatively straight course towards the surface, in comparison to the more tortuous dendrites of basket and, especially, of stellate cells. The thicker dendrites have fairly smooth contours, although some of them appear to be rough in their proximal segments (Figs. 88–90). As they become thinner with each successive branching in the molecular layer, they also become slightly varicose, and spicular appendages project from them irregularly at right angles. They are, however, very much less bristly than the dendrites of stellate and basket cells.

Probably the most distinctive feature of the Golgi cells is their axons. From one to three axons extend from the cell body, usually its deeper surface, or from one of the major dendrites close to the cell body. When there are several axonal trunks, one of them tends to be more contorted and to branch more profusely than the others (Figs. 88 and 90). The axon starts out as a fairly straight, smooth, and thin process. It can describe a bend or two, and then a short distance from its origin, without any perceptible tapering, it branches dichotomously at right angles. This mode of division, noted by Golgi in his original descriptions, is strikingly different from the forking at acute angles characteristic of the dendrites. The axons divide again and again, each branch always projecting straight out at right angles from the stem, and then, after a sharp bend or a gentle curve or two, dividing once more. The repeated dichotomies at right angles and the arching, recurving habit of the collaterals quickly result in the construction of a dense plexus, in which the individual fibers can be followed only with great difficulty even at high magnifications. Ramón y Cajal writes (1890a, p. 17):

> "When one of these axons appears completely impregnated in a (Golgi) preparation, it is almost impossible to follow its complete arborization, so abundant, sinuous, and contorted are its ramifications; it is only in the incomplete impregnations of adult animals and mainly in those of young mammals, in which the fibers are stouter and less numerous, that one can study the course and divisions of the axon."[6]

6 «Lorsque dans une préparation apparait complètement imprégné un de ces cylindres-axes, il est presque impossible de suivre la totalité de l'arborisation, tant sont abondantes, flexueuses et entortillées ses ramifications; c'est seulement dans les imprégnations incomplètes des animaux adultes, et principalement dans celles des jeunes mammifères où les fibres sont plus fortes et moins nombreuses, qu'on peut étudier la marche et les divisions du cylindre» (Ramón y Cajal, 1890a, p. 17).

Fig. 88. A large Golgi cell: horizontal view. This Golgi cell has dendrites that radiate from the cell body outwards into the granular layer. The axons of this cell consist of thin threads (*axon*) that arise from dendrites (*arrows*). These stalks then emit many fine, beaded collaterals that ramify within and beyond the territory of the dendritic expansion. Rapid Golgi, 90 μm, adult rat. Camera lucida drawing

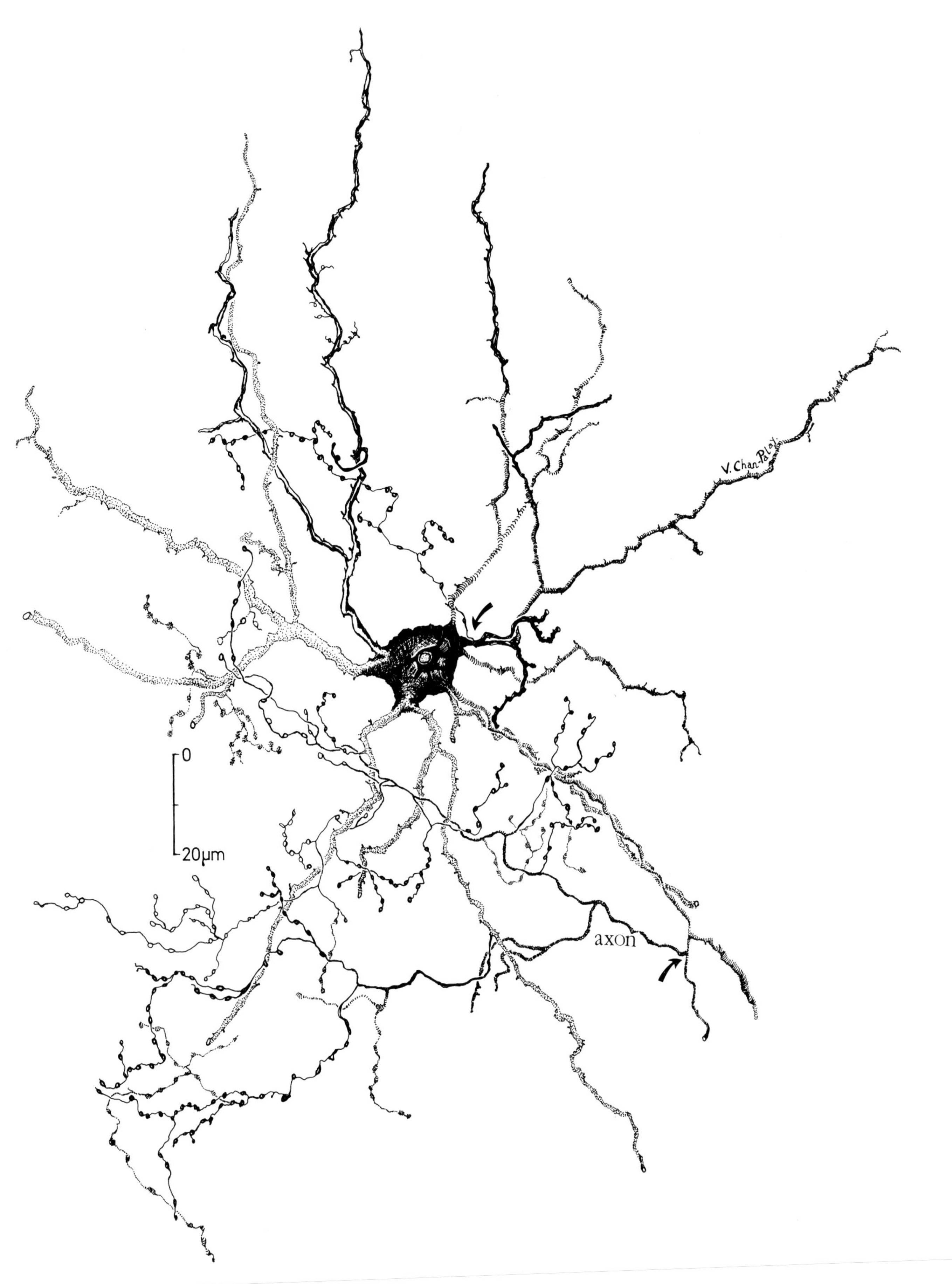
V. Chan-Palay
0
20µm
axon

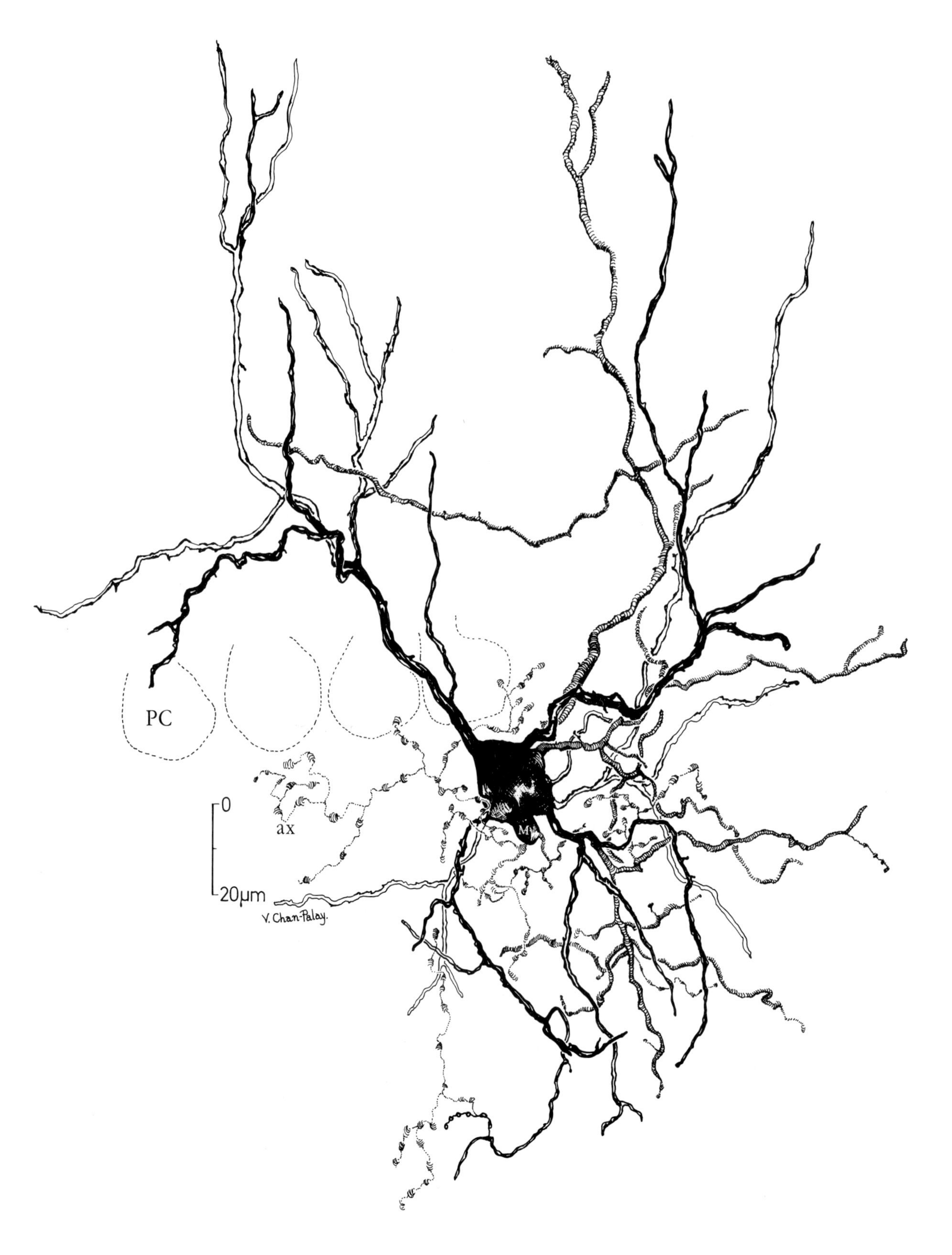

Fig. 89. A large Golgi cell: sagittal view. This Golgi cell emits a large number of long, branching dendrites that invade both the molecular layer and the granular layer. The latter set of dendrites, however, have a shorter excursion. The boundary between these two layers is indicated by the layer of Purkinje cells (*PC*). A peculiar protrusion of the cell body (*M*) suggests that this may be a possible site for the extensive synapse *en marron*. From this hillock issue the beaded axons of this cell (*ax*). Rapid Golgi, 90 µm, adult rat. Camera lucida drawing

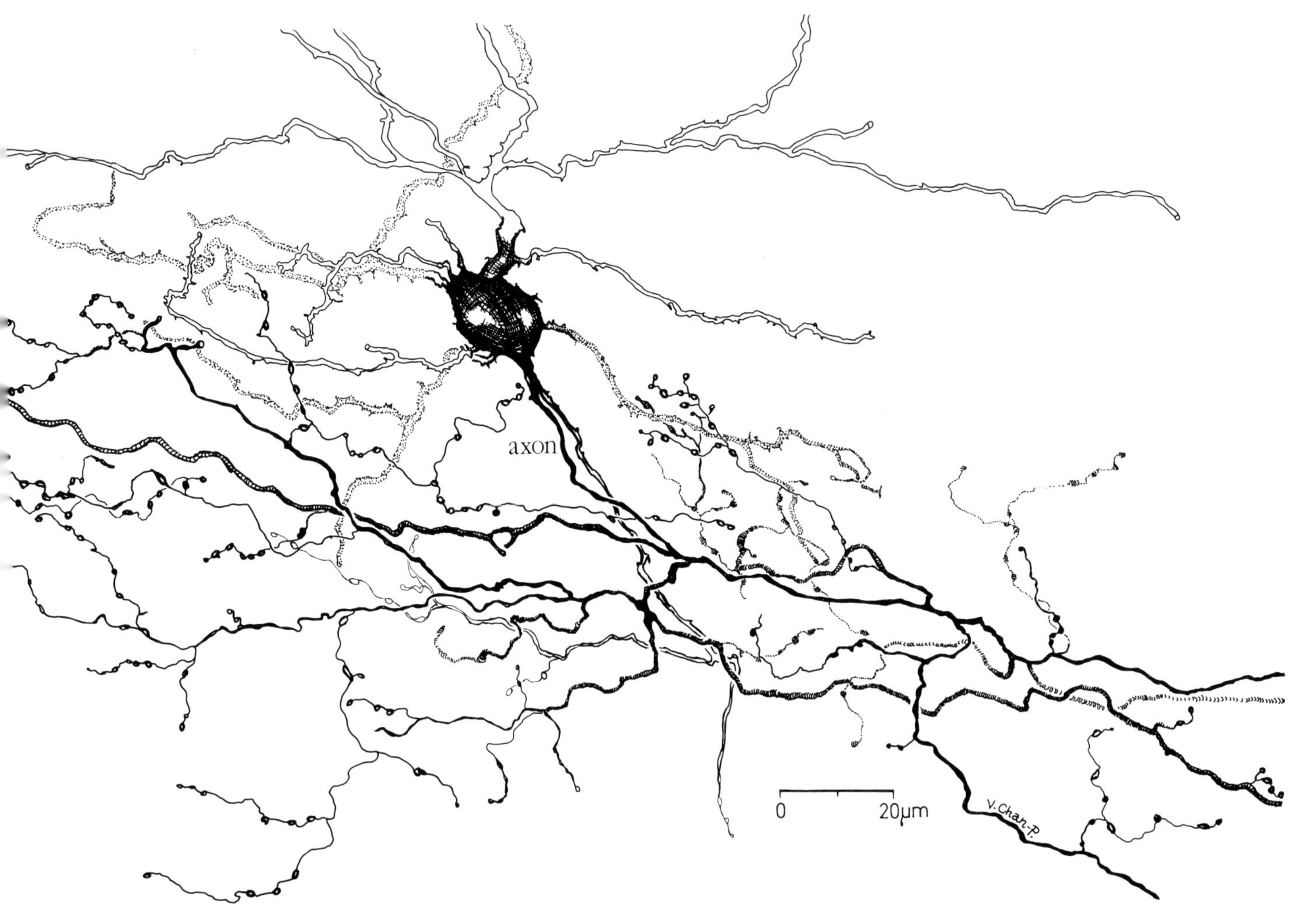

Fig. 90. The axonal system of a Golgi cell. This large Golgi cell has several small dendrites and one large dendritic trunk from which subsidiary branches issue. The axon emerges from the basal pole as a smooth, slender thread. The axon then branches into other smooth stems, which then break up into varicose beaded collaterals. All of these branches, however, arise from only one axonal stem. Rapid Golgi, 90 μm, sagittal, adult rat. Camera lucida drawing

An idea of the complexity of this arborization can be gained by examining Fig. 90, which shows a portion of the ramifications from only one of the several axonal trunks of a large Golgi cell. With successive branchings the axonal twigs become thin, varicose strands, resembling a string of beads (Figs. 88–90). The varicosities accentuate twists, kinks, or crimps in the course of the axon with ellipsoidal or spherical expansions, and often they form terminal nubbins on short side branches.

Even though the axonal arborization of the Golgi cell is unusually rich and intricate, its density is not everywhere the same. In well-impregnated specimens two degrees of compactness are evident, as may be seen in the drawings of Ramón y Cajal (1911, Figs. 30 and 31). These illustrations show a relatively loose ground plexus in which foci of considerably greater density occur here and there. The openings in the ground plexus are large enough to contain several perikarya of granule cells. When the denser foci are examined with high magnifications (100× oil immersion objectives), it can be seen that they consist of intertwined varicose fibers arranged like a cage around the periphery of an unimpregnated center. Although many of the fibers appear to end freely in these dense zones, they seldom penetrate into or cross the unstained central mass. Ramón y Cajal (1911) discovered that the fibers of the Golgi cell axonal plexus terminate in these dense zones by synapsing with the dendrites of granule cells. Each dense zone therefore corresponds to the periphery of a glomerulus, which is the unstained mass within it.

b) The Fine Structure of Large Golgi Cells

Thus far we have been concerned primarily with the size, shape, and distribution of the large cells in the granular layer, characteristics that are the fundamental criteria for their identification first in light microscopic preparations and then in electron micrographs. Relatively little attention is given in the literature to the internal structure of these cells. JAKOB (1928) provided one of the rare accounts of their appearance in Nissl preparations. He noted that with the Nissl stain three kinds of cells could be distinguished: (1) a large cell with a large nucleus and voluminous cytoplasm containing coarse Nissl bodies; (2) a less common and slightly smaller cell with an elongated nucleus and scanty cytoplasm containing small Nissl bodies; and (3) a smaller cell with a distinct nucleus and little cytoplasm containing punctate Nissl bodies. These cell forms can be correlated with those found in Golgi preparations and can be equated with the cell types described by other authors. The first and third types clearly correspond to two of RAMÓN Y CAJAL's subcategories of Golgi cells, the large and the small, while the second includes the Lugaro cell and the various much less numerous atypical forms mentioned earlier.

Electron microscopy also reveals the existence of two distinct kinds of Golgi cells in the granular layer of the rat, and they will be described here for the first time. Whether these two kinds of cells correspond precisely to the large and small Golgi cells found in impregnated specimens by light microscopy cannot yet be determined with the available evidence. The two cell types distinguished in electron micrographs are, however, different in size as well as in fine structure, and it is therefore convenient to designate them as large and small. As measured in electron micrographs, the large cells vary from about 9 to 16 μm in diameter and the smaller cells range from 6 to 11 μm. Since both cells are multipolar and often tend to be elongated, the dimensions given are only approximations, as the overlap of ranges indicates. Both cells can be found throughout the granular layer from the margin of the white matter, where they may be surrounded by myelinated fibers, to the infraganglionic plexus, where they must be distinguished from Lugaro cells. Neither Golgi preparations nor electron micrographs give a very precise idea of their number.

The Perikaryon of Large Golgi Cells. Since in addition to their size differences, the somata of the two cell types have distinctive internal structures, they will be described separately. The nucleus of the larger cell is usually eccentric and is roughly spherical or ovoid (Figs. 91–93). Extensive infoldings, however, give the nucleus of this cell the most irregular shape in the cerebellar cortex. Its dimensions are therefore difficult to ascertain; in overall measurements, disregarding the invaginations, the diameter varies between 9 and 15 μm in different cells. The nuclear content is rather uniformly dispersed so that the nucleus appears pallid in contrast to the nuclei of the surrounding granule cells. In some cells chromatin is condensed in small heaps at the nuclear envelope, especially along the invaginations. These cells have single large, rounded nucleoli, 2–3 μm in diameter, usually located eccentrically in the nucleus. The nucleolus has the usual composite structure found in large neurons, with anastomosing strands of the pars granulosa interspersed with globular masses of the pars fibrosa.

The nucleus also contains a rodlet or circlet of fine periodic filaments (Figs. 93 and 94), similar to, but less complex than that found in the nuclei of various other neurons in the central nervous system (SIEGESMUND *et al.*, 1964; CHANDLER and WILLIS, 1966; SOTELO and PALAY, 1968) and in sympathetic ganglia (MASUROVSKY *et al.*, 1970; SEITE, 1970; SEITE, ESCAIG, and COUINEAU, 1971; SEITE, MEI, and COUINEAU, 1971). These are to be equated with the bâtonnets intranucléaires discovered by RONCORONI (1895); although the bâtonnets are supposed to occur in other cerebellar cells, especially in granule cells (RAMÓN Y CAJAL, 1909), we have not seen the rodlets of filaments in any cerebellar neuron other than the large Golgi cells.

The nuclear envelope often approaches close to the surface membrane of the cell because of the eccentric position of the nucleus (Fig. 91). The envelope is fitted with the usual array of pores, and sends out tubular streamers that fuse with the granular endoplasmic retic-

Fig. 91. Perikaryon of a large Golgi cell. This large Golgi cell has a characteristically lobulated nucleus (*l Go*) and a prominent nucleolus with pars fibrosa (*pF*) and pars granulosa (*pG*) evident even at this magnification. The cell also has a well-developed Golgi apparatus (*Go*) that rings a large portion of the nucleus, and large Nissl bodies (*NB*). Synaptic terminals (*arrows*) contact the cell soma and emerging dendrite (*Go d*). Two granule cells (*gr*) lie adjacent to the cell as well as to an unusually large clump of axons belonging to a Golgi cell (*Go ax*, ▲▲). Lobule IV. ×12000

Fig. 92. Large Golgi cell with initial segment. This Golgi cell has a lobulated nucleus (*l Go*) and several small Nissl bodies (*NB*). The initial segment (*IS*, ▲▲) emerges from one pole of the cell as a fairly fleshy process before it moves out of the plane of this section. A large dendritic process (*Go d*) emerges from the cell body at the top right. Lobule X. ×15000

Fig. 93. A large Golgi cell. This cell has a very lobulated nucleus (*l Go*), and the perikaryon is filled with many small Nissl bodies (*NB*) among the other organelles. Several granule cells (*gr*) in a cluster abut against the cell soma. A nuclear rodlet is included (*rectangle*) in the nucleus of the Golgi cell. Lobule IV. ×13000

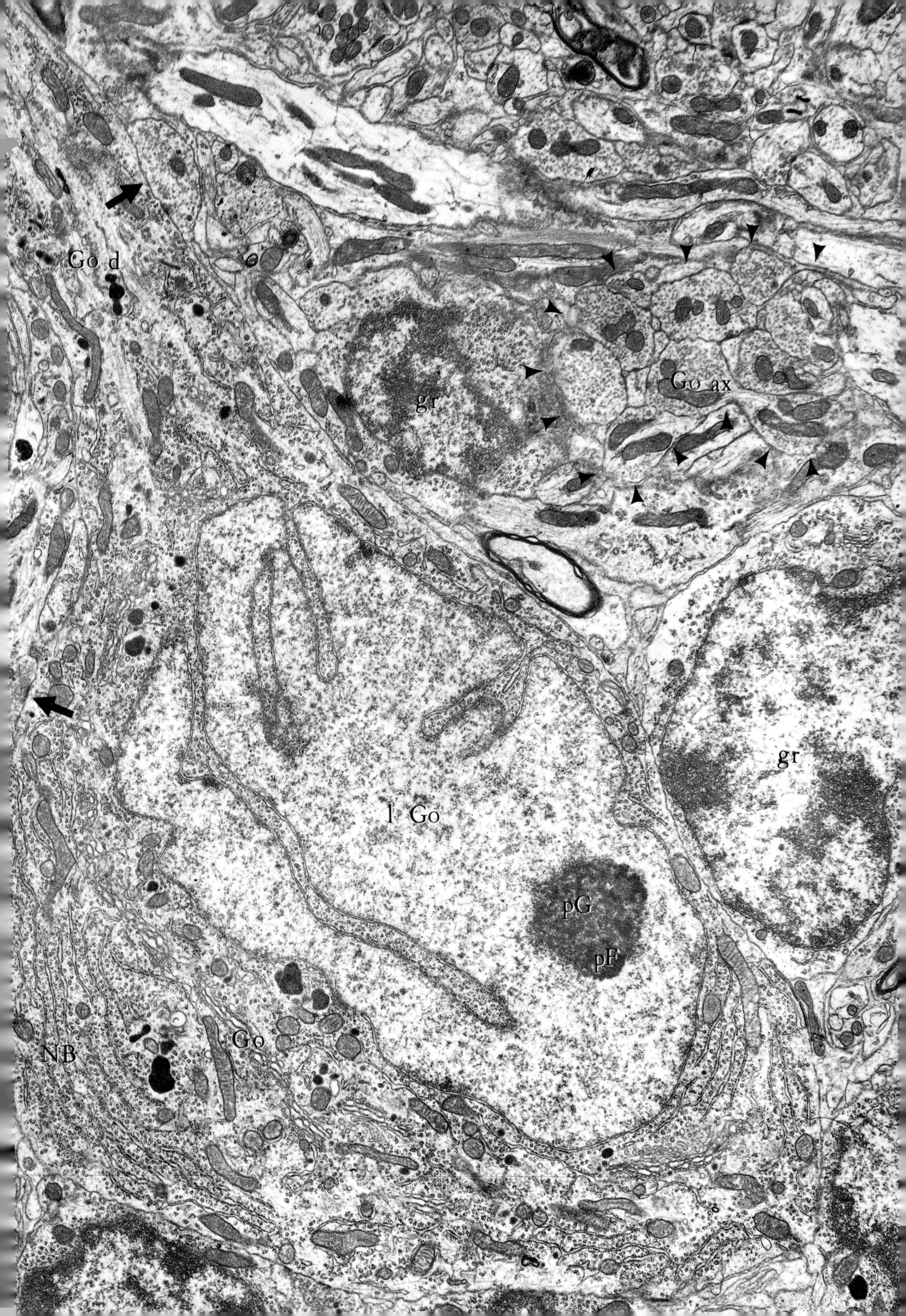
Go d
gr
Go ax
gr
l Go
pG
pF
NB
Go

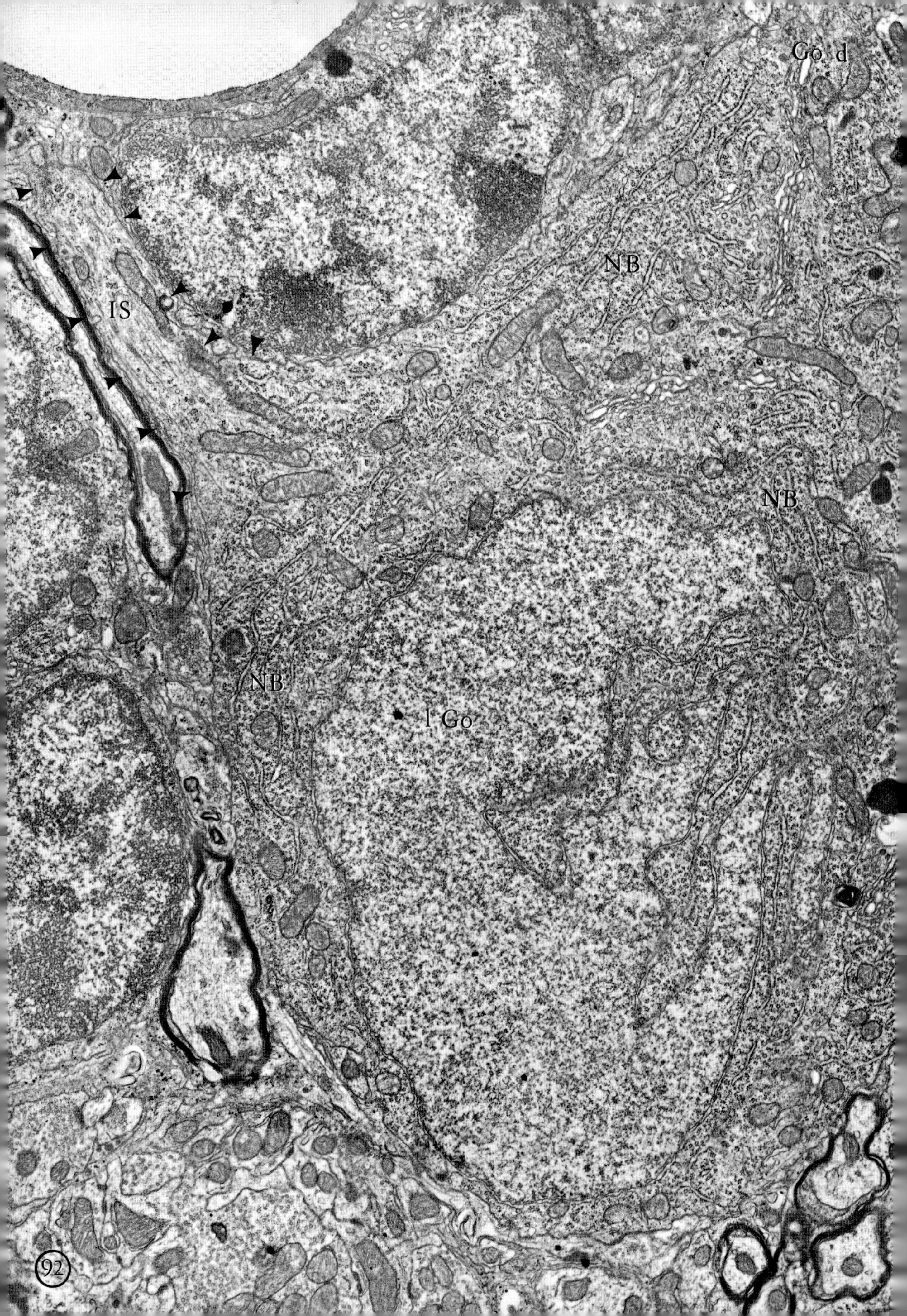
Go d
IS
NB
NB
NB
l Go
92

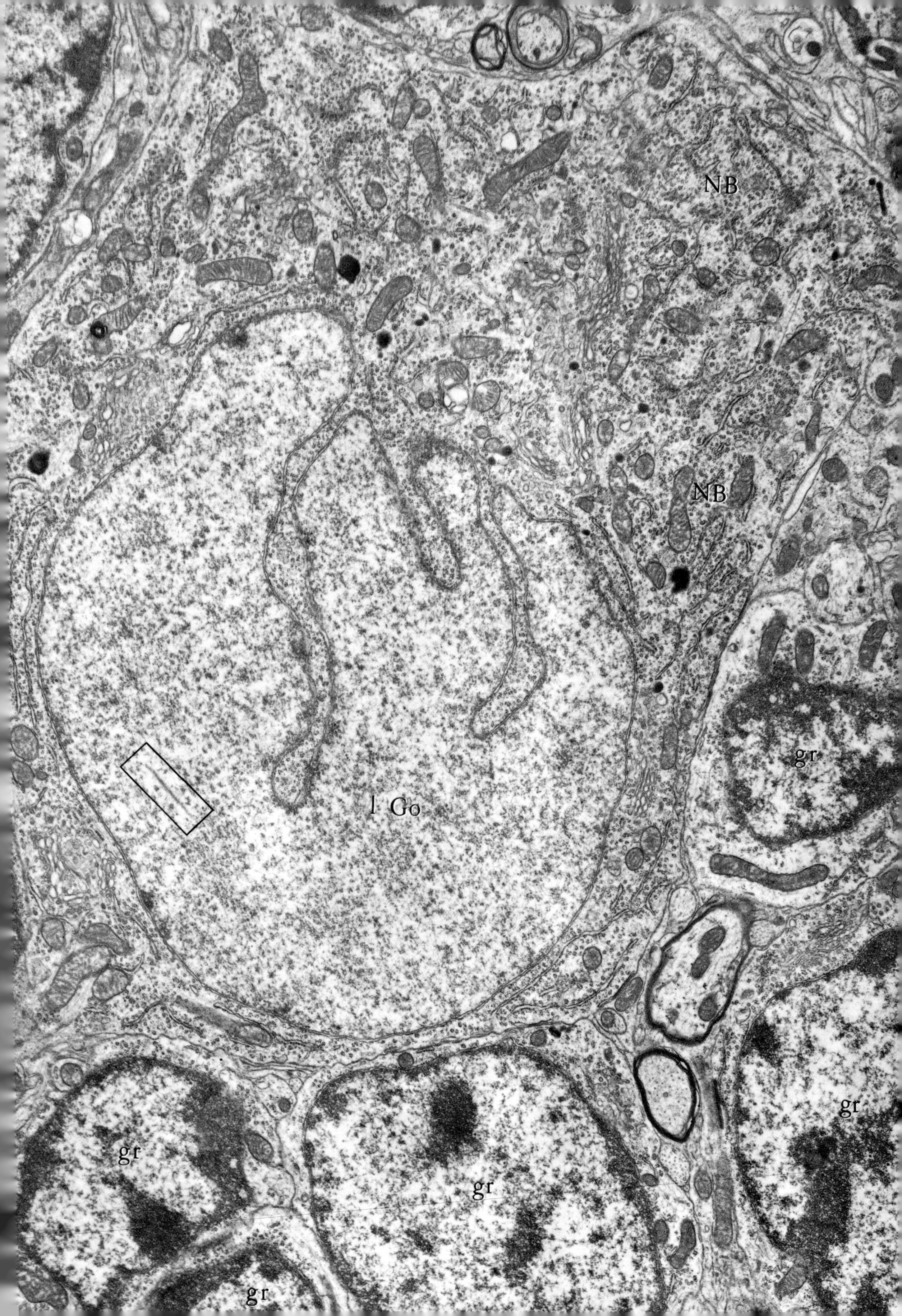
NB
NB
gr
l Go
gr
gr
gr
gr

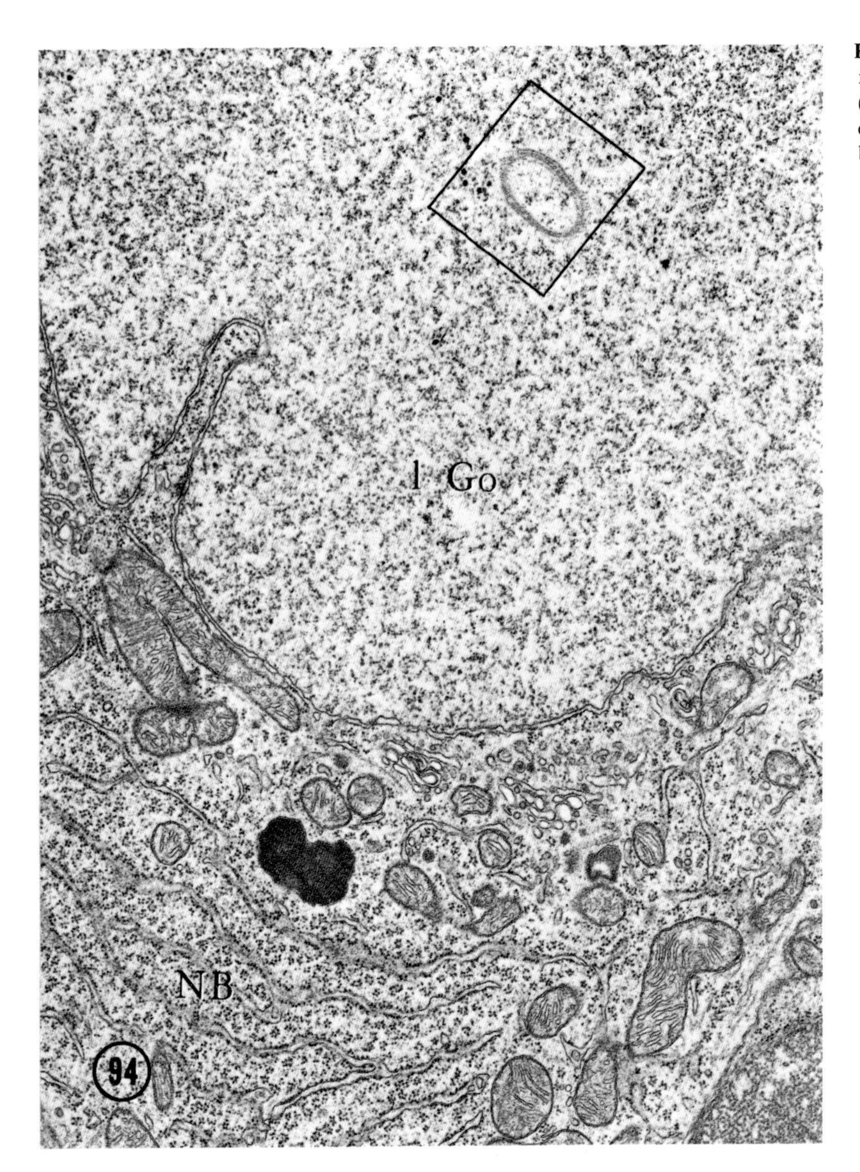

Fig. 94. Nuclear inclusion in a large Golgi cell. An inclusion consisting of a ring or circlet of filaments (*rectangle*) is shown in the nucleus of this large Golgi cell. The cell contains a large, well-developed Nissl body characteristic of this cell type. Lobule VI. × 5600

ulum in the cytoplasm. The outer surface of the envelope is itself studded with rows and clusters of ribosomes only at large intervals.

As was noted by optical microscopy, the cytoplasm of the large Golgi cell is relatively voluminous and its constituents display a distinctive pattern. In the electron micrographs of thin sections it can be seen that the organelles tend to fall into a circular arrangement, with the nucleus lying inside the circle but off center. The Nissl substance is arranged in two eccentric zones, one immediately about the nucleus and the other in the peripheral cytoplasm. The Nissl substance of the inner zone not only encloses the nucleus but also fills the deep cytoplasmic invaginations that penetrate into it. This distribution of the Nissl substance explains why the nucleus appears so well demarcated from the cytoplasm in Nissl preparations. In this perinuclear zone the Nissl substance consists largely of clustered ribosomes dispersed in the matrix surrounding the nucleus. Except for a few more highly ordered places where one may speak of a Nissl body, the granular endoplasmic reticulum is loosely disposed around the nucleus as an erratic, branching, and contorted network of channels. Frequent confluences with the nuclear envelope occur. In some cells the Nissl substance is heaped up at one pole of the nucleus to form a thin nuclear cap (Fig. 92). Such masses appear in the electron micrographs as highly ordered arrays of three or four parallel cisternae of the granular endoplasmic reticulum with clouds of clustered ribosomes suspended in between. Both membranous and ribosomal components of the Nissl substance extend out across an intermediate zone to the peripheral cytoplasm, where they combine to form typical large Nissl bodies (Figs. 91–93). In this outer zone the branching, anastomosing cisternae of the endoplasmic reticulum are spread out roughly parallel to the surface membrane of the cell in imbricated arrays of seven or eight members. Large expanses of the cisternae are devoid of granules, and the free ribosomes appear almost entirely as rosettes. In between the large Nissl bodies groups of free ribosomes are dispersed irregularly together with isolated tubules or cisternae of the endoplasmic reticulum. The distribution and massing of the Nissl substance in the electron micrographs of these cells correspond quite accurately to the Nissl pattern of the large Golgi cells as described and figured by JAKOB (1928) and as seen in our own preparations for the optical microscope. At this level of magnification the large cells display a thin perinuclear layer of basophilic cytoplasm accentuating the nucleus and a peripheral border of heavy Nissl bodies, with a sprinkling of fine basophilic particles in an intermediate pale zone.

The most conspicuous structure lying in the intermediate zone proves to be the Golgi apparatus (Figs. 93 to 95). Consisting of a compact set of four to six overlapping cisternae and associated vesicles, it partially or completely encloses the nucleus, as if in a cage. The isolated curved stacks of cisternae which appear in single electron micrographs of thin sections are parts of a continuous open reticulum (Figs. 91–93). Sometimes these stacks extend halfway around the nucleus. In other instances they are small islands arranged like the links of a chain in the intermediate zone.

The mitochondria are generally short, thin rods with their cristae oriented transversely or obliquely across the inner compartment. Because they are excluded from the main mass of the Nissl bodies, they tend to collect in the intermediate zone along with the Golgi apparatus and also to be arranged in a perinuclear orbit. In a thin section the profiles of the mitochondria are often strikingly clustered together in tight groups of two to four or more, a combination that suggests that each cluster represents a section through a single mitochondrion of complex, racemose shape. Lysosomes, granules of lipofuscin pigment, multivesicular bodies, and vagrant tubules of the agranular endoplasmic reticulum are found in the intermediate zone, although they also occur in the rest of the cytoplasm in lesser numbers. Centrioles, individually or paired, are also encountered in this region in fortunate sections. Microtubules course through the intermediate zone, again in a roughly perinuclear trajectory. They also are visible in the open spaces (*Straßen*) between the Nissl bodies and in the wider spaces at the bases of dendrites where they collect as they pass out of the cell body. Individual neurofilaments are encountered infrequently. The matrix of the perikaryon contains a fine, fluffy material that generally imparts a cloudy background to the electron micrographs of this cell.

It is important to point out, in view of the possible confusion of the large Golgi cell with the Purkinje cell, that the hypolemmal cisterna is very poorly represented in the Golgi cell and that even subsurface cisternae are infrequent. The close association of a mitochondrion with a subsurface cisterna, so characteristic of the Purkinje cell, almost never occurs in the Golgi cell.

The soma of the large Golgi cell often lies immediately adjacent to granule cell perikarya, dendrites, or axons without any intervening neuroglial cell processes. Occasionally subsurface cisternae in the two apposed cells are matched in the same way as subsurface cisternae in neighboring granule cells. Although thin lamellae of neuroglial cytoplasm clasp the surface here and there, especially in the region around synapses, the cell body is not invested with a neuroglial capsule like that enclosing the Purkinje cell.

The soma of the large Golgi cell receives very few synaptic endings. The vast majority of the axo-somatic synapses are contributed by terminal boutons and varicosities in the fibers of the deep plexus woven by the recurrent collaterals of the Purkinje cell axon as they pass through the granular layer. Profiles of these boutons or *terminaisons en passant* appear singly or in groups of two or three in electron micrographs (Figs. 91 and 95). The ending is applied flat against the surface of the cell, forming a rather extensive junction, but the synaptic complex is developed at only one or two small spots along this interface (Figs. 95 and 96). Each macula is characterized by a widened synaptic cleft crossed by slender filaments. Thin, symmetrical densities are attached to the cytoplasmic or internal surfaces of the apposed plasmalemmas. The presynaptic terminal is nearly filled with oblate vesicles and a cluster of mitochondria, or a single branching mitochondrion, all suspended in a cottony matrix. The axon terminals shown in Figs. 95 and 96 probably belong to recurrent collaterals of Purkinje cell axons.

The somata of some large Golgi cells engage in a most remarkable articulation with either climbing or mossy fibers, which has been given the name *synapse en marron* (CHAN-PALAY and PALAY, 1971a and b). As the small Golgi cells also participate in the same kind of junction, the description of the synapse *en marron* will be deferred until after the small cells have been considered.

The Dendrites of Large Golgi Cells. The fleshy dendrites of the large Golgi cell appear as rapidly tapering prolongations of the soma, filled with the same organelles, more or less linearly arranged (Fig. 91). At the start the

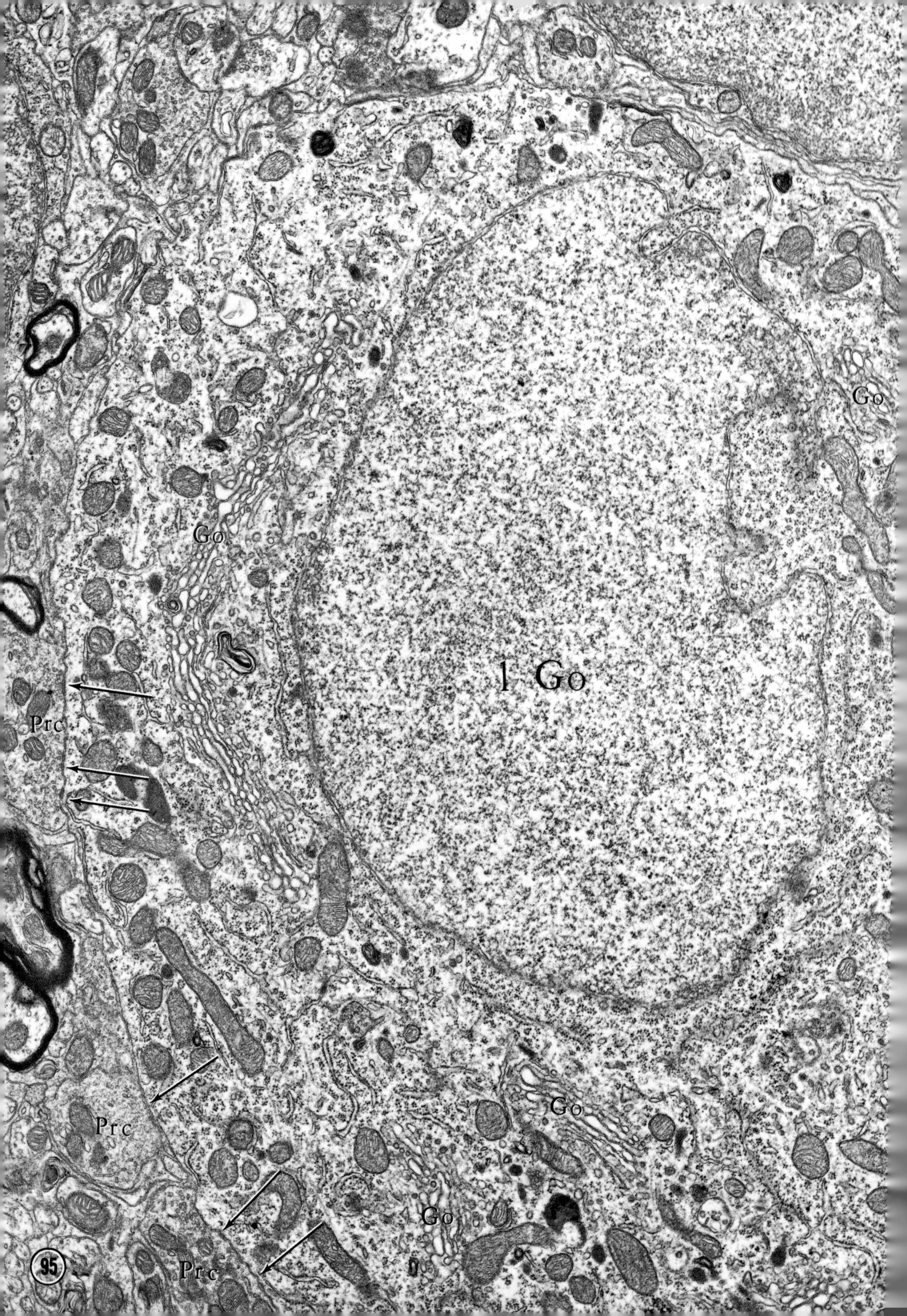

Go
Go
Prc
1 Go
Prc
Go
Go
Prc
95

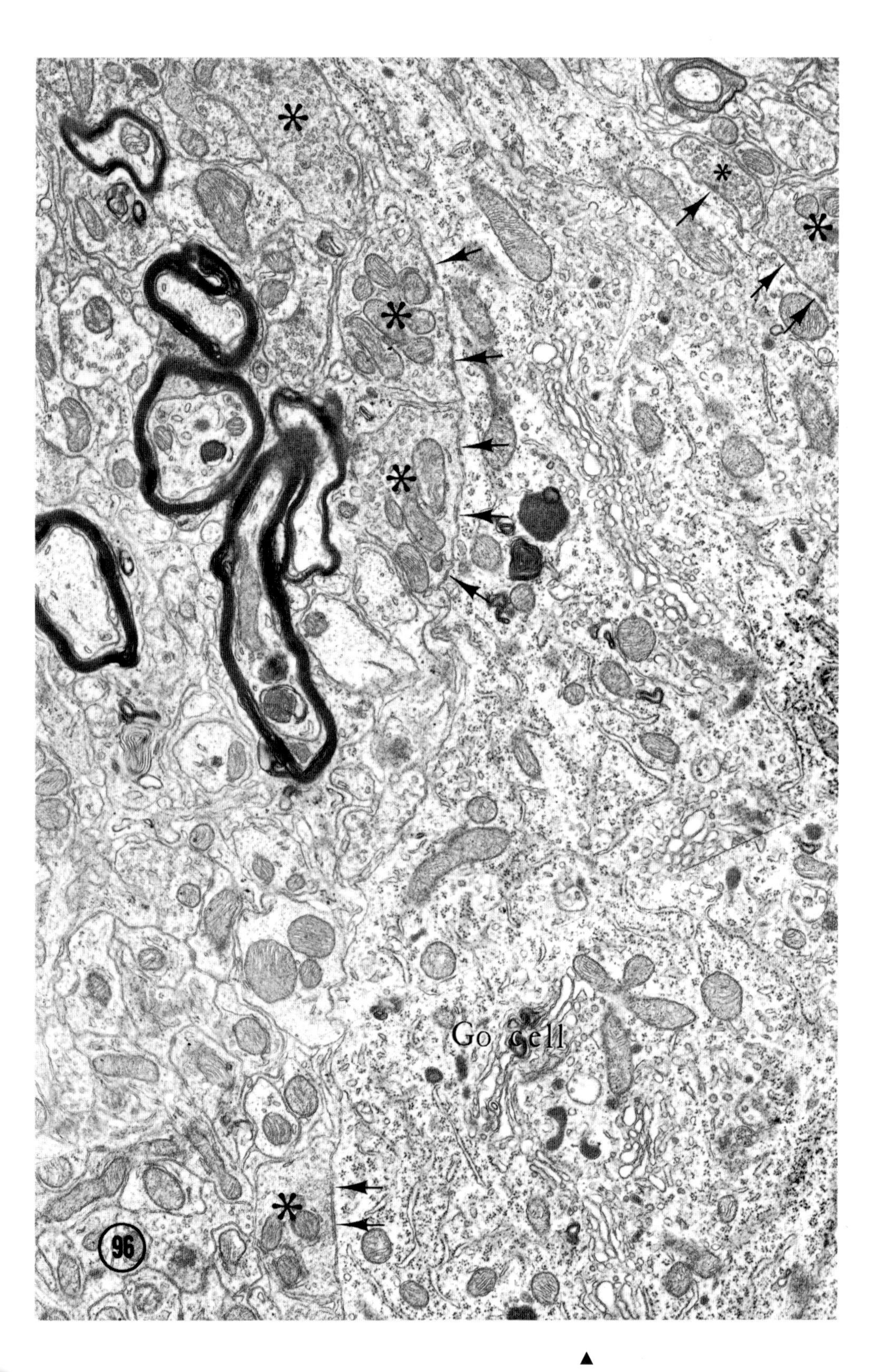

◀ **Fig. 95.** A large Golgi cell. The nucleus of this cell (*l Go*) is encircled by a garland of profiles of the Golgi apparatus (*Go*), which winds in and out of the section. The somatic surface of this cell receives synaptic contacts (*arrows*) from three terminals, probably belonging to recurrent collaterals of Purkinje axons (*Prc*). Lobule X. × 18000

▲
Fig. 96. A cluster of boutons of recurrent collaterals (∗) in synapse with the soma and dendrite of a deep large Golgi cell (*Go Cell*). These boutons form many small synaptic junctions with the surface of the Golgi cell (→). (Lobule X, vermis.) × 15000

mitochondria and the microtubules tend to take up a central position while the granular endoplasmic reticulum drifts into the periphery. The Nissl bodies, at first rather large, gradually dissipate. Beyond the primary branch points the Nissl substance becomes reduced to small assemblages of ribosomes with or without bits of the granular endoplasmic reticulum. The Golgi apparatus stretches a short distance into the root of the dendrite, often beginning in the center and then shifting to the periphery with the Nissl bodies. Like them, it soon fragments and then disappears altogether. Lysosomes and multivesicular bodies are dispersed without a preferential location. The smooth endoplasmic reticulum, which starts out as numerous twisted, branching tubules and elongated cisternae, becomes reduced to a few longitudinally oriented tubules. As the dendrite narrows and branches and with increasing distance from the cell body, all of the organelles except the microtubules deviate to the periphery and scatter so that the persistent microtubules remain as the most distinctive components (Fig. 97). As can be seen in Golgi preparations (Fig. 88), the first segment of a large dendritic trunk bears irregular projections of various lengths and shapes. Electron micrographs show that these appendages are simple or complex fungoid processes. Some of them have several branches, and they do not usually have synaptic complexes, even when they protrude into an axonal terminal. Such processes were first pointed out by HÁMORI and SZENTÁGOTHAI (1966b; see also ECCLES, ITO, and SZENTÁGOTHAI, 1967), and have been described in more detail by MUGNAINI (1972). Some of the dendritic trunks—"hairy" dendrites, as MUGNAINI (1972) calls them—participate in a distinctive synaptic formation with the terminal expansions of mossy or climbing fibers. The elongated terminal, running alongside the shaft of the dendrite, engulfs the projecting appendages but develops synaptic complexes only with its face opposite the dendritic shaft. The junction is somewhat reminiscent of both the glomerulus and the synapse *en marron*, which will be described later.

The large dendritic trunks and their primary branches in the granular layer also articulate with the varicose terminals of the recurrent collaterals from the Purkinje cell axons (Fig. 98). Axo-dendritic synapses are much more common on these cells than axo-somatic synapses. They occur in clusters, each terminal encapsulated on its rounded free surface by a thin neuroglial sheet. As on the cell body, the junctional interface is a broad, flat zone of apposition with the synaptic junctions marked either by small macular densities underlining a clump of flattened vesicles or by a long, narrow, ribbon-like density. These junctions have the same features as the less numerous junctions of recurrent collaterals on the cell body: a widened synaptic cleft flanked by thin pre- and postsynaptic densities. Some of the axo-dendritic junctions on Golgi dendrites in the granular layer are characterized by a secondary row of discrete, round densities lying beneath the postsynaptic membrane density (Fig. 99). This row resembles the postsynaptic organelle described by MILHAUD and PAPPAS (1966) in the habenula and interpeduncular nuclei and by AKERT, PFENNINGER, and SANDRI (1967) in the subfornical organ of the cat. We have not seen this organelle in any other neurons of the rat.

It must be pointed out that in electron micrographs of thin sections we cannot distinguish the dendrites of large Golgi cells from those of small Golgi cells unless they are found in continuity with their parent cell bodies. Nevertheless, the identification of a dendritic profile as belonging to one of the Golgi cells does not present any great difficulty in the granular layer. It is necessary to distinguish them only from the dendrites of granule cells and Lugaro cells. The latter can be the source of some confusion, as they are of about the same caliber and have the same connections as the dendrites of the Golgi cells, but they lie almost exclusively in the horizontal plane, parallel to and just beneath the Purkinje cell bodies. At this level most Golgi cell dendrites are already ascending vertically toward the brain surface. The granule cell dendrites are much smaller and have no synapses on their surfaces until they reach their terminal arborizations in the glomeruli.

As has already been indicated, the dendritic trunks of the Golgi cells give off relatively few branches in the granular layer and these eventually turn upward toward the surface of the folium. In electron micrographs of the molecular layer they have to be distinguished from the dendrites of Purkinje cells, basket cells, and stellate cells. These other dendrites are much more numerous than the Golgi cell dendrites in any section of the molecular layer. The Golgi preparations (Figs. 87–90) show that Golgi cell dendritic stems dichotomize only a few times during their course through this layer, giving rise to stout branches that pursue a nearly straight vertical path toward the surface of the brain. Most other dendrites are more tortuous in their ascent. The ordinary Purkinje cell den-

▶

Fig. 97. The Golgi cell dendrite in the granular layer. A portion of a more distal dendrite of a Golgi cell courses through the center of this field (▲▲) and divides. It is flanked on the left by several profiles of a mossy fiber with loosely dispersed synaptic vesicles (*MFd*) around which dendrites of granule cells swirl. To the right of the dendrite, three granule cells lie in apposition with its surface. Lobule IV. × 11 000

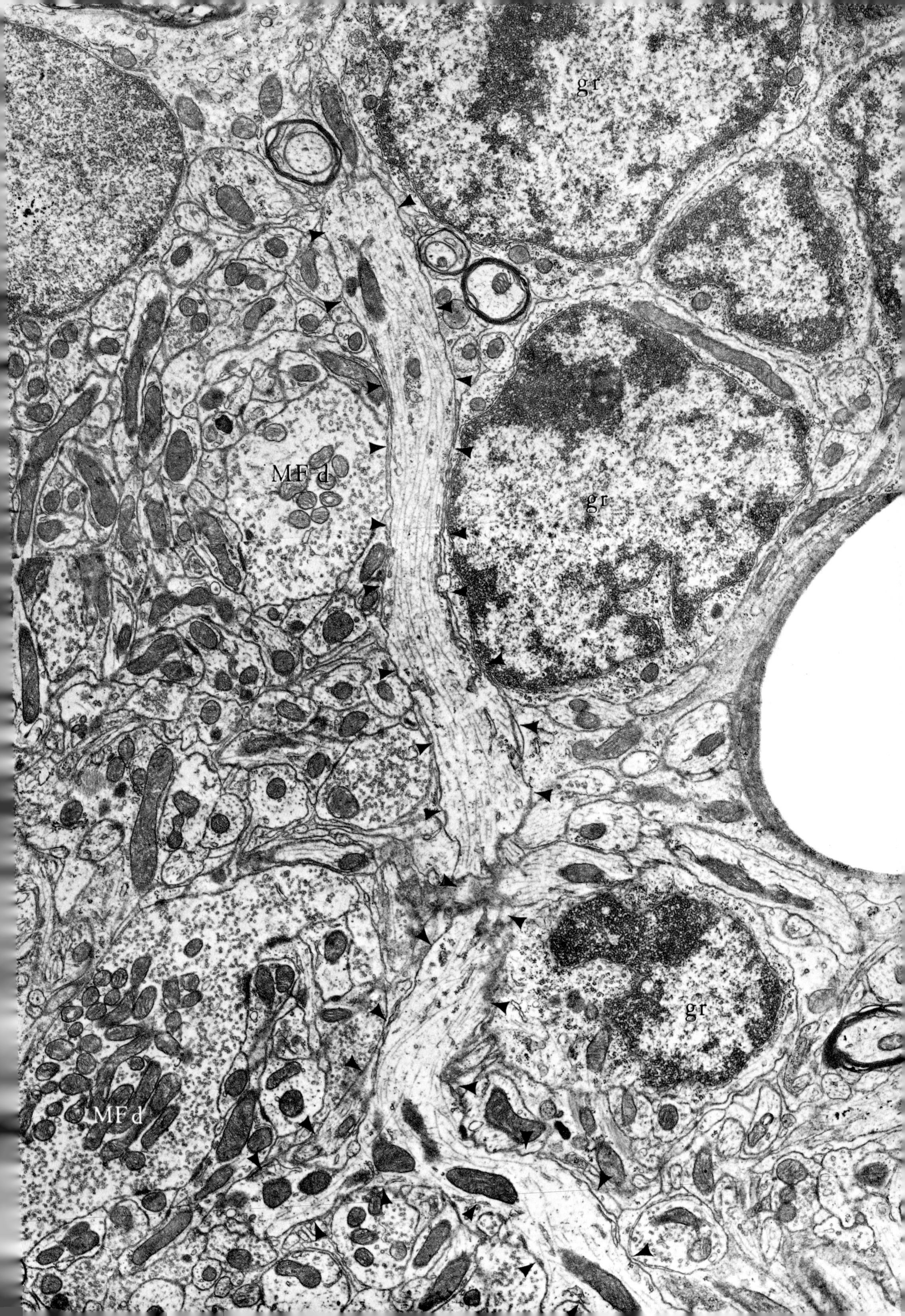
gr
MF d
gr
gr
MFd

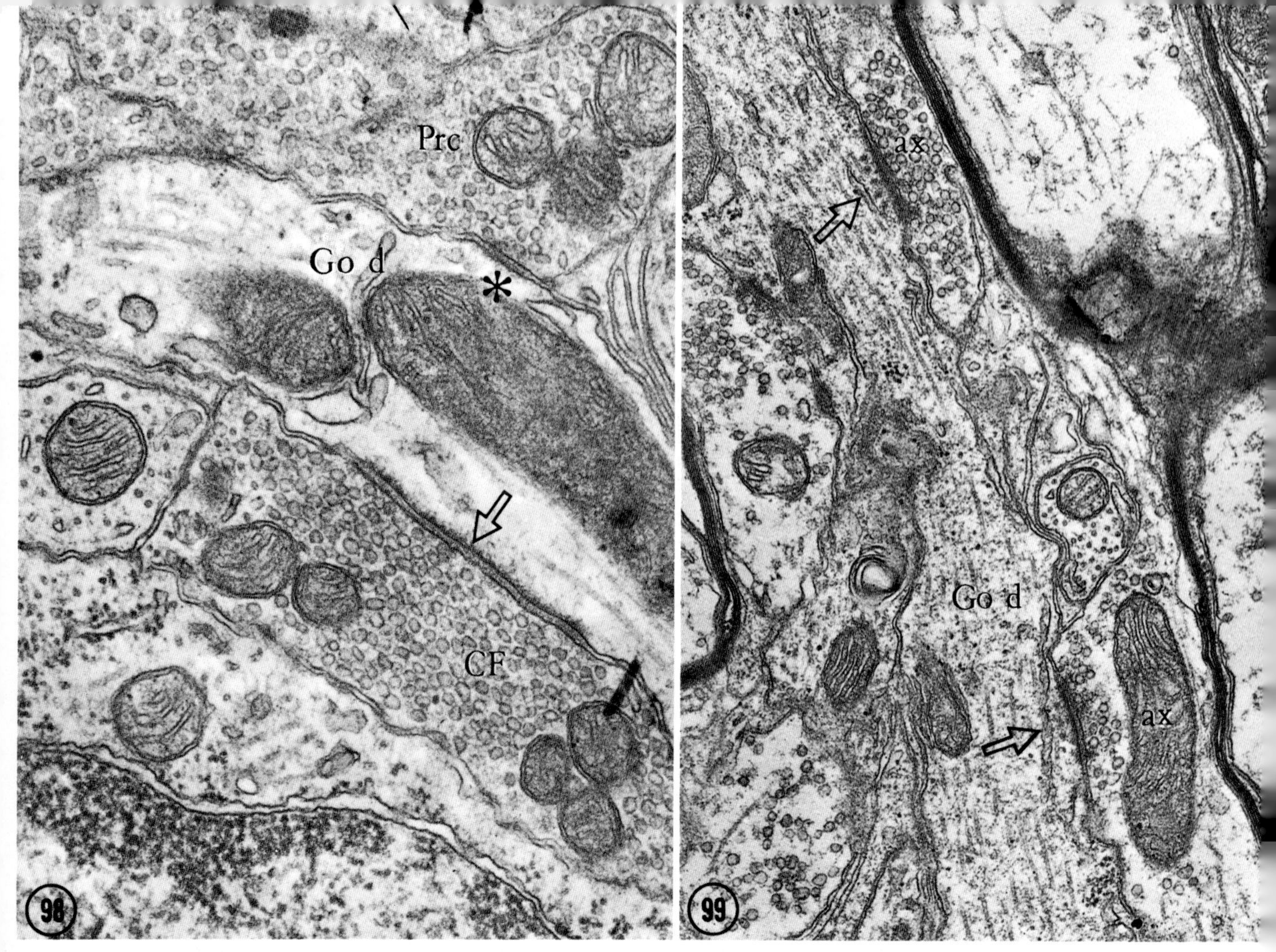

Fig. 98. Synapses on a Golgi dendrite in the granular layer. A terminal of a Purkinje cell recurrent collateral (*Prc*) synapses upon a Golgi cell dendrite (*Go d*). The junction exhibits a widened synaptic cleft and almost symmetrical membrane densities (*asterisk*). A terminal, possibly belonging to a tendril collateral of a climbing fiber, also synapses upon the Golgi dendrite. This synapse exhibits a Gray's type 1 junction (*arrow*). Lobule VI. ×15000

Fig. 99. Subsynaptic specializations. The Golgi dendrite (*Go d*) in the granular layer receives two synapses (*arrows*) from axons (*ax*) possibly belonging to granule cells. The postsynaptic region deep to the junction is specialized; a row of granules about 300 Å in diameter lies about 300 Å away from the plasmalemma. Lobule IV. ×23000

drite with its numerous branchings and its distinctive thorns and internal fine structure is not mistakable for the Golgi cell dendrite. Basket and stellate cell dendrites, however, may sometimes be confused with it in transverse sections. Because the Golgi cell dendrites are relatively stout and straight, they are most readily recognized in longitudinal section. Long stretches of these processes can be captured in a single thin section. The configuration seen in these images matches the pattern seen in Golgi preparations. Therefore, they provide a promising starting point for characterizing the Golgi cell dendrites in the molecular layer.

Figs. 100 and 101 show portions of distal Golgi dendrites in longitudinal section as they course through the molecular layer. The dendrite follows a gentle curve towards the pial surface. It is about 1 μm thick, and it contains long, slender mitochondria and the usual complement of microtubules, both running longitudinally. Here and there are minute bits of granular endoplasmic reticulum and occasional tubules of the agranular reticulum. The smooth contours and the poorly developed hypolemmal cisterna distinguish this dendrite from the occasional Purkinje cell dendrite that is also straight. Furthermore, it is not separated from the surrounding axons by a neuroglial sheath, as the Purkinje cell dendrites usually are. The surface of the dendrite is in contact with a large number of synaptic terminals, most of them small varicosities in parallel fibers (Figs. 100 and 101). These can be identified by their small size and the spherical synaptic vesicles clustered at the interface. The synaptic contact formed between the parallel fiber and the shaft of the Golgi cell dendrite is similar to that formed between the parallel fiber and the thorns of Purkinje cell dendrites, but not exactly the same. The synaptic cleft is widened and the dense plaque attached to the inner side of the dendritic, or postsynaptic, plasmalemma is thicker than the presynaptic densities in the axon. This asymmetry is, however, not always so marked as it is in the junction between the parallel fiber and the thorns on the Purkinje cell dendrite. Axo-dendritic synapses are also made with the axons and axon collaterals of basket cells and stellate cells (Fig. 100). These junctions show slight or no widening of the synaptic cleft. The synaptic interface often appears rigidly straight, and the synaptic densities are nearly symmetrical. In most cases the dense material hanging from the postsynaptic membrane is barely discernible. The presynaptic terminals contain loosely aggregated, ellipsoidal or polygonal synaptic vesicles. Aside from these simple synaptic contacts the Golgi dendrite articulates with the horizontal axon of the basket cell in the so-called girdle synapse (Fig. 169), already described (CHAN-PALAY and PALAY, 1970, Figs. 5, 6).

The characteristics of the Golgi cell dendrite as determined in longitudinal section provide the criteria for identifying it in transverse section (Figs. 102 and 103). These are relatively large caliber (about 1–1.5 μm across), smooth profile, absence of the hypolemmal cisterna, a meager complement of mitochondria, both microtubules and neurofilaments, and, perhaps most important, a high ratio of synaptic junctions to surface area. Most of the synapses are made by parallel fibers, which cluster about the circumference of the dendrite like petals in a flower. Occasionally a basket or stellate cell terminal is mingled with them. Neuroglial processes can also lie against its surface, but it is not protected by a neuroglial sheath from the axons or dendrites that may contact it without synapsing.

Occasionally slender spicules project from the surface of the Golgi cell dendrite. These are encountered rarely in electron micrographs, not surprisingly since the samples are so small. These spicules contain only a finely filamentous matrix, like the spines and filopodia of other nerve cells. We have seen very few of these appendages in electron micrographs of Golgi cell dendrites. Some have had granule cell varicosities in synaptic contact on their surfaces.

The Axonal Plexus of the Golgi Cell. The axons of the Golgi cell are almost impossible to follow in random sections at the electron microscope level. Only short lengths appear in the micrographs. The reasons for this inconvenience are clear when the axons are traced with high optical magnifications in Golgi preparations. Beyond the initial segment, which is often fairly straight, they become highly tortuous and branch profusely in their characteristic right-angled manner. The three-dimensional plexus that is generated in this way almost precludes finding a long stretch of the axon in any thin section used for electron microscopy. Nevertheless, the fragments that are encountered in the situations prescribed by the description of Golgi preparations have the form and dimensions expected, and therefore can be identified with little fear of contradiction. FOX, DUTTA, HILLMAN, and SIEGESMUND (1965) were the first to show by experimental means and electron microscopy that Golgi axons end in the periphery of the glomerulus. They undercut the cerebellar cortex and allowed the extrinsic afferents, the mossy fibers, to degenerate. The healthy axons remaining in the glomeruli were identified as the Golgi axons. These results have been amplified by FOX, HILLMAN, SIEGESMUND, and DUTTA in 1967 and confirmed independently by HÁMORI (1964), HÁMORI and SZENTÁGOTHAI (1966b), and MUGNAINI (1972) (also see Chap. VI). The detailed structure of the glomerulus will be described in a separate section when the mossy fiber is considered. For the present purpose a brief account of the Golgi axon itself will be given.

The initial segment of the Golgi cell is necessarily encountered in thin sections only infrequently (Fig. 92). It emerges with hardly any cytoplasmic differentiation from a barely perceptible pucker in the soma or a dendritic trunk. It is a rather stout fiber, 0.8–1.0 μm in diameter, but we have been unable to trace it further in thin sections. At its origin the axon has the usual internal structure of initial segments (CONRADI, 1966, 1969; PALAY, SOTELO, PETERS, and ORKAND, 1968; PETERS, PROSKAUER, and KAISERMAN-ABRAMOF, 1968; CHAN-PALAY, 1973a). Apparently it becomes very thin after repeated branchings, for when it is next picked up, in the margin of the glomeruli, it is barely 0.2 μm thick. Here it appears as a highly varicose terminal or preterminal fiber woven over the granule cell dendrites that surround the central mossy fiber rosette in the glomerulus (Figs. 87–90). High voltage electron micrographs (Fig. 104), especially stereoscopic pairs, give a vivid impression of the dense plexus woven by these varicose fibers in the outskirts of the glomerulus. Some of the varicosities reach into the depths of the glomerulus, where they come into apposition with the mossy fiber without establishing synaptic contact. The tenuous Golgi axon contains both microtubules and neurofilaments and little else between varicosities (Figs. 143 to 145, and 152). These expansions vary from 1 to 4 μm in diameter, and succeed one another at intervals varying from tenths of a micron to several microns. They are generally spheroidal or elliptical enlargements in the course of the fiber, filled with small polygonal, flattened vesicles and a cluster of mitochondrial profiles (Figs. 91

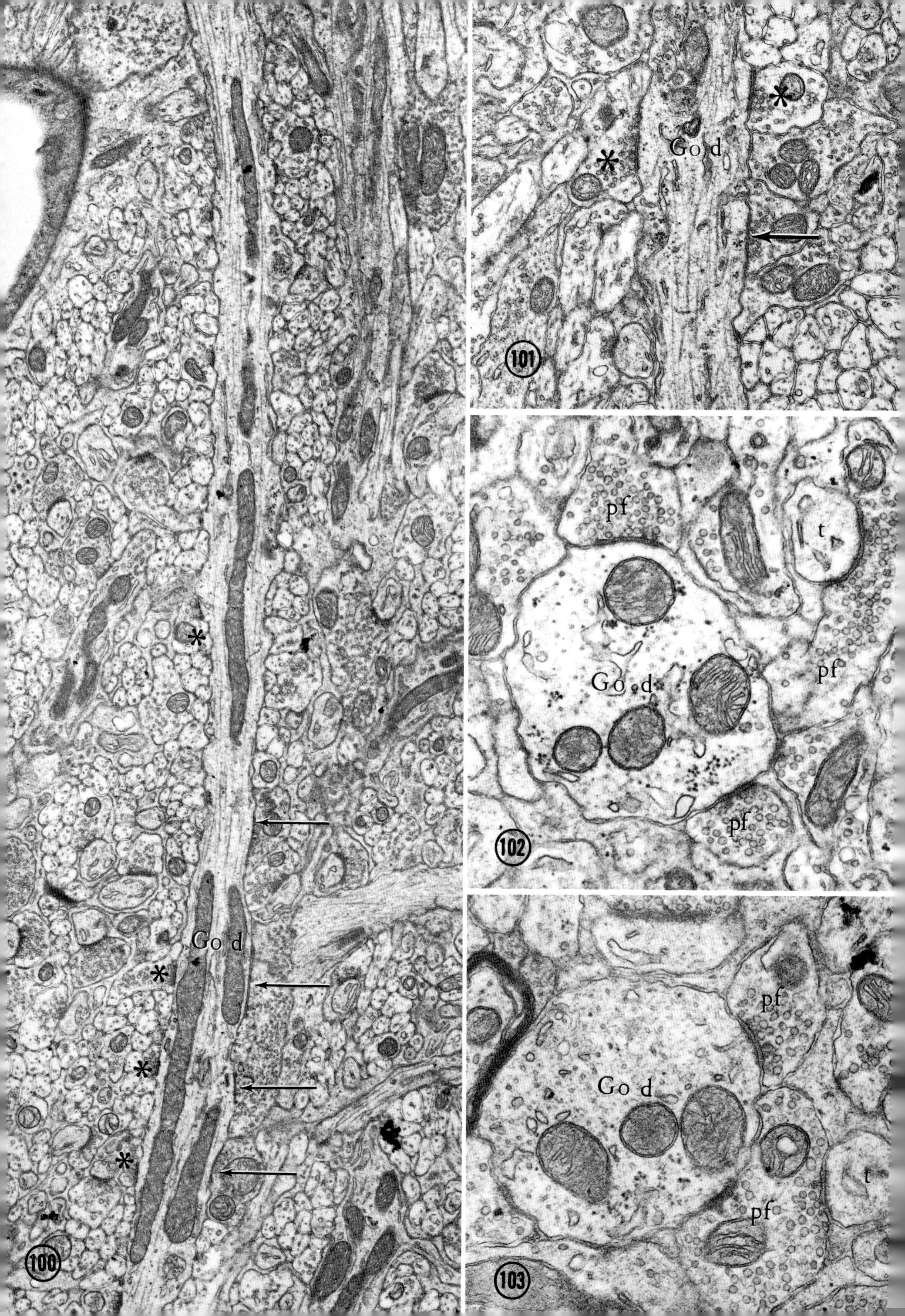
Go d
*
*
*
*
*
Go d
*
*
101
100
pf
t
pf
Go d
pf
102
pf
Go d
t
pf
103

and 105). The microtubules and neurofilaments pass right through the centers of these expansions. In axial sections the vesicles appear swept to each side and sometimes a long, slender mitochondrion extends from one varicosity to the next as if slithering between them. But most sections display the varicosities as roughly circular independent profiles crowded with their small, flattened vesicles and a few mitochondria (Fig. 91).

They are found most commonly around the periphery of the glomerulus where the Golgi axons articulate with the granule cell dendrites.

Although the junctions bring a broad, flattened axonal surface into apposition with the granule cell dendrite, synaptic specializations at the interface are only infrequently encountered. They are small macular complexes, circular or elliptical in outline, from 0.1 to 0.3 μm in diameter, in which the synaptic cleft is widened and filled with fine, filamentous material (Fig. 105). A similar interstitial material appears in patches between adjacent Golgi terminals, as well as between granule cell dendrites, all over the glomerulus, without widening of the interstitial cleft. But in addition, the synaptic complex is characterized by a mat of filamentous material attached to the inner side of the dendritic (postsynaptic) membrane, which is balanced by a thinner, much less dense plaque on the axonal (presynaptic) side. A small collection of ovoid or polygonal synaptic vesicles often presents itself in the axon opposite the synaptic interface. These junctions are, except for the shapes of the vesicles, not much different from the synapses effected between the mossy fiber and the same dendrites. This observation is interesting in view of the general principle that in the cerebellar cortex the type 1 synapses are uniformly excitatory junctions, whereas the Golgi cell is thought to have an inhibitory effect on the granule cells with which it synapses. Further consideration of these interesting synaptic terminals will be given in the section on the glomerulus.

◀

Figs. 100 and 101. Synapses on Golgi dendrites in the molecular layer: longitudinal section. As the Golgi dendrite (*Go d*) courses radially through the molecular layer, synapses occur on its surface. The dendrite receives small punctate synapses from parallel fiber varicosities (*asterisks*) and broader synapses from basket or stellate cell axons (*arrows*), Fig. 100, lobule VI. ×15000; Fig. 101, lobule V. ×23000

Figs. 102 and 103. Synapses on Golgi dendrites in the molecular layer: cross section. Such profiles are typical of these dendrites in the molecular layer, and they are easily recognized. The profile is generally large and lacks the hypolemmal cisterna. The dendritic surface (*Go d*) also bears several synapses. These two figures show synapses with parallel fiber varicosities (*pf*). Thorns of Purkinje dendrites (*t*) also synapse upon these varicosities, Fig. 102, lobule X. ×37000; Fig. 103, lobule X. ×41500

3. The Small Golgi Cell

In its overall form and in the configuration of its processes the small Golgi cell resembles the large Golgi cell, but it has a smaller cell body and its dendritic tree arises from several more or less equal trunks radiating outward from the cell body and branching only a few times (Fig. 87). The perikaryon is occupied almost entirely by an eccentric nucleus. In toluidine blue preparations it appears as a pale, creased nucleus with finely delineated contours lying among the only slightly smaller and much more intensely colored nuclei of the granule cells. The nucleolus is usually conspicuous and eccentric. Groups of two or three small Golgi cells clustered close together are not infrequent. As is true of the large Golgi cell, the small Golgi cell has not been described as a distinct cytological entity before. In preparations for the optical microscope, it may have been confused with the astrocyte, which possesses a nucleus of about the same size and shape, and, like the small Golgi cell, has little perikaryal basophilia. For example, Palkovits *et al.* (1971 b) considered all nuclei in the granular layer larger than 10 μm in diameter as belonging to Golgi cells, but they placed all nuclei less than 10 μm in diameter and clearly not those of granule cells in a category named astrocytes. Their counts of Golgi cells in the cat cerebellar cortex may therefore be too low, and their estimate that 89% of all Golgi cells are located in the upper half of the granular layer may be excessive in view of the fact that most of the small Golgi cells reside in the deeper half of the granular layer. We have not attempted to make counts of the Golgi cells in the rat.

a) The Fine Structure of Small Golgi Cells

In the electron micrographs the nucleus of the small Golgi cell proves to be an elongated ellipsoid, 5–10 μm across and deeply incised by thin laminae of ribosome-laden cytoplasm (Figs. 106 and 107). Although most of the chromatin is homogeneously distributed, as is consistent with the pallor of the nucleus in the light microscope, small condensations are marginated against the nuclear envelope and often the eccentric nucleolus is associated with one of them. The nuclear envelope throws out streamers that continue into the granular endoplasmic reticulum, and the outer surface of the envelope itself is studded irregularly with rows of ribosomes.

Compared to the large Golgi cell with its prominent, large Nissl bodies, the small Golgi cell has relatively

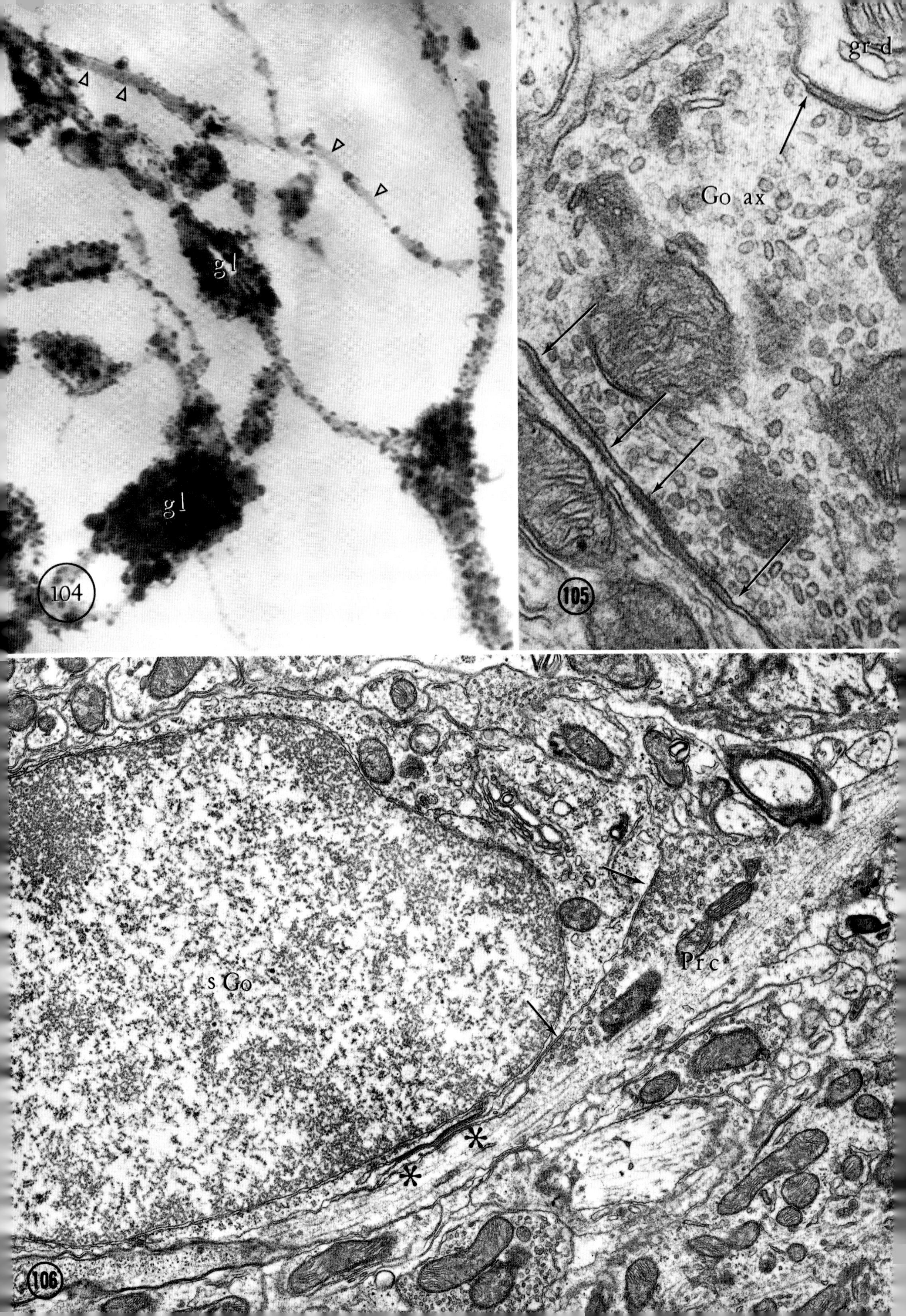
gl
gl
104
gr d
Go ax
105
Prc
s Go
*
*
106

little Nissl substance. It resembles the large cell stripped of the outer cytoplasmic zone (Fig. 107). The granular endoplasmic reticulum is rather poorly represented, being confined to small aggregates widely distributed over the cytoplasm. Small polysomal arrays of free ribosomes are strewn throughout the perikaryon, but near the nucleus they condense into a cloud that invests the nucleus, extending into the folds of the nuclear envelope and excluding virtually all other organelles.

The perikaryon of the small Golgi cell is characterized by unusually large numbers of small mitochondria (Fig. 107). Most of them appear in electron micrographs as slender rods, 0.2 μm in diameter, with oblique or longitudinal cristae. The high proportion of circular profiles and short arcs suggests that the mitochondria are probably short, bacillary forms. Occasionally, long profiles, 2–4 μm long, are encountered. The presence of a few mitochondrial profiles 0.4 μm across or larger, and of branching forms among clusters of circular profiles, indicates that at least some of the mitochondria are digitiform. The mitochondrial aggregates are especially striking in the cytoplasm subjacent to the large axo-somatic *synapse en marron* that will be described further on (CHAN-PALAY and PALAY, 1971b).

The Golgi apparatus is also less elaborate in the small Golgi cells than it is in the large ones (Figs. 106 and 107). The islands of imbricated cisternae and associated vesicles are smaller and farther apart; at least, fewer of them are met with in any random section. It is rare to find the long, arciform aggregates of cisternae that are typical of the large cells. Lysosomes, multivesicular bodies, and dense core vesicles are also rather scarce in these cells.

Fig. 104. High voltage electron micrograph of Golgi axons after rapid Golgi impregnation. The two components of the Golgi reaction, background density defining the fibers and surface incrustation, are shown. Note the variation in sizes of the globular crystals on some of the axons. Portions of the Golgi axons in this field have few or no globules (*arrowheads* Δ). Other axons are well incrusted with globular particles (*gl*). 3 μm, lobule X. ×15000

Fig. 105. Synapses of the Golgi cell axon. Terminals of the Golgi cell axon (*Go ax*) generally have a dark axoplasmic matrix containing polymorphic synaptic vesicles. The axon synapses with dendrites of granule cells (*gr d*) through atypical junctions (*arrows*) showing widened synaptic clefts and inconspicuous pre- and postsynaptic membrane densities. Lobule VI. ×76000

Fig. 106. Synapse upon the soma of a small Golgi cell. The soma of this cell (*s Go*) receives a terminal of a Purkinje cell axon (*Prc*) bearing two small synaptic junctions (*arrow*). A subsurface cisterna appears between the asterisks. As in most small Golgi cells, the Nissl substance is not organized into distinct bodies. Lobule X. ×23000

The matrix of the cytoplasm (Fig. 107) is permeated by a fine, filamentous material through which numerous microtubules course in a perinuclear trajectory. These fine filaments give the matrix of the cell a dense, fibrillar appearance in electron micrographs. In the subsynaptic region of the synapse *en marron* this filamentous material forms a thick matting, which seems to exclude all other organelles except for a few tubules of the agranular endoplasmic reticulum. A similar filamentous matrix occurs in dendritic thorns, spines, and filopodia of all sorts of neurons. It probably represents a gelated state of the cytoplasm, which is responsible for maintaining the geometric stability of the surface projections, in this case the wrinkled subsynaptic surface.

Few synaptic terminals are attached to the surface of the small Golgi cell, and the majority of these are terminals of recurrent collaterals of Purkinje cell axons (Fig. 106). In addition, many small Golgi cells display an extensive synapse *en marron*, which will now be described.

b) The Synapse en Marron

Both large and small Golgi cells have a peculiar, irregularly corrugated surface, which gives them an unusually ruffled profile in electron micrographs. This rugose surface is sculptured to receive the processes of other cells, neurons or neuroglia, that course through the adjoining neuropil. In the regions where a mossy fiber rosette or a climbing fiber terminal abuts against the soma, the corrugated surface bears a further series of small, uneven ridges and furrows. Because this part of the cell resembles a Spanish chestnut (Fig. 108), the articulation found here between the expanded axon terminal and the soma of the Golgi cell has been termed the synapse *en marron* (CHAN-PALAY and PALAY, 1971a and b). The articulating surface is shown in Fig. 109a and b, which are photomicrographs of a large Golgi cell taken at two levels of focus. The lower, hemispherical surface of this cell is marked by deep pits or furrows that correspond to the wrinkles seen in thin sections by conventional electron microscopy (Fig. 110). A similar region is seen in the large Golgi cell in Fig. 87. The junctions can also be recognized in 1 μm plastic sections stained with toluidine blue by virtue of the dense aggregation of mitochondria in the large axonal terminal abutting against the Golgi cell. An electron micrograph of the synapse *en marron* was published by HÁMORI and SZENTÁGOTHAI (1966b) and by ECCLES, ITO, and SZENTÁGOTHAI (1967) but not recognized as such. These authors described the picture as simply a synapse between a mossy fiber and a dendrite of a Golgi cell. The

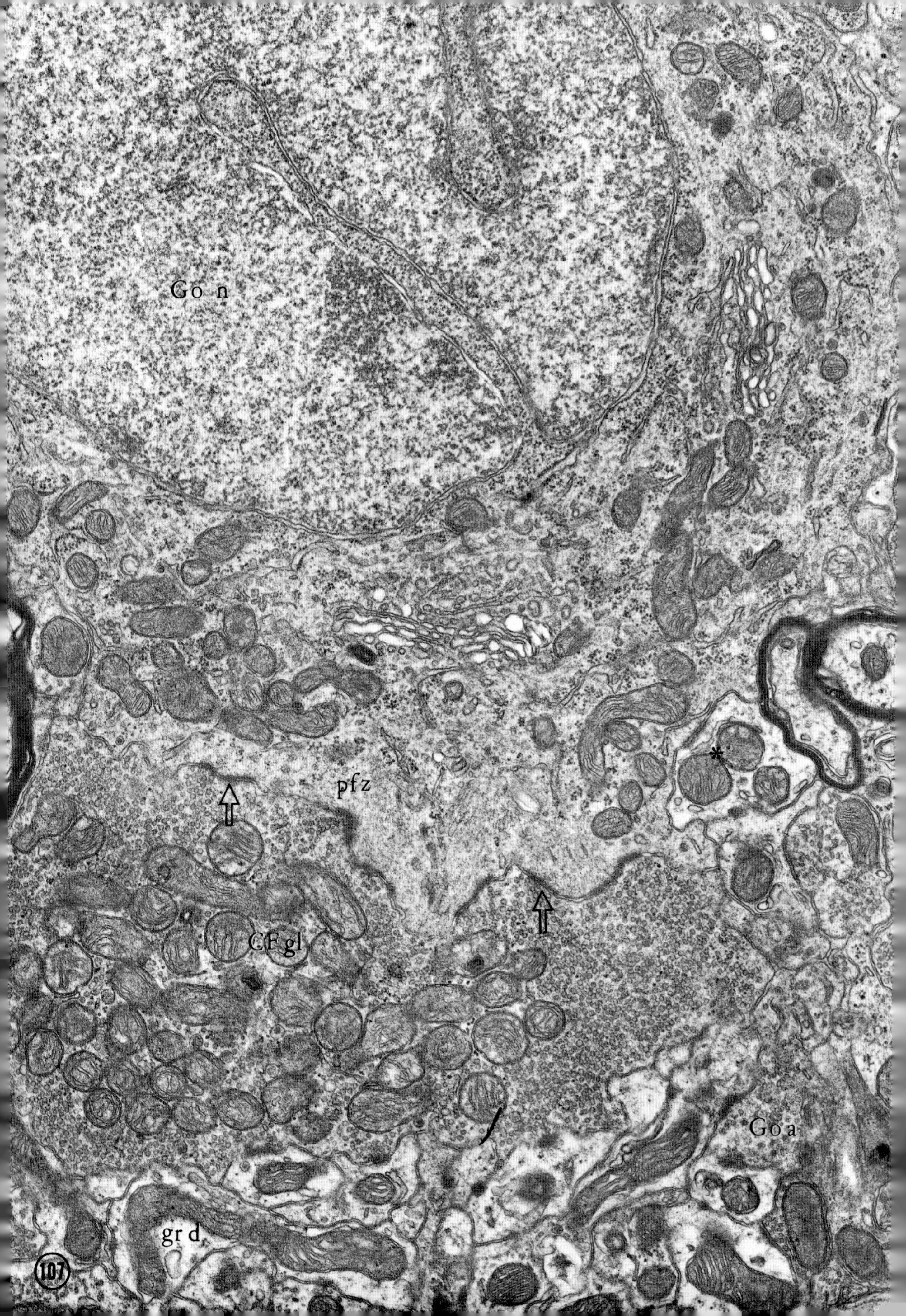
Go n
pfz
*
CF gl
Go a
gr d
107

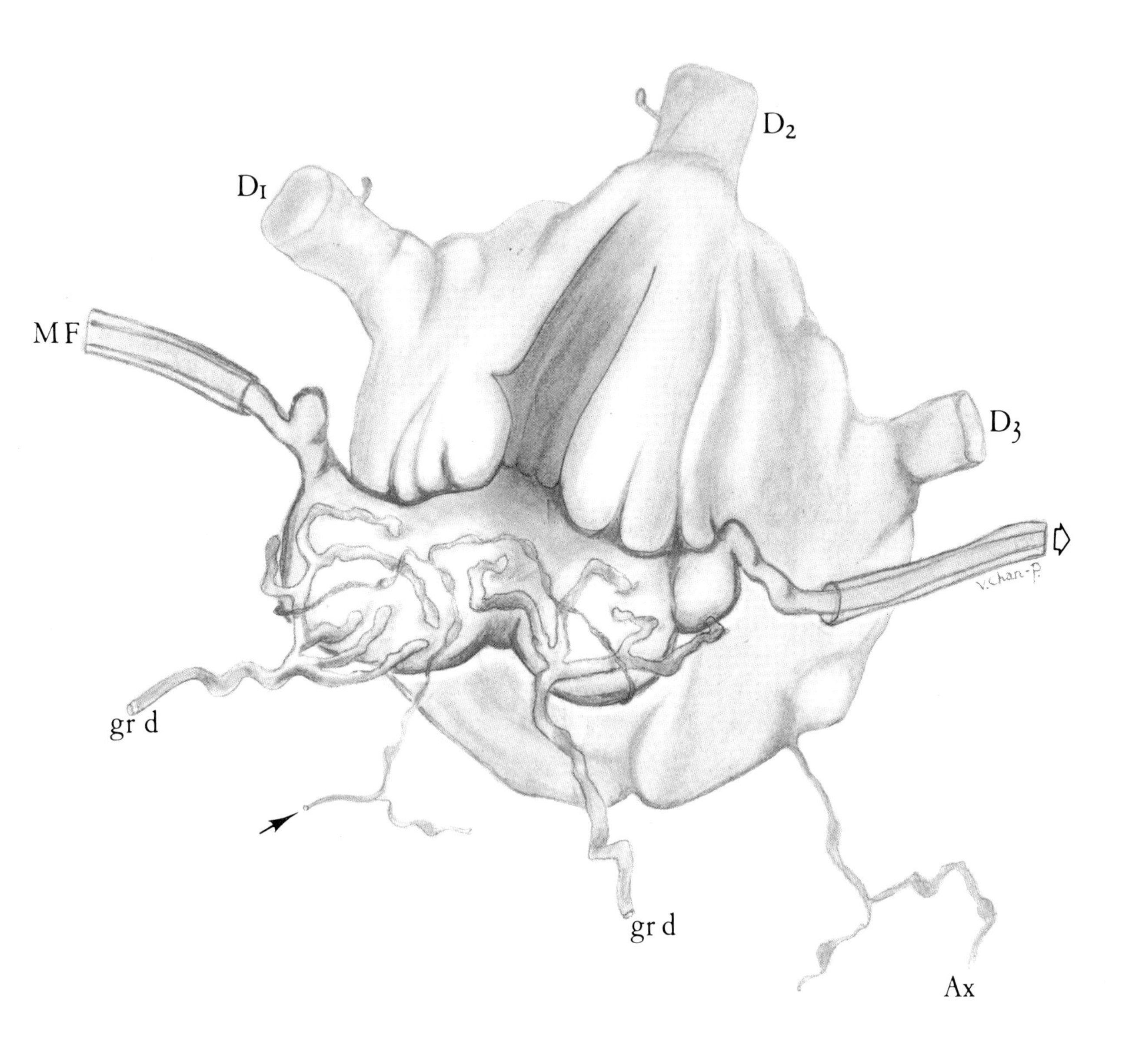

◀

Fig. 107. The synapse *en marron* on the soma of a small Golgi cell. The perikaryal surface of the Golgi cell lying adjacent to this climbing fiber glomerulus is wrinkled like a Spanish chestnut. The terminal of the climbing fiber (*CF gl*) that is pressed against it is reciprocally ridged and furrowed. Extensive synapses typical of Gray's type 1 occur in the depths of these shallow furrows (⇐). Because the wrinkled Golgi cell surface resembles that of a chestnut, this form of axo-somatic junction has been called the synapse *en marron*. A broad zone of cytoplasm, about 0.3–1 μm deep beneath the junctional complex, lacks organelles but is filled with a dense fibrillar matrix. The component filaments of this postsynaptic fibrillar zone (*pf z*) are oriented radially, spreading outward toward the synaptic surface. Deep to this zone the usual cytoplasmic organelles begin to reappear, first tubules of the agranular endoplasmic reticulum, then rough endoplasmic reticulum, mitochondria and all other organelles as the nucleus of the Golgi cell (*Go n*) is approached. Surfaces of the climbing fiber terminal not in contact with the Golgi cell are entwined by dendrites of granule cells (*gr d*) and occasional Golgi axons (*G a*). A punctum adhaerens between a granule cell dendrite and the Golgi cell body is indicated by (*). The cytoplasm of this Golgi cell lacks well-defined Nissl bodies. Lobule V. ×17000

▲

Fig. 108. The synapse *en marron* between the Golgi cell body and a mossy fiber rosette. The cell body of this Golgi neuron displays broad, rounded projections abutting against a mossy fiber *terminaison en passant*. These projections are further wrinkled into a series of uneven longitudinal ridges and furrows. Electron microscopy has shown that the sides and depths of these furrows bear extensive synapses with the mossy fiber. The mossy fiber (*MF*) enters this drawing from the left as a myelinated fiber. It loses its myelin sheath and distends into a terminal formation that is pressed against the Golgi cell. The fiber leaves the region of the cell body to enter the next segment of its myelin sheath at the right (⇒). Dendritic claws of granule cells (*gr d*) in company with a Golgi axon (→) clasp the surface of the mossy fiber that is not in contact with the Golgi neuron. The initial portions of three dendrites (D_1, D_2, D_3) emerge from the Golgi cell body, and its fine, thinly beaded axon leaves at the bottom right hand corner (*Ax*)

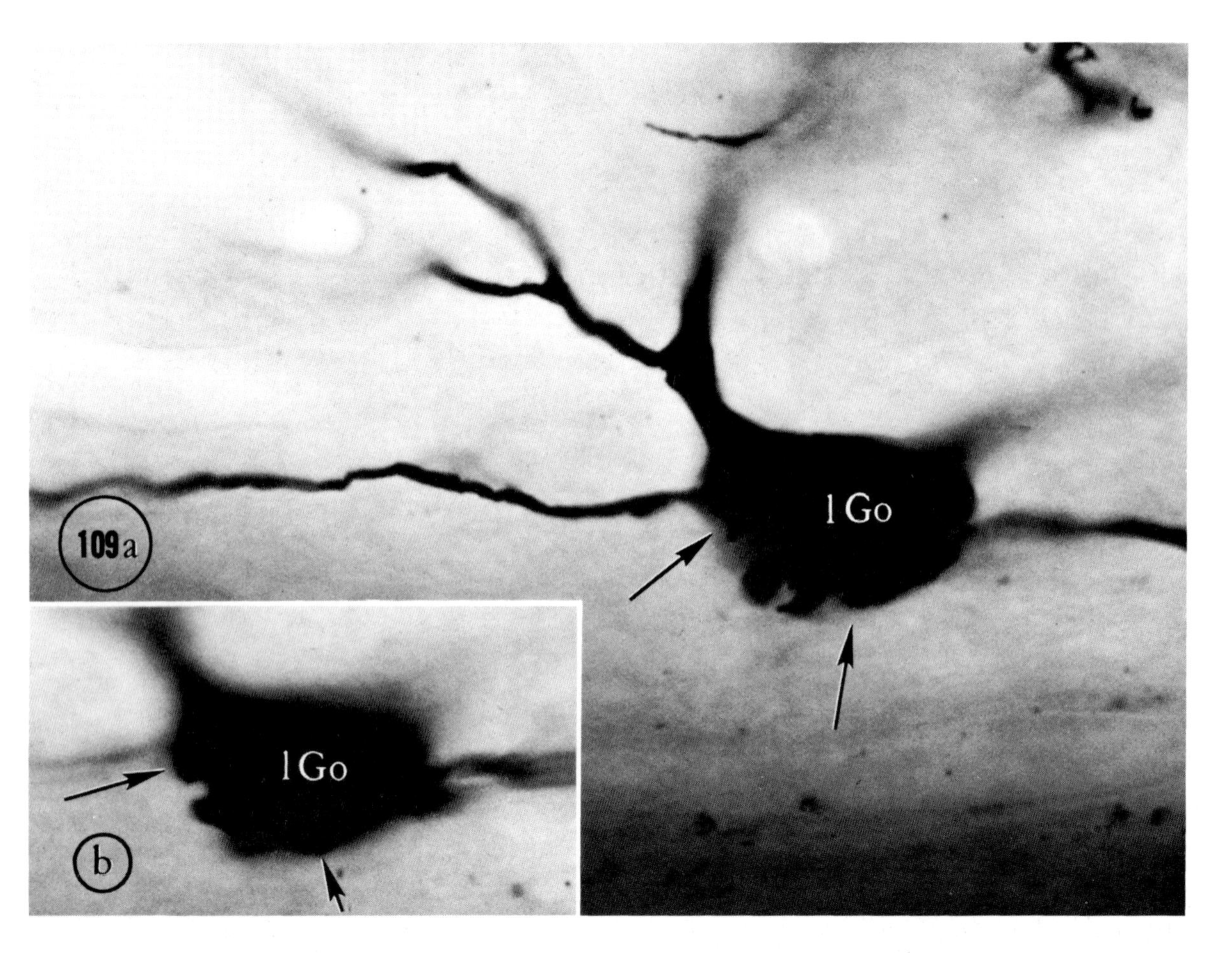

Fig. 109a and b. A large Golgi cell: the synapse *en marron* postsynaptic surface. These two photomicrographs show a single large Golgi cell (*l Go*) at two levels of focus in (a) and (b). The surface of the cell body bears several projections (*between arrows*). Each projection probably represents a hillock of the ruffled postsynaptic cytoplasm bearing the synapse *en marron* as seen in Figs. 107 and 108. Rapid Golgi, 90 µm, adult rat. ×1400

extraordinary features of this axo-somatic synapse were not analyzed until 1970, when we were fortunate in obtaining sections of the junctional interface that included the nucleus and soma of the Golgi cell (Chan-Palay and Palay, 1971a and b). We were thus able to demonstrate unequivocally that the postsynaptic component is the cell body of the Golgi cell.

The synapse *en marron* is an extensive axo-somatic synapse. It encompasses a good part of the somatic surface of the Golgi cell. For this reason, electron micrographs of thin sections usually include only fragments of it, and these have rather puzzling contours (Figs. 110–113). These isolated profiles, enclosing a fibrillar matrix and a few tubules, set in fields of synaptic vesicles and mitochondria, could be understood only after a study of sections that include the nucleated portion of the Golgi cell. Then it becomes clear that they derive from a region of highly specialized structure subjacent to the synaptic interface, the postsynaptic fibrillar zone.

As may be seen in Figs. 110 and 117, within the Golgi cell just beneath the junction there is a broad zone 0.3 to 1.0 µm deep that is devoid of the usual cytoplasmic organelles. Instead, it is filled with the dense, fibrillar matrix that forms the background to the rest of the cell. The component filaments of the matrix tend to radiate outward toward the wrinkled synaptic interface. Deep to this postsynaptic fibrillar zone the usual cytoplasmic organelles begin to reappear, first the tubules of the agranular endoplasmic reticulum, some of which venture

Fig. 110. The synapse *en marron* between mossy fiber and Golgi neuron. A rounded projection of Golgi cell cytoplasm (*Go*) is shown with its surface thrown into ridges and furrows. The furrows are the sites of extensive ribbon-like synapses (*) with a mossy fiber rosette (*MF*), and the ridges are bare of synapses (→). Underlying the synaptic interface is a broad postsynaptic fibrillar zone (*pfz*) with its typical, dense filamentous matrix. A few tubules of agranular endoplasmic reticulum extend into this zone from a larger accumulation deep to it. Other cellular organelles increase in number as one proceeds away from the cell surface. The rest of the mossy fiber profile is entwined with dendrites of granule cells (*gr d*) and the axons of Golgi cells (*Go a*). Lobule X. ×24000

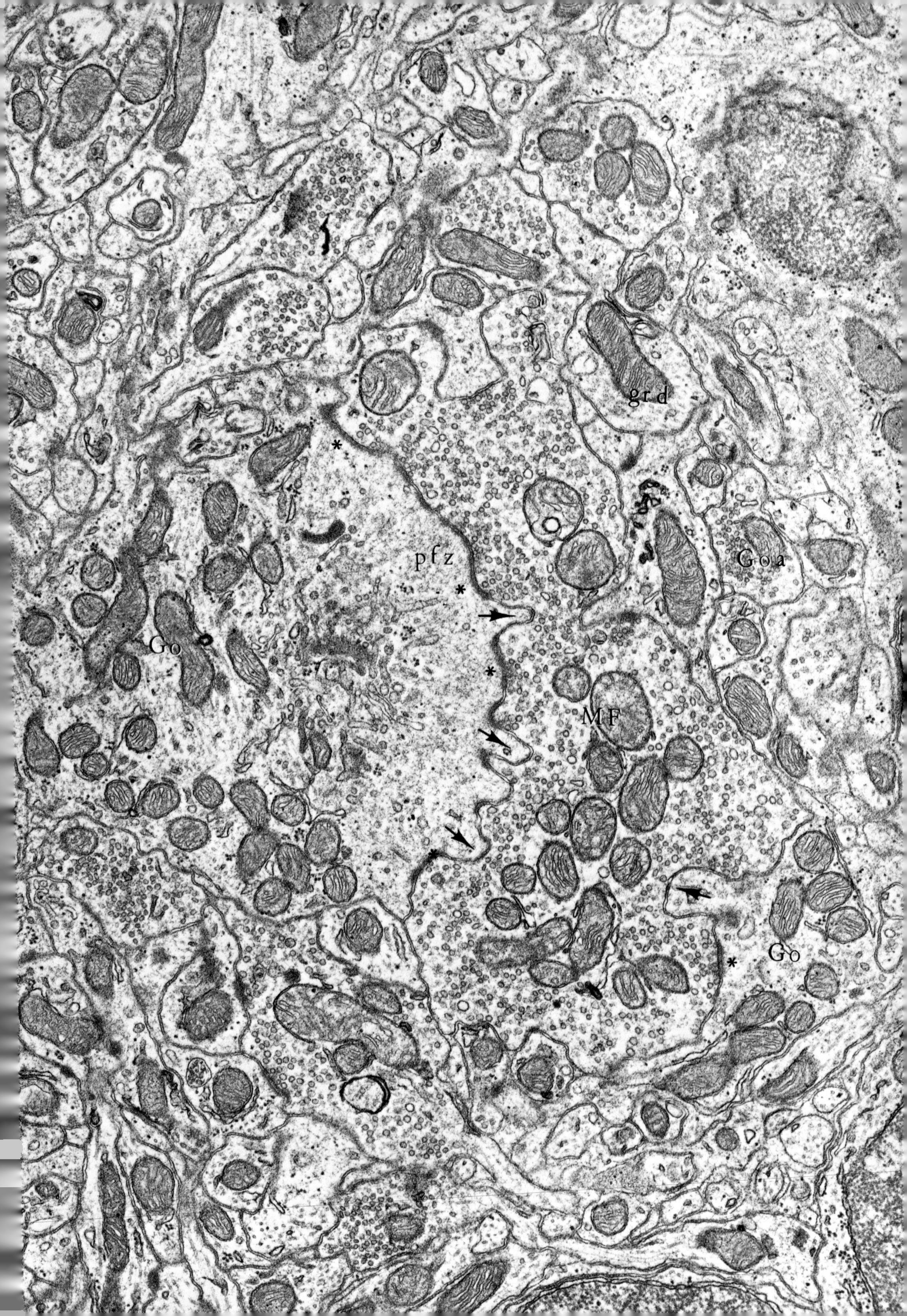
grd
pfz
Goa
Go
MF
Go

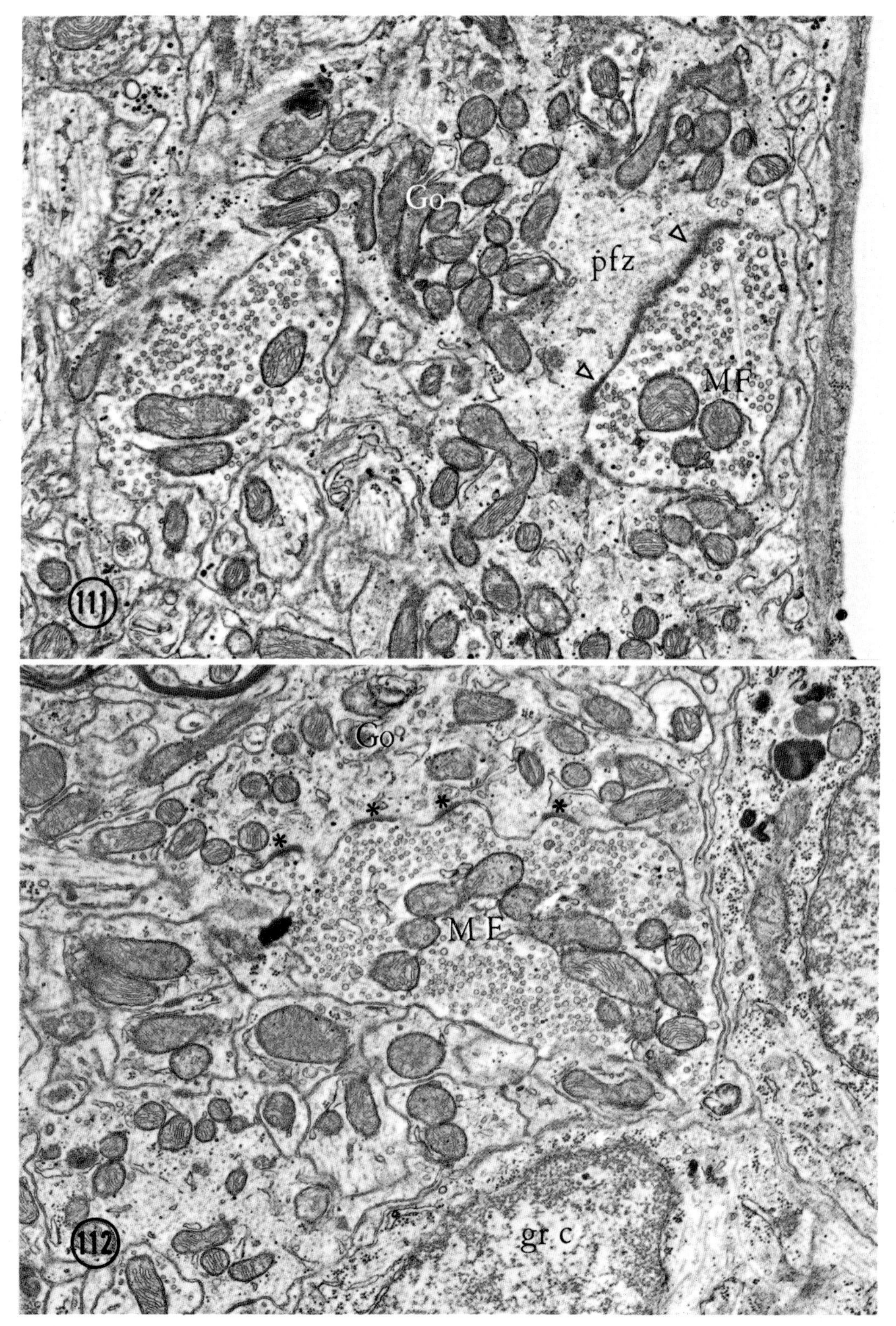

Fig. 111. Longitudinal section through a synapse *en marron.* This electron micrograph passes longitudinally through a furrow in the Golgi cell (*Go*) that contains a synapse *en marron* (▿) with a mossy fiber (*MF*). A long synaptic complex lies in the furrow. The postsynaptic fibrillar zone (*pfz*) displays the dense matrix typical of these synapses. Lobule IX. × 18000

Fig. 112. Cross section through the synapse *en marron.* The cytoplasm of the Golgi cell (*Go*) is thrown into a series of wrinkles with ridges and furrows that are cut transversely. Four saucer-shaped synapses (*) made with an abutting mossy fiber (*MF*) are separated from each other by the low ridges of Golgi cytoplasm. Profiles of granule cells (*gr c*) and other axons and dendrites surround these structures. Lobule IX. × 18000

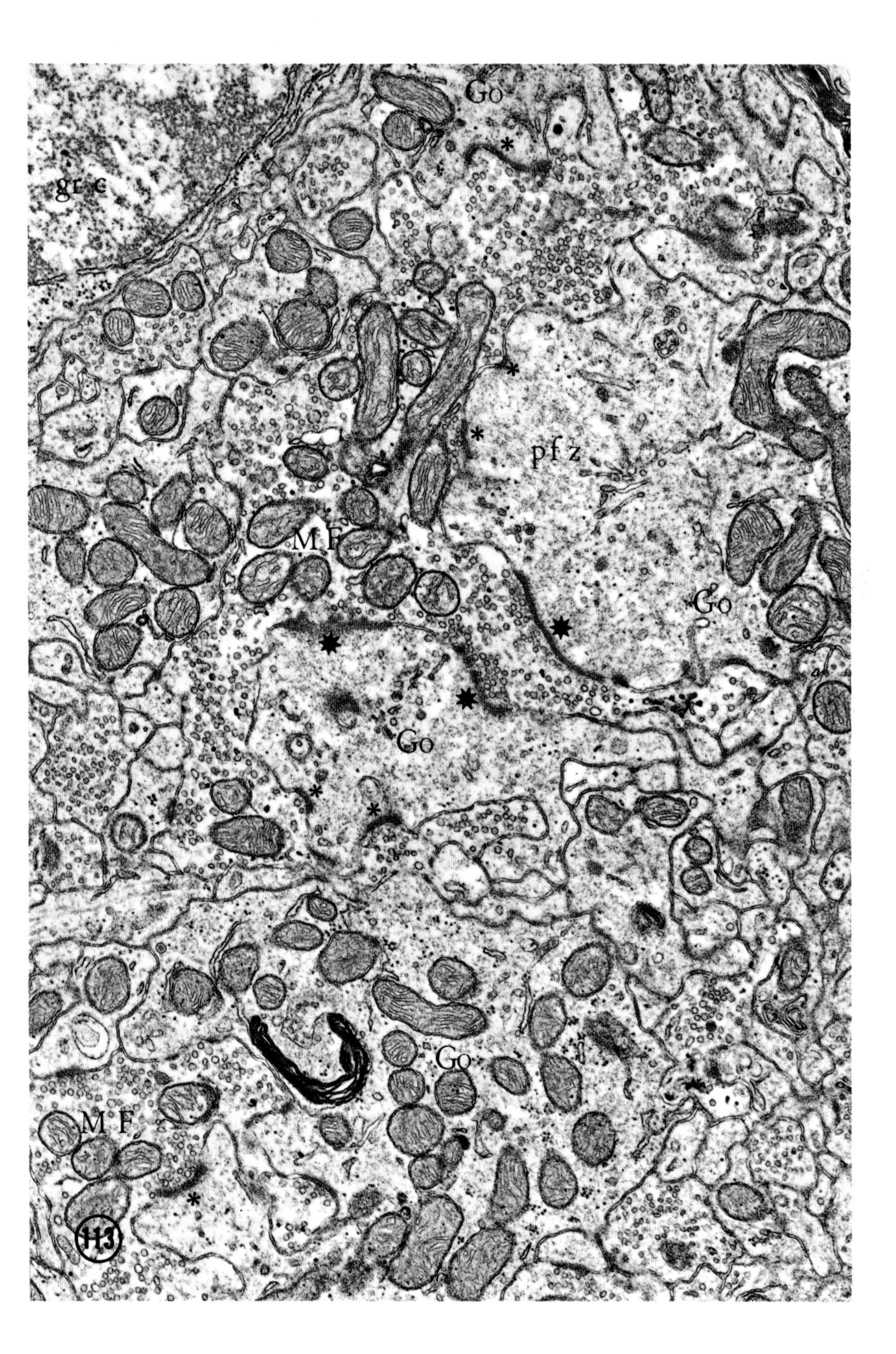

Fig. 113. The synapse *en marron* between mossy fiber and Golgi neuron. This illustration shows many synaptic complexes in the furrows of the Golgi cell. Several rounded projections of the Golgi perikaryon are indicated (*Go*) and all show the thick postsynaptic fibrillar zone (*pfz*) typical of these synapses. In those regions where the section passes at right angles to the long axis of the furrows, short, saucer-shaped synaptic complexes are seen (*). Where the section passes longitudinally through the furrows, long, straight, ribbon-like synapses are shown (✱). The mossy fiber rosette (*MF*) is surrounded by axons, dendrites, and granule cells (*gr c*). ×24000

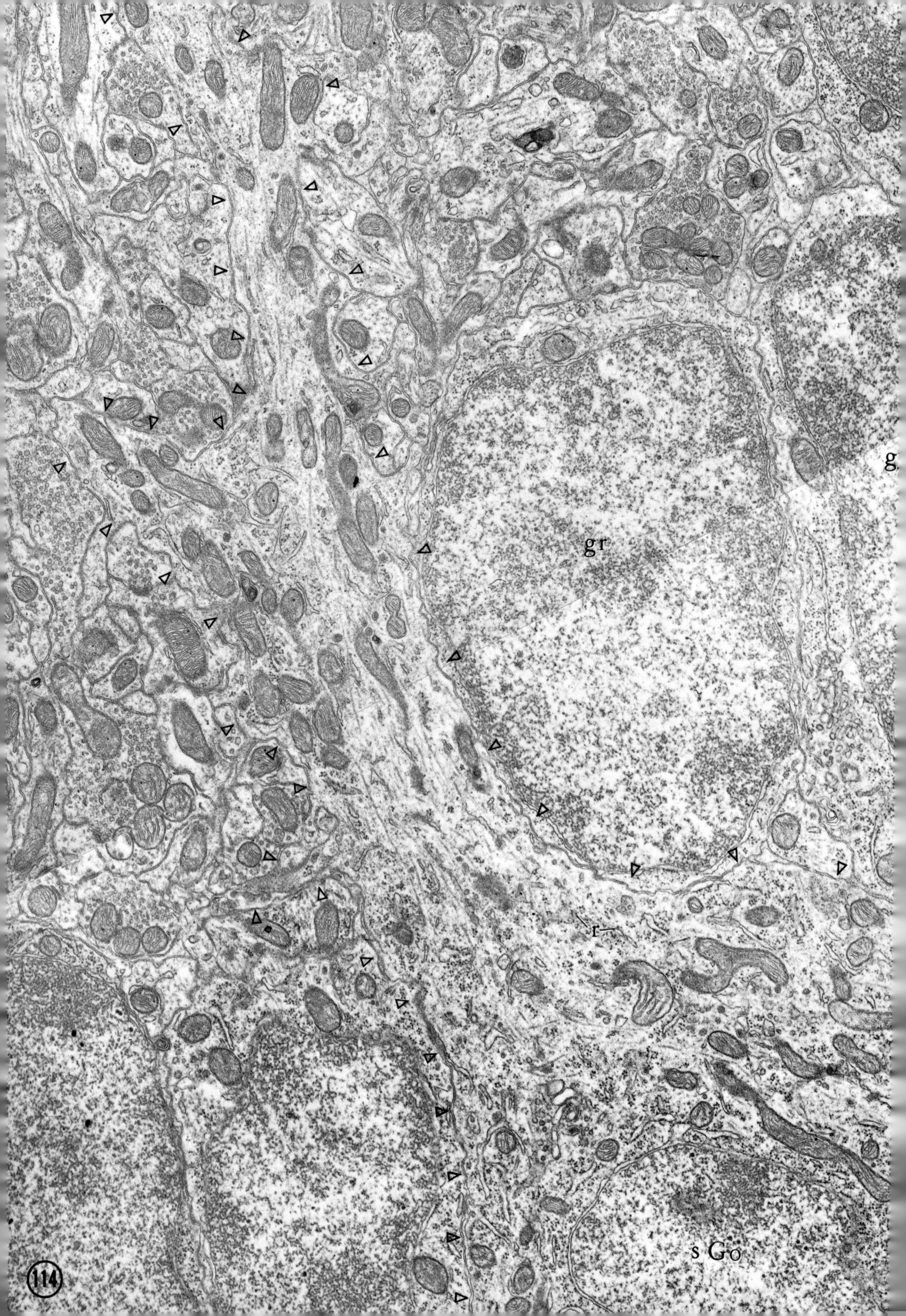
g
gr
r
s Go
114

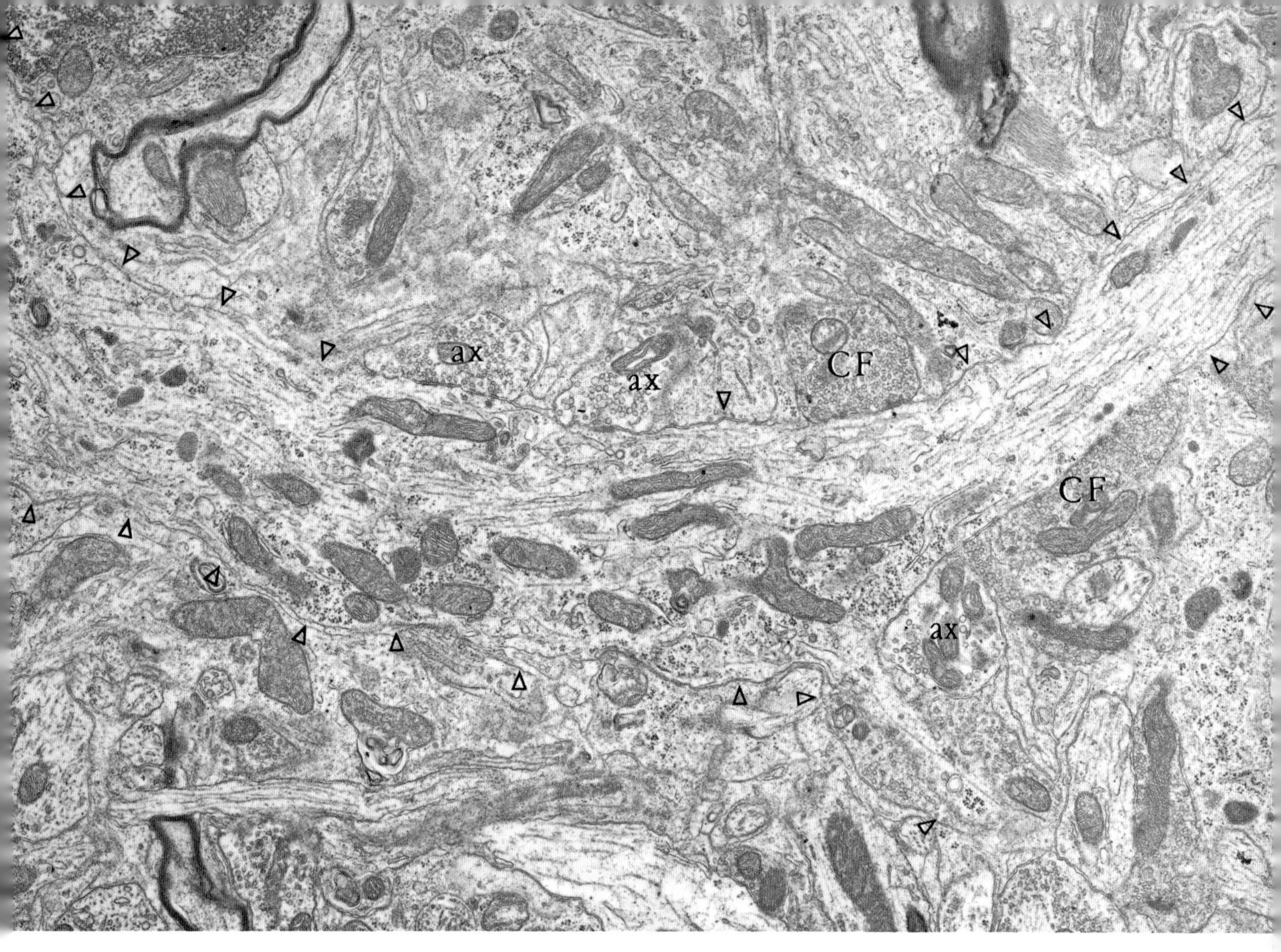

Fig. 115. Proximal dendrite of a small Golgi cell. This dendrite (ΔΔ) is also somewhat shaggy in outline, although less so than the one in Fig. 114. It bears two types of synapses on its surface. There are synapses with two terminals of the climbing fiber (*CF*), as well as synapses with three other terminals (*ax*) probably belonging to axons of granule cells. Lobule V. × 15000

into the fibrillar zone, then successively the granular endoplasmic reticulum, the clustered mitochondria, and finally the Golgi apparatus and all of the other organelles. Mitochondria often accumulate like a fringe at the margin of the postsynaptic zone. This arrangement of organelles does not occur in dendrites, and although the postsynaptic fibrillar zone does resemble the matrix of dendritic thorns and spicules, it is, of course, much more extensive in the cell body than in these appendages. The meandering outlines of the isolated profiles that were so baffling at first arise simply from the irregularly undulating and ridged surface that has given the synapse its name.

The junctional interface consists of the apposed plasmalemma of the Golgi cell body and either a mossy fiber or a climbing fiber. The presynaptic surface of the terminal is reciprocally folded to match the rugosities in the surface of the cell body. The furrows in the cell are broad and flat and are separated by low, narrow ridges. Long ribbons of synaptic complexes run in the furrows, while the ridges lack synaptic specializations. The synaptic complexes are characterized by slightly widened synaptic clefts traversed by dense filaments and by thick accumulations of filamentous material asymmetrically disposed on the postsynaptic cytoplasmic surface. Because the synaptic complexes run longitudinally in the furrows, they appear like shallow saucers between ridges in thin sections passing across their long axes. Thin sections passing parallel to the furrows (Figs. 111 and 113) show long, flat synaptic complexes, interrupted only by short barren intervals or low ridges. Since, however, the corrugations on the body of the Golgi cell run in all directions, a thin section of the synapse *en marron* often shows islands or peninsulas of perikaryon bearing several long, flat synaptic complexes with short, saucer-shaped junctions around them. The

◄

Fig. 114. Proximal dendrite of a small Golgi cell. This small Golgi cell (*s Go*) has many small clusters of ribosomes (*r*) dispersed in the cytoplasm. A large dendrite emerges from the cell soma (ΔΔ). The surface of the dendrite is somewhat shaggy in outline. Lobule IV. × 19500

more detached examples of these islands can be taken as profiles of dendrites, but the presence of the postsynaptic fibrillar zone and the absence of microtubules indicate that they are profiles of protuberances separated from the perikaryon by the section.

Since the synapse *en marron* is only one part of a very complex articulation involving Golgi cells, granule cells, and mossy or climbing fibers, further description of the axonal members will be deferred to their respective chapters.

c) The Dendrites and Axons

The dendrites of the small Golgi cells begin as prolongations of the cell body, and, like those of the large Golgi cell, they have highly irregular contours (Fig. 114). Because of the shaggy outline of both soma and dendrites, it is often difficult to decide whether a generally circular profile belongs to a dendrite or the cell body. The rafts of slender mitochondria (Fig. 115) that often flow out into the dendrites do not assist in the identification, since they also occur in the soma, like the Golgi apparatus and the small clusters of ribosomes. With increasing distance from the perikaryon, however, the microtubules become the predominant organelle (Fig. 97). Aligned parallel with the axis of the dendrite, they provide the most consistent identifying feature, as in other types of nerve cell. The more distal parts of the dendrites cannot be distinguished from those of the large Golgi cells, except that they lie in the granular layer for the most part, whereas the dendrites of large Golgi cells also extend into the molecular layer.

As seen in electron micrographs, the proximal parts of these dendrites are frequently setulose (HÁMORI and SZENTÁGOTHAI, 1966b; MUGNAINI, 1972). Few of these appendages, however, appear in Golgi preparations even when examined with high magnification lenses. Only rarely can one see a slender filiform process or a bulbous projection sticking out from the dendrite. At the electron microscope level it can be seen that these appendages are ectoplasmic extensions occupied by a fine, fibrillar matrix in which are suspended a few contorted tubules or cisternae of the agranular endoplasmic reticulum. Occasionally a mitochondrion finds its way into one of them. Close to the cell body the dendrites pass through glomerular fields in which they articulate with either climbing fibers (Fig. 115) or mossy fiber profiles. Synaptic complexes appear along the shaft of the dendrite or at the bases of the appendages. The dendritic trunks of the large Golgi cells form similar synapses, as has already been mentioned.

The axons of the small Golgi cells cannot be distinguished from those of the large Golgi cells.

4. Summary of Synaptic Connections of Golgi Cells

Afferents	mossy fiber	————	Golgi cell
	climbing fiber	————	Golgi cell
	Purkinje cell	————	Golgi cell
	granule cell	————	Golgi cell
	basket cell	————	Golgi cell
	stellate cell	————	Golgi cell
Efferent	Golgi cell	————	granule cell

Chapter V

The Lugaro Cell

1. A Little History

In 1894 LUGARO described a large cell lying in the upper granular layer of the cat and generating a peculiar axonal arborization that runs longitudinally, with the parallel fibers in the lower molecular layer. Because the axonal plexus was located more in the molecular than in the granular layer, LUGARO named this neuron the intermediate cell. Despite many attempts, RAMÓN Y CAJAL (1911) never succeeded in impregnating such an axon in his own Golgi preparations. Nevertheless, he considered Lugaro's cell to be one of those horizontal fusiform cells that he classed among the large stellate (Golgi) cells of the granular layer. Now, as mentioned in the previous chapter, GOLGI (1874 and 1883) had originally described and figured these horizontal cells, but he did not differentiate their axonal pattern from that of the more common globular or polygonal cells that have become known by his name. His illustration does not depict the axon beyond its origin. RAMÓN Y CAJAL (1911) traced the axon of the horizontal cell downward through the granular layer to the white matter. In some instances ascending collaterals were given off to the molecular layer. LUGARO described neither the cell body nor the dendrites of his intermediate cell. His figures all show the cell only in the plane of the axonal arborization, that is, in the longitudinal plane. They display a triangular cell body with rather short, radiating dendrites, some of which ascend into the molecular layer. Hence, it was merely a presumption on the part of RAMÓN Y CAJAL to equate Lugaro's cell with the horizontal fusiform cells. This is all the more evident when one notices that LUGARO specifically included a description of the horizontal fusiform cell in his discussion of the large cells in the granular layer and introduced the description of his intermediate cell with the remark that it is an additional, previously unknown cell type. Neglecting these considerations, most later authors have accepted Cajal's classification of Lugaro's cell.[7]

Probably no student of cerebellar cytoarchitecture saw Lugaro's cell again until FOX and BERTRAM in 1954 found a cell in the cerebellar cortex of the monkey that emitted an axon matching Lugaro's description and figures. In 1959 FOX presented a fuller account of his re-examination of the intermediate cells of Lugaro. He found them to be elongated, fusiform cells, oriented transversely in the folium and located immediately beneath the Purkinje cell layer. Their long dendrites issue from opposite poles of the cells, and their axons ascend obliquely into the molecular layer where they give rise to extensive collaterals, which course longitudinally among the parallel fibers in the lower part of this layer. Although FOX could not follow their axons for more than a short distance in transverse (parasagittal) sections, in no case did he find horizontal cells with axons directed downward toward the white matter as RAMÓN Y CAJAL had described. FOX, therefore, left open the question of whether all fusiform horizontal cells are intermediate cells of Lugaro or whether there are several varieties of horizontal cells (according to the distribution of their axonal plexus) among which one is the Lugaro cell. For our present purpose it is only necessary in this regard to record that Fox's observations provide the evidence that the Lugaro cell (recognized by its location and axonal arborization) is a fusiform horizontal cell.[8]

FOX also had the good fortune to make observations concerning the synaptic connections engaged in by the fusiform horizontal cells. On the one hand he found that their horizontally directed dendrites are in contact with the pinceaux formed by basket cell axon collaterals beneath the Purkinje cell bodies, whereas their descending dendritic branches come into synaptic relations with both mossy fiber rosettes and Golgi cell axons. On the other hand he found that collaterals from the long axonal

7 ESTABLE (1923) expressed the opinion that Lugaro's cell was nothing but a slightly displaced basket cell, and Fox (1959) thought that certain of Lugaro's figures were indeed incompletely impregnated, fusiform basket cells.

8 We are grateful to Dr. Fox, who kindly provided an opportunity for one of us (SLP) to examine the original sections on which his account was based. They showed precisely what he described and illustrated, although, of course, the interpretation of synaptic relationships of the dendrites must be revised in the light of our electron microscopic findings.

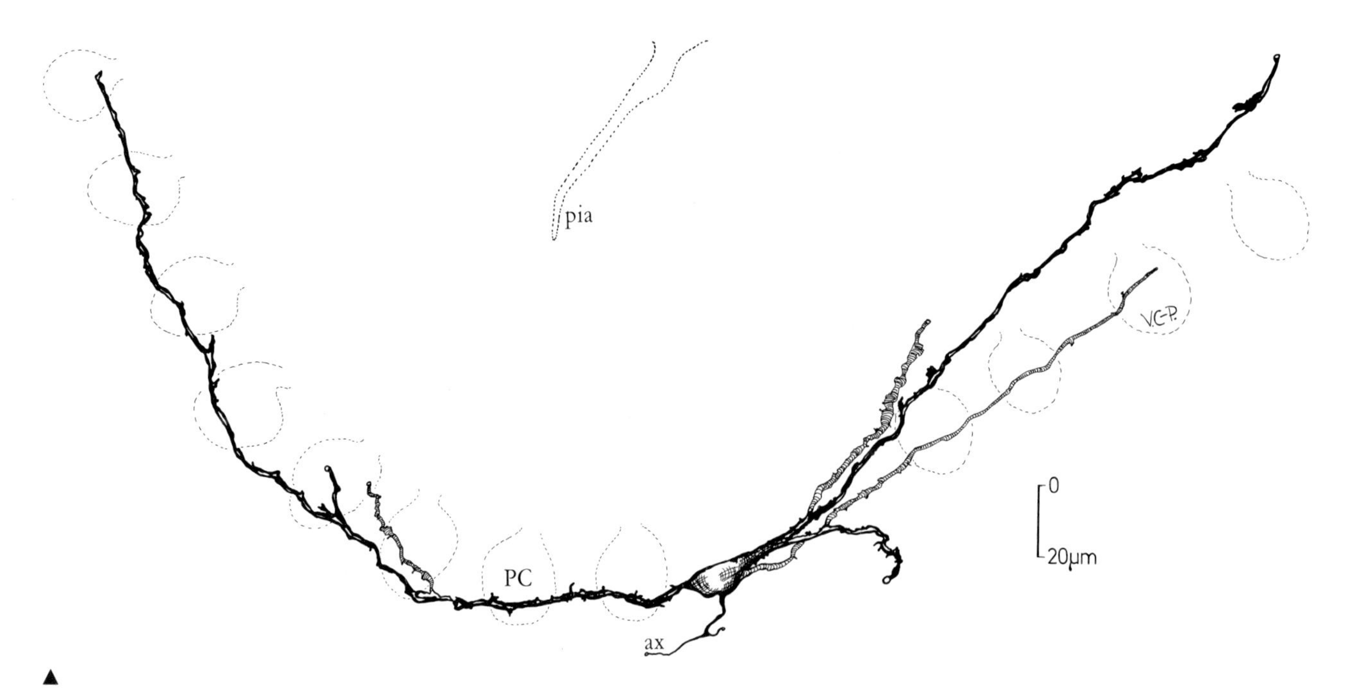

▲

Fig. 116. Lugaro cell at the base of a folium. The long, fleshy dendrites of this cell issue from both poles of the elliptical cell body. They follow the curve of the folium as indicated by the layer of Purkinje cells (*PC*). The axon (*ax*) of this cell descends into the granular layer and is lost to view. The pial surface of the folium is shown (*pia*). Rapid Golgi, 90 µm, sagittal, adult rat. Camera lucida drawing

Fig. 117. Lugaro cell with dendrites and descending axon. The dendrites issue from the fusiform cell body to course in the sagittal plane of the folium at the level of the Purkinje cell layer (*PC*). The axon descends into the granular layer. Rapid Golgi, 90 µm, sagittal, adult rat. Camera lucida drawing

▼

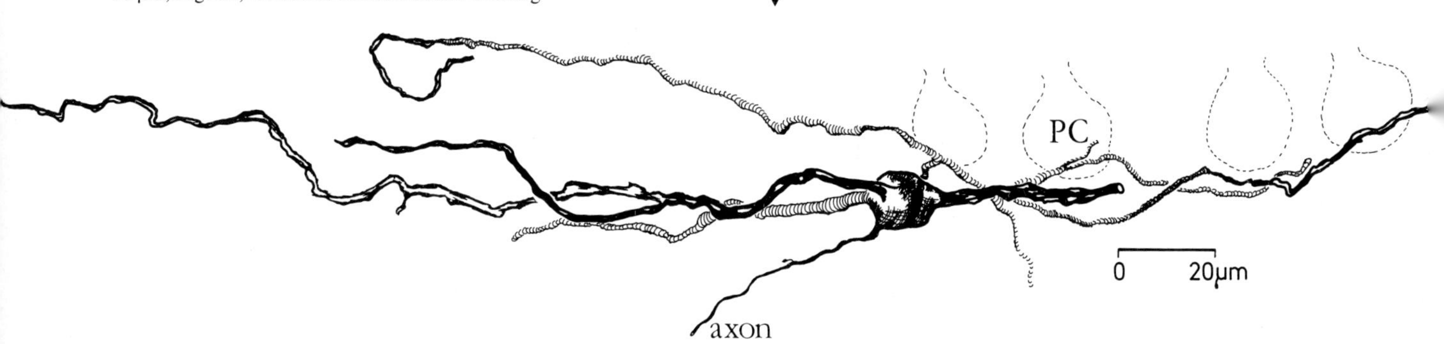

arborization contacted the cell bodies of basket cells in the lower molecular layer. He considered it likely that the Lugaro cell should receive afferents from the recurrent collaterals of Purkinje cell axons since they ramify in the infraganglionic level that the Lugaro cell dendrites occupy.

In our own Golgi preparations we have not yet encountered the axonal arborization described by LUGARO and by FOX. This failure may be a consequence of an unfavorable plane of section. As FOX has remarked, the axons usually can be followed only for a short distance. We have seen, however, a few examples of horizontal fusiform cells, the axons of which ascend into the deepest molecular layer and ramify among the perikarya of the Purkinje cells and their primary dendrites (Figs. 119 and 124). These obviously incomplete arborizations are cut off by the plane of section. They might be the stems of LUGARO'S axonal plexus. In addition, we have seen horizontal fusiform cells that give rise to axons directed downwards through the granular layer and reaching the white matter in the center of the folium, just as RAMÓN Y CAJAL (1911) described and figured them (Figs. 116–118). O'LEARY *et al.* (1968) described and pictured a number of Lugaro cells with descending axons. FOX (1959) did not encounter such axons. There are, therefore, at least two kinds of horizontal fusiform cells, indistinguishable as far as their perikarya and dendrites are concerned, but possessing different axonal patterns and distributions. Since the axons run in opposite directions, the cells may well have quite different connections. A more thorough

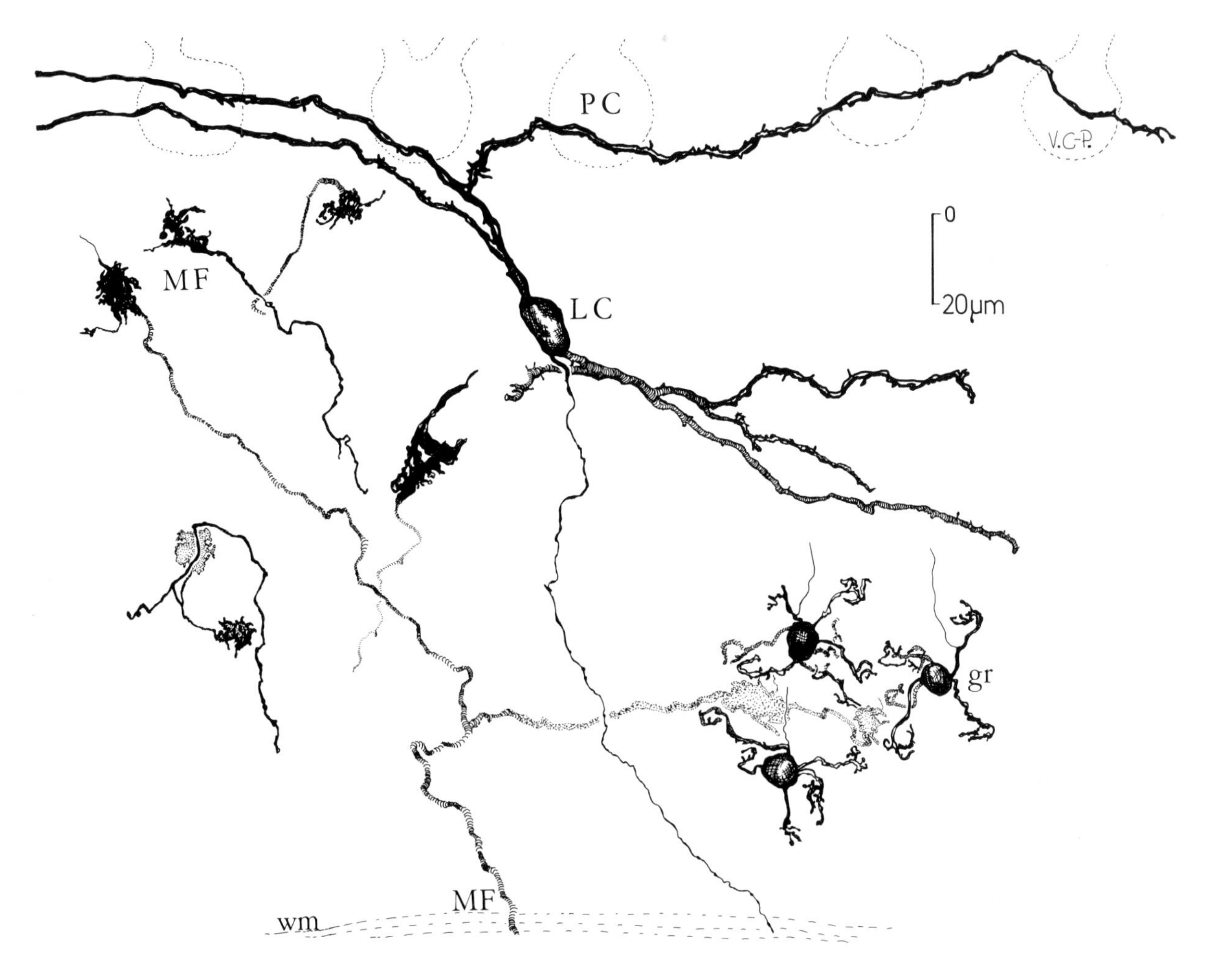

Fig. 118. Lugaro cell with long descending axon. This Lugaro cell lies with its elliptical cell body (*LC*) athwart the upper granular layer. The long fleshy dendrites rise to the level of the Purkinje cell layer (*PC*) and travel in the sagittal plane. Other dendrites reach into the depths of the granular layer. The axon courses through the granular layer to reach the white matter. Several mossy fibers (*MF*) are drawn, one rosette of which is shown in relation with granule cells (*gr*). Rapid Golgi sagittal, 90 μm, adult rat. Camera lucida drawing

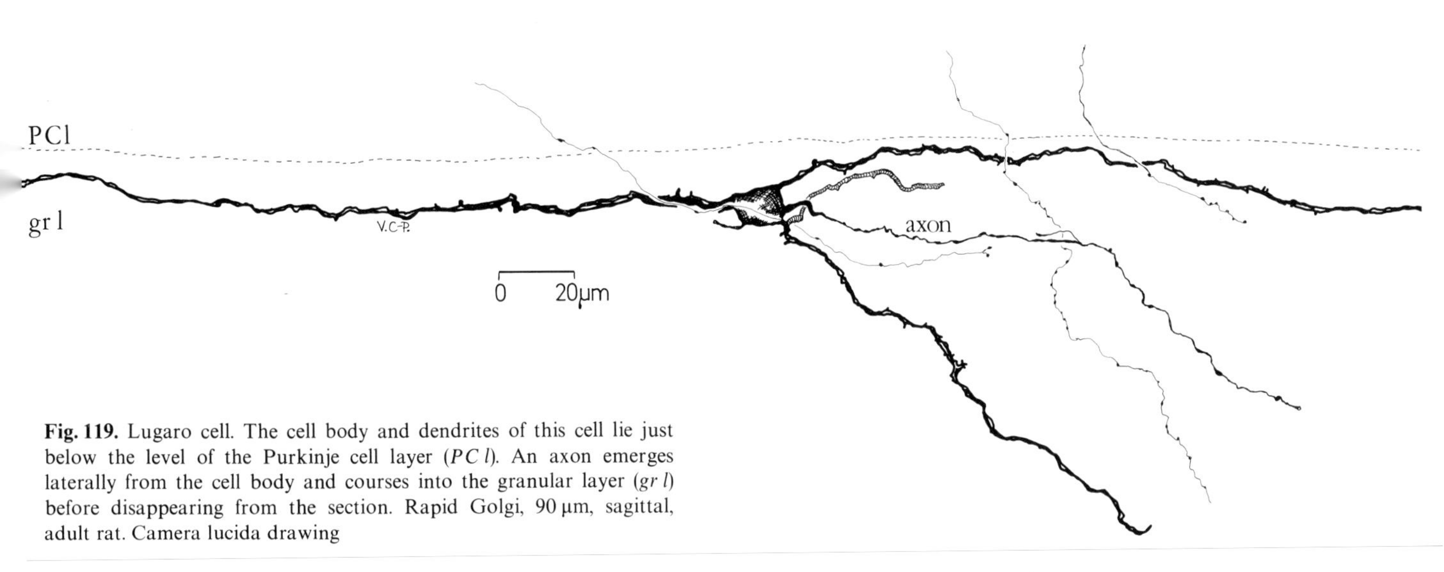

Fig. 119. Lugaro cell. The cell body and dendrites of this cell lie just below the level of the Purkinje cell layer (*PC l*). An axon emerges laterally from the cell body and courses into the granular layer (*gr l*) before disappearing from the section. Rapid Golgi, 90 μm, sagittal, adult rat. Camera lucida drawing

study of Golgi preparations is, however, necessary in order to characterize the axons of both cell types. Until that has been done, these cells must all be considered together in this description, since neither by impregnation techniques nor by electron microscopy have we found any way to distinguish between the perikarya or dendrites of the horizontal fusiform cells that lie beneath the Purkinje cell layer.

2. The Lugaro Cell in the Light Microscope

In sections stained with toluidine blue the horizontal cell appears as a creased, ellipsoidal nucleus with its major axis lying parallel to the pial surface. The cells are located in the outermost third of the granular layer, usually just underneath the sheet of Purkinje cell bodies. Little cytoplasm is discernible, and the Nissl substance is scanty and diffuse.

The cell is much more impressive in Golgi preparations. As seen in fortunate parasagittal sections (Figs. 116–119), across the axis of the folium, the cell extends its dendrites horizontally for seemingly vast distances an either side of its fusiform perikaryon. In some instances the cell body lies obliquely or almost vertically athwart the granular layer. Its dendrites still originate from the two poles of the cell and rise toward the level of the Purkinje cell bodies, where they spread out. Fox (1959) illustrates such a cell, as do O'Leary *et al.* (1968), and one is shown in Fig. 118. As Golgi (1883) mentioned, it is difficult to ascertain the length of the cell body because the thick dendrites emerge from its gradually tapering extremities. The perikaryon is 9 or 10 μm in diameter at its thickest, and about 25 or 30 μm in length. The dendrites run in an arc that follows the curve of the folial surface (Fig. 116). They are remarkably straight, considering the distance they span, but they have small crooks and twists along their course. Although the whole dendritic expansion cannot be contained within a single section, it is not unusual to find examples extending for 300 to 600 μm from one end to the other without reaching their true terminations. One major dendrite is given off from each end of the perikaryon, and other smaller dendrites can also arise from the middle portions of it. Most of the branches originate from bifurcations at a considerable distance from the cell body, and they generally run in the same horizontal plane as the major dendrites, rising only gently into the molecular layer. Both trunks and branches are fitted with bent or twisted thread-like appendages or with thin spines terminating in a small bulb. The dendrites thus lie in the infraganglionic plexus in the midst of the terminal arborization generated by the recurrent collaterals of the Purkinje cell axons (Figs. 43 and 124). As Fox has pointed out, there should be connections between these collaterals and the horizontal cell dendrites, and electron microscopy demonstrates that these are the most numerous connections formed by these cells.

The dendrites also are in a position to contact collaterals of basket cell axons, especially the distal tips that make up the pinceau formation surrounding the initial segment of the Purkinje cell axon. Fox (1959) found that the descending collaterals of basket axons converge on the horizontal cells and their dendrites. In this case, however, electron microscopy shows that the contacts are only apparent, as the dendrites pass through the pinceaux or their periphery without engaging in synaptic articulation with these axons. Finally, the mossy fibers, especially those terminating in the upper third of the granular layer, are in a position to contact the horizontal cell body or its dendrites, and Golgi preparations can be cited to substantiate the possibility (Fox, 1959). But this articulation is also not borne out by electron microscopy.

The axons of the horizontal cells take origin from the cell body (Figs. 116–119) or from a large dendrite near the cell body (Fig. 124). Those that pass into the molecular layer ascend obliquely among Purkinje cell bodies and soon ramify in all directions, but with a distinct preference for the horizontal plane within the supraganglionic plexus. Our observations are inadequate to make com-

▶

Fig. 120. The Lugaro cell body. This electron micrograph shows the Lugaro cell with a typical lobulated nucleus (*LC*) and a prominent nucleolus (*ncl*). The cell bears a synapse (*arrows*) with a terminal of a Purkinje cell recurrent collateral. A Purkinje cell (*PC*), a Golgi epithelial cell (*Go ep*), and basket axons share the field with the Lugaro cell. Lobule IV. ×14000

Fig. 121. The dendrite of a Lugaro cell synapsing with two boutons of the recurrent collateral plexus. The bouton Prc_1 shows a single, fairly long synaptic complex (←). The bouton Prc_2 shows a series of three short synaptic complexes (→). These junctions are typical of those made by recurrent collateral terminals, with the widened interstitial cleft, containing a dark material, and shallow, symmetrical pre- and postsynaptic densities. Bouton Prc_2 contains a dark proteinaceous clump (⇐). In the lower right hand corner a recurrent collateral bouton synapses on the soma of a Purkinje cell (*PC*). Lobule V, vermis. ×25000

Fig. 122. Cytoplasmic inclusion in a Lugaro cell. The perikaryon of this Lugaro cell contains a cytoplasmic filamentous inclusion (*arrow*), and bears two large terminals of Purkinje axon recurrent collaterals (*Prc*) and a parallel fiber (*asterisk*) on its surface. Lobule X. ×25000

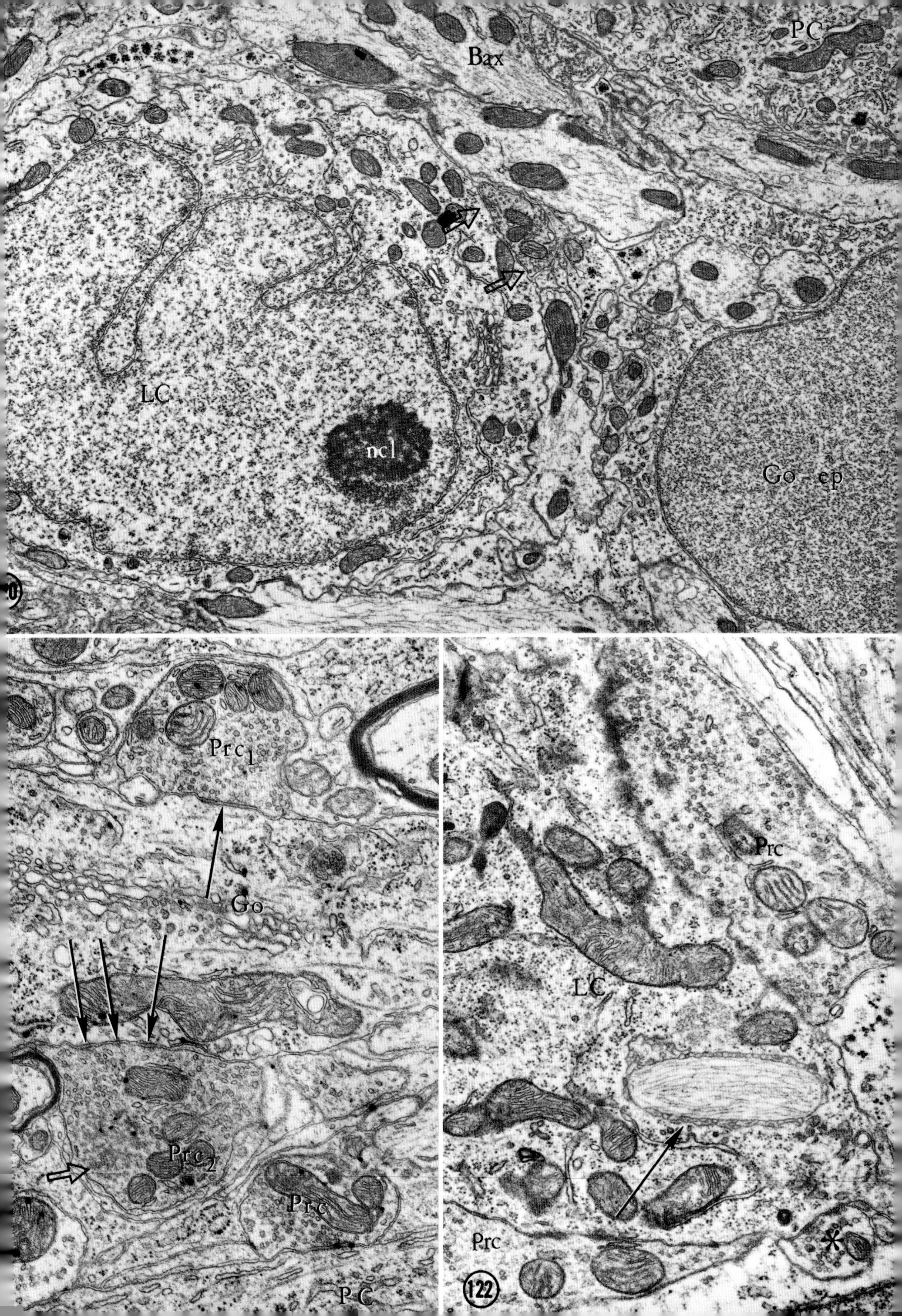

Bax
PC
LC
ncl
Go - ep
Prc1
Go
Prc2
Prc
PC
Prc
LC
Prc
*
122

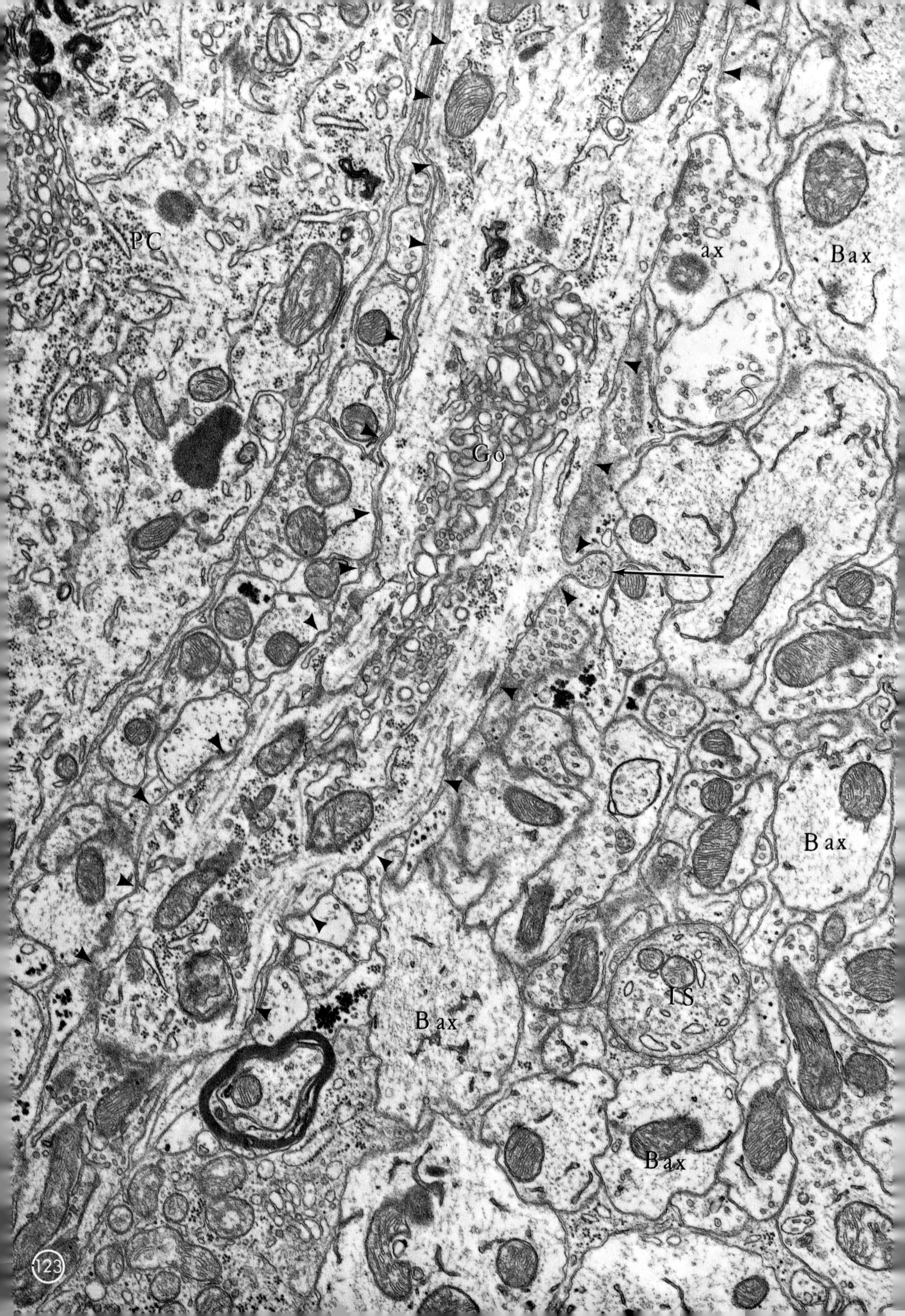
PC
ax
Bax
Go
Bax
Bax
IS
Bax
123

parisons with either LUGARO's or FOX's descriptions and figures, except to say that in the usual preparations the axons cannot be followed for more than a few tens of microns before they go out of the plane of section. (The Lugaro cell in Fig. 118 is unusual in our material, as its axon descends for 150 μm before passing out of the section.) The branches that are emitted in these short stretches appear tenuous and varicose. According to FOX (1959), these ascending axons send varicose offshoots to end on the perikarya of basket cells. As we have not yet been successful in identifying the axons of the horizontal cells in electron micrographs, we cannot verify this description or add any new information. The descending axons of horizontal cells appear much smoother and more robust than the ascending ones (Figs. 116–118). They descend in an arc through the granular layer, giving off few collaterals, and disappear in the white matter. Again, we have not been able to find the destination of these fibers.

3. Fine Structure of the Lugaro Cell

In electron micrographs of parasagittal sections, the horizontal cell can be recognized as an ellipsoidal perikaryon lying beneath the Purkinje cell layer. The perikaryon measures 7 or 8 μm across, and is occupied almost completely by an elongated nucleus, which contains a large, prominent nucleolus [1.5–2.0 μm in diameter (Fig. 120)]. The chromatin is homogeneously dispersed except for some small marginal masses near the nuclear envelope and around the nucleolus. Characteristically, the nuclear envelope is deeply indented and creased. These invaginations are filled with cytoplasm bearing free ribosomes in rosettes, helical rows, and other polysomal arrays. Ribosomes are also attached to the cytoplasmic, outer surface of the nuclear envelope, forming long strings at wide intervals.

◄

Fig. 123. Dendrite of a Lugaro cell. This fleshy dendrite of a Lugaro cell (▲▲) contains all of the usual cytoplasmic organelles as well as a well-developed Golgi apparatus (*Go*). This is invariably found in proximal dendrites of Lugaro cells. The dendrite emits a spiny projection (*arrow*), which exhibits no synaptic contacts in this section. The dendrite is flanked by the soma of a Purkinje cell on the left (*PC*) and the axons of basket cells (*B ax*) on the right. The pinceau formed by basket axons encircles the initial segment of the Purkinje axon (*IS*), seen here in cross section. Lobule V. ×26500

The perikaryal cytoplasm, typically more voluminous at both poles of the nucleus than at its equator, contains all of the organelles usual for neurons. The granular endoplasmic reticulum appears as stringy profiles ramifying between the nuclear envelope and the plasmalemma. Occasionally, ordered arrays coalesce out of the usual random arrangements, forming distinct Nissl bodies composed of three to five imbricated cisternae that are stretched out parallel to the plasmalemma. Free ribosomes are dispersed in clusters and rows all over the cytoplasm. The Golgi apparatus encircles the nucleus with an open meshwork, the nodal points of which comprise four or five curved, fenestrated, and closely apposed cisternae surrounded by clouds of small vesicles. Round, irregular, or sausage-shaped lysosomes together with multivesicular bodies are also found in the vicinity of these nodes. A few dense-core vesicles occur in the periphery of the Golgi complexes. The mitochondria are generally short and plump with either transverse or oblique cristae. They and all of the other cytoplasmic organelles stream out into the dendrites, which originate from the tapering ends of the perikaryon (Figs. 121 and 123). It is difficult to decide where the dendrite, strictly speaking, begins, since the first part of the process has the same composition as the perikaryon. Gradually, however, as the process becomes narrower, its cytoplasm begins to clear. The longitudinally arranged microtubules, which are inconspicuous and widely separated in the cell body, become more prominent and assume a regular arrangement. The Golgi apparatus disappears, the Nissl substance becomes fragmentary, reduced to discrete collections of free ribosomal clusters around branching cisternae of the endoplasmic reticulum, and the mitochondria become sparser and longitudinally oriented. The core of the dendrite is filled with longitudinal microtubules, fairly regularly spaced. At irregular intervals the dendrites give off small, digitiform processes filled with a fibrous matrix (Fig. 123).

Both the cell body and the dendrites are partially ensheathed by a highly perforate neuroglial envelope, which the preterminal axons and synaptic terminals must penetrate in order to reach the postsynaptic surface (Fig. 120–123). The vast majority of the terminals are varicosities and synaptic end bulbs contributed by the recurrent collaterals of Purkinje cell axons ramifying in the infraganglionic plexus (Figs. 120–124). The terminals cluster against the surface of the horizontal cell perikaryon, each presenting a rather flattened zone of apposition in which a set of small, discrete, macular synaptic complexes occurs. These synaptic junctions are intermediate between Gray's type 1 and type 2. The apposed surface membranes

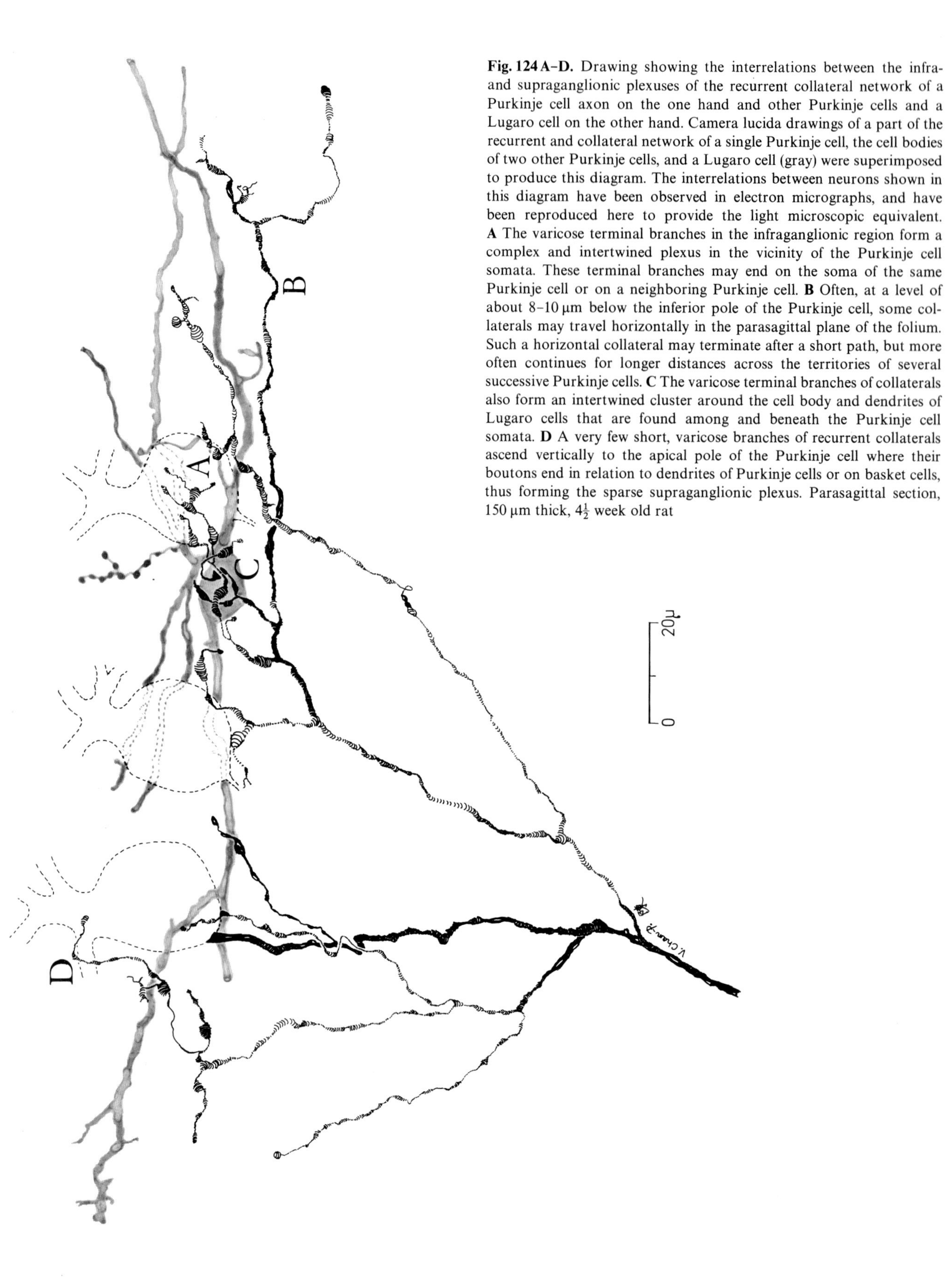

Fig. 124 A–D. Drawing showing the interrelations between the infra- and supraganglionic plexuses of the recurrent collateral network of a Purkinje cell axon on the one hand and other Purkinje cells and a Lugaro cell on the other hand. Camera lucida drawings of a part of the recurrent and collateral network of a single Purkinje cell, the cell bodies of two other Purkinje cells, and a Lugaro cell (gray) were superimposed to produce this diagram. The interrelations between neurons shown in this diagram have been observed in electron micrographs, and have been reproduced here to provide the light microscopic equivalent. **A** The varicose terminal branches in the infraganglionic region form a complex and intertwined plexus in the vicinity of the Purkinje cell somata. These terminal branches may end on the soma of the same Purkinje cell or on a neighboring Purkinje cell. **B** Often, at a level of about 8–10 μm below the inferior pole of the Purkinje cell, some collaterals may travel horizontally in the parasagittal plane of the folium. Such a horizontal collateral may terminate after a short path, but more often continues for longer distances across the territories of several successive Purkinje cells. **C** The varicose terminal branches of collaterals also form an intertwined cluster around the cell body and dendrites of Lugaro cells that are found among and beneath the Purkinje cell somata. **D** A very few short, varicose branches of recurrent collaterals ascend vertically to the apical pole of the Purkinje cell where their boutons end in relation to dendrites of Purkinje cells or on basket cells, thus forming the sparse supraganglionic plexus. Parasagittal section, 150 μm thick, $4\frac{1}{2}$ week old rat

are usually parallel with each other, but the interstitial cleft is widened and shallow densities are arranged symmetrically on the pre- and postsynaptic sides of the junction. Similar synapses occur on the long, fleshy dendrites issuing from the perikaryon. Synapses with the axons of granule cells or parallel fibers on Lugaro cells have also been observed (Figs. 122 and 123).

We have not been able to recognize profiles of the Lugaro cell axon in electron micrographs. Judging from the form of the fibers in Golgi preparations, they should be slender, relatively straight profiles with varicose collaterals in the deep molecular and Purkinje cell layers. These have not yet come to light. Alternatively, the Lugaro cell axons should be coarse fibers proceeding through the granular layer into the white matter. But since we have recognized them only in a few Golgi preparations, we cannot even say whether they are myelinated or not.

4. Summary of Synaptic Connections of Lugaro Cells

Afferents	Purkinje cell	———	Lugaro cell
	granule cell	———	Lugaro cell
Efferents			unknown

Chapter VI

The Mossy Fibers

1. A Little History

RAMÓN Y CAJAL reported his discovery of the mossy fibers in May 1888 in his first paper on the cerebellum of birds. In this paper he referred to these fibers as *fibras nudosas* and described them as follows:

> "These are the most numerous of all those fibers which course through the granular layer. They are characterized by showing from place to place nodular expansions which appear to consist of an irregular accumulation of silver precipitate. When these nodosities are examined in the short, most delicate impregnations, it can be seen that they are true arborizations, short and varicose, which decorate certain parts of the fibers in the manner of moss or brambles on a wall. In many places this granular arborescence is supported by a short, delicate stem like a flower."[9]

Further on in the same paragraph he uses the phrase "*arborizaciones musgosas*" in reference to the collaterals of these fibers. Almost a year later (1889a and 1890a) in a paper on the granular layer of the mammalian cerebellum, he appears to have settled on the name "mossy fibers" («Les branches, que nous appellons mousseuses [sic] à cause de leur ressemblance avec la mousse qui tapisse les arbres ...»—RAMÓN Y CAJAL, 1890a, p. 13), a term that was quickly taken up by KÖLLIKER (1890), VAN GEHUCHTEN (1891), RETZIUS (1892a and b), and DOGIEL (1896), all of whom confirmed his findings. Apparently RAMÓN Y CAJAL did not feel that the name was self-explanatory, as in 1893 in a general account of the newer discoveries in the nervous system he again comments that the mode of formation of these fibers resembles that of the mossy covering of trees.[10] In 1911 he explains once more in detail:

> "But these fibers display a much more interesting detail, one that has earned them the name 'mossy' by which we made them known. Along their trajectory and at their extremity a sort of thickening appears that is serrated, or rather, bristling with radiating branchlets, short, thick, varicose, and moderately ramified. The whole gives the impression of a fragment of moss."[11]

The image remains a bit abstruse, even today after more than 80 years of general acceptance and unquestioned usage. Careful examination of methylene blue and silver-stained preparations (DOGIEL, 1896; RAMÓN Y CAJAL, 1911, 1926; CRAIGIE, 1926) does not suggest the velvety covering of moss-grown trees so much as coiled, unravelled loops of yarn. KÖLLIKER (1890) regarded the "mossy" thickenings and appendages as probable artifacts of the Golgi method, and DOGIEL (1896) was of the opinion that the reagents used in the Golgi method cause precipitates to form in the interstices between the thread-like rami of the fiber and thus thicken the fiber so that rosettes appear along its course. Undoubtedly, over-impregnation with silver salts does contribute to the mossy appearance of the fibers. Nevertheless, the rosettes

9 «Fibras nudosas. — Son las más recias de todas las que marchan por la capa granulosa y se caracterizan por presentar, de trecho en trecho, unos abultamientos nudosos que se diría estén constituídos por un acúmulo irregular de plata precipitada. Examinadas dichas nudosidades (Fig. 2a), en los cortex más finamente impregnados, se echa de ver que son verdaderas arborizaciones, cortas y varicosas, que guarnacen ciertos parajes de las fibras a la manera de un musgo o maleza de revestimiento. En muchos sitios, esta arborizatión granulosa está sostenida por un tallo corto y delgado que tiene el aspecto de una flor (Fig. 2a, F).» Estructura de los centros nerviosos de las aves. Revista Trimestral de Histología Normal y Patológica, No. 1, 1888a. p. 312 in Trabajos escogidos, vol. 1. Madrid, JIMÉNEZ Y MOLINA, 1924.

10 „Sie (mossy fibers) sind dadurch in charakteristischer Weise ausgezeichnet, daß ihre Zweige von Zeit zu Zeit bolbenartig, nach Art einer Rosenknospe, anschwellen, von welcher Stelle dann kurze rauhe Zweige abgehen, eine Bildungsweise, die der Moosbekleidung der Bäume ähnlich erscheint." RAMÓN Y CAJAL, 1893, Arch. f. Anatomie und Entwickelungsgeschichte, p. 345. This paper was a review of Cajal's work originally published in Spanish, and translated into German by HANS HELD at the request of WILHELM HIS.

11 «Mais ces fibres présentent un détail bien plus intéressant, celui-là même qui leur a valu le nom de *moussues*, sous lequel nous les avons fait connaître. On voit, en effet, sur leur trajet et à leur extrémité, des espèces d'épaississements découpés en dentellures, ou plutôt hérissés de ramuscules épais, courts, divergents, variqueux et ramifiés modérément. Le tout donne l'impression d'un lambeau de mousse» (RAMÓN Y CAJAL, 1911, p. 57).

or mossy appendages represent real specializations at the extremities and along the course of these fibers, and nothing is to be gained by altering their time-honored name.

2. The Mossy Fiber in the Light Microscope

Mossy fibers are thick, heavily mylinated nerve fibers that enter the granular layer from the white matter in the center of each folium. During its course through the white matter, each mossy fiber divides many times, giving off as many as 20 or 30 collaterals to the granular layer as it ascends towards the summit of the folium (RAMÓN Y CAJAL, 1889a, 1890a, 1911). Some fibers bifurcate deep in the white matter, and their resultant branches serve two or more adjacent folia. In this way each mossy fiber distributes to a large field of cortex which overlaps extensively with that of its neighbors. As can be seen in well-impregnated Golgi preparations (Figs. 125a, b, and 126), the preterminal segments of the mossy fibers radiate from the central sheet of white matter into the granular layer, each fiber crooking its way among the granule cells as it approaches its terminal field. After entering the granular layer, the collaterals and the distal portion of the parent trunk itself bifurcate a few times, forming a loose terminal arbor.

Like their parent trunks in the white matter, these preterminal portions of the mossy fibers within the granular layer are also myelinated. This characteristic was first recognized by DOGIEL (1896), who saw and illustrated the crosses of Ranvier, which mark the nodes of myelinated fibers in methylene blue preparations. Originally RAMÓN Y CAJAL (1889a and 1890a) thought it possible that a great many of the myelinated fibers in the granular layer were provided by the mossy fibers, but by 1911 he had arrived at the erroneous conclusion, which he expresses categorically, that none of the branches of the mossy fibers is myelinated and that the myelin sheath ceases soon after the main trunk leaves the white matter. He had been unable to repeat DOGIEL's observations with the methylene blue method, and he thought that DOGIEL had made an error, since mossy fibers are easy to impregnate by the rapid Golgi method, even in adult animals, and such a result never occurs with myelinated fibers. By 1926, however, RAMÓN Y CAJAL had changed his mind again, and in his last paper on cerebellar histology (p. 222) he simply affirmed what DOGIEL had stated 30 years before. The presence of the myelin sheath covering the preterminal portions of the mossy fibers is readily confirmed by electron microscopy.

The terminal or synaptic portion of the mossy fiber is precisely that formation which gives the fiber its name. In keeping with his image, RAMÓN Y CAJAL referred to them as arborizations, seeing the mossy region not simply as an irregular swelling but as a set of discrete processes, emitted from the fiber along its course and at its tip (Fig. 125c). He discriminated three locations for these arborizations: (1) along the course of the fiber in the granular layer (*arborisations collatérales*), (2) at branching points (*arborisations de bifurcation*), and (3) at the actual terminals of the fiber (*arborisations terminales*) (Fig. 126). The arborizations in the course of the fiber are the most frequent, as might be expected. The myelin sheath is interrupted at the sites of these formations, and in many instances it begins again beyond them. In effect, the formation is developed in an extended node of Ranvier. This arrangement can be seen in thin plastic-embedded sections stained with toluidine blue, as well as in more conventional preparations for the light microscope, and it has been confirmed by electron microscopy. In all locations the form of these arborizations is more or less the same. According to RAMÓN Y CAJAL (1911), the fiber thickens and sends out a small number of thick, varicose, rarely branched processes, each terminating in a large, round granule. Sometimes the formation is limited to a few bulging varicosities; in other instances it consists of multiple processes from one or both sides of the fiber. One of the most interesting forms of the arborization occurs at the terminals of the mossy fibers under the layer of Purkinje cell bodies. Here, as RAMÓN Y CAJAL (1911) remarks, the fiber turns parallel to or away from the surface of the folium and bursts into a number of long processes ending in a bouquet of small spherules.

Our own rapid Golgi preparations of the rat's cerebellar cortex show the mossy fibers as thick, somewhat contorted fibers running through the granular layer. Entering the cortex from the central white matter, they branch off along the sides of the folium and fan out at its summit (Figs. 125a, b, and 126). As they ascend through the granular layer, the fibers ramify extensively. Although the thickness of individual fibers varies considerably along their course, in general two classes of fibers can be distinguished, one about 1.5 µm and the other about 0.4 µm in diameter. It is possible that at least some of the thinner fibers are terminal twigs of the larger ones. Both thick and thin fibers give rise to the same varieties of mossy excrescences (Figs. 126 and 127). These characteristic formations, to which the fibers owe their name, occur at the ends and along the course of

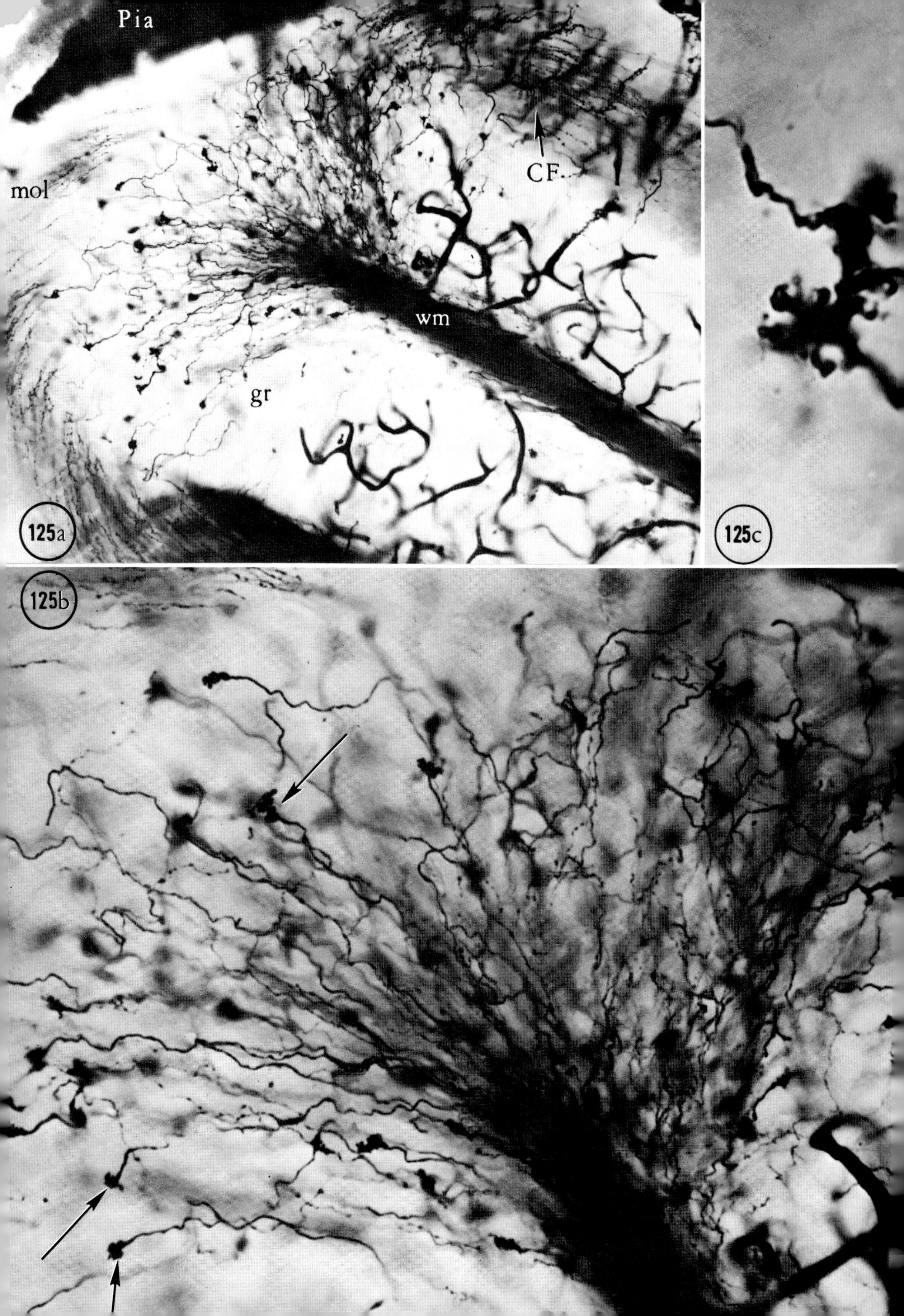
Pia
mol
CF
wm
gr
125a
125c
125b

the fiber at intervals of 20 to 80 μm or more, and involve sections of axon from 7 to 22 μm long. Most of them are in the neighborhood of 10–15 μm long. As the fiber enters the mossy section, it gradually distends to form an irregularly spindle-shaped, tuberous enlargement from which numerous short, bent, and sometimes branching processes extend in all directions. The axis of the fiber passing through these enlargements is sometimes fairly straight and sometimes uncinate or more contorted.

Although the variety displayed by the mossy formations is so enormous that classification can be no more than a descriptive guide, three kinds can be recognized. The first kind is a fairly simple fusiform enlargement with globular contours and a few short, finger-like projections (Fig. 127). The second has a more complicated shape with a massive central expansion out of which several twisted and angular, tapering, or filiform appendages project (Figs. 125c, and 126). Both of these are most common along the course of the fiber, but both, especially the second, also occur at terminals. The third variety is a much more open and graceful formation. It occurs particularly at terminals, most commonly in the outer third of the granular layer and just beneath the Purkinje cell layer (Fig. 126). The central portion of this formation is an elongated cone or tuber that expands into a curiously flower-like figure by sending out recurving tubular and filamentous processes. These appendages are often thin threads terminating in a spherical knob. Terminal inflorescences like these deserve the name "rosettes" much more than the majority of the formations to which that name is fancifully applied in the literature. It is worth noting that, as can be seen in the drawings of Golgi preparations (Figs. 126 and 127), any particular mossy fiber exhibits either simple or complex mossy formations. It would, therefore, be conceivable that fibers from a particular source, such as spinal cord or pons, might have a specific pattern. We have attempted to pursue this idea by some experimentation (see section on experimental lesions: Figs. 150–157), but the results obtained have unfortunately been only limited.

So far as we have been able to discern, the three varieties of mossy fiber terminal occur with about equal frequency in all lobules of the cerebellar cortex. Brodal and Drabløs (1963), in a frequently cited paper, reported that a large proportion of the mossy fiber terminals in the posterior lobules appear to be different from those in the rest of the cerebellum. They studied material from adult rats and cats prepared by two silver impregnation methods, and material from very young rats (2–6 days) prepared by the rapid Golgi method. In the nodulus, flocculus, ventral uvula, and ventral paraflocculus, they found most mossy fiber terminals to be more massive and coarser than they were in the other lobules. We have not seen anything like this in our material, taken from adult rats, but we have noted that mossy fibers during the first postnatal week are poorly developed throughout the cerebellum, as few granule cells have descended into the internal granular layer, while the axonal plexus of the Golgi cells is already quite elaborate (see Fig. 206 for an illustration of mossy fibers at the ninth postnatal day).

The characteristics of the Golgi impregnation may result in a somewhat coarser appearance than these structures should have. As mentioned earlier, Dogiel (1896) thought that the metallic precipitate filled in the interstices and obliterated the delicate details that are visible in methylene blue preparations. Recent studies of Golgi preparations in the high voltage electron microscope indicate that the superimposition of silver globules attached to the surfaces of nerve processes and appendages can distort their true shape, but that impregnation with red silver chromate follows quite closely the configuration of the cells and processes (Chan-Palay and Palay, 1972b).

Fig. 125. a Mossy fibers in a folium. A spray of mossy fibers emerges from the white matter (*wm*) to enter and ramify in the granular layer (*gr*). Several climbing fibers (*CF*) are also visible in the molecular layer (*mol*). The pial surface is indicated (*pia*). Rapid Golgi, adult rat, sagittal, 90 μm, dorsal paraflocculus. ×225. **b** A spray of complex mossy fibers. This photomicrograph shows the same field as Fig. 125a at a higher magnification and a different focal level. The mossy fibers here give rise to many rosettes, all of which are of the complex variety (*arrows*). Rapid Golgi, adult rat, sagittal, 90 μm, dorsal paraflocculus. ×530. **c** A complex mossy fiber rosette. The fiber enters the illustration in the lower right hand corner, enlarges into a complicated rosette before continuing out of the field at the upper left. Rapid Golgi, adult rat, 90 μm, sagittal view. ×2700

Fig. 126. The complex mossy fiber. These mossy fibers have been drawn from a single section of a folium in a rapid Golgi preparation. The fibers issue from the white matter (*wm*) and reach out through the granular layer, giving rise to varicosities and rosettes along their length until the Purkinje cell layer (*PC*) is reached. Each complex mossy fiber arcs gently in its course and gives rise to increasingly more complex rosettes. The terminal rosettes (*arrows*) are usually the most complicated and the largest and have been termed the filigree type. Rosettes are also formed at the bifurcation points (*asterisks*) of mossy fibers. Only one truly simple mossy fiber (*crossed arrow*, left hand panel) is shown in this drawing. Rapid Golgi, lobule IX, 90 μm, sagittal, adult rat. Camera lucida drawing

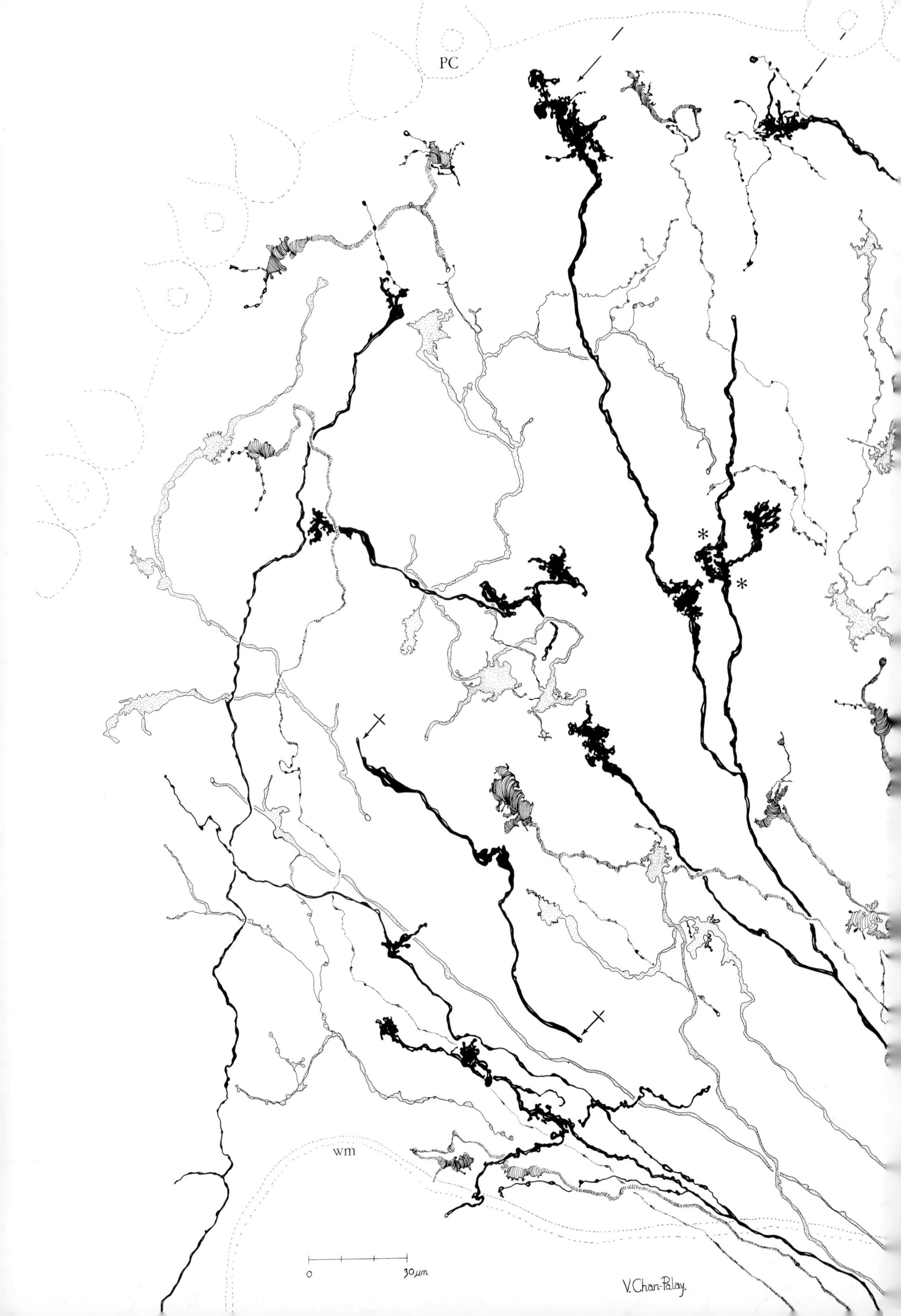

PC
wm
0
30 µm
V. Chan-Palay.

PC
*
*
*
wm

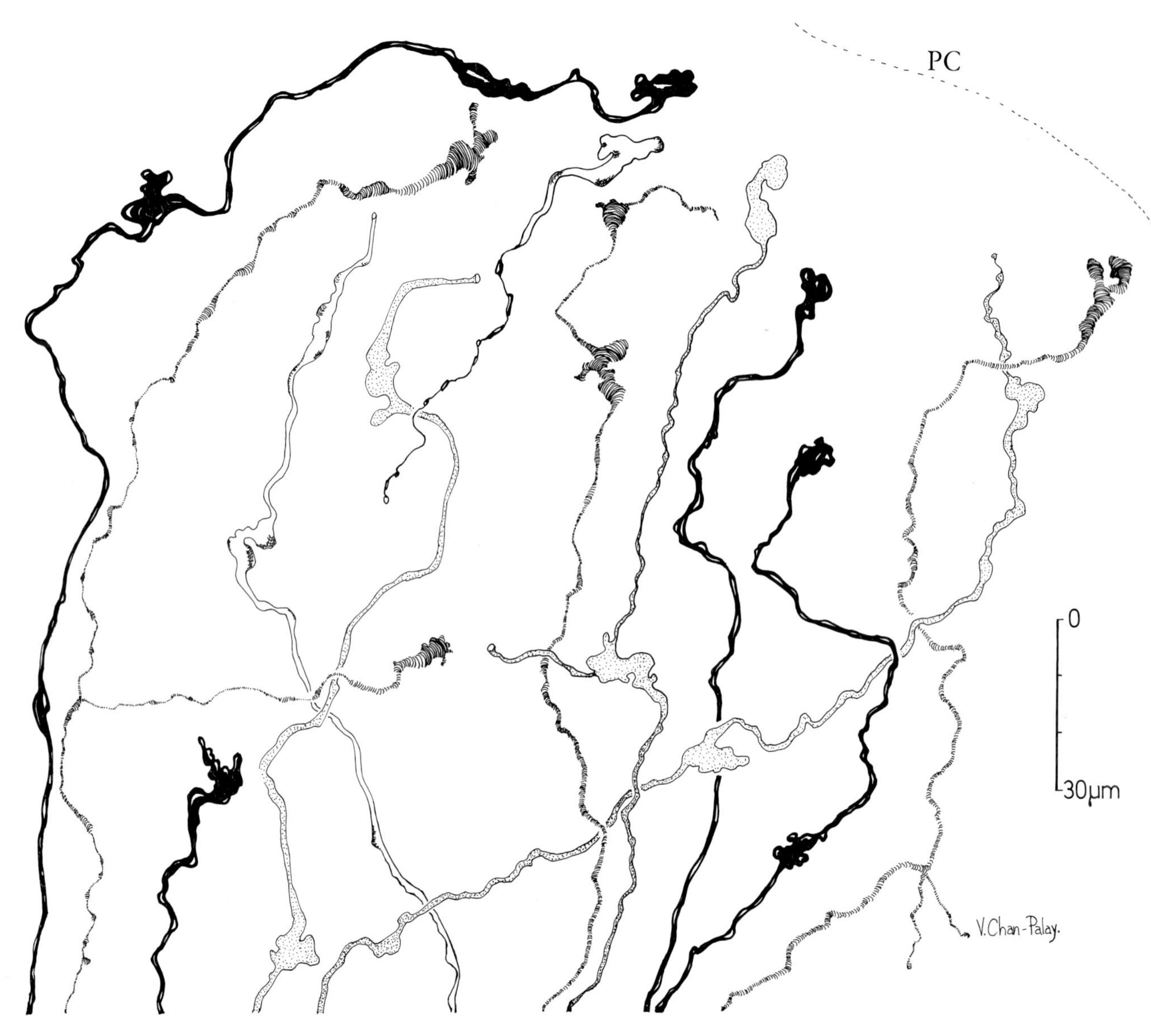

Fig. 127. Simple mossy fibers. Mossy fibers with simple varicosities or rosettes are illustrated here. Besides simple outlines in their varicose portions, these mossy fibers also generally exhibit less complex branching patterns. Simple mossy fibers like these usually give rise only to simple varicosities along their length. Compare this with Fig. 126. Rapid Golgi, composite camera lucida drawing, 90 μm, sagittal, adult rat

With neurofibrillary stains still another appearance is manifested. In these preparations (Ramón y Cajal, 1908, 1911, 1926) the mossy fiber appears as a thick cable of more or less parallel neurofibrillary strands which, after several twists and turns, enters the mossy section. Here it tends to spread out somewhat, or unravel, and various nubbins and short collateral branches are produced in the cable by the folding and duplication of a few strands. Also, longer branches are given off from the axial cable that are filled with a fibrillary network and terminate in a ring of neurofibrils. As Craigie (1926) pointed out, when the mossy fibers in methylene blue preparations and neurofibrillary stains are compared, it is seen that mossy fibers in the silver preparations are more slender, more convoluted, and less branched. The same conclusion results from a comparison of Golgi preparations with neurofibrillary stains (for a more modern comparison see Brodal and Drabløs, 1963). The difference is due to the facts that the neurofibrillary cable lies approximately in the axis of the fiber, that it tends to be less taut when it passes through the mossy expansions,[12] and finally, that it sends branches only into a few of the

12 The convolutions of this neurofibrillary cable have been erroneously represented in diagrams as the form of the whole mossy fiber in a recent publication (Llinás and Hillman, 1969).

characteristic processes that constitute the rosette. Therefore, the neurofibrillary stain gives only an incomplete rendition of the mossy fiber.

3. The Glomerulus

a) The History of a Concept

RAMÓN Y CAJAL (1905, 1908, 1911, 1926) and his followers (CRAIGIE, 1926) recognized that the neurofibrils are embedded in and enclosed by a sheath of axoplasm, which intervenes between the fibrils and the granule cell dendrites with which the mossy fiber articulates. This sheath, unstained by neurofibrillary methods, can be colored to some extent by plasma staining methods, for example, mitochondrial stains. Applying such methods, HELD (1897) discovered that nerve endings everywhere are exceedingly rich in mitochondria (which he called neurosomes). Among the granule cells in the cerebellum he described and figured dense collections of mitochondria in a lightly staining matrix, which he recognized as the rosettes of the mossy fibers, discovered only a few years earlier by RAMÓN Y CAJAL. These protoplasmic islets had been noticed, but erroneously interpreted, twenty years before by DENISSENKO (1877), who described the arrangements of the granule cells in ring-like clusters around irregular clearings.[13] Drawing on the information provided by RAMÓN Y CAJAL's most recent studies (1894, 1896), HELD appreciated that the protoplasmic islets were composed of mossy fiber rosettes interwoven with the claw-like terminals of granule cell dendrites and the terminal branchlets of the Golgi cell axons. He named these complex formations *glomeruli cerebellares* (sic) in order to indicate their similarity to the glomeruli in the olfactory bulb.

13 Most authors (e.g., JAKOB, 1928) have agreed with RAMÓN Y CAJAL (1911, 1926) in stating that Denissenko's description of tightly packed groups of eosinophil cells refers to the parenchymatous or protoplasmic islets of twentieth century authors. But Denissenko's text and especially his pictures show rather clearly that he was dealing with cell bodies that resemble ordinary granule cells in size, shape, and arrangement. These *Eosinzellen* would appear to be granule cells giving a variant staining reaction in tissue that has undergone delayed fixation. On the other hand, Denissenko did describe clearly the clustering pattern of the regular granule cells (which he called *Haematoxylinzellen* because their nuclei stained intensely with haematoxylin): „... diese Zellen nicht selten rings um Öffnungen angebracht sind, die eine unregelmäßige, eckige Form besitzen. An dickeren Präparaten konnte man bei verschiedener Einstellungen sehen, daß von diesen Öffnungen, Kanäle abgehen, deren Wände auch von diesen Zellen belegt sind." Such openings seem to correspond more closely to the protoplasmic islets than do the Eosinzellen, which, in any case, Denissenko thought were nerve cells.

HELD is usually given the credit for discerning the composition of the cerebellar glomeruli, probably because he was the first to recognize them as the sites of nerve endings and because he adopted the reticularist point of view, which was then in the ascendant. Actually, it was RAMÓN Y CAJAL (1894, 1896) who had analyzed this complex synapse in earlier publications, and he did not fail to claim his priority (RAMÓN Y CAJAL and ILLERA, 1907; RAMÓN Y CAJAL, 1911). Nevertheless, it must be pointed out that RAMÓN Y CAJAL was not always certain about the intimate relations between the various elements in the glomerulus, and he had changed his mind about them more than once. The uncertainty arose from the well-known difficulty of visualizing clearly the members of a synapse in Golgi preparations. The selectivity of the method usually precludes the impregnation of both pre- and postsynaptic members at the same spot. Consequently it was impossible to locate the site of the articulation between the axons and dendrites in the granular layer. At first (1889a, 1893), RAMÓN Y CAJAL reported that the dendrites of the granule cells embrace the cell bodies of other granule cells. After further investigation (1894, 1896, 1911), he deduced that the granule cell dendrites articulate with the mossy fiber rosettes in the protoplasmic islets. His reasoning was exemplary. Dendrites from several granule cells converge on the same spot. In Golgi preparations that have been counterstained with thionin, it can be seen that nearly all of the mossy fiber rosettes extend between the granule cells. The open spaces apparently enclosed by the claw-like terminals of the granule cell dendrites match the dimensions and shapes of the mossy fiber rosettes, rather than the spherical shapes of the neuronal perikarya. Finally, the articulation between the mossy fiber rosette and the granule cell dendrites can be seen, though rarely, in Golgi preparations. These observations convinced RAMÓN Y CAJAL of the correct mode of termination of the mossy fibers. Thus, when HELD showed that the nerve endings in the granular layer did not form pericellular nests and that they were confined to the cell-free areas lying among the granule cells, he confirmed RAMÓN Y CAJAL's interpretation and eliminated some earlier proposals such as DOGIEL's (1896) that the mossy fiber rosettes surrounded granule cells. Neurofibrillary methods, which were capable of demonstrating both axons and dendrites, later corroborated this analysis of the composition of the glomeruli (BIELSCHOWSKY and WOLFF, 1904; RAMÓN Y CAJAL and ILLERA, 1907; RAMÓN Y CAJAL, 1911).

LUGARO (1894) suggested that the mossy fibers articulate with the terminals of the Golgi cell axons, which pervade the granular layer. RAMÓN Y CAJAL (1911) expressed considerable doubt that such a connection could occur. Still, he recognized that the terminal ramifications of the Golgi axons are components of the protoplasmic islets. In a later paper (1926) he had the remarkable prescience to state in a footnote (p. 225) that in keeping with the law of dynamic polarization he would expect that "certain dendrites of the granule cells enter into connection with the mossy fibers while others are subject to the influence of the axonal twigs of the Golgi cells."

Thus, the glomerulus was recognized as a complicated structure at the very beginning of its history. It should be clear from the brief review of its discovery given above that the nature of the articulation between its components was an insoluble problem with the techniques then available. It is to RAMÓN Y CAJAL's credit that with only fragmentary evidence he was able to discern the correct answers so early. Nevertheless, the constitution of the glomerulus continued to be a highly vexed question, even in his own laboratory. A great deal of importance was attached to it because it was one of the few sites in the central nervous system where a nerve fiber could be seen to terminate in relation to the processes of another nerve cell. The nature of this articulation quickly became a focus in the contest between the reticularists and the neuronists. As the history of this conflict is not pertinent to our purpose here, it will not be summarized (see RAMÓN Y CAJAL, 1934, and JANSEN and BRODAL, 1958, for reviews of the issues and data).

In 1954 the SCHEIBELS undertook a fresh examination of the glomerulus by means of a modified rapid Golgi method. Their results were reported in a comprehensive review by JANSEN and BRODAL (1958), but never independently published by the SCHEIBELS. They made a careful study of the mossy fiber rosettes and granule cell dendrites in order to analyze the relationship between them and the other components of the glomerulus. They found that the claw-like arborizations of the granule cell dendrites frequently nestled against the globular cell bodies of other granule cells. Such dendro-somatic contacts could be reciprocal; that is, a dendrite of one granule cell could be in contact with another granule cell, which in turn sends a dendrite to the first granule cell. Apparent dendro-dendritic contacts made by converging dendrites could be explained if the granule cell body that they contact failed to be impregnated or to be included within the thickness of the section. Mossy fiber rosettes were also seen to be closely applied to the somata of granule cells. The SCHEIBELS never saw convincing examples of articulations between the mossy fiber rosettes and the dendritic claws of granule cells. Furthermore, they found that the axons of Golgi cells also contact the somata of granule cells by means of terminal boutons or boutons en passant.

On the basis of these findings, JANSEN and BRODAL indicated that the traditional conception of the glomerulus was in need of revision. It was no longer appropriate to consider it as a cell-free protoplasmic islet. Instead, it was a complex synaptic tangle of axons and dendrites centered upon the cell bodies of granule cells. In this tangle the dendrites of granule cells, the rosettes of mossy fibers, and the axons of Golgi cells all impinged upon the granule cells and perhaps upon one another. Mossy fibers were also thought to affect the Golgi cells and the Purkinje cells by means of direct axo-somatic contacts. The difficulties of reliably identifying synaptic junctions in Golgi preparations left these interesting proposals without a strong foundation. Not only did the suggestion that granule cells are linked together by dendro-somatic synapses conflict with the prevailing conception of dendritic function, but the observations also recalled a number of previously discredited descriptions and interpretations. For example, RAMÓN Y CAJAL (1889a) had originally thought that the claw-like arborizations of the granule cell dendrites grasped the cell bodies of other granule cells, and BECHTEREW (1896) had proposed that granule cells were connected to one another through an interweaving of their dendrites. Then, it will be remembered that DOGIEL (1896) believed that he saw granule cell bodies enclosed within the coils of the mossy fiber rosette, and PENSA (1931) declared that the glomerulus included small cell bodies, which, he was convinced, were those of oligodendrogliocytes. Nevertheless, the new interpretations were received with considerable interest and approval by established authorities like JANSEN and BRODAL (1958), who attempted to revise the contemporary model of the functional circuitry of the cerebellar cortex in the light of their discussions with

Fig. 128. A mossy fiber apposition with a granule cell soma. The mossy fiber enlargement is simple, with a population of dispersed synaptic vesicles (*sMFd*). The profile abuts against the soma of a granule cell (*gr*). Intermittently, the interstitial space between the axon and cell membrane is filled with a dense material (*fine arrows*). The mossy fiber terminal is also surrounded by many dendrites of granule cells with which it synapses (*arrowheads*). Axons of Golgi cells lie in the periphery of this glomerulus (*asterisk*). Lobule V. × 14000

Fig. 129. Simple mossy fiber adjacent to granule cells. This mossy fiber has a more clustered population of synaptic vesicles (*sMFc*). The axon abuts against two granule cells (*gr*). Intermittently the interstitial space is occupied by a dense material (*arrows*). Lobule X. × 18500

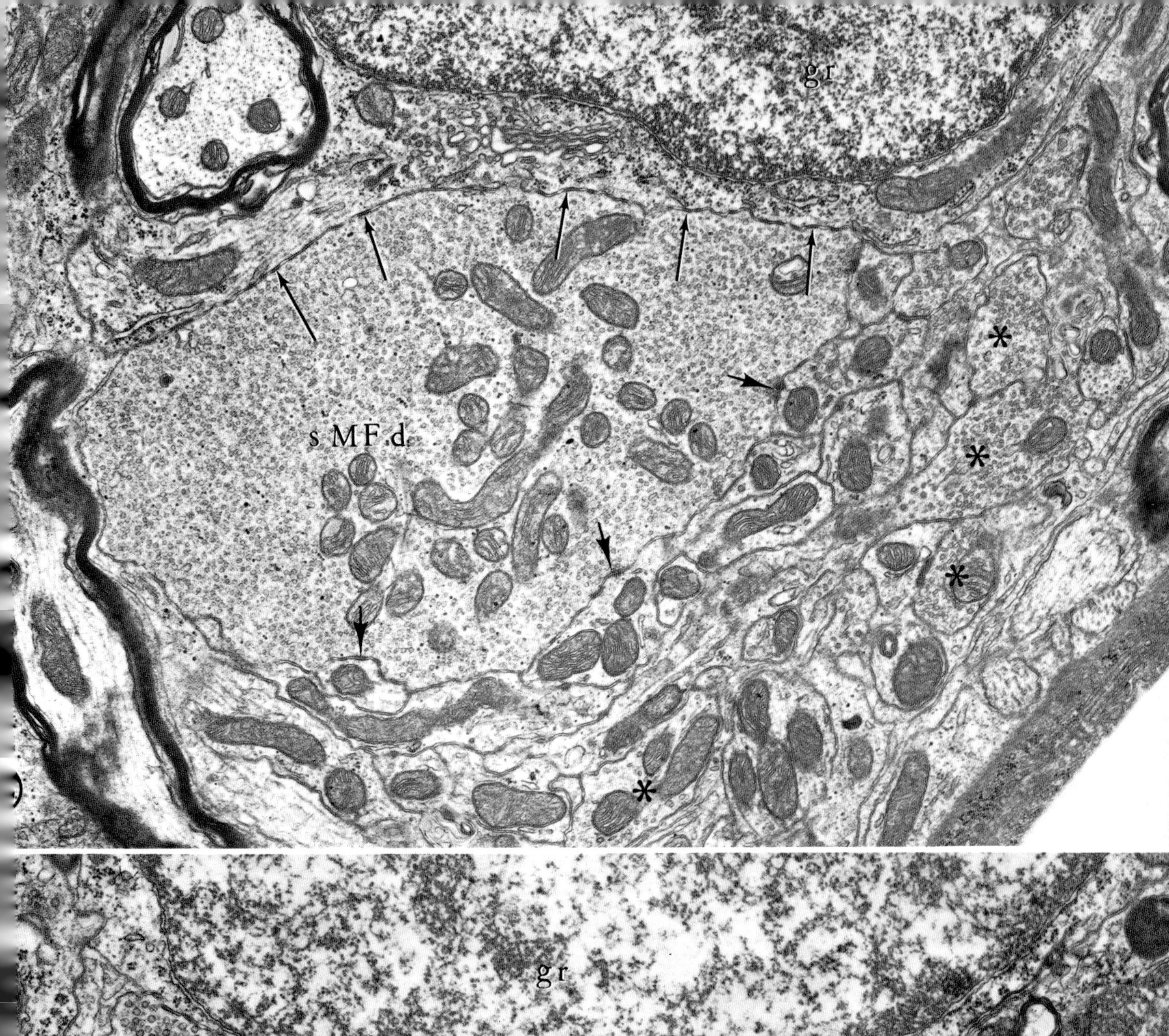

gr
s MF d

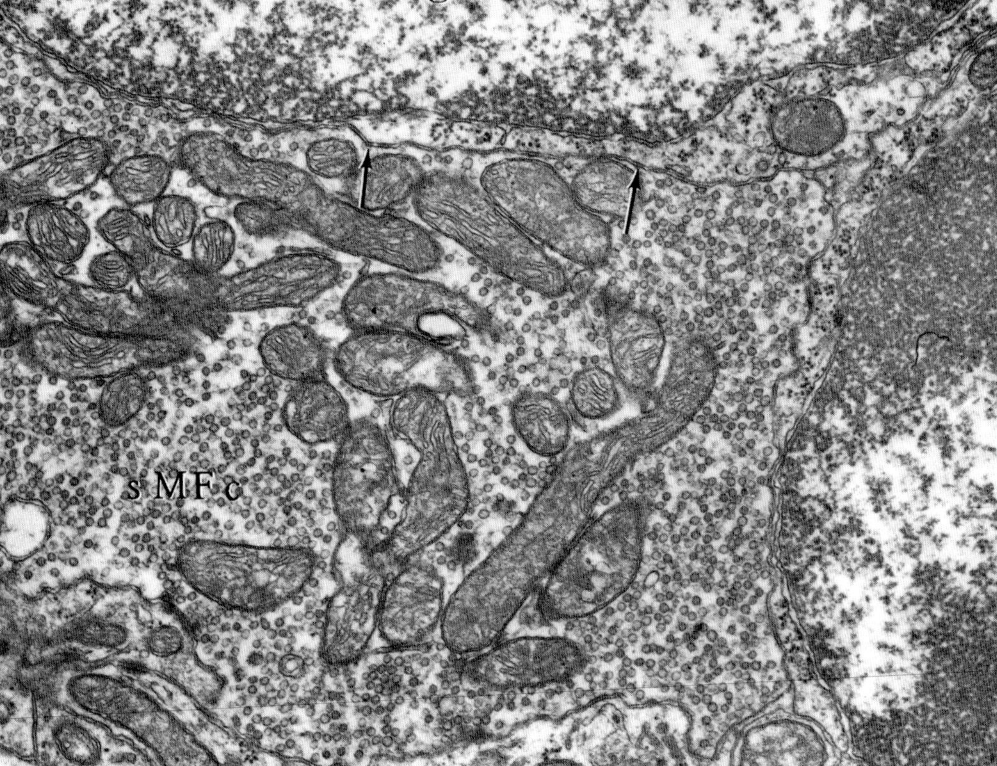

gr
s MF c
gr

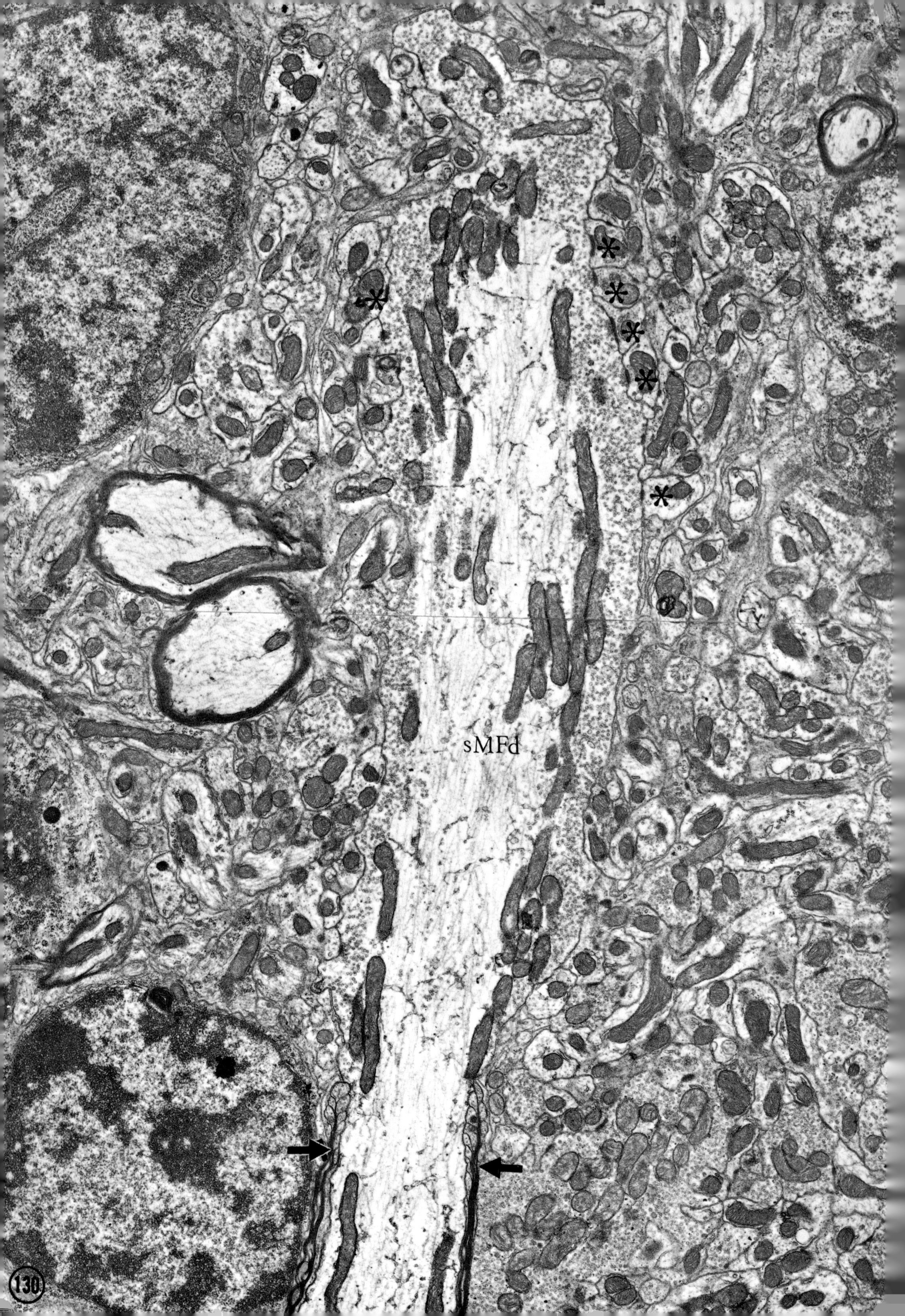
sMFd
130

the SCHEIBELS. That sympathetic interest discloses how tenuous is the evidence from optical microscopy for the traditional conception of the glomerulus.

The preparations studied by the SCHEIBELS showed nothing different from those of RAMÓN Y CAJAL and many other competent investigators using the Golgi method. Only the interpretation of the preparations was different. RAMÓN Y CAJAL arrived at his conclusions on the basis of an argument, already summarized above, that proceeded from one fundamental observation: both granule cell dendrites and mossy fiber rosettes converge on apparent clearings among the granule cells. Using a similar finding but paying more attention to the cell bodies around these clearings, the SCHEIBELS arrived at a different conclusion. Actually, it was not technically possible for any of these investigators to distinguish between mere proximity and a true synapse, nor could they detect which face of a fiber, dendritic claw, or an axonal rosette was the properly synaptic one. Only a different method, capable of differentiating the possible varieties of interfacial contact, could settle the problem of the glomerulus.

b) The Fine Structure of the Glomerulus

Although the cerebellar glomerulus was among the first examples of the interneuronal synapse[14] to be examined in the electron microscope (PALAY, 1956, 1958), the first detailed studies of this formation were reported by GRAY and by PALAY independently in 1961. These two papers and many subsequent ones have corroborated completely the analysis given by RAMÓN Y CAJAL (BIRCH-ANDERSEN *et al.*, 1962; DAHL *et al.*, 1962; HÁMORI, 1964; HÁMORI and SZENTÁGOTHAI, 1966b; FOX, HILLMAN, and SIEGESMUND, 1967; LARRAMENDI, 1969a and b; LLINÁS and HILLMAN, 1969; MUGNAINI, 1972). The electron micrographs show that the mossy fiber rosette is the central component of the glomerulus and that it synapses with the dendrites of granule cells, not with the cell bodies. Although granule cell dendrites can pass close to the somata of granule cells and puncta adhaerentia can develop between them along the zone of contact, the morphological features of the synapse never appear at these sites. Furthermore, the granule cell body never appears in synapse with any kind of axonal terminal even when, as occasionally happens, a mossy fiber rosette (Figs. 128 and 129) or a climbing fiber varicosity, a Purkinje cell recurrent collateral, or a Golgi cell axon (Fig. 149) comes into contact with it. Images like those in Figs. 128 and 129 may be taken, however, as a vindication of the SCHEIBELS, who observed articulations between mossy fiber rosettes and the somata of granule cells in their Golgi preparations. The two mossy fiber terminals in these figures are clearly in direct apposition to the surfaces of one or two granule cells. But these junctions do not display the cytological characteristics of synapses, even though the interstitial space between these elements is typically occupied by a discontinuous dense material that does not appear generally in the interstitial space. The presence of this material may suggest that some specific role should be sought for this junction. Granule cell bodies are also frequently in direct apposition to one another, as the clustering pattern seen in the optical microscope indicates, but they never display any synaptic junctions along the extensive contact zone. The filamentous and granular densities that intermittently occur in the interstitial space are, like those just mentioned, not associated with any specializations along the inner faces of the apposed membranes. Thus, electron microscopy clearly returns the glomerulus to its original status as a complex articulation between mossy fiber rosettes and the terminal dendritic arborizations of granule cell dendrites. The nature of this articulation will now be described as it appears in electron micrographs of thin sections.

The Form of the Mossy Fiber Terminal. The terminal branches of the mossy fiber can be recognized in the micrographs as large, thinly myelinated axons traversing the granular layer. They characteristically contain numerous neurofilaments and a few microtubules, long bacillary mitochondria, and tubular elements of the agranular endoplasmic reticulum—all running parallel with the longitudinal axis of the fiber. As the mossy fiber approaches the territory of the glomerulus, its myelin sheath terminates in the manner usually associated with the node of RANVIER. In fact, as mentioned earlier in the light microscopic description, the rosettes are frequently developed in nodes of RANVIER (Fig. 126), and the myelin sheath takes up again when the axon leaves the glomerulus.

14 Quite good electron micrographs of cerebellar glomeruli from rat were made by GEORGE PALADE at the Rockefeller Institute at the end of 1952.

Fig. 130. Simple mossy fiber emerging from myelin sheath. This mossy fiber emerges from its myelin sheath (*arrows*) and flows out as a simple enlargement containing a dispersed population of synaptic vesicles (*sMFd*). Many synapses (*asterisks*) are made with dendrites of granule cells that abut against it. Lobule IV. ×13000

Scarcely has the last myelin lamella terminated when the axoplasm flows out like a stream into a lake at the bottom of a valley. A simple mossy fiber is shown emerging from its myelin sheath in Fig. 130. In complex mossy fibers, the fiber becomes irregularly distended, as much as two or three times its previous diameter, and forms a tapering enlargement, 10 to 20 microns long, that corresponds to the rosette of light microscopy. Rounded protuberances and projecting arms distort its outline. The more salient of these appendages are often thin, tubular stalks, which terminate in an expanded knob at some distance from the main mass (Fig. 131) and which match quite precisely a formation commonly seen in the Golgi preparations.

In an earlier section of this chapter the mossy fiber terminals were described as appearing in Golgi preparations in one of three forms with different degrees of complexity. It is difficult to identify these forms with assurance in electron micrographs of random thin sections through the glomeruli. A section can completely miss the twists and offshoots that are critical for classification in the optical microscope. Besides, the profile of a mossy fiber has an irregular shape as a rule, and so we are demanding a classification of profiles according to degrees of irregularity. As MUGNAINI (1972) has written, most of the published electron micrographs of glomeruli show mossy fibers like those in Figs. 130 and 132, simple in outline with a few bulbous protuberances. The resemblance between these profiles and the simple mossy fiber terminals of the Golgi preparations (Fig. 127) is easy to accept. But more complicated rosettes with lateral branches and a filigree shape (Figs. 126 and 125c) are also very numerous. The reason for their infrequent appearance in publications has more to do with the exigencies of the printer's format than with the specimens studied. Electron micrographs of complex mossy fiber terminals are shown in Figs. 131 and 133–135. These have highly branched profiles; offshoots are rarely captured in continuity with the axial mass (Fig. 131), and usually appear as detached islands surrounded by granule cell dendrites. Some sections, missing the central mass, present only a collection of such islands, an image that can be understood readily if the configuration in Golgi preparations is kept in mind.

The Core of the Mossy Fiber. The fine structure of the mossy fiber terminals is conveniently described from the inside out. All of them, whether simple or complex in shape, have basically the same internal architecture, consisting of three overlapping arrays of organelles. The neurofilaments in the myelinated portion of the mossy fiber continue through the rosette into the next myelinated portion. As can be seen in both longitudinal (Figs. 130, 132, 144a, and 146) and transverse sections of the fiber (Fig. 147), they form a loose bundle in the core of the rosette as they maintain the axial direction of the fiber and do not reflect its surface convolutions. The neurofilamentous bundle can twist and buckle as it passes through the rosette, but it does not usually contribute twigs to the appendages branching from the central varicosity. They often contain a few microtubules. The appearance of the bundle of neurofilaments is, in general, compatible with the form of the rosettes as displayed in material stained with the neurofibrillary methods (compare, for example, the drawings in RAMÓN Y CAJAL, 1911, 1926 or CRAIGIE, 1926).

The core of neurofilaments is surrounded by a cylinder composed of vermiform mitochondria, arranged in parallel with the filaments and one another. This concentrated stream of mitochondria accounts for HELD's description of the mossy fibers in the glomerulus. These organelles generally display longitudinally oriented cristae and have a dark matrix. They vary from 1 to 3 μm in length, and are about 0.25 μm in diameter. They can lie two or three deep in their clusters with little space between them. Nevertheless, both tubular elements of the agranular reticulum and occasional microtubules manage to be insinuated among them.

The Synaptic Vesicles. From light microscopy we have learned that the neurofibrillary cable in the center of the rosette is surrounded by a thick cortex or sheath of "neuroplasm," which intervenes between the fibrils and the dendrites of the granule cells. This "neuroplasm" appeared faintly granular in reduced silver or in thionin-

▶

Fig. 131. A knob from a complex mossy fiber rosette. This complex mossy fiber rosette extends a long process, which ends in a knob (▲▲). Round synaptic vesicles are dispersed in the axoplasm (*cMFd*). Osmium tetroxide fixation. × 39000

Fig. 132. Mossy fiber glomerulus. This mossy fiber rosette is relatively simple, although its outline (▲▲) is not as smooth as that of most simple mossy fibers. The synaptic vesicles are dispersed in the axoplasm (*MFd*). Lobule VII. × 17000

Fig. 133. A complex mossy fiber glomerulus. The mossy fiber rosette in the center of the glomerulus has a complex shape and contains many synaptic vesicles (*cMFc*) arranged in a homogeneous cluster in the axoplasm (ΔΔ). The rosette is surrounded by axons of Golgi cells (*) and by dendrites of granule cells. Lobule IX. × 19000

Fig. 134. Complex mossy fiber glomerulus. Several profiles of the same complex mossy fiber rosette can be seen in this section (*cMFc*). The profiles contain many synaptic vesicles in large clusters. The rosette is surrounded by swirls of granule cell dendrites (*gr d*) and axons of Golgi cells (*). Lobule IX. × 18000

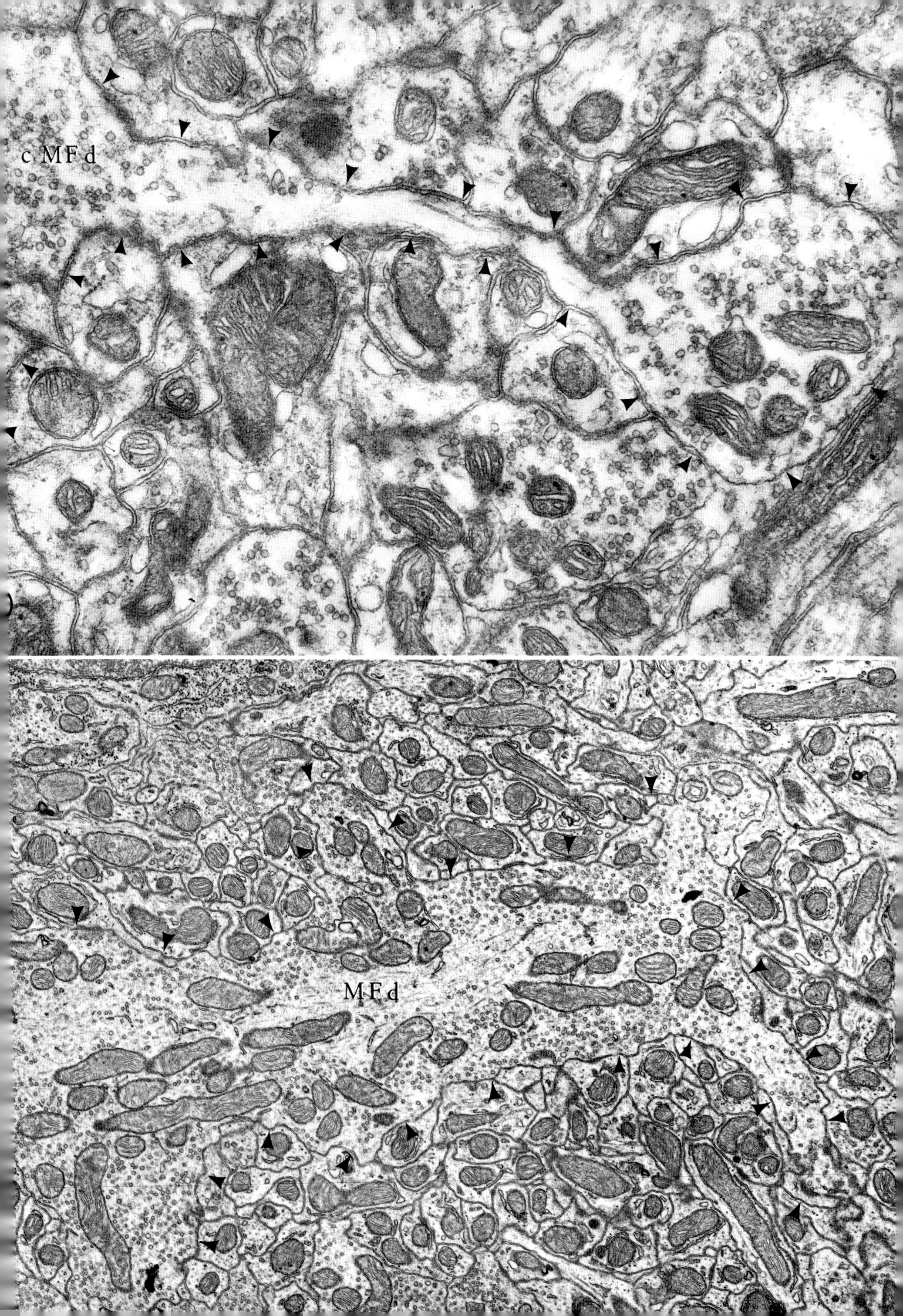
c MFd
MFd

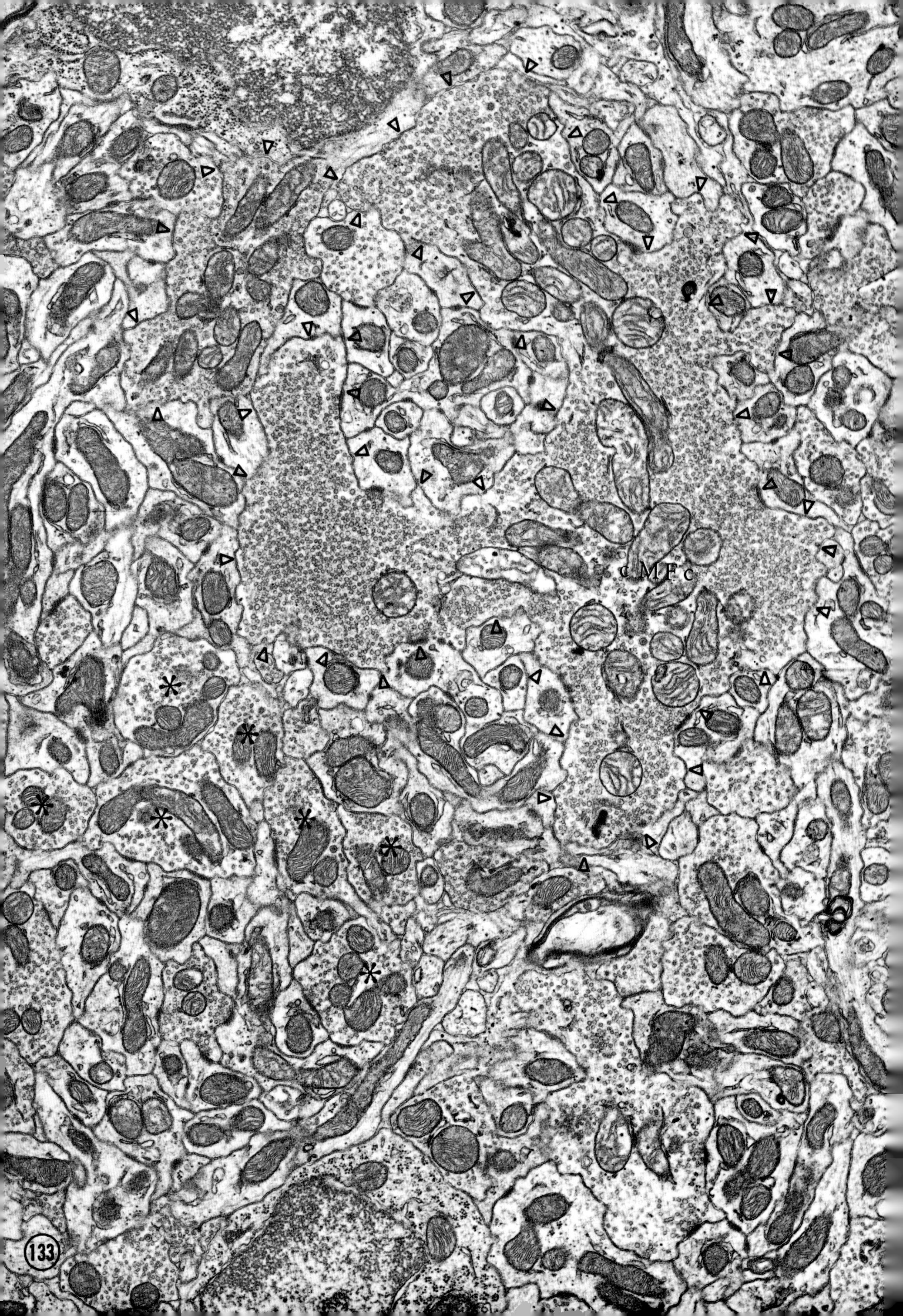

c MF c
133

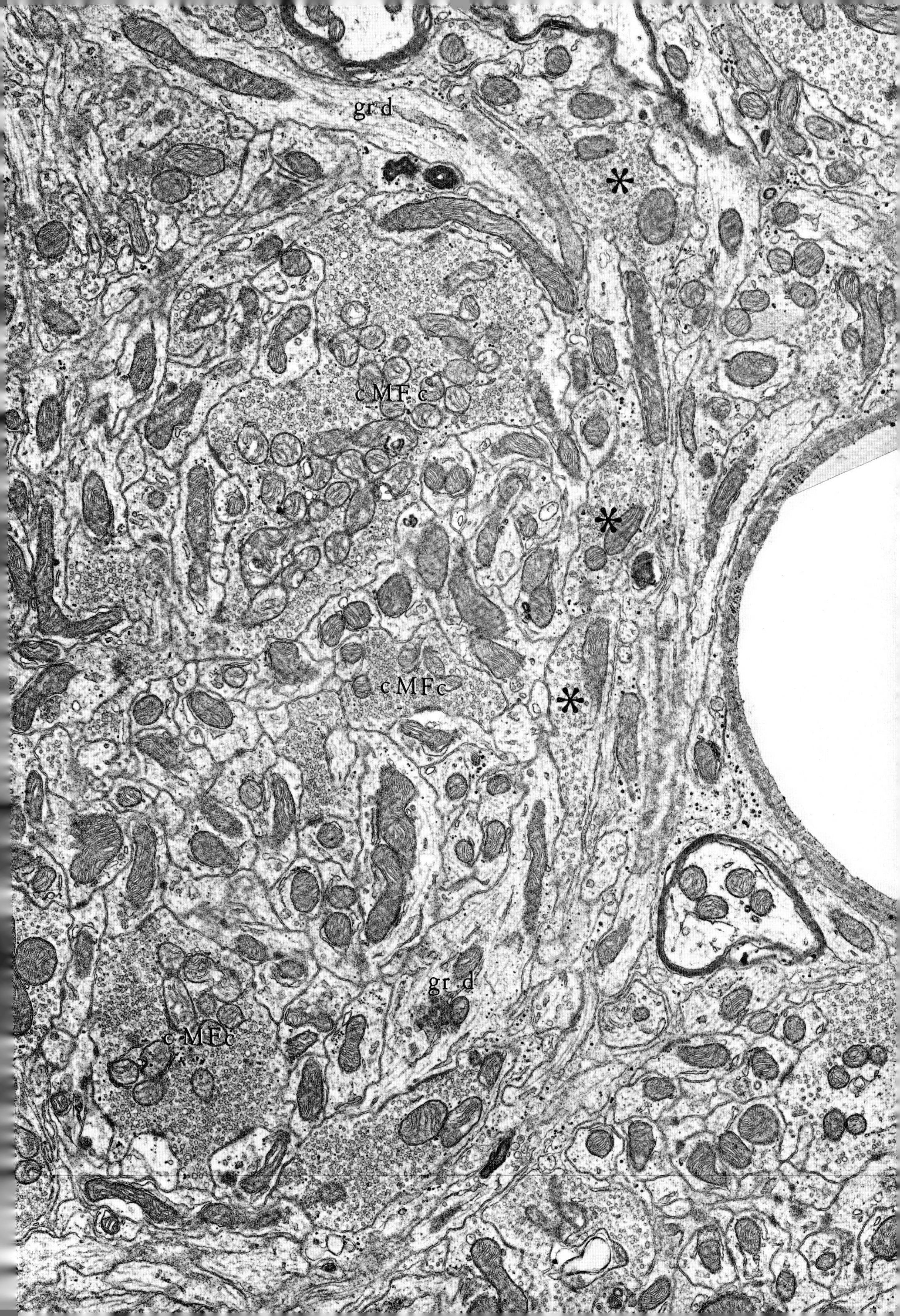
gr d
*
cMF c
*
cMFc
*
gr d
cMFc

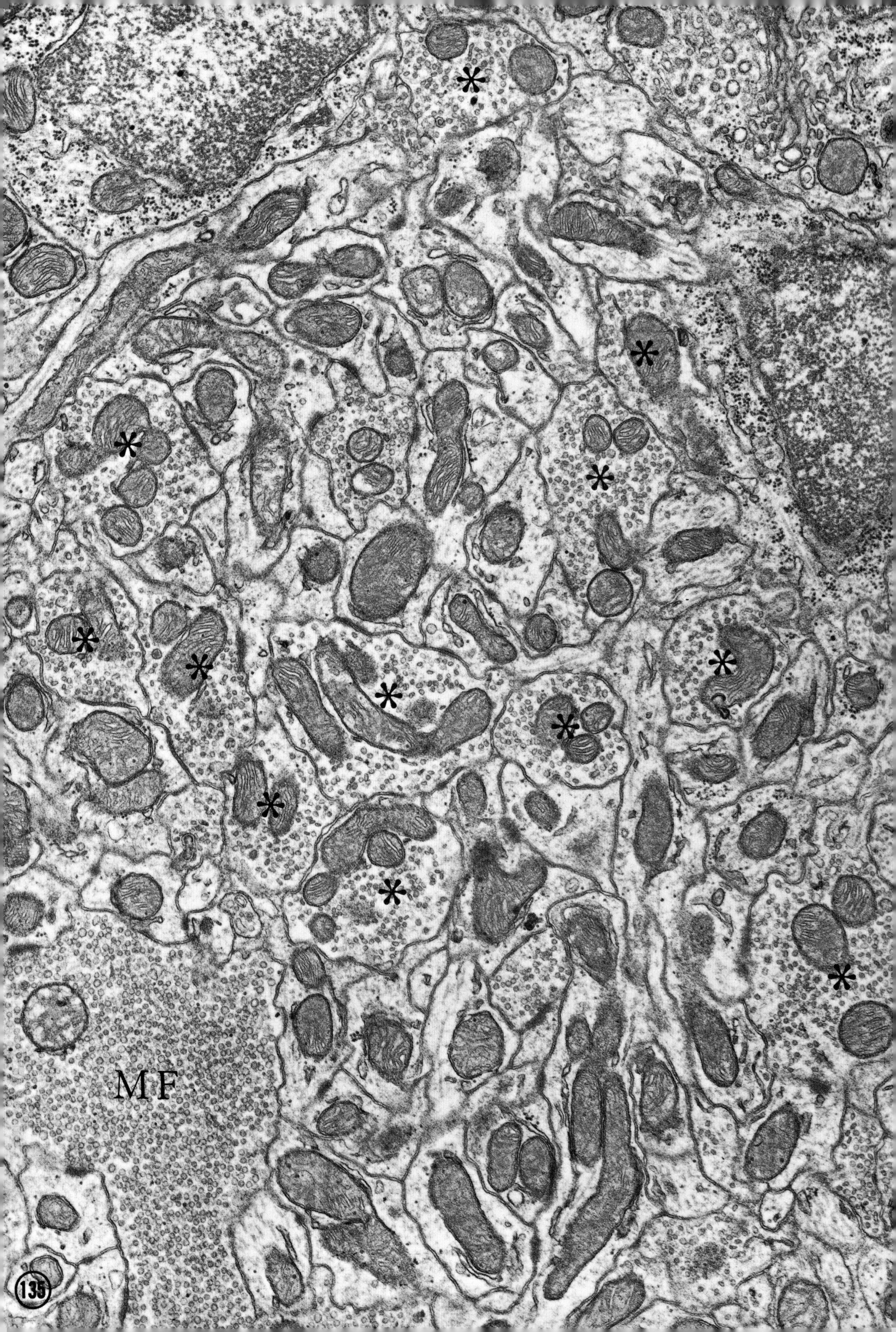
MF
135

Fig. 136. Histogram of the diameters of synaptic vesicle profiles in two types of mossy fibers, dispersed and clustered. The ranges of vesicle sizes are similar in the two varieties of mossy fibers, from 300–560 Å. The actual diameter of the synaptic vesicles in both the dispersed and the clustered mossy fibers is about 480 Å

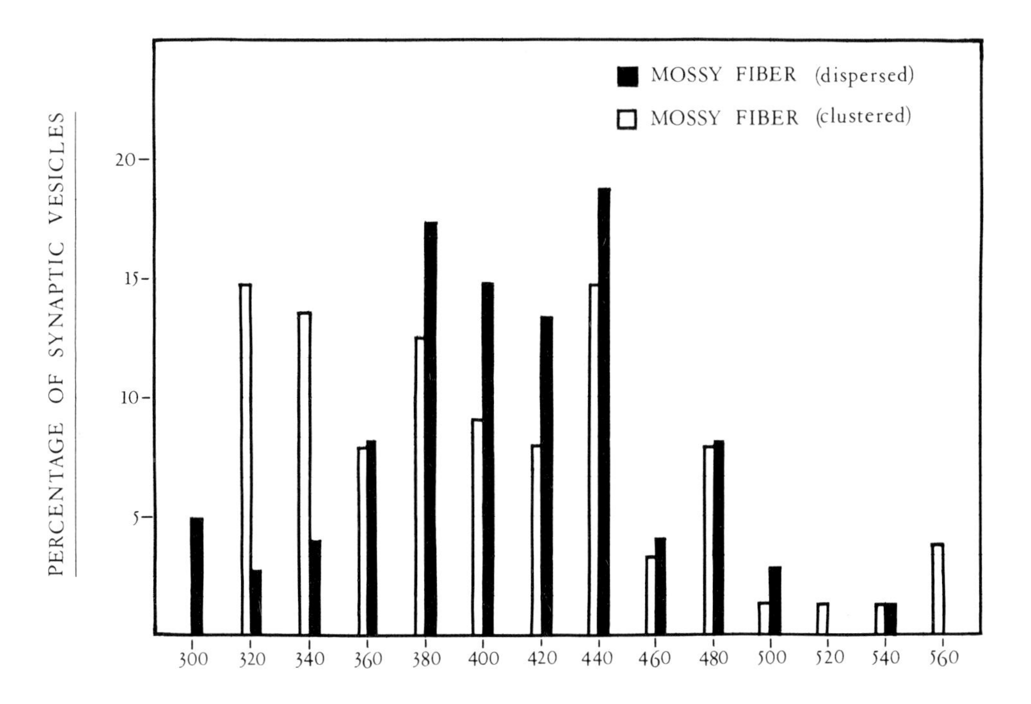

stained preparations, and its nature could not be discerned. The electron micrographs of thin sections show that this cortical zone is occupied by throngs of spherical synaptic vesicles (Figs. 128–132). The profiles of these vesicles range from 300 to 560 Å in diameter (measurements on 160 vesicles), and the diameter of the average vesicle should be around 440–460 Å (Fig. 136). They are all circular in outline and have clear centers.

Inspection of the illustrations (Figs. 128, 129, 137, and 138) will indicate that there is considerable variation in the concentration of vesicles in different mossy fiber terminals. Study of random thin sections does not allow us to ascertain whether the concentration of vesicles is consistent in all terminals of the same fiber. Furthermore, both simple and complex terminals have the same variety of concentrations. The terminals can be arranged in at least two classes according to the crowding of the vesicle population. In one class the vesicles are loosely scattered over the cortical axoplasm surrounding the neurofibrillary core (Figs. 128, 130–132, 137, and 138). In the second class they are fairly tightly packed (Figs. 129, 133–135, 137, and 138). We have named these two classes respectively "dispersed" and "clustered," merely for convenience in description. In the dispersed terminals the vesicles tend to be scattered in small groups here and there within the terminal, especially opposite the synaptic junctions with granule cell dendrites. In the clustered variety, the ending tends to be uniformly filled with vesicles as if in one great cluster. The size range of the vesicles is similar in these two classes (Fig. 136). Representative counts were made to determine the population density of the vesicles in the cortical axoplasm of both kinds of mossy fiber terminal, using random thin sections. A total of 1935 vesicles were counted in a total area 42.8 μm^2 of axoplasm at 26000× magnification. The results are displayed as a histogram in Fig. 139. Dispersed terminals have a mean concentration of 116 vesicles per μm^2, while clustered terminals have a mean concentration of almost twice that, 222 per μm^2. These should be compared with the extraordinary density of the vesicle population in the various terminals

Fig. 137. A comparison of mossy fibers containing dispersed synaptic vesicles with mossy fibers containing clustered vesicles. The rosette in the upper half of the figure (*MF d*) has synaptic vesicles dispersed in the axoplasm. The rosette in the lower half of the figure (*MF c*) has many more synaptic vesicles clustered together. Compare this with Fig. 138, a climbing fiber glomerulus. Profiles of Golgi cell axons are distributed in the periphery of the glomeruli (*asterisks*). Lobule V. ×25000

Fig. 138. Adjacent climbing fiber and mossy fiber glomeruli and a myelinated granule cell. A climbing fiber efflorescence (*CF*) and a mossy fiber rosette (*MF d*) with a dispersed population of synaptic vesicles lie adjacent to one another. A neighboring granule cell (*gr*) is surrounded by a myelin sheath (*my*). (Compare this with Fig. 137.) Lobule V. ×21000

Fig. 135. A mossy fiber glomerulus. The mossy fiber profile (*MF*) is surrounded by many axons of Golgi cells (*asterisks*). These axons synapse upon numerous profiles of granule cell dendrites in the illustration. Lobule IX. ×26000

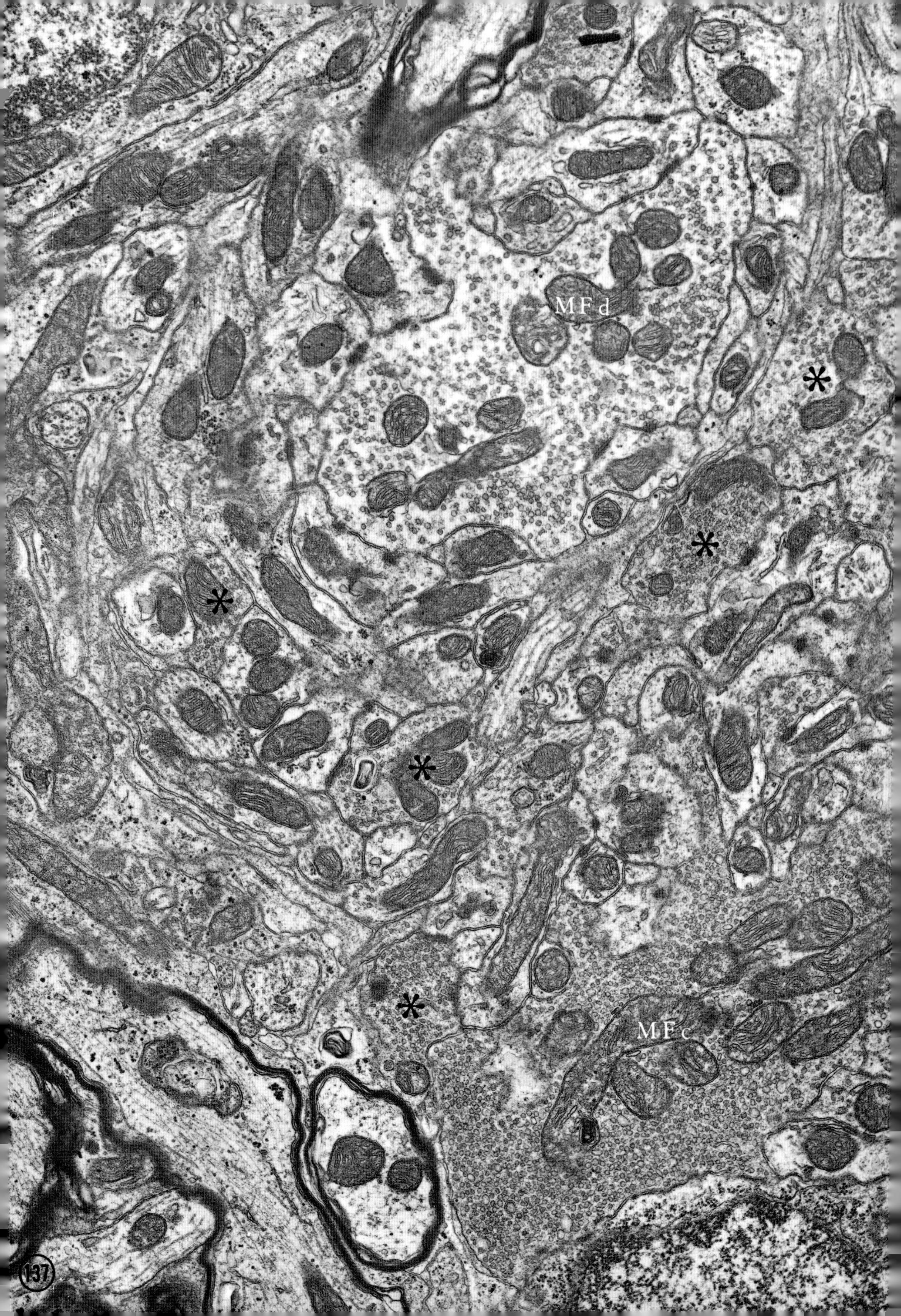
MFd
MFc
137

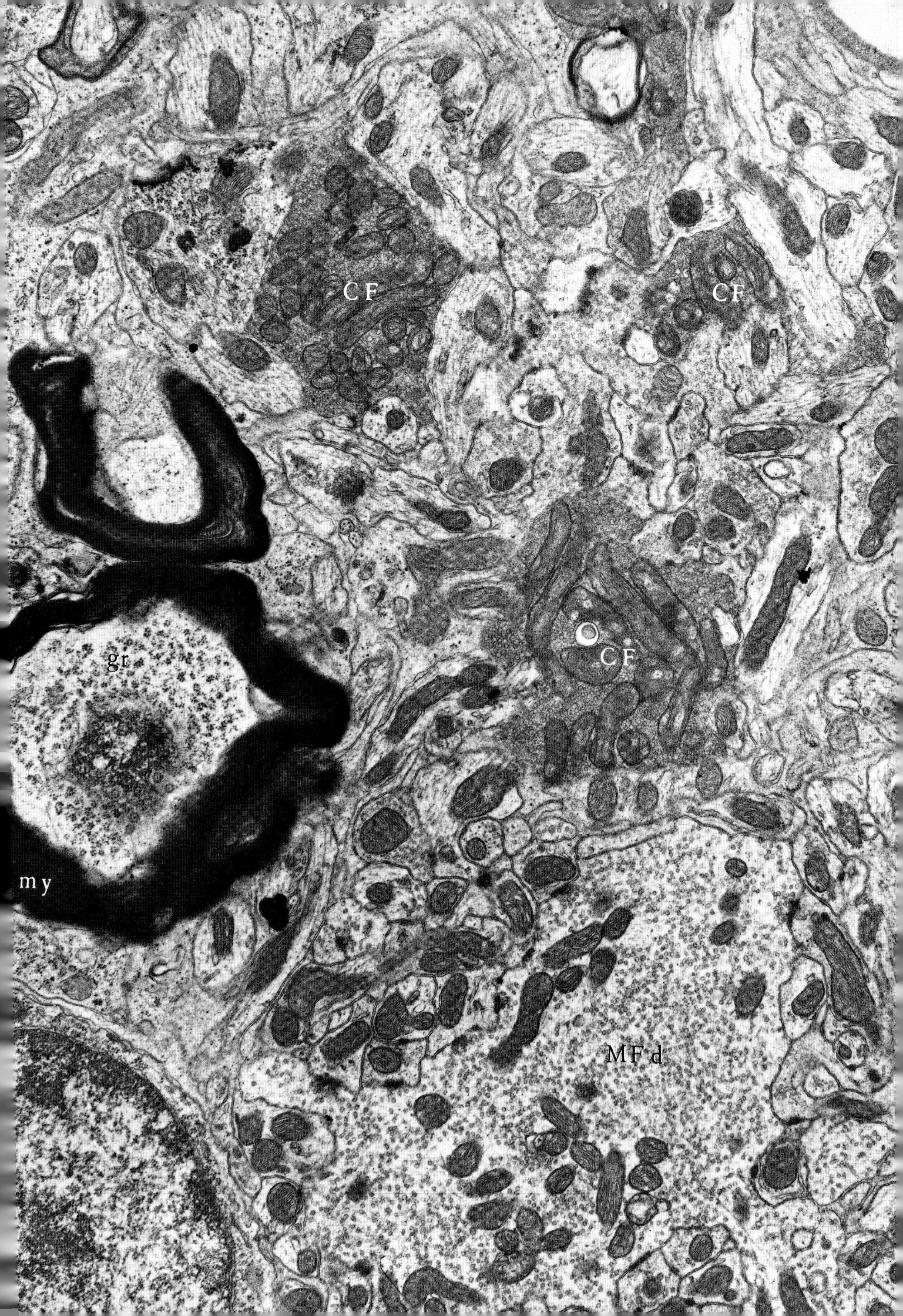
CF
CF
gr
CF
my
MFd

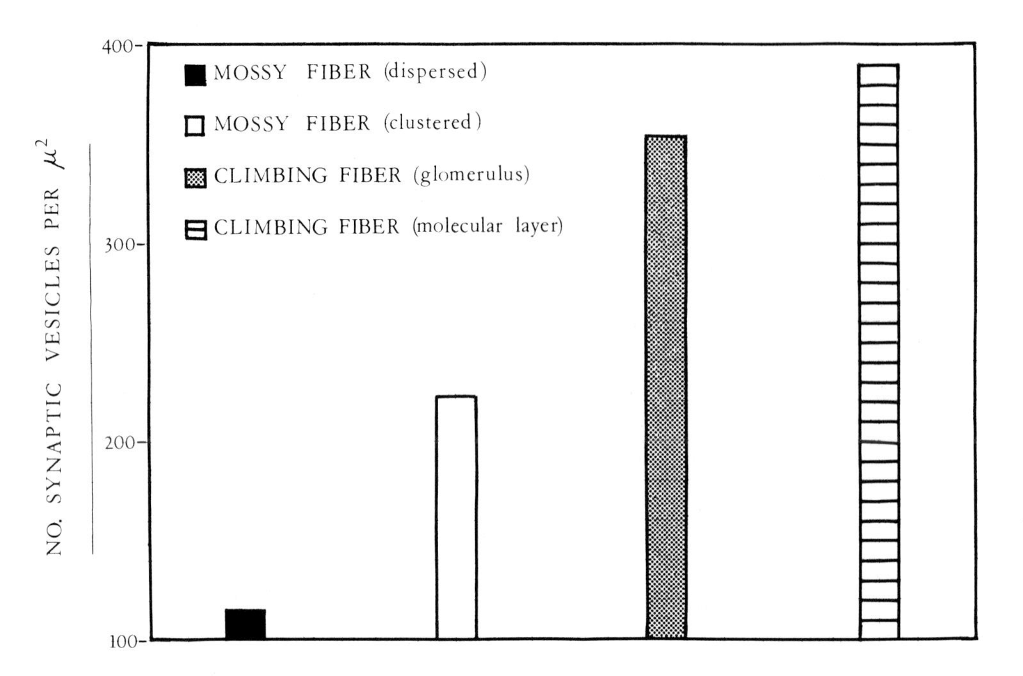

Fig. 139. Histogram showing the mean concentration of synaptic vesicles in the cortical axoplasm of mossy fiber rosettes and climbing fiber terminals. Climbing fiber axoplasm in both the granular (*stipple*) and molecular (*stripe*) layers has a high packing density of 340–390 synaptic vesicles per μm^2. Mossy fibers with clustered synaptic vesicles (white) have 222 per μm^2, and mossy fibers with dispersed synaptic vesicles (black) have half that amount (116 per μm^2)

of climbing fibers, which ranges between 340 and 390 vesicles per μm^2 (Fig. 138).

With these morphological distinctions in mind, it is possible to classify mossy fiber rosettes into four groups according to their external form and the disposition of their synaptic vesicles. Simple mossy fiber rosettes can have either dispersed synaptic vesicles or synaptic vesicles that are clustered. By the same token, complex mossy fiber rosettes can have synaptic vesicles that are either dispersed or arranged in denser clusters. Our experience has shown that most electron micrographs of mossy fiber glomeruli can be classified in this way (see figure legends). Climbing fiber glomeruli are not easily confused with any of these profiles.

There is another kind of vesicle, also circular in outline but somewhat larger, up to 1000 Å in diameter, and with a dense, granular center. In most mossy fiber rosettes, such vesicles are rather rare, at most only one or two appearing in a section of a rosette. None can be found in the majority of random sections. However, in a few rare profiles, a rosette may be seen with a fair number of such vesicles (Fig. 140). These may be indicative of a special physiological state.

Gray (1961) described still another kind of vesicle, about 1000 Å in diameter, which he tentatively interpreted as a central sphere, 600–800 Å in diameter, surrounded by a shell of tightly packed vesicles, 150–200 Å in diameter. Such "complex vesicles," as he called them, seem to have been numerous in his material. Although they appear to be much less abundant in the mossy fiber rosettes of our preparations, they are easily found. They correspond to the "coated," "spiny," or "alveolate" vesicles that have been described in a wide variety of cells, both nervous and non-nervous. They are commonly found in nerve endings of all types, both in the central and peripheral nervous systems. A careful three-dimensional analysis of these complex vesicles by Kanaseki and Kadota (1969) reveals that they are simple vesicles surrounded by a framework composed of hexagonal elements. A similar alveolate coating appears on the inner face of the plasmalemma lining pits in the nonsynaptic surface of the mossy fiber rosettes and other nerve terminals. In secretory cells, for example, in the epithelium of the vas deferens (Friend and Farquhar, 1967), similar coated vesicles and pits are evidently involved in ferrying substances picked up from the extracellular phase to the multivesicular bodies and the lysosomes deeper in the cytoplasm. Similar phenomena have been shown to exist in neuronal perikarya (see Novikoff, 1967a and b). In recent experiments on neuromuscular junctions in the frog, Heuser and Reese (1972) have demonstrated that tracer substances, for example, horseradish peroxidase, are taken into coated pits in the junctional surface of the nerve terminal, sequestered within coated vesicles, and transported to certain vacuoles with which the coated vesicles merge. Their results were interpreted as indicating that the coated vesicles provide the vehicles for (a) the re-uptake of transmitter released into the synaptic cleft and (b) the return of the membrane added to the cell surface by the fusion of synaptic vesicles with the plasmalemma. Gray and Willis (1970) have discussed a similar proposal in regard to synaptic terminals in the central nervous system. This mercantile hypothesis of the role of the coated vesicles in nerve terminals is strongly supported by the evidence adduced by PALADE and his colleagues (JAMIESON and PALADE, 1967a and b, 1971a and b; MELDOLESI, JAMIESON, and PALADE, 1971) for the circulation of specific membrane fragments among the granular endoplasmic reticulum, the Golgi apparatus, the secretory vesicles, and the surface membrane in gland cells.

Finally, in this connection it may be noted here that CECCARELLI, HURLBUT, and MAURO (1972) have provided very strong evidence that synaptic vesicles in frog neuromuscular junctions are recycled after discharging their burden of transmitter. They showed that in a neuromuscular preparation bathed in a Ringer solution containing peroxidase, tetanic stimulation for three hours resulted in the appearance of many peroxidase-containing vesicles in the nerve terminal, whereas in the controls, bathed in curare without stimulation, only a few such vesicles appeared. They also succeeded in depleting the terminals of their vesicles by tetanic stimulation lasting more than four hours.

The Granule Cell Dendrites. The multilobulate contour of the mossy fiber rosette provides numerous coves and bays in which the dendrites of granule cells are lodged. It will be remembered that, as may be seen in Golgi preparations (Figs. 53 and 54), the terminal arborization of these dendrites is a digitate spray of short, varicose fibers that curl about the mossy fiber rosette. Dendrites from many different granule cells contribute to each glomerulus. In electron micrographs (Figs. 141 and 142) of sections passing obliquely through the periphery of the glomerulus, these dendrites can be seen running more or less parallel with one another and forming a sheath about the mossy fiber terminal. Each finger-like dendritic branch contains a single long mitochondrion partially wrapped in a flattened cisterna of endoplasmic reticulum. At irregular intervals neighboring dendrites are bound together by puncta adhaerentia, small spots that are characterized by a slight widening of the interstitial cleft and the symmetrical disposition of dense, filamentous plaques on the cytoplasmic sides of the apposed surface membranes.

Sections passing across the axis of the dendrites show that they run in deep or shallow depressions or grooves let into the surface of the mossy fiber rosette. The rosette is thus almost completely encircled by a palisade of round or polygonal profiles representing the granule cell dendrites (Figs. 141 and 142). Each profile is nearly filled by a single cross section of a mitochondrion and a cisterna of the endoplasmic reticulum wrapped partially around it. The profiles of dendrites are often two and three deep, packed into bays between the twisted appendages of the rosette. Puncta adhaerentia occur at the mid-points of the surfaces apposing one dendrite to another, and also frequently along the contact between dendrite and mossy fiber.

In three dimensions the articulation between the granule cell dendrites and the mossy fiber rosettes is very much like that of the fingers of a hand grasping a soft plastic object and molding it. This image has much to commend it. In some recent publications, drawings of three-dimensional models are presented which show the dendrites with subspherical tips fitting into excavations in the surface of the rosette. The latter is given a pitted surface like that of a golf ball. Sections of the glomerulus in all planes fail to support this model, as the dendrites run along the surface of the rosettes in furrows making synaptic contacts *en passant* as well as at their tips.

The synaptic junction between the granule cell dendrite and the mossy fiber terminal is characterized by a widened synaptic cleft produced by a slight concavity or lifting away of the dendritic plasmalemma. The cleft is occupied by fine, filamentous material, and frequently a thin line can be seen halfway between the apposed membranes. A thick plaque of filamentous material is attached like a beard to the cytoplasmic side of the dendritic surface membrane, and on the axonal side tufts of similar material project into the mossy fiber rosette. A small cluster of round synaptic vesicles is usually applied to this density. The junctional complex between granule cell dendrites and the mossy fiber rosette thus corresponds closely to Gray's type 1. It should be noted that the synaptic complex is limited to a small macula only about one third as wide as the contact zone between the two nerve processes. The number of these maculae along the extent of any dendrite is unknown.

A different kind of information about the glomerulus is obtainable from replicas of freeze-fractured tissue (PALAY, RAVIOLA, and CHAN-PALAY, 1973). Low power electron micrographs of the granular layer in such preparations show the components of the glomerulus in relief and give an almost three dimensional impression of the topography. In Fig. 142A, a glomerulus high in the granular layer appears as a whorl of branching processes, the claw-like terminals of the granule cell dendrites, disposed about and upon an extensive, folded or ridged background, which represents the surface of the central mossy fiber rosette. At higher magnifications (Figs. 142B and C), the broken projections of the mossy fiber are easily identifiable as they are filled with throngs of synaptic vesicles and mitochondria. Surrounding them

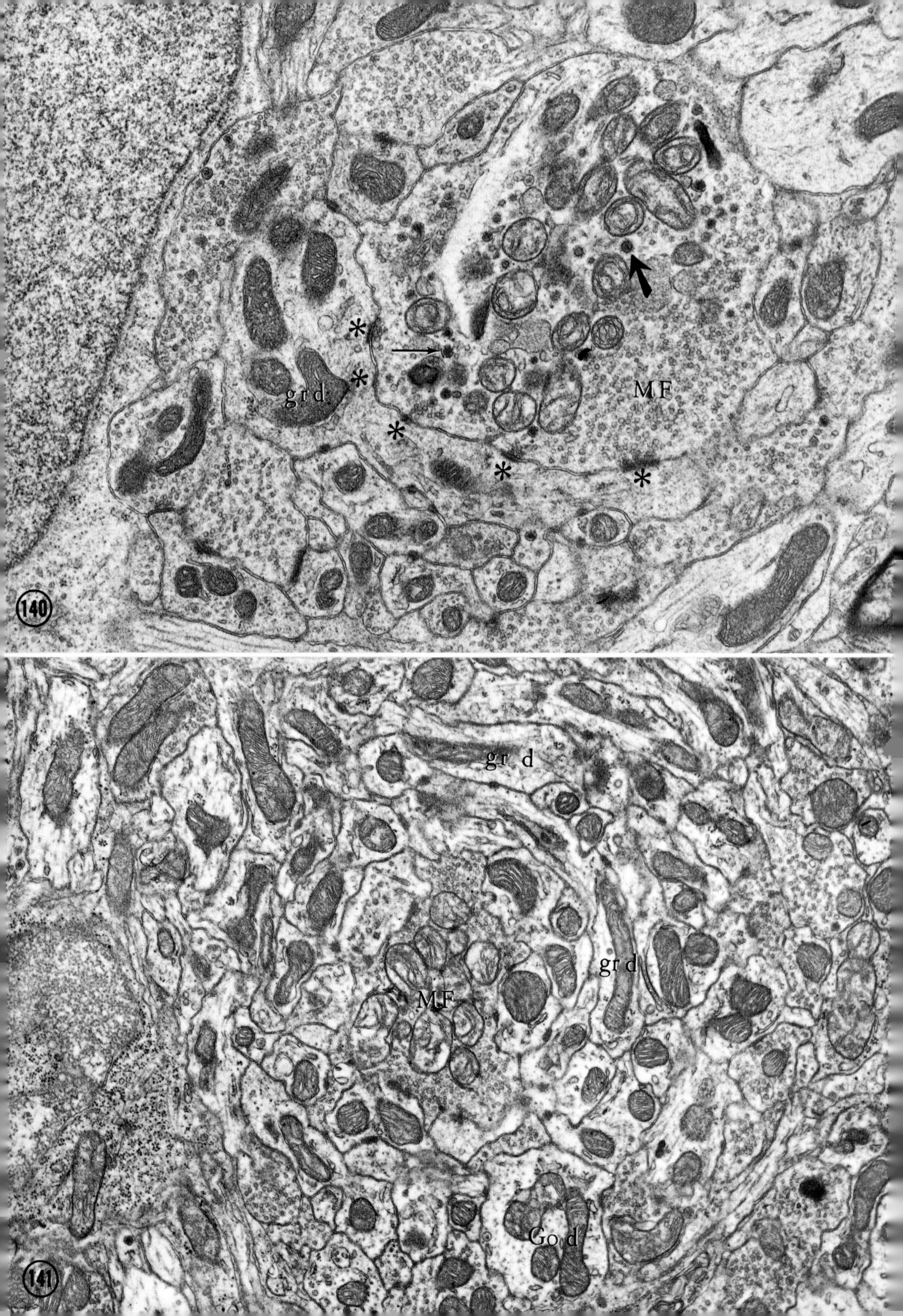

140
grd
MF
gr d
gr d
MF
Go d
141

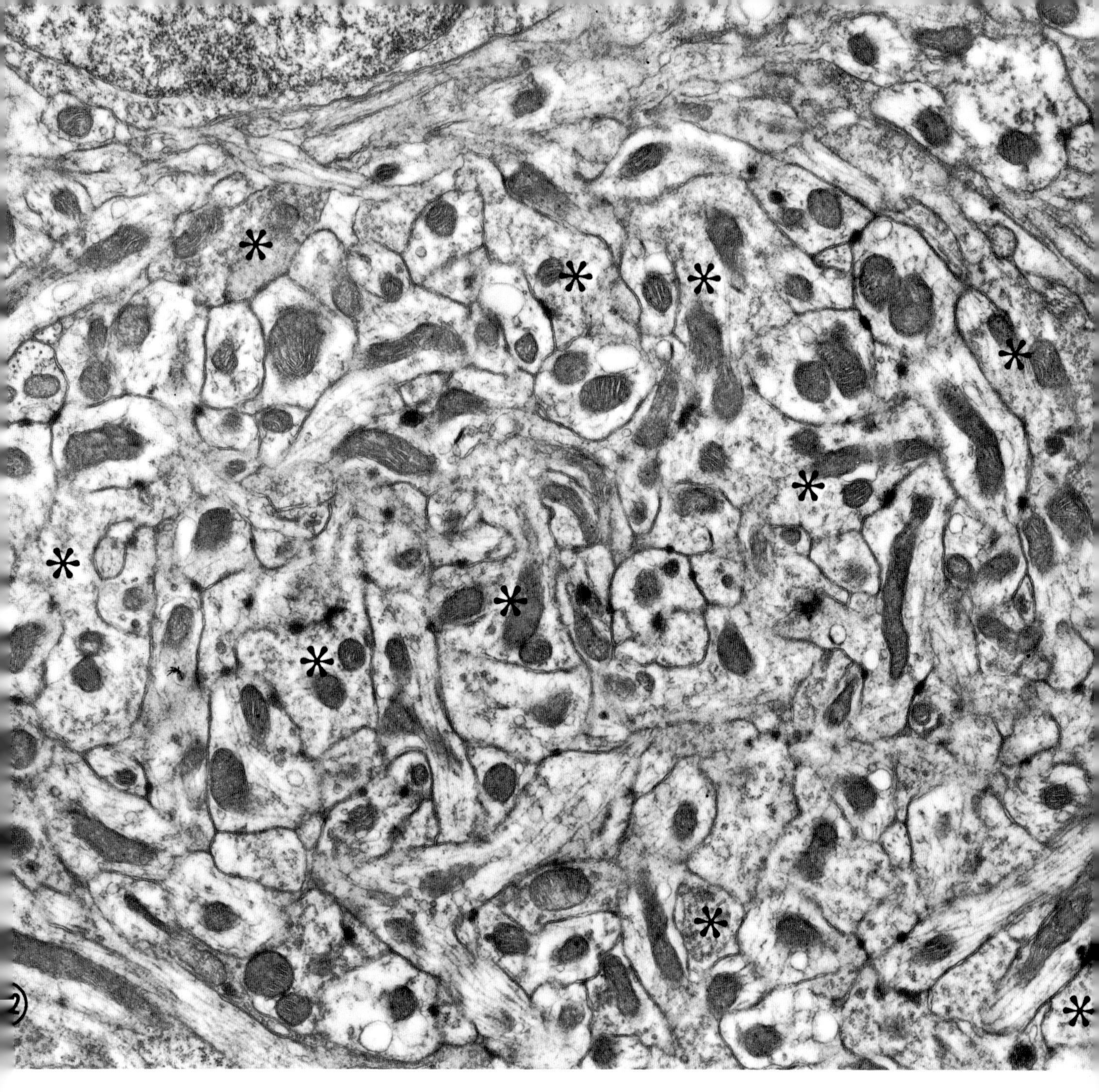

Fig. 142. A section through the periphery of a glomerulus. A tangle of granule cell dendrites and axons of Golgi cells (*asterisks*) form the center of this illustration. Osmium tetroxide fixation. ×15000

◄

Fig. 140. Dense core vesicles in mossy fiber rosette. This mossy fiber rosette has many large dense core vesicles (*small arrow*) among the usual small clear synaptic vesicles. Several "spiny" vesicles are also present in this profile (*large arrow*). A granule cell dendrite wraps around the axon and makes several small synapses along the interface (*asterisks*). Lobule VIII. ×26000

Fig. 141. Swirl of granule cell dendrites in a mossy fiber glomerulus. The central mossy fiber terminal (*MF*) is surrounded by a swirl of dendrites belonging to granule cells (*gr d*). A Golgi dendrite (*Go d*) occurs in the periphery of the glomerulus. Lobule V. ×14000

►

Fig. 142A. Low power view of the upper granular layer. A Purkinje cell (*PC*) appears at the upper left and three granule cells (*gr c*) are seen at the right. The lower half of the field is occupied by a large glomerulus in which the claw-like terminal branches of granule cell dendrites (*gc d*) and numerous fragments of mossy fibers (*mf*) can be made out. Neuroglial processes (*ng*) are also evident. The large arrow indicates the direction of shadowing. Replica, freeze fracture. ×14000

Fig. 142B. Mossy fiber rosette. The left upper part of the field is occupied by the broken axoplasm of a mossy fiber terminal in which synaptic vesicles and mitochondria are embedded. The B face of the plasmalemma of this axon extends toward the center of the field from the mass of axoplasm. The B face is marked by several sets of small papules (*arrows*) each associated with a cluster of particles. These arrays correspond to the sites of the presynaptic specializations. The large arrow indicates the direction of shadowing. Replica, freeze fracture. ×78000

Fig. 142C. Mossy fiber rosette and granule cell dendrites. The background of this field is provided by the A face of a mossy fiber rosette, covered with small particles of different sizes. Ridges and grooves in the surface result in the appearance of a mountainous landscape, that is the complement of the B face in Fig. 142B. Shallow depressions of oval and roughly circular outline are relatively clear of A face particles, but exhibit small pits around which particles are clustered (*arrows*). Covering parts of the mossy fiber are the B faces of several granule cell dendrites. These faces are generally smooth except for small aggregates of particles (*circles*, *pa*) which correspond to the puncta adhaerentia seen in conventional thin section electron microscopy. The large arrow indicates the direction of shadowing. Replica, freeze fracture. ×77000

PC
gr c
ng
gcd
gcd
mf
mf
mf
142 A

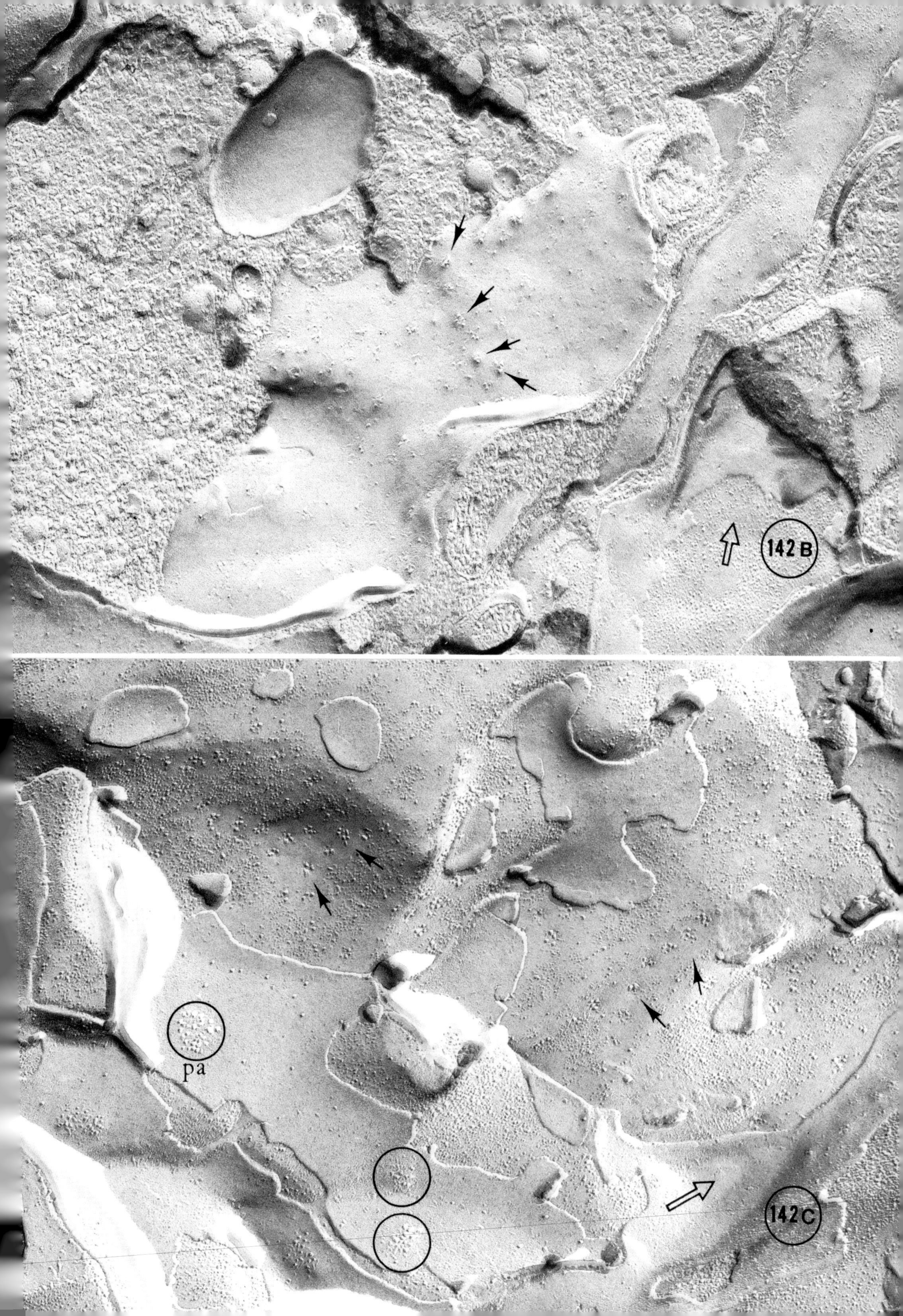
142 B
pa
142 C

is often a palisade of granule cell dendrites, each containing a central mitochondrion within a circle of the endoplasmic reticulum, much as in conventional thin sections.

The mossy fiber rosette presents a broad surface for synaptic junctions to these surrounding dendrites. This surface is grooved and facetted to receive them much as expected from a study of thin sections. The intramembranous synaptic specializations appear as linear or macular arrays about 0.3–0.5 μm in largest dimension. In the B face of the plasmalemma (Fig. 142 B) these arrays consist of chains or rounded collections of small papules, each 250–300 Å across, upon which is seated a cluster of 4–5 intramembranous particles (*arrows*). In some instances the entire array rests on an elevated macular zone of the membrane. The complementary appearance is presented by the A face (Fig. 142 C). The raised maculae of the B face are represented here by flattened or shallow depressions of irregular outline and cleared of the usual A face particles. These depressions are marked with small pits arrayed in a single row (*arrows*) or as irregular collections (↑). A cluster of 3–6 particles marks the mouth of each pit. According to AKERT *et al.* (1972) and PFENNINGER *et al.* (1972) such B face projections and their complementary A face pits represent the points of attachment of synaptic vesicles to the presynaptic membrane. Thus, freeze-fracture preparations of the mossy fiber rosette demonstrate the same kind of intramembranous structure in the axonal synaptic surface as appears in the axonal membrane of the parallel fiber (see Chap. III).

Fig. 142 C shows still another intramembranous specialization. This is a group of particles (*circles*, *pa*) aggregated on the B face of the granule cell dendrite. The clusters are 800–1 300 Å across and are spotted here and there over the dendritic membrane. These clusters are much smaller than the presynaptic maculae and arrays of pits that are found on the mossy fiber rosette. They also occur on the B faces of dendrites running along or crossing over other granule cell dendrites. These clusters, therefore, must correspond to the puncta adhaerentia that appear commonly in thin sections along the interfaces of adjacent granule cell dendrites as well as along the junctions of mossy fiber rosettes with these dendrites. The complementary A face specialization of the punctum adhaerens has not yet been recognized; it may be represented by circular clearings in the randomly dispersed A face particles. Although puncta adhaerentia are very common on the dendritic membranes, as the field in Fig. 142 C indicates, we have not yet found any membranous specialization or array of particles in these freeze fractured preparations that can correspond to the postsynaptic site. Except for the circular aggregates representing the puncta, the B face of the dendritic membrane appears smooth and undifferentiated. Clearly the subject needs further investigation.

The Golgi Cell Axonal Plexus. The terminal axonal arborization of the Golgi cell also contributes to the glomerulus by interweaving with the dendritic claws of the granule cells (Figs. 143 and 144). Although these axons occasionally come into apposition with the mossy fiber, they have not been seen in synapse with it (see Figs. 144 and 145 for a possible exception). Usually Golgi axons appear on the outskirts of the glomerulus, where their varicosities synapse with the dendrites of the granule cells. These contacts are on the side of the dendrite opposite to that synapsing with the mossy fiber. The dendrite is therefore subject to afferent input from two different sources, one excitatory and the other inhibitory. These synapses have already been described (see Chap. III). Golgi axons and their synapses with granule cell dendrites are best demonstrated in glomeruli where the central axon terminal has degenerated, such as after a spinocerebellar lesion (see Fig. 152).

Golgi cell dendrites are not commonly entangled in the glomerulus, but they must pass through at least the edge of glomerular territories in order to reach the molecular layer. Besides, mossy fiber rosettes form a distinctive synapse *en marron* against the corrugated surface of Golgi cell perikarya and the bases of their dendrites. These synapses, already described fully in the section on the Golgi cell, bring the Golgi cell into the glomerulus and, by explaining some of the observations of the Scheibels, justify to some extent their interpretations. Those mossy fibers that engage in *marron* synapses with Golgi cells are therefore capable of activating not only a large number of granule cells through their axo-dendritic synapses, but also at the same time, a Golgi cell through an extensive axo-somatic synapse. The effect of these two varieties of articulation is to activate at the same time both inhibitory and excitatory limbs of a circuit involving the Golgi cell, the granule cells, and the Purkinje cell.

The Protoplasmic Islet. Finally, some comments should be made concerning the organization of the glomeruli in the protoplasmic islets. RAMÓN Y CAJAL (1911, 1926) and other investigators have noted that a number of mossy fiber rosettes can participate in the construction of a single islet. These rosettes can originate from different mossy fibers or from collaterals of the same fiber. As many as five or six mossy fibers can contribute to an islet. These rosettes are not, however, in axo-axonic synapse with one another, or even usually in contact, despite the appearance given by GOLGI and reduced silver

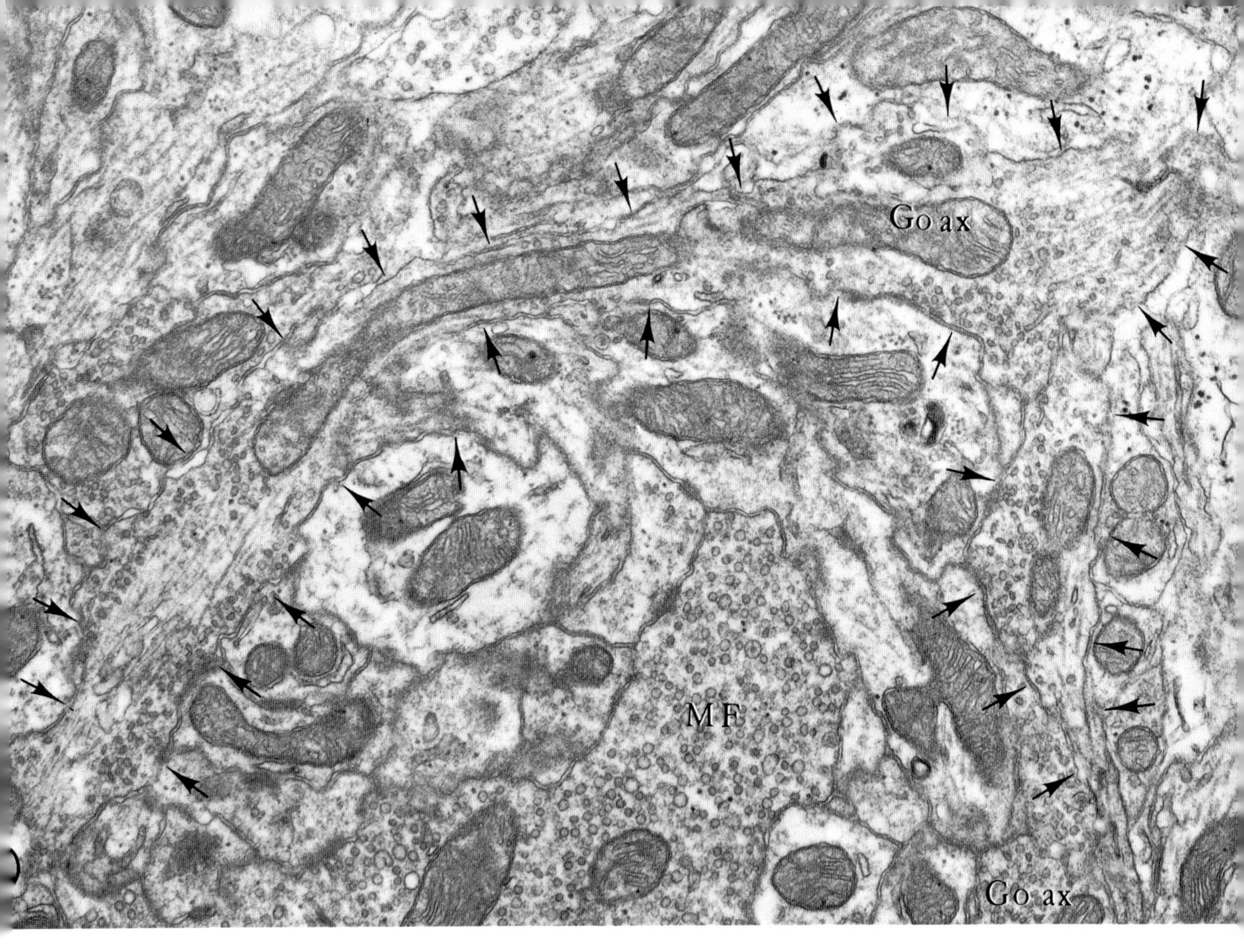

Fig. 143. The Golgi cell axon. This stretch of Golgi axon (*Go ax, arrows*) surrounds the periphery of a mossy fiber glomerulus (*MF*). Lobule X. × 30000

preparations. Electron micrographs of thin sections show that each rosette is surrounded by its own palisade of granule cell dendrites. It is quite unusual to see dendrites shared between two mossy fibers. Frequently, tenuous neuroglial sheets divide the fields of nerve fibers so that each mossy fiber rosette and its attendant granule cell dendrites is demarcated from its neighbors by a perforated sheath (CHAN-PALAY and PALAY, 1972c). The detailed description of the neuroglia will be given in Chap. XI. In considering this arrangement of the rosettes, dendrites, and neuroglia, it becomes desirable, then, to distinguish between the glomerulus and the protoplasmic islet. The islet is regarded as a cell-free zone[15] of neuropil containing one or more glomeruli, each of which consists of a central mossy fiber rosette and its surrounding dendrites. According to this conception, each glomerulus would contain only one mossy fiber rosette. This definition has some bearing on the estimates of the amount of divergence effected in the glomerulus, for if a protoplasmic islet contains more than one glomerulus, the basic assumption for calculations in the literature may have to be revised. For example, FOX, HILLMAN, SIEGESMUND, and DUTTA (1967) counted up to 15 granule cell dendrites entering a single glomerulus in Golgi preparations, and MUGNAINI (1972) shows a photomicrograph of a granule cell in a Golgi preparation that apparently sends two dendrites into the same glomerulus. But since the boundaries of a glomerulus within the protoplasmic islet cannot be distinguished in Golgi preparations, there is no assurance that these dendrites are all terminating in relation to the same mossy fiber. Similarly, ECCLES, ITO, and SZENTÁGOTHAI (1967) made a simple calculation of the divergence provided by each glomerulus in the cat by proceeding from the reasonable assumption that each granule cell dendrite ends in a different glomerulus. Then they calculated that the number of granule cells served by any one mossy fiber rosette would equal the ratio of granule cells to glomeruli in the granular layer

15 No special distinctions need to be made for the rosettes synapsing *en marron*, since the Golgi cells, like the neighboring granule cells, lie around the periphery of the protoplasmic islet.

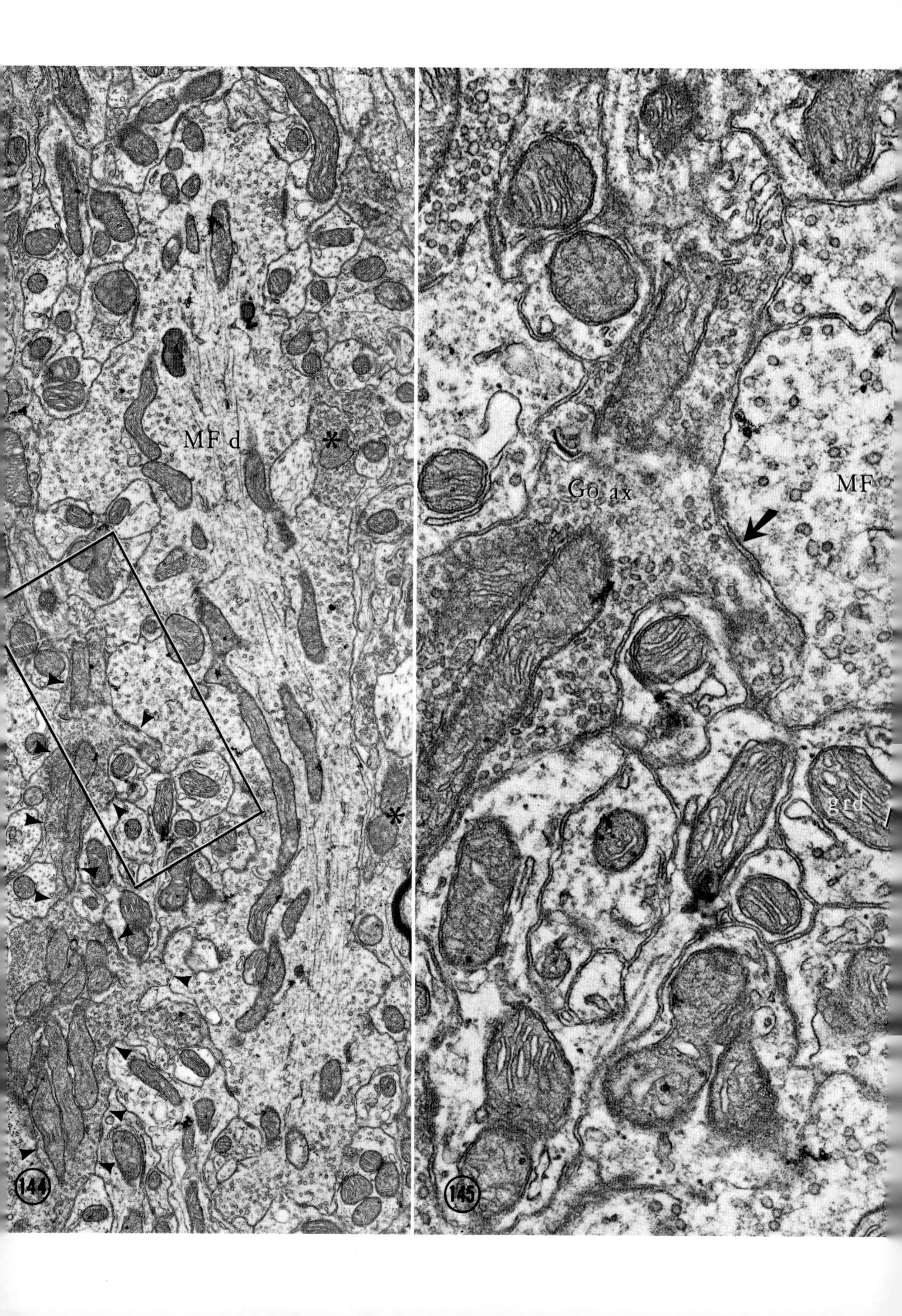
MF d
Go ax
MF
grd
144
145

(4.5) multiplied by the average number of dendrites per granule cell (4.5). This yields a product of 20 granule cells in synapse with a mossy fiber rosette, a value that is in satisfying agreement with the number of granule cell dendrites entering a glomerulus according to FOX, HILLMAN, SIEGESMUND, and DUTTA (1967). However, what was meant by "glomeruli" is really protoplasmic islets. Therefore, the number of glomeruli was underestimated by an unknown factor. Similar considerations might apply to the recent revision of this estimate by PALKOVITS *et al.* (1972). To our knowledge, the number of glomeruli per islet has not been determined in any animal. It is, therefore, not clear how much divergence actually occurs in the glomerulus. One suspects that it is less than has been thought heretofore.

We agree, however, with MUGNAINI (1972) that the axonal plexuses of Golgi cells overlap or interlace to some degree and that any particular glomerulus may be served by more than one Golgi cell. Consequently, the territories occupied by the axonal plexuses of Golgi cells are not necessarily exclusive. Thus it may happen that neighboring granule cells could be under the control of different Golgi cells, or that a single granule cell might respond to different Golgi cells from one moment to the next.

4. The Identification of Different Kinds of Mossy Fibers

In Chap. I it was mentioned that afferent fibers from diverse sources terminate in the cerebellar cortex as mossy fibers. In the present chapter we have indicated that only a few different kinds of mossy fibers can be distinguished by their size or by the form and content of the rosettes. For example, there are simple and complex rosettes, dispersed and clustered patterns of synaptic vesicles. There have also been suggestions that mossy fibers from different sources have different forms (BRODAL and DRABLØS, 1963) or different territories of distribution in the granular layer (ECCLES, ITO, and SZENTÁGOTHAI, 1967). Although the evidence is not unequivocal and may even be of doubtful validity, the suspicion persists that the widely different sources of fibers ending as mossy fibers should be reflected in their form, content, distribution, or pharmacology.

MUGNAINI (1972) has suggested that the varied glycogen content of mossy fiber rosettes might be a useful marker for tracing fibers to their origins. In the normal rat only a few mossy fiber terminals contain glycogen. Another feature that might prove useful in this connection is the presence of acetylcholinesterase. Histochemical studies at the light microscope level (HEBB, 1959; CSILLIK *et al.*, 1963; SHUTE and LEWIS, 1965) have demonstrated that acetylcholinesterase activity is differentially distributed in the lobules of the cerebellar cortex in such a way that the anterior lobe and the posterior vermis, i.e., the spinal parts of the cerebellum, exhibit much more activity than the rest of the cerebellum. Most of the activity is localized in the granular layer, in the glomeruli (HEBB, 1959; CSILLIK *et al.*, 1963; SILVER, 1967). Examination of the reaction at the electron microscope level (SHIMIZU and ISHII, 1966; BROWN and PALAY, 1972) shows that the

◄

Fig. 144. A mossy fiber/Golgi axon apposition. A mossy fiber rosette with dispersed synaptic vesicles is flanked on the right by two small Golgi axon terminals (*asterisks*). On the left a large segment of a Golgi axon approaches and abuts against the mossy fiber. The region of this axo-axonal apposition (*rectangle*) is enlarged in the next figure. Lobule X. ×15000

Fig. 145. The Golgi axon/mossy fiber junction. The Golgi axon (*Go ax*) abuts against the mossy fiber surface, and an axo-axonal junction (*large arrow*) is formed. This type of synapse is very rare. The common form of synaptic contact is one made by a granule cell dendrite (*gr d*) with the mossy fiber (*fine arrow*). Lobule X. ×47000

►

Fig. 146. Mossy fiber with small amounts of glycogen. This mossy fiber (*MF*) rosette shows a distinct neurofilamentous core (*n*) that is typical of mossy fibers. Glycogen particles are strewn in the axoplasm (*arrow*). These particles may indicate an early stage in the degeneration of the mossy fibers belonging to the spinocerebellar tracts. Hemisection of spinal cord at C4, severed spinocerebellar tracts, 10 hours after surgery. Lobule II. ×22000

Fig. 147. Simple mossy fiber with clumps of glycogen. This mossy fiber (*MF*) has many fairly large clusters of glycogen particles (*arrow*) in the axoplasm. The axon still retains normal synapses with granule cell dendrites. The surrounding neuropil also is apparently healthy and well preserved. Hemisection of spinal cord at C4, severed spinocerebellar tracts; 10 hours after surgery. Lobule IV. ×15000

Fig. 148. Simple mossy fiber with large collections of glycogen. This mossy fiber rosette (*MF*) has many large clusters of glycogen particles scattered among neurofilaments (*n*), synaptic vesicles, and mitochondria. Hemisection of spinal cord at C4, severed spinocerebellar tracts; 18 hours after surgery. Lobule IV. ×17000

Fig. 149. Complex mossy fiber and Golgi axon adjacent to granule cell. The mossy fiber rosette (*MF*) has many complicated projections (*asterisks*) that protrude into the periphery of the glomerulus. The axoplasm of this mossy fiber also shows large clumps of glycogen. A portion of a Golgi axon (*Go ax, arrows*) abuts against the soma of a granule cell (*gr*). Hemisection of spinal cord at C4, severed spinocerebellar tracts; 18 hours after surgery. Lobule IV. ×19000

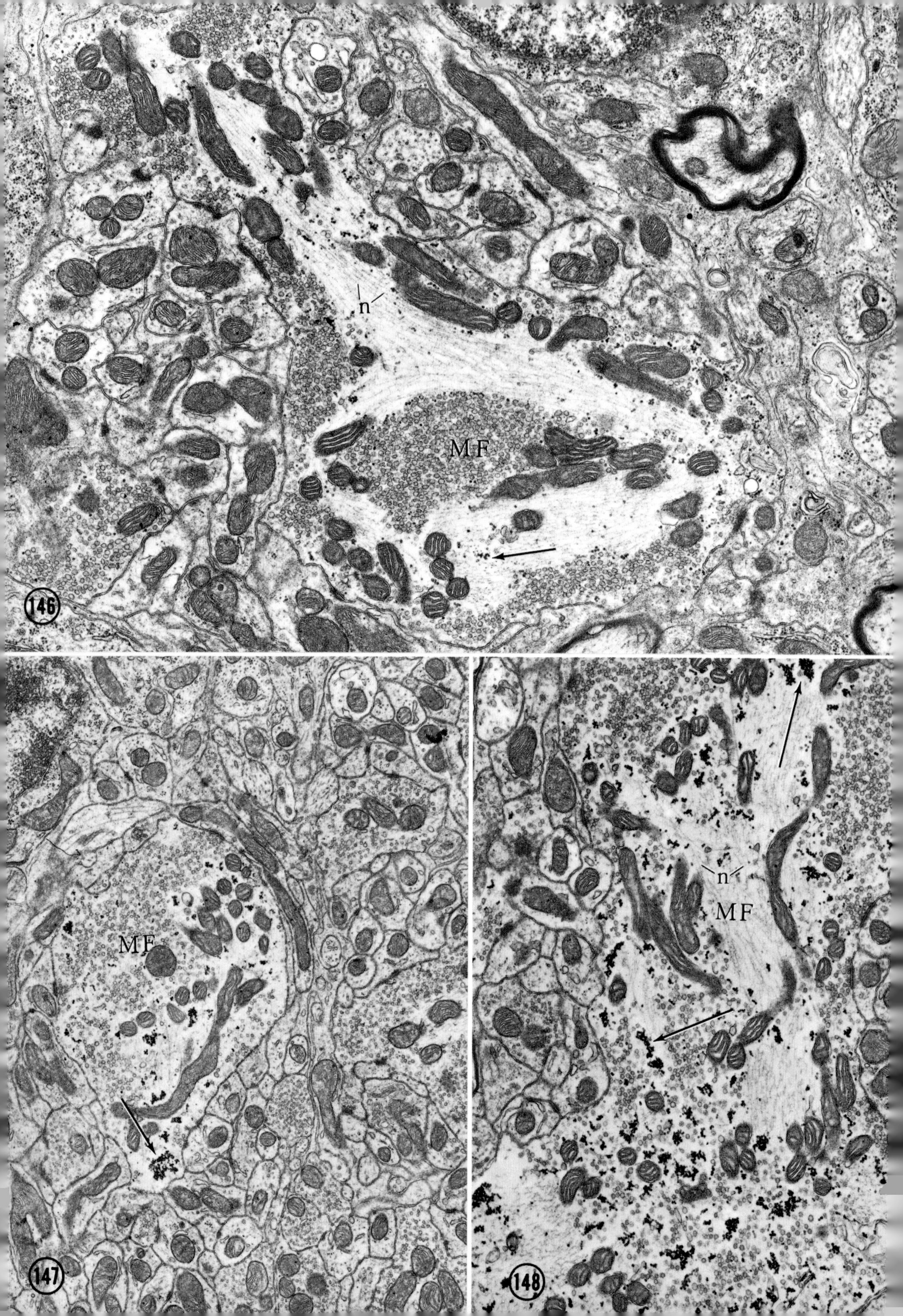
n
MF
146
MF
147
n
MF
148

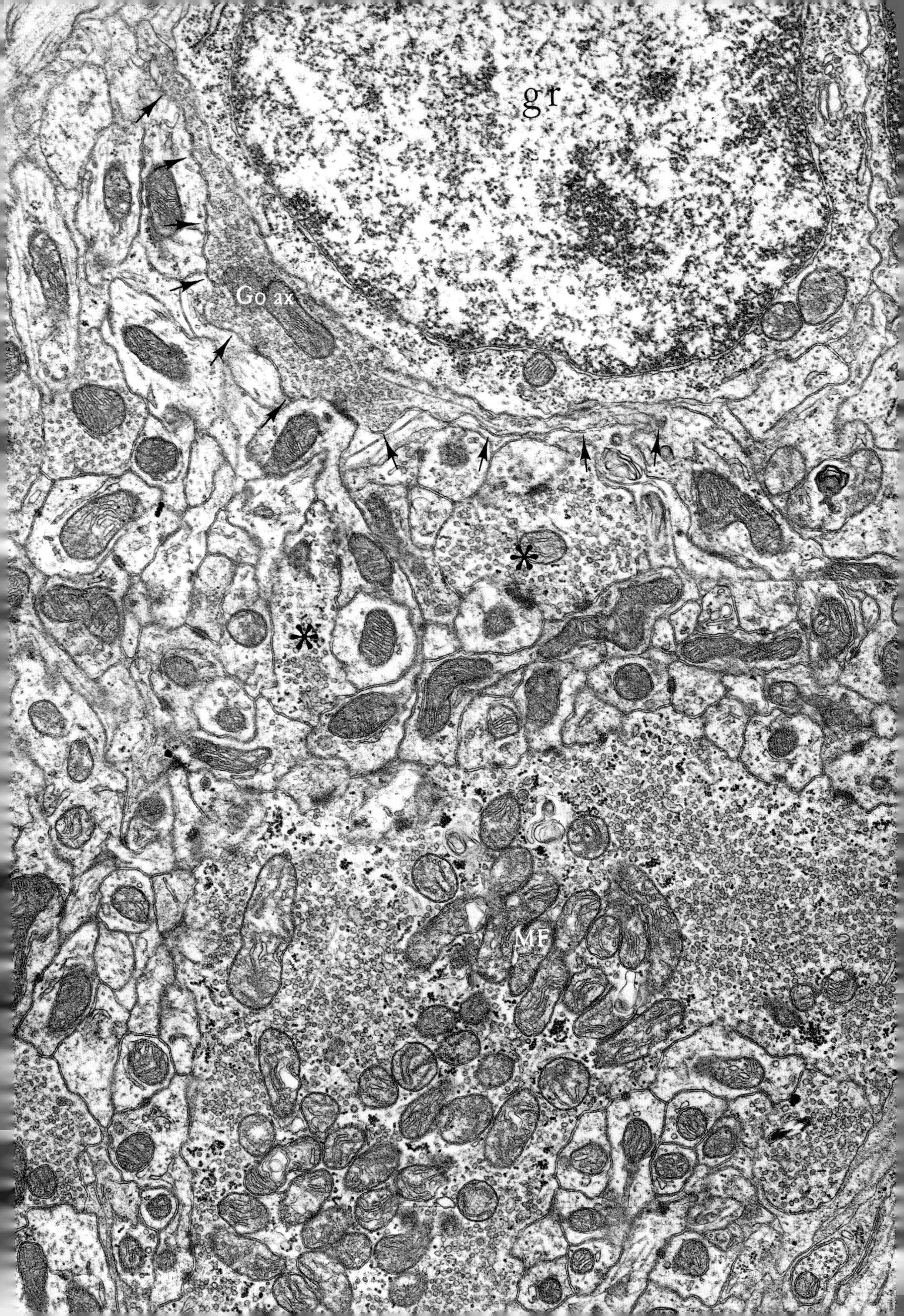
gr
Go ax
MF

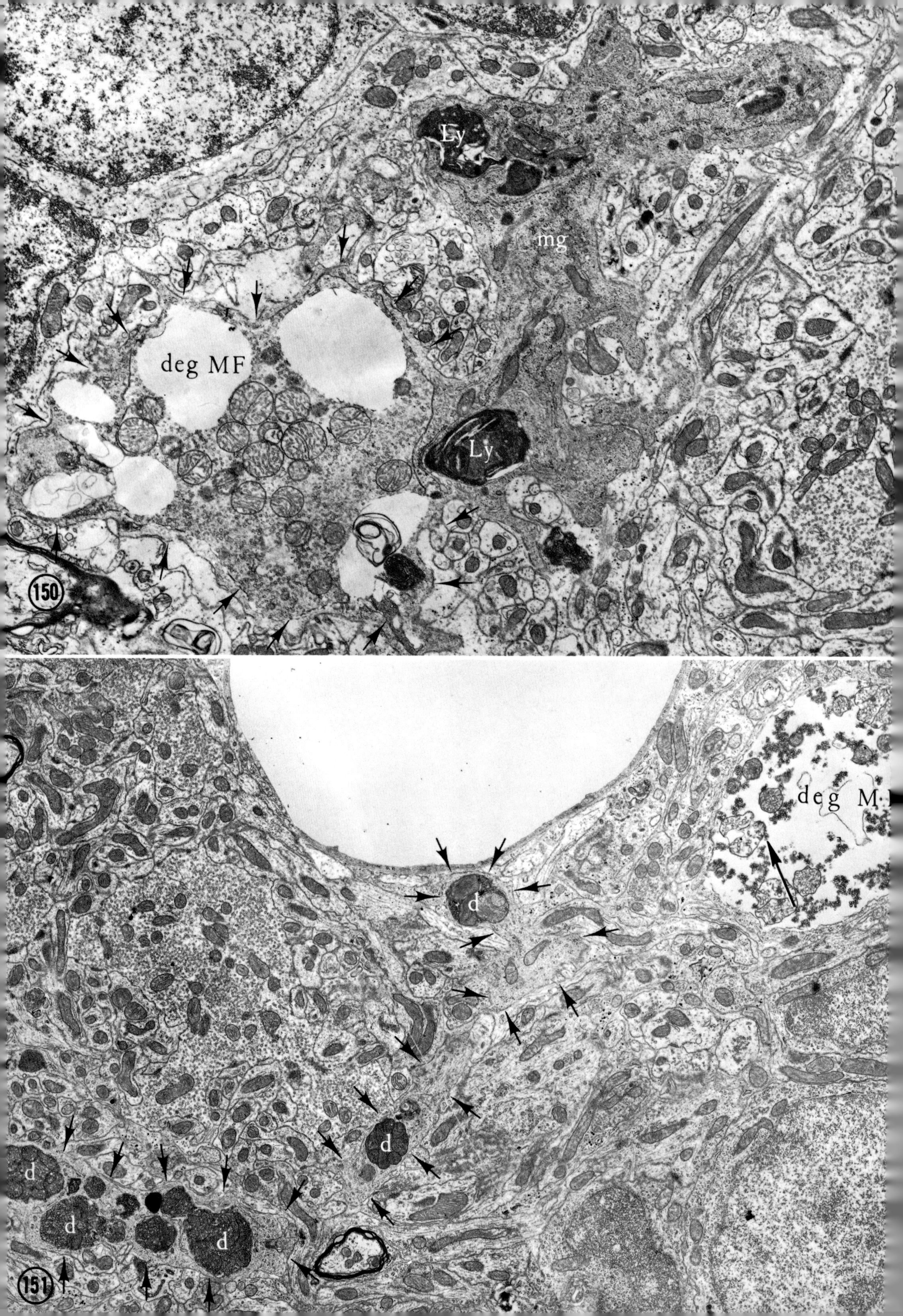
Ly
mg
deg MF
Ly
150
deg M
d
d
d
d
d
151

reaction product is very delicately distributed in the interstitial spaces of some glomeruli, especially the spaces between the mossy fiber ending and the granule cell dendrites. The distribution of esterase activity is not uniform in those lobules where it is most intense. Some glomeruli exhibit activity, while others in the vicinity show none at all. These differences might reflect differences in the cells of origin of the fibers, for example, those coming from spinal centers might be active while those of pontine or reticular origin might be inactive. But no morphological feature of either the rosettes or the rest of the glomeruli could be correlated with the presence or absence of acetylcholinesterase. Terminals as different as climbing and mossy fiber were both found with and without enzyme activity (BROWN and PALAY, 1972). Unfortunately, the test for enzyme was not combined with an experimental interruption of the spinocerebellar or any other afferent pathway.

MUGNAINI (1972) tried to distinguish spinocerebellar fibers from other afferents in an electron microscopic study of degenerating mossy fibers after low cervical hemisection of the cord in the cat. He was unable to demonstrate any consistent distribution of the spinal afferents within the granular layer of the appropriate lobules. They were dispersed all over the layer, as indeed he predicted from the study of Golgi preparations and as SASAKI and STRATA (1967) found by electrophysiological recording. MUGNAINI, however based his study on the time table of degeneration extracted from silver impregnation procedures (BRODAL and GRANT, 1962). Signs of degeneration in these preparations began to appear after several days and were hardly detectable before three days. Consequently, in MUGNAINI's electron micrographs the degenerating mossy fibers are already quite far advanced on their path to destruction, and it is no longer possible to recognize their original shape or the pattern of the organelles in the rosettes. In order to do this, it would be necessary to detect mossy fibers in the earliest stages of degeneration. Consequently, a similar experiment was carried out on adult rats, and emphasis was laid on the earliest intervals after the lesions (CHAN-PALAY, 1973b).

The spinal cord was hemisected at midcervical levels (C4 or C5), and the degenerative events in lobules I to V in the cortex of the ipsilateral vermis were studied by both light and electron microscopy. The animals were perfused with fixative 3, 6, 10, 12, 18, 24, 36, 48, and 72 hours and 7 days after the surgical lesions.

Fig. 150. Degenerating spinocerebellar mossy fiber. This degenerating mossy fiber has swollen mitochondria, a dark, patchy axoplasmic matrix, and scattered synaptic vesicles. Large pools have appeared in the axoplasm. The entire rosette (*arrows*) is already engulfed by the cytoplasm of a microglial cell (*mg*); hence no synaptic junctions are evident. The microglial cell cytoplasm includes several large masses of debris in lysosomes (*Ly*), probably the remains of other spinocerebellar glomeruli. Hemisection of spinal cord at C4; 24 hours after surgery. Lobule IV. ×11500

Fig. 151. Degenerating spinocerebellar mossy fibers. In the upper right hand corner the axoplasm of a mossy fiber undergoes lysis (*deg MF*). Mitochondria are blown up (*small arrow*), and synaptic vesicles are packed into dark pyknotic clumps (*large arrow*). On the left a trail of dark pyknotic clumps (*d*) containing debris from synaptic vesicles and mitochondria can be seen within the cytoplasm of a phagocytotic neuroglial cell (*arrows*). The surrounding neuropil is apparently healthy and well preserved. Hemisection at C4, severed spinocerebellar tracts; 24 hours after surgery. Lobule IV. ×18000

Fig. 152. Crenated remains from dying spinocerebellar mossy fiber. The axoplasmic debris is a pyknotic mass in which crushed and crumpled mitochondria (*mit*) and collapsed synaptic vesicles (*large arrow*) can be discerned. This dark mass is surrounded by a thin shell of neuroglia (*gl*) whose presence is indicated by glycogen particles. The loss of the central mossy fiber terminal renders the surrounding granule cell dendrites and Golgi cell axon (*asterisks*) in the glomerulus more conspicuous. The small arrows indicate synapses between granule cell dendrites and the thin portions of the beaded Golgi axon. Hemisection of spinal cord at C4, severed spinocerebellar tracts; 36 hours after surgery. Lobule IV. ×23000

Fig. 153. Remains of spinocerebellar mossy fiber within a microglial cell. The cytoplasm of this microglial cell (ΔΔ) is replete with large clumps of crenated mitochondria and synaptic vesicles (*d*). Large lipid droplets (*l*) are present as well. Hemisection of spinal cord at C4; 48 hours after surgery. Lobule V. ×21000

Fig. 154. Replete microglial cell. This microglial cell (*mg*) is in a late stage in the digestion of the debris from spinocerebellar mossy fibers. Only numerous lysosomes (*Ly*) remain to tell the tale. The cell lies adjacent to a Purkinje cell soma. Hemisection of spinal cord at C4; 72 hours after surgery. Lobule IV. ×18000

Fig. 155. Replete microglial cell resting beside a blood capillary. This microglial cell (*mg*), replete with the lysosomal remains of spinocerebellar mossy fibers, comes to rest upon a blood capillary in the molecular layer. Hemisection of spinal cord at C4; 72 hours after surgery. Lobule IV. ×16000

Fig. 156. Removal of debris from within a myelinated axon. A tongue of cytoplasm from a microglial cell (*mg, arrows*) invades a myelinated spinocerebellar mossy fiber in order to remove the debris (*d*) within. The mass consists of crumpled up mitochondria and vesicles. Hemisection of spinal cord at C4; 48 hours after surgery. Lobule IV. ×27000

Fig. 157. Degenerated segments of spinocerebellar mossy fibers in the white matter. Two myelinated spinocerebellar mossy fibers (*large white arrows*) with degenerating axoplasm are seen at the right and left of the figure. A large myelinated axon, probably a mossy fiber, courses through the center of the field (*mMF*). Two Purkinje cell axons (*PC ax*) flank it on either side. The identifying characteristic of these axons is the presence of the hypolemmal cisterna (*small arrow*). Hemisection of spinal cord at C4; 48 hours after surgery. Lobule V. ×13000

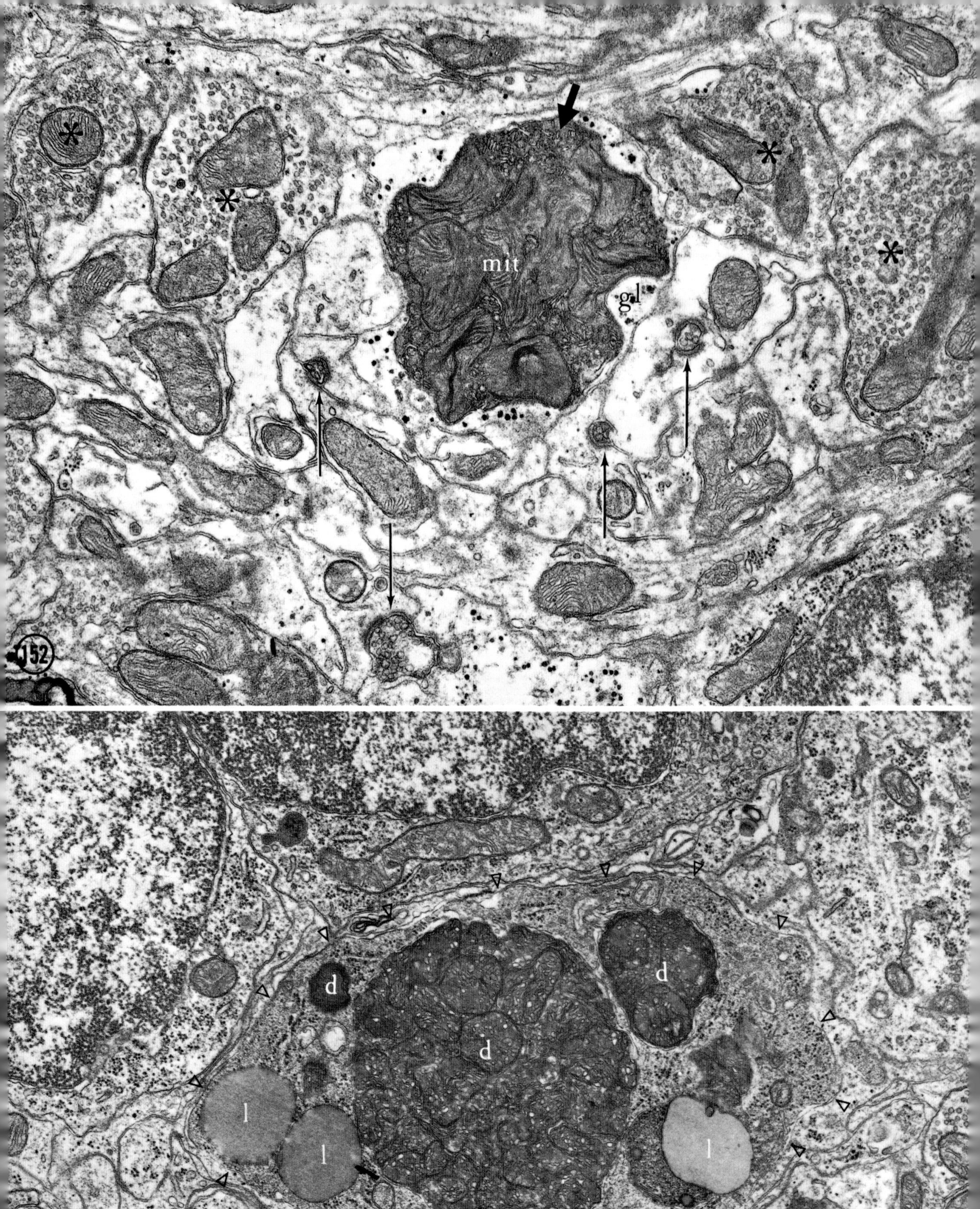
mit
gl
152

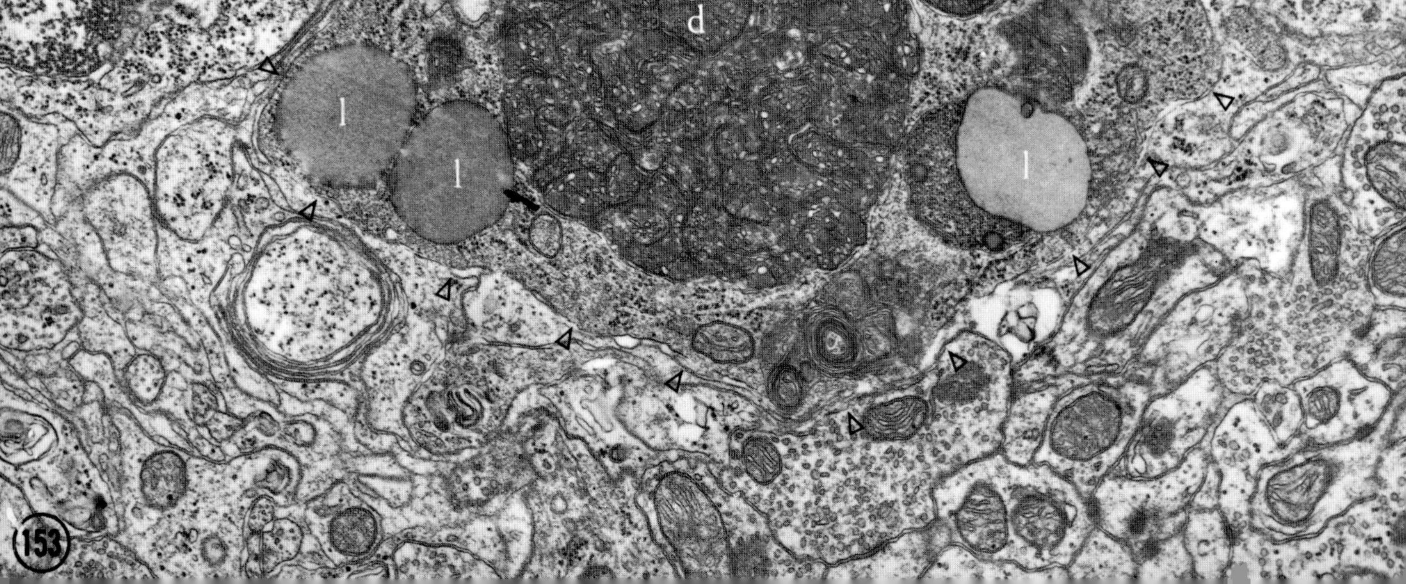
d
d
d
l
l
l
153

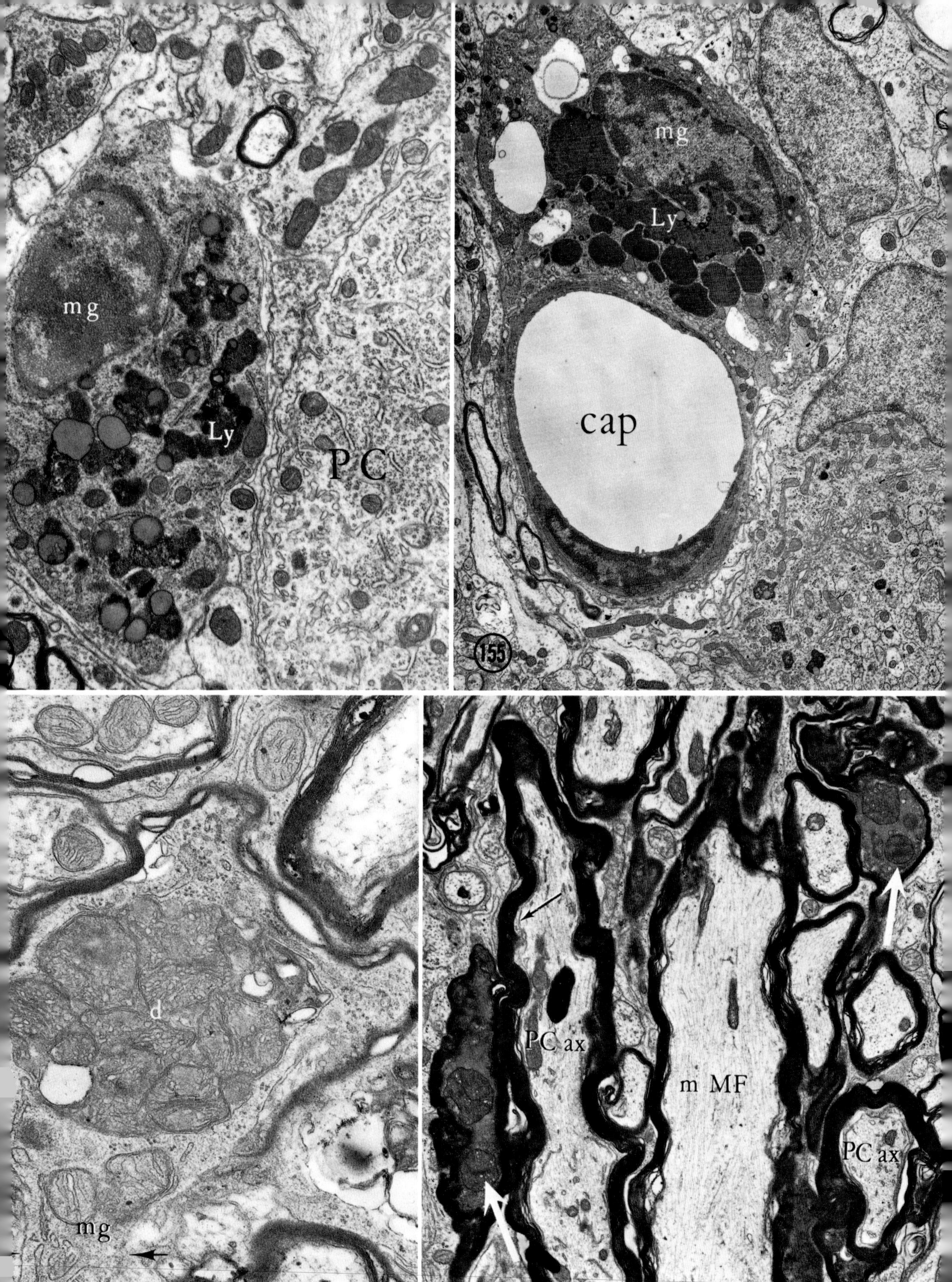
mg
Ly
PC
mg
Ly
cap
155
d
mg
PC ax
m MF
PC ax
157

The earliest sign of impending degeneration in certain mossy fiber terminals is the accumulation of small clumps of glycogen. These appear as early as 10–12 hours after the lesion. Although single β particles are normally present in the central axons of some glomeruli, larger clumps or aggregations have been seen only in these experimental animals. The clumps grow progressively larger with time, and sometimes form multiple α glycogen foci (see Figs. 146 and 149). These foci can be found in many glomeruli, especially in lobules I–III, approximately 18 hours after the lesion. They are never found in the normal animal. Concurrently with the clumping of glycogen, the mitochondria in these terminals also begin to change. They become darker, and the cristae swell somewhat and acquire distorted shapes (Fig. 148).

By 24 hours the axon terminal can already be undergoing lysis. Mitochondria become swollen and very round. Large clear pools appear in the terminal, and some synaptic vesicles become flattened and elongated. The remainder of the background matrix may be even more flocculent than in the normal state (Fig. 150). Synaptic junctions, however, may still persist. In Fig. 151 the terminal in the right hand corner shows a late stage in lysis and clumping of synaptic vesicles. The mitochondria are almost completely swollen, and the clumped synaptic vesicles are flattened and pushed to one side as if in a discarded heap. No synaptic junctions are evident.

Fig. 152 shows a later stage (about 36 hours) in degeneration. Here the axon terminal has shrunken into a distorted sac in which the crenated remains of synaptic vesicles and crumpled mitochondria are evident. Such a terminal is still surrounded by a healthy tangle of granule cell dendrites that form synapses with equally robust Golgi cell axons. Insinuated between these are slips of normal neuroglial cytoplasm containing glycogen particles.

Early in the developmental sequence of degeneration, even before 24 hours, small cells with dark basophilic nuclei can be seen strewn throughout the granular and molecular layers. The origin of these cells is unknown. In the electron microscope they have the cytological characteristics common to microglial cells. On their arrival, these cells are generally not distended with debris or lysosomes. In due course their conspicuous dark cytoplasmic pseudopodia insinuate themselves into and between the glomeruli that are early in their degenerating sequence (see Figs. 150 and 151). In the normal rat, the microglia are rarely encountered in the cerebellar cortex. Thus, in the experimental animal their presence is a clear flag marking degenerating glomeruli.

By 48 hours, it is common to find profiles of microglial cells containing many clumps of crenated synaptic vesicles and mitochondria that are barely recognizable. These profiles are usually tucked in between the remaining neuropil, which can be remarkably healthy in appearance (Figs. 151 and 153). Later stages (72 hours onwards) in the digestion of such debris within the microglial cells are signaled by the formation of large, homogeneous, dark lysosomes; lipid droplets, and the disappearance of recognizable axonal debris. Often microglia laden with such inclusions can be found to have wandered even into the molecular layer (Figs. 154 and 155). Here they may come to rest next to a blood capillary for the remainder of the observation period, i.e., 1 week. It is as if they were arrested here until their load of debris has been digested.

Fig. 157 shows that the processes of a microglial cell can even invade the confines of the myelin sheath in order to evacuate pyknotic clumps of mitochondria and other cellular debris from within it. Profiles of degenerating axoplasm within myelin sheaths are rather common in the later survival periods, e.g., from 2 days onwards (Fig. 157). In fact, after the third day, nearly all degenerating glomeruli have been cleared away and only myelinated portions of degenerating axons are left. These experiments suggest that the cytological events following cord transection reach a climax very early for unmyelinated sections of axons. The myelinated preterminal parts of degenerating axons can persist for a longer period of time. Generally by the end of a week only the myelin sheaths are left to be removed.

In studying these degenerating fibers it was found that the actual time sequence of degeneration varied from one axon to another. Some axons showed a rapid onset of degeneration; others were much slower. This means, therefore, that especially in the intervals from 2 days onwards one is likely to find not only late degeneration in some glomeruli but also some early degenerative signs in others (Fig. 153). The overall developmental sequence of degeneration, however, becomes clear upon study of a very large sample of material at all stages beginning shortly after the lesion.

Up to the present time, however, these observations on the sequence of degenerative changes following spinal cord section have not given an unambiguous answer to the question of whether the mossy fibers originating from the various afferent tracts, i.e., vestibular, spinal, and pontine, are distinguishable cytologically from one another. The observations gathered from the early stages of degenerating mossy fiber glomeruli suggested the following conclusions. Firstly, only glomeruli that were clearly supplied by mossy fibers degenerated after spinal cord hemisection. The glomeruli described as belonging to climbing fibers did not fit anywhere into the scheme of degeneration (CHAN-PALAY and PALAY, 1971a). Secondly, the mossy fibers that were observed in their earliest stages of degeneration were generally simple rosettes, not those of the more complicated variety. This is shown in Figs. 146–148, where the clumps of glycogen in glomeruli are confined to the central axonal profile and are not found in smaller axonal profiles around the central one. These simple mossy fibers always had only dispersed synaptic vesicles. Perhaps, then, the cerebellar afferents of spinal origin are distributed in the vermis as mossy fibers with simple rosettes containing dispersed vesicles. This is, however, only an empirical correlation. Since the spinocerebellar tracts distribute bilaterally, it was not possible to obtain a clear control to show that only simple mossy fibers originate from these tracts. Rarely, however, a more complex mossy fiber rosette with clustered synaptic vesicles was found to be undergoing an early stage of degeneration (see Fig. 149). Thus, a categorical statement cannot be made correlating the spinal origin of mossy fibers with simple rosettes and dispersed synaptic vesicles.

5. Summary of Intracortical Synaptic Connections of Mossy Fibers

mossy fiber —— granule cell —— Purkinje cell
mossy fiber — Golgi cell — granule cell — Purkinje cell
mossy fiber — granule cell — basket cell — Purkinje cell
mossy fiber — granule cell — stellate cell — Purkinje cell

Chapter VII

The Basket Cell

Compared with the granule cell, the basket cell is quite complicated. It receives synapses from parallel fibers and to a limited extent from climbing fibers, but it devotes its entire axonal output to the Purkinje cells with the possible exception of a few contacts on other basket cells and Golgi cells. A large number of widely dispersed parallel fibers converge on its dendrites, but its axon sends divergent impulses to only a small number of Purkinje cells. In a recent study PALKOVITS *et al.* (1971 c) estimated that in the cat the axon of the basket cell makes contact with only 9 Purkinje cells. The number may be somewhat smaller in the rat. The same authors obtained densities of 6577 basket cells per mm^3 of molecular layer in the cat. About 95% of the basket cells were located in the deeper half of the layer. These values are to be compared with densities of 18695 per mm^3 for stellate cells and 18433 for neuroglial cells. These results indicate that basket fibers are much more numerous than had been previously supposed. Unfortunately, equivalent figures for the rat are not available.

1. A Little History

Although basket cells and their axons were noted by a number of earlier workers, it was again the merit of GOLGI (1883) and especially of RAMÓN Y CAJAL (1888 a and b) to provide the definitive description of them at the level of the optical microscope. RAMÓN Y CAJAL (1888 a and b) was the first to discover the characteristic terminal plexus elaborated around the Purkinje cell body by the axons of these cells. He named this formation the pericellular nest (*nid péricellulaire*). His finding was quickly confirmed by KÖLLIKER (1890), who referred to the same structure as a "fiber basket" (*Faserkorb*), and to the cells that gave rise to it as "basket cells" (*Korbzellen*). These terms have superseded those originated by RAMÓN Y CAJAL, who called the cells "large or deep stellate cells" (*grandes cellules étoilées, cellules étoilées profondes*), and even in his great book on the structure of the nervous system, CAJAL (1911) refers to them interchangeably as *grandes cellules étoilées* and *cellules à corbeilles*.

In the history of neuroanatomy the discovery of the pericellular basket surrounding the Purkinje cell is of the greatest importance, equal in significance to the discovery of the nerve cell itself. It was the first clear observation of an axonal terminal in the central nervous system, and RAMÓN Y CAJAL immediately recognized its significance. From his study of this terminal plexus he grasped two important principles: that nerve cells need only be in contact, not in continuity with one another in order for a nerve impulse to be transmitted, and that the flow of the impulse is directed from the axon of one cell to the cell body of another (1911, p. 27). These ideas later became incorporated in a general statement proclaiming the integrity of the nerve cell and its processes—the Neuron Doctrine, which is the basis of modern neuroanatomy and neurology.

2. The Form of the Basket Cell and Its Processes

The basket cell has a roughly pyramidal or ovoid shape, and it lies in the lower third of the molecular layer with its long axis parallel to the Purkinje cell layer in the sagittal plane. In Nissl preparations little more can be seen of it than its triangular or oval cell body, about 10 μm long, and in this it is the nucleus and nucleolus that attract attention. The scanty cytoplasm contains only small, granular Nissl bodies.

a) The Dendrites

In Golgi preparations (Figs. 158 and 187), however, the cell is much more impressive than one would have imagined from its Nissl pattern, for the squat, pyramidal cell body throws out long dendrites and an axon of extra-

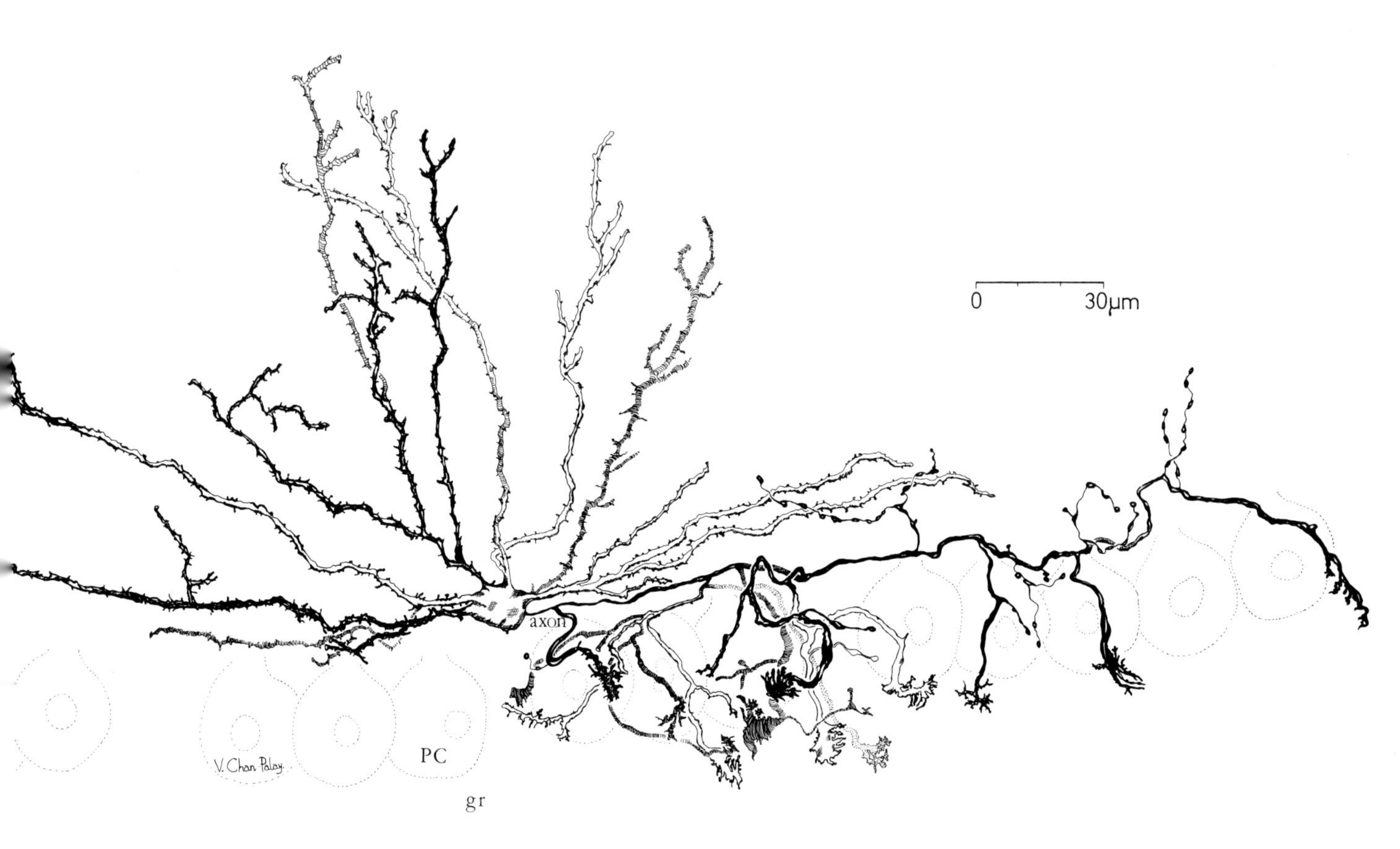

Fig. 158. The dendrites and axon of a basket cell. The dendrites of this cell extend from the cell body, radiating into the molecular layer along a fairly straight course. Spines project from the dendritic surface. The axon leaves the cell as a thin initial segment and travels horizontally through the territories of several Purkinje cell bodies (*PC*). Thick descending collaterals reach down to form the pericellular basket and pinceau around the Purkinje cell body and its initial segment. Along the course of the axon several thin beaded ascending collaterals emerge and reach into the molecular layer. The granular layer (*gr*) is indicated. Rapid Golgi, 90 μm, sagittal, adult rat. Camera lucida drawing

ordinary complexity. The dendrites spread out widely in a vertical direction from the points and upper sides of the cell body, and thus produce a fan-shaped field in the parasagittal plane. Many of them extend horizontally for some distance above the Purkinje cells before turning upwards. Their course is moderately straight though sometimes contorted. Although they gradually taper towards their extremities, they have an irregular form with variations in diameter. It is pertinent to point out here that in Golgi preparations, basket cell dendrites tend to be less twisted than dendrites of stellate cells, though not as straight as those of Golgi cells. The dendrites give off relatively few branches in their trajectory, and both trunks and branches are fitted with sparse filopodia or spines. The number of these appendages varies greatly from cell to cell and in different Golgi preparations. Usually in any particular preparation the dendrites of basket cells display more spines than those of stellate cells. The dendrites of basket cells that project deeply, towards the granular layer, are sparse and fairly short, whereas most of the dendrites ascend upwards even to the pial surface, where their terminal branches run horizontally or recurve downward.

It should be noted that the dendritic tree of the basket cell expands in the parasagittal plane and is in a position to receive the same inputs as the Purkinje cell dendritic tree. The planar disposition of the dendritic tree appears clearly in sections passing through the transverse plane, i.e., parallel with the longitudinal axis of the folium, as in Fig. 159. Ramón y Cajal (1911) pointed out that the dendritic trees of the two kinds of cells alternate, with the basket cell dendrites occupying the lamellar space left open between successive ranks of the Purkinje cells. He observed, however, that the dendritic arborization of the basket cell is not so flattened as that of the Purkinje cell and that some branches or even the cell body itself could depart from the general plan in order to go around a Purkinje cell dendrite. Actually, there is a good deal of interdigitation of the two dendritic trees, which appears in electron micrographs of some regions as a regular alternation of the dendrites from the two cell types. But since the Purkinje cells themselves are not arrayed in

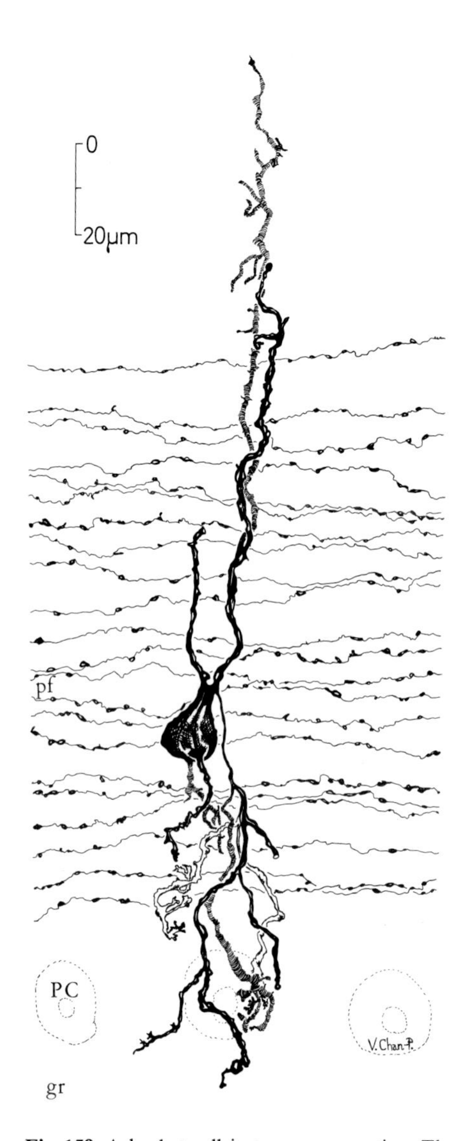

Fig. 159. A basket cell in transverse section. The cell and its processes are shown as they appear in the sagittal plane. The descending collaterals of the basket cell axon wrap around the cell body of a Purkinje cell and the initial segment of its axon. The width of the axonal expansion of this basket cell is approximately 30 µm. The Purkinje cell layer (*PC*), granular layer (*gr*), and parallel fibers (*pf*) in the molecular layer are indicated. Rapid Golgi, 90 µm, transverse, adult rat. Camera lucida drawing

linear ranks (contrary to the usual diagrams of cerebellar circuitry), it is very unlikely that an alternation of whole dendritic trees could occur consistently in this cortex (BRAITENBERG and ATWOOD, 1958; ARMSTRONG and SCHILD, 1970). In fact, the branches of the basket cell dendrites enter into complex relations with the Purkinje cell dendrites and with climbing fibers, Bergmann fibers, and even basket cell axons (CHAN-PALAY and PALAY, 1970). These intricacies are not well displayed in Golgi preparations, and require a study of the electron micrographs of the molecular layer for their elucidation.

b) The Axon and Its Collaterals

The axon of the basket cell emerges either from the perikaryon or from one of the major dendrites, and quickly assumes a horizontal course in the parasagittal plane, weaving through one Purkinje cell territory after another (Figs. 158 and 160). Usually the basket cell axon runs above the perikarya of the Purkinje cells at the level of the first dendritic branching. The initial segment of the axon is extremely slender and smooth, in contrast to the rest of its course. Generally it is fairly straight, but sometimes it arcs and loops apparently around other cells or large dendrites before it arrives at its definitive level. At the site of its first bifurcation the axon abruptly doubles or triples in caliber. From this point onward, as the main horizontal stem of the axon crooks its way among the lower dendrites of the Purkinje cell trees, it emits a succession of descending, ascending, and transverse collaterals. The basket cell axon is, therefore, in a position to effect synapses with the Purkinje cells lying in a roughly rectangular or oval field. FOX, HILLMAN, SIEGESMUND, and DUTTA (1967) calculated that each basket cell could synapse with 216 Purkinje cells in the monkey's cerebellum. SZENTÁGOTHAI (1965; ECCLES, ITO, and SZENTÁGOTHAI, 1967) estimated that in the cat the field is somewhat smaller, comprising only 50–70 Purkinje cells, and he comments in the later citation (ECCLES, ITO, and SZENTÁGOTHAI, 1967) that this estimate is probably too high. In a more recent re-examination of this question, PALKOVITS *et al.* (1971c) revised these figures drastically downward. On the basis of the ratio of Purkinje cells to basket cells (1:6) and the number of basket axon collaterals descending around the average Purkinje cell body (40–60), and considering the potential number of available Purkinje cells in the field (70), they estimated that each basket cell could supply only an average of 9 Purkinje cells with preterminal collaterals.

Each descending collateral (Figs. 158 and 160) is a thick, sinuous process that branches and rebranches as it descends in a cascade over the primary dendrites of the Purkinje cell or their immediate vicinity to reach the soma, where it enters into numerous synaptic contacts all over its surface. The branches give off short, varicose twigs that terminate in small bulbs on the surface of the Purkinje cell soma. Descending collaterals from several basket cells converge on the soma of a single Purkinje cell, and their branches interlace to form the pericellular basket that gives the parent cells their name. SZENTÁGOTHAI (1965; and ECCLES, ITO, and SZENTÁGOTHAI, 1967) estimated that each basket represents the contributions of 20–30 basket cells (cat). Newer counts in Bielschowsky-

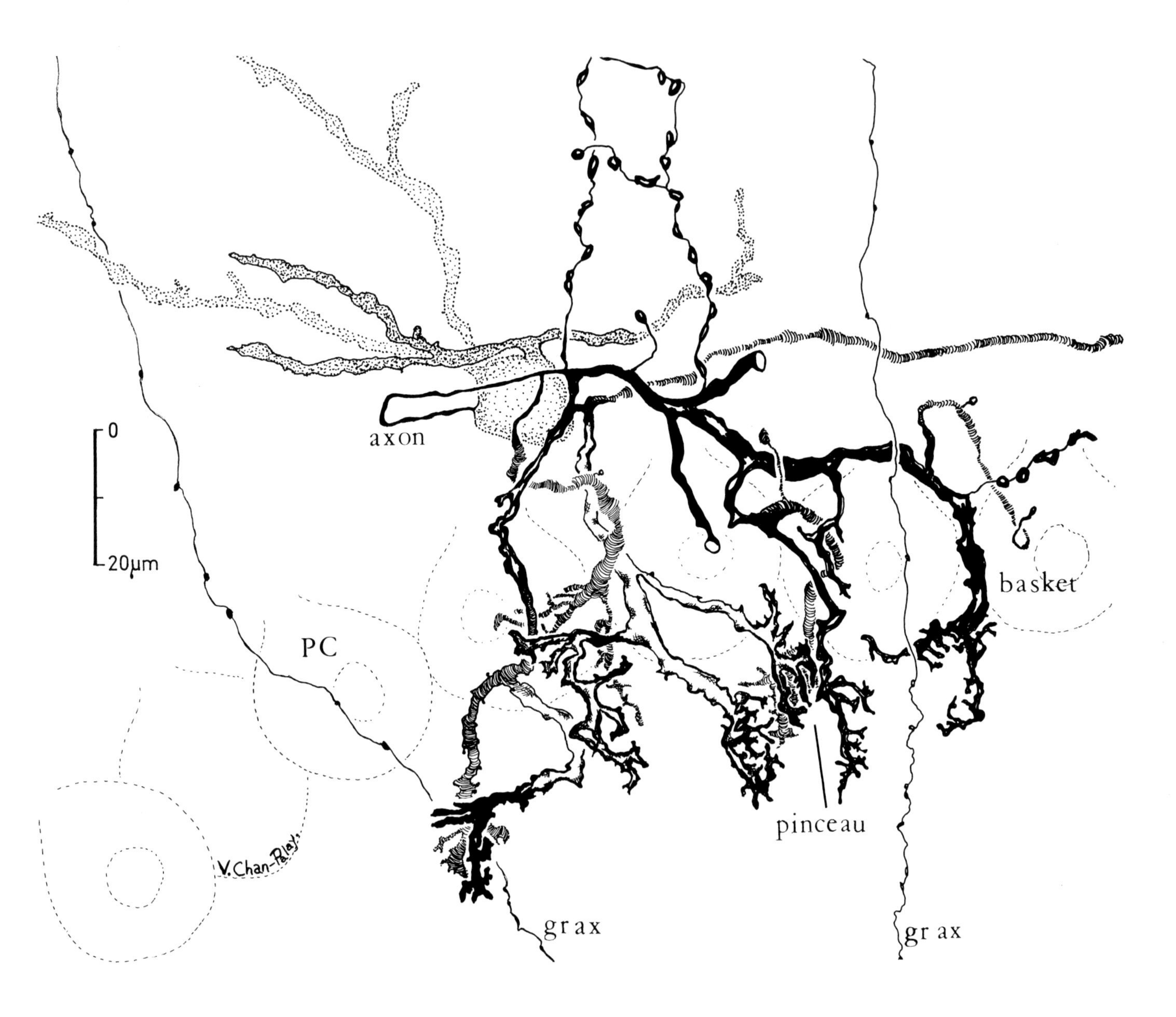

Fig. 160. The descending collaterals of a basket cell. The axon emerges from the left of the cell body, makes a hairpin bend, and runs horizontally in the field. Only a portion of the axon is shown in this illustration. In this field alone, five descending collaterals issue to contribute to the pericellular basket and the pinceau around the Purkinje cell. The Purkinje cell layer is indicated (*PC*) as well as ascending axons of granule cells. Camera lucida drawing

silver preparations by PALKOVITS *et al.* (1971 c) indicate that 40–60 preterminal branches enter into the typical pericellular basket. This number seems to be excessive for the rat, in which only 3–7 afferent axons compose the usual basket. In addition, it should be noted that a particular horizontal axon can contribute several collaterals to the same basket, often from a considerable distance, the descending branches inclining backward to join their fellows emitted nearer the cell body.

The descending collateral and its arborization do not terminate on the pericellular basket but continue beyond the Purkinje cell body into a periaxonal plexus surrounding the initial segment of the Purkinje cell axon. This pinceau, as RAMÓN Y CAJAL called it, is a peculiar axo-axonic synapse almost unparalleled in the rest of the vertebrate nervous system. In Golgi preparations (Figs. 158 and 160) it appears as a tangle of thick, branching axonal twigs hanging below the Purkinje cell body, and in neurofibrillary silver stains it looks much more like the paintbrush called for by CAJAL's name. The Purkinje cell axon enters into this plexus immediately on leaving the cell body, and when it emerges it slips into its myelin sheath. The mode of termination of the basket cell axons in this plexus was a mystery to the earlier neuroanatomists (see RAMÓN Y CAJAL, 1911, and JAKOB,

1928). The interpretation of the pinceau became a focal point in the endless controversy between the neuronists and the reticularists. With the methods at their disposal neither of these schools could produce evidence that would satisfy the other. Only in recent years has it been possible with electron microscopy to see the actual junctions in the axo-axonic synapse and to analyze its extraordinary intricacy.

Ascending collaterals are also given off at irregular intervals from the main horizontal stem of the basket cell axon (Fig. 158). These branches ascend into the middle third of the molecular layer, penetrating the Purkinje cell dendritic trees more or less at right angles to the major dendrites (which often run somewhat horizontally at first) and then parallel with many of the smaller dendritic branches. In their paths the ascending collaterals encounter secondary and tertiary dendrites as well as spiny branchlets, all of which they can accompany for short distances. The ascending collaterals are generally much more slender than the horizontal stem and the descending collaterals, and they travel along a fairly straight axis, although with crooks and bends around it. These divagations can be associated with slight distensions in the caliber of the axon.

The transverse collaterals are thick, and many of them simply bend down to join baskets lying lateral to the main stem of the axon. They give off both descending collaterals that contribute to neighboring pericellular baskets and ascending collaterals that contact the Purkinje cell dendritic tree. According to ECCLES, ITO, and SZENTÁGOTHAI (1967) and PALKOVITS *et al.* (1971 c), the transverse collaterals in the cat usually extend a distance equivalent to three Purkinje cells to either side of the main horizontal axon. As may be seen in Fig. 159, the lateral extent in the rat is considerably smaller than it is in the cat. The collaterals hardly ever extend beyond the Purkinje cells in the next row. Thus the territory into which the basket cell axon ramifies is narrow and involves fewer Purkinje cells than in the cat.

The pattern of synapses made by all of these axonal branches is not very clear in Golgi preparations. Obviously a major articulation with the cell body, the axon initial segment, and the dendrites of the Purkinje cell is indicated by these observations, but the nature and extent of the contacts become evident only with a careful electron microscopic study of the axonal arborization.

Before leaving the optical microscopy of the basket cells we must note that the identification of those cells is, as their name implies, dependent upon tracing their axon collaterals into the pericellular baskets. Since this is possible only in successful Golgi impregnations and certain neurofibrillary stains, other criteria must be applied in other kinds of preparations, especially in electron microscopy, for the identification of the cell bodies, dendrites, and initial, nonsynaptic part of the axon. Fortunately, the Golgi preparations provide these criteria: the intermediate size of the cell body, its shape, and its almost exclusive location in the lower third of the molecular layer at the level of the major Purkinje cell dendrites and their first branches; the horizontal course and tenuity of the axon initial segment and the robustness of the remainder; the relatively straight ascending course of the dendrites (compared with those of stellate cells, which are very contorted). Usually there should be no difficulty in identifying the neurons in the deeper molecular layer as basket cells. Unfortunately, there is a gradation in the form of the cells, and the extent to which they contribute collaterals to the pericellular baskets diminishes with the distance of the cell body from the Purkinje cell layer. RAMÓN Y CAJAL (1911, p. 25) remarks that two kinds of basket cells can be distinguished: the one with an axon the descending collaterals of which all go into baskets, even its terminal arborization, and the other with an axon which gives off descending collaterals only in the beginning of its course and which during the rest of its course gives off only thin ascending collaterals to the molecular layer. *In our Golgi preparations there seems to be a continuous spectrum of patterns extending from those deeper basket cells whose axons emit many descending collaterals and few ascending ones* (see Fig. 158) *to those more superficial cells whose axons give rise principally to ascending collaterals and very few descending ones.* Furthermore, the thickness of the horizontal axon and its descending branches also diminishes as the cells are located at successively higher levels in the molecular layer. The gradually shifting form of the basket cell and its axon from deeper to more superficial levels of the molecular layer do not constitute a progressive series leading to the typical stellate cell (see Chap. VIII), contrary to the implications of the literature from RAMÓN Y CAJAL (1889 b) and SMIRNOW (1897) onwards to RAKIC (1972). Our studies (CHAN-PALAY and PALAY, 1972 a) show that the basket cell and the stellate cell are distinct entities, distinguishable on the basis of their dendritic and axonal arborizations. These differences indicate that the two cell types have different roles in the circuitry of the cerebellar cortex.

3. The Fine Structure of the Basket Cell

a) The Perikaryon

In electron micrographs as in Nissl preparations the basket cell soma (Fig. 161) is occupied almost entirely by a large oval nucleus lying with its major axis parallel to the Purkinje cell layer. What is not usually noticeable in the Nissl preparations is that the nucleus can be deeply creased (Figs. 162 and 163). Often a thin knife of cytoplasm slices more than halfway through the nucleus, carrying with it a few mitochondria, the endoplasmic reticulum, and hordes of ribosomes in clusters. The nucleus itself has irregular contours. It is of the open type, usual for neurons, with widely dispersed chromatin (Fig. 161). Sometimes blocks of condensed chromatin are located near the nuclear envelope. The nucleolus, 1–2 μm in diameter and consisting of the usual partes fibrosa and granulosa, is frequently in a peripheral position and may be embedded in one of these blocks of chromatin (Fig. 162). Occasionally the nucleolus is divided into two bodies, of about equal size. Aside from its irregular contour the nuclear envelope is not remarkable. It throws out frequent streamers that are confluent with the granular endoplasmic reticulum and even with the plasmalemma. Here and there ribosomes in rows and spirals are attached to its outer surface, but most of its surface is bare.

Although the cytoplasm is scanty, the eccentricity of the nucleus and the rather thick stems of some of the dendrites provide space at one pole of the nucleus where the perikaryal organelles collect (Figs. 161 and 163). The Golgi apparatus is found here along with lysosomes, multivesicular bodies, and assorted vesicles in various stages of their life history. These in turn are surrounded by small mitochondria and cisternae of the granular endoplasmic reticulum. Centrioles can also be found here (Fig. 163). Except for the region immediately occupied by the Golgi apparatus, the whole perikaryon is pervaded by small clusters of free ribosomes. The granular endoplasmic reticulum is spread out as single undulating, branching cisternae roughly parallel with the nuclear envelope; only rarely are two or three cisternae arrayed in a stack to form a conventional Nissl body. The agranular endoplasmic reticulum is rather deficient and is represented almost entirely by a few subsurface cisternae scattered sparsely under the plasmalemma (Fig. 161). Since even many of these have ribosomes attached to their deep surface, they are perhaps more properly considered part of the granular reticulum of this cell. The mitochondria, varying from 1 to 3 μm long and 0.1 to 0.3 μm wide, arch round the nucleus to fit into the thin shell of perinuclear cytoplasm. In the more ample parts of the cell near the roots of dendrites, the mitochondria may be larger, may branch or bend back on themselves. Secondary Golgi complexes also occur in these places.

Small numbers of microtubules trace arcs around the nucleus. At the roots of the dendrites they tend to gather in parallel array and pass out of the cell body into the processes (Fig. 161).

b) The Dendrites

As in the dendrites of other neurons, parallel microtubules, arranged in a gentle helicoid, are conspicuous organelles, but neurofilaments are also present in greater numbers than in most dendrites. The Golgi apparatus and the Nissl substance similarly continue out of the cell body into the first part of the dendrite (Fig. 163). The Golgi apparatus extends for only a few microns, and the Nissl substance soon fragments into progressively smaller masses, but these can be found quite far out in the secondary and tertiary branches. In the dendrites the agranular endoplasmic reticulum, which was not a salient feature of the perikaryon, forms a continuous loose network of predominantly longitudinal tubular elements. With increasing distance from the cell body the mitochondria become elongated and farther apart, measuring as much as 6 or 7 μm long. Branching forms are not unusual in the thicker parts of the dendritic branches.

As is shown by the Golgi preparations, the dendrites of the basket cell usually have somewhat irregular contours and follow a generally upward course toward the pial surface. Transverse sections are much more common than longitudinal sections, no matter which plane is taken. Profiles in transverse section are rounded or oval, sometimes facetted, and contain one or two mitochondria and several microtubules, also in transverse section. A short cisterna of the smooth endoplasmic reticulum and a few ribosomes complete the picture.

Fig. 161. Soma of a basket cell. The nucleus of this basket cell (*B Nuc*) contains a mass of nucleolus-associated chromatin (*rectangle*). Three small subsurface cisternae (*asterisks*) occur under the perikaryal surface. Two synapses (*arrows*) with parallel fibers can be seen on the somatic and the dendritic surface. Lobule III. × 15000

Fig. 162. Soma and dendrites of a basket cell. This section passes through a lobulated basket cell nucleus (*B Nuc*) and two dendrites (ΔΔ, *Bcd*). The somatic surface of the cell exhibits two synapses with parallel fibers (*arrows*) and a junction (*asterisk*) with an axon of a stellate or basket cell. Lobule VIII. × 13000

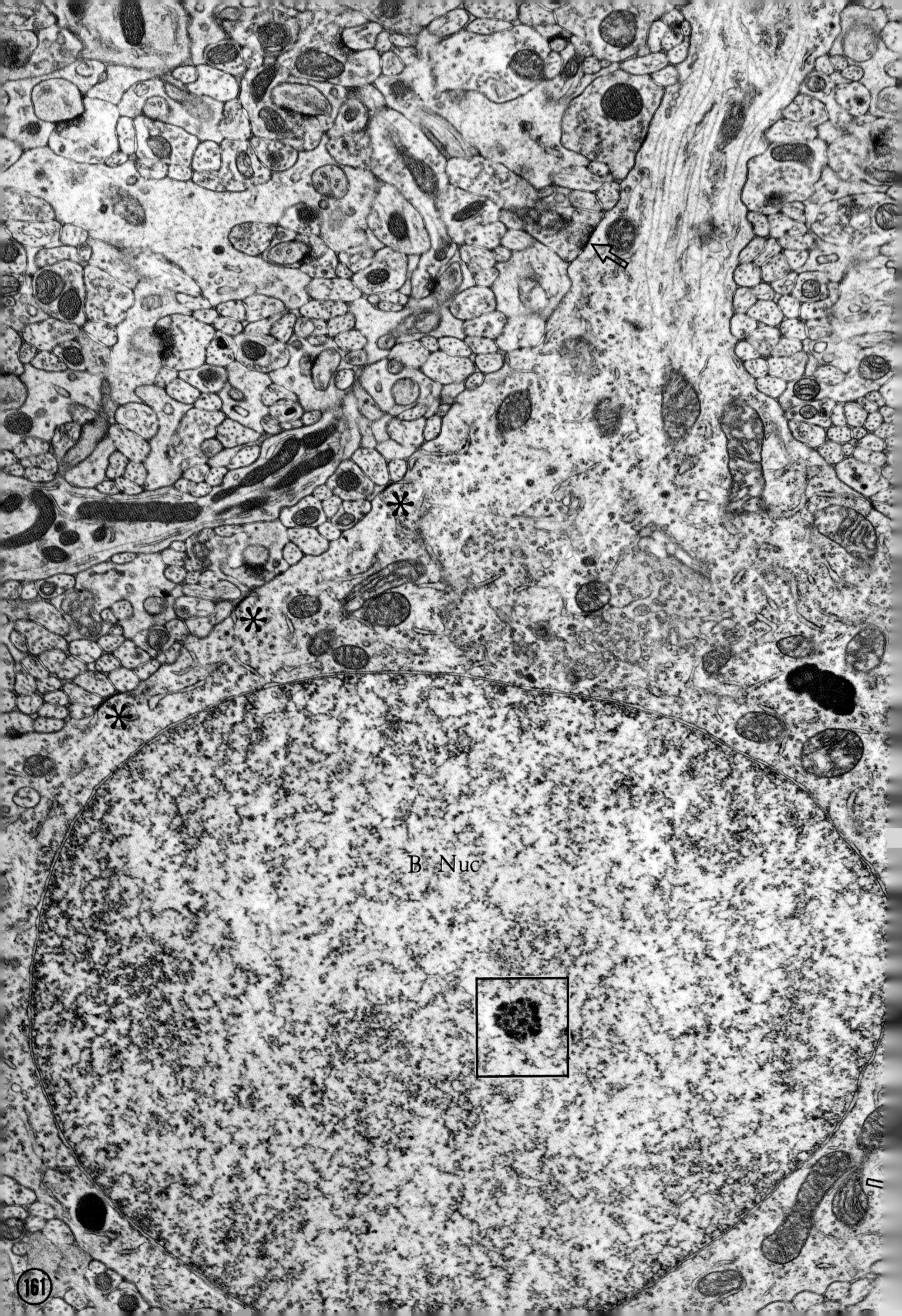
B Nuc
161

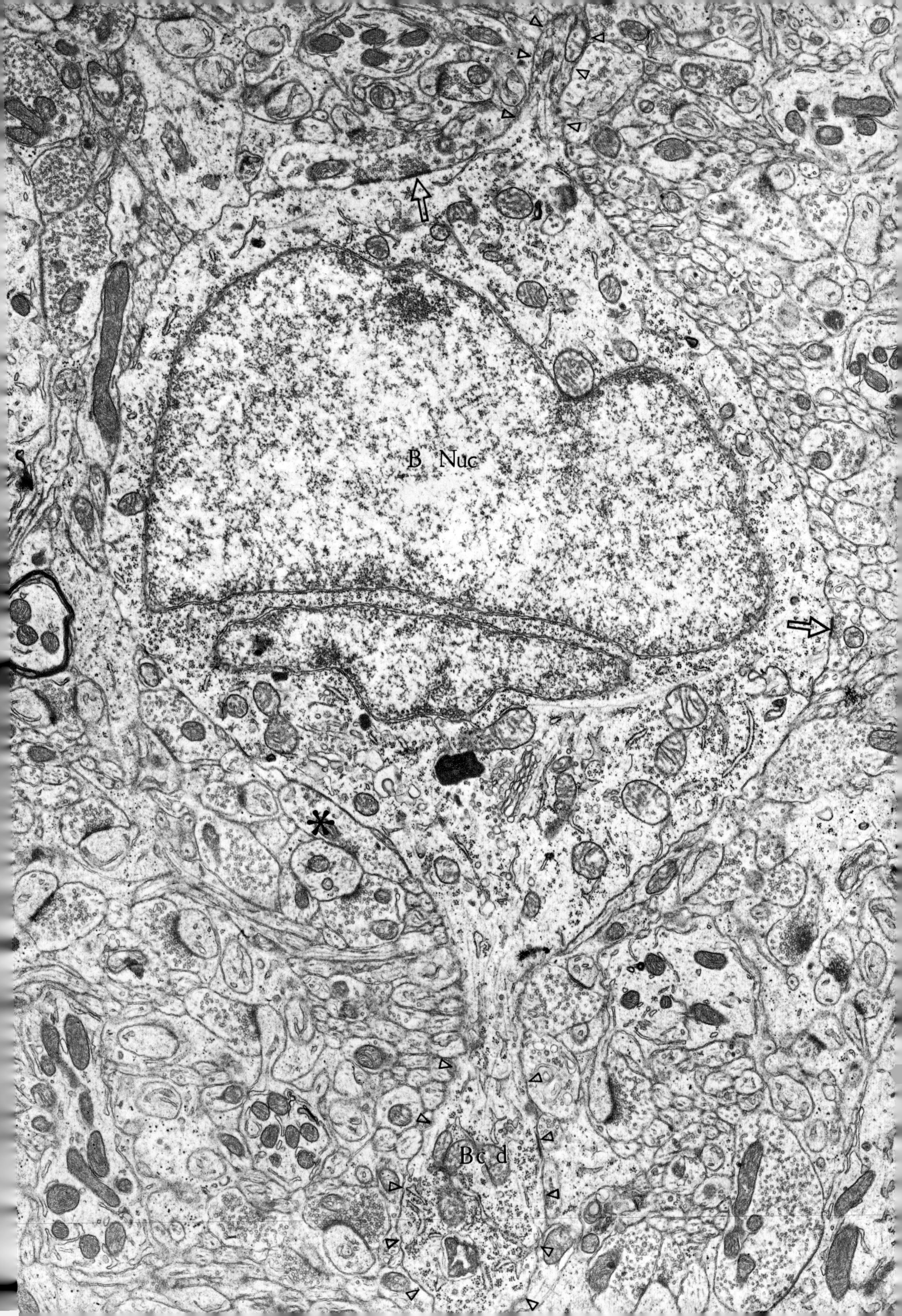
B Nuc
*
Bc d

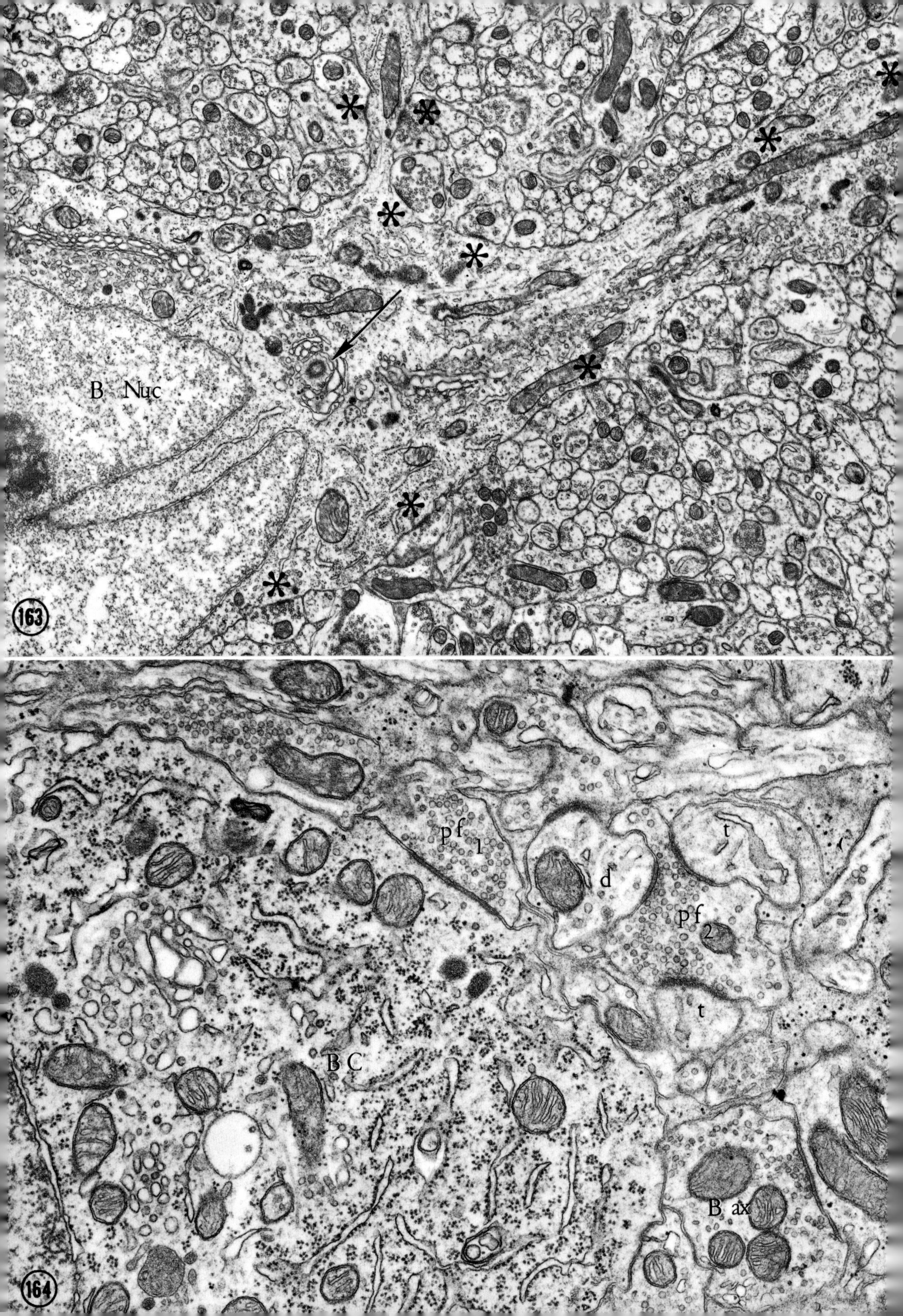
B Nuc
163
pf 1
d
t
pf 2
t
BC
B ax
164

Usually identification is aided by the presence of one or two preterminal axons synapsing in a characteristic fashion on the surface of this dendrite (Fig. 164).

Longitudinal sections of basket cell dendrites can sometimes be traced to the cell of origin, and they substantiate the characteristics that have been listed. In particular, they show (Fig. 165) that the shaft of the basket dendrite is studded with clustered synaptic junctions made by parallel fibers coursing usually at right angles through the dendritic trees of the Purkinje and basket cells. Each synaptic junction is marked by a small dark spot atop a flat plateau projecting slightly from the shaft of the dendrite (Figs. 165, 166, and 167).

Since these mesa-like junctions are characteristic of the basket cell, they will be described here in detail (Figs. 164–167). At the site of the synapse the parallel fiber distends slightly to accommodate a small collection of round vesicles and sometimes a mitochondrion. The parallel fiber can also produce a larger varicosity or a pedunculated excrescence that partially surrounds the slender dendrite. In any case the synaptic specialization is a small dark macula, about 0.5 μm in diameter, that occupies only part of the junctional interface, sometimes only a small part of it. The synaptic cleft is slightly larger than the ordinary interstitial space, and it is crossed by fine filaments extending between the apposed plasmalemmas. A thin and irregular presynaptic density is matched by a thicker and darker postsynaptic density. The pronounced asymmetry and the narrow cleft are characteristics of Gray's type 1 and type 2 synapses respectively (GRAY, 1959). These junctions are therefore of an intermediate type, and contrast with the typical type 1 junctions made by the same parallel fibers with Purkinje cell dendritic thorns.

There is also another type of junction on the shafts of the basket cell dendrites (Figs. 164–166). This type is probably produced by axons of other basket and stellate cells. They are characterized by flat macular densities, approximately symmetrical on both sides of the synaptic cleft, which is hardly enlarged. The synaptic vesicles in the axonal terminal are elliptical or flattened, and are dispersed through the presynaptic ending rather than clustered as in parallel fibers. These endings are less common than those formed by the parallel fibers, and they correspond more closely to Gray's type 2.

The rough and irregular silhouettes of basket cell dendrites are partly produced by the spicules of filopodia projecting from them. These spines are encountered rather infrequently in electron micrographs on account of the thinness of the sections and the improbability of including a slender appendage still attached to a dendrite (Fig. 166). The longer spines usually appear in electron micrographs as detached, narrow, elliptical, or small, round profiles only about 0.2 μm in diameter and encrusted with considerably larger terminals of parallel fibers. In longitudinal section these spines are long, thin, tapering projections covered with synaptic terminals. Others are merely simple flat-topped elevations in the surface of the dendrite or blunt, stubby processes less than a micron high. The synaptic junctions on these spines resemble the junctions of intermediate type that have just been described on the shaft of the basket cell dendrites. Spines of all these various shapes contain a fine, filamentous matrix and a few tubules or vesicles of the agranular endoplasmic reticulum.

The electron micrographs show that in their ascending course basket cell dendrites come into close association with both Purkinje cell dendrites and Bergmann fibers. They can cross paths with these structures or run along parallel with them for some distance. Unlike the Purkinje cell dendrites, however, they are not ensheathed in neuroglial processes and they are always separated from the Purkinje cell dendrites by the thin sheets of neuroglial cytoplasm that compose the sheath of that cell. No dendro-dendritic junctions of any kind have ever been seen between these two nerve cells. As will be seen more fully below, basket cell dendrites also can run alongside basket cell axon collaterals and even climbing fibers clambering over the surface of Purkinje cell dendrites. Synaptic junctions have not been seen between these axons and the basket cell dendrites in such situations. We have also not seen the gap junctions found by SOTELO and LLINÁS (1972) between dendrites of basket cells in the cat.

The foregoing remark concerning the neuroglia should not be taken to signify that neuroglial processes never touch the surface of basket cells. LEMKEY-JOHNSTON and LARRAMENDI (1968a) measured the outlines of basket cells and stellate cells in electron micrographs from mouse

◄

Fig. 163. Soma and dendrites of a basket cell. This basket cell has a lobulated nucleus (*B Nuc*), and in the perikaryon a centriole surrounded by the Golgi apparatus is present (*arrow*). Two dendrites emerge from the soma. In this section the somatic and dendritic surfaces of this cell bear a total of ten synapses with parallel fibers. Lobule II. ×14000

Fig. 164. Synapses on a basket cell body. The somatic surface of this basket cell (*BC*) carries a large synapse with a parallel fiber varicosity (pf_1). A second parallel fiber varicosity (pf_2) effects three synapses, two with Purkinje cell dendritic thorns (*t*) and one with a dendrite, probably of a basket cell (*d*). A second such dendrite (d_1) synapses with a basket axon profile (*B ax*). Lobule X. ×30000

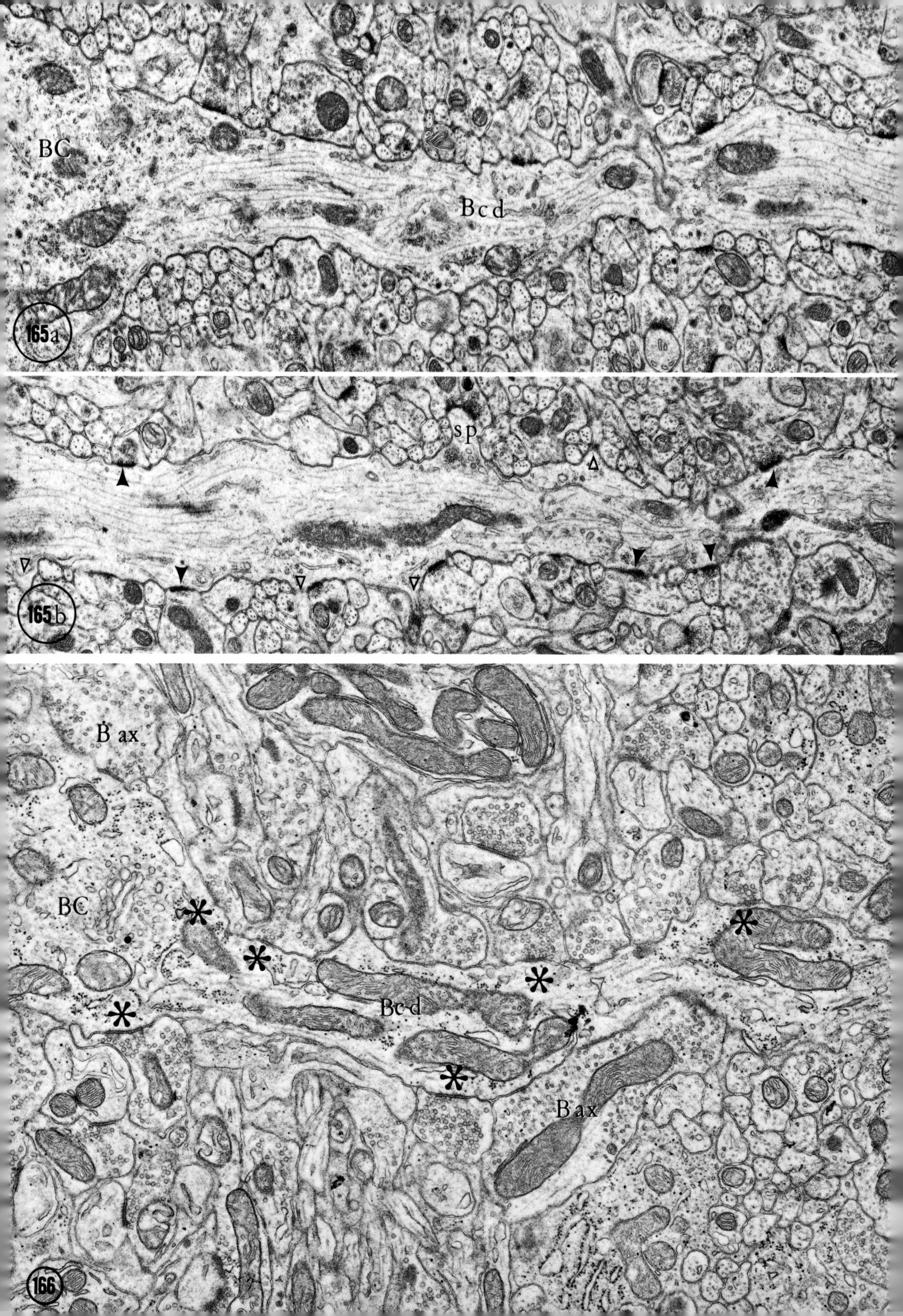

BC
Bcd
165a
sp
165b
B ax
BC
Bcd
B ax
166

cerebellum, and calculated the percentages of the linear profiles that are occupied by neuroglial processes. They found that only a small proportion of the somatic perimeter is in contact with the neuroglia (basket cell, 14.4%; stellate cell, 11.8%), and the condition of the dendritic profiles is not much different (basket cell, 14%; stellate cell, 22%). Although such small proportions hardly amount to proper clothing, these authors take issue with an earlier statement by one of us (PALAY, 1966) to the effect that basket and stellate cells are not ensheathed in neuroglia. Such questions are, of course, relative, and the contrast with the Purkinje cell is obvious, since that cell is almost completely enshrouded with neuroglial processes, except where synaptic junctions occur. In fact, LEMKEY-JOHNSTON and LARRAMENDI (1968a) refer several times to the fact that the neuroglial processes on the basket and stellate cells are usually offshoots from the sheath of a neighboring Purkinje cell.

A glance at Figs. 161–163 and 165 will show that the soma and dendrites of the basket cell are immersed in a sea of nerve fibers, the unmyelinated parallel fibers, few of which even pause in their trajectory over the surface to articulate with it. Despite the paucity of the neuroglia, synapses are infrequent on the soma and only a little more numerous on the dendrites. Only a minute fraction of the fibers passing through the territory of the basket or stellate cell dendrites come close enough to make contact, and very few of them actually synapse with them. Quantitative data concerning this relationship would be just as interesting as those concerning the number of Purkinje cells contacted by a single parallel fiber or basket cell axon. In the absence of such data, the estimates made in the mouse by LEMKEY-JOHNSTON and LARRAMENDI (1968a and b), on the distribution of synapses over the surfaces of basket and stellate cells, take on considerable interest. Taking up the somata of these cells first, they found that only 15% of the basket cell and 4.5% of the stellate cell perimeter was in contact with axon terminals having so-called "active zones." But if to these figures are added the terminals in which the "active zone" was not in the plane of section, the proportions become 30% and 18% respectively. Some 56% of the basket cell soma and 70% of the stellate cell soma is covered by axons that are merely passing by. Thus the cell body of the basket cell has nearly 50% more synaptic interface than that of the smaller stellate cell. When the sources of the terminals were considered (LEMKEY-JOHNSTON and LARRAMENDI, 1968b), it turned out that parallel fibers account for 74% of the total number of synaptic junctions on the soma of the basket cell while climbing fibers account for 6%, other basket and stellate cell axons for 11%, and recurrent collaterals of Purkinje cell axons for 9%. (LEMKEY-JOHNSTON and LARRAMENDI also considered a category of "very low basket cells" which are almost entirely preempted by recurrent collaterals of Purkinje cell axons. It seems clear from their description that these cells are Lugaro cells.) In the somata of stellate cells the corresponding proportions are 92% for parallel fibers and 8% for the axons of other basket and stellate cells. Thus the somata of both of these cells are overwhelmingly under the influence of parallel fibers. Certainly, however, the lower incidence of other afferents does not necessarily indicate a lower efficiency of synaptic action, since each fiber has no more than one or two junctions on any particular cell.

LEMKEY-JOHNSTON and LARRAMENDI (1968a and b) obtained similar results when they studied the dendrites of these cells. Again, the basket cell dendrites had more synaptic articulations than stellate cell dendrites, 34% compared to 29%. At this level of the cells about 50% of their surface is covered with axons passing by and 14% and 22% respectively covered by neuroglia. However, the kind of axon engaging in axo-dendritic synapses proved to be almost entirely the parallel fiber, 100% for the stellate cell dendrites and 89% for the basket cell dendrites. Climbing fibers synapsing with basket dendrites accounted for less than 1% of the synapses counted, and 4% were Purkinje cell recurrents and 6% other basket or stellate cell axons. Thus the conclusion is quite clear that, in the mouse, parallel fibers constitute the most important input to basket and stellate cells, so far as numbers are concerned. There is no reason to think that the situation is different in other mammals. This input is

◄

Fig. 165a and b. Basket cell dendrite. **a** The dendrite of a basket cell (*Bcd*) emerges from the cell body (left, *BC*) and extends radially towards the cortical surface. The process is large, about 4 μm across, and contains slender, long mitochondria, fragments of granular and agranular endoplasmic reticulum, and abundant microtubules. The two distinguishing characteristics of the basket cell dendrite are its origin from cells in the low molecular layer and the pattern of synapses on its surface. ×19000. **b** is a continuation of the dendrite illustrated in a. In this stretch synapses with parallel fibers are more numerous. The junctional interfaces (►) resemble a series of low plateaus, contributing to the irregular contours of the dendrite. Spines of various sizes project from the surface of the fiber intermittently. Some spines (▷) are small and lack synapses; others (*sp*) are blunt and stubby, topped with two or more synapses with parallel fibers. Lobule III. ×19000

Fig. 166. Synapses on a basket cell dendrite. The dendrite of this basket cell (*Bcd*) emerges from the cell body (*BC*) and weaves gently across the field. Nine sets of synapses are made in this short stretch. The majority (seven) are made with parallel fibers (*asterisks*), and the rest are made with axons of other basket cells (*B ax*). Lobule X. ×25500

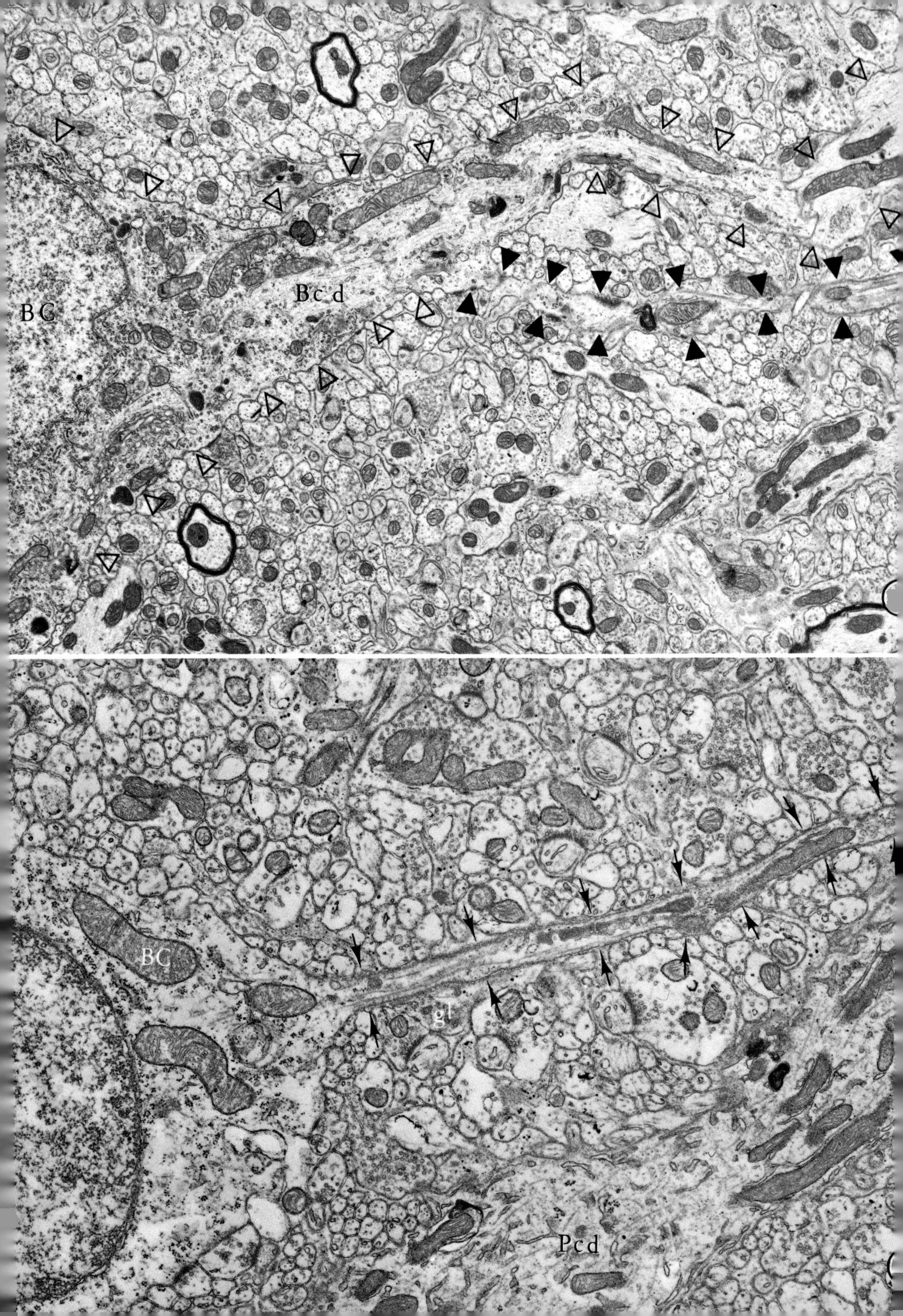
BC
Bc d
BC
gl
Pcd

excitatory, and accounts for the discharges of basket and stellate cells when the mossy fiber-granule cell relay is activated (ANDERSON, ECCLES, and VOORHOEVE, 1964; ECCLES, LLINÁS, and SASAKI, 1966a, b, c; ECCLES, ITO, and SZENTÁGOTHAI, 1967). Inhibitory inputs to these cells are numerically considerably less important, but since they are relatively more frequent on the somata and proximal dendrites than they are distally (LEMKEY-JOHNSTON and LARRAMENDI, 1968b), they may be individually just as efficient in suppressing basket or stellate cell discharges as the more numerous parallel fiber synapses are in activating them. The difficulty with interpreting these statistical data in functional terms is that they give no information about the individual cell and its particular balance of inputs.

One cannot but be impressed with the wide spread of basket cell dendrites and by the paucity of parallel fiber synapses on their surface. Considering these features in relation to the design of the cerebellar cortex, we may hypothesize that *the basket cell takes small samples of the activity in a wide variety of parallel fibers throughout the thickness of the molecular layer. We suggest that the basket cell is apparently not concerned with the details of the information carried by individual parallel fibers from specific glomerular inputs, but rather with the overall balance of activity impinging on the Purkinje cells corresponding to its dendritic field.* It is like an automaton that puts up an umbrella when it rains in a neighboring county.

c) The Axon

The initial segment of the basket cell axon is only about 0.5 µm in diameter. It arises from either the cell body or one of the major dendrites without a distinct axon hillock (Fig. 168). A few microtubules gather into fascicles and run into the axon along with tubular agranular endoplasmic reticulum and an elongated mitochondrion or two. Neurofilaments are not prominent. Clusters of free ribosomes occur in diminishing numbers and in decreasing association with the endoplasmic reticulum with increasing distance from the cell body. Ribosomes tend also to become dispersed singly at the distal end of the initial segment. The plasmalemma is underlain by the typical tripartite undercoating of the initial segment, but it is less conspicuous than in the larger initial segment of the Purkinje cell axon. Unlike all other parts of the basket cell, the initial segment is surrounded by an incomplete neuroglial sheath derived from the laminar expansions of the Golgi epithelial cells. The sheath varies in thickness from a few hundred Ångstroms to several microns, depending upon the proximity of the neuroglial cell body. Curiously, this sheath does not prevent parallel fibers, basket cell axons, and axons of other types from coming into direct apposition with the initial segment. We have not, however, seen axo-axonic synapses on the initial segment of basket cells.

Beyond the initial segment, profiles of the basket fibers are easy to recognize in electron micrographs because of their location, course, and fine structure. The main horizontal axon and all of its branches tend to be rather fleshy and of uniform caliber (Fig. 169). Ordinarily they are the largest axons in the molecular layer. They vary from 1 to 3 µm in diameter, the smallest branches being the ascending and the largest being the descending collaterals. Not infrequently the descending collaterals are thicker than the parent axon. The axoplasm is characteristically light because of the paucity of organelles. The agranular endoplasmic reticulum is drawn out into a longitudinal, fine, loose meshwork of branching and flattened tubules with occasional cisternal dilatations at points of anastomosis. Some tubules cross from one side of the fiber to the other, producing a distinctive appearance in longitudinal sections. These cross links are not, however, peculiar to the basket fibers. The mitochondria are

Fig. 167. The basket cell dendrite and axon initial segment. The cell body (*BC*) of this basket cell gives rise to a dendrite (*Bcd*, ΔΔ). From the dendrite, an initial segment issues, coursing first out of, then within the plane of this section (▲▲). Lobule V. ×5000

Fig. 168. The basket cell initial segment. This axon (*arrows*) arises directly from the basket cell body (*BC*). In its first portion, the spiral lamina of the undercoat can be discerned, as well as a partial neuroglial covering (*gl*). A Purkinje cell dendrite (*Pc d*) is at the bottom of the field. Lobule IV. ×8500

Fig. 169. The horizontal axon of the basket cell. The horizontal basket axon (*h Bax*) traverses the lower molecular layer perpendicular to the ascending columns of parallel fibers (bottom left corner). This axon is about 3 µm in diameter and has gentle distentions along its length, which are the sites of synapses. The axoplasm is light, with numerous neurofilaments and an open meshwork of the agranular endoplasmic reticulum (*SER*); synaptic vesicles are found only at synaptic interfaces. As this horizontal axon weaves among the branches of the Purkinje cell tree, it makes a long synaptic contact (*arrow*) with the shaft of a spiny branchlet (sb_1) when the two meet at an acute angle. Girdle synapses (⇒) are formed when the horizontal axon meets a dendrite at right angles. Such synapses are shown three times: with a tertiary Purkinje dendrite (*Pc d*) in the center, a Golgi dendrite (*Go*) at the right, and with the shaft of a spiny branchlet (sb_2) at the extreme right. In the left lower corner a parallel fiber synapses (ΔΔ) with two thorns from Purkinje cell spiny branchlets. Lobule IV. ×14500

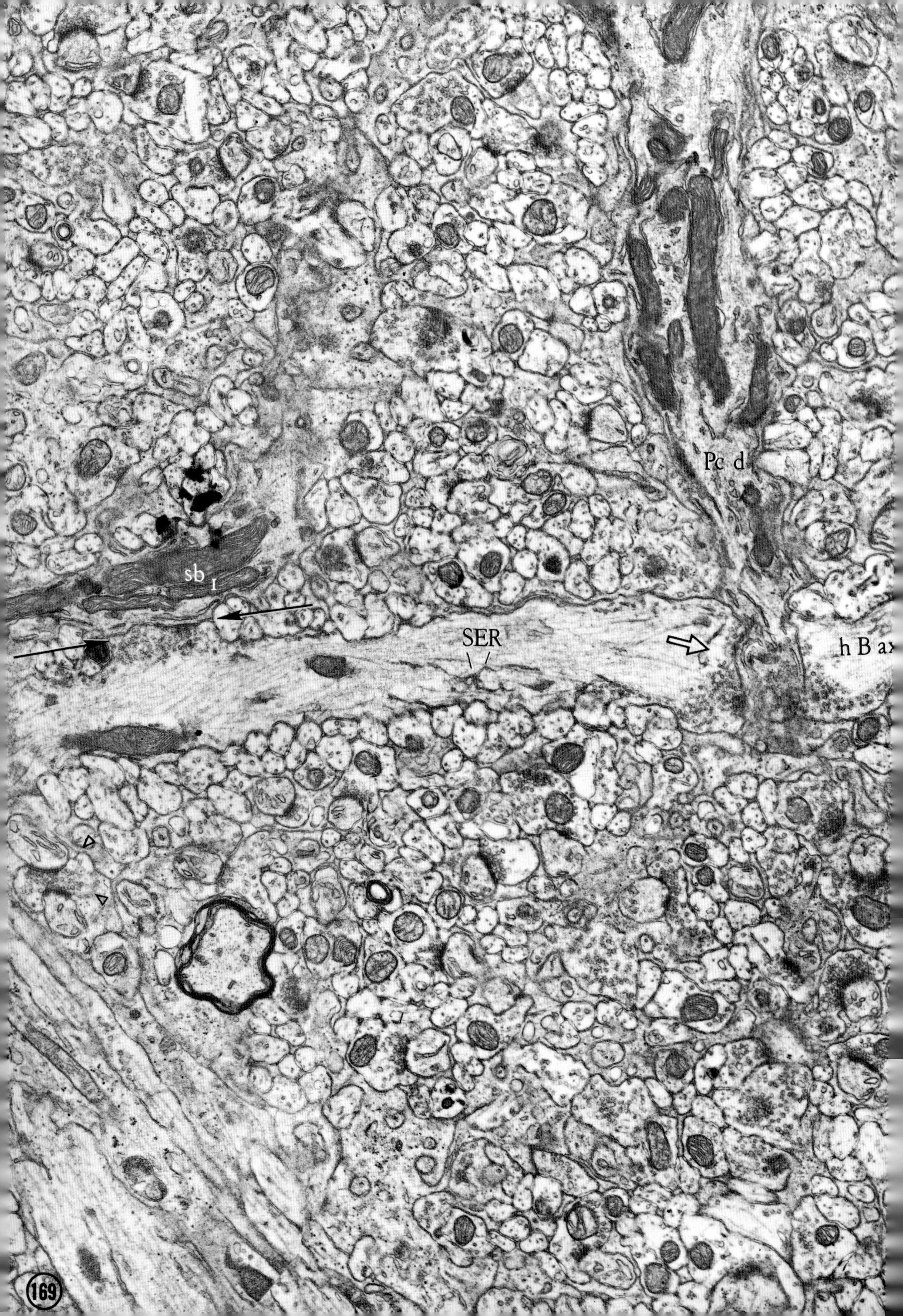
Pc d
sb
SER
h B ax
169

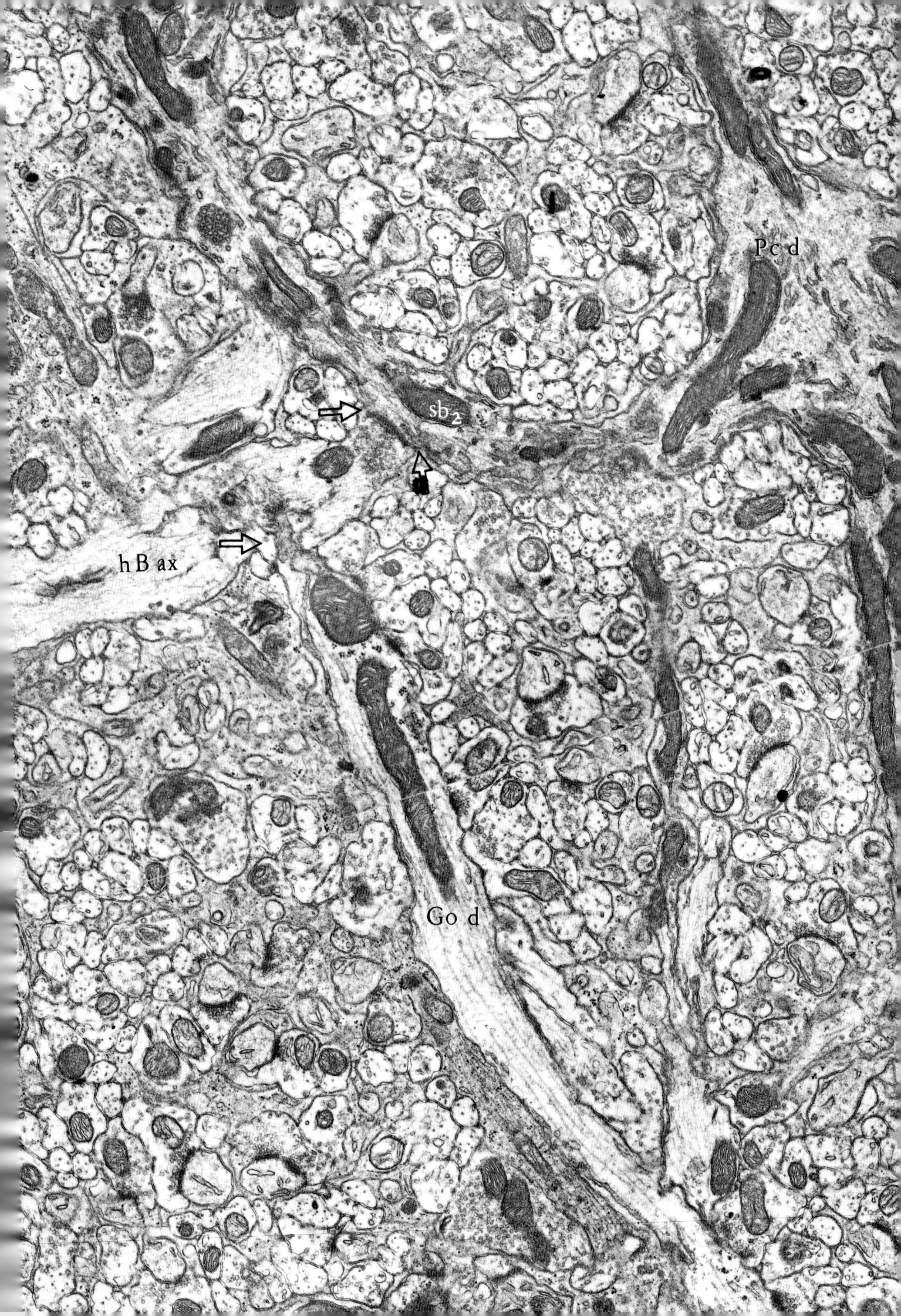
Pc d
sb 2
hB ax
Go d

sparse but very long, 0.3–1 μm in diameter, and usually oriented parallel to the major axis of the fiber. Microtubules are infrequent and vesicles are found only in presynaptic positions. The neurofilaments are therefore the most abundant axoplasmic constituents. Among unmyelinated nerve fibers, basket fibers are unusual for the profusion of their neurofilaments and the scantiness of their microtubules. The filaments run longitudinally but with a loosely twisting course. Swinging around mitochondria and other organelles, they fill up the entire cross-sectional profile of the axon.

Because it is not ensheathed by neuroglia, the axon of the basket cell comes into direct apposition with the perikarya, dendrites, and other axons in its vicinity. It can run alongside the perikaryon of another basket cell for 6 or 7 μm without any intervening cellular element and without any morphological sign of synaptic interaction between them. Occasional puncta adhaerentia occur along these interfaces. The axon courses generally at right angles through extensive fields of parallel fibers without displaying any surface specializations in relation to them. The horizontal stem of the basket cell axon does make *en passant* synapses with the dendrites and perikarya in its path. These take different forms depending upon the angle at which the pre- and postsynaptic structures meet. When the fiber meets a dendrite at less than a right angle (Fig. 169), it runs along with the dendrite for some distance, making a broad synaptic contact with the shaft before turning off onto its own trajectory again. These synaptic junctions are very common. The curling form of the spiny branchlets of the Purkinje cell dendrites makes them extremely favorable sites for the formation of such glancing synaptic junctions. In Golgi preparations these sites are indicated by slight rounded or fusiform enlargements of the basket cell axon. In electron micrographs it can be seen that the synaptic vesicles are loosely aggregated to one side next to the bulging synaptic interface, while the neurofilaments and microtubules continue longitudinally through the dilated region (Figs. 169 and 170). Occasionally the dilatation contains one or two long, thick mitochondria. The synaptic interface is flat, with the pre- and postsynaptic membranes parallel to each other or very slightly bowed. The synaptic cleft is about as wide as other interstitial spaces, and thin, shaggy layers of dense material are symmetrically attached to the cytoplasmic faces of both membranes. These synapses are similar to those formed by basket fibers on the perikaryon of the Purkinje cell and occasionally on the perikarya of other basket cells. They conform to the second type of synapse in Gray's classification (Gray, 1959). Other dendrites such as those of Golgi cells or even other basket cells can also be contacted in the same way.

In those places where the horizontal axon of the basket cell meets a dendritic trunk of the Purkinje cell at right angles (Fig. 169), it forms an interesting girdle synapse. Here the basket fiber wraps around the shaft of the dendrite and continues on its way without further diversion. In Golgi preparations these junctions merely appear as tight loops or hooks in the course of the fiber. But in electron micrographs it can be seen that these formations are synaptic junctions. The axon flows around the dendrite with hardly any change in its dimensions, and synaptic vesicles pile up in loose aggregates around the curving presynaptic surface. As the apposed plasma membranes are nearly flat, parallel planes, the whole formation resembles a stiff belt or girdle around the dendrite. The resultant junction conforms to Gray's type 2 with thin, symmetrical densities on either side of the synaptic cleft. Although these girdle synapses usually involve a Purkinje cell dendrite, they can also involve an ascending dendrite of a Golgi cell. One can imagine that such synapses must be generated during the development of the molecular layer when the growing basket cell axon encounters a suitable Purkinje cell or other dendrite in its path and is obliged to detour around it.

The synaptic vesicles in basket fibers usually have a more or less circular profile. A few vesicles in each presynaptic terminal formation are ellipsoidal or irregular in outline, but most can be interpreted as being spheres. The profiles vary in diameter from 400 to 540 Å with a few still smaller (340 Å) profiles that are attributable to the short axes of ellipsoids. Dense-cored vesicles have not been seen in basket fiber endings. The vesicles are generally loosely aggregated, and seldom fill the profile

Fig. 170. Synapse between basket cell axon and Purkinje cell dendrite. This field cuts through the Purkinje cell dendrite in cross section. It is seen that the dendrite (*Pc d*) is enwrapped by two axons of basket cells ($B\,ax_1$ and $B\,ax_2$). The axons are of large caliber, with many neurofilaments and synaptic vesicles collected at the junctional interface (*arrows*). $B\,ax_1$ is seen in longitudinal section, $B\,ax_2$ in cross section. The synaptic complex is that of Gray's type 2. Lobule X. ×40000

Fig. 171. The pericellular basket and pinceau. The three descending collaterals of a basket cell axon wrap around the body of the Purkinje cell (*large arrow*), and then burst into many fine processes which are disposed around the initial segment of the Purkinje cell axon. This is the pinceau (*small arrow*). The pinceaux around three other Purkinje cell axons are also seen. The dotted lines indicate the position of Purkinje cell somata (*PC*). The molecular layer (*mol*) and the granular layer (*gr*) are also indicated. Rapid Golgi, 90 μm, sagittal, adult rat. Lobule X. ×500

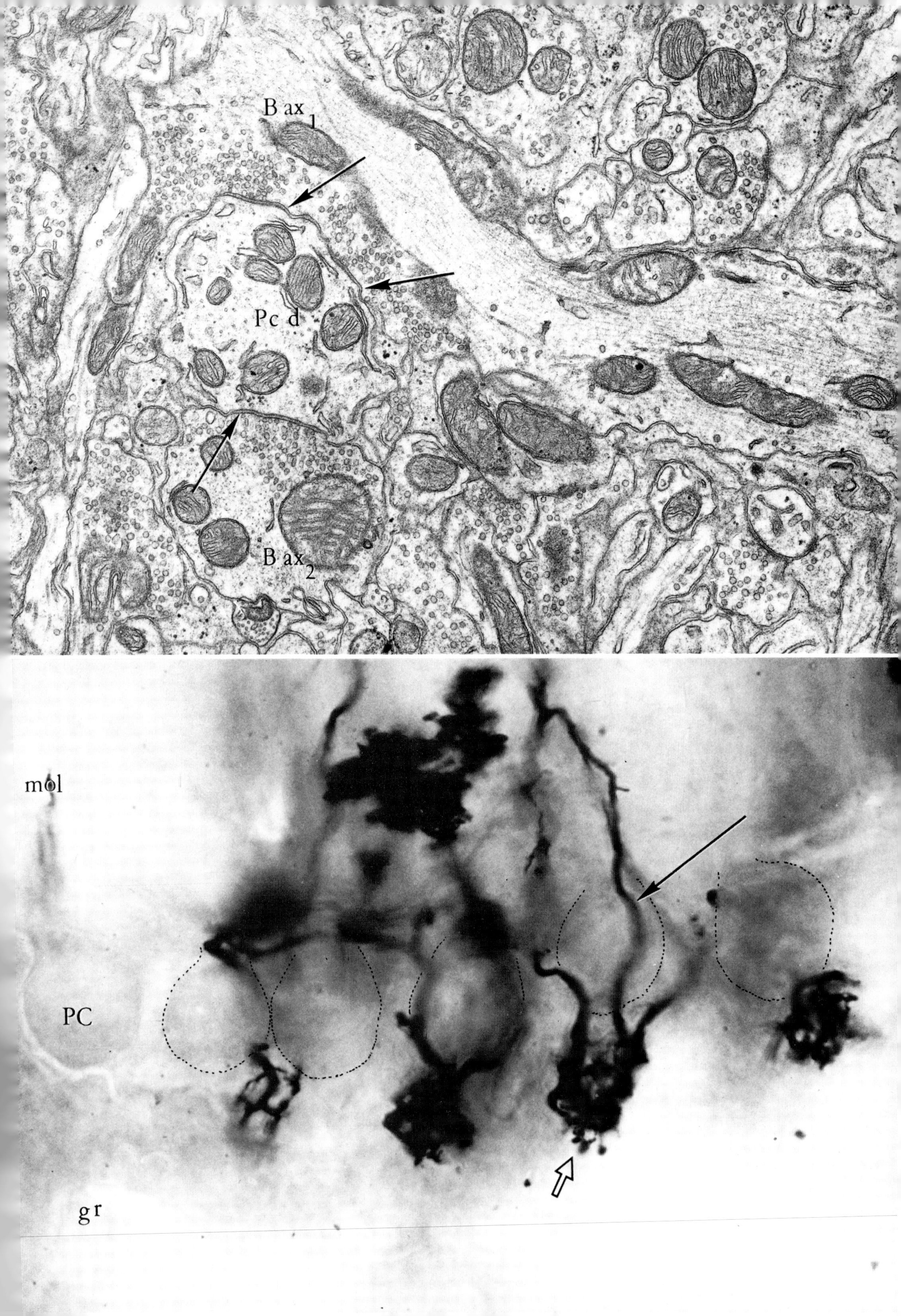
B ax1
Pc d
B ax2
mol
PC
gr

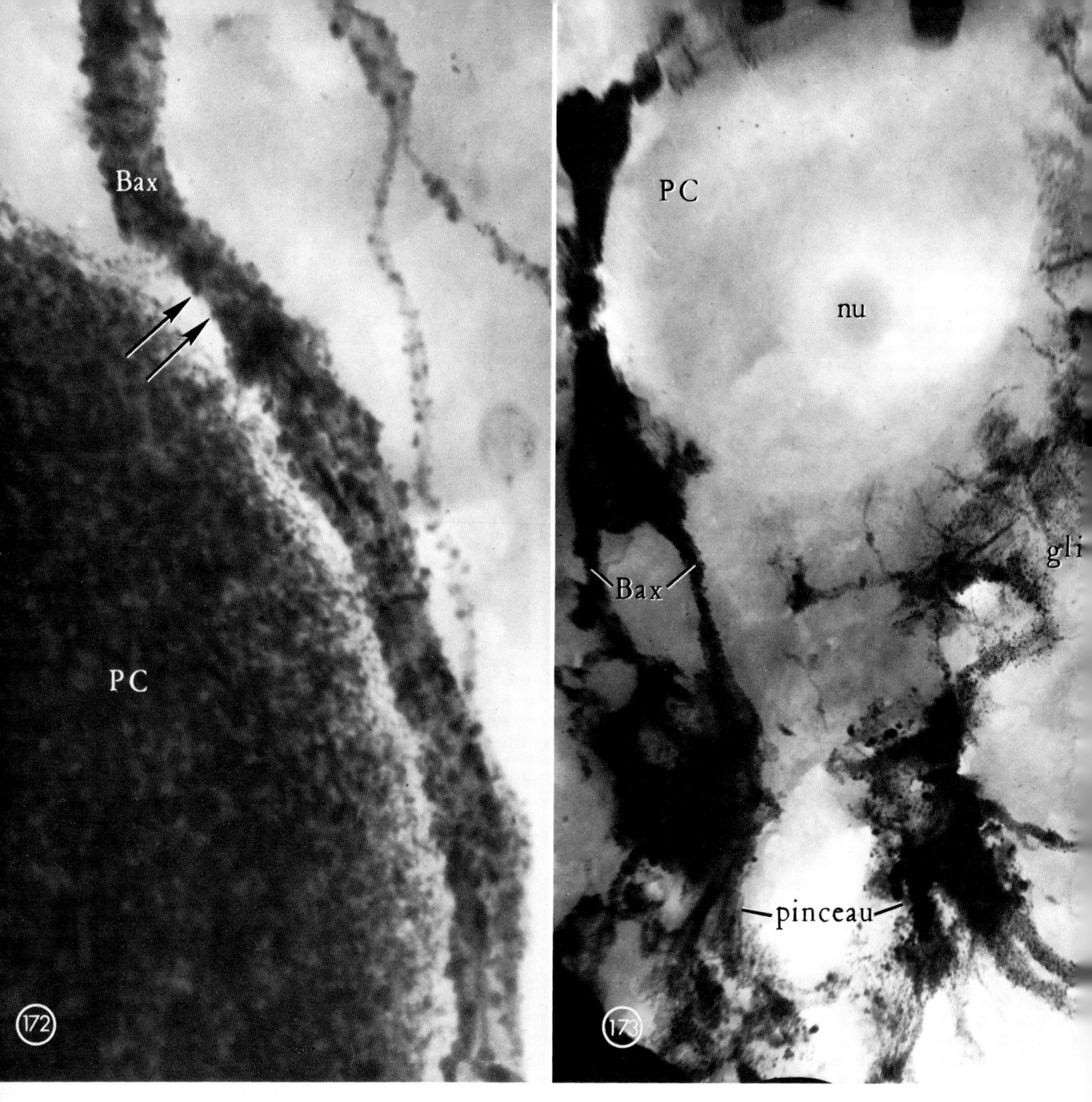

Fig. 172. High voltage electron micrograph of a descending collateral of a basket cell axon and a Purkinje cell body after rapid Golgi impregnation. The pericellular basket around the Purkinje cell soma is formed by many descending collaterals of the basket cell. One such collateral is illustrated here (*B ax*). The axon comes into close juxtaposition, possibly in synaptic contact, with the Purkinje cell body (*PC*). Intermittently along the extent of this apposition, the interstice between them widens, leaving spaces possibly large enough to accommodate neighboring nerve or neuroglia processes (*arrows*). The electron micrograph equivalent of this in thin sections is seen in Figs. 175 and 176. Rhesus monkey cerebellar cortex, 5 μm. Lobule X. × 13000

Fig. 173. High voltage electron micrograph of the pericellular basket, pinceau, and neuroglial sheath around the Purkinje cell after the rapid Golgi method. The unstained Purkinje cell body (*PC*), nucleus and nucleolus (*nu*) are indicated by their shadows. The cell body is circled on the left by the descending collaterals of a basket axon, which break up at their tips to form the pinceau, or paintbrush, around the initial segment of the Purkinje cell axon. On the right half of this cell, portions of a neuroglial sheet (*glia*) appear as impregnated undulating films that generally completely shroud the nonsynaptic surface of the Purkinje cell perikaryon, its dendrites, and the initial segment. Rhesus monkey cerebellar cortex, 5 μm. Lobule X. × 2500

of a basket fiber terminal. They usually are confined to the axoplasm immediately adjacent to the synaptic interfaces.

Ascending collaterals of the basket cell axon often can be found running alongside Purkinje cell dendrites or climbing over them for long distances, such as 10 μm or more. Usually they are separated by thin slips of neuroglial cytoplasm, which ensheath the Purkinje cell and its processes. Here and there, through fenestrae in this sheath, the basket axon and the shaft of the dendrite come into apposition (Fig. 170). The synapses at these sites resemble those described above as glancing synapses between the main axon of the basket cell and various dendrites. Ascending collaterals also enter into complex relations with dendrites of basket cells and with climbing fibers. Such formations will be described in a later section.

The Pericellular Basket. Descending collaterals of the basket cell axon surround the perikaryon of the Purkinje cell with a loose wickerwork of thick, branching fibers of irregular caliber, from 0.5 to 3 μm in diameter. A close correlation of electron micrographs with Golgi preparations is necessary in order to understand the construction of this pericellular plexus. A Golgi preparation like that illustrated in Fig. 171, in which the descending collaterals are completely impregnated, clearly proves the aptness of Kölliker's name for this formation. In this figure it can be seen that three transparent Purkinje cell bodies are embraced, like three bottles of Italian wine, by the descending fibers as they curve round and interlace in a complicated tangle beneath.

But the Golgi preparations only indicate the presence of an articulation between the descending axons and the perikarya. Even at higher magnifications, as in the high voltage electron microscope, the positions of synaptic junctions along these fibers cannot be made out with any certainty. As shown in Fig. 172, the descending fibers display slight distentions as they slip around the Purkinje cell body, but a thin clear crevice intervenes between the two, and this is apparently filled with a tenuous sheet of neuroglial cytoplasm, which can also be seen in some Golgi preparations (Fig. 173). In such material, the soma of the Purkinje cell also appears transparent or is vaguely suggested by light shadows, but it is fitted out in a thin, ragged integument composed of one or more velamentous processes of Golgi epithelial cells.

A corresponding image is usually encountered in thin sections examined by conventional electron microscopy. Fig. 174 shows a quadrant of a Purkinje cell which is entirely encompassed by imbricated slips of neuroglial cytoplasm, containing little more than β granules of glycogen. The basket fibers and their terminals are kept at a slight distance, except where perforations in the sheath permit them to come into apposition. At such fenestrae in the neuroglial sheath, a descending collateral can form either short or extensive junctions with the underlying perikaryon. In Fig. 175 a basket fiber can be seen touching down through an interruption in the sheath to synapse *en passant* with the perikaryon after having first contacted a small dendrite.

Usually the thin sections permit us only such glimpses of the basket fibers as they shift in and out of the plane of the section. Occasionally, however, a fortunate section provides a more panoramic view corresponding closely to the images of Golgi preparations (Figs. 171 and 172). In Fig. 176 a descending basket axon collateral is shown coursing alongside a Purkinje cell for about 15 μm. For the best part of that distance the fiber is in immediate apposition with the surface of the Purkinje cell. In the rest of its course a tenuous layer of neuroglia is interposed between it and the cell. Although this fiber is not very thick, it displays the typical characteristics of the basket cell axon—a core of neurofilaments, long, sinuous mitochondria, scattered groups of round and elliptical synaptic vesicles, and macular synaptic complexes. It is interesting to note that although the contact zone is quite long and synaptic vesicles are loosely strewn throughout almost the whole length of this axon, the synaptic specializations are limited to macular areas about 1 μm or less in diameter. In Fig. 177 a passing basket fiber articulates with a Purkinje cell briefly, but develops two small synaptic complexes in the junctional zone. The synaptic vesicles are only slightly closer to the axolemma in the region of the synaptic complexes than elsewhere in the fiber. The two synaptic junctions are marked by a widened synaptic cleft, which is occupied by a thin, discontinuous plate. The filamentous material adherent to the postsynaptic surface is not very dense, and that on the presynaptic surface is scanty and interrupted. The most conspicuous characteristic of the synaptic junction is the nearly strict parallelism and apparent rigidity of the two apposed membranes. These junctions are similar to Gray's type 2. As the basket axon collateral sweeps down around the Purkinje cell, they are spotted at intervals along the contact zone between axon and perikaryon. In between, the apposed membranes exhibit no morphological specializations, and shoots of neuroglial cytoplasm can separate the two neural structures so that the descending collateral alternately constricts when it passes over the neuroglia and distends opposite each successive fenestra in the sheath like a string of sausages.

In addition to these synapses *en passant*, the descending collaterals also produce true *boutons terminaux* at the ends of stubby branchlets that poke finger-like through the neuroglial sheath to touch the perikaryon of the Purkinje cell. These boutons are nearly filled with round or ellipsoidal synaptic vesicles, and are characterized by macular synaptic complexes identical to those just

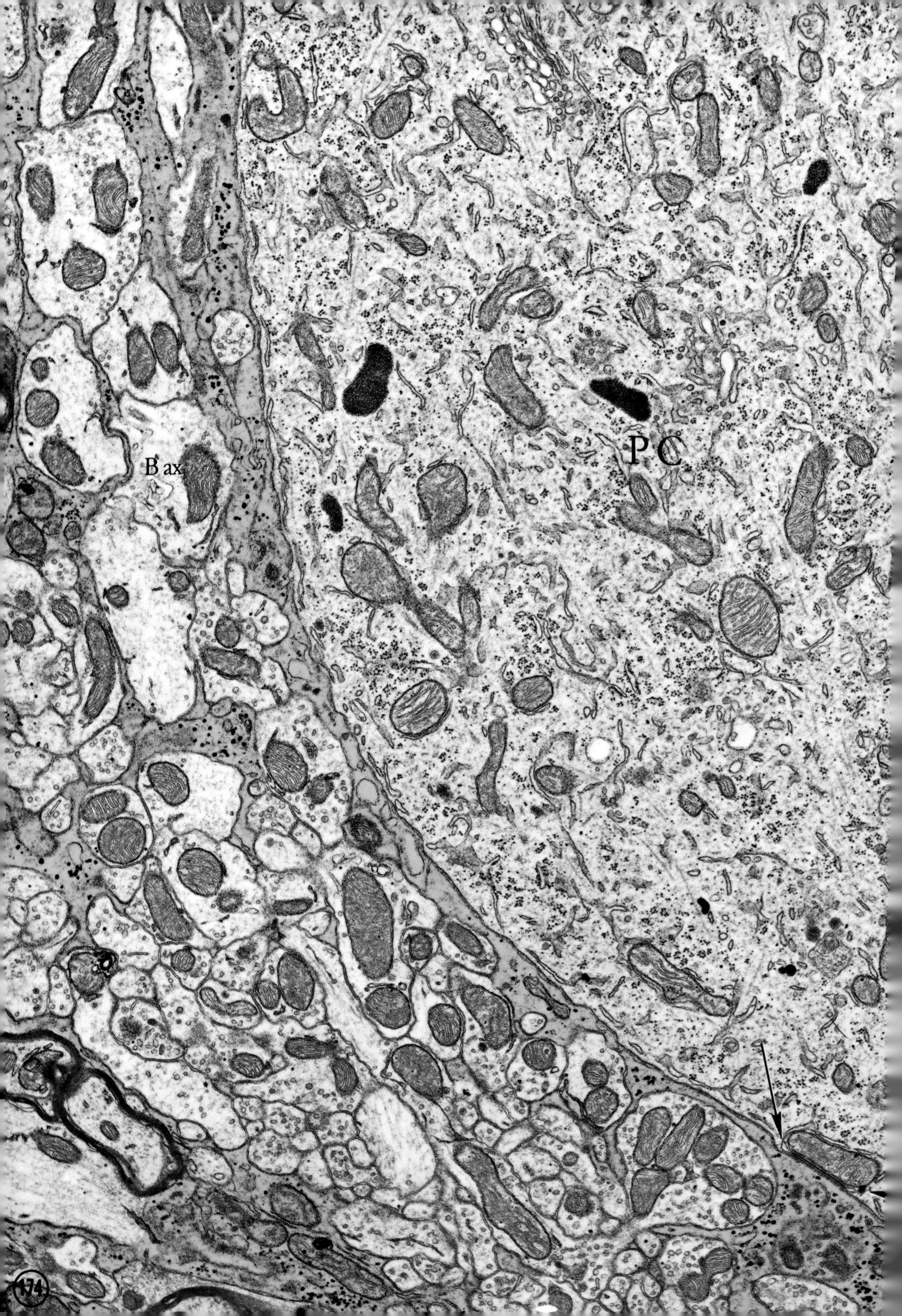

B ax
PC
174

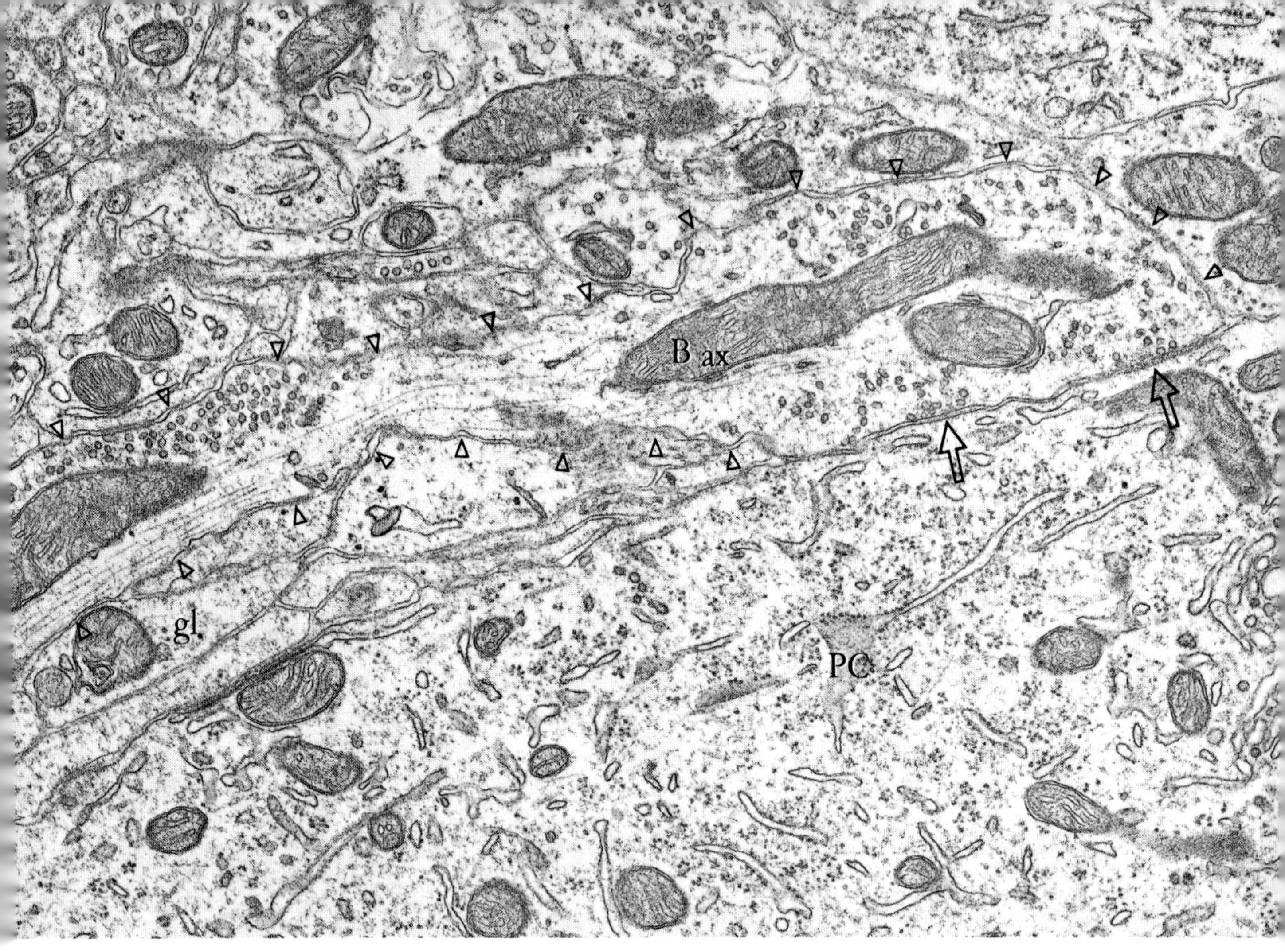

Fig. 175. The neuroglial sheath between basket cell axon and Purkinje cell soma. The basket axon (*B ax* ΔΔ) is separated from the soma of the Purkinje cell (*PC*) by the cytoplasm of neuroglial processes (*gl*). In order to effect a synapse (*arrows*), the axon has to perforate this sheath, so that it can come into direct apposition with the Purkinje cell. Compare this with Fig. 172. Lobule X. ×29000

described. Neurofilaments enter into the boutons only in small numbers, but one or two mitochondrial profiles are commonly seen.

Terminaisons en passant and *boutons terminaux* belonging to the basket cell axon can occur anywhere on the surface of the globular Purkinje cell body, but they occupy only a small proportion of that surface. A similarly small proportion is occupied by endings of the recurrent collaterals from Purkinje cell axons, and the vast majority of it is covered by neuroglial processes and is not synaptic.

The study of freeze-fractured preparations of the pericellular basket provides an opportunity to compare the presynaptic intramembranous structure of the basket axon with that of the parallel fiber. Although these two axons exert effects of opposite sign on the Purkinje cell, the structure of their presynaptic membranes is quite similar. Fig. 177A shows the B faces of several basket fibers crossing the membrane of a Purkinje cell body. The surface of the Purkinje cell exhibits shallow grooves and depressions in which the basket fibers and terminals are seated. The usual A face particles are strewn over the higher planes of this surface and appear in reduced numbers along the depressions. The B faces of the basket fibers are generally smooth, except for circular raised areas of a cobbly texture in which a few papules with clustered particles can be seen. These B face features are similar to those exhibited by the presynaptic terminals of parallel fibers (Fig. 85B). The cobbly areas of basket fiber terminals, however, appear to be less elevated than those of the parallel fibers. The low relief of the presynaptic areas in the freeze-fractured preparation probably corresponds to the flatness of the presynaptic membrane that is characteristic of the Gray's type 2

◄

Fig. 174. The pericellular basket in cross section. The neuroglial processes ensheathing the Purkinje cell body (*PC*) and inserted among the basket cell axons (*B ax*) have been colored in yellow. The basket axons have been cut in cross section. Besides the larger axonal profiles of the pericellular basket, many fine terminals of the pinceau are also in the field. A subsurface cisterna of the Purkinje cell is indicated between arrows. Lobule IX. ×20000

synapses between the basket axon and the Purkinje cell, as seen in conventional thin section electron microscopy.

It is interesting and disappointing that the freeze fracturing technique does not reveal a more conspicous difference between the membrane structure of this inhibitory synapse and that of the excitatory synapses impinging on other regions of the same cell. These preparations show even less difference between excitatory and inhibitory synapses than conventional thin section electron microscopy. It should be recalled, however, that the freeze-fracture method displays only the assemblages and aggregations of intramembranous particles and the irregularities of the surface on which they are distributed. It does not differentiate the chemical specificities of the membrane components.

The Pinceau. The descending collaterals, as mentioned earlier, continue beyond the cell body to form a characteristic plexus, the pinceau, surrounding the initial segment of the Purkinje cell axon. This plexus is too dense and complicated to be resolved at the light microscope level. It was among the first complex synapses to be investigated with the electron microscope (PALAY, 1964a and b). The articulation between the fibers of this plexus and the Purkinje cell axon is apparently the unique representative of its type in the mammalian nervous system. Its nearest analogue is the axon cap found upon the initial segment of the Mauthner cell axon in teleost fishes. Whereas this latter synapse has received some neurophysiological study (FURUKAWA and FURSHPAN, 1963), the pinceau has unfortunately not been examined by similar methods, although its structure has been more clearly analyzed than that of the Mauthner cell axon cap.

The pinceau, as its name suggests, is shaped like a miniature globular paintbrush with its broad or thick base attached to the Purkinje cell body and its pointed tip ending at the onset of the myelin sheath on the Purkinje cell axon. In actuality it is a rather unkempt brush, more like a dense thicket of convoluted and branching fibers (Figs. 158, 160, and 171). The large axons of the pericellular basket collect, several ranks deep, around the base of the Purkinje cell body (Figs. 170 and 178), the innermost ones engaging in synapses *en passant* around the exit point of the Purkinje cell axon. Beneath this level the basket fibers descend into the pinceau, the largest fibers making up a kind of palisade around the periphery of the formation. Electron micrographs of longitudinal sections, i.e., parallel with the Purkinje cell axon, show that the thick, fleshy axons are arrayed vertically under the cell body, clasping the initial segment like long fingers (Figs. 179 and 180). They run along very close together, separated in some places by velamentous neuroglial slips. Certain specialized junctions that occur between these axons will be described below. In transverse sections the structure appears as a whorl of axons 15–20 μm or more across, with the initial segment of the Purkinje cell at its approximate center (Figs. 123 and 181).

The large axons, 2 or 3 μm in diameter, are characterized by a light cytoplasm filled with longitudinal neurofilaments and parallel mitochondria. Seven to ten mitochondrial profiles can appear within a single axonal outline, but many of these represent sections of an individual mitochondrion of complex shape. The cristae in these organelles are generally longitudinally oriented, and the matrix is dark. Scattered, highly branched tubules and cisternae of the agranular reticulum form a cribriform network extending along and across these large axons, but they contain few synaptic vesicles (Figs. 179 and 180). Microtubules are also infrequent. The large axons give off stout branches of similar or smaller caliber that extend inward across the pinceau and subdivide it into enclaves containing still smaller branches and terminal twigs (Fig. 178). The middle-sized branches, 1–2 μm in diameter, are usually short and contain 3–5 profiles of thin mitochondria, or a single large mitochondrion, as well as many neurofilaments. Some of the smaller branches contain swarms of round and ellipsoidal vesicles. The terminal twigs are blunt, vermiform processes, 0.1–0.6 μm in diameter, coiled about one another in a dense tangle. They contain large numbers of round, ellipsoidal, or discoidal vesicles, loosely packed in a pallid matrix. A very small proportion of these vesicles contain dense centers. Sample counts of vesicles in adjacent axonal profiles showed that in 304 vesicles only 6, or 2%, had dense cores. Although neurofilaments seem not to extend into these terminal processes, the axoplasmic matrix has a fine, cottony texture. Occasionally a slender mitochondrion fills most of the process, and microtubules are rarely seen.

All of these processes—large axons, middle-sized branches, and terminal twigs—are bound together oc-

Fig. 176. The pericellular basket around a Purkinje cell. A descending collateral of a basket cell axon (*B ax, black arrowheads*) approaches the Purkinje cell somatic surface (*PC*) from the upper right hand corner. In the initial part of this trajectory the axon is separated from the somatic surface (*between asterisks*) by a thin neuroglial sheath. However, the axon is in direct apposition with the Purkinje cell body for the next 15 μm before leaving the field. More than fourteen small synaptic junctions (*arrows*) occur along this extensive axo-somatic apposition. Lobule V. × 17000

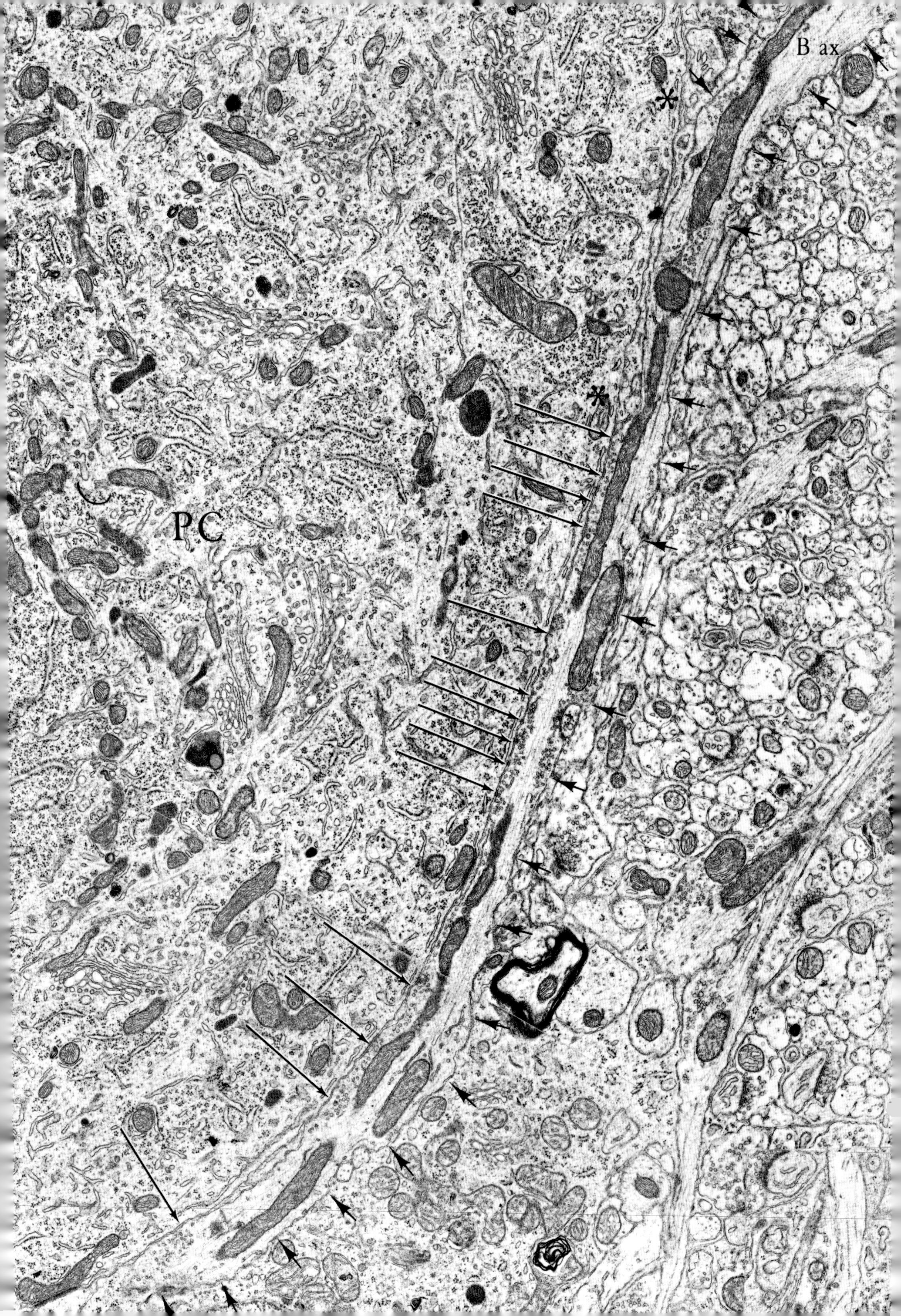
B ax
*
*
PC

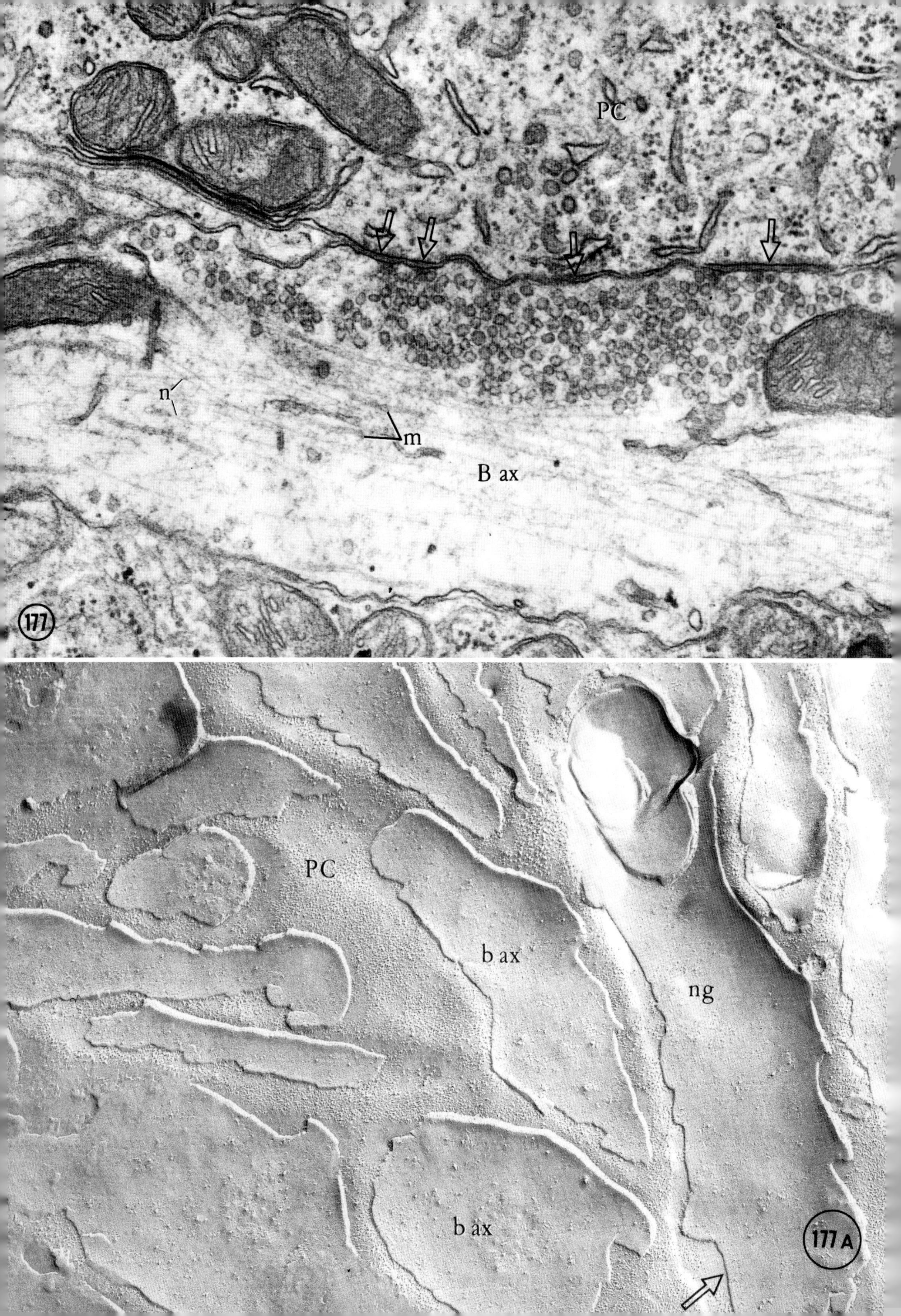
PC
n
m
B ax
177
PC
b ax
ng
b ax
177 A

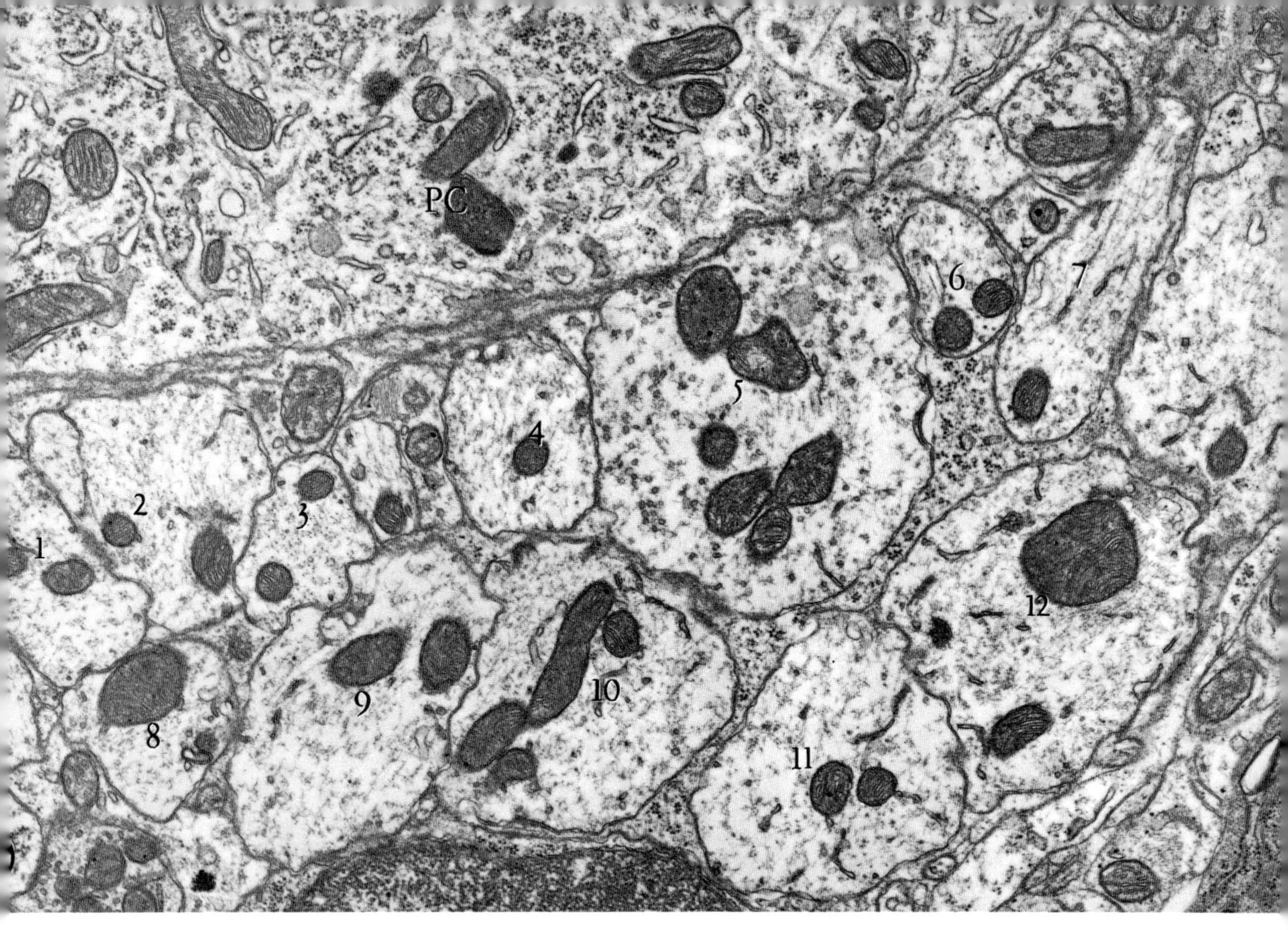

Fig. 178. Neuroglial, basket axon, and Purkinje cell relationships in the pericellular basket. The Purkinje cell body (*PC*) is surrounded by a thin neuroglial sheath. The sheath insinuates itself among the many basket axons in the field (1–12) and other small terminals. The axon terminals must penetrate the neuroglial sheath in order to come into direct apposition with the postsynaptic surface (No. 5). Lobule VII, Crus I. ×17000

Fig. 177. The basket axon, Purkinje cell soma synapse. The basket axon (*B ax*) with its many neurofilaments (*n*) and occasional microtubules (*m*) exhibits three synaptic junctions with the Purkinje cell (*PC*) body (*arrows*). The synapses are of an intermediate variety with a slightly widened synaptic cleft. Synaptic vesicles are collected only near the synaptic interface. Lobule X. ×85000

Fig. 177A. Basket axon terminals on the surface of a Purkinje cell. The background of this field consists of the A face of the Purkinje cell body (*PC*). The A face is covered with the usual particles. Coapted to this surface are fragments of the plasmalemmas of several basket axons (*b ax*) displaying their B faces. These faces exhibit circular patches of cobbly texture and some fine papules associated with clustered particles. The patches correspond to the small macular synaptic complexes of the basket axon-Purkinje cell body synaptic junction. The B faces of neuroglial processes (*ng*) are also seen. The large arrow indicates the direction of shadowing. Replica, freeze fracture. ×64000

casionally by puncta adhaerentia and much more frequently by macular or ribbon-like patches of peculiar construction. The latter junctions have been recently described in some detail first by GOBEL (1971) in the cat and by SOTELO and LLINÁS (1972) in the cat and rat. MUGNAINI (1972) also mentions them briefly as a variety of gap junction. They are characterized by the presence of transverse septa stretching at regular intervals across the interstice between axons, like a ladder (Figs. 182a, b, and 183a, b). The apposed axonal membranes become almost rigidly parallel and straight without changing the average distance between them. The septa consist of dense particles, about 80 Å across, arrayed within the interstice in a honeycomb pattern with a center-to-center spacing of approximately 200 Å. The honeycomb pattern becomes evident in *en face* views of the junction (Fig. 183a, b). The particles vary somewhat in shape, and are often irregular, suggesting a Greek cross, as SOTELO and LLINÁS (1972) have noted (Fig. 183a). Their form can also be described as Z- and Y-shaped (Fig. 183b), depending upon the plane of section, and such forms are in harmony with the analysis of the septate junction in an invertebrate (*Hydra*) given by HAND and GOBEL (1972). Indeed, GOBEL (1971) considered the interaxonal junctions in the pinceau to be septate junctions, even though

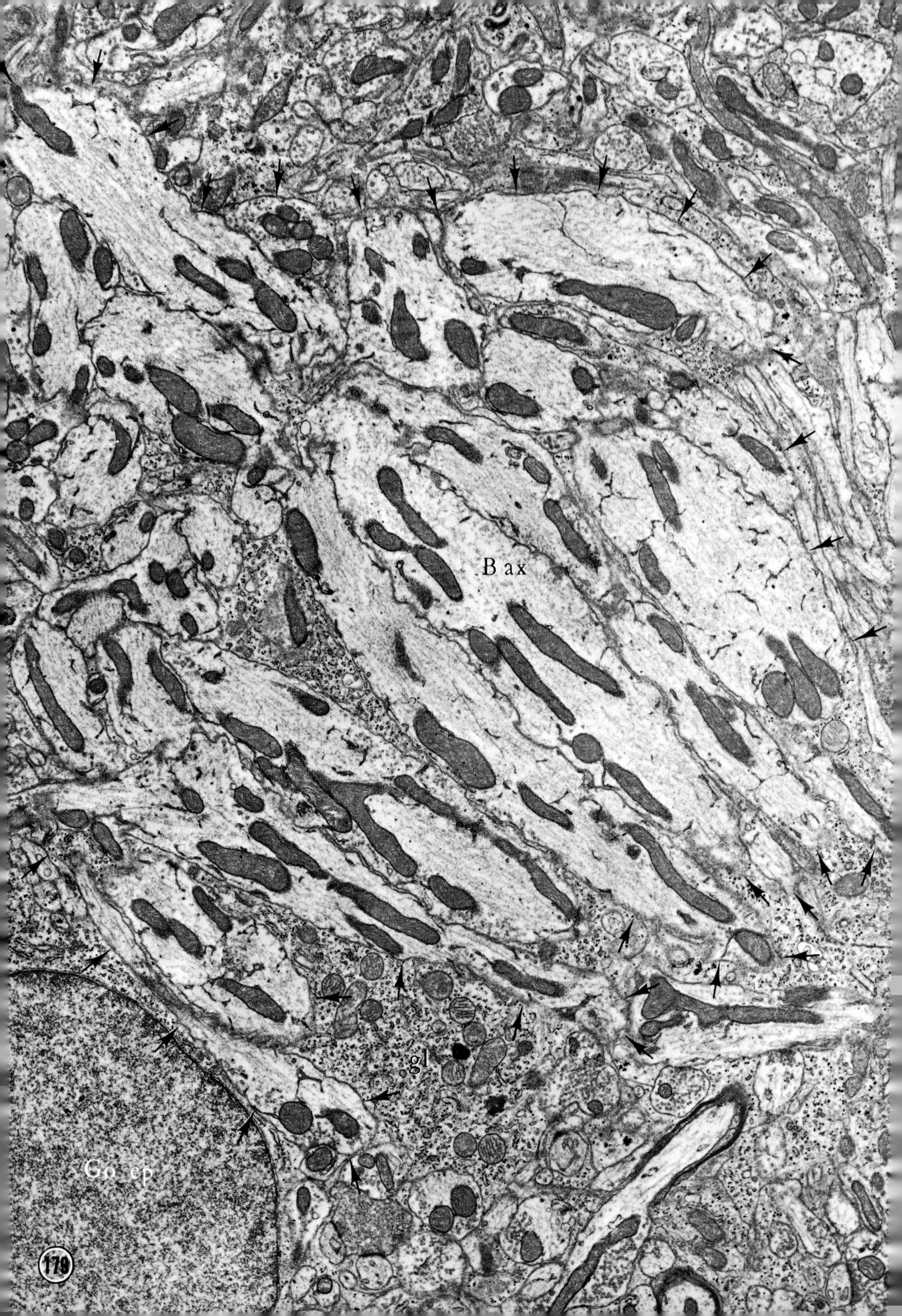
B ax
gl
Go ep
179

he found certain differences between them and the typical septate junctions of invertebrates (WOOD, 1959; GILULA, BRANTON, and SATIR, 1970; HAND and GOBEL, 1972). FRIEND and GILULA (1972) described similar formations in the interstitial spaces of the rat's adrenal cortex, which they studied by a variety of techniques. But unlike GOBEL, they emphasized the differences between these and the invertebrate septate junction. Therefore, they referred to them as "septate-like," and suggested that further investigation, especially with freeze-fracturing methods, would be necessary before the structure of the interaxonal junctions in the pinceau could be identified. SOTELO and LLINÁS (1972) were also more cautious than GOBEL, although they proposed that the junctions are modified forms of septate junctions.

The differences between these interaxonal junctions and the typical septate desmosomes of invertebrates were summarized by GOBEL (1971) as follows: (a) they tend to be shorter, (b) they are bounded by focal constrictions of the interstitial space around the perimeter of the junctions, (c) the septa tend to stretch diagonally between the apposed axonal membranes. In addition, according to SOTELO and LLINÁS (1972), the septa are not so "prismatic" as they are in the true septate junction. In our opinion, none of these differences is of any great importance. The length of the septate junctions in invertebrate epithelia varies a good deal, and gaps in the ranks of the septa are frequent. In our material (rat) the constriction of the interstitial space at the margin of the junction is not a consistent feature, and the septa are usually arrayed perpendicular to the apposed membranes. The "prismatic" or cylindrical form of the septa in invertebrates seems, furthermore, to be somewhat exaggerated in the minds of investigators, especially in view of the model reconstructions published by HAND and GOBEL (1972).

Since these differences in form, disclosed by conventional electron microscopy of thin sections, seem to be trivial, the objections of FRIEND and GILULA (1972) to the identification of the interaxonal junctions with septate junctions must be considered more seriously, even though their own investigation dealt with a different organ. In their material, an intercellular junction in the adrenal cortex, they found that the septa were often cylindrical particles that lay in the center of the interstitial space and did not always bridge the whole gap. The contact zone was permeable to lanthanum nitrate, and in freeze-fracture preparations failed to display the characteristic membrane structure of septate junctions. They considered this last difference of the greatest significance if, as they believe, the interaxonal junction is the same as their adrenal junction. It is exactly for this reason that the investigation of the interaxonal junctions in the pinceau should be pursued with a variety of techniques that can characterize their structure. For it is entirely possible that FRIEND and GILULA have misjudged the situation: their adrenocortical junctions do not appear exactly the same as the interaxonal junctions in conventional thin sections, and therefore the results of their study may not apply to the latter.

In these circumstances functional interpretations of the interaxonal junctions would seem to be at least premature. Nevertheless, all of the above authors have considered the possibility that these junctions act as electrotonic coupling agents between cells. GOBEL suggested that they held the axonal components of the pinceau together against the dissecting pressure of the velate astrocytes. FRIEND and GILULA made a related proposal, that the interstitial particles acted as struts holding the extracellular spaces open. SOTELO and LLINÁS suggested that the junctions might be electrotonic coupling devices or, alternatively, that they are high-resistance elements capable of channeling the extracellular flow of current.

Fig. 179. Basket axons of the Purkinje cell pericellular basket. The thick descending collaterals of basket cell axons (*B ax*) fall in a cascade over the cell body of a Purkinje cell. The collaterals are conspicuous because of their large caliber and their many longitudinally oriented mitochondria and neurofilaments (*black arrowheads*). The axons are packed tightly together in an orderly manner. The rounded contours of the group, however, give the clue that the soma of a Purkinje cell lies deep to the basket that the axons form. The axonal profiles are here and there interspersed with slips of neuroglial cytoplasm (*gl*). A Golgi epithelial cell is present in the lower left corner (*Go ep*). Lobule V. ×13500

Fig. 180. The pericellular basket and the neuroglial sheath. This view of the pericellular basket complements that shown in the previous figure (179). Here the axons of the pericellular basket (*black arrowheads*) cascade down around the belly of a Purkinje cell body that is just deep to this section. The axons are packed closely together. In the center of the field, the section plane cuts through the layers of basket axons (*asterisks*) and passes tangentially through the neuroglial sheath and the plasmalemma around the Purkinje cell. In the surround, slips of neuroglial cytoplasm (*gl*) pervade the neuropil. A Golgi epithelial cell is indicated (*Go ep*). The large arrow points to a process full of microtubules that courses past the pericellular basket. This is probably a portion of the climbing fiber stalk. Lobule V. ×12000

Fig. 181. The pinceau around the Purkinje axon initial segment. This cross section of the Purkinje axon initial segment (*arrowheads*) is surrounded by a neuroglial sheath and then enwrapped in a thick layer of terminal basket axons. These terminals make up the pinceau. Three granule cells (*gr*) and a Golgi epithelial cell (*Go ep*) flank this field. Crus I, VII. ×15000

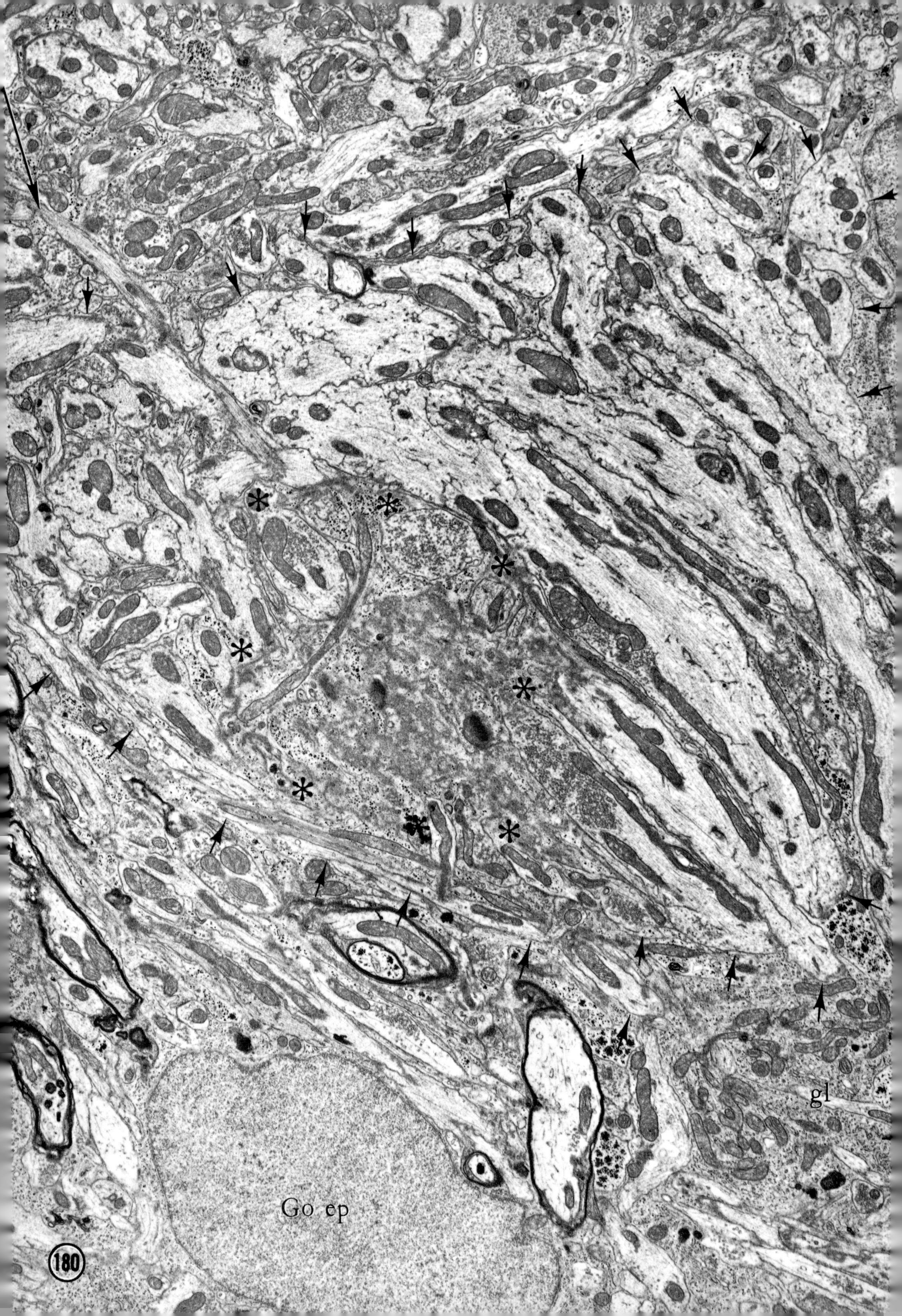
gl
Go ep
180

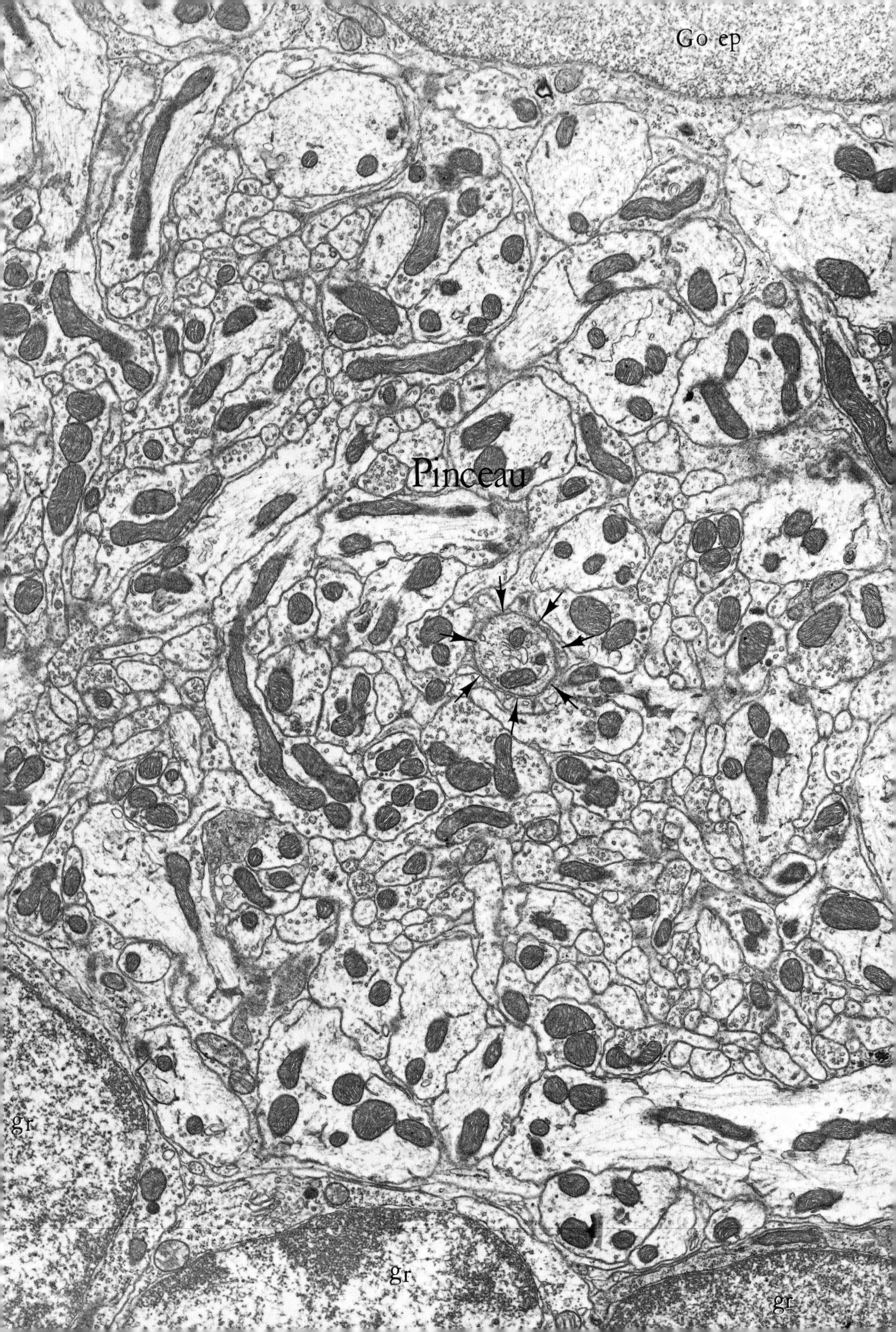
Go ep
Pinceau
gr
gr
gr

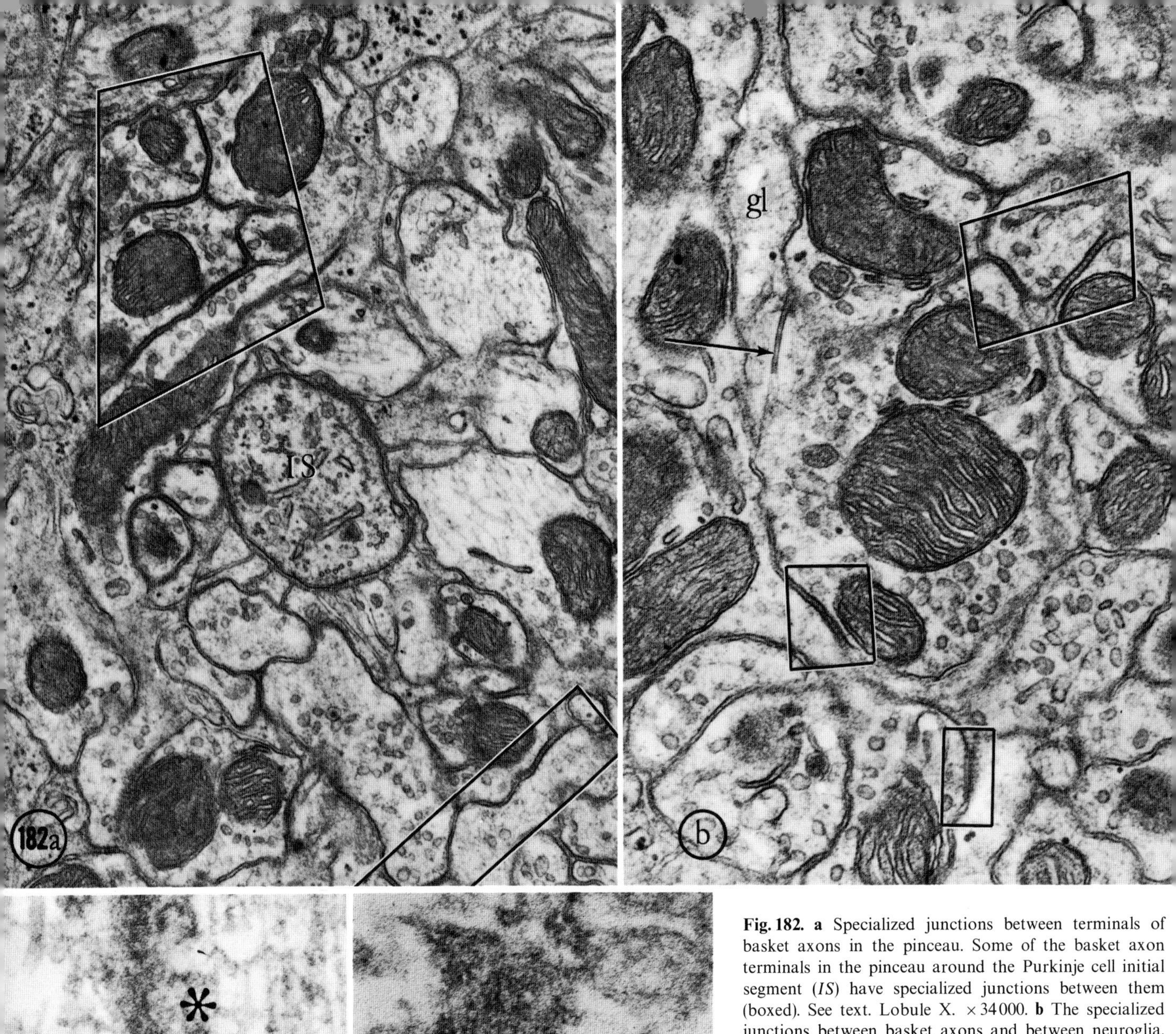

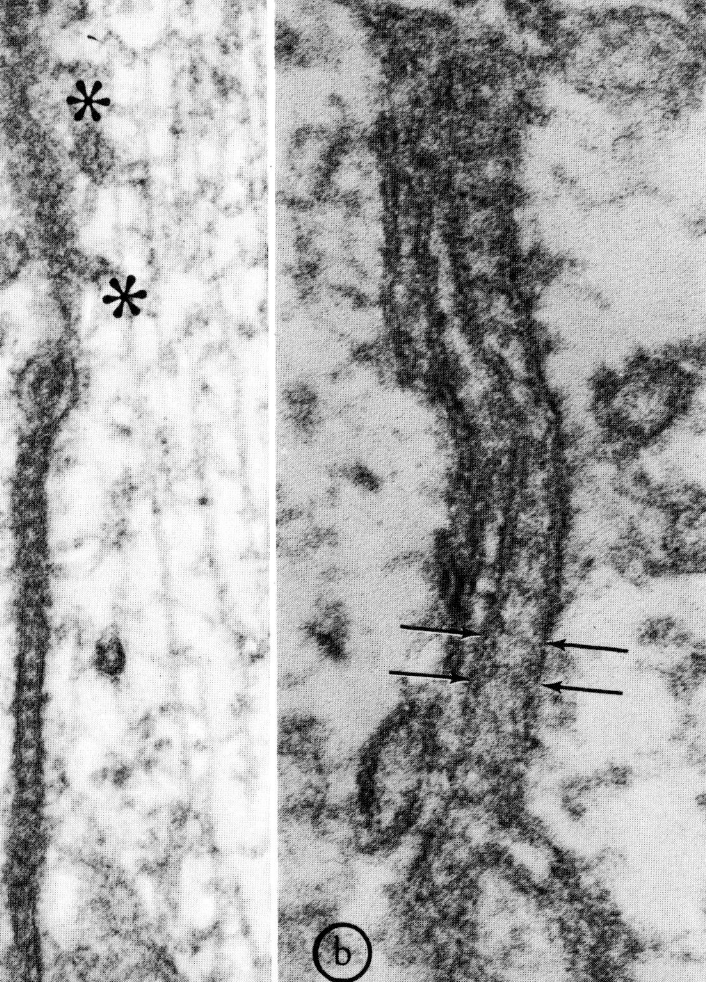

Fig. 182. a Specialized junctions between terminals of basket axons in the pinceau. Some of the basket axon terminals in the pinceau around the Purkinje cell initial segment (*IS*) have specialized junctions between them (boxed). See text. Lobule X. ×34000. **b** The specialized junctions between basket axons and between neuroglia. In this field of several basket axon terminals in the pinceau, the specialized junctions between basket axons are enclosed in boxes. However, another kind of specialized junction (*arrow*) can be made out between leaflets of neuroglial cytoplasm. (Refer to text and Fig. 185.) Lobule X. ×47500

Fig. 183a and b. Specialized junctions between basket axon terminals—high magnifications. These junctions have transverse septa (*arrows*) between the apposed plasmalemmas. The septa consist of dense particles, about 80 Å across, arrayed in the interstices in a honeycomb pattern (between *asterisks*). The center-to-center spacing between septa is about 200 Å. (See text.) a Lobule II. ×261000; b Lobule VI. ×600000

In this connection it may be worth while to note that FRIEND and GILULA state that the particles in their "septate-like" adrenal junctions had a strong affinity for potassium pyroantimonate. Since the binding of this precipitate indicates the presence of calcium and other cations, the particles in the interstitial space might well contain deposits of cations. If this finding should prove to be true of the junctions on the pinceau as well, then they may have a role in the ionic balance and, therefore, the flow of current in the interstitial fluid in the pinceau, and thus an important influence on the inhibitory efficiency of the basket cell axon synapse.

It is only fair to point out in relation to these interaxonal synapses that interstitial densities occur throughout the granular layer between granule cells (see Figs. 61 and 62), between granule cells and various axons (Figs. 128 and 129) or dendrites, and between neuroglial processes (Fig. 182). In most cases the material is not very well defined, perhaps not well enough preserved. In some it appears vaguely filamentous, as if fine hairs were spanning the extracellular space. In others it is more organized and consists of periodic densities. An example of this type occurs in the clefts between the neuroglial processes in the pinceau, especially around the Purkinje cell axon (Figs. 182b and 185). In these contacts, thin filaments traverse the intercellular clefts just as in the septate-like junctions between the basket axons, but the repeat period is about half that of the latter. In this respect the contacts resemble the "gap" junctions elsewhere, but in this case the intercellular cleft is not so narrow as it is in the usual "gap" junction. The total thickness of the contact measures about 180 Å across, with an interval of about 70 Å between the membranes. Such junctions may be modified gap junctions or merely incompletely preserved ones. More likely, they may be a different type of junctional specialization. The variety of interstitial arrangements suggests that the pinceau would be a difficult but challenging field for the cytologist who would interpret freeze-fractured specimens.

The Neuroglial Sheath. It was mentioned earlier that the pinceau is subdivided into enclaves of terminal twigs by swirls of fleshy axons. Some of these large fibers reach into the center of the pinceau where they approach the initial segment of the Purkinje cell axon. This latter fiber, however, is enveloped in a remarkable neuroglial sheath that only rarely gives access to the axonal surface beneath it. The sheath consists of thin prisms, bars, or sheets of neuroglial cytoplasm, often so thin as to be composed of nothing more than surface membrane and enclosed cytoplasmic matrix (Figs. 40, 182a, 184a, b, c, and 185). The cytoplasmic processes are usually laid on edge radially all around the surface of the initial segment, so that in transverse sections it appears to be centered in a floret. Sections in the longitudinal plane show that the neuroglial processes run for long distances alongside the initial segment without interruption. These processes are actually the edges of extremely tenuous broad sheets that permeate the whole pinceau from its outer surface. They originate from velate protoplasmic astrocytes or Golgi epithelial cells in the immediate vicinity of the Purkinje cell body, and they invade the pinceau from all directions, slipping between the larger basket cell axons, about which they weave a ragged and defective cover (Figs. 179, 180, and 184a). Some of the processes contain long, slender mitochondria, granular endoplasmic reticulum, and fascicles of gliofilaments. In some preparations they are jammed with β glycogen particles. Usually they appear as long sheets consisting only of the surface membrane of the neuroglial cell reduplicated over itself and enclosing no discernible intracellular organelles. The neuroglial sheets tend to circumvent the tangled, worm-like terminal branches of the axons, so that these are usually in direct apposition with one another over most of their surfaces. The pattern of the neuroglia (Fig. 181) exaggerates the subdivision of the pinceau into enclaves, and suggests that each enclave of small branches and terminal twigs may represent the terminal arborization of a single descending collateral. Although reconstructions from serial sections would be necessary to prove this suggestion, the appearance of the basket fibers in Golgi preparations is consonant with it.

Moreover, the entire pinceau is surrounded by a highly perforate neuroglial sheath of the same veil-like nature. Since no cell bodies, neuronal or neuroglial, lie within the precincts of the pinceau, the adjacent cells can only arrange themselves around the field of axons. The laminar processes of the neuroglia partially circumscribe this field as they insinuate themselves between some of its constituents. In the cat the pinceau has been described and diagrammed as completely ensheathed by astrocytic processes (HÁMORI and SZENTÁGOTHAI, 1965), and it has been suggested (ECCLES, ITO, and SZENTÁGOTHAI, 1967) that certain electrical properties of the Purkinje cell response to stimulation of basket cells reflect the slow leakage of transmitter from this encapsulated field. Whatever the merits of these speculations, the pinceau in the rat is very incompletely ensheathed by neuroglia.

The openness of the neuroglial barrier is further attested by the invasion of the pinceau by an assortment of other neural processes. These include the dendrites of Lugaro cells and the preterminal branches of recurrent collaterals of Purkinje cell axons, which, coming together

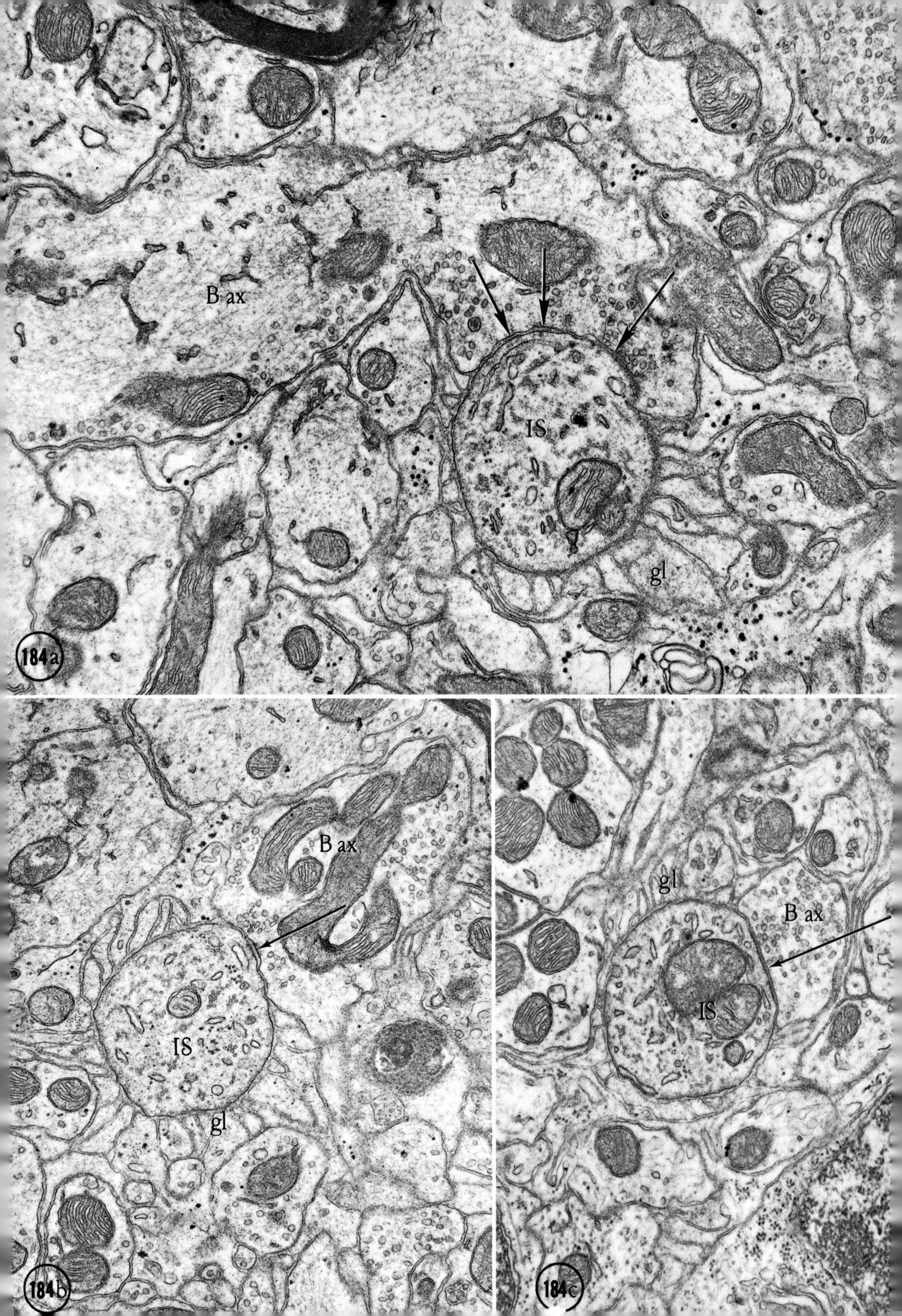
B ax
IS
gl
184a
B ax
IS
gl
184b
gl
B ax
IS
184c

in synapse, can appear among the peripheral fibers of the pinceau. Dendrites of granule cells and Golgi cells also occur there. In addition, the ascending stem of the climbing fiber, with or without its myelin sheath, can pass right through the pinceau on its way to the molecular layer (Fig. 180). The dendrites are usually accompanied by neuroglial laminae, but the axons are often bare. None of these intruders pass close to the initial segment of the Purkinje cell axon, and it is evident that they are not an integral part of the pinceau. Yet one wonders what effect the neural activity of their environment has upon them.

The Axo-Axonal Synapse. Although the overwhelming majority of the basket cell terminal axonal arborization is comprised in the pinceau, some of the large basket cell collaterals penetrate through this thicket and breach the neuroglial barrier around the Purkinje cell initial segment to synapse with the axon within. Such axo-axonal synapses occur regularly in the upper third of the initial segment in the rat. In hundreds of electron micrographs of random sections including the central axon in the pinceau we have encountered only 43, of which 35 were found in neighboring pinceaux in two horizontal sections from two different specimens. The latter circumstance suggests that there may be a small number of synaptic junctions, perhaps one or two, on each initial segment and that these are all disposed at approximately the same distance from the cell body, perhaps in the proximal quarter to third of the axon. Only reconstruction from serial sections could answer this question. Most of our examples are clearly synapses *en passant*, and three of them are possibly *boutons terminaux*. Several examples had two synaptic terminals each. The junction takes the form of the axo-dendritic and axo-somatic junctions described earlier in this section. The basket cell axon flattens against or partially encompasses the Purkinje cell initial segment (Figs. 184a, b, c, and 185). Oval or round synaptic vesicles aggregate rather loosely at the interface. Increased density of the cytoplasm on either side of the junction produces a symmetrical or nearly symmetrical accentuation of the apposed axonal membranes. In some of our examples the postsynaptic membrane is clearly preponderant (Fig. 184); in others there is no difference between them (Figs. 184b and 185). The synaptic cleft is somewhat wider than the interfacial space beyond the synaptic complex; it is not, however, wider than the interstitial space in many nonsynaptic regions. Fine filaments can usually be seen crossing the cleft. This junction therefore appears like a chemical synapse, and its inclusion in the field of the pinceau points up further the interesting peculiarity of this complicated axo-axonal articulation.

The foregoing description differs in some important respects from previous accounts of the pinceau. In earlier publications the rarity of direct axo-axonal contacts was emphasized (Palay, 1964b, 1967; Peters, Palay, and Webster, 1970). At the time of those reports, only three axo-axonal synapses between basket cell terminals and the initial segment of the Purkinje cell axon had been found in our material, despite surveys of nearly a hundred different initial segments in random sections through the cerebellar cortex. Hámori and Szentágothai (1965), however, had declared that in the cat basket axons form numerous true synaptic junctions in the pinceau, particularly with the initial segment of the Purkinje cell axon, and they underscored this finding by presenting a schematic drawing of the pinceau at electron microscopic magnification in which the initial segment was covered with synaptic junctions of Gray's second type (also reproduced in Eccles, Ito, and Szentágothai, 1967). This discordance could easily reflect a species difference. Mugnaini (1972), however, reported that in the cats he examined, synaptic junctions on the initial segment were just as rare as they are in the rat, and Sotelo and Llinás (1972) had a similar impression. During the past three years a more intensive study of the emerging axon (Chan-Palay, 1972a) has greatly increased the number of pinceaux that we have examined, and consequently we have found many more examples of these axo-axonal junctions. Nevertheless, we cannot confirm the descriptions of Hámori and Szentágothai (1965; repeated in Eccles, Ito, and Szentágothai, 1967; and in Bell and Dow, 1967). Although we can now say that one or two synaptic junctions occur regularly on the axon of each Purkinje cell in the rat, there is still no suggestion that the initial segment is covered with them as in the drawing of Hámori and Szentágothai. These authors also described, and indicated in their drawing, an unusual amount of extracellular space surrounding the finger-like terminal branches of the basket fibers and a distinct neuroglial capsule enwrapping the whole complex. Neither

Fig. 184. a Axo-axonal synapse between basket cell axon and Purkinje axon initial segment. The initial segment (*IS*) of this Purkinje cell axon is surrounded by neuroglial processes (*gl*). The axon receives a set of three synapses (*arrows*) from a basket cell axon (*B ax*). The basket cell axon is typically large and full of neurofilaments. It wraps partially around the initial segment. The small synaptic complexes are characterized by a widened synaptic cleft and sparse postsynaptic densities. Lobule X. × 39000. **b and c** The axo-axonal synapse. Two other initial segments (*IS*) of Purkinje cell axons are shown here in synaptic contact (*arrow*) with axons of basket cells (*B ax*). The neuroglial sheaths around the Purkinje cell axon are well displayed (*gl*). Lobule X. × 31000

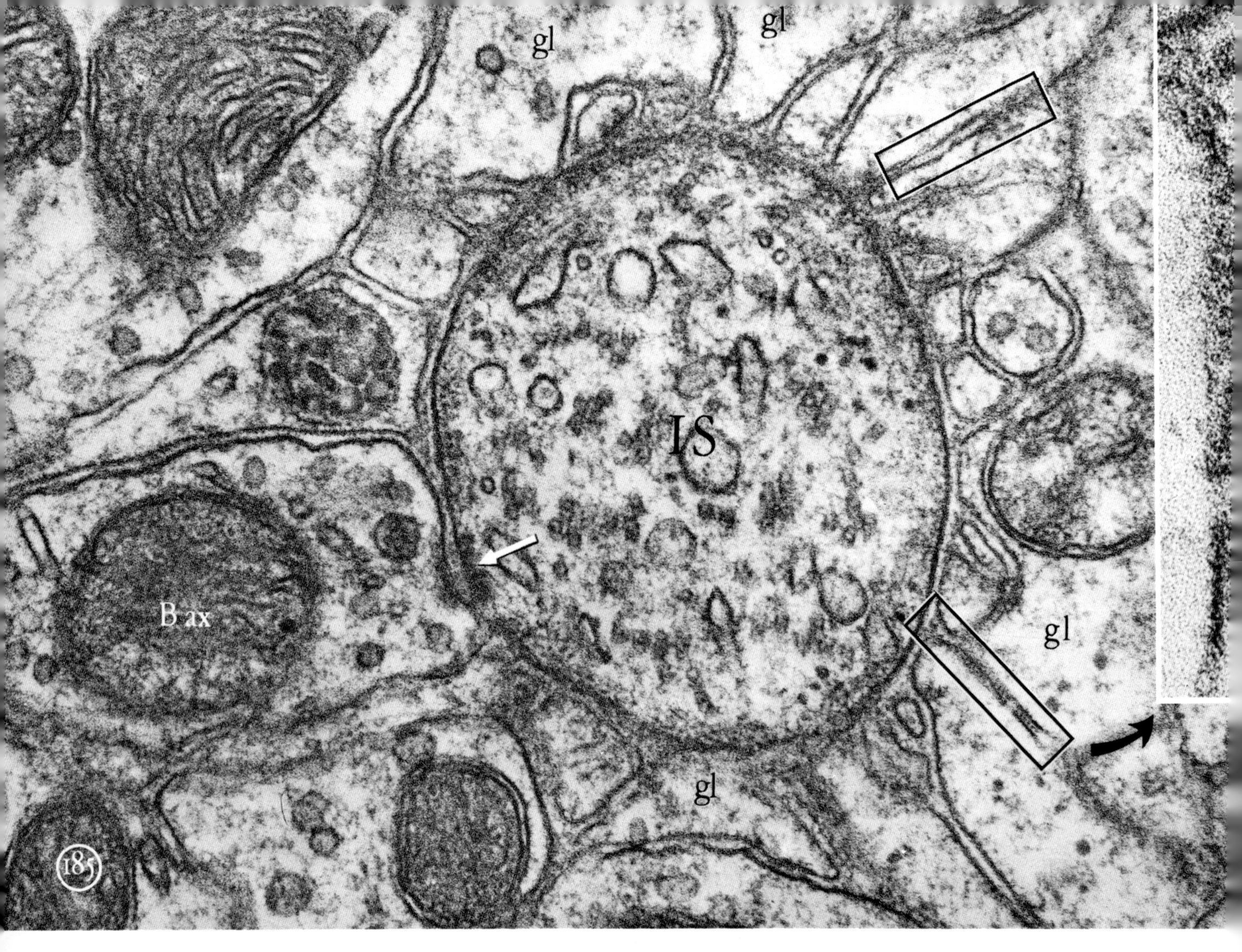

Fig. 185. The specialized neuroglial junction. This Purkinje cell axon initial segment (*IS*) is almost completely surrounded by the processes of neuroglial cells (*gl*). The cell-to-cell apposition between neuroglia often shows junctional specializations (boxed, also see text and Fig. 182b, *arrow*). The inset shows one of these junctions at higher magnification. The interstitial cleft is traversed regularly by short densities. On the initial segment, a terminal of a basket axon (*B ax*) has penetrated the neuroglia to make a synaptic contact. This synaptic complex is not seen in true cross section (*white arrow*). Lobule VI. × 102000. Inset × 386000

of these features has been confirmed by subsequent observations in the cat (GOBEL, 1971; MUGNAINI, 1972; SOTELO and LLINÁS, 1972), and both are absent in the rat (PALAY, 1964, 1967) and monkey (FOX, HILLMAN, SIEGESMUND, and DUTTA, 1967). Consequently, they should not be used for erecting hypotheses to explain the slow build-up and decay of inhibition by basket cell axons (ECCLES, ITO, and SZENTÁGOTHAI, 1967).[16]

16 ECCLES, ITO, and SZENTÁGOTHAI (1967) attribute to PALAY (1964b) the idea that the action of the pinceau should be similar to that of the axon cap on the Mauthner cell initial segment because of the neuroglial capsule around this peculiar formation. While that author did recognize the similarity between the pinceau and the axon cap, he also pointed out that the pinceau is *not* encapsulated by neuroglia. On the contrary, it is the Purkinje cell and its axon that are encapsulated.

In summary, the descending collaterals of the basket cell axon converging on the Purkinje cell initial segment terminate almost entirely in panicles of rounded and digitiform endings, only a few of which actually contact the surface of the Purkinje cell axon. These few synapses are located at the border between the first and second thirds of the initial segment, and bear the morphological features of a chemical synapse. The rest of the surface of the initial segment is ensheathed in neuroglial processes. The small number of synaptic terminals signifies that at most only one or two of the numerous collaterals in the pinceau can end in this fashion; all others must end freely.

The pinceau is a most unusual interneuronal articulation. In the first place, it is an axo-axonal synapse, which is an uncommon type. Second, it includes not only a simple chemical synaptic junction but also a voluminous and complicated synapse *à distance*. No other example of this type is known in the mammalian nervous system. The only analogous structures known are the axon cap of the Mauthner cell in teleosts and the spiral axo-axonal synapse in the sympathetic ganglia of amphibia. In principle, however, the innervation of glands and smooth muscle also could be considered analogous.

What could be the function of this peculiar articulation? At first sight, its location appears to be ideal for an inhibitory synapse, since by surrounding the initial segment it is in a favorable position to control the flow of current from the cell body. Because of the structural similarity between the axon cap of the Mauthner cell (ROBERTSON, BODENHEIMER, and STAGE, 1963) and the pinceau, PALAY (1964b and 1967) put forward the suggestion that they might have analogous modes of operation. The studies of FURUKAWA and FURSHPAN (1963) on the Mauthner cell indicated that the afferent fibers composing the axon cap produce an electric field which inhibits spike discharges in the axon. ECCLES, ITO, and SZENTÁGOTHAI (1967), however, found no evidence for fast electrical inhibition in the Purkinje cell, and they dismissed the analogy with the Mauthner cell. They supposed that the prolonged inhibitory action of the basket cells on the Purkinje cell might be explained by the accumulation of the transmitter around the initial segment followed by its slow dissipation after the cessation of basket cell activity. But such a hypothesis, even if descriptively correct, does not take into account the architecture of basket cell axon collaterals and their numerous synapses on the dendrites and soma of the Purkinje cell. The slow build-up and decline of basket cell inhibition might not be a characteristic of function in the pinceau itself. Based on a faulty concept of the pinceau, this hypothesis does not consider the small postsynaptic surface available to the periaxonal fibers and the thorough encapsulation of the initial segment with neuroglial processes. SOTELO and LLINÁS (1972) have recently revived PALAY'S earlier suggestion, explaining that the peculiar arrangements of the pinceau would "distribute the outward action currents of the basket cell end feet as a vertical sleeve around the Purkinje cell axon," thus blocking the latter. But they presented no new evidence. A direct investigation of the pinceau by intracellular electrodes has not been reported as yet. Such a study is probably the only way to solve the puzzle of the pinceau.

However, even if the pinceau should prove to be an electrical inhibitor, the action of the chemical synaptic junctions directly on the initial segment is not explained by such models. One might assume that a synaptic junction on the first third of the initial segment would have a critical effect upon the firing of that axon. Since there are only one or two of such junctions, there can be only one or at most two parent basket cells that are given the privilege of exerting such a critical effect. The other fibers are collaterals of basket cells that have their privileged status elsewhere, i.e., on other Purkinje cell axons. We may therefore propose the hypothesis that a peculiar combination of discharge patterns would be necessary in order to display the potential effect of the two modes of articulation. For example, the discharges of the terminals *à distance* might damp the Purkinje cell and slow its frequency, but a coincidence of discharge with that of the chemical axo-axonal synapse might be necessary in order to silence the Purkinje cell altogether. The privileged basket cell would determine the effect on its Purkinje cell, but would merely contribute to the periaxonal field around the other Purkinje cells in its territory. This scheme could operate whether or not the pinceau proves to have an electrotonic mode of action. Not only transmitter but also potassium ions could be expected to accumulate in the periaxonal interstitial spaces, and these should have a significant effect on the electrogenic activity of the initial segment. The thorough encapsulation of the initial segment with neuroglial processes perhaps takes on additional significance in view of the outpouring of potassium ions that would accompany prolonged activation of the periaxonal basket fibers.

4. Summary of Synaptic Connections of Basket Cells

Afferents	granule cells	————	basket cell
	climbing fiber	————	basket cell
	basket cell	————	basket cell
	stellate cell	————	basket cell
Efferents	basket cell	————	Purkinje cell
	basket cell	————	basket cell
	basket cell	————	stellate cell

Chapter VIII

The Stellate Cell

1. A Little History

The stellate cells compose a class of small polymorphous neurons lying in the outer two thirds of the molecular layer. They were described by a number of early authors, FUSARI (1883), PONTI (1897), SMIRNOW (1897), in addition to RAMÓN Y CAJAL (1889b), who in his great book on the nervous system (1911) gave SMIRNOW the credit for writing the most detailed and exact account of the different kinds of stellate cells.[17] SMIRNOW distinguished two kinds of stellate cells according to the configuration of their axons: (1) cells with long horizontal axons running in the parasagittal plane and (2) cells with highly arborized short axons. RAMÓN Y CAJAL reversed the order of these two classes and, furthermore, divided the short-axoned cells into two subclasses. His scheme, which has been followed by most subsequent investigators (e.g., ECCLES, SZENTÁGOTHAI, and ITO, 1967) may be paraphrased thus:

(1a) Dwarf cells with more or less fusiform cell bodies, possessing few dendrites and furnished with a thin, short axon, which ramifies into a simple arborization.

(1b) Somewhat larger cells with stellate, ovoid, or fusiform cell bodies, possessing numerous diverging dendrites and an axon which spreads out into an extensive and complicated plexus of ascending and descending collaterals. This plexus extends through a large part of the molecular layer, reaching even into the lower third, which is usually reserved for the basket cells. Some of these stellate cells lie deep in the molecular layer themselves and contribute small collaterals to the baskets.

(2) Larger stellate, fusiform, or triangular cells, provided with long dendrites and a long axon running transversely across the folium and terminating in a sparse ramification distributed in the outer half of the molecular layer. In its course the axon gives off a few short ascending or descending collaterals, which terminate in the outer molecular layer.

It should be noticed that the small stellate cells of type 1a are concentrated in the outer third of the molecular layer and that, although they occur in small numbers elsewhere in the molecular layer, they are the only neuronal cell bodies in the outer third. In contrast, the larger stellate cells tend to lie in the middle and lower thirds of the molecular layer, where they are mingled with the upper basket cells. The distribution of stellate cells makes it difficult to distinguish the two cell types in electron micrographs of the middle molecular layer, while any neuronal cell body in the outer molecular layer can be identified with assurance as that of a stellate cell.

Most authors have regarded the stellate cells as related to basket cells, and have followed RAMÓN Y CAJAL in thinking that there is an uninterrupted chain of transitional forms between them. Both basket and stellate cells are oriented in the same plane athwart the longitudinal axis of the folium. Both also occupy similar positions in the circuitry of the cerebellar cortex, for they both receive synapses principally from the parallel fibers and they direct their output principally to the Purkinje cells. There is, however, a very important difference between the two cell types in so far as the distribution of their axonal arborizations is concerned. In large measure the axonal collaterals of the stellate cell impinge upon the shafts of the Purkinje cell dendrites, whereas the basket cell axon, in addition to having this disposition, participates in the construction of an elaborate plexus about the perikaryon of the Purkinje cell and in the pinceau beneath it. Once this fundamental difference is noted, other differences between the two cell types become evident. As a result of an intensive study of the stellate cells (CHAN-PALAY and PALAY, 1972a), we have come to the conclusion that they are genuinely a different cell type from the basket cells and therefore merit separate treatment. This conclusion is substantiated by the fact that stellate cells occur in the cerebella of all classes of vertebrates that have been examined, whereas basket cells appear only in birds and mammals. In view of the morphological differences, it may be predicted that the functional roles of these two cells in the operation of the cerebellar cortex may be quite different.

17 SMIRNOW, writing from Tomsk in Western Siberia, had proposed in his paper that the basket cells should be named Cajal cells.

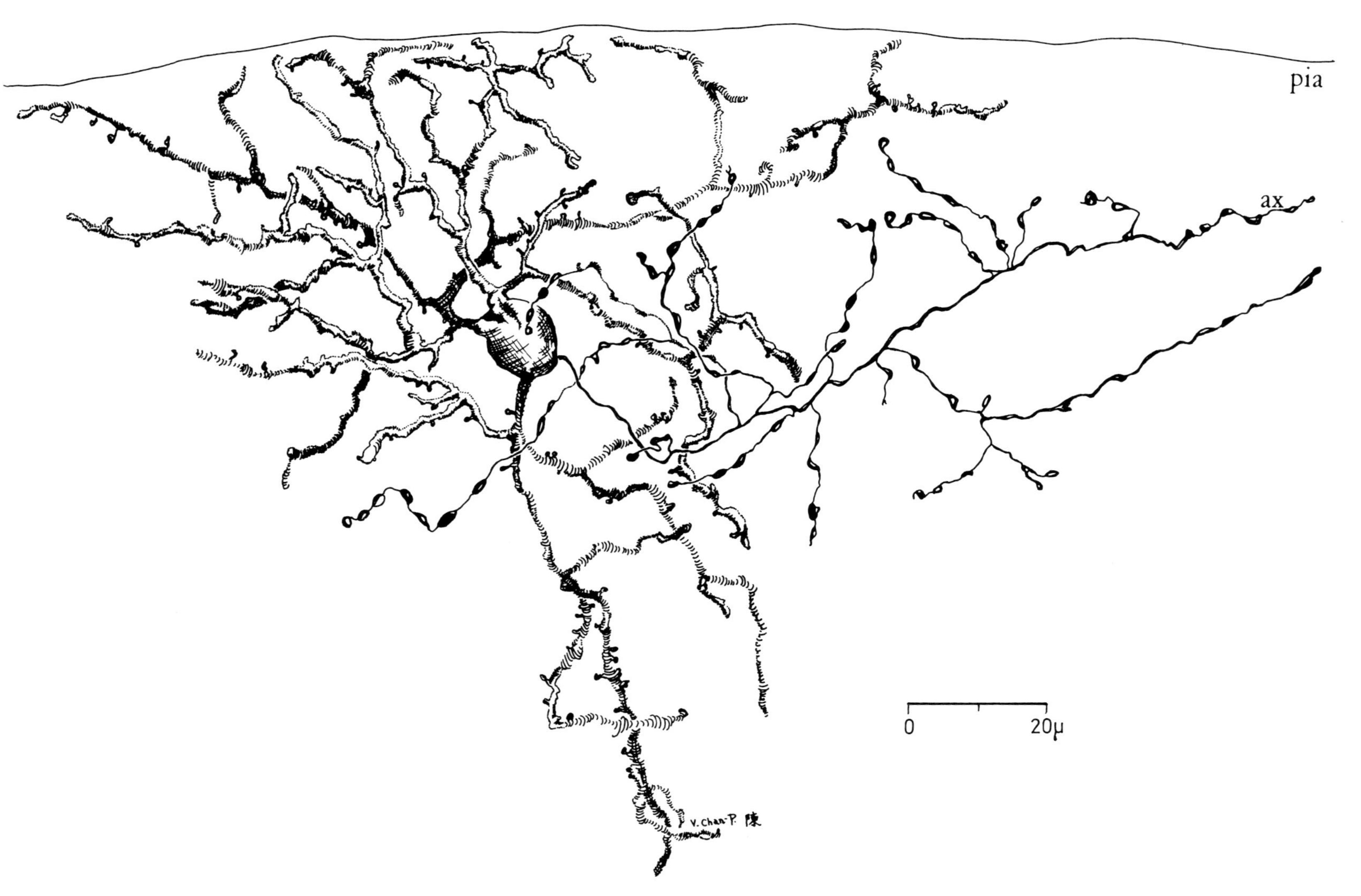

Fig. 186. A superficial stellate cell in a parasagittal section of rat cerebellar cortex. This small stellate cell lies in the upper third of the molecular layer near the pial surface. The many contorted branches of its dendritic tree arise from three main dendrites which radiate from the cell body. The dendritic surfaces display a few short spicules or tiny round bulbs on thin threads. The axon (*ax*) begins as a thin, smooth initial segment which soon meanders from its original course, breaking up into many fine, beaded collaterals. The entire plexus arborizes within the field of the dendrites. Rapid Golgi preparation. Camera lucida drawing. Section 90 μm thick. 100 × oil immersion objective

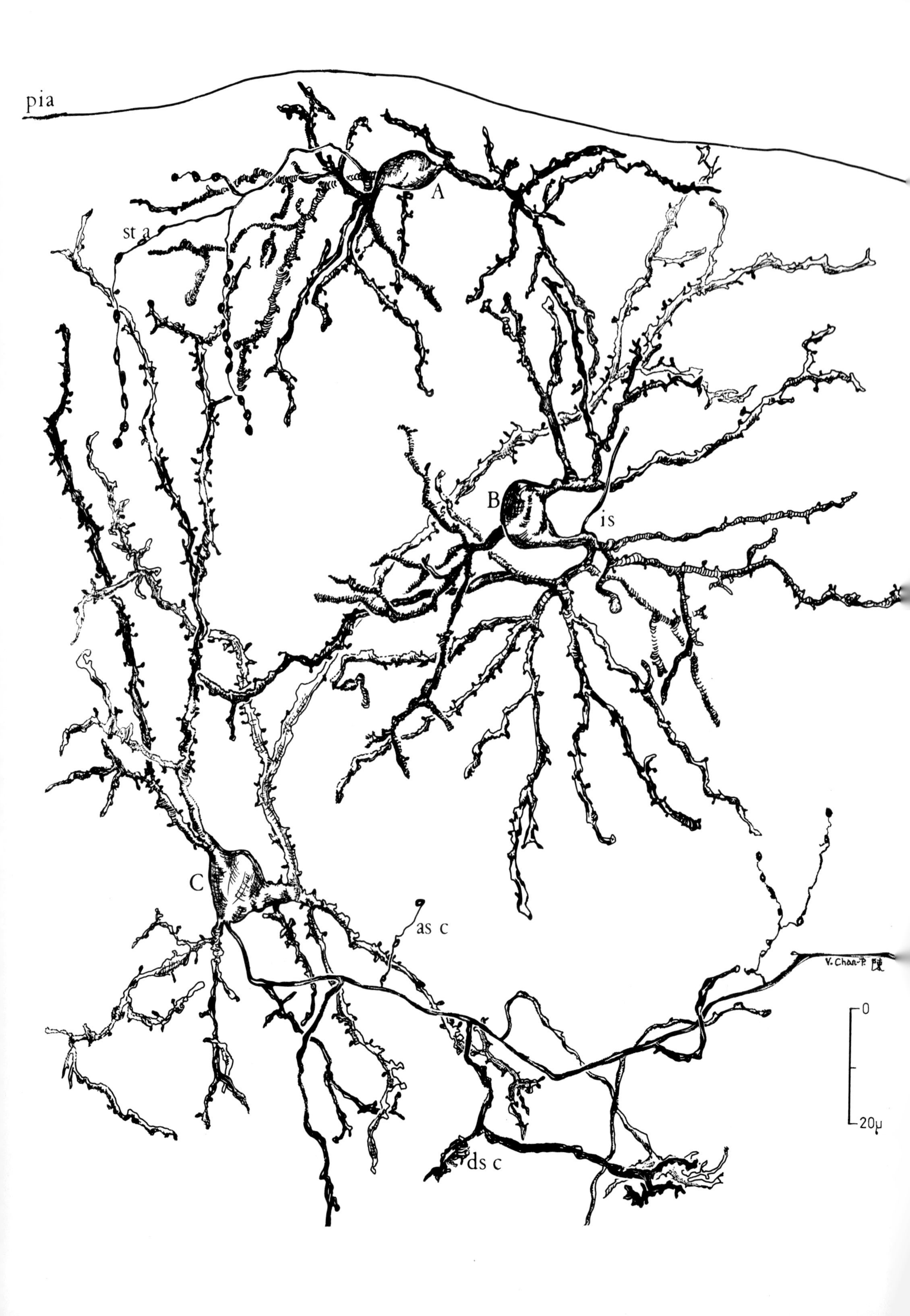

pia
A
st a
B
is
C
as c
ds c
V. Chan-P.
0
20μ

2. The Stellate Cell in the Light Microscope

a) The Superficial Short Axon Cell

It is easier to begin the description of the stellate cells with those that reside in the outer third of the molecular layer near the pial surface, because, as was mentioned above, they are the only neuronal cell bodies present there. They can be quite superficial, lying just within the neuroglial end feet that form the limiting membrane under the pia. The superficial stellate cells have small somata about 5–10 μm in diameter. The appearance of their dendrites in Golgi preparations is typical of all stellate cells. They are irregular in caliber and very contorted, with many abrupt changes in direction, as if twisting and hooking around invisible obstacles in their course. These characteristics can be seen in Figs. 186–189. The dendrites are relatively short, and branch profusely, ending in twigs that run horizontally under the pial surface (Fig. 186). They are fitted with a few appendages in a variety of forms: lumpy, warty excrescences, short spicules, or tiny, round bulbs tethered to the dendrite by thin strands. RAMÓN Y CAJAL (1911) described the dendrites as "finely varicose." The dendritic trees of the stellate cells near the pial surface fall into two patterns. First, major dendrites can originate from opposite sides of the cell body, giving the cell a somewhat bipolar appearance. These dendrites can extend either horizontally, that is, parallel with the pial surface (Fig. 187A) or vertically (Fig. 188A, B). Secondly, a cell can have dendrites that ascend toward the pial surface and then cascade downwards, often reaching below the level of the cell body (Fig. 188C).

Fig. 187. A comparison of the dendritic arrangements of a superficial and a deep stellate cell with that of a basket cell in a parasagittal section of the molecular layer. The superficial stellate cell (*A*), lying just underneath the pial surface, gives rise to short, crooked dendrites that issue from either pole of the cell body and run horizontally for a limited distance. The axon ramifies simply into three beaded collaterals (*st a*). The deep stellate cell (*B*) lies in the middle third of the molecular layer. Its relatively long, contorted dendrites radiate from the cell body and bear a few spines. Except for the initial segment (*is*), the axon has not been illustrated here. The basket cell (*C*) lies in the lower third of the molecular layer, and its long, relatively straight dendrites extend across the entire layer to reach up towards the pial surface. The dendrites of this cell are particularly thorny, and they branch at their tips to run for short distances sometimes just under the pia. The smooth, thick axon (*b ax*) gives rise to many thick descending collaterals (*ds c*) that form the pericellular basket and pinceau. A few thin, beaded ascending collaterals (*as c*) are also illustrated. Rapid Golgi preparation. Camera lucida drawing. Section 120 μm thick. 100× oil immersion objective

The axon of the stellate cell originates from the cell body (Fig. 186) or, less often, from a major dendrite (Figs. 187A, B, 188A–C, and 190), by way of a barely perceptible axon hillock. The initial segment is usually thin, straight, and smooth for about 10 μm, but it soon gives way to a meandering, branching axon. Thin, crooked collaterals arise from it at approximately right angles or somewhat less. RAMÓN Y CAJAL (1911) described the axons of stellate cells as covered with varicosities. In our preparations they are short, tenuous, varicose threads resembling strands of loosely strung, imperfect pearls (Figs. 186 to 191). As can be seen in Figs. 186–188, the axon of the superficial stellate cell is usually fairly short, and it is confined to the same restricted field as its dendrites. The axon can loop back and forth through a depth of about 30–40 μm in this field (Fig. 188A), giving off its varicose collaterals at irregular intervals. Both axons and dendrites of most superficial stellate cells generate fairly simple arborizations, as is illustrated in Figs. 187A and 188A, B, C. However, some stellate cells in the outer third of the molecular layer have been encountered with complicated, highly branched dendrites and more extensive axonal arborizations (Fig. 186). These cells need not be considered as another class of stellate cells, as they are merely more complicated versions of the superficial short axon cells.

Like the basket cells, all stellate cells extend their dendrites only in the plane that lies transverse to the axis of the folium (Fig. 189). Since this plane is the same as that of the Purkinje cell dendritic tree, the two are interleaved, sometimes in an alternating fashion, the parallel fibers threading their way through both arborizations. The axonal plexus of the stellate cell deviates, however, from the strict flat plane of the dendritic tree and extends on either side. When viewed in Golgi preparations sectioned parallel with the axis of the folium, the profiles of the dendritic tree are confined to a thin plane 15–20 μm thick, whereas the axons spread out over 40 μm. Such a spread is necessary if the axons are to synapse with the dendrites of Purkinje cells on either side. The dimensions of the axonal plexus generated by the short axon superficial cells suggest that they contact only a few Purkinje cells.

b) The Deeper Long Axon Stellate Cell

In the middle third of the molecular layer and deeper, stellate cells with very long axons are encountered. These are distinct from the superficial short axon cells, and constitute a second type (Figs. 187C, 188D, 190, and 191). Their major dendrites emerge from the cell body and

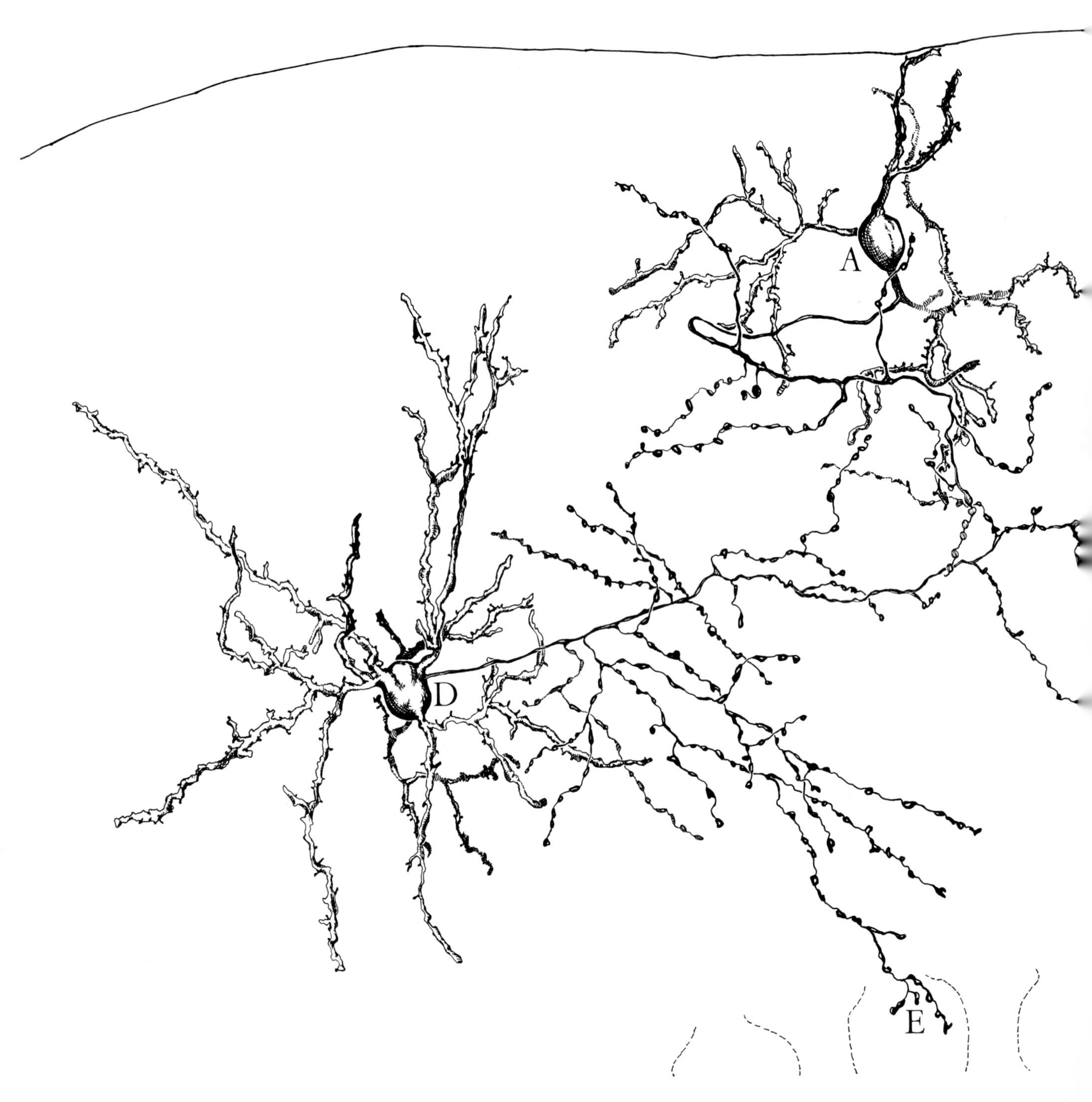

Fig. 188. Three superficial stellate cells and a deep long axon stellate cell in the molecular layer. These three examples of short axon superficial stellate cells (*A*, *B*, *C*) are simpler than that illustrated in Fig. 1. Cell *C* has dendrites that issue from the cell body and ascend nearly to the pial surface, then cascade downwards sometimes to reach below the level of the cell body. The axonal arborization is distributed in a tangle of beaded threads around these dendrites. Ascending axons of granule cells (*ga*) come into relation with the dendrites and cell body of stellate cells *B* and *C*, respectively. The long axon stellate cell (*D*) has a typical axonal plexus. The initial part of the axon gives rise to a thin bush of beaded collaterals, some ascending, but most descending. One of the descending collaterals (*E*) reaches down to branch in the region of the pericellular basket of a Purkinje cell soma. The remainder of this axon (*d s ax*) continues across the folium for about 450 µm, giving off occasional collaterals in its traverse. Camera lucida drawing, sagittal section. Section 120 µm thick, 100 × oil immersion objective

give rise to a large number of contorted branches that radiate outwards into the molecular layer, dividing frequently. Resembling the dendrites of the more superficial stellate cells, they also have few spiny appendages.

The axon can extend for lengths of up to 450 µm (Fig. 188 D), always running in the parasagittal plane and

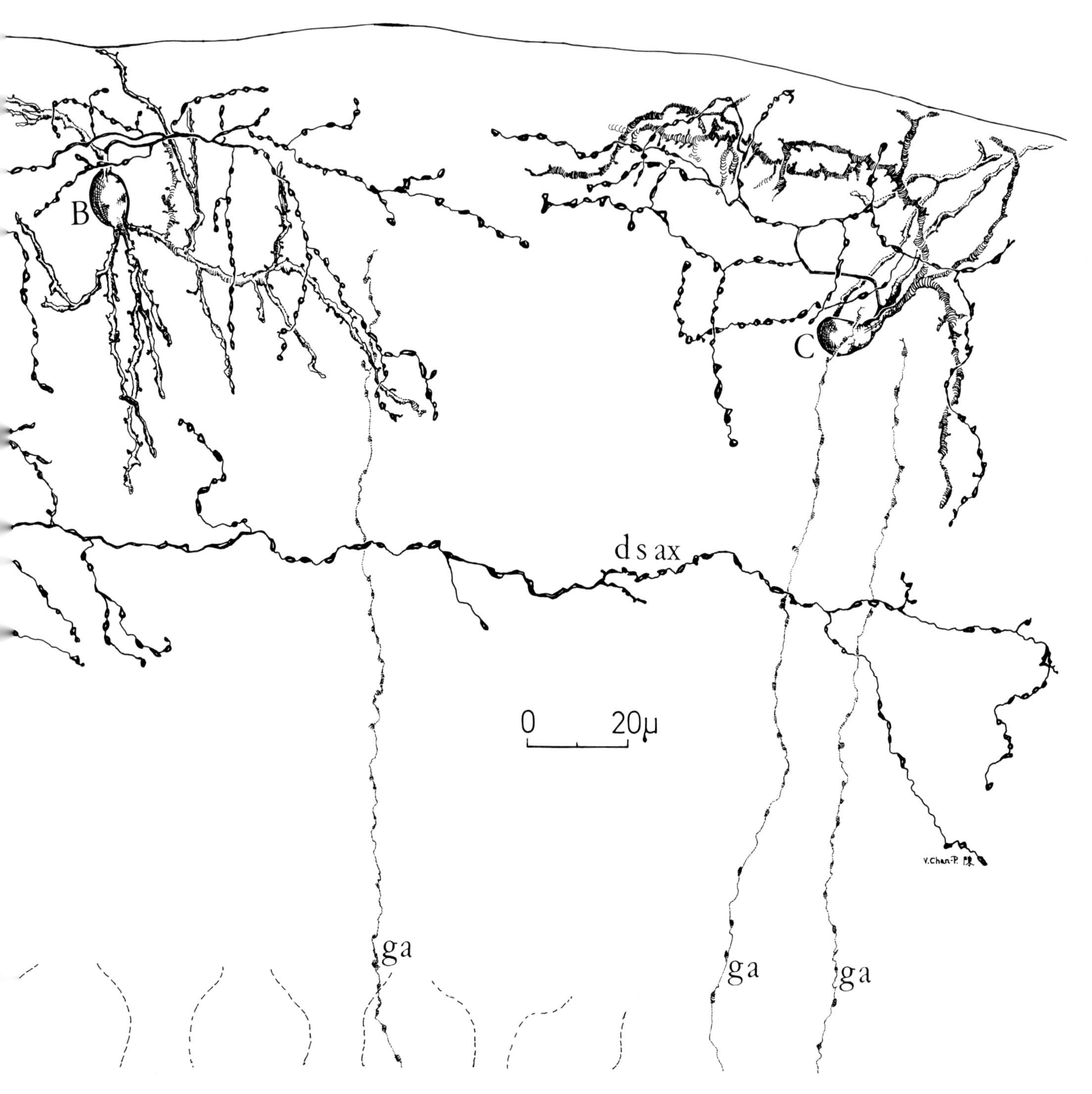

slipping through the fan-like arborescence of the Purkinje cell dendrites. A typical example of such a deep stellate cell is shown in Fig. 191. The main stem of the axon emerges from the cell body and remains at approximately the same level in the molecular layer as it occupied at its beginning. In the first third of its traverse after the initial segment it gives rise to a number of ascending and descending, tenuous, and varicose collaterals. The main stem does not thicken appreciably after the initial segment and all of the branches are slender. The ascending collaterals sometimes branch again to produce a simple plexus. The descending collaterals always branch repeatedly, giving the resulting arborization the appearance of a thin, short beard. The number of collaterals emitted from the remaining two-thirds of the axonal stem decreases as the last varicosity that signals the tip of the axon is approached (Fig. 188 D).

Rarely, one of the descending collaterals issuing from the axon of a deep stellate cell gives off a branch that continues as a long, thin varicose thread down to the

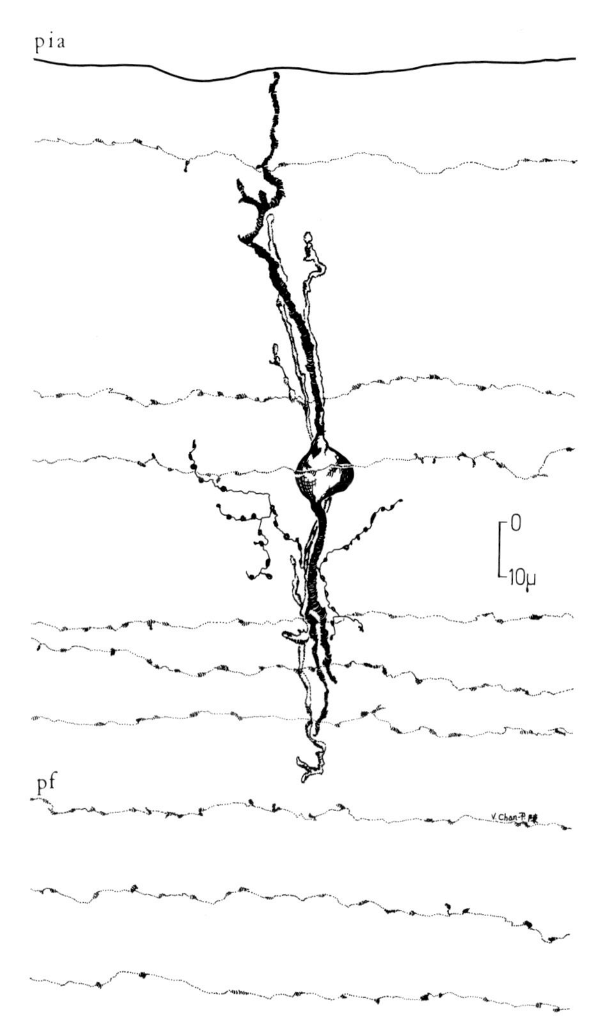

Fig. 189. Stellate cell in a section parallel to the longitudinal axis of the folium. The axons and dendrites of this stellate cell project a narrow profile when viewed in this plane of section. Parallel fibers (*pf*) running along the axis of the folium come into relation with the dendrites and soma of this cell. Camera lucida drawing. Section 90 μm thick. 100 × oil immersion objective

level of the Purkinje cell bodies (Fig. 188E). Here it may divide into three or four shorter branches. One cannot decide on the basis of Golgi preparations alone whether these stellate axon collaterals actually come into synaptic contact with the Purkinje cell body or only contribute to the pericellular basket around it and its initial segment.

c) The Difference between Stellate and Basket Cells

The superficial stellate cells are easy to recognize and identify, since they are the only neurons in the upper molecular layer. But the existence of the deep long axon stellate cells with their descending axon collaterals invites comparison and confusion with the basket cells, which they may seem to resemble in some degree. The forms of all of these cells should be compared in Figs. 158–160, 187, 188, and 191. Particular attention should be given to the dendritic and axonal arborizations. Although the dendrites of both cell types extend in the parasagittal plane, those of the basket cell spread upwards in a fan-shaped field, unlike the dendrites of stellate cells, which tend to radiate in all directions within the plane. Compared to the highly contorted stellate cell dendrites, basket cell dendrites are relatively straight, reaching to the pial surface, and are beset with many more spicules or other small appendages. These differences seem less imposing, however, when the axons are compared. In the basket cell the *defining* characteristic is, of course, the generation of basket collaterals and particularly the terminal formation of the pinceau. A neuron in the molecular layer whose axon does not contribute to the formation of the pericellular basket and the pinceau cannot, by definition, be a basket cell. Some ambiguity, however, arises when the descending collaterals of the long axon stellate cell are seen to enter the layer of Purkinje cell bodies and even, in some more complete impregnations, are seen to mingle with the pericellular plexus. Here the form of the axon and its collaterals should provide the clue to differentiating these two cells. Descending basket axon collaterals are thick, fleshy fibers; they continue beyond the base of the Purkinje cell body to terminate in a burst of short, finger-like branches. These fibers differ markedly from the tenuous, varicose threads of the deep stellate cell axon collaterals that contribute to the pericellular basket. Furthermore, it should be noted in the deep stellate cells of Figs. 188 and 191 that the long axon of the deep stellate cell gives off most of its branches close to its origin, leaving the remainder rather barren. The basket cell axon, in contrast, tends to give off its descending branches only at some distance from the cell body.

In a recent paper on the development of neurons in the molecular layer of the monkey's cerebellar cortex, RAKIC (1972) observed that the small neurons of this layer display different dendritic patterns according to their position in the grid of parallel fibers. The dendrites of deeper cells generally radiate upwards, those of cells at intermediate depths radiate in all directions, and those of more superficial cells tend to radiate downwards. The patterns were thought to reflect the possibilities for directional growth through the maturing lattice of the molecular layer at the time when the cells in question were laid down. Aside from being a somewhat rigid conception, this hypothesis does not consider the possibility of genetic differences between cells. As RAKIC did not take into account the axonal arborizations of these cells, he could not differentiate between stellate and basket cells. Our

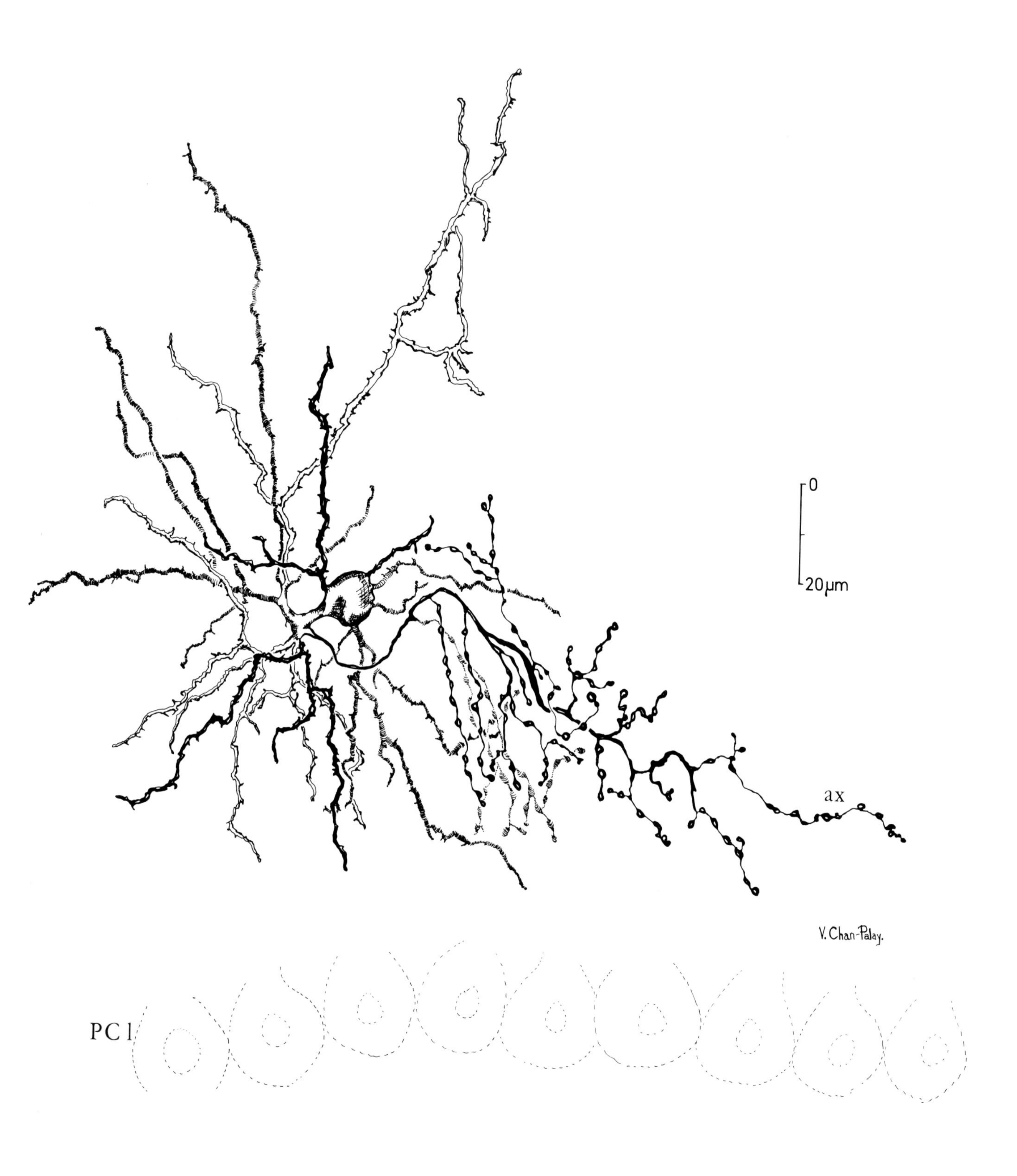

Fig. 190. The deep long axon stellate cell. This long axon stellate cell has a typical radiating spray of contorted, spiny dendrites that reach into the neighboring molecular layer. The axon of this cell (*ax*) arises from the main dendritic trunk. It courses along horizontally in the sagittal plane. Many varicose descending collaterals issue from the first third of its course. Thereafter both ascending and descending collaterals occur. The Purkinje cell layer (*PC l*) is indicated. Rapid Golgi, sagittal, 90 µm, adult rat. Camera lucida drawing. 100× oil immersion objective

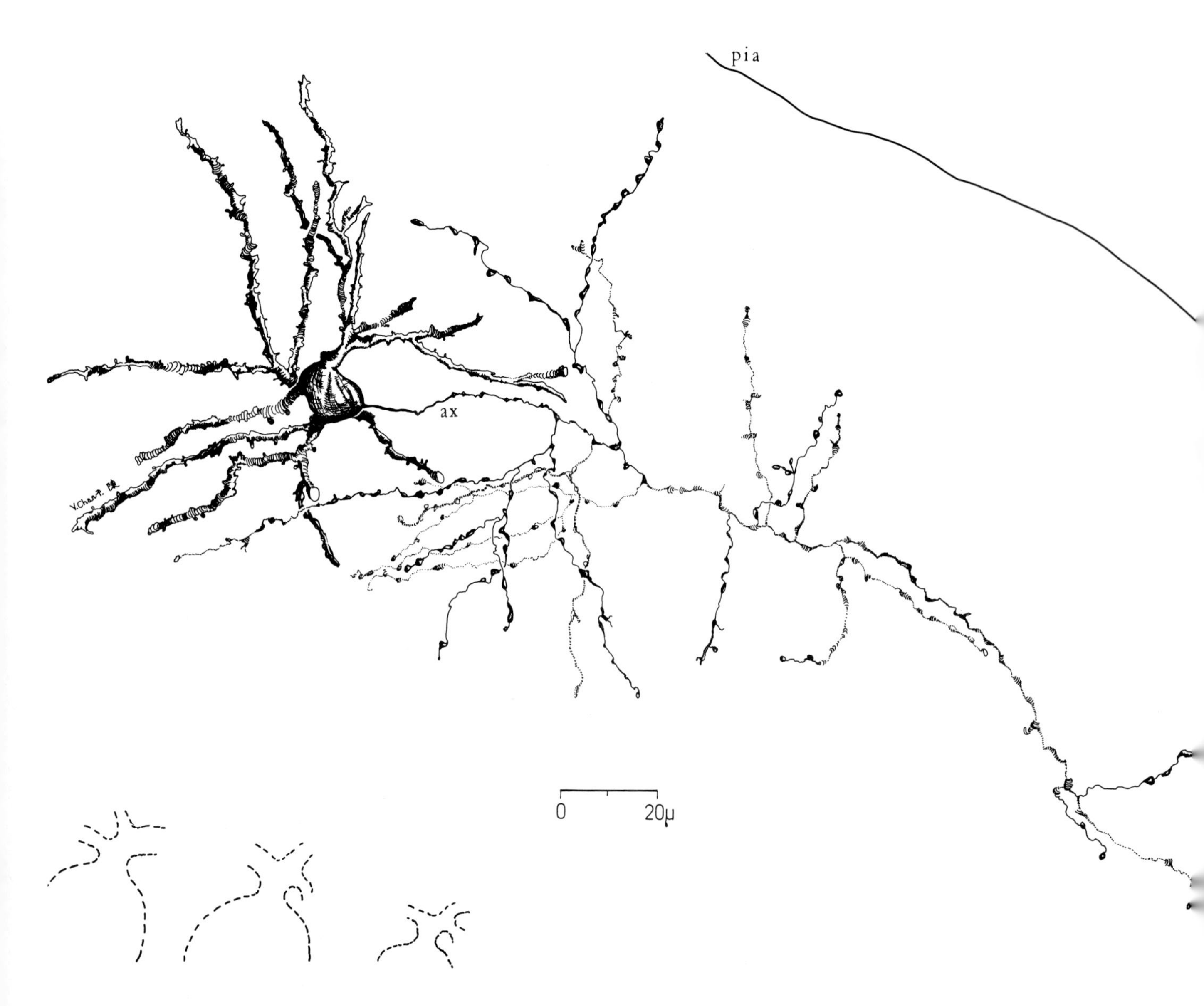

Fig. 191. Deep long axon stellate cell. This cell lies in the molecular layer midway between the pial surface and the Purkinje cell layer. It is a typical deep stellate cell, as its axon (*ax*), which courses about 320 μm across the folium, gives off numerous descending, and a few ascending, beaded collaterals in its initial portion. Farther out along the axon, the number of collaterals is small. Parasagittal section. 90 μm thick. Camera lucida drawing. 100 × oil immersion objective

own observations on the rat's cerebellar cortex generally agree with those reported by RAKIC so far as the dendritic patterns are concerned, but our studies indicate that the two cells cannot be assimilated into a single cell type, since their axonal trees are markedly different.

Additional evidence for the distinction between these two classes of cells is found in the quantitative analysis of the neurons in the molecular layer of cats by PALKOVITS *et al.* (1971 c). By comparing the ratio of the lengths and breadths of cell somata with their volume, these authors show that neurons of the molecular layer separate into two classes. The stellate cells are distinctly smaller and more ellipsoidal, while the basket cells tend to be larger and rounder. By these criteria, numerous stellate cells and few basket cells were recorded in the upper half of the molecular layer, whereas many more basket cells and fewer stellate cells were found in the lower half of the layer.

In summary, the stellate cells of the molecular layer can be classified according to their dendritic and axonal patterns as follows:

1. Superficial cells with short, circumscribed, contorted dendrites and limited axonal fields (upper third of the molecular layer);

2. Deep stellate cells with radiating, contorted dendrites and long axons with thin, varicose collaterals (middle third of the molecular layer);

3. Deep stellate cells with radiating, contorted dendrites, and long axons with varicose collaterals that contribute meagerly to the pericellular basket (middle third of the molecular layer).

3. The Fine Structure of the Stellate Cell

Because of the large variety of basket and stellate cells in the deeper half of the molecular layer, it is best to begin a description of the fine structure of the stellate cells with the more superficial group, which, as was pointed out above, can be securely identified simply by their location. The following account is adapted from our recent study (CHAN-PALAY and PALAY, 1972a). It is based on electron micrographs of the superficial stellate cells primarily and, where indicated, on micrographs of deeper stellate cells when they could be distinguished from basket cells.

a) The Cell Body

Superficial stellate cells generally have small elliptical perikaryal profiles, 5–9 μm in diameter, that are occupied almost entirely by their nuclei. Their nuclei are characteristically rounded, with one or several shallow indentations (Figs. 192 and 195). The deeper cell bodies can be larger, up to 12 μm in diameter, and they have a more voluminous cytoplasm. Their shape tends to be more ellipsoidal than that of the outer stellate cells, and can even be multipolar. The deeper cells are more likely to have deep and complicated nuclear indentations than the superficial cells (Fig. 193). In all, the chromatin is usually homogeneously dispersed throughout the karyoplasm except for a thin, irregular marginal condensation. In a few cells one or two large masses of heterochromatin occur in this peripheral zone or nearer the center. The nucleolus is also frequently marginal in position (Figs. 193 and 194).

The Cytoplasm. As in most small neurons, the cytoplasm of the stellate cell is rather scanty and simple (Figs. 192–195). The nuclear envelope often throws out irregular streamers that join with the granular endoplasmic reticulum (Figs. 192 and 194). This latter organelle is represented by a few ramifying tubules or cisternae sometimes very long and stringy, but usually randomly dispersed. Occasionally three or four cisternae are assembled into a small Nissl body (Fig. 192). Although its outer surface is studded with ribosomes arranged here and there in single and double rows, surprisingly long stretches of the cisternal membrane are free of any granules. In the cytoplasmic matrix, ribosomes are arranged in rosettes of four to six members, loosely scattered about, not only in the vicinity of the granular endoplasmic reticulum but also in the membrane-free zones and the nuclear indentations. It should be noted that ribosomes are frequently attached to the outer wall of the nuclear envelope. The agranular reticulum is confined to a few vagrant tubules and occasional subsurface cisternae.

The Golgi apparatus, consisting of stacked agranular, fenestrated cisternae and associated vesicles, presents a highly variable configuration from one cell to another. In some stellate cells it is fragmented into small aggregates of cisternae together with clustered vesicles (Figs. 193 and 194). These are dispersed over the perikaryon and appear in the roots of the major dendrites, apparently as disconnected bodies. In other stellate cells, the Golgi apparatus is organized into a coherent shell of overlapping cisternae entraining swarms of vesicles of different sizes. This shell is applied like a cap to the pole of the nucleus nearest the origin of one of the dendrites (Fig. 192), and it can extend out into the dendrite for five or ten microns. Sometimes, in fortunate sections, a single centriole or a pair is encountered in the vicinity of the Golgi apparatus. A few lysosomes and multivesicular bodies are strewn about the cytoplasm, but are generally inconspicuous.

In the cat a peculiar laminated inclusion body occurs in the cytoplasm of stellate cells (MORALES and DUNCAN, 1966; UCHIZONO, 1969; MUGNAINI, 1972). This structure consists of sheets of parallel, closely spaced tubules alternating with layers of fine parallel filaments, the whole formation arranged in a whorl. The tubules have been variously interpreted as microtubules (PETERS, PALAY, and WEBSTER, 1970) and as derivatives of the endoplasmic reticulum (MUGNAINI, 1972). The latter interpretation would seem to be correct, as MUGNAINI (1972) has demonstrated continuity between the tubules in the inclusion and elements of the endoplasmic reticulum at its periphery. Similar structures were originally discovered by MORALES, DUNCAN, and REHMET (1964) in the lateral geniculate body of the cat, and have since been seen there by many authors (see PETERS, PALAY, and WEBSTER, 1970) as well as in the striate cortex (KRUGER and MAXWELL, 1969). They have not yet appeared in the rat, either in the visual system or in the stellate cells of the cerebellar cortex.

In low magnification micrographs the most prominent organelles in the cytoplasm of stellate cells are the mitochondria. Although these structures are sparse, their size

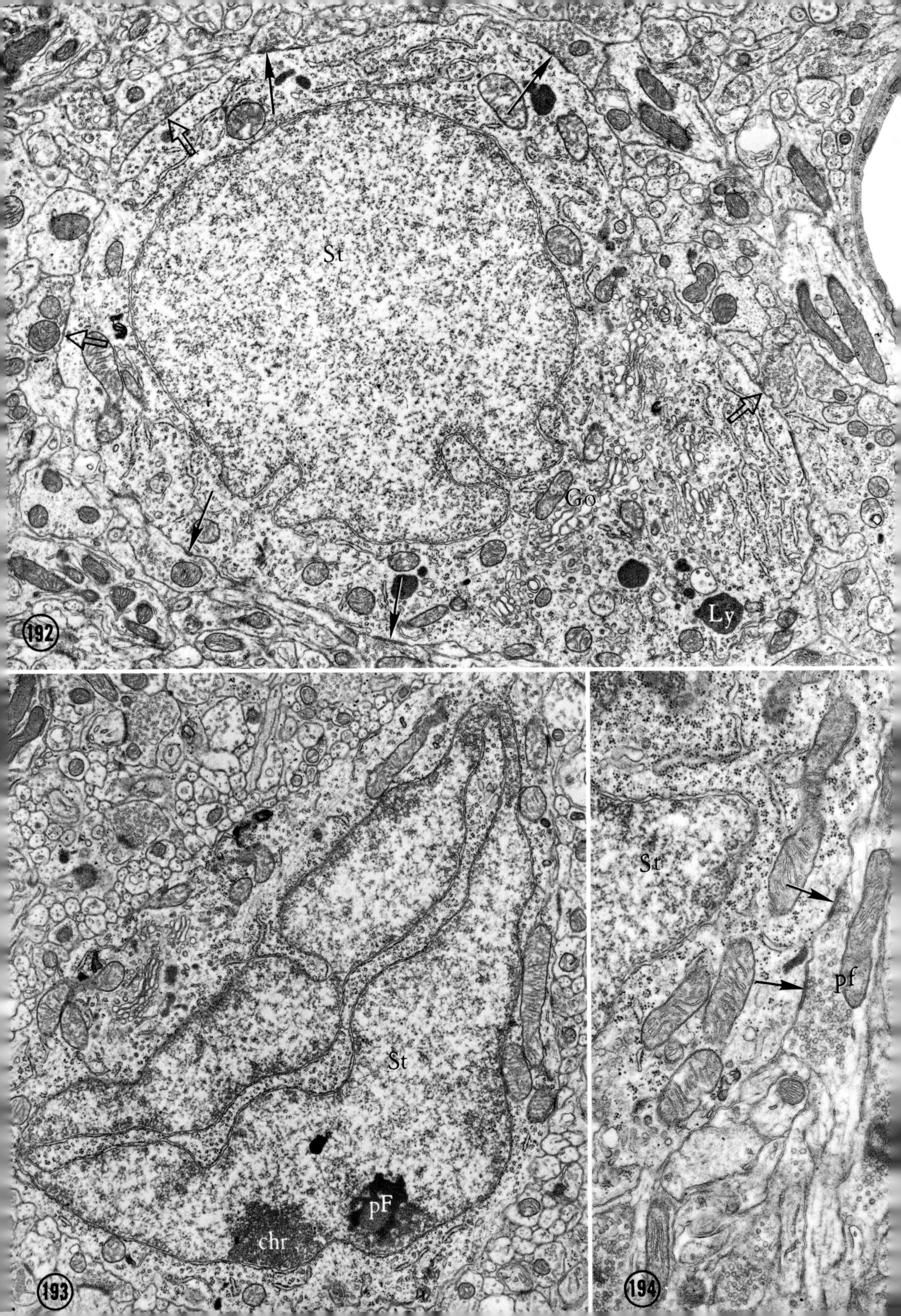
St
Go
Ly
192
St
pF
chr
193
St
pf
194

and construction makes them stand out in the meager perikaryon (Fig. 194). They tend to be small rods or globules, ranging from 0.2 to 0.4 μm in diameter and 0.8 to 1.6 μm in length. Occasionally they can be up to 9 μm long, forming a branching, curved rodlet, lying parallel to the nuclear envelope and looking as if a number of small mitochondria had coalesced. The mitochondrial matrix is dense and the cristae are highly pleomorphic, varying from the ordinary transverse plates in small mitochondria to undulating oblique, longitudinal, or even tubular cristae in the larger organelles.

Microtubules and neurofilaments are also scarce, often hidden from view in low power micrographs by the crowding of polysomes and other organelles into the thin rim of perinuclear cytoplasm. Microtubules appear singly, running around the nucleus and assembling into groups at the bases of the dendrites or the axon.

b) The Dendrites

The larger dendrites begin as conical extensions of the perikaryon with all of the cytoplasmic organelles funneling through the narrow orifice and aligning themselves parallel to the longitudinal axis of the dendrite. The Golgi apparatus is frequently a major component of the proximal segment of the dendrite, nearly filling it with a longitudinal stream of fenestrated cisternae, tubules, and vesicles. Mitochondria are aligned along the edges of this stream, and it is surrounded by a sleeve of matrix in which microtubules run longitudinally and tags of the rough endoplasmic reticulum or groups of polysomes appear. Farther out, the cytoplasm of the dendrite clears as the Golgi apparatus drops out. Now the most conspicuous components are the mitochondria, sinuous, long, and surrounded by a leash of parallel microtubules and occasional neurofilaments (Figs. 196 and 197). The agranular endoplasmic reticulum appears as sparse tubules in the thicker parts of the dendrite, and polysomes are infrequent. The cytoplasm of the smaller dendrites has this same composition from their origins onwards.

Because of the irregular caliber and crooked, ramifying course of the dendrites, they cannot be followed continuously for long distances in the thin sections used in electron microscopy. Stretches 6 to 10 μm long are frequently encountered, running athwart the grid formed by the parallel fibers (Figs. 196 and 197). These correspond to the profiles expected from the Golgi preparations. They have irregular contours, frequent bifurcations, closely spaced varicosities, and thin, tapering appendages filled with a fibrous matrix that match the spicules of the Golgi preparations. The stellate cell receives most of its synapses on these more peripheral dendritic branches.

As neither the cell body nor its processes are ensheathed by neuroglial cells, the dendrites traverse a sea of parallel fibers, which, lying in direct apposition to them, are their principal afferents. Many of these fibers simply pass by without developing any synaptic specializations (Fig. 197); others expand into synaptic varicosities that impinge upon the shaft or the appendages of the dendrite (Figs. 196 and 197). Synapses occur singly or in clusters, with lengthy bare intervals in between. The parallel fiber varicosity encloses a loose aggregate of round synaptic vesicles. Only one third to one half of the junctional interface is involved in the synaptic complex (Fig. 196). In this zone, the interspace or synaptic cleft is enlarged by a slight concavity of the presynaptic membrane, while the postsynaptic membrane remains flat. Dense, filamentous material crossing the synaptic cleft, together with asymmetrical pre- and postsynaptic cytoplasmic densities, makes the whole complex dark and conspicuous. The synapse resembles those made by the parallel fiber on the thorns of Purkinje cell dendrites, except that the postsynaptic dense plaque is usually not so thick in the stellate cell dendrite as in the Purkinje cell thorn. Nevertheless, the appearance of the synaptic junctions made by the parallel fiber on these two very different sites is very nearly the same.

Synapses of the parallel fibers on the somata of stellate cells, however, tend to be somewhat different from those on the dendrites (Figs. 194 and 195). There is a greater variation in the depth of the postsynaptic density, and the synaptic cleft is not usually widened. Such differences are not consistent from cell to cell in the same section, and are probably not significant. Nearly all of the parallel fiber endings are synapses *en passant*; the fibers synapsing

◄

Fig. 192. The stellate cell perikaryon. This stellate cell has a slightly lobulated nucleus (*st*) and the organelles usual for neurons. The Golgi apparatus (*Go*) is well developed, and lysosomes (*Ly*) are numerous. The somatic surface bears four synapses with parallel fibers (*black arrows*) and three synapses with other stellate axons (*open arrows*). Lobule VII, Crus I. × 14000

Fig. 193. The stellate cell nucleus. This very lobulated nucleus of a stellate cell (*St*) has the usual marginated chromatin as well as a clump (*chr*) situated next to the nucleolus. The pars fibrosa (*pF*) of the nucleolus is amalgamated with another clump of chromatin. Lobule III. × 14000

Fig. 194. Synapse on a stellate cell soma. The somatic surface of this stellate cell (*St*) bears two synaptic junctions (*arrows*) with a parallel fiber varicosity (*pf*). Lobule X. × 30500

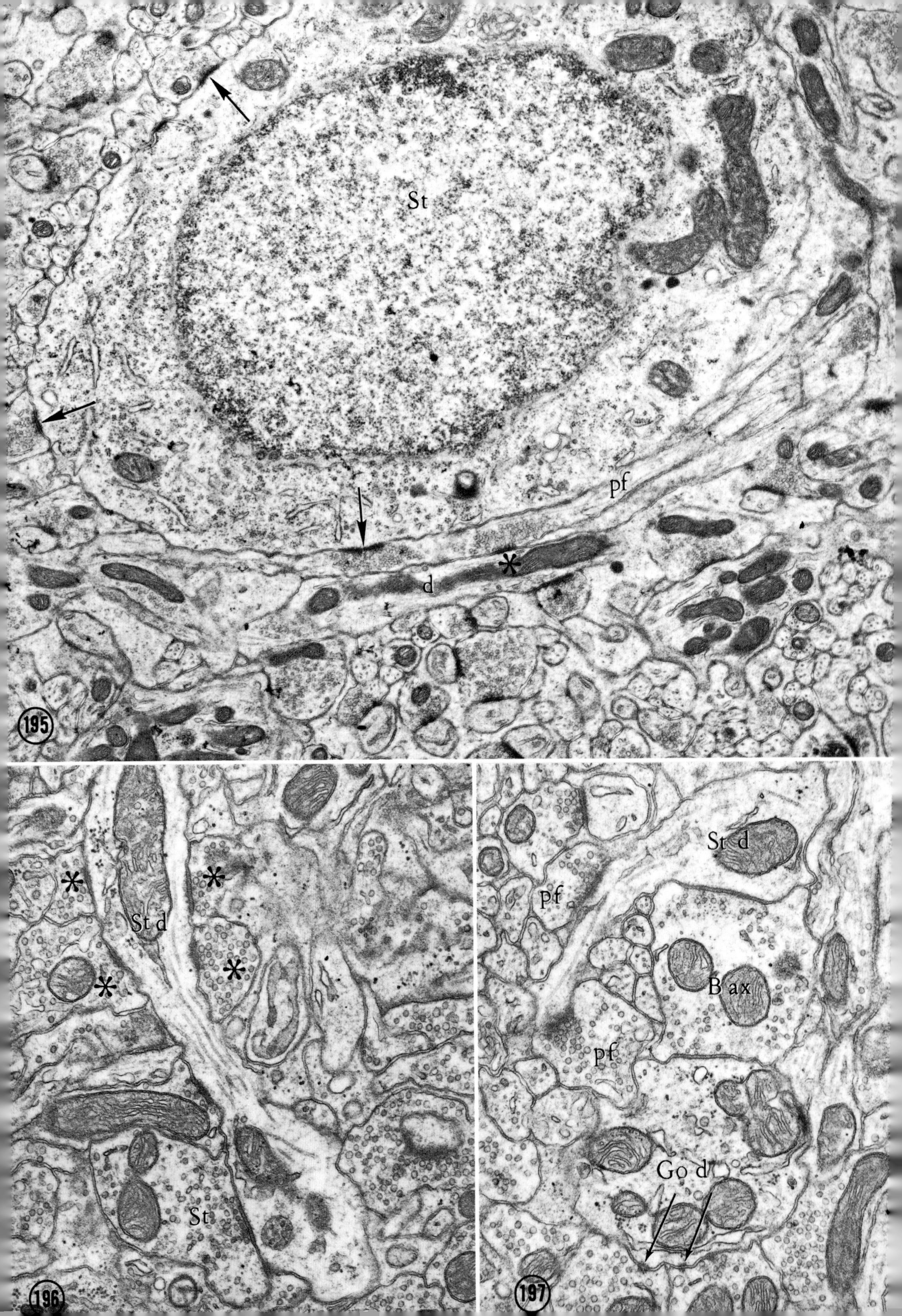

St
pf
*
d
195
*
*
St d
*
*
St
196
St d
pf
B ax
pf
Go d
197

on the cell body can contact neighboring dendrites in the same way (Fig. 195). Although the cell body and the proximal segments of the dendrites are surrounded by parallel fibers, very few of them synapse with these portions of the cell. Profiles of the entire cell body display only from one to five synaptic endings, although the whole of the cell outline can be in direct contact with parallel fibers (Figs. 192 and 195). These observations agree with the findings of LEMKEY-JOHNSTON and LARRAMENDI (1968b) in the mouse and those of MUGNAINI (1972) in the cat.

Besides the varicosities of parallel fibers, another type of axon terminal synapses with the somata and dendrites of stellate cells, but is encountered much less frequently. This type is represented by the endings of axons from other stellate cells and rarely the endings of ascending collaterals from basket cells. Such contacts were first brought to notice in a paper of the SCHEIBELS (1954) dealing with the climbing fiber as seen in Golgi preparations, and they were further described in a summary of their observations in JANSEN and BRODAL'S review of cerebellar structure (1958). Stellate cell terminals effect *bouton* or *en passant* synapses with the shafts of the dendrites (Fig. 195) and with the cell bodies (Fig. 196). In neither site are they numerous. The presynaptic ending contains elliptical and round synaptic vesicles, some of which are collected near the synaptic interface. The synaptic cleft is wider than the usual interstitial space and is traversed by fine filaments. On the cytoplasmic sides of the apposed surface membranes a thin, delicate fringe is attached more or less symmetrically. The junction is intermediate between GRAY'S two types. Basket cell endings are very similar except that the ending includes scattered elliptical vesicles and a complement of neurofilaments, which is the mark of basket axon collaterals (Fig. 197). The SCHEIBELS (1954) also described junctions between climbing fibers and stellate cells or stellate cell dendrites. We have seen these junctions only a few times, and usually on the cell bodies of deeper stellate cells and basket cells (Fig. 218). They will be described in the section on the climbing fiber.

Fig. 195. Parallel fiber synapses on the stellate cell soma. This stellate cell (*St*) bears three synapses (*arrows*) with parallel fibers on its surface. One of these axons (*pf*) also synapses (*asterisk*) with another dendrite (*d*). Lobule III. × 17000

Fig. 196. Synapses on a stellate cell dendrite. This dendrite of a stellate cell (*St d*) bears four synapses with parallel fibers on its surface (*asterisks*). Further along, an axon of a stellate cell (*St*) synapses upon it. Lobule X. × 30500

Fig. 197. Synapses on a stellate cell dendrite. This assemblage includes two parallel fibers (*pf*) synapsing on a stellate cell dendrite (*St d*), as well as a basket axon (*B ax*) synapsing upon a Golgi cell dendrite (*Go d*). The Golgi cell dendrite receives several synapses from other parallel fibers (*arrows*). Lobule X. × 31000

c) The Axon

The initial segment of the axon originates from a funnel-shaped protuberance on the cell body or one of the dendrites. A few terminals of other stellate cell axons are located upon this small axon hillock. The initial segment, from 0.45 to 0.6 μm across, contains the usual fasciculi of microtubules that have streamed into it from the cell body, as well as fragments of the rough endoplasmic reticulum, isolated ribosomes, and small lysosomes (Fig. 198). The characteristic undercoating of the plasmalemma begins beyond the point where the axon has assumed its definitive caliber. We have not had the good fortune to follow the initial segment of a stellate cell axon to its end. We have found four examples of initial segments that are continuous with the stellate cell body or one of its dendrites. All four are enclosed in thin neuroglial sleeves.[18] It is not clear whether this sheath is characteristic of all stellate cell initial segments or is merely a fortuitous finding, for in each case a spiny branchlet of the Purkinje cell crosses the trajectory of the initial segment. The sheath about the axon may be only an offset from that around the Purkinje cell dendrite. This observation is mentioned for two reasons. First, the remainder of the axonal arborization is not ensheathed in neuroglia and second, initial segments of basket and Purkinje cell axons are similarly ensheathed in the neuroglia.

The collaterals of the stellate cell axon can be identified in electron micrographs on the basis of their similarity to the axons seen in the Golgi preparations. It should be remembered that in the upper molecular layer parallel fibers and stellate cell axons are the only axonal units present. The impregnated axon is a thin, beaded thread with relatively few branches but pursuing a tortuous, highly irregular course across the path of the parallel

18 MUGNAINI (1972) published two electron micrographs (Figs. 35 and 36) of an axon emerging from a dendrite, and another (Fig. 34) of an axon emerging from the cell body of stellate cells. Both examples are comparable to ours, and both are accompanied by neuroglial processes, which begin at the level of onset of the initial segment undercoat. UCHIZONO (1969) published a micrograph (Fig. 19) purported to be an axon emerging from a dendrite of a stellate cell, but this profile is clearly that of a Purkinje cell dendrite, and the so-called axon is only a branch that passes obliquely out of the section plane.

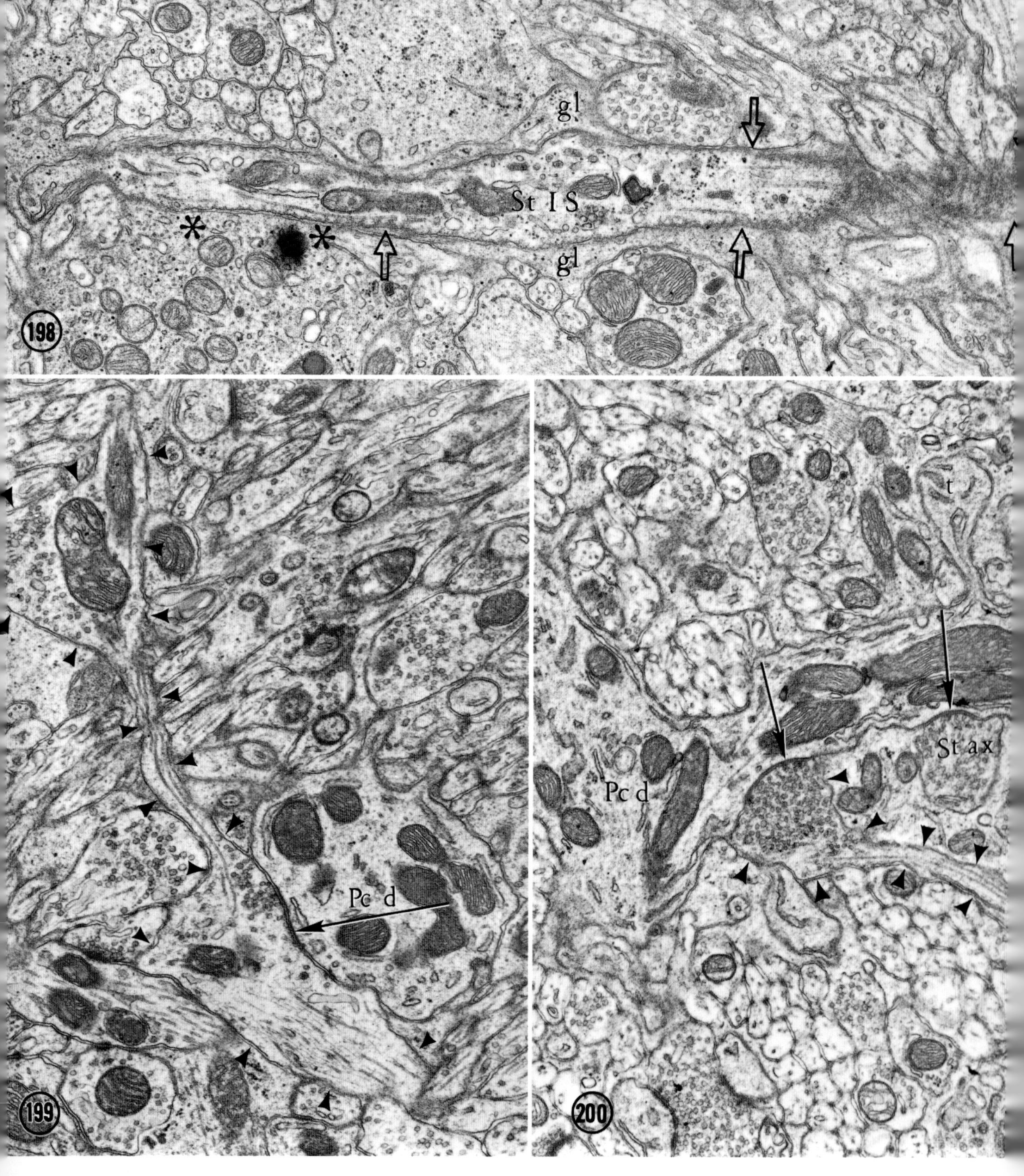

Fig. 198. Initial segment of stellate cell axon. The initial segment of this axon (*St I S*) is partially covered with neuroglial cytoplasm (*gl*). The undercoat of this axon shows a tripartite structure in longitudinal section (*between asterisks*). At and between sets of arrows, where the section grazes the surface of the axon, the structure of the spiral lamina, or the innermost layer of the undercoat can be discerned. (Compare this with Figs. 35–40.) Lobule X. ×25000

Fig. 199. The stellate cell axon. The varicose nature of the stellate axon is evident in this illustration (▲▲). The axon synapses (*large arrow*) upon the shaft of a Purkinje cell dendrite (*Pcd*). Lobule X. ×25000

Fig. 200. The stellate cell axon. The axon of the stellate cell (*St ax*) impinges upon and synapses (*arrows*) with the shaft of a Purkinje cell spiny branchlet (*Pc d*). A thorn (*t*) emerges from the upper aspect of this dendrite. Lobule V. ×21000

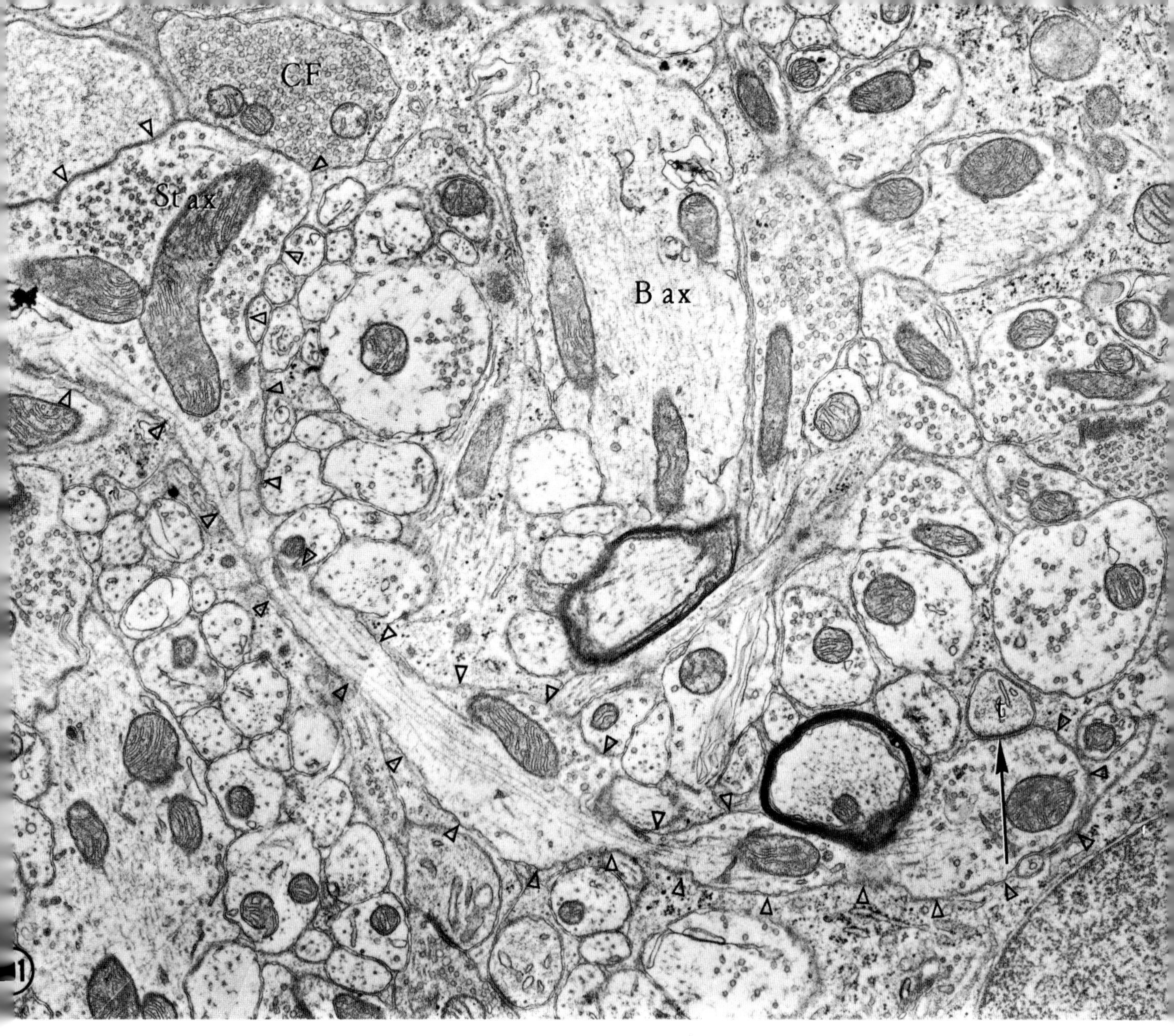

Fig. 201. The stellate cell axon. The stellate axon (*St ax*, ΔΔ) is a beaded fiber that can be readily distinguished from a neighboring climbing fiber terminal (*CF*) and a basket axon terminal (*B ax*). The stellate axon rarely may synapse upon a Purkinje dendritic thorn (*t*) by means of a Gray's type 2 synaptic junction (*arrow*). Lobule II. × 14500

fibers. In electron micrographs of the outer third of the molecular layer numerous axonal profiles are encountered that answer to this description (Figs. 199–201). These profiles are of narrow fibers, about 0.1 μm in diameter, that distend at closely spaced intervals into wide bays, 1.5 to 2.0 μm across. The fibers twist and turn, bending into a different direction with each expansion. Consequently, they cannot be followed in thin sections for more than a few microns; however, as they twist in and out of the plane of the section their trail can often be picked out in a micrograph as a disconnected chain of tailed circular or elliptical profiles. The fine, thread-like connecting links are, of course, often missed by the thin sections. They contain one or two microtubules, and occasionally a long mitochondrion, which run from one varicosity to the next. Small, slightly flattened or ovoid synaptic vesicles occupy these varicosities along with one or two elongated or branched mitochondria.

The varicosities make synaptic contact with the shafts of dendrites belonging to Purkinje cells (Fig. 200), Golgi cells, basket cells, and stellate cells (Fig. 196). They also contact the somata of stellate cells (Fig. 192). In all cases the structure of the synaptic junction is the same. The synaptic complex occupies only about a third of the junctional interface, and thin plaques of filamentous densities are nearly symmetrically arranged on the cytoplasmic sides of the apposed surface membranes. The synaptic cleft is slightly larger than the surrounding interstitial spaces and is bridged by crossing filaments. Rarely, one of these varicosities is encountered synapsing upon a thorn of a Purkinje cell dendrite (Fig. 201). These junctions are not different from those of parallel fibers on thorns, except for the flattened vesicles in the axon.

We have collected only two instances of axonal profiles similar to those of the stellate cell and located in the region of the basket surrounding the Purkinje cell. In one the profile comprises two elongated varicosities connected by a thin thread containing one or two microtubules, a couple of neurofilaments, and a large mitochondrion that extends through both varicosities. There are flattened synaptic vesicles in these enlargements, and the junctional interface of one varicosity has three short synaptic junctions with the soma of the Purkinje cell against which it abuts. The synaptic junctions are similar to those described above. These axons may represent contributions from deep long axon stellate cells to the pericellular basket similar to that illustrated in Fig. 188D, E. We have not yet discovered any axonal profile that fits the description of stellate axons in the pinceau or in the surrounding infraganglionic plexus.

4. Some Physiological Considerations

Like the Golgi and basket cells, the stellate cells have been implicated in the inhibitory control of the Purkinje cell. Inhibition by way of stellate cells has been documented in mammals (Anderson *et al.*, 1964; Eccles, Llinás, and Sasaki, 1966a, b, c, e; Eccles, Ito, and Szentágothai, 1967), reptiles (Llinás and Nicholson, 1969; Kitai, Shimono, and Kennedy, 1969), amphibia (Rushmer and Woodward, 1971), and cartilaginous fishes (Nicholson, Llinás, and Precht, 1969; Eccles, Táboriková, and Tsukahara, 1970). At first sight basket cells and deep stellate cells, with their long axons running in the parasagittal plane, appear to be quite similar and might be considered as having more or less the same function. More careful study of their form, however, indicates that these two cells incorporate different principles and that their roles in the circuitry and physiology of the cerebellar cortex cannot be the same. Both cells have dendritic trees spread out in the parasagittal plane, and, like the Purkinje cell, they present an extremely thin profile broadside to the stream of parallel fibers impinging on them in the longitudinal axis of the folium. Both cells have axons that, coursing in the parasagittal plane, synapse on a succession of Purkinje cells. But these similarities should not obscure the great differences between them. The basket cells have widespread dendrites that receive inputs from all levels of the molecular layer over a fan-like expanse. The stellate cell dendrites are much more circumscribed, and they receive inputs from a comparatively restricted stream of parallel fibers. The basket axon projects over a long distance, and, though it may supply basket collaterals to only a few Purkinje cells, it usually skips the one or two Purkinje cells in the vicinity of its perikaryon. In contrast, the axon of the stellate cell ramifies profusely close to the cell body, synapsing mainly with the Purkinje cells in the immediate vicinity, and then continues onward, giving off only a few additional twigs. This mode of arborization is epitomized still further by the axons of the more superficial stellate cells, which ramify entirely in the vicinity of the cell body. Furthermore, in addition to making scattered synaptic contacts with the shafts of the Purkinje cell dendrites, the elaborate axonal plexus of the basket cell impinges on the Purkinje cell in three critical locations: the branch points of the major dendrites, the perikaryon, and the initial segment of the emerging axon. The stellate cell is much more modest; it deploys its axonal terminals over a limited arena in the distal branches and spiny branchlets of the Purkinje cell dendritic tree. Finally, even though some collaterals of the deeper stellate cell axons contribute to the pericellular basket, they have a form completely different from that of the basket axon collaterals. Their sparseness, their simple varicose form, and their failure to join in the pinceau all indicate that they cannot have the efficacy of the basket cell fibers.

The fine anatomy, therefore, suggests that the inhibition resulting from the activation of a stellate cell is confined to the Purkinje cells and dendrites in the region immediately around the cell body of the stellate cell. Thus a parallel fiber volley would elicit principally local inhibitory effects from both superficial and deep stellate cells and, through them, a progressively weaker inhibitory effect on successive Purkinje cells lying "off beam" in the parasagittal plane. In terms of the analysis proposed by Eccles, Ito, and Szentágothai (1967, p. 208, Fig. 114), this would mean that the inhibition evoked by stellate cell activity would be largely "on beam" rather than "off beam." Because of its limited axonal arborization, the stellate cell, unlike the basket cell, could not be expected to have strong inhibiting effects on the discharges of the Purkinje cell, unless a large number of stellate cells were activated in concert. Stellate cells might, however, be important in resetting the membrane potential of the Purkinje cell dendrites after the passage of parallel fiber excitation. On the basis of these local effects, we have suggested (Chan-Palay and Palay, 1972a) that in physiological studies, the stellate cells should be considered as distinct from basket cells. The fact that stellate cells are phylogenetically older supports this view.

5. Summary of Synaptic Connections of Stellate Cells

Afferents	granule cells	———	stellate cell
	climbing fiber	———	stellate cell
	basket cell	———	stellate cell
	stellate cell	———	stellate cell

Efferents	stellate cell	———	Purkinje cell
	stellate cell	———	basket cell
	stellate cell	———	stellate cell

Chapter IX

Functional Architectonics without Numbers

In previous chapters we have described all of the neuronal types in the cerebellar cortex and one of the principal fiber pathways coming into it, the mossy fibers. We therefore have in hand the components of several important circuits through which the mossy fibers can exert an influence on the Purkinje cells, which provide the only outflow from this cortex. We shall not attempt to review the physiological evidence that has been adduced concerning the activity of these circuits. Neither shall we attempt to construct any comprehensive theories of how the cerebellar cortex operates. Our purpose is much more modest. We mean only to discuss some of the functional implications of the morphology. Fig. 202 presents a diagram of the interneuronal connections that have been confirmed by electron microscopy of the mammalian cerebellar cortex. As the climbing fiber will be deferred for a later chapter (Fig. 230), only the mossy fiber input is included in this epitome.

It is evident from the morphology summarized in Fig. 202 that the great incoming system of mossy fibers cannot exert any direct effect on the Purkinje cells, as they have no connections with them. They can exert an influence only through the granule cells and the Golgi cells with which they synapse. The basic plan of the mossy fiber system is to bring impulses into a folium along a highly divergent line, which disperses the input to a very large number of interneurons (granule cells). These in turn transfer the input to a large number of Purkinje cells and a smaller number of other interneurons (basket, stellate, Lugaro, and Golgi cells). Some idea of this dispersion is given in a recent study by PALKOVITS *et al.* (1972) in the cat. These authors calculated that within any particular folium an average mossy fiber generates 16 rosettes, each of which receives a dendrite from 112 granule cells, and each of these granule cells, receiving excitation from four different mossy fibers, conveys its impulses to 45 out of the 225 Purkinje cell dendritic trees through which it passes. Since all granule cells associated with a particular rosette are located beneath the territory of a single Purkinje cell, their parallel fibers constitute a family that establishes synapses with overlapping combinations of 45 Purkinje cells in the same row of 225 dendritic trees through which they pass. In other words, the average mossy fiber input sustains a divergence to 1 792 granule cells (16 rosettes × 112 granule cells), which then contact 45 Purkinje cells each, or 225 Purkinje cells all together. These figures do not include the further divergence produced by the ramification of mossy fibers in the white matter before they enter the cortex.

The excitatory effect of mossy fibers on granule cells and through them on Purkinje cells is well documented by much physiological evidence (ECCLES, LLINÁS, and SASAKI, 1966b; SASAKI and STRATA, 1967; ECCLES, ITO, and SZENTÁGOTHAI, 1967). Although a particular mossy fiber articulates with a large number of granule cells, its discharges are not sufficient to activate any one of them, unless these discharges are coordinated with those of other mossy fibers ending on the other dendrites of the same cell. Consequently, the discharges of any one mossy fiber result in the activation of only a few of the hundreds of granule cells with which it establishes contact. It is unlikely that a granule cell would articulate more than once with any particular mossy fiber, for (1) granule cells usually do not send more than one dendrite into a glomerulus, and (2) the distance between rosettes on the average mossy fiber is more than twice the average distance between glomeruli or the average length of granule cell dendrites (PALKOVITS *et al.*, 1972). The result of these conditions is that the granule cells activated by any individual mossy fiber are likely to be widely dispersed in the granular layer within the distribution territory of that mossy fiber. By the same token, the parallel fibers of these granule cells will course at different depths in the molecular layer, impinging upon different branches of the dendritic tree and possibly on the trees of completely different Purkinje cells. In other words, the Purkinje cell is presented with samples of the same mossy fiber discharge in several different parts of its dendritic tree at once.

MARR (1969) has pointed out that the number of separate events that a Purkinje cell can process is vastly

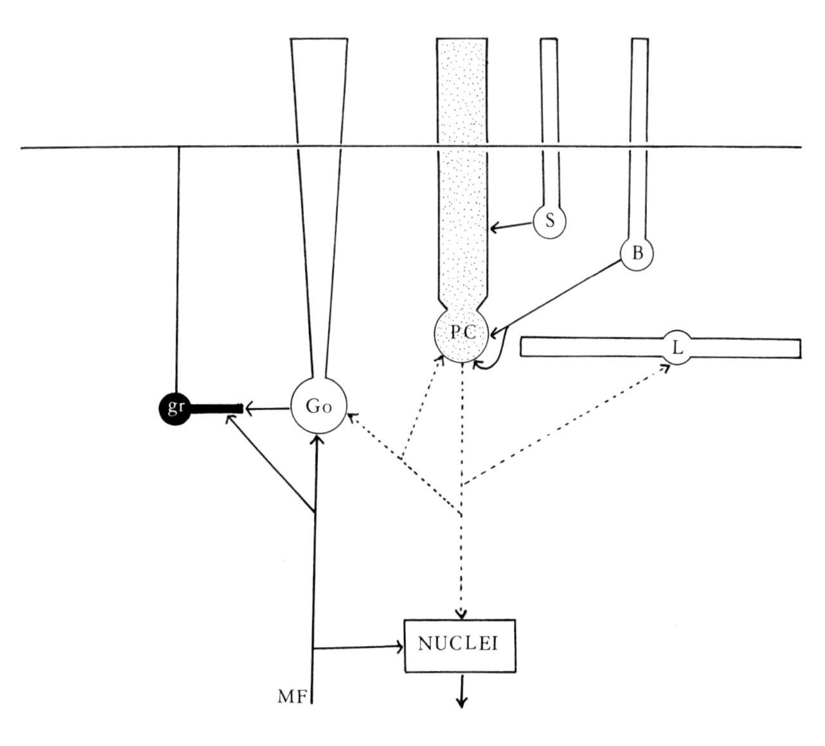

Fig. 202. Diagram of the mossy fiber-Purkinje cell circuit and interneuronal relays. The connections shown in this diagram are drawn from synaptic contacts that have been confirmed by electron microscopy. The mossy fiber (*MF*) contacts granule cells (*gr*), Golgi cells (*Go*), and the cerebellar nuclei. Granule cell axons synapse with Golgi cells, Purkinje cells (*PC*), basket cells (*B*), stellate cells (*S*), and Lugaro cells (*L*) (connection not shown). Stellate cell and basket cell axons contact the Purkinje cell; Golgi cells synapse with granule cells, and the Lugaro cell output is yet unknown. The only outflow of the cortex is through the Purkinje axon (*dotted lines*), which goes to the cerebellar nuclei. Recurrent collaterals in the cortex contact Golgi cells, Lugaro cells, and other Purkinje cells (not necessarily itself as shown in this diagram)

increased as the number of parallel fibers (or granule cells) occupied by the average event decreases. Of the thousands of parallel fibers traversing the dendritic tree of a Purkinje cell, only a very small proportion need be conceived of as discharging simultaneously. Probably most of them are silent at any particular instant. Since the active ones are distributed throughout the dendritic tree, ending on different branches, their activating effects will be subject to further filtration by the inhibitory action of other interneurons, stellate and basket cells, which impinge on the shafts of the dendrites, especially on the branching points of the major dendrites (CHAN-PALAY and PALAY, 1970). Therefore, in order for a small number of parallel fibers to be effective in activating the Purkinje cell, either they would have to operate against a background of high excitability, perhaps maintained by a large number of parallel fibers firing asynchronously all over the dendritic tree, or they would have to be concentrated on a particular subdivision of the dendritic tree so that their summated action could overcome the inhibitory effects of the interneurons. For the latter condition to exist, it is required that the topographical representation of the mossy fibers should be conserved in the distribution of the parallel fibers over the Purkinje cell dendritic tree. It would be surprising if this condition were not met, although its existence would be extremely difficult to prove by experiment.

1. The Uses of Inhibition

The abundance and elaborateness of the interneurons in the cerebellar cortex baffled RAMÓN Y CAJAL. He could see how the granule cells dispersed the incoming nervous currents like telegraph lines to numerous substations, but he could only imagine that the circuits in which the Golgi, stellate, and basket cells participated somehow reinforced the basic mossy fiber-granule cell-Purkinje cell system (RAMÓN Y CAJAL, 1911). His notion was similar to what modern engineers call positive feedback. Of course, RAMÓN Y CAJAL was not aware of inhibition, which was not discovered until after his death.

a) Basket Cells

As reference to Fig. 202 will show, the stellate and basket cells also receive inputs from the mossy fiber system via the granule cells and their parallel fibers. Since their dendrites extend in the same plane as the Purkinje cell dendritic tree and their axons terminate on the Purkinje cells, they are connected in parallel with the primary mossy fiber-granule cell-Purkinje cell circuit. The same considerations regarding the distribution of the parallel fibers in the dendritic tree of the Purkinje cell that were discussed in the previous section could be applied to the dendrites of these cells. It is important to note, however, that because of its sparse, diverging dendritic arborescence, the basket cell is much less likely to be activated by any particular parallel fiber passing through its territory than is the Purkinje cell. Most of the parallel fibers, by far, pass through without making contact on a basket cell. As mentioned earlier, the basket cell is constructed to sample activity in a wide expanse of the molecular layer, representing the afferent excitation in diverse parts of the Purkinje cell dendritic field. It receives relatively few inhibitory inputs from other basket cells (LEMKEY-JOHNSTON and LARRAMENDI, 1968b). The basket cell would therefore seem to be less discriminating than the stellate cell regarding the informational content of the fibers to which it responds.

The action of the basket axon on the Purkinje cell is clearly inhibitory (ANDERSON, ECCLES, and VOORHOEVE, 1964; ECCLES, ITO, and SZENTÁGOTHAI, 1967), like that of the other interneurons, but it must be noted specifically that its effects are exerted upon the Purkinje cells lateral (in the parasagittal plane) to the territory of its dendritic tree. Thus the inhibition thrust upon these Purkinje cells is unrelated to the parallel fiber activity within their dendritic trees. The basket cell axons seem to provide a mechanism for sharpening the boundary between activated and inactive Purkinje cells. Pathways bringing excitation into the dendritic trees not only confront this "overload" of inhibition, but also contribute to the depression of the Purkinje cells on either side of their trajectory. But this lateral inhibitory effect can be overemphasized. In view of the considerations discussed above, the behavior of the parallel fiber system is likely to be quite different under natural conditions than under the massive "on beam" or "LOC" stimulation of physiological experiments. Since, as mentioned earlier, the basket cell dendritic trees are widespread and sparse, most parallel fibers are likely to pass through their territories without synapsing, whereas they will establish synapses with the Purkinje cell. Even with the ratio of six basket cells for every Purkinje cell (PALKOVITS *et al.*, 1971c), most parallel fibers can be expected to activate Purkinje cells without meeting a basket cell dendrite. It may therefore be necessary to entertain the idea that some parallel fibers and their parent granule cells are specifically designed to convey to the basket cells a sample of the activity going on in the corresponding parts of the granular layer.

b) Stellate Cells

In contrast, the stellate cells, while contributing to some extent to lateral inhibition, are principally concerned with the inhibition of the particular dendritic tree that they inhabit. Although there is some evidence that they receive synapses from parallel fibers that also articulate with Purkinje cell thorns, it seems reasonable to expect that they contact at least neighbors among the parallel fibers and that these represent neighboring granule cells. But that condition provides no assurance that these granule cells are associated with the same constellations of mossy fibers or that they are even activated coherently. Therefore, the stellate cells, like the basket cells, may be responding to parallel fibers that are unrelated to the Purkinje cell dendritic tree through which they pass.

We have seen that the axonal plexus of stellate cells has mainly a local distribution on the Purkinje cell dendritic tree. Its inhibitory influence (ECCLES, ITO, and SZENTÁGOTHAI, 1967; RUSHMER and WOODWARD, 1971) is therefore restricted to the more distal branches and spiny branchlets of the Purkinje cell dendrites. The utility of this local and peripheral inhibition is not immediately obvious, especially in view of the strategically located pericellular and periaxonal plexus of the more forceful basket cell. But the basket cell axon, it will be remembered, also gives rise to similar ascending collaterals, which clamber over dendrites and coil about branch points. The role of these collaterals is as obscure as the strong inhibitory power of the pinceau is obvious. Considering the similarities between the accessory collaterals of the basket cell axons and the arborization of the stellate cell axons, it may be suggested that whatever the role of one, the other has the same function. We have recently proposed (CHAN-PALAY and PALAY, 1970) that the ascending basket cell collaterals form part of a control mechanism that can switch in or out whole sectors of the Purkinje cell dendritic tree and can thus determine which part of its afferent field will be allowed to figure in the integrations performed by the Purkinje cell. This hypothesis will be discussed further after the climbing fiber has been considered.

MARR (1969) has propounded a related hypothesis to explain the intermingling of excitatory and inhibitory inputs on the peripheral dendrites. MARR considered that the stellate cell axons would parcel the dendritic tree into small units of excitation and inhibition. Each unit, behaving like a small computer, would summate the postsynaptic effects of opposite sign and would pass the effect along to the next larger dendrites, where further summation would occur, and so on. This tessellated design would permit a large number of events to be processed without occlusion in the major dendrites. Its operation would be even more effective if the branch points are capable of generating dendritic spike potentials

(FUJITA, 1968). This scheme has the attraction of providing a mechanism for fine discriminations at the most distal parts of the dendritic tree. The structural design of the stellate cell would readily fit it for such a role.

c) Golgi Cells

The Golgi cell is unusual among the interneurons in that it receives direct input from the mossy fibers as well as synapses from the parallel fibers. Since its axons articulate with the granule cell dendrites, it is connected in parallel with the main mossy fiber-granule cell-Purkinje cell circuit but bypasses the Purkinje cell (see Fig. 202). It is not the only interneuron to receive primary afferents, as the basket cells also receive climbing fiber synapses, but at least some Golgi cells have a massive and important axo-somatic synapse that must subject it to powerful excitation by either mossy or climbing fibers, as the case may be. They also receive inputs from a widespread dendritic tree in the molecular and granular layers, which, like the arborization of the basket cell, samples activity in parallel fibers throughout the molecular layer. Like the dendrites of the basket cell, this dendritic tree cannot be a device for fine discriminations, and since it spans the territories of numerous Purkinje cells in both the longitudinal and transverse axes of the folium, it cannot be very selective in regard to topography either. The indiscriminate dendritic arborization seems to be matched by an equally global axonal plexus in the granular layer, which seems to be designed to touch every dendrite of the millions of granule cells within its territory. The Golgi cell would therefore be expected to operate in a totalitarian way, suppressing all granule cells in its territory that are weakly excited by mossy or climbing fiber activity (ECCLES, SASAKI, and STRATA, 1967a and b). Such a global inhibition suggests that the Golgi cell axonal plexus provides a device for clearing the lines and resetting the circuits for fresh volleys of incoming impulses.

But, as has been indicated above, the Golgi cell has access both to the impulses impinging upon the granule cells (via mossy fibers) and to activity evoked in the granule cells (via parallel fiber synapses on its dendrites). And it should be added, as reference to Fig. 202 will confirm, that it also has access to the final resultant evoked in the Purkinje cells (via recurrent collaterals). What can be the value of these afferents, bearing information from the main circuit at different points in its processing? Referring only to the mossy fiber and parallel fiber innervation, MARR (1969) has suggested that the apparent duplication is a device for monitoring the effects of Golgi cell inhibition on the granule cell dendrites in the glomerulus. In effect, the Golgi cell would be adjusting the threshold of the granule cells in response to the difference between the mossy fiber input and parallel fiber activity. Of course, the influence of the Purkinje cell recurrent collaterals would allow even finer tuning of this device than Marr suspected. Since the Golgi cell and granule cells would be activated at about the same time by a barrage of mossy fiber discharges, the Golgi cell would begin to exert its inhibitory effect after the granule cells have sent off their first impulses. These, passing along the parallel fibers, would excite the Golgi cell to continue or even increase its inhibitory discharges, thus depressing or extinguishing the granule cell, upon which the parallel fiber impulses would subside and the Golgi cell would no longer be facilitated. If the parallel fibers were successful in exciting the Purkinje cell, the impulses returning to the granular layer would inhibit the Golgi cell, thus releasing granule cells for the next mossy fiber volley. At each synaptic area there is opportunity for interaction of all the other effects depending upon timing, frequency, after-hyperpolarization, and other factors. The dendritic arborizations, according to this concept, would act as sensing devices for the efficacy of inhibition produced by the Golgi cell. The role of the Golgi cell would be to reset the threshold of the granule cells according to the conditions of input.

In order for this scheme to operate, it would be necessary that the parallel fibers impinging on the Golgi cell dendrites in the molecular layer should originate from the same granule cells (or a suitably representative sample of them) as are under the influence of its axons in the granular layer. It must be pointed out that, in any particular event, a relatively small number of parallel fibers responding to impulses from several mossy fibers could activate a Golgi cell through its dendrites in the molecular layer and evoke global suppression of granule cells throughout the territory of the Golgi cell, not just in the particular set activated by the mossy fiber in contact with that Golgi cell. Presumably some pattern of synchrony in the mossy fiber input representing certain regions or certain movements is necessary in order for granule cells to escape from the pervading and probably continuous inhibitory atmosphere of the Golgi cells in the granular layer. We have already suggested that the glomerulus, to which both mossy fibers and Golgi axons contribute, would act as a noise filter in the transmission of signals from the mossy fiber to the Purkinje cell (CHAN-PALAY and PALAY, 1971b).

d) Purkinje Cells

The Purkinje cell itself has a pathway for inhibiting its neighbors and certain interneurons. We have already seen that by way of its recurrent collaterals ending on Golgi cells it can disinhibit the granule cells, thus releasing them to respond to fresh stimuli. Each discharge from the Purkinje cell would wash the slate clean so that it is ready for a new message. This provision should be quite important, as the granule cells are the gateway for a major source of impulses to the Purkinje cell. Recurrent collaterals also pass to neighboring Purkinje cells, producing a weak inhibitory effect that should reinforce the lateral inhibitory effects of the basket cells, perhaps developing at the same time or a little later.

A puzzling feature of the Purkinje cell recurrent collaterals is their rich articulation with the Lugaro cell. These cells seem designed to intercept the inhibitory output of dozens of Purkinje cells, but their role in the circuitry of the cerebellar cortex is still a mystery. If LUGARO (1894) and FOX (1959) are correct in their description of the axons as ramifying in the longitudinal direction, these fibers could provide an inhibitory output to basket cells, which otherwise lack the inhibitory feedback that all other cells in the cerebellar cortex receive. In this case, inhibition of the Lugaro cells by the Purkinje cell collaterals would result in disinhibition of the basket cells. Alternatively, they could have excitatory effects, and then inhibition by Purkinje cell recurrents would result in suppression of basket cells by the process of disfacilitation. In either case they would be interesting cells to study. It is unfortunate that the physiologists have not given them special attention.

2. The Shapes of Synaptic Vesicles

It must be evident from the foregoing chapters that one of the prime criteria for the identification of a nerve ending is the shape of its synaptic vesicles. Under the conditions of fixation generally used, including those followed in the present account, both spherical and ellipsoidal, or flattened, vesicles occur to some extent in all fibers, but VALDIVIA (1971) has documented the already well-accepted observation (UCHIZONO, 1965) that each fiber type in the cerebellar cortex has a characteristic ratio of flattened and spherical vesicles. For example, mossy, climbing, and parallel fibers contain almost exclusively spherical synaptic vesicles, each fiber deviating from exclusiveness by a different percentage. Axons of basket, stellate, Golgi, and Purkinje cells contain pleomorphic vesicles, both spherical and more or less flattened, in different proportions. Valdivia showed that these proportions can be altered by changing the osmotic pressure of the buffer in the fixative or in the rinsing solutions before dehydration and embedding. Because of this variability according to the method of fixation, the shape of the vesicles alone is not a sufficient criterion for identification; it must be considered in the context of other features, such as form and location of the ending, concentration or dispersal of the vesicles, presence of neurofilaments and microtubules, specializations of the contact zone, and other details relevant to the region.

UCHIZONO (1965, 1967a and b, 1969) has carried this correlation between vesicle shape and identity of the nerve ending one step further. He has proposed that terminals containing flattened vesicles are inhibitory, whereas those containing spherical vesicles are excitatory in nature. This correlation is based on observations in the mammalian cerebellar cortex and in the crayfish stretch receptor neuron. In both of those sites physiological evidence concerning the differential action of the various cell and fiber types is both voluminous and concordant. BODIAN (1966) made a similar correlation for the endings of afferents to the motor neuron in the spinal cord. These correlations have elicited a great deal of discussion (see PALAY, 1967; PETERS, PALAY, and WEBSTER, 1970; and discussion following UCHIZONO's 1969 paper). In the cerebellar cortex, in which both morphological and physiological analyses have reached an advanced state, the correlation between vesicle shape and postsynaptic action of a terminal seems to be nearly perfect. But it is still debatable whether this example can be extended unambiguously to other parts of the central nervous system. Since excitation or inhibition is a response of the *postsynaptic* element, it might be expected that the shape of the synaptic vesicles represents some parallel but not necessarily relevant function of the presynaptic fiber, such as the osmolarity of the axoplasmic matrix or its protein content or the binding of specific transmitter chemicals. Some have suggested that the shape of the vesicles is related to the transmitter secreted by the nerve ending. If that hypothesis should turn out to be true, then the correlation between vesicle shape and synaptic action seen in the cerebellar cortex may not be generalizable, since there are examples, in invertebrates, of contrary responses to activation of the same terminals under different conditions or to the same transmitter in different cells. Consequently, it is all the more remarkable that the correlation between vesicle shape and postsynaptic effect is so consistent in the cerebellar cortex. It may be that the cells and fibers of the cerebellar cortex deal in a currency of only two transmitters, one for inhibition and another for excitation, and that this is the reason why the correlation is so definite.

Another early correlation between form and function at the synapse has fared almost as well. ANDERSON *et al.* (1963) suggested that synapses with Gray's type 1 junctions were excitatory, whereas those with Gray's type 2 junctions were inhibitory. When the distribution of type 1 junctions is examined, it turns out that they are always associated with terminals containing spherical vesicles. In the cerebellum these are characteristics of synapses that have been shown to be excitatory by physiological experiment, the synapses made by mossy fibers, climbing fibers, and parallel fibers. But difficulties arise in connection with the definition of Gray's second type. In the cerebellum this category is not as distinct as could be desired, since there are many examples of intermediate or, rather, indeterminate junctions, which cannot be classed with either type and which are known to be inhibitory. As a result of such findings, this correlation might be recast so that junctions of Gray's type 1 are excitatory and all others are inhibitory. This slight modification of the original hypothesis vitiates any residua of the earlier criticisms of this correlation based on an incomplete study of the cerebellar cortex (PALAY, 1967; PETERS, PALAY, and WEBSTER, 1970) and already conceded to be wrong (CHAN-PALAY and PALAY, 1970). Nevertheless, its applicability to centers outside the cerebellar cortex remains to be proved by coordinated cytological and physiological observation.

A reservation should, however, be entered into the record here. These hypothetical correlations refer to the present divisions of all synaptic activity into excitation and inhibition, and they depend upon the accuracy of the sometimes highly presumptuous identification of the cells or fibers in question by electrophysiological means. As HÁMORI has amusingly brought out (see discussion of UCHIZONO'S 1969 paper), the physiologists may change their minds, and then where shall we be? More seriously, it should be understood that the anatomical discriminations are based on visible evidence that can be reproduced, criticized, and re-evaluated in the light of further study. The physiological identifications are usually based on transients that cannot be recaptured for later evaluation, such as the depth of the recording or stimulating electrode; histological, or better still, cytological identification of the points studied electrically is rarely carried out. Since this confirmation of presumptions would slow the pace of investigation, the reluctance of physiologists to undertake it is understandable, especially since the flow of consistent results from several laboratories and several different species makes the identifications of the various types of responses seem reasonable. But to the morphologist the entire edifice has an air of tentativeness, because an essential step is missing in the train of the argument. Besides, continued progress in the refinement of physiological technique might result in the discovery of more subtle and abstruse synaptic actions and thus provide functional correlates for some of the variations in junctional morphology that we now try to fit into a simple bifid scheme.

3. A Hitherto Unrecognized Fiber System

By means of fluorescence microscopy, HÖKFELT and FUXE (1969) discovered a previously unrecognized class of fibers in the cerebellar cortex. These fibers, containing norepinephrine and detectable because of their green fluorescence, extend from the white matter in the center of the folia into all three layers of the cortex, where they generate a sparse plexus. Although the fibers are not restricted to any particular plane, the majority ascend vertically through the granular layer. Some run horizontally in the parasagittal plane, and tangential sections demonstrate that they are moderately concentrated in the layer of the Purkinje cells. Since none of the cell bodies in the cerebellar cortex are fluorescent, HÖKFELT and FUXE concluded that the fibers must originate from cells lying outside the cerebellar cortex. At first they proposed the lateral reticular nucleus as a likely source of the fibers, but more recent experiments (OLSON and FUXE, 1971) have disclosed that they arise mainly from the locus coeruleus, which also innervates the cerebral cortex in a similar manner. BLOOM, HOFFER, and SIGGINS (1971) demonstrated similar fibers containing norepinephrine in the deeper molecular layer, which they suggested synapse with Purkinje cell dendrites and thorns, since microiontophoretic application of norepinephrine depresses the firing of Purkinje cells (HOFFER, SIGGINS, and BLOOM, 1971). They injected norepinephrine labeled with tritium and processed the tissue for autoradiography. At the electron microscope level, they found the label in scanty nerve endings, which contained large granular vesicles. They were not able to find the label in mossy or climbing fibers or in terminals showing the synaptic specializations. Their illustrations show autoradiographic grains over neuroglial processes and parallel fibers, as well as profiles that may be terminals.

Hökfelt and Fuxe (1969) also found another group of fibers that was disclosed by their fluorescence only after incubation of brain slices in 6-hydroxytryptamine. These fibers run horizontally mainly in the parasagittal plane of the molecular layer. It was thought that these fibers represent axons of 5-hydroxytryptamine neurons, since the 6-hydroxy analogue is known to accumulate in them. However, no fluorescent cell bodies were found in the cerebellar cortex with this technique, and it was therefore concluded that the cell bodies must lie in the raphe nuclei where most 5-HT nerve cells are congregated.

It is not easy to assimilate the systems of fibers revealed by fluorescence techniques into the categories of fibers in the traditional roster. All previously recognized afferents to the cerebellar cortex follow a different trajectory and either end in the granular layer or ascend along the Purkinje cell dendrites in a specific and distinct arborescence. Although Hökfelt and Fuxe (1969) described some fibers ending in formations resembling glomeruli, the majority of the fibers seem not to follow the familiar paths. It must be admitted, however, that with their low magnification studies the various authors have not demonstrated a *continuous* trajectory of fibers passing from the white matter into the granular and thence into the molecular layer. Their illustrations only show fluorescent beaded fibers in all of these places, a few indeed crossing from the granular layer into the molecular. Some of them might be mossy fibers ending in glomeruli. Some of the horizontal fibers in the molecular layer could well be basket axons or deep stellate cell axons, the absence of somatic fluorescence notwithstanding.

Nevertheless, it would not be too surprising if a third category of afferents, in addition to climbing and mossy fibers, had been missed in Golgi preparations in spite of intensive study during the past century. The number of fibers involved is small; they may have failed to impregnate for some reason. Besides, the fragments of fibers seen in the fluorescent sections resemble fragments of fibers already recognized in Golgi preparations, as indicated above. A large number of the fibers resemble the vertical stems of granule cell axons; others look like the horizontal stems of basket fibers. Some could be climbing fibers. Consequently, without the fluorescent marker they cannot be discerned as different from other fibers of the same size and course.

It is even less surprising that they have not been recognized by electron microscopy. Their number is small, especially in the cerebella of adult animals, which provide the vast majority of the specimens examined. Their irregular, beaded course would make them elusive indeed, as our experience with the stellate cell axon indicates. Besides, we do not yet have a good method for selectively identifying or marking aminergic fibers and endings in the central nervous system. Although immersion in potassium permanganate might preserve their fine, dense synaptic granules (Hökfelt, 1967a and b), the method is a poor fixative for central nervous system. A more intensive study by means of autoradiography with labeled norepinephrine might provide some clues to the identity of these fibers.

4. The Inhibitory Transmitter

Meanwhile evidence has been accumulating that the principal inhibitory transmitter in the cerebellar cortex is γ-aminobutyric acid (GABA). The Purkinje cell layer contains higher concentrations of GABA and associated enzymes than the molecular and granular layers separated by dissection from the cerebellar cortex (Kuriyama *et al.*, 1966). Hökfelt and Ljungdahl (1970) showed that in brain slices incubated with labeled GABA the transmitter was taken up by small cells in the cerebellar cortex, presumably basket and stellate cells, but not by Purkinje cells. This result was unexpected, since there is considerable strong evidence indicating that the transmitter released from Purkinje cell axon terminals is GABA (Obata, Ito, Ochi, and Sato, 1967; Fonnum *et al.*, 1970; Obata, Otsuka, and Tanaka, 1970). Recently Hökfelt and Ljungdahl (1972) extended their study to living rats injected stereotactically with tritiated GABA, and examined their autoradiograms in the electron microscope. Because of much improved fication and resolution, they were able to show that the transmitter was taken up and concentrated in basket and stellate cells and their axons, particularly in the pericellular baskets and the pinceaux. A few Golgi cells may also have been labeled, but the Purkinje cells were conspicuous for their lack of label. The discrepancy between the autoradiographic and biochemical and pharmacological results is not necessarily irreconcilable. The Purkinje cell body may not be a site of uptake for this transmitter. One is reminded that the Purkinje cell is almost totally enclosed in neuroglial processes, which may have some effect upon this result.

The autoradiographic results of Hökfelt and Ljungdahl (1970, 1972) nicely corroborate a study of inhibition of Purkinje cells in the frog by Woodward, Hoffer, Siggins, and Oliver (1971). These authors applied GABA by microiontophoresis to the surface of Purkinje cell bodies, and found that it produced a prolonged hyperpolarization in these cells, accompanied by reduction in the amplitude of action potentials and cessation of spontaneous firing, an action closely mimicking the results of activating the stellate cells through parallel fiber stimulation. Somewhat similar results in mammals with topically applied norepinephrine (HOFFER, SIGGINS, and BLOOM, 1971) indicate that these may be two parallel systems for inhibitory control of the Purkinje cell. Again, autoradiography at the electron microscope level may be necessary to localize these two inhibitory pathways.

Chapter X

The Climbing Fiber

1. A Little History

RAMÓN Y CAJAL (1888b) discovered the climbing fiber during his study of the cerebellar cortex of birds. In view of the difficulties electron microscopists had in identifying this nerve fiber nearly a century later, his first account is interesting.

> "In conclusion we will also mention the existence of certain vertical, or more or less oblique, fibers that contribute to the intricacy of the molecular layer. They come from the granular layer, penetrate obliquely into the molecular layer, following a somewhat vertical course, and at various levels terminate in a varicose, radiating arborization, disposed in the form of an irregular star. The most notable feature presented by these stellate figures is that in several places they appear to be formed of two fibers, one thick and continuous with the stem, the other thin and weakly impregnated. We have not been able to determine the origin of this second fiber, which is sometimes intimately apposed to the first and ramifies along with it; nevertheless, in several cases it was possible to observe that it arises from the principal stem or one of its thick branches." [19]

Almost all of the principal characteristics of the climbing fiber are mentioned in this first description—the spreading arborization, the varicosities, and particularly the separation into a thick main stem and almost parallel, twisting tendrils. At first RAMÓN Y CAJAL was puzzled by these stellate figures. He wondered whether they were continuous with some nerve or neuroglial cell in the granular layer or with some myelinated fiber in the white matter. A few months later, in a French version of this paper (RAMÓN Y CAJAL, 1889b), he reported that since he had been able to follow the stem to the white matter he tended to think that the stellate figures were the terminal ramifications of nerve fibers. In another paper in the same year (RAMÓN Y CAJAL, 1889a; 1890a, in French) he noted that during its tortuous course through the granular layer, the stem of the stellate figures displayed neither thickenings nor efflorescences, in marked contrast to the mossy fibers. Furthermore, he remarked that the terminal arborization in the molecular layer extends in a single flat plane like that of the Purkinje cell dendritic tree. He also convinced himself that the secondary fibers in this arborization all arise from the principal fiber or its main branches. In this rapid sequence of papers RAMÓN Y CAJAL disclosed the essential features of the climbing fibers without quite understanding them.

In the next year, however, everything fell into place. The key lay in the embryology of the cerebellar cortex. In the previous year RAMÓN Y CAJAL (1889a and 1890a) had discovered, in newborn and very young animals, a new class of nerve fiber originating from the white matter and ascending through the granular layer. When these fibers reached the layer of Purkinje cells, they burst into a dense arborescence that surrounded the Purkinje cell bodies. Fancying that these formations resembled birds' nests, RAMÓN Y CAJAL called them cerebellar nests (*nidos cerebelosos, nids cérébelleux*). At the time of this discovery, he entertained the idea that these pericellular plexuses might persist into adulthood, forming with the pericellular basket two overlapping pericellular nests. But he never found such forms in the adult cerebellum. By 1890, he had completed an intensive study of numerous specimens taken from rat, cat, dog, and chick at different ages, both very young and adult. He now recognized

19 «Para terminar, citaremos también la existencia de ciertas fibras verticales o más o menos oblicuas, que contribuyen a complicar el entrelazamiento de la zona molecular. Provienen de la zona granulosa, penetran oblicuamente en la molecular, siguen un curso casi vertical, y, a nivel variable, rematan por una arborización varicosa, divergente, dispuesta en forma de estrella irregular (Fig. 3a, lám. XI). Lo más notable que tales estrellas presentan, es que en muchos puntos aparecen formadas por dos fibras, una gruesa (*b*), continua con el tallo, y otra delgada (*c*) y débilmente impregnada. El origen de esta segunda fibra, que algunas veces se adosa intimamente a la otra, ramificándose como ella, no puede determinarse; no obstante, en algún caso hemos podido notar que nacía del mismo tallo principal o de una de sus gruesas ramificaciones» (RAMÓN Y CAJAL, 1888b, p. 39).

(RAMÓN Y CAJAL, 1890b) that the cerebellar nests do not persist into adulthood but progressively transform into the stellate figures that he had discovered earlier. Therefore, the cerebellar nests and the stellate figures were different stages in the maturation of the same fiber. The parallel maturation of these fibers and the dendritic trees of the Purkinje cells, also described for the first time in this paper, led RAMÓN Y CAJAL to understand the significance of this arrangement.

"The study of the transition forms and comparison of them with the disposition (of the fibers) in the adult permitted us in addition to discover a very singular fact, which perhaps throws a vivid light upon the manner of connection between the nervous elements, to wit: that the ramified plexiform arborization formed by the fibers of the fourth kind (those ending in pericellular nests), in the depths of the molecular layer, twists and twines along the ascending main stem dendrite of the Purkinje cell, as well as about its principal branches, in the fashion of tropical vines or ivy climbing along the trunk and large branches of a tree. That is why we call these terminal plexuses *climbing arborizations or plexuses*."[20]

Although he had some difficulty in obtaining good preparations of climbing fibers in adult animals, his studies of immature animals, in which the impregnation of climbing fibers was much more common, convinced him that each Purkinje cell was furnished with a climbing plexus generated by a single myelinated fiber. He perceived, in addition, that the intimate and exclusive connection between the climbing fiber plexus and the dendritic tree of the Purkinje cell had for its purpose the transmission to the Purkinje cell of the impulse that excites the climbing fiber, in other words, that it was what we now call a synapse.[21]

These observations and interpretations of RAMÓN Y CAJAL were quickly confirmed by VAN GEHUCHTEN (1891) and RETZIUS (1892a and b). VAN GEHUCHTEN published a beautiful figure showing six climbing fibers copied from RAMÓN Y CAJAL's papers (1889a, 1890a and b), but his drawing of a climbing fiber from his own preparations is not very convincing. RETZIUS, however, corroborated every detail, including the developmental sequence, in rabbit, cat, dog, and man. LUGARO (1894) provided a similar confirmation. KÖLLIKER, the dean of nineteenth century histology, apparently did not succeed at first in impregnating the climbing fibers. As he could not distinguish between climbing fibers and mossy fibers, whose characteristic efflorescences he considered artifactitious, he (1890) thought that both fibers were the same. Even as late as 1896 he still thought that they were closely related, and was not convinced of the necessary correlation between the disposition of the climbing fiber arborization and the form of the Purkinje cell dendritic tree. Nevertheless, he provided three drawings of climbing fibers in a young kitten that agreed with RAMÓN Y CAJAL's.

The discovery of the climbing fiber and its articulation with the Purkinje cell dendrites was an important link in the chain of evidence that led RAMÓN Y CAJAL to formulate the neuron doctrine and the law of polarization in their most general forms (RAMÓN Y CAJAL, 1890b, 1894). Here was an extensive axonal arborization, clearly terminal and clearly ending on the surface of another cell, exactly like the ending of a motor nerve on the surface of a skeletal muscle cell. The analogy carried an important double message. It not only disclosed the direction of conduction and transmission in the axon, but it also revealed the direction of conduction in the dendritic tree (toward the cell body) and confirmed his suspicion that dendrites, like the perikaryon, constituted a receptive apparatus. Furthermore, by completing the list of the direct afferents to the Purkinje cell, his discovery of the climbing fiber also led him to perceive the principle that when a nerve cell establishes multiple contacts with a variety of afferents, each kind occupies a different and specific territory on the surface of the postsynaptic cell. For example, in the Purkinje cell, the perikaryon is supplied by the basket fibers, the dendritic trunks by the climbing fiber, and the dendritic thorns by the parallel fibers. RAMÓN Y CAJAL was not yet in a position to conclude that the nerve cell is an integrating unit, or that the location of efferent terminals on its surface is related to the specific role of those afferents in the response pattern of the cell, but he thought that the disposition of the afferent terminals on different parts of the cell served

20 «L'étude des formes de transition mentionnées et leur comparaison avec les dispositions adultes, nous ont de plus permis de découvrir un fait très singulier qui est peut-être appelé à jeter une vive lumière sur le mode de connexion des éléments nerveux, à savoir: que l'arborisation plexiforme ramifiée, formée par les fibres de la 4me espèce au sein de la couche moléculaire, s'enroule et serpente en quelque sorte tout le long de la tige protoplasmique ascendante de la cellule de Purkinje ainsi qu'autour des branches principales de celle-ci, à la façon des lianes ou du lierre qui grimpent le long du tronc et des grosses branches d'un arbre. C'est pourquoi nous appelons ces plexus terminaux *arborisations* ou *plexus grimpants*» (RAMÓN Y CAJAL, 1890b, p. 459).

21 «Il est évident que l'intime et exclusive connexion que les plexus grimpants tiennent avec l'arborisation protoplasmique des cellules de Purkinje a pour objet de transmettre à celles-ci le mouvement que les anime; ...» (RAMÓN Y CAJAL, 1890b, p. 462).

to keep separate the impulses arriving over different pathways.[22]

RAMÓN Y CAJAL's classic description of the climbing fiber, reaffirmed in repeated publications (e.g., RAMÓN Y CAJAL and ILLERA, 1907; RAMÓN Y CAJAL, 1909, 1911, 1934; ESTABLE, 1923), remained without serious modification until relatively recently. In his comprehensive review of the cerebellum, JAKOB (1928) could only confirm the older observations and re-emphasize the established one-to-one relation between the climbing fiber and the Purkinje cell. In 1954, however, the Scheibels published an important study with the Golgi method, which laid the groundwork for a considerable expansion of our concept of the climbing fiber. They examined a large collection of material from rat, cat, dog, monkey, chimpanzee, and man, both immature and adult. They applied a number of variants of the Golgi method, among which the most useful was one that involved only brief exposures to the usual reagents. This variant produced an increased incidence of red impregnations of both cells and fibers (see Chap. II). With this procedure the SCHEIBELS were able to make a much more detailed examination of the climbing fiber plexus than had been done previously. They found that the climbing fiber makes contact with nearly every kind of cell and fiber in the cerebellar cortex. At the level of the Purkinje cell bodies, small collaterals left the mainstem of the climbing fiber and re-entered the granular layer where after a few bifurcations they were lost to view. In a few instances the SCHEIBELS were able to follow them to their apparent ends as single large nodules. Some 25 years earlier PENSA (1931) had described and figured similar collaterals, not only retrograde, but also emerging from the climbing fiber during its trajectory through the granular layer; his description and pictures were, however, so thoroughly reticularist that they elicited little notice. More recently SZENTÁGOTHAI and RAJKOVITS (1959) demonstrated such collaterals by degenerative techniques, and FOX, ANDRADE, and SCHWYN (1969) have figured fine thread-like collaterals emitted by the climbing fiber in the granular layer in a Golgi preparation. Of such collaterals the SCHEIBELS (1954) wrote: "We can only speculate that they come in contact with the mossy fiber-granule cell complex, or the large stellate cells (Golgi cells) of the granular layer, or both, since our evidence is inconclusive on this point" (p. 737). The actual distribution and form of the endings of those collaterals were described for the first time by CHAN-PALAY and PALAY (1971), who carried out an intensive light and electron microscope study of them. Their results will be related below.

In addition to the collaterals going to the granular layer, the Scheibels also reported the existence of a second type of retrograde collateral that descended from the stem of the climbing fiber to participate in the formation of the basket around the Purkinje cell. Besides, they found collaterals that ended on the soma, neck, or primary dendrites of adjacent Purkinje cells. Such collaterals may contact as many as five or ten neighboring Purkinje cells. They suggested reasonably that these climbing fiber collaterals could account for some of the ring endings that RAMÓN Y CAJAL and ILLERA (1907) saw on the soma and major dendrites of Purkinje cells and sought to attribute to the endings of recurrent Purkinje cell axons. Furthermore, the Scheibels showed that, in addition to the extensive contact with the Purkinje cell dendritic tree, the climbing fiber also articulated with cell bodies and dendrites of basket and stellate cells by means of *boutons terminaux* and *boutons en passant*. Long contacts between climbing fiber collaterals and ascending collaterals of the basket fibers were reported, and the Scheibels postulated that these represented true axo-axonal connections rather than merely axo-dendritic relations to the underlying Purkinje cell dendrite.

The observations of the SCHEIBELS signified that the climbing fiber has a much wider distribution than it had been thought to have. By means of its collaterals, its influence extended not only to more than one Purkinje cell, but also to other kinds of nerve cells in both the molecular and granular layers. Furthermore, because of the lateral spread of the axonal plexuses belonging to stellate, basket, and Golgi cells, the action of the climbing fiber in contact with these cells through its collaterals was no longer confined to the flat parasagittal plane of the Purkinje cell, but was diffused through a three-dimensional space shaped like an inverted cone. The SCHEIBELS proposed that by means of all these collateral connections the climbing fiber participated in a number of feed-back circuits that could regulate the flow of impulses in the cerebellar cortex. Some of the functional implications set forth by the SCHEIBELS on the basis of

---

22 «Quand diverses cellules doivent établir une connexion avec une seule, les ramifications de leur expansions nerveuses se mêlent, en s'appliquant autour de celle-ci, les contacts pouvant s'opérer en differents endroits, soit du corps soit de l'arborisation protoplasmique, de manière à ce que les transmissions ne se confondent pas» (RAMÓN Y CAJAL, 1890b, p. 467). «La differentiation compliquée de l'appareil protoplasmique — expansions basilaires, tige protoplasmique radiale, panache terminal, etc. — que nous presentent les pyramides du cerveau et en partie aussi les corpuscules de Purkinje du cervelet, paraît avoir pour but de permettre que chacune des cellules de cette espèce puisse établir des contacts distincts avec diverses catégories de fibres nerveuses» (RAMÓN Y CAJAL, 1894, p. 465).

purely morphological observations have actually been borne out by subsequent electrophysiological investigations (see for example: ECCLES, ITO, and SZENTÁGOTHAI, 1967). It must be remembered, however, that apparent contacts seen in Golgi preparations do not necessarily signify the existence of synapses at these points. The Golgi method is simply not capable of defining the position of synapses, or disclosing the nature of apparent contacts, or even ascertaining whether such contacts are real or mere approximations. Unimpregnated neural or neuroglial processes can be interposed in such contacts without being detectable at the light microscope level. While it is difficult to imagine what kind of relation other than synapsis could be involved in the long articulation between the climbing fiber and the Purkinje cell dendritic tree, some of the other contacts recorded by the SCHEIBELS seem more open to question. For example, they described axo-axonal contacts between climbing and parallel fibers and between climbing fibers and the axons of stellate and basket cells. The only way to decide whether such apparent contacts are synapses is to examine them in the electron microscope.

The resolution of this question awaited the development of electron microscopy of the central nervous system to the point where neurocytologists could profitably attack such problems. For, in order to recognize the synapses made by the climbing fiber arborization, it was necessary to identify in electron micrographs all of the components of the cerebellar cortex. The perikarya of the various cell types were rather quickly recognized (BAUD, 1959; GRAY, 1961; HERNDON, 1963, 1964; PALAY, 1964a and b; FOX, HILLMAN, SIEGESMUND, and DUTTA, 1967); their axons and dendrites were gradually characterized (GRAY, 1961; PALAY, 1961; FOX, SIEGESMUND, and DUTTA, 1964; HÁMORI, 1964; HÁMORI and SZENTÁGOTHAI, 1965, 1966b, 1968; FOX, HILLMAN, SIEGESMUND, and DUTTA, 1967), but the climbing fiber proved to be elusive. It is strange that an axonal plexus as elaborate and singular in appearance as the climbing fiber should have occasioned so much confusion among students of cerebellar fine structure. Several authors apparently missed finding it altogether. Instead, the axons of basket cells, especially the ascending collaterals, were mistakenly identified as climbing fibers (KIRSCHE *et al.*, 1964; PALAY, 1964b, 1967; HÁMORI and SZENTÁGOTHAI, 1966a; FOX, HILLMAN, SIEGESMUND, and DUTTA, 1967). Certain authors (e.g., GOBEL, 1967) were unable to distinguish among the several larger axons in the molecular layer, and therefore evaded the question of identification. It was not until LARRAMENDI and VICTOR (1967) approached the fine structure of the cerebellar cortex from the standpoint of embryology that the climbing fiber was recognized in electron micrographs and properly identified. Their observations have been amply confirmed in a number of species, including mammals, birds, reptiles, amphibians, and teleosts (KORNGUTH *et al.*, 1968; HILLMAN, 1969; KAISERMAN-ABRAMOF and PALAY, 1969; LLINÁS and HILLMAN, 1969; MUGNAINI, 1969, 1972; SOTELO, 1969; UCHIZONO, 1969; CHAN-PALAY and PALAY, 1970).

LARRAMENDI and VICTOR began with the knowledge that the immature Purkinje cell has a very irregular shape, its surface thrown up into multiple complex cytoplasmic processes (RAMÓN Y CAJAL, 1911). LARRAMENDI (1965) and MUGNAINI (1966, 1969; MUGNAINI and FORSTRØNEN, 1967) showed by electron microscopy that many of these processes are in synaptic contact with various axons. During subsequent development most of the cytoplasmic processes are resorbed by the perikaryon as the dendritic tree becomes elaborated. As is well known, the cell body and primary dendrites of the Purkinje cell have a smooth round surface. LARRAMENDI and VICTOR (1967), however, found that in the young adult mouse some of the processes persisted as thorns projecting from the surface of the large dendrites and the apical part of the cell body. These thorns had a shorter stem and more bulbous head than those attached to the more distal spiny branchlets, and they were in synaptic contact with a distinctive axonal terminal packed with round synaptic vesicles. In a few instances in which the parent axon could be found connected to such terminals, the axoplasm appeared to be full of microtubules. Similar terminals seldom contacted the smooth surface of the perikaryon or shafts of the dendrites, and when they did no synaptic specializations occurred at the apposition sites. The synapsing terminals found in these positions had a different composition and could be traced either to the myelinated axons of Purkinje cell recurrent collaterals or to the neurofilamentous axons of basket cells. By comparing the distribution of the distinctive terminals synapsing on somatic and dendritic spines with the classical course of the climbing fiber in Golgi preparations, Larramendi and Victor concluded that they belonged to the climbing fibers.

These investigators had made two important discoveries about the climbing fiber. They distinguished the terminals of the climbing fiber from those of the other axons in the lower molecular layer, and they established that the climbing fiber synapses exclusively on thorns emitted from the soma and larger dendrites of the Purkinje cell. The existence of these thorns had been wholly unsuspected by light microscopists, except in the immature Purkinje cell, and the first photomicrographs of them in

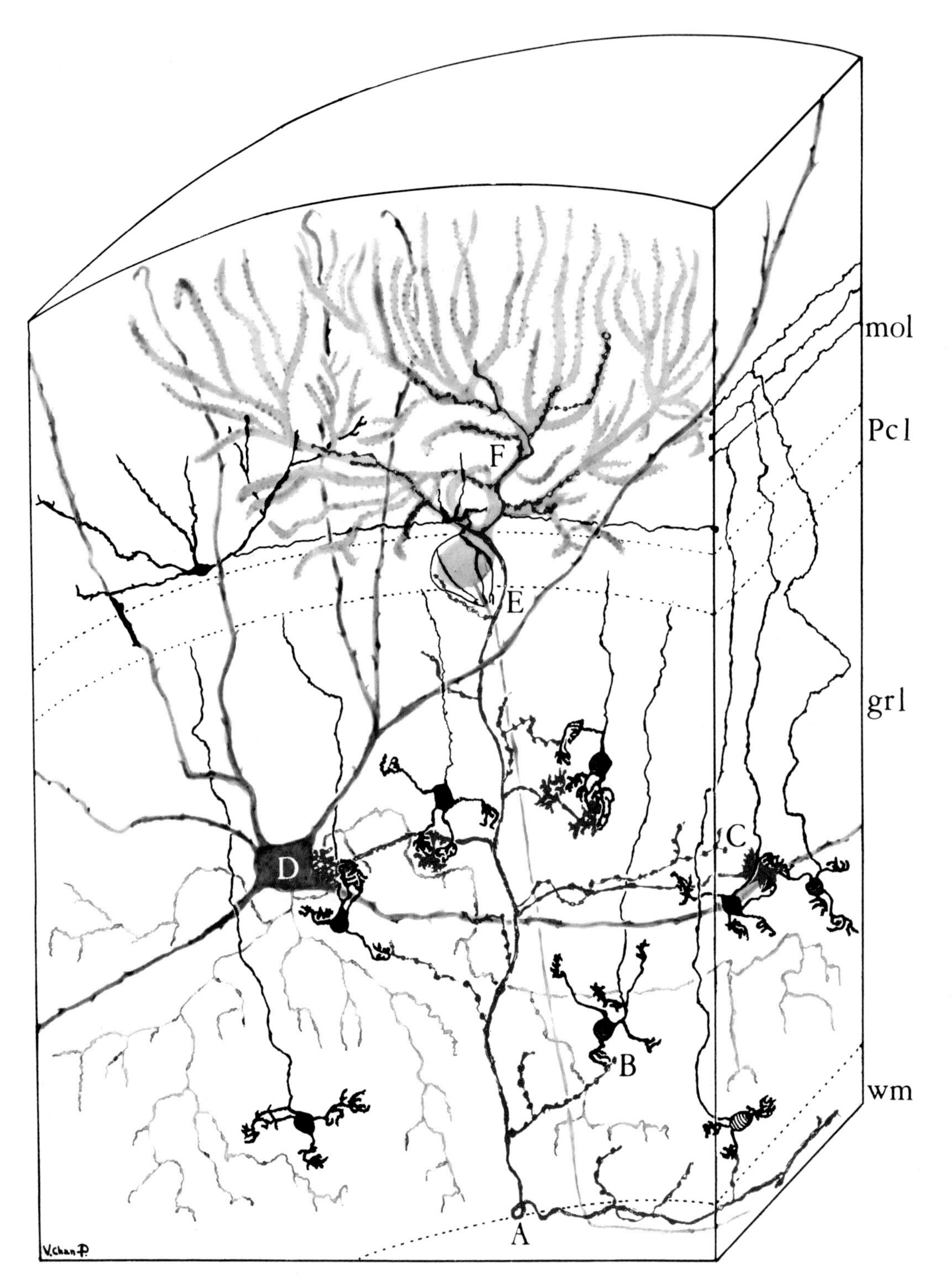

Fig. 203. Diagram showing relations of the climbing fiber in the granular and molecular layers. Camera lucida drawings of climbing fibers (red), a Purkinje cell (gray), granule cells (black), a basket cell (black), and a Golgi cell (dark gray) from rapid Golgi preparations were superimposed to produce this diagram. The interrelations of the processes of these cells have been observed in electron micrographs, and are reproduced in this figure in order to provide the light microscope equivalent. (*A*) The parent stalk of the climbing fiber traverses the white matter (*w m*), bending at a right angle to enter the granular layer (*gr l*). Fine tendril collaterals bearing globose varicosities are emitted at irregular intervals from it and synapse with dendrites of granule cells (*B*), and the somata of Purkinje cells and Golgi cells. Glomerular collaterals of the climbing fiber end in large efflorescences that are the central structures in glomeruli (*C*). The dendrites of granule cells and axons of Golgi cells participate in the formation of these glomeruli (the axonal plexus of the Golgi cell has been simplified in this illustration). An efflorescent terminal of the climbing fiber forms an extensive synapse *en marron* with the wrinkled surface of the Golgi cell soma (*D*). The claw-like dendrites of granule cells complete the glomerulus on the sides of this climbing fiber core not in contact with the Golgi cell. Tendrils (*E*) issue from the parent stalk in the region of the basket around the Purkinje cell axon. The climbing fiber continues through the Purkinje cell layer (*Pc l*). Here it branches into two smooth stems which continue to ascend into the molecular layer (*mol*) on the Purkinje cell dendrites. From these stems beaded tendrils (*F*) are emitted that come into relation with ascending collaterals of the basket cell on dendrites of the Purkinje cell. Rapid Golgi, 120 μm, camera lucida. About ×200

Golgi preparations were not published until after the work of LARRAMENDI and VICTOR was generally accepted (SOTELO, 1969; MUGNAINI, 1970, 1972). Indeed, no one else had mentioned them in electron microscopic studies, either. Since they had not been reported before in work on the cat, LARRAMENDI and VICTOR suggested that the persistence of thorns on the soma and large dendrites might be a species characteristic of the mouse. Subsequent work, however, has demonstrated the general occurrence of these spines and their exclusive association with the climbing fiber.

In our paper on the interrelation between climbing fibers and basket cell axons (CHAN-PALAY and PALAY, 1970), we summarized the reasons why these two fibers, now obviously distinct, were at first confused by electron microscopists of the cerebellar cortex. We considered that the major obstacle was not inadequate technique or insufficient diligence on the part of electron microscopists, but rather an incomplete understanding of the climbing fiber at the level of the light microscope. As indicated above, lacking knowledge that the principal dendrites project thorns, electron microscopists expected that the climbing fiber should appear as a commonly occurring axon attached to the smooth trunks of the Purkinje cell dendritic tree. Actually this location is favored by the ascending collaterals of the basket cell axon, which are much more frequent than had been supposed (CHAN-PALAY and PALAY, 1970). These collaterals also twine round the larger dendrites and cling to their smooth surface, with which they synapse. LARRAMENDI and VICTOR (1967) found that in the lower reaches of the Purkinje cell dendritic tree (dendrites 2 µm or more in diameter), basket axon synapses were two to four times as numerous as climbing fiber synapses. Furthermore, climbing fiber terminals are mostly varicosities *en passant*, in the course of the fine, tortuous secondary branches; consequently, profiles of any length are not frequent in electron micrographs of thin sections. All of these factors combine to make it difficult to recognize the climbing fiber in electron micrographs. In view of this awkward history, the following account of the climbing fiber in the rat starts with a comprehensive description of its appearance in Golgi preparations and then proceeds to its fine structure. Most of the information included derives from two earlier papers (CHAN-PALAY and PALAY, 1970, 1971a), but numerous details, previously unpublished, have been added.

2. The Climbing Fiber in the Optical Microscope

In the white matter at the center of a cerebellar folium, the climbing fiber can be recognized as a fairly thick (2–3 µm, as measured in Golgi preparations), gently contorted, myelinated axon (Fig. 203). In its course through the white matter it gives off several fine collaterals that radiate into the granular layer. RAMÓN Y CAJAL (1890b, 1911) found similar but more robust collaterals, which, he said, join the terminal arborization of the parent fiber in the molecular layer. The main stem pursues a more or less straight trajectory through the white matter, but eventually bends nearly at right angles to enter the granular layer. The fiber ascends obliquely or vertically through this layer, twisting and bending around presumed

Fig. 204 A–E. The climbing fiber arborization in the granular, Purkinje cell, and molecular layers. **A** This climbing fiber enters from the white matter (*wm*) as an unbranched, twisted stalk, and courses through the granular layer. In the upper third of the granular layer, a branching collateral plexus of beaded tendrils diverges from the parent fiber (*arrow*). At the level of the Purkinje cell body (*PC*) the axon breaks into its terminal arborization to climb upon the dendrites of a Purkinje cell. Rapid Golgi, 90 µm, sagittal, adult rat. **B** This second climbing fiber emits a glomerular collateral with an efflorescence (*arrow*) in the granular layer before continuing into its terminal arborization in the molecular layer. At the level of the Purkinje cell body, several thin, beaded tendrils issue and turn away from the parent fiber. Rapid Golgi, 90 µm, sagittal, adult rat. **C** The stalk of this climbing fiber leaves the white matter (*wm*) and emits two collateral tendrils in the low granular layer. As is typical of such branches, they extend almost at right angles away from the parent stalk. The main fiber continues through the granular layer to the Purkinje cell layer (*PC*). Here, at the apical pole of a Purkinje cell, the fiber describes a series of tortuous hairpin bends before continuing upwards to form its terminal arborization. Rapid Golgi, sagittal, 90 µm, adult rat. **D** The climbing fiber seen in the transverse plane. This climbing fiber is seen in profile, in a section taken at right angles to its fan-shaped arborization. The parent fiber leaves the white matter (*wm*) and gives off several long, beaded collaterals in the granular layer. These collaterals course laterally, away from the parent fiber, for about 150 µm (*arrow*), giving the climbing fiber arborization a dimension in the transverse plane of the brain. These collaterals may also ascend to the level of the Purkinje cell body (*PC*). The parent fiber twists in a tortuous manner, making several loops at the level of the Purkinje cell, before continuing into the molecular layer. Here, the characteristic fan-like arborization is seen in profile, and the tendrils can be followed for 30–40 µm at their widest extent. Rapid Golgi, 90 µm, transverse, adult rat. **E** This climbing fiber leaves the white matter (*wm*) and describes a curved course through the granular layer, following the gentle curve imposed on it by the arrangement of cells at the summit of the folium. Along this curve, the fiber gives rise to several relatively short, varicose collaterals that ascend to the upper granular layer. Above the Purkinje cell layer (*PC*) the parent fiber issues many beaded branches that droop downwards towards the Purkinje cell body. The fiber then continues upward to generate its characteristic terminal arborization. Rapid Golgi, 90 µm, sagittal, summit of folium in the cerebellar cortex, lobule IX. Camera lucida drawing

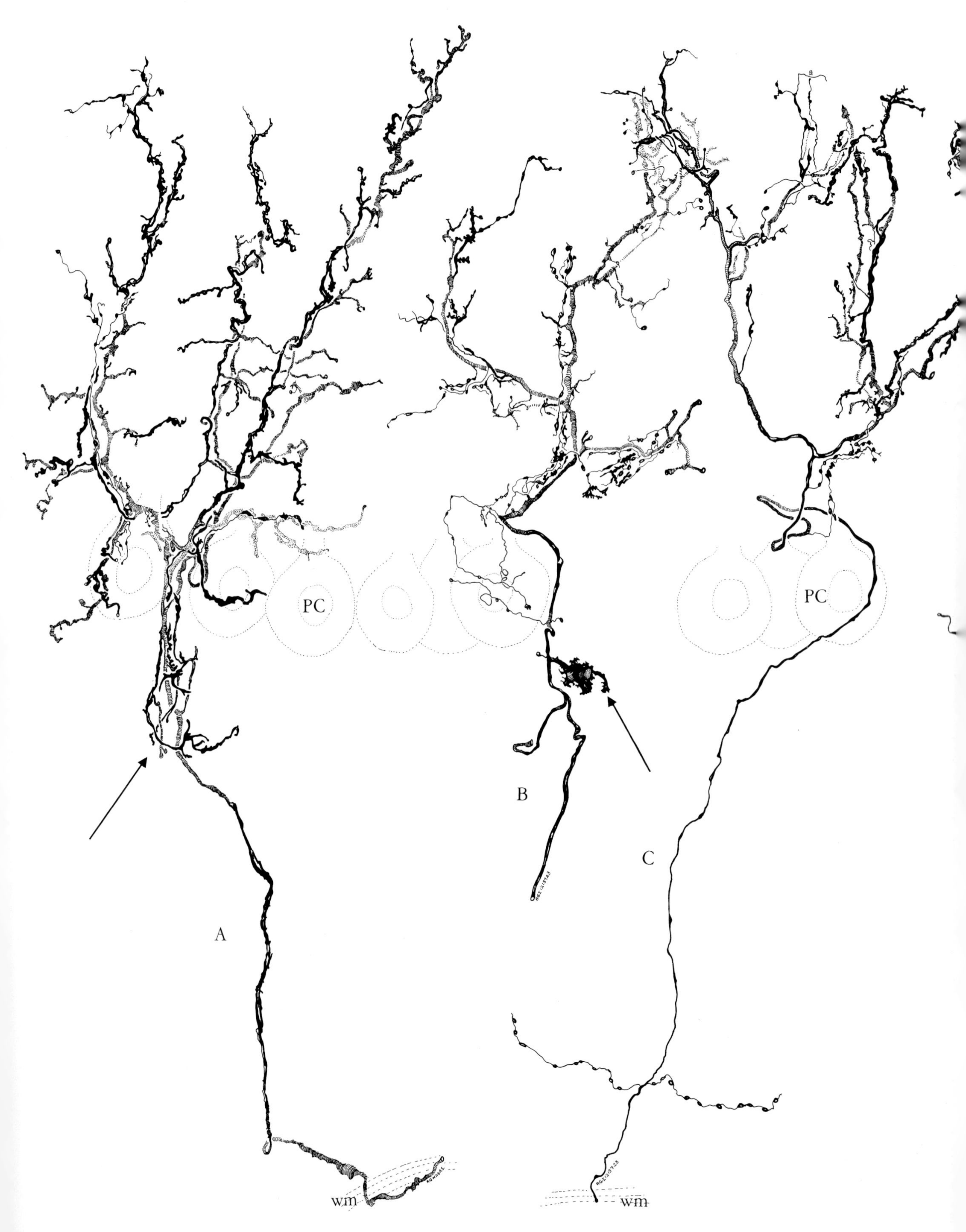
PC
PC
A
B
C
wm
wm

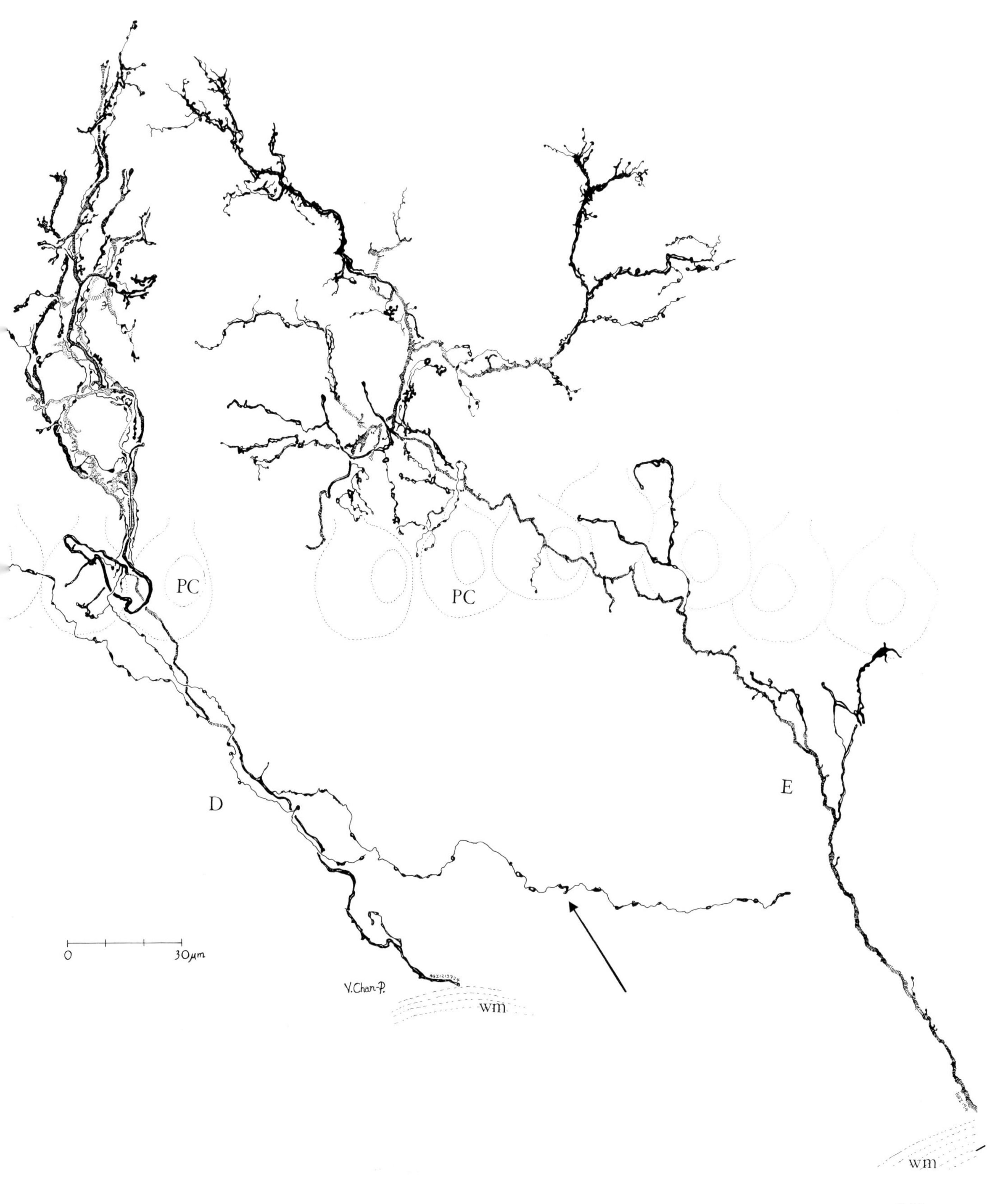
PC
PC
D
E
0
30μm
V. Chan-P.
wm
wm

obstacles in its path until it reaches the Purkinje cell layer.

Two distinct types of collaterals are emitted from the climbing fiber during its course through the granular layer (Fig. 204 A–E). The first type is a fine, beaded collateral consisting of several globular or bulbous dilatations, about 2 μm thick, connected by a fine thread (Fig. 204 B–E). These tendrils, resembling the common collateral in the terminal arborization to be described in the molecular layer, remain close to the parent stem and twine around it (Fig. 204 C2 d, e), or they may pursue a somewhat tortuous course, meandering away from it for a short distance (Fig. 204 C, D). Similar collaterals were shown in a drawing of a Golgi preparation by Fox, Andrade, and Schwyn (1969). The second kind of collateral begins as a contorted, thick branch (about 2 μm) that maintains approximately the same dimensions as the main stem from which it springs, and abruptly terminates in a truncated spray or large efflorescence about 12–15 μm long (Fig. 204 B). A collateral of this kind may have a single efflorescence or may branch again to end in another one. Since the spray terminating this kind of collateral forms the central structure in a glomerulus, it has been given the name glomerular collateral. In the Golgi preparations such a spray has a characteristic stellate or ragged appearance caused by the presence of fairly thick, stubby projections. Neither of these collaterals in the granular layer had been described before (Chan-Palay and Palay, 1971 a), although some of those figured by Pensa (1931) and by Fox, Andrade, and Schwyn (1969) presumably correspond to them.

Unlike the terminal ramifications of the climbing fiber in the molecular layer, which, as Ramón y Cajal (1889 a) noted, spread out within a more or less narrow parasagittal plane, the collaterals in the granular layer are less confined. Since the tendril and glomerular collaterals spread out in any one of the three cardinal planes, climbing fibers in the granular layer extend not only parasagittally but also transversely, along the folium. The three-dimensional orientation of the ramifications in the granular layer may explain why they escaped the attention of earlier investigators. It is difficult enough to find the terminal arborization in the molecular layer in continuity with the parent fiber in the granular layer (witness Ramón y Cajal's perplexity about the sources of his stellate figures), and its collaterals, spreading through the granular layer in a plane roughly perpendicular to that of the terminal arborization, are often severed by the plane of section.

At about the level of the Purkinje cell bodies the climbing fiber takes leave of its myelin sheath and begins its characteristic climbing course. By the time the fibers have reached the tops of the Purkinje cell bodies, very few of them are still myelinated (see Figs. 212 and 213). As it emerges from its sheath the main stalk of the climbing fiber is about 2 μm thick and has smooth contours. It runs alongside the Purkinje cell body and its main dendrite, apparently closely applied to them at least for stretches several microns long. Electron microscopy shows that the fiber is, on the contrary, separated from them by the neuroglial sheath of the Purkinje cell and variable amounts of neuropil. In this initial unmyelinated portion of its course the climbing fiber also gives off collaterals that return to the granular layer (Figs. 204 A, B, D, E, and 205). These are the retrograde collaterals that were first described by the Scheibels (1954), who were sometimes able to trace them to terminations in large single nodules. In our own material (Figs. 204 and 205) they appear as tendril collaterals, which we have frequently encountered emerging from either the parent stem of the climbing fiber or from one of its first branches above the Purkinje cell bodies. Recurving downwards, the tendrils ramify sparsely and interlace with the parent fiber in a complicated fashion, as they descend into the infraganglionic plexus. Here they end in a meager raceme of varicose branchlets or in an irregular talon-like figure.

At the level of the first bifurcation of the Purkinje cell dendrite, the climbing fiber also divides and so begins its terminal arborization. Its branches follow the secondary ramifications of the dendritic tree with further subdivisions, but without maintaining a close correspondence to the finer branching pattern of the dendrites. Most of the climbing fiber arborization is confined to the lower two thirds of the molecular layer. After the first division or two the parent fiber begins to show small variations in diameter and occasional varicosities as it twists and turns over the surface of the Purkinje cell dendrites. Its most remarkable characteristic is that it emits fine, beaded

Fig. 205. Terminal arborization of the climbing fiber in the molecular layer. At the level of the Purkinje cell body (*PC*), the parent stalk of the climbing fiber gives rise to a number of fine, beaded collaterals, which run either away from or up along, in parallel with the parent stalk. The main fiber then divides into four main branches at the level of the primary Purkinje cell dendrite. As these stems climb the Purkinje cell dendritic tree, they become more gnarled and irregular. Along their lengths, fine tendrils are emitted. Each tendril consists of a fine thread connecting a series of varicosities. Some of these tendrils run parallel with the stem of the climbing fiber and ascend on the secondary Purkinje cell dendrites. Others branch laterally, away from the climbing fiber stem to form an exuberant plexus on tertiary Purkinje cell dendrites. The entire terminal arborization of the climbing fiber does not extend beyond the lower $\frac{2}{3}$–$\frac{3}{4}$ of the molecular layer. The pial surface (*pia*) is indicated. Rapid Golgi, 90 μm, sagittal, adult rat. Camera lucida drawing

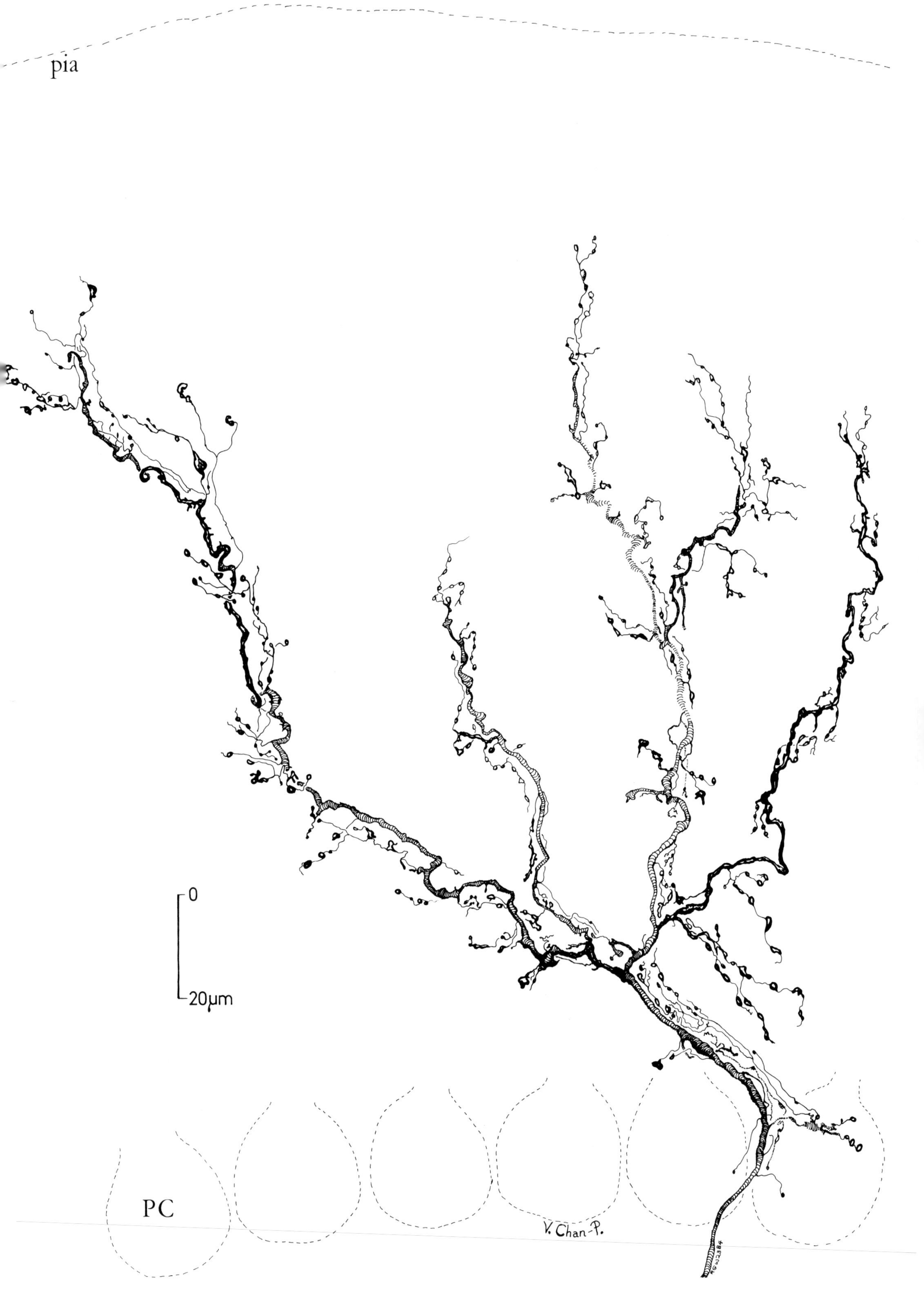

pia

0

20µm

PC

V. Chan-P.

collaterals that run along more or less parallel with it, twining like tendrils round the same dendrite or clinging to its smaller branches (Figs. 204 A–E, and 205). These tendrils constitute the second fiber that puzzled RAMÓN Y CAJAL (1888b) in his original observations. They were figured in RAMÓN Y CAJAL's drawings (for example, Fig. 22 in CAJAL, 1909) as forming a sort of plexus overlying the principal smooth dendrites of the Purkinje cell. They are better shown by the SCHEIBELS (1954) in the drawings of their paper on the climbing fiber and more recently by FOX, ANDRADE, and SCHWYN (1969) in a magnificent photomicrographic montage. The thick stem of the climbing fiber and its delicate, beaded collaterals, meandering over the lattice provided by the Purkinje cell, ensheathe the larger dendrites in a reticular sleeve resembling filigree. Thus an extensive interface is made available for synaptic contact between the Purkinje cell dendrites and the climbing fiber. It is all the more remarkable, then, that as electron microscopy shows, most of this surface is not used for synaptic contact; instead, synaptic junctions occur only between the beads and varicosities in the course of the fiber and the thorns projecting from the dendrites.

The thin tendrils, however, are not restricted to the trunk and major branches of the Purkinje cell dendritic tree and do not even follow it faithfully. Some of them extend across voids in the dendritic tree to end on spiny branchlets, or on basket or stellate cell bodies and their dendrites. Some of them also run along in parallel with ascending stems of granule cell axons.

a) The Immature Climbing Fiber Plexus

In the historical prelude to the present chapter, we saw how an understanding of development played an important part in the recognition of the climbing fiber in the adult cerebellum. In our studies on the cerebellar cortex of the adult rat we have found, as indicated above, evidence in both Golgi preparations and in electron micrographs of a rich collateral plexus in the granular layer. Since this plexus is even more exuberant in the infant rat than it is in the adult, it must undergo considerable remodeling and regression during the first three weeks after birth. Nevertheless, a strong contingent of collaterals in the granular layer persists into adulthood. We cannot, however, maintain that every climbing fiber in the adult gives rise to such collaterals, although probably many of them emit at least a few tendril collaterals. There is a great deal of variation in the details and profuseness of the collateral plexus in the immature state, and one would expect some variation in the more skeletonized plexus of the adult.

Fig. 206 presents a drawing of a Golgi preparation from a nine day old rat. In addition to the climbing fibers, granule cells and mossy fibers were impregnated in this region. The climbing fibers are in early stages of what RAMÓN Y CAJAL called "nid" formation, that is, they have generated around the immature Purkinje cells the transitory pericellular nests that are characteristic of climbing fibers at this age. It is noteworthy, in view of the later one-to-one ratio of climbing fibers to Purkinje cells, that no more than one pericellular nest is elaborated by each fiber (in this field not all fibers are shown to have even one). For our present purpose, however, we wish to direct attention to the profuse branching of these fibers within the granular layer and just beneath the layer of Purkinje cell bodies. The two fibers in the right hand panel, which are the least highly ramified of the set, give rise to a sparse plexus of varicose collaterals that closely resemble the tendril collaterals of the adult. The fibers in the left hand panel are remarkable for their extended rami coursing along beneath the row of Purkinje cells and projecting a variety of irregular ascending collaterals that apparently terminate on the Purkinje cell bodies.

This exuberant plexus in the infraganglionic layer disappears with further development. Although the SCHEIBELS (1954) described collaterals that stretch from the territory of one Purkinje cell to another, they are certainly uncommon in adult animals. The meanderings of the main stem pictured by FOX, ANDRADE, and SCHWYN (1969) and the retrograde collaterals described by the SCHEIBELS (1954) could be residua of this immature arborescence. Most of the other branches are resorbed, withdrawn, or rejected during the later postnatal development. These considerations agree with the interpretations proposed by O'LEARY, INUKAI, and SMITH (1971) in a recent paper on the embryogenesis of climbing fibers in the rat. While we can only substantiate the conclusion that the majority of these branches are removed during maturation of the climbing fiber, the number still persisting in the young adult, whose cerebellar cortex is apparently completed, is not negligible. On the contrary, our preparations at light and electron microscope levels indicate that a good many of the collaterals survive and establish synapses with cells in the granular layer.

It is interesting in this connection that WOODWARD, HOFFER, SIGGINS, and BLOOM (1971) have recorded climbing fiber responses from immature Purkinje cells in the rat as early as the third postnatal day, and more securely on the fourth postnatal day. At this time climbing fibers should still be quite primitive, but apparently they are already capable of firing the Purkinje cell in their characteristic fashion. It might be speculated that the collaterals from such fibers are also capable of forming successful synapses with other cells in the internal granular layer.

3. The Climbing Fiber in the Electron Microscope

As the actual articulations of the climbing fiber with the various postsynaptic elements cannot be visualized in Golgi preparations by optical microscopy, it is necessary to turn to electron microscopy. When studying electron micrographs, however, it must be remembered that climbing fibers are "not always parallel to the Purkinje cell dendrite, but on the contrary they often coil in spirals, or better, they make elaborate zigzags, sometimes in front, sometimes behind the dendrites"[23] (RAMÓN Y CAJAL and ILLERA, 1907). Furthermore, the fine tendril collaterals are not only contorted but are also very slender, so that the probability of capturing long stretches of them in a single thin section is very small. Therefore in electron micrographs one would expect to see, accompanying the large Purkinje cell dendrites, a succession of detached profiles like islands in an archipelago, and would only occasionally encounter the threads bridging between them. One would also expect to find completely isolated profiles at some distance from the main arborization, which should represent sections through the more distal varicosities of tendril collaterals.

As it courses through the granular and Purkinje cell layers, the main stem of the climbing fiber is recognizable as a slender, thinly myelinated fiber, about 1 or 2 µm across (Figs. 212 and 213). It might be confused with the myelinated ascending stems of granule cell axons, which are, however, somewhat thinner. But these are uncommon in the rat, unlike the cat (MUGNAINI, 1972). The main characteristic of the myelinated portion of the climbing fiber is its large quantity of microtubules, which distinguishes it from Purkinje cell axons and mossy fibers, both of which are also thicker.

23 «Un de ces détails consiste en ce que le trajet de ces fibres n'est pas toujours parallel aux dendrites de Purkinje, mais au contraire, elles s'enroulent souvent en spiral ou, encore mieux, elles font des zig-zags à grand developpement; les tours de spirale se trouvent les uns devant, les autres derrière les tiges Purkinje» (RAMÓN Y CAJAL and ILLERA, 1907, p. 13).

a) The Terminal Arborization in the Molecular Layer

In the molecular layer the climbing fiber is displayed to best advantage in thin sections that graze the surface of the large Purkinje cell dendrites. The fiber now released from its myelin sheath is characterized by its smooth contours and its large complement of microtubules running longitudinally. The mitochondria are long and slender, the endoplasmic reticulum is tubular and sparse, and synaptic vesicles are absent. The parent fiber, from 1 to 2 µm in diameter, emits very fine branches, about 0.1 µm in diameter, which lead to a string of globular or bulbous varicosities filled with spherical synaptic vesicles. As indicated, these tendrils cannot be followed for any great distance in thin sections. Their train is marked out by a series of circular or elliptical profiles jammed with round synaptic vesicles and occasionally containing a small mitochondrion. The microtubules, endoplasmic reticulum, and mitochondria pass through the axis of the

→

Fig. 206. Developing climbing fibers, mossy fibers, and granule cells in the ninth postnatal day. The various elements in this illustration have been selected to show the characteristics of mossy fibers and climbing fibers at one stage in the development of the cerebellar cortex. The pial surface (*pia*), the limits of the external granular layer (*ext gr*), Purkinje cell layer (*PC*), internal granular layer (*int gr*), and white matter (*wm*) are all indicated. Beginning from the left, two immature granule cells (*gr c*) with their shaggy dendritic processes are shown in their descent into the internal granular layer, with their axons trailing. Two mossy fibers are shown (*MF*) reaching into the granular layer from the white matter. These mossy fibers are quite distinct. They have highly irregular, contorted filopodia projecting in bouquets from the growing tips. A few nearly mature rosettes are also present. Developing climbing fibers. The climbing fiber on the far right (CF_1) already has many of the distinguishing characteristics of the adult climbing fiber. It leaves the white matter (*wm*) and courses through the granular layer. From the main stalk several thin, beaded collaterals are emitted, as well as a few thicker, more varicose ones. At the level of the Purkinje cell body, these thicker collaterals branch profusely to form a *nid*, or nest (1), of beaded terminals around the cell body. A second climbing fiber (CF_2) emerges from the white matter, branches in the granular layer and also forms a *nid* or nest (2) around another Purkinje cell body. The third climbing fiber (CF_3) pursues a far more complicated course. The fiber leaves the white matter, courses through the granular layer. In the upper granular layer it divides into several thick branches that travel horizontally under the layer of Purkinje cells. The main branch forms a nest (3) around one Purkinje cell. Other branches of the same climbing fiber reach into and contribute to the nest (2) of the Purkinje cell situated about 125 µm away and already supplied by the terminals of CF_2. Still other branches of CF_3 course for 200 µm to the right, at the level of the Purkinje cell layer, and contribute small branches to three or four nests along the way. These branches may be only exploratory collaterals that are later resorbed by the axon. Certainly, evidence of persistent climbing fiber collaterals in the granular layer resembling those emitted by CF_1 and CF_2 has been documented in the adult rat (see text). Rapid Golgi, sagittal, 120 µm, rat cerebellar cortex, 9th day postnatal. Composite camera lucida drawing

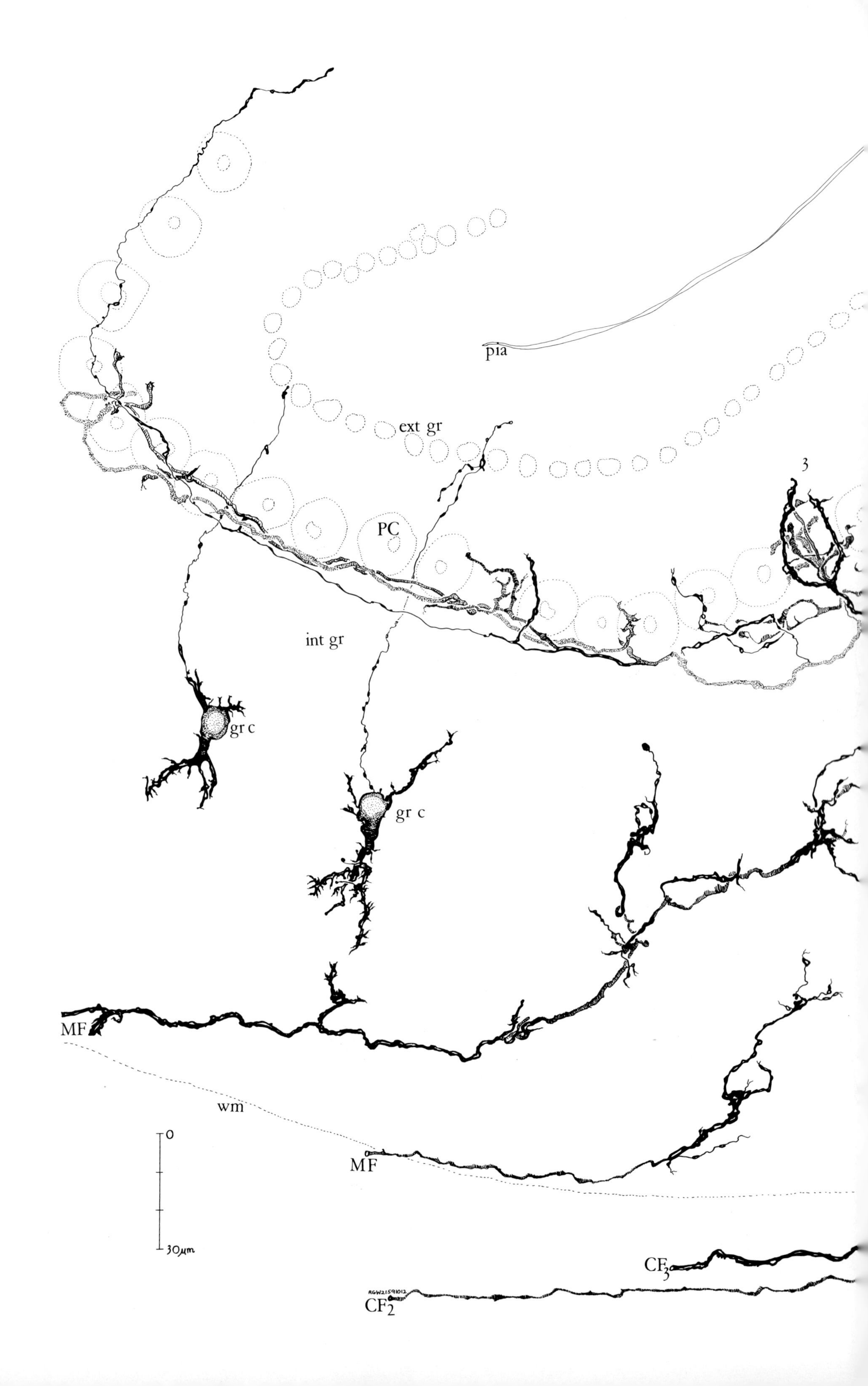

pia
ext gr
3
PC
int gr
gr c
gr c
MF
wm
0
30µm
MF
CF3
RGW21591012
CF2

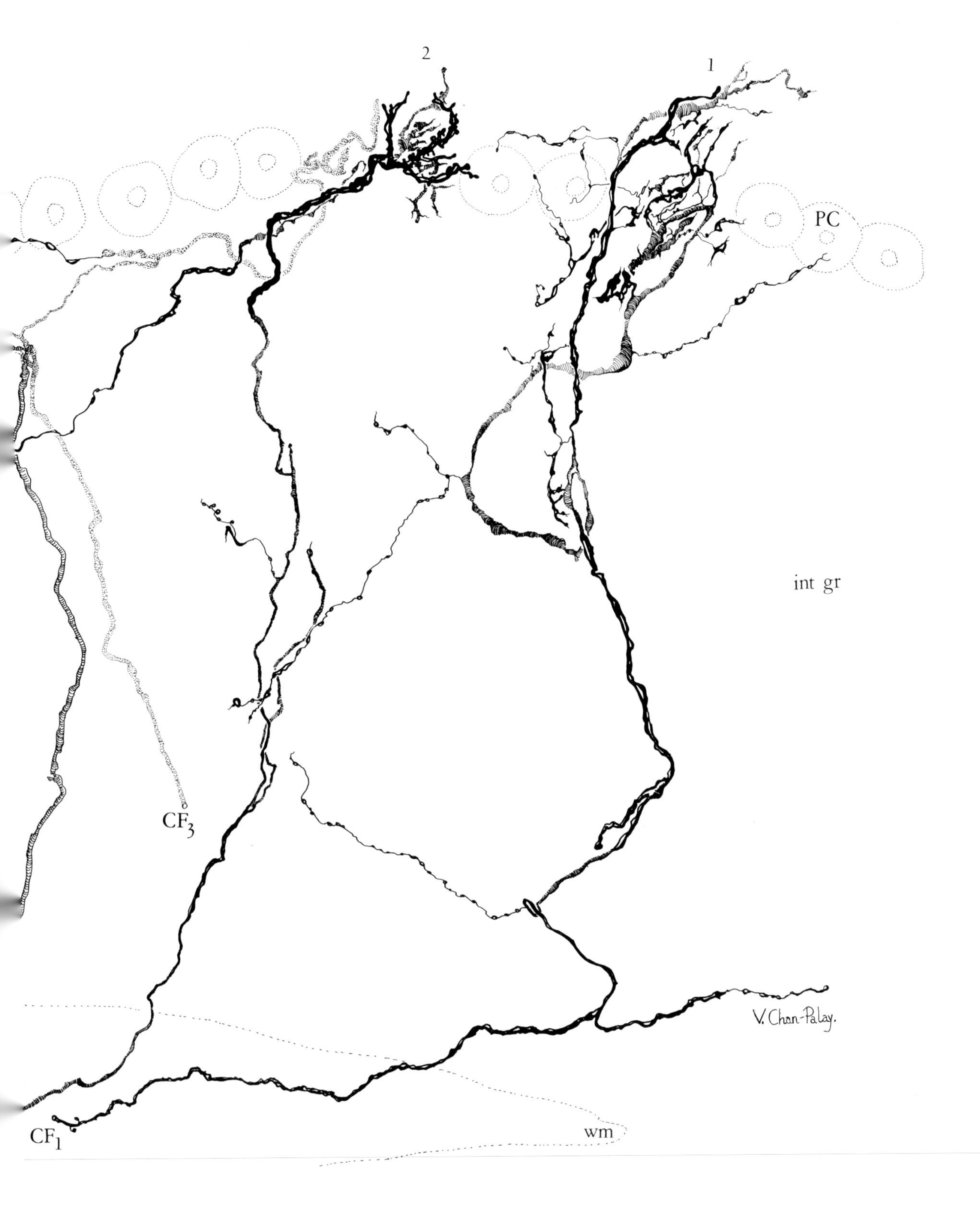
pia
ext gr
2
1
PC
int gr
CF_3
V. Chan-Palay.
CF_1
wm

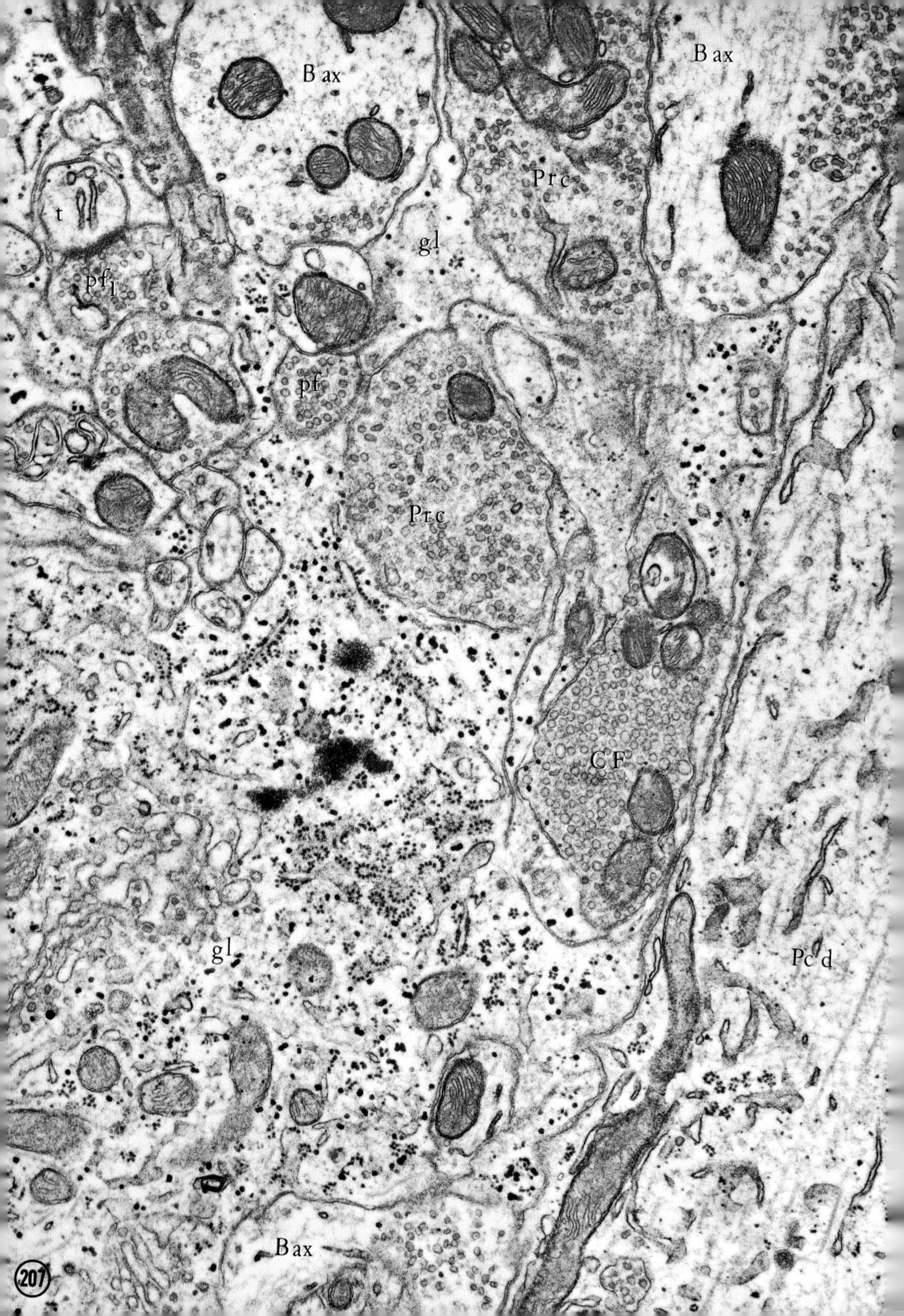
B ax
B ax
t
Pr c
gl
pf1
pf
Pr c
CF
gl
Pc d
B ax
207

tendril, while the synaptic vesicles aggregate to one side or another in the varicosities.

The profiles of synaptic vesicles in the climbing fiber varicosities are all circular, ranging from 440 to 590 Å in diameter. Occasionally one or two larger vesicles, about 1 000 Å in diameter, are encountered, and they are filled with a dense central granule. The most characteristic feature of the climbing fiber varicosities is the dense packing of the synaptic vesicles. In micrographs of the axoplasm close to the synaptic interface, there are at least four times as many profiles of synaptic vesicles per unit area as there are in a similar region in basket fibers (Fig. 207). This characteristic feature serves to identify climbing fiber synapses, and it will be re-emphasized in the section dealing with the connections of the climbing fiber in the granular layer.

Each tendril varicosity is in contact with from one to six short, stubby thorns projecting from the perikaryon or major dendrites of the Purkinje cell. This is the part of the cell and dendritic tree that had formerly been considered smooth. As has already been described (Chap. II), these thorns differ in a number of ways from those found on the spiny branchlets in contact with parallel fibers (see Fig. 23). They are shorter than those on spiny branchlets and are more clavate in form, having a shorter, more sharply tapering stalk. The thorns contain a few cisternae and tubules of the endoplasmic reticulum suspended in a finely filamentous matrix. As can be seen in Figs. 208 to 210, the thorns tend to be arranged in groups encircling the varicosities in the climbing fiber and impressing them with either shallow or deep, rounded indentations. The synaptic interface is usually on the side of the head of the thorn, not on its tip. The synaptic cleft is a widened interstitial space (240 Å compared with the usual 125 Å), and the apposed surface membranes are accentuated by dense, filamentous material adherent to their cytoplasmic faces, slightly more on the postsynaptic side. These synapses are therefore similar but not identical to those between parallel fibers and Purkinje cell dendritic thorns (Fig. 210). They are closest to Gray's type 1 junctions, but the asymmetry of that type is not so well displayed by climbing fiber synapses as by parallel fiber synapses. Rarely the climbing fiber varicosity articulates not only with one or more thorns but also with the shaft of the dendrite. This contact is more like Gray's type 2 junction, with straight, parallel apposed membranes, almost symmetrical densities, and a shallow synaptic cleft.

Large varicosities with the contours to be expected from climbing fiber tendril terminals can be found beside the Purkinje cell dendrites in material prepared by the freeze-fracture technique. Two such terminals display the A faces of their surface membranes in Fig. 210 A. The synaptic junction with a Purkinje cell dendritic thorn is signalled in this kind of terminal by a fairly deep depression of circular outline on the A surface. The floor of the depression is marked with small, shallow pits among which clumps of A face particles are strewn without noticeable pattern. The structure of the presynaptic membrane does not seem to differ essentially from that of the parallel fiber and mossy fiber examined by the same method. The depth of the depression in the presynaptic A face corresponds to the wide synaptic cleft seen in conventional thin section electron microscopy.

The neuroglial sheath surrounding the Purkinje cell dendrite also partially ensheathes the climbing fiber and its varicosities, thus separating them from adjacent fibers but also to some extent from the dendrite. This neuroglial sheath is undoubtedly the "granular layer" described by Ramón y Cajal (1890 b and 1911) as a conductile material lying between the nerve fibers on the hand and the perikaryon and main stem dendrites of the Purkinje cell on the other.[24] In micrographs of sections grazing the surface

◄

Fig. 207. Terminal boutons of axons in the molecular layer. Two profiles of recurrent collaterals (*Prc*) can be recognized by their dark axoplasmic matrix and their population of flat, elliptical, and round synaptic vesicles. In this same field, axonal profiles belonging to basket fibers (*B ax*), a climbing fiber (*CF*), and parallel fibers (*pf*) can be seen in the vicinity of a Purkinje cell dendrite (*Pcd*). A parallel fiber varicosity synapses with a Purkinje cell thorn (*t*). Extensive sheets of neuroglial cytoplasm (*gl*) containing glycogen are insinuated between the axons and dendrites in the field. Lobule X, flocculus. × 34 000

24 «Quant à la manière dont la connexion s'établit, nous pensons qu'elle se réalise par des contacts multipliés et souvent rendus plus intimes au moyen d'entrelacements et de véritables engrenages. Les arborisations grimpantes du cervelet viennent particulièrement à l'appui de cette manière de voir. Peut-être, comme incline à admettre His, il existe aussi entre les parties nerveuses en contact une matière conductrice comparable à la substance granuleuse des plaques motrices. Nous croyons avoir aperçu quelque chose de pareil autour du corps et de la tige ascendante des cellules de Purkinje. Il s'agit d'une couche granuleuse que se colore en brun ou en jaune par le chromate d'argent, restant indépendante des cellules et des fibres. Cette couche constitue une bourse très inégale montrant des lignes et des impressions dues probablement à la présence des pinceaux descendants» (Ramón y Cajal, 1890 b, p. 466). «La *substance granuleuse* entoure le tronc et les grosses branches protoplasmiques des cellules de Purkinje; elle englobe par conséquent l'arborisation grimpante et la soude pour ainsi dire aux dendrites. Lorsque ce ciment s'imprègne seul, on constate, ... qu'il se continue avec l'enveloppe cimentaire du corps des cellules de Purkinje et qu'il est sillonné par des stries longitudinales; ces dernières sont vraisemblablement les empreintes de l'arborisation grimpante. Mais bien souvent arborisation et ciment sont imprégnés à la fois; il est alors difficile de distinguer les fibrilles grimpantes immergées dans une masse granuleuse marron clair» (Ramón y Cajal, 1911, p. 65).

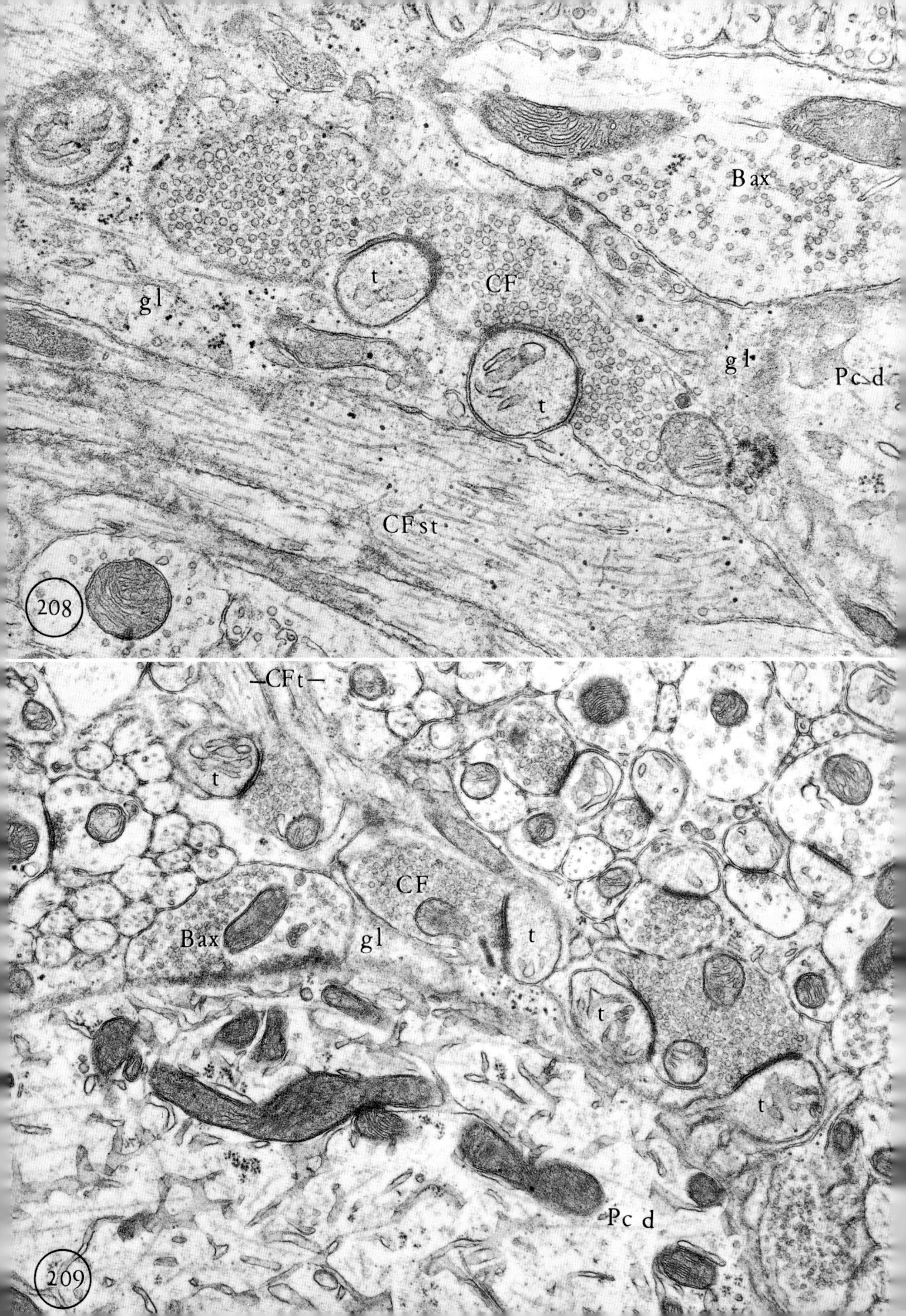

B ax
t
CF
g l
g l
Pc d
t
CF st
208
-CF t-
t
CF
Bax
g l
t
t
t
Pc d
209

of the dendrite and passing through the neuroglial sheath, the necks of the thorns can be seen poking through in order to present their synaptic faces to the overlying varicosity of the climbing fiber. The varicosity is, so to speak, grasped by the fingers provided by the cluster of thorns and held away from the shaft of the dendrite, which is enshrouded in neuroglial processes (Fig. 211). Thus, the main stalk or a tendril of the climbing fiber is suspended like a vine from a succession of fixed points marked by the varicosities in its course and the clustered thorns on the dendrite. In many cases, however, the belly of the varicosity rests directly upon the shaft of the dendrite without intervening neuroglia and without any synaptic membrane specializations in the region of the apposition. The scarcity of synaptic contacts between climbing fiber varicosities and the shafts of Purkinje cell dendrites emphasizes the prevailing rule that climbing fibers synapse with thorns.

The Functional Significance of the Climbing Fiber Arborization. In concluding this description of the climbing fiber in the molecular layer, the form of its terminal arborization deserves special emphasis. When RAMÓN Y CAJAL compared it to the terminal ramifications of a motor nerve on a striated muscle cell, he already recognized that this extensive synaptic apparatus should have an important activating effect upon the Purkinje cell. That intuition has been borne out by modern electrophysiological experiments (ECCLES, LLINÁS, and SASAKI, 1966d; ECCLES, ITO, and SZENTÁGOTHAI, 1967). ECCLES, ITO, and SZENTÁGOTHAI (1967) have remarked on the strong correlation between "the powerful all-or-nothing synaptic activation of Purkinje cells and the classical histological findings of the very extensive synaptic contact that each Purkinje cell receives from a single climbing fiber." Modern histological and electron microscopic observations also support this correlation (LARRAMENDI and VICTOR, 1967; CHAN-PALAY and PALAY, 1970). It must be pointed out, however, that it is not merely the extent of the synaptic contact between the climbing fiber and the Purkinje cell dendritic tree that requires notice, but also its disposition and its detailed structure, for these may have interesting physiological correlates, too. After all, only a small part of the putative contacting surface is concerned with synaptic transmission—the articulation between the varicosities and the special dendritic thorns. As Figs. 203–205 show, the main stem of the climbing fiber subdivides only a few times as it follows the Purkinje cell dendrites, and each subdivision gives off many fine, beaded tendrils that twist and turn round the same dendritic branch as their parent fibers. Nearly all of the synaptic contacts made by climbing fibers are formed on the surfaces of the globose varicosities belonging to these fine tendrils. This plexiform mode of arborization can be abstracted to a long, thick, fast-conducting axial fiber with many short, thin, slow-conducting offsets, which make practically all of the terminal connections. This form appears favorable for the rapid conduction of an impulse over a long distance with almost simultaneous dispersal to multiple sites of transmission. The small caliber of the lateral offshoots or tendrils with their relatively slow rate of conduction would ensure that the activation of synaptic junctions is nearly simultaneous all over the arborization. Since a single Purkinje cell receives nearly the whole of this synaptic input, the arrangement of the terminal climbing fiber arborization appears to be a device for overwhelming the entire dendritic tree and thus for achieving very high efficiency of activation. These characteristics agree with the properties of the climbing fiber-Purkinje cell system as described in electrophysiological experiments (ECCLES, ITO, and SZENTÁGOTHAI, 1967).

Not less remarkable than the overall form of the terminal arborization is the structure of the articulation between the climbing fiber and the major dendrites of the Purkinje cells. As the figures show, the delicate tendrils

◄

Fig. 208. Stem and tendril of a climbing fiber with a basket cell axon on the surface of a Purkinje cell dendrite. A climbing fiber (stem and synaptic varicosity) and a basket axon are seen running parallel to each other on the surface of a Purkinje cell dendrite (*Pc d*), which comes into the plane of section at the right of the figure. The basket axon (*B ax*) contains neurofilaments and a loose cluster of more or less round synaptic vesicles (diameter 400–540 Å). The climbing fiber stem (*CF st*) is recognized by its diameter (2 μm), and numerous microtubules. An elongated varicosity of the climbing fiber tendril (*CF*) with its dense population of round synaptic vesicles (440–590 Å) lies in the center of the field and synapses with two Purkinje cell dendritic thorns (*t*). The neuroglial sheath (*gl*) of the Purkinje cell dendrite envelops both the thorns and the climbing fiber and sends delicate slips around the basket axon. Lobule IV. ×39000

Fig. 209. The climbing fiber tendril with synapses on thorns of the Purkinje cell dendrite. The connecting thread (*CF t*) of a climbing fiber tendril enters this field from the upper left. This fiber is distinguished by its small diameter (0.75 μm) and abundant microtubules. Three profiles of varicosities (*CF*) of the climbing fiber are found in the immediate vicinity, all partially surrounded by slips of neuroglia (*gl*), which are extensions from the sheath around the Purkinje cell dendrite (*Pc d*). These varicosities show the typical dense populations of round synaptic vesicles and synaptic junctions with thorns (*t*) of the Purkinje cell dendrite. One of these thorns is seen in perfect profile; it has a short neck with a clavate head bearing a synapse on one side. Other profiles of thorns in this figure do not include the narrow necks; each thorn contains a few cisterns or tubules of endoplasmic reticulum embedded in a fine filamentous matrix. In contrast, a basket cell axon (*B ax*), distinguished by its loose population of synaptic vesicles, synapses on the shaft of this dendrite. Lobule IV. ×27000

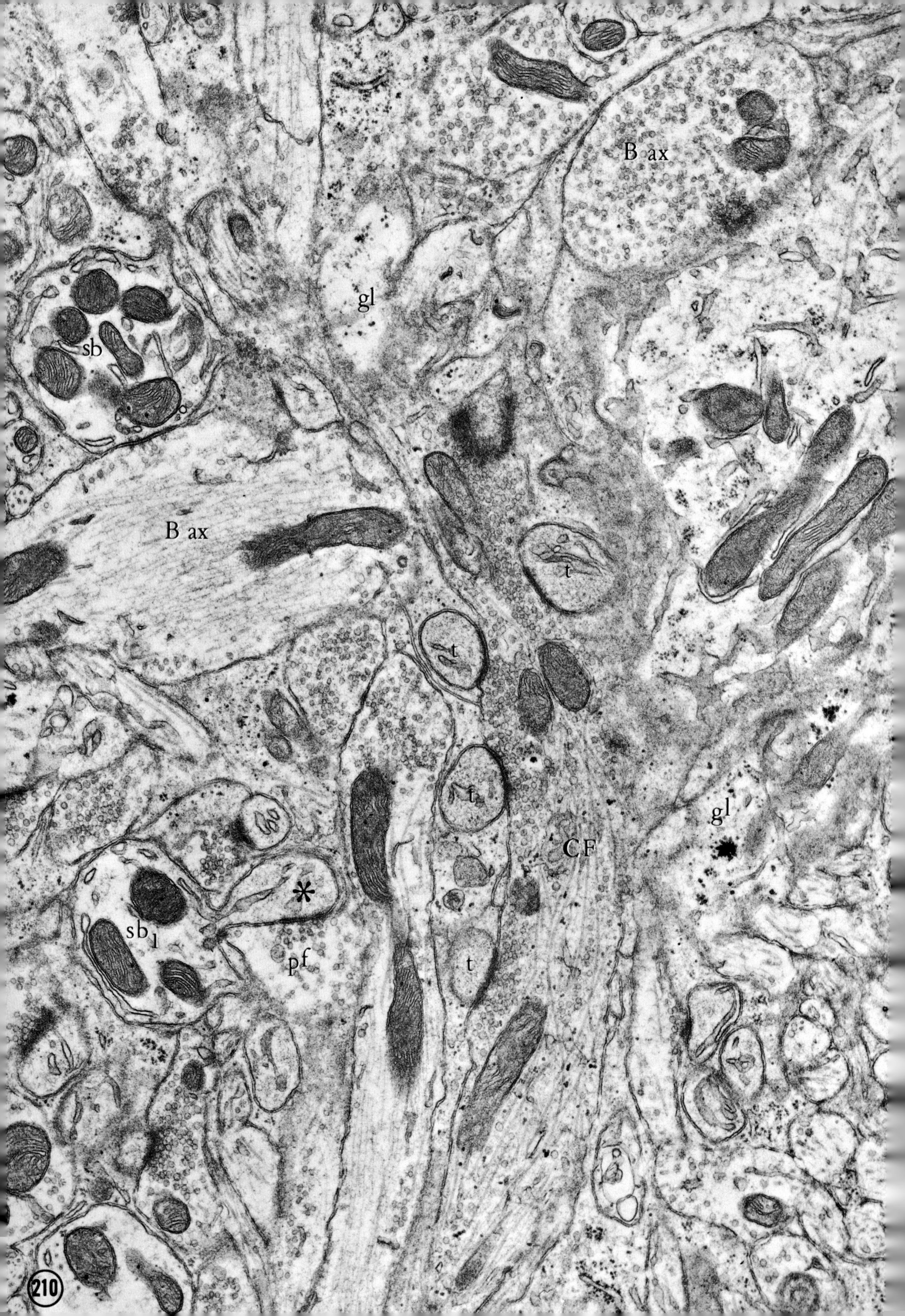
B ax
gl
sb
B ax
t
t
t
gl
CF
*
sb1
pf
t
210

Fig. 210A. Two terminals of climbing fiber tendril collaterals. Each terminal is represented by the A face of a large varicosity, recessed to receive the head of a stubby dendritic thorn. The depths of the recesses are cobbly and pitted, and particles are arranged in ill-defined clusters. A Purkinje cell dendrite appears in the lower margin of the picture. The large arrow indicates the direction of shadowing. Replica, freeze fracture. × 79000

twining round the dendrites here and there expand into varicosities of various sizes, each of which is grasped by a cluster of thorns projecting from the dendrite. Although most of the deep surface of the varicosity can be applied to the shaft of the dendrite and the stems of the thorns, and although the whole varicosity can be jammed with synaptic vesicles, the junctional interface displays synaptic specializations only at the sides of thorns not at their extremities (Figs. 208–210). The basket fiber and its collaterals also coil about the dendrite and dip down through the neuroglial sheath to contact the shaft of the dendrite. But in contrast to the climbing fiber, they make plane synaptic contacts on the surface of the shaft without the intermediation of thorns. The juxtaposition of these two different styles of articulation on the same postsynaptic surface—the one on thorns and the other on the shaft—and their known opposite physiological effects indicate that some fundamental morphological correlate of synaptic action is illustrated here. Moreover, if we compare the two kinds of excitatory synapse on thorns in the Purkinje cell dendritic tree, we see that each varicosity in a climbing fiber usually articulates with several thorns, all arising from the same dendritic surface, whereas each varicosity in a parallel fiber, with comparable synapses on more distal dendrites, usually contacts only one or two thorns. Although examples of parallel fiber varicosities in contact with two or even three thorns are not rare (Figs. 83–85; see also GRAY, 1961; GOBEL, 1967; UCHIZONO, 1969), they may be presumed to project from different branches of the dendritic tree, if not from different Purkinje cells.

Fig. 210. The climbing fiber in the molecular layer. A climbing fiber varicosity (*CF*) courses through the center of this field. Many microtubules and round synaptic vesicles are present in the axoplasm. Four dendritic thorns (*t*) of a Purkinje cell are in synaptic contact with its surface. Two spiny branchlets of Purkinje cell dendrites (*sb* and sb_1) are present in the field. One of these (sb_1) has a thorn (*t*) which synapses with a parallel fiber (*pf*). Profiles of basket axons (*B ax*) and neuroglial cytoplasm containing glycogen (*gl*) are indicated. Lobule V. × 27000

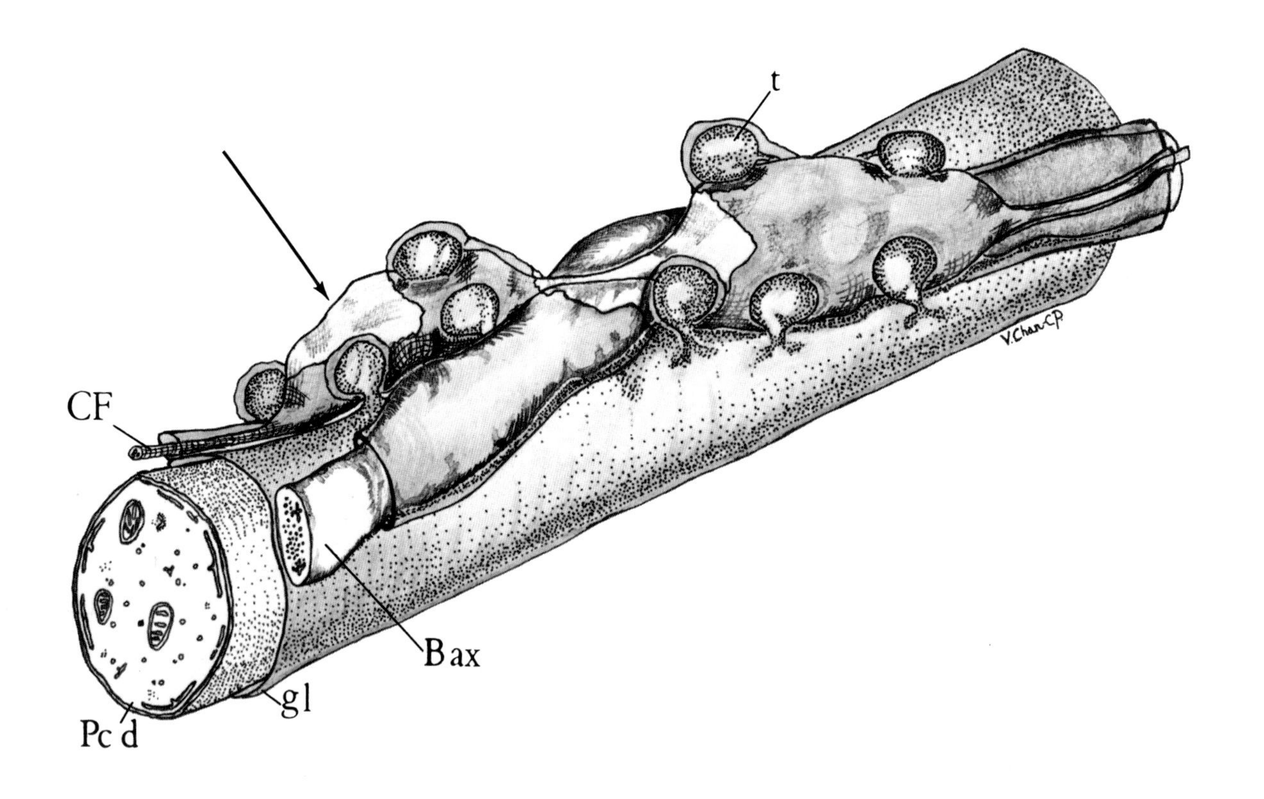

Fig. 211. Diagram of a climbing fiber tendril and a basket axon on a Purkinje cell dendrite. This drawing illustrates a segment of a Purkinje cell dendrite (*Pc d*) on the surface of which portions of a climbing fiber tendril (*CF*) and an ascending basket axon collateral (*B ax*) are entwined. The tendril of the climbing fiber has two synaptic varicosities connected by a fine thread. The surface of each varicosity is impressed by a cluster of dendritic thorns (*t*). The dendritic shaft is covered by a veil of neuroglia (*gl*), and thorns emerging from it are also ensheathed, like fingers in a glove, on all surfaces except at the synapse. The basket axon is a smooth fiber with gentle distentions where synapses occur *en passant* with the shaft of the Purkinje cell dendrite. This axon may be covered by a sleeve of neuroglia extending from the same sheath that covers the dendrite. The climbing fiber tendril and basket axon weave along, first together then apart, on the surface of the Purkinje cell dendrite. Although neuroglia covers the dendrite and the thorns, and may extend over part of the basket axon, it may also be incomplete, leaving other surfaces of the climbing fiber and the basket axon bare (*arrow*).

The Advantages of Thorn Synapses. What might be the functional significance of synapses on thorns? At first sight it might appear that since the junctional interface is greatly increased by the distension of the axon to form a varicosity on the one hand and by the thorns protruding into it from the dendrite on the other, the result would be a great increase in the area of apposed surface membrane that is available for synaptic transmission of the signal. But the areas of structural specialization, the so-called "active zones," occupy only a small proportion of the junctional interface. Furthermore, most of the dendritic surface is not used for synaptic junctions of any kind, but is covered with neuroglia. Thus, it seems that the advantage of increased surface area is not put to use. Therefore, thorns must have some specific function.

Another attractive possibility is that the thorns are simply small extensions of the dendritic surface, thrust out in order to capture the appropriate afferents among the thousands of fibers coursing through the dendritic tree (see also PETERS and KAISERMAN-ABRAMOF, 1970). According to this idea, thorns would be one of the devices for selecting out of a crowded environment the precise array of connections that is characteristic of a particular nerve cell. The ontogeny of the Purkinje cell, indeed, suggests some such idea, although the operation of this principle would not explain why thorns persist into maturity in connection with climbing fibers and not with basket fibers (LARRAMENDI and VICTOR, 1967). Even if this simple notion is correct to the extent that thorns enable dendrites to make selective synaptic contacts with axons, it is not sufficient to explain the other features of thorns.

Recently several more elaborate but related hypotheses have been put forward. These take into consideration not only the location of the thorns on the dendritic tree, but also the electrical properties implicit in their form. The SCHEIBELS (1968), in a summary of a conference on dendritic thorns, ascribed an integrative role to them. They were impressed by the pairing of presumed inhibitory and excitatory terminals on the thorns of neocortical pyramidal cells (COLONNIER, 1968). They suggested that

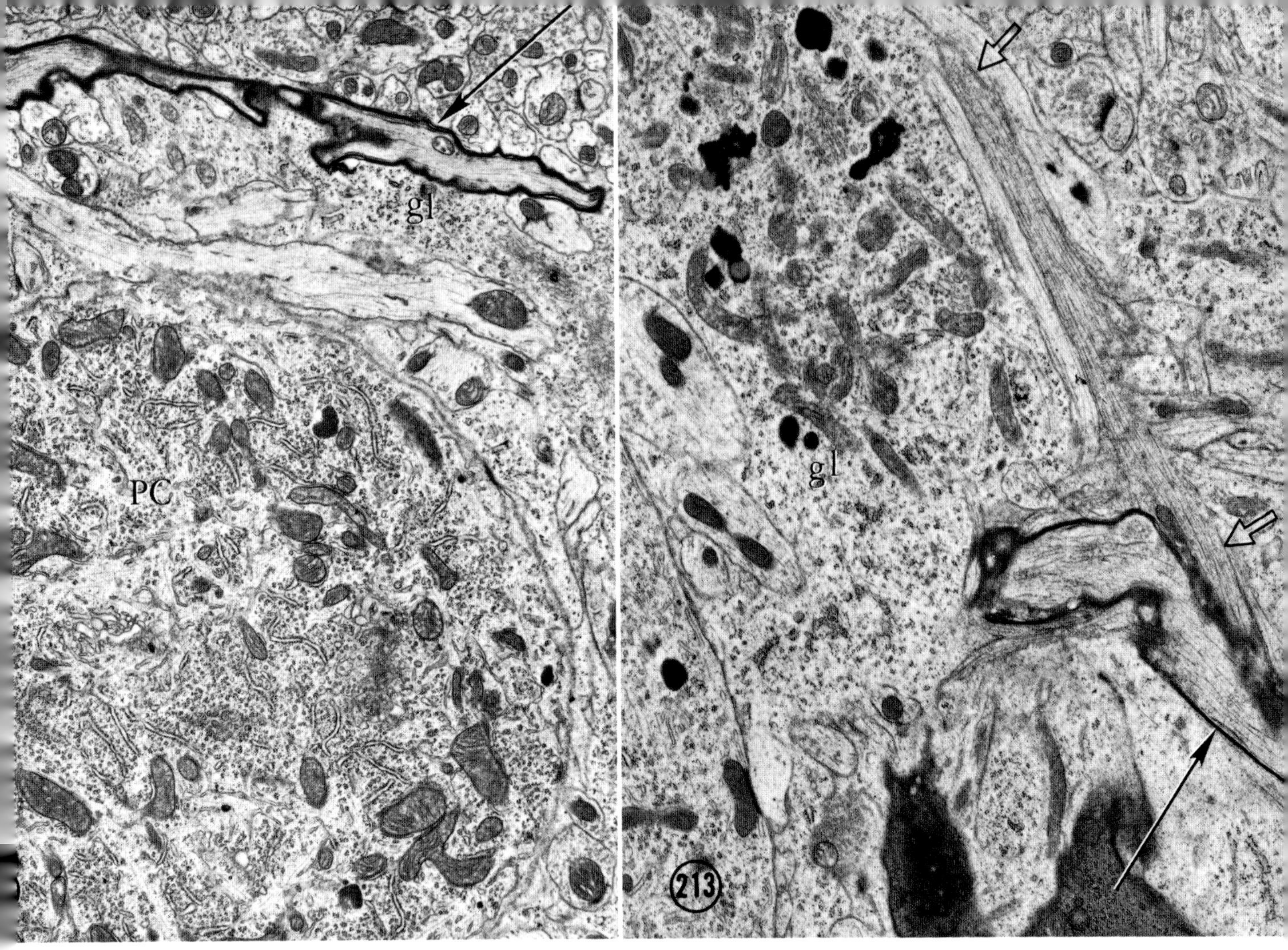

Fig. 212 and 213. The climbing fiber stalk; myelinated and unmyelinated portions. Portions of the climbing fiber stalk can be seen in sections of cortex at the level of the Purkinje cell bodies (*PC*). These axons are usually filled with microtubules in a dark axoplasmic matrix, and may be myelinated (*large arrows*) or unmyelinated (*small arrows*). Processes of neuroglial cytoplasm (*gl*) occupy some of the surrounding neuropil. The stalk of the climbing fiber is distinctive and can be easily distinguished from the myelinated axons of Purkinje cells. See Figs. 50 and 157. Fig. 212, lobule III. ×9000; Fig. 213, lobule III. ×15000

thorns are sites for complex suppressive modulations of postsynaptic potentials in the dendrite. This idea also receives some support from the work of JONES and POWELL (1969), who observed that in the small number of neocortical dendritic thorns beset with two afferent terminals, the presumably excitatory terminal was always seated on the tip of the thorn, while the presumably inhibitory one was attached to the stem. In this position the inhibitory terminal might be capable, on appropriate timing of its discharges, of suppressing or at least of interacting with the postsynaptic potentials elicited by the more distal excitatory terminal.

A somewhat simpler idea has been presented by DIAMOND, GRAY, and YASARGIL (1969, 1970) in their speculations on the articulation between the Mauthner cell axon and spinal motor neurons in teleost fishes. They proposed that "the dendritic spine provides a postsynaptic region which is effectively isolated from other synapses in the neurone, in such a way that the immediate and long-term effects of pre-synaptic activity at the spine occur with little or no interference from synaptic activity generated elsewhere in the cell. The activity generated in the spine, when transmitted to the parent dendrite, though attenuated, can summate there with other responses and so contribute to dendritic integration." This important suggestion deserves very serious consideration by neurophysiologists concerned with rationalizing the functional architecture of the neuron. LLINÁS and HILLMAN (1969) have published a related hypothesis in which they regard the spine as a current-injecting device. An extreme instance where this isolating function would be expected to come into play is given by WESTRUM (1970), who found spines projecting from the initial segments of neocortical pyramidal cells. These spines were contacted by one or two terminals of presumably inhibitory nature. In view of the depolarizing events believed to occur in the initial segment, the isolation of such a strategically placed inhibitory input would seem most desirable.

If these notions have any general relevance, they would

have to be incorporated in the dendritic tree of the Purkinje cell, which has stood for decades as the prototype of thorn-encrusted dendrites. They would have to explain the operation of both the individualized thorn synapses with parallel fibers and the clustered thorn synapses with climbing fibers. These aspects were examined by CHAN-PALAY and PALAY (1970), who attempted to derive a comprehensive theory of the operation of thorns, which, although arrived at independently, is based upon considerations similar to those of DIAMOND *et al.* (1969, 1970) and LLINÁS and HILLMAN (1969). The following account is taken from their paper, with slight modifications.

It is rather striking that as a general rule inhibitory inputs to the Purkinje cell dendritic tree end upon the smooth shafts of the dendrites, while excitatory inputs end upon the thorns. A few exceptions to this rule have been recorded—terminals of stellate or basket cell axons synapsing with thorns (MUGNAINI, 1969, 1970, 1972, and in our own material)—but they are decidedly rare. One might infer from this differential distribution that the role of the thorns is to segregate functionally different parts of the postsynaptic membrane. Thus not only would the synapses of specific afferent fibers be distributed in a precise pattern over the surfaces of the dendrites, but the synaptic loci of opposite sign would also be isolated from each other by virtue of the electrical properties of thorns relative to the dendritic shaft. The narrow neck of the thorn can be expected to have a high longitudinal resistance which would tend to damp any fluctuations in current spreading from the dendritic shaft. By the same token, only a small amount of current could be introduced into the dendrite by synaptic action on a thorn (LLINÁS and HILLMAN, 1969). In this light, the thorn may be viewed as a device for evoking in the dendrite only a weak postsynaptic potential from the activation of a single afferent fiber. Synapses on thorns seem to be ideally designed for a system that involves vast numbers of similar convergent fibers, whose excitation evokes a fluctuating background of depolarization in the postsynaptic cell. In such a system the individual fibers can be expected to count for very little. The form of the synaptic contact, therefore, is one way of regulating and weighting the effectiveness of a synapse. For example, in the Purkinje cell-parallel fiber system the form of the thorn synapses would tend to restrict all of the parallel fibers to a narrow range of potency.

What then could be the value of using this restrictive device in a system like the climbing fiber synapse, which has a high degree of effectiveness? Here the thorns are grouped together, and each cluster projects into a varicosity linked by short lengths of slender axon to other similar varicosities. As the varicosities are packed with synaptic vesicles, the excitation of each one in rapid sequence discharges a great deal of transmitter into the synaptic cleft and activates all of the attached thorns simultaneously. Since the thorns are clustered together, the small amount of current injected into the dendrite from each thorn would produce a small postsynaptic potential that would summate first with the small postsynaptic potentials arising from the other thorns in the cluster. Spreading electrotonically, this potential would then summate with others arising from similar thorn clusters that were activated at almost the same moment along the path of the climbing fiber. Articulating a varicosity so that it synapses on a cluster of thorns thus could restore the synapse to the potency it would otherwise have if it were located simply on the dendritic shaft. But this result would be achieved without losing the great advantage inherent in the geometry of the thorn—the relative isolation of the subsynaptic membrane from electrical events in the rest of the dendrite. Therefore, clustering would permit an escape from the principal disadvantage of the thorn—the attenuation of the conducted signal—and ensure a high rate of success for the synaptic action of the efferent fiber.

Recent electrophysiological work throws an interesting sidelight on the form of this synaptic junction. Repetitive activation of climbing fibers results in a diminution of their power to excite the Purkinje cell, and with frequencies of stimulation above 100 per second, the Purkinje cell fails to discharge (ECCLES, LLINÁS, SASAKI, and VOORHOEVE, 1966; extending earlier work of GRANIT and PHILLIPS, 1956). The failure of the Purkinje cell to follow the tetanic stimulation is not due to a deficiency in the excitatory postsynaptic mechanisms but to excessive depolarization resulting from summation of successive excitatory postsynaptic potentials and the suppression of the spike generating mechanism. When the tetanic stimulation of the climbing fiber is discontinued, the level of depolarization in the Purkinje cell slowly declines and the spike generating mechanism slowly recovers with progressively increasing amplitude and decreasing frequency of the spike potentials. This after-discharge is interpreted as indicating the continuing action of accumulated transmitter. "Evidently there can be a large accumulation of the excitatory transmitter at climbing fiber synapses, and its dissipation can be extremely slow" (ECCLES, ITO, and SZENTÁGOTHAI, 1967, p. 171).

These experimental results correlate closely with the functional morphology of the climbing fiber explained above. Each varicosity is jammed with synaptic vesicles all near one "active zone" or another. Even if each action

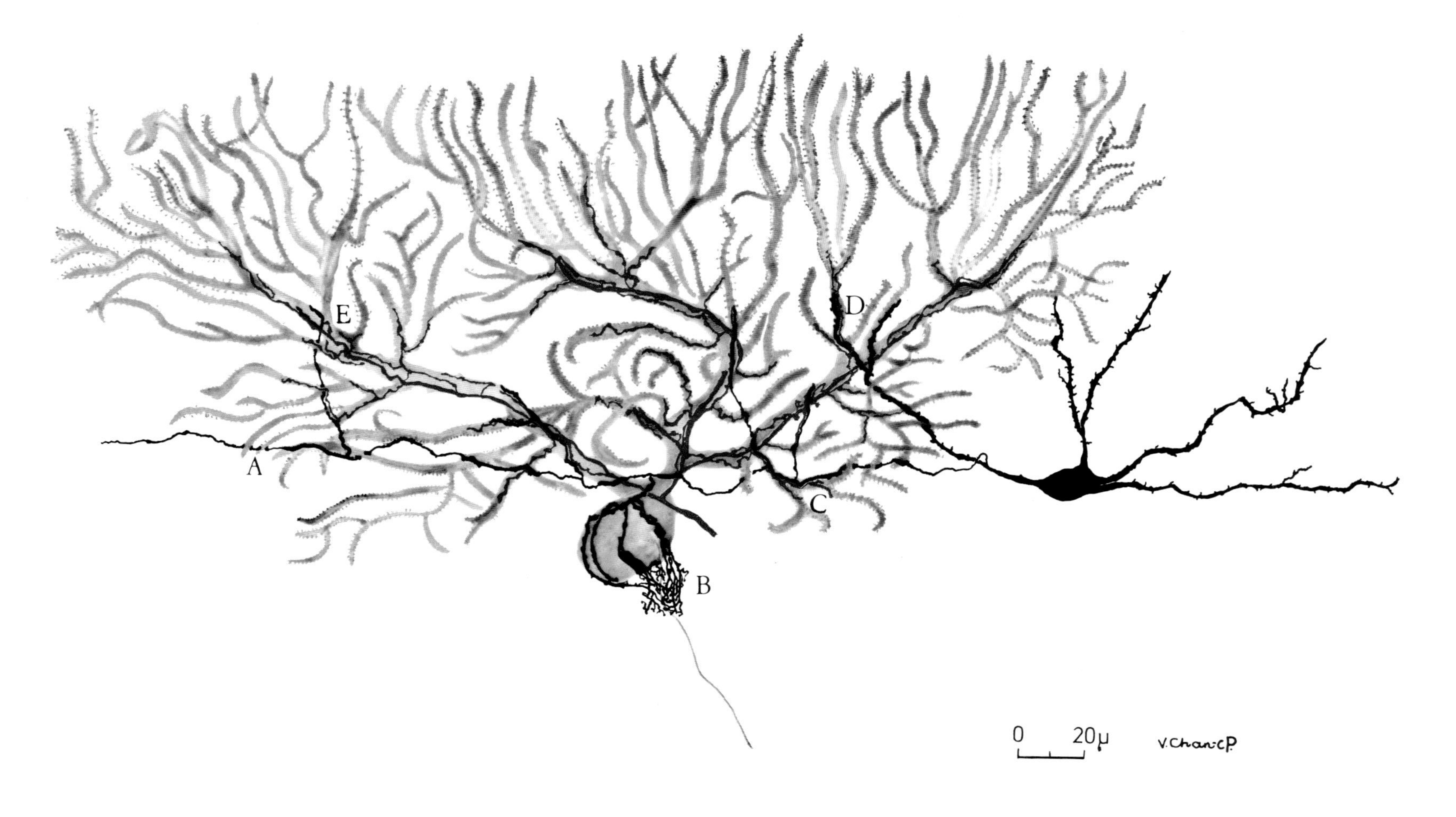

Fig. 214 A–E. Diagram showing the interrelations of basket cell, Purkinje cell, and climbing fiber. Camera lucida drawings of a basket cell (black), a Purkinje cell (gray), and a climbing fiber (red) from rapid Golgi preparations were superimposed to produce this diagram. The interrelations of the processes of these three cells have been observed in the electron micrographs, and are reproduced in this figure in order to provide the light microscope equivalent. **A** The horizontal axon of the basket cell issues from the cell body, and in its trajectory across the Purkinje cell territory weaves in and out among the branches of the dendritic tree. Long synaptic contacts are formed when the axon meets a dendrite at an acute angle, and girdle synapses are formed when it meets a dendrite at right angles. **B** Descending collaterals of the basket axon run down the Purkinje cell soma to form the distinctive basket around it and the initial segment. The stalk of the climbing fiber (red) enters the low molecular layer in the vicinity of the main dendrite of the Purkinje cell. Here it branches into two smooth stems which continue to ascend into the molecular layer on the Purkinje cell dendrites. From these stems fine tendrils bearing large varicosities are emitted. **C** Ascending collaterals of the basket cell axon are given off at various intervals along its length. These branches ascend the Purkinje cell dendritic tree on secondary and tertiary dendrites, entwined with tendrils of the climbing fiber. **D** Dendrites of the basket cell spray out from the cell body, and several of them come into close relation with tendrils of the climbing fiber on dendrites of the Purkinje cell. **E** At the bifurcations of secondary Purkinje cell dendrites, climbing fiber and basket axons gather to form an extensive plexus over the branch points. Rapid Golgi, 100 μm, camera lucida. × 500

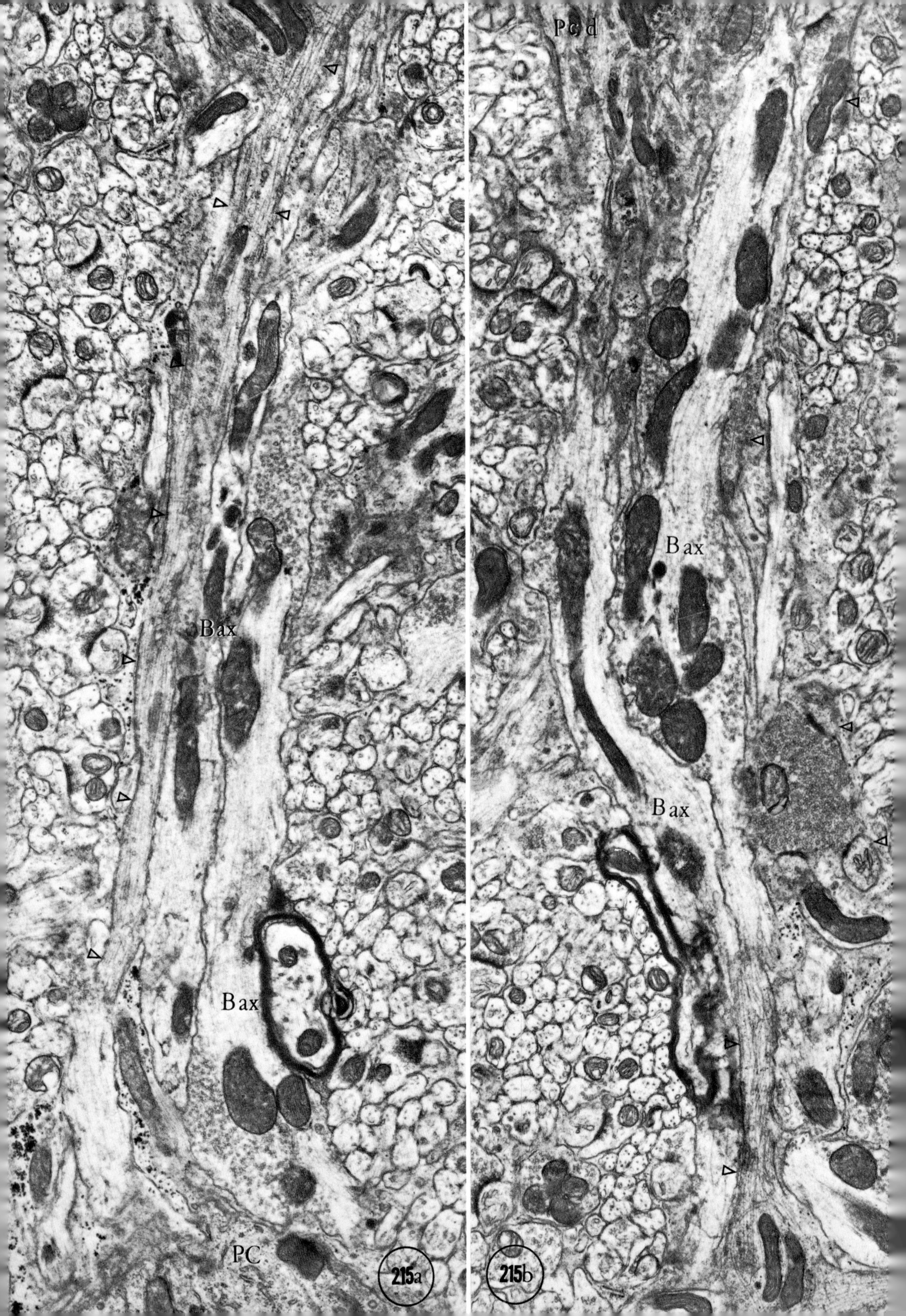
Pcd
Bax
Bax
Bax
Bax
PC
215a
215b

potential invading the varicosity releases only a small proportion of its vesicles, the amount of transmitter discharged from such a large ending is likely to be great. Recent electrophysiological studies on synaptic efficacy in the spinal cord very strongly suggest that the excitation of a large presynaptic terminal releases a larger number of quanta of transmitter than the excitation of a small terminal (EIDE *et al.*, 1967; KUNO and MIYAHARA, 1968, 1969). Repetitive stimulation of the climbing fiber at high frequencies would result in a considerable accumulation of transmitter in the narrow, convoluted synaptic cleft between the varicosity and the clustered thorns. Under these circumstances dissipation of the transmitter by uptake, diffusion, or enzymatic destruction would be too slow to keep up with the rate of secretion. Furthermore, the relatively small proportion of the interface that is occupied by the subsynaptic membrane would also contribute to the slow removal of the transmitter.

Thus, the terminal arborization of the climbing fiber incorporates in a single fiber a large number of morphological devices that could enhance the effectiveness of a synapse. As has already been remarked, it has an unusually extensive area of synaptic contact with its target, the Purkinje cell. The synaptic interface is, however, greatly fragmented and widely dispersed over the surface of the major dendrites. The difference in caliber between the thick main stem of the fiber and the fine preterminal tendrils, which effect virtually all of the synapses, ensures that all of the terminals are activated practically simultaneously, thus engendering a widespread, multicentric, postsynaptic potential in the dendrite. Many of the terminals are large varicosities, and all are replete with synaptic vesicles. These would discharge their content of transmitter into a narrow, complicated synaptic cleft where it would dissipate slowly. The terminals are strategically located upon thorns which, because they are marshalled into groups, provide for efficient transmission of the afferent impulse while segregating the receptive surface from electrical events in the rest of the dendrite. Probably no more successful example of integrated structure and function can be found in the central nervous system.

Relationships with Basket and Stellate Cells. In their 1954 paper, the Scheibels reported that collaterals of basket and stellate cell axons enter into contact with the climbing fiber by running along together or intertwining with it or even by forming typical end feet on it. They suggested that these contacts were axo-axonal synapses, "rather than axo-dendritic relationships to the underlying Purkinje cell dendrite" (p. 743). Since the proximity of axons in Golgi preparations can be exaggerated and since axo-axonal synapses had not yet been seen in electron micrographs of the molecular layer, we made a close study of the fibers running along the Purkinje cell dendtritic tree in order to re-examine the question (CHAN-PALAY and PALAY, 1970).

As Fig. 214 shows, the ascending collaterals of the basket cell axon encounter the climbing fiber and its plexiform collaterals as they run vertically through the Purkinje cell dendritic tree. They can accompany one another for short, or sometimes long distances, both climbing up the Purkinje cell dendrites and synapsing in their characteristic fashions with the shaft or the thorns respectively. In electron micrographs of sections that graze the surface of the principal dendrites but provide longitudinal profiles of the fibers clinging to them (Figs. 215 and 216), the basket axon and the climbing fiber can be seen running along together, weaving in and out, first on one side of the Purkinje cell dendrite, then on the other, forming an exuberant plexus over the surface of the dendrite and then parting. The two axons are sharply different in structure, and the contrast is very evident in micrographs where they appear together (Figs. 207, 215, and 216). The basket fiber is thick, with stout mitochondria and numerous neurofilaments, whereas the climbing fiber is thin, with slender mitochondria and abundant microtubules. Here and there the basket fiber dilates somewhat to accommodate loose collections of synaptic vesicles; the climbing fiber distends widely to form distinct vari-

Fig. 215. a The bifurcation of the main stalk of the climbing fiber on the primary Purkinje cell dendrite. The main stalk of the climbing fiber (Δ) is a smooth, thin (2 μm) axon containing longitudinally oriented microtubules and a few slender mitochondria. It ascends on the surface of the primary Purkinje cell dendrite that emerges from the cell body (*PC*) at the bottom of the figure but is out of the plane of section. The climbing fiber is accompanied by basket axons (*B ax*) distinguished by their abundance of neurofilaments and loose population of synaptic vesicles. The climbing fiber stalk bifurcates into left and right stems (Δ) at the point where the trunk of the Purkinje cell dendrite also divides into its secondary dendrites. × 15000. **b** The stem of the climbing fiber and the trail of a tendril on the Purkinje cell dendrite. This figure is a continuation of the field shown in Fig. 215a. The bifurcation of the main stalk of the climbing fiber (Δ) is repeated at the bottom of this illustration. The right stem of this axon leaves the field near the bottom right hand corner. The left stem of the climbing fiber continues upward and soon weaves out of the plane of section. The trail of the climbing fiber tendril (Δ) is indicated first by a varicosity densely packed with synaptic vesicles and in contact with three dendritic thorns, and then by profiles of two other varicosities in the middle and near the top of the figure. The last varicosity is connected to a thin thread which leaves the field at the upper right hand corner. This tendril is probably a branch of the left stem of the climbing fiber that runs on the Purkinje cell dendrite (*Pc d*) appearing at the top left. Ascending and descending collaterals of basket cell axons (*B ax*) complete the drape of fibers on the surface of this dendrite. Lobule V. × 15000

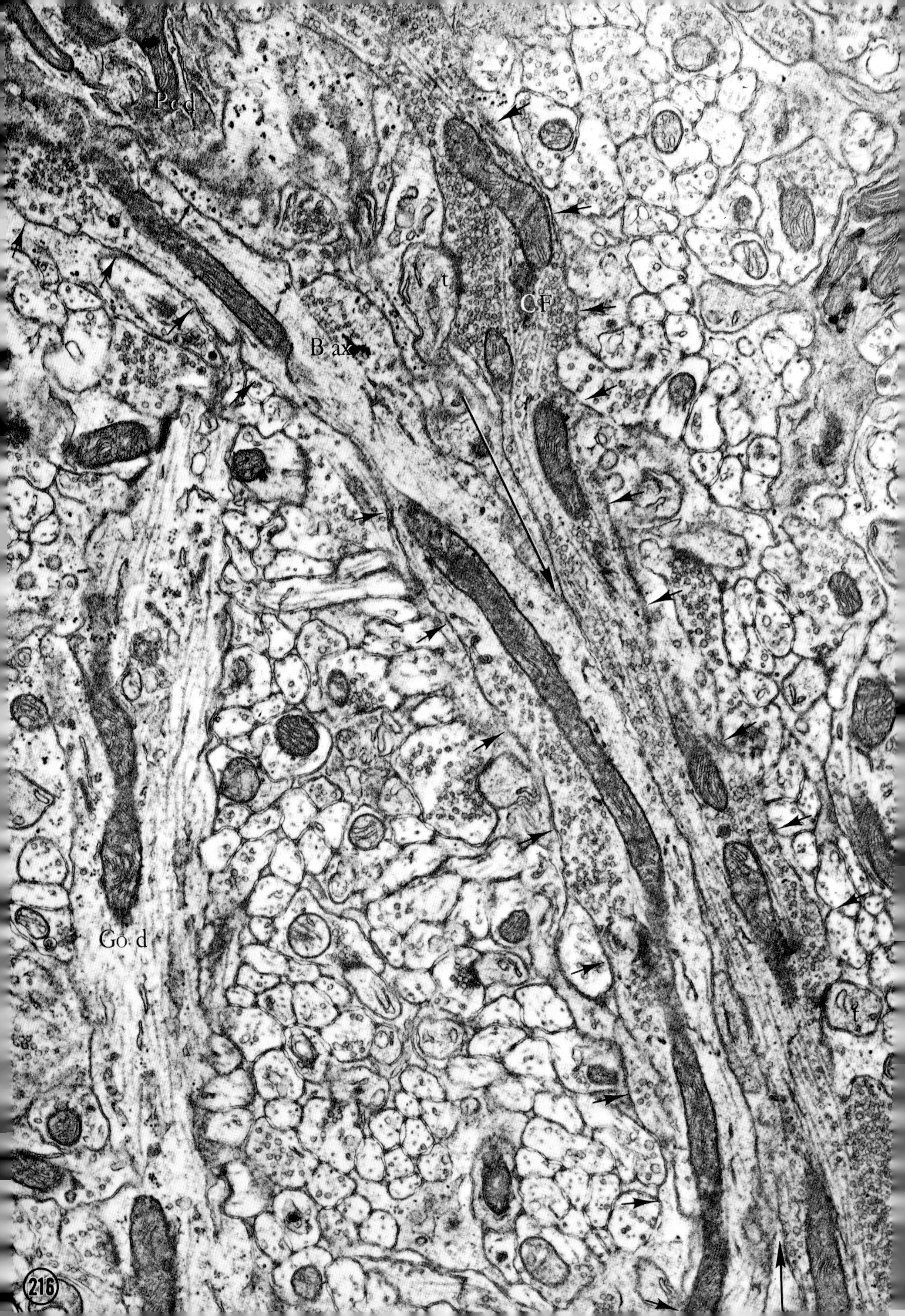

Pc d
t
CF
B ax
Go d
t
216

cosities densely packed with synaptic vesicles. The basket fiber contains somewhat flattened or ovoid vesicles; the climbing fiber contains only spherical vesicles. Each has its characteristic synapses *en passant* on the underlying dendrite, the basket fiber forming type 2 synapses on the shaft and the climbing fiber forming a modified type 1 synapse on the thorns.

Usually the two fibers are separated from each other by thin slips of the neuroglial sheath that surrounds the Purkinje cell and all of its processes. Frequently, however, the two fibers come into immediate apposition for long stretches, several microns long in many instances (Fig. 216). These are apparently the axo-axonal contacts that were detected by the SCHEIBELS in their study of Golgi material. Careful examination of these contacts in electron micrographs has failed to disclose any specializations of the apposed surfaces, either adherent plaques or synaptic complexes. Indeed, where the adjacent regions of both fibers contain synaptic organelles, such as vesicles and presynaptic densities, these organelles are located on opposite sides of the fibers, away from the coapted surfaces (Fig. 216). The electron micrographic evidence indicates that each fiber comes into synaptic relation only with the Purkinje cell dendrite underlying it and not with the other axon.

Sometimes a dendrite of a basket cell or a Golgi cell accompanies the climbing fiber or one of its beaded tendril collaterals. Or a basket cell dendrite can be bundled together with a basket axon collateral and the climbing fiber—all three apparently twining around the Purkinje cell dendrite together. In these instances also there is no evidence of synaptic articulation between the basket cell processes and the climbing fiber. As may be seen in Fig. 217, the basket cell dendrite makes numerous synaptic contacts with the parallel fibers touching its free surface and the climbing fiber synapses with numerous Purkinje cell thorns protruding into its free surface. But

◄———

Fig. 216. Interrelation between a climbing fiber and an ascending collateral of a basket axon. These two axons are running together upon the surface of a Purkinje cell dendrite (*Pc d*). The dendrite can be seen at the top left of this illustration and then passes just out of the plane of this section in the rest of the illustration. The basket axon (*B ax*) and the tendril of the climbing fiber (*CF*) accompany each other along the dendrite (*arrowheads*). In their common trajectory, the two axons are in direct apposition for about 6 µm without the formation of any axo-axonal synaptic junctions (*between large arrows*). In fact, the synaptic vesicle populations in both the climbing fiber and the basket axon are located away from their apposed surfaces. Two thorns of the Purkinje cell dendrite (*t*) are seen synapsing upon the climbing fiber. The dendrite of a Golgi cell (*Go d*) and parallel fibers fill the neuropil in the field. Lobule V. ×27000

the long interface between the two processes is free of synaptic specializations, even though neuroglia is not interposed between them. In contrast, basket cell axons frequently can be found in electron micrographs ending *en passant* upon dendrites of the same or another basket cell or Golgi cell, thus confirming one of the SCHEIBELS' observations in Golgi preparations. Our observations do not exclude the possibility that climbing fibers synapse with basket or stellate cells in other parts of their course, as was reported by the SCHEIBELS. We have, indeed, seen climbing fiber endings on the perikarya of stellate and basket cells in the manner noted by the SCHEIBELS (Fig. 218). It appears, however, that the long parallel paths of climbing fibers and basket cell axons or dendrites are not synaptic devices, even when they are immediate neighbors.

The plexus of climbing fibers and basket cell axons is especially rich at the branching points of the major Purkinje cell dendrites. As has already been mentioned, the ramifications of the dendritic trunks beyond the first one or two subdivisions are not usually simple bifurcations. Commonly the dendrite expands and gives off several branches of different calibers in different directions. Climbing fibers and basket axons seem to find a particularly favorable foothold in the crotches thus formed (Fig. 219). Consequently the branching trunks in the lower part of the dendritic tree are draped in clinging axons. In these zones, the extensive neuroglial sheath around the dendrites is highly perforate in order to give access to the broad synaptic complexes of the basket fibers and to let the thorns pass through on their way to contact the climbing fiber. The dendritic shaft is not, however, completely clothed in axons. They form a partially unraveled skein as they climb and coil about the dendrite. Transverse sections prove that the surface is mostly free of them. At branching points, however, the effect produced by these festoons of fibers is a high concentration of basket axon and climbing fiber synapses at strategic places in the complex dendritic tree of the Purkinje cell.

Although these fibers do not synapse with each other, they are in a position to affect the same dendrite. *Their common disposition suggests that by their opposing actions —the one inhibitory, the other excitatory—they regulate the flow of information along the dendrite.* The draping of major branching points in the dendritic tree with festoons of basket cell axons intertwined with the climbing fiber should be especially effective in this regard. Convergence of inhibition and excitation at these critical points might provide a mechanism for controlling the spread of current from the more distal parts of the

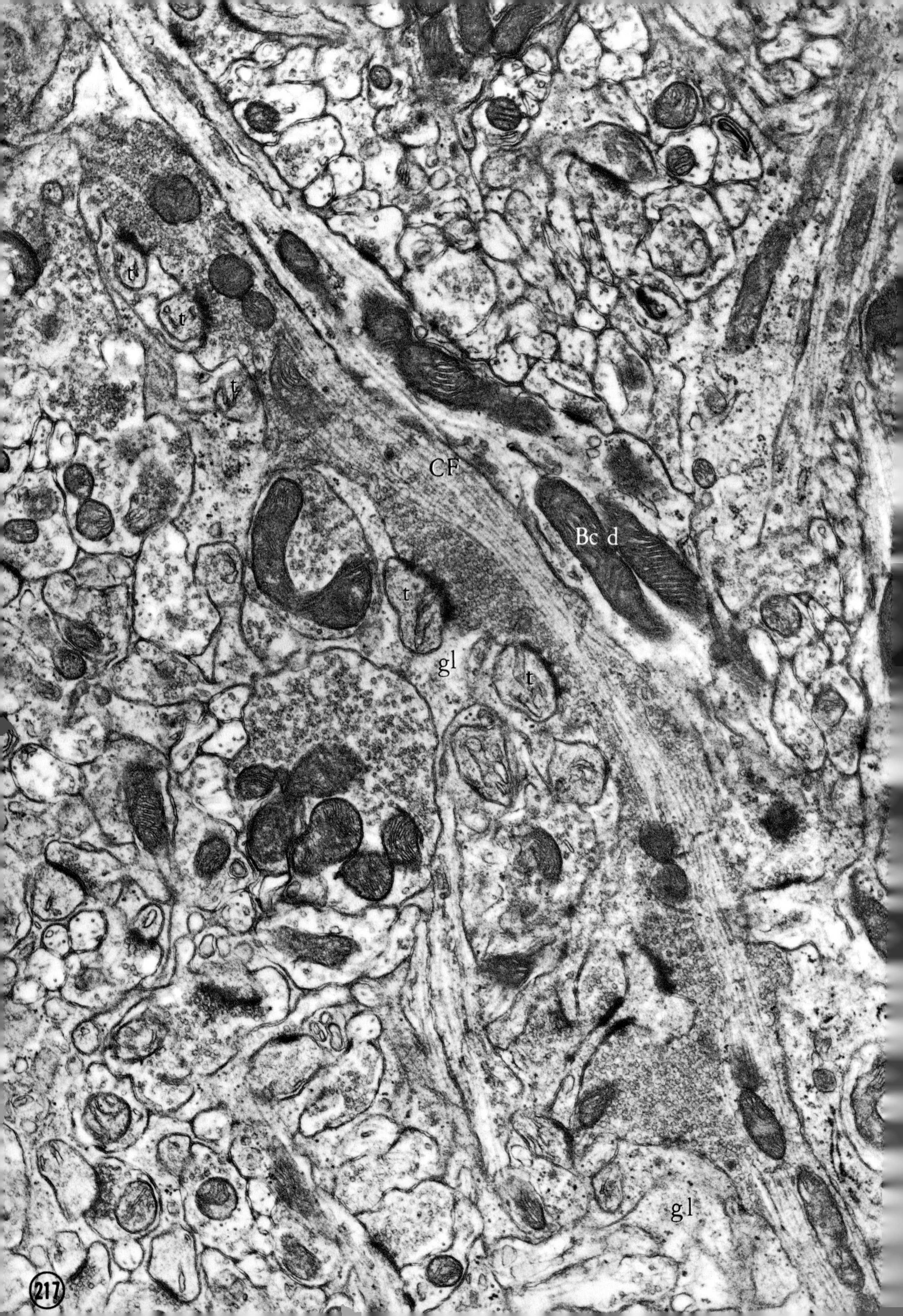
t
t
t
CF
Bc
d
t
gl
t
gl
217

Fig. 217. Climbing fiber tendril and basket cell dendrite in apposition. A basket cell dendrite (*Bc d*) and a climbing fiber tendril (*CF*) run in apposition for a considerable distance (8 μm). The climbing fiber is distinguished by a core of microtubules with dense collections of synaptic vesicles against the surface away from its coaptation with the basket cell dendrite. Several Purkinje cell dendritic thorns (*t*) synapse upon the climbing fiber, and neuroglia (*gl*) partially covers it. Lobule V. × 27000

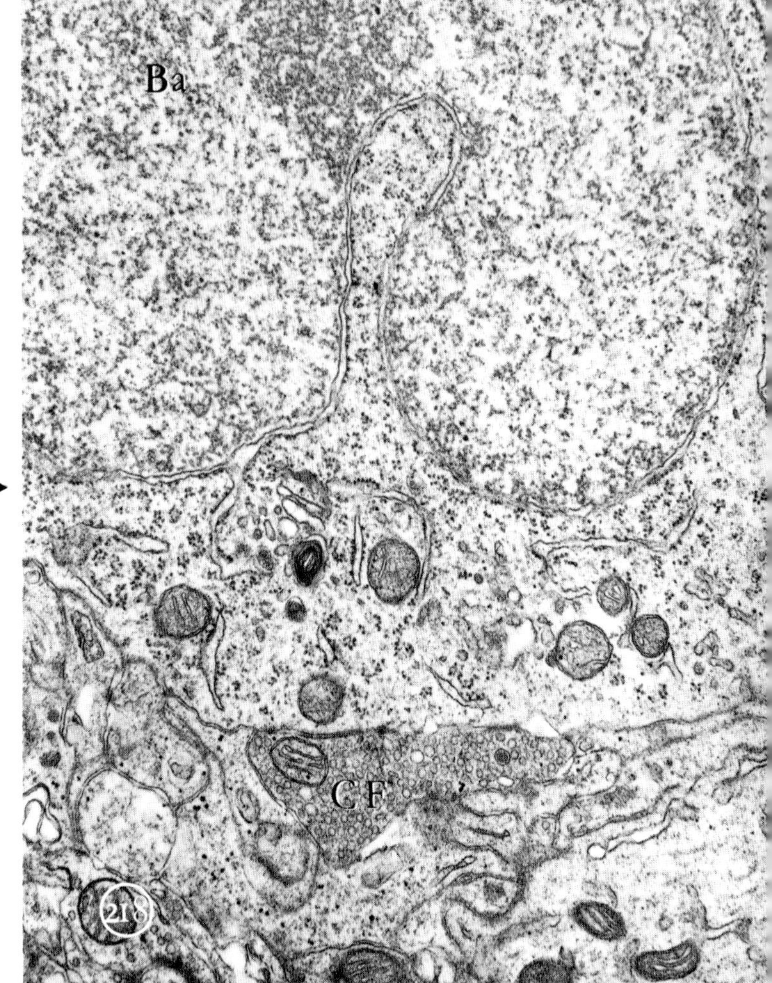

Fig. 218. The climbing fiber termination on a basket cell. A climbing fiber varicosity (*CF*) contacts the somatic surface of a basket cell (*Ba*) and probably synapses with it just deep to this section. The hint of a synaptic junction is visible here. Lobule I. × 23000

Fig. 219. The Purkinje cell dendritic bifurcation. This montage shows the complex festoon of axons draped over a major branching point in the Purkinje cell dendritic tree (*Pc d*). The dendrite crosses the illustration and divides just beneath the plane of the section. A sheet of neuroglia (*gl*) invests the dendrite and dissects between it and the axonal profiles abutting on it. A climbing fiber varicosity (*cf*) bearing three dendritic thorns is encircled by various profiles of basket axons (*B ax*). One of these basket axons (extreme left) synapses on the shaft of a Purkinje cell spiny branchlet (*sb*). Profiles of parallel fibers fill the surrounding field, in which there is also a distal synapse between a climbing fiber and a thorn (*arrow*). Lobule IV. × 15000

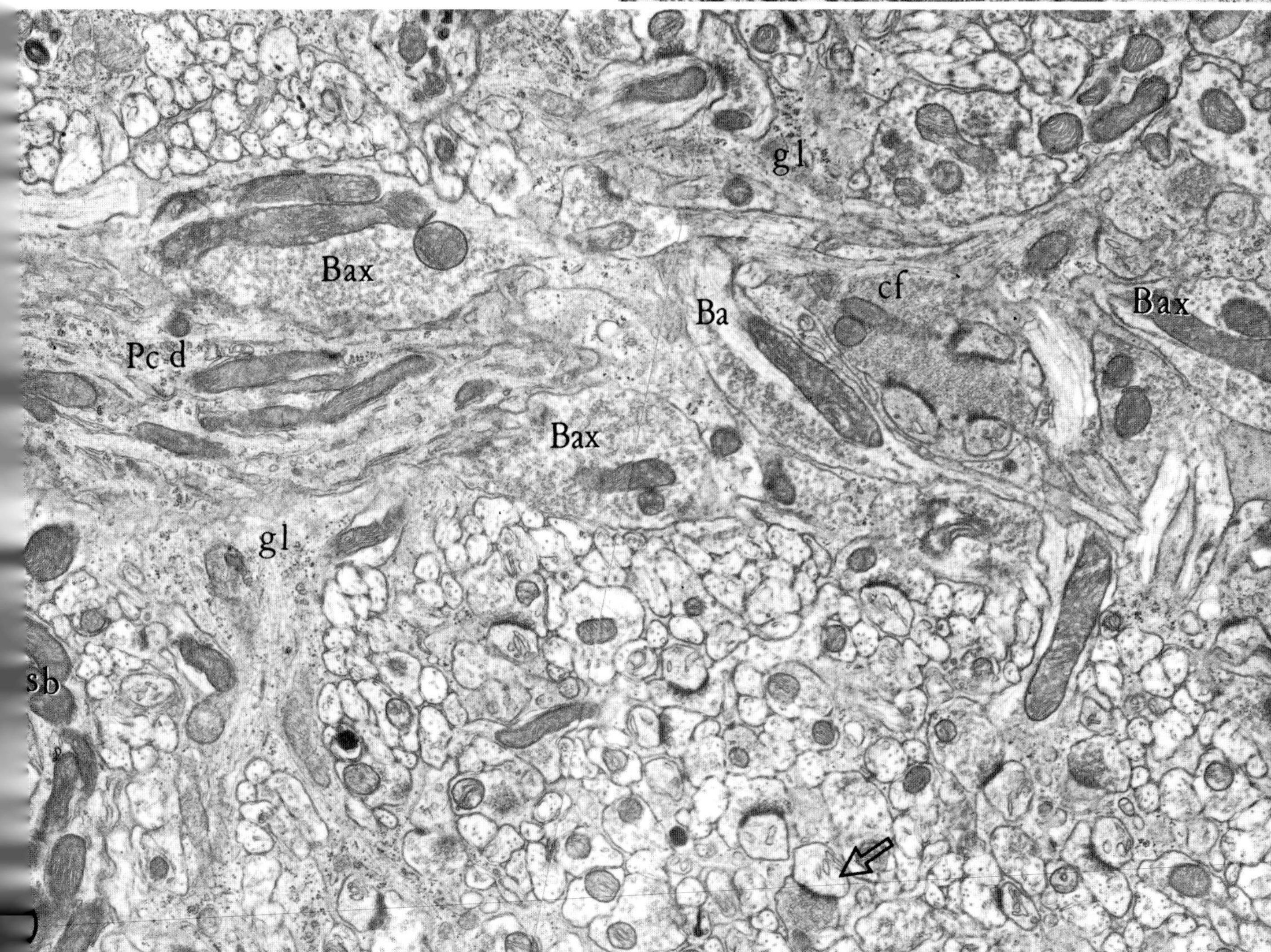

arborization through the mainstem dendrite and into the cell body and axon initial segment. By this means, local integration of the opposing influences of these two powerful afferents could selectively tune in or out whole sectors of the dendritic tree distal to the branch point involved, thus modulating the sources of effective input into the Purkinje cell. Vast numbers of parallel fiber synapses might be temporarily shunted out, while the effectiveness of others in another part of the dendritic arborization might be enhanced. Moreover, the arrangement of climbing and basket fibers at dendritic branching points also increases the dependence of any one Purkinje cell on the activity in the afferents to neighboring Purkinje cells, for although each Purkinje cell is in general influenced by only a single climbing fiber, it is influenced by several basket cell axons, the territories of which overlap to various extents. These factors greatly complicate the analysis of the responses of Purkinje cells to natural stimuli. Finally, it may not be superfluous to mention that the skeins of basket and climbing fibers on branching points of the major dendrites are situated in particularly favorable positions to control the development of dendritic spikes (FUJITA, 1968; LLINÁS, NICHOLSON, FREEMAN, and HILLMAN, 1968; LLINÁS, NICHOLSON, and PRECHT, 1969; LLINÁS and NICHOLSON, 1969, 1971; MARTINEZ, CRILL, and KENNEDY, 1971).

b) The Climbing Fiber and Its Collaterals in the Granular Layer

The recognition and identification of the climbing fiber and its terminal formations in electron micrographs of the granular layer was considerably simplified by their prior characterization in the molecular layer. The principle followed was that *the internal fine structure of a type of nerve fiber, like that of neuronal types, remains constant and characteristic no matter where it may be found or what its external shape may be.* For climbing fibers, the high concentrations of microtubules in the parent fiber and its collaterals and the extraordinarily high concentration of round synaptic vesicles in its presynaptic formations proved to be consistent in both layers of the cerebellar cortex (CHAN-PALAY and PALAY, 1971a and b). As described above, two kinds of collaterals and terminal formations were discovered in Golgi preparations of the granular layer: tendril collaterals, resembling those in the molecular layer, and glomerular collaterals, specific to the granular layer. Since the synaptic relations of these collaterals could be elucidated only by means of electron microscopy, it was necessary to begin with a painstaking examination of the granular layer in extensive montages of electron micrographs (see Figs. 267h, i). The entire thickness of the layer was covered by overlapping micrographic fields so that the trajectory of fibers ascending through the layer might be detected. In fact, the glomerular collaterals were first recognized in electron micrographs, then in 1 μm thick plastic sections stained with toluidine blue, and finally in Golgi preparations, once their probable shapes had been predicted. This reciprocating interplay between the Golgi preparations and the electron micrographs has borne fruit on many occasions in our study of the cerebellum, and illustrates the value of using both approaches in any investigation of synaptic relations. Our findings on the climbing fiber in the granular layer were reported in a communication of 1971 (a), from which the following account is adapted. More recent additions to the original report have been incorporated as indicated.

The Climbing Fiber Glomerulus. The glomerular collateral of the climbing fiber terminates in a special variety of glomerulus, distinguishable from the mossy fiber glomerulus even in the light microscope. In 1 μm Epon sections stained with toluidine blue (RICHARDSON *et al.*, 1960), these glomeruli are very large, at least 15 μm long. They contain one or more basophilic masses which send out short, irregular projections and are encircled by the lightly stained dendrites of granule cells (Fig. 220a). The mossy fiber glomerulus (Fig. 220b) displays a generally weakly basophilic core surrounded by similar lightly stained granule cell dendrites. In sections that pass longitudinally through the central axis of the glomerulus, the core often contains many long, thread-like mitochondria. In electron micrographs this pale axial structure proves to be the mossy fiber terminal itself, which contains loosely arranged synaptic vesicles embedded in a matrix characterized by a predominance of neurofilaments rather than microtubules (Figs. 130, 132, and 221). In contrast, the core of the climbing fiber glomerulus consists of several irregular axonal profiles jammed with synaptic vesicles in a dark finely filamentous matrix. The dense packing of the vesicles accounts for the basophilia of the glomeruli in the toluidine blue-stained sections. Longitudinally oriented mitochondria and some microtubules occupy the central axis of the axon terminal. The profiles of the synaptic vesicles are all circular, and range from 420 to 600 Å in diameter. There are also a few larger vesicles, about 1000 Å in diameter and filled with dense centers. Small amounts of particulate glycogen are scattered among the synaptic vesicles. In a few fortunate sections (Figs. 222 and 223) the stem of the climbing fiber collateral can be traced as it makes its tortuous

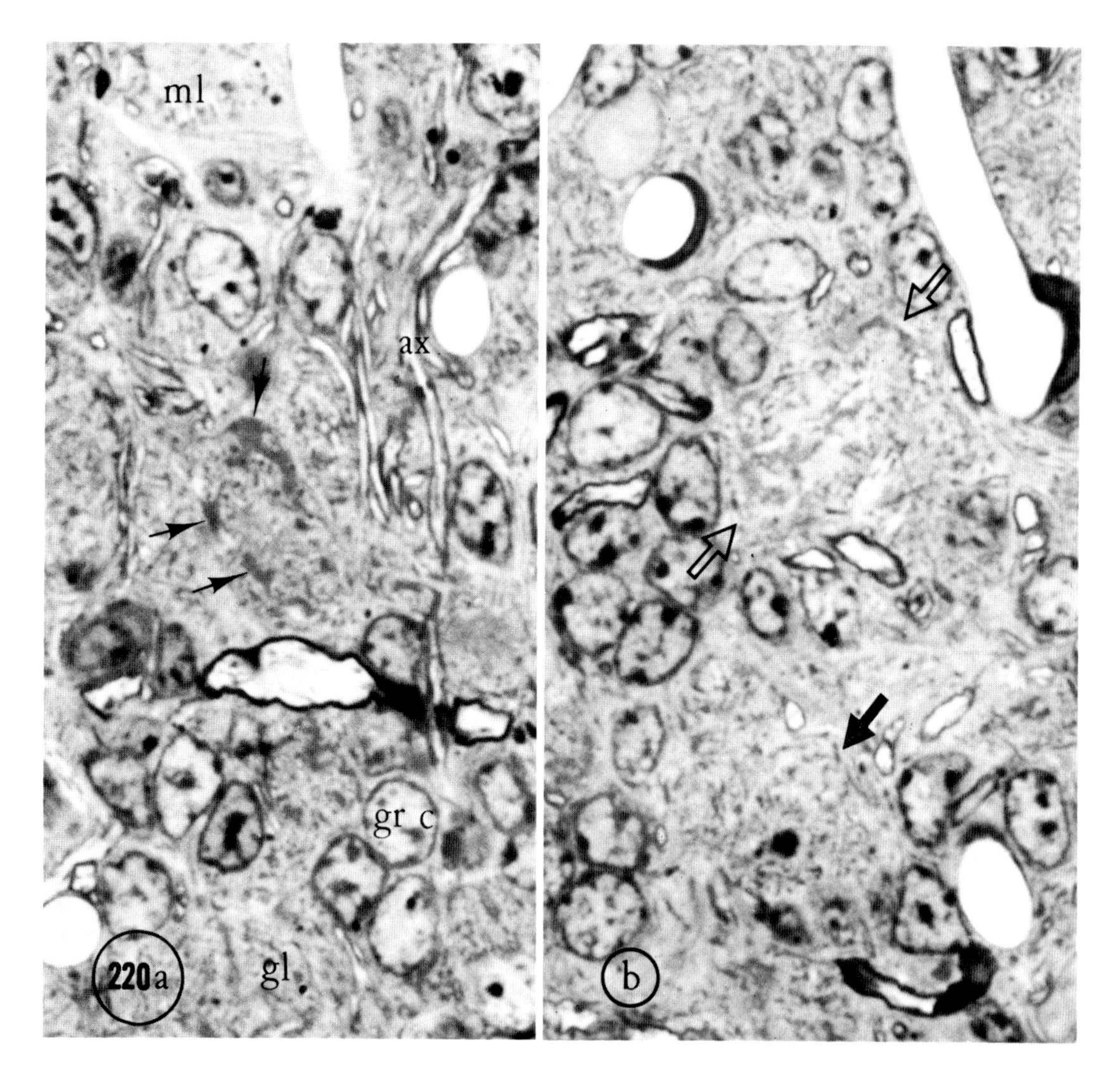

Fig. 220. a Light micrograph of the climbing fiber glomerulus. The large, elongated glomerulus of the climbing fiber can be readily distinguished in the granular layer among the complex network of neurons, axons, and dendrites. This glomerulus is elliptical, about 15 µm long, and contains five or six irregular, stellate profiles (*arrow*) of the basophilic core that represents the climbing fiber termination. The core is encircled by numerous lightly stained profiles of granule cell dendrites and Golgi axons. The rest of the granular layer shows columns of finely myelinated axons (*ax*), various other glomeruli (*gl*), and the nuclei of granule cells (*gr c*). The molecular layer (*m l*) is seen at the top of the figure. 1 µm epoxy section, toluidine blue. × 1800. **b** Light micrograph showing a mossy fiber glomerulus and a degenerating glomerulus in the granular layer. This material was taken from an animal with an experimental hemisection of the cervical spinal cord. Two neighboring glomeruli are indicated in the granular layer. The mossy fiber glomerulus (top, *open arrows*), apparently healthy, is shown in a section that passes near the central axis of the glomerulus. There is a large, weakly stained core structure with many thread-like mitochondria disposed toward the center of it. This terminal is surrounded by lightly stained dendrites of granule cells. The glomerulus with the degenerating terminal (bottom, *arrow*) has a single dark, intensely basophilic center surrounded by the lightly stained granule cell dendrites. A comparison of these two types of glomeruli with the climbing fiber glomerulus (Fig. 220 **a**) shows that the three species of glomerulus can be readily distinguished from one another on the basis of their appearance in the light microscope. 1 µm epoxy section, toluidine blue. × 1800

approach into the glomerulus. It is between 1 and 2 µm in diameter, has smooth contours, and contains longitudinally oriented microtubules in a dense matrix. Synaptic vesicles occur in the stem either singly or in sets of four or five.

Electron micrographs of the climbing fiber glomerulus show that the central structure or axonal efflorescence is surrounded by numerous other axons and dendrites as in most cerebellar glomeruli (Fig. 223). The dendrites of granule cells predominate in the immediate periphery of the central terminal. Their profiles contain long mitochondria, tubules of the granular endoplasmic reticulum, and microtubules embedded in a light matrix. Typical profiles show transverse sections of the dendrites as rounded or polygonal outlines, each enclosing a single robust mitochondrion in cross section. These represent long, finger-like dendritic terminals which interlace with one another. Often, sections passing through other planes relative to the long axis of the central fiber show longitudinal profiles of the granule cell dendrites swirled

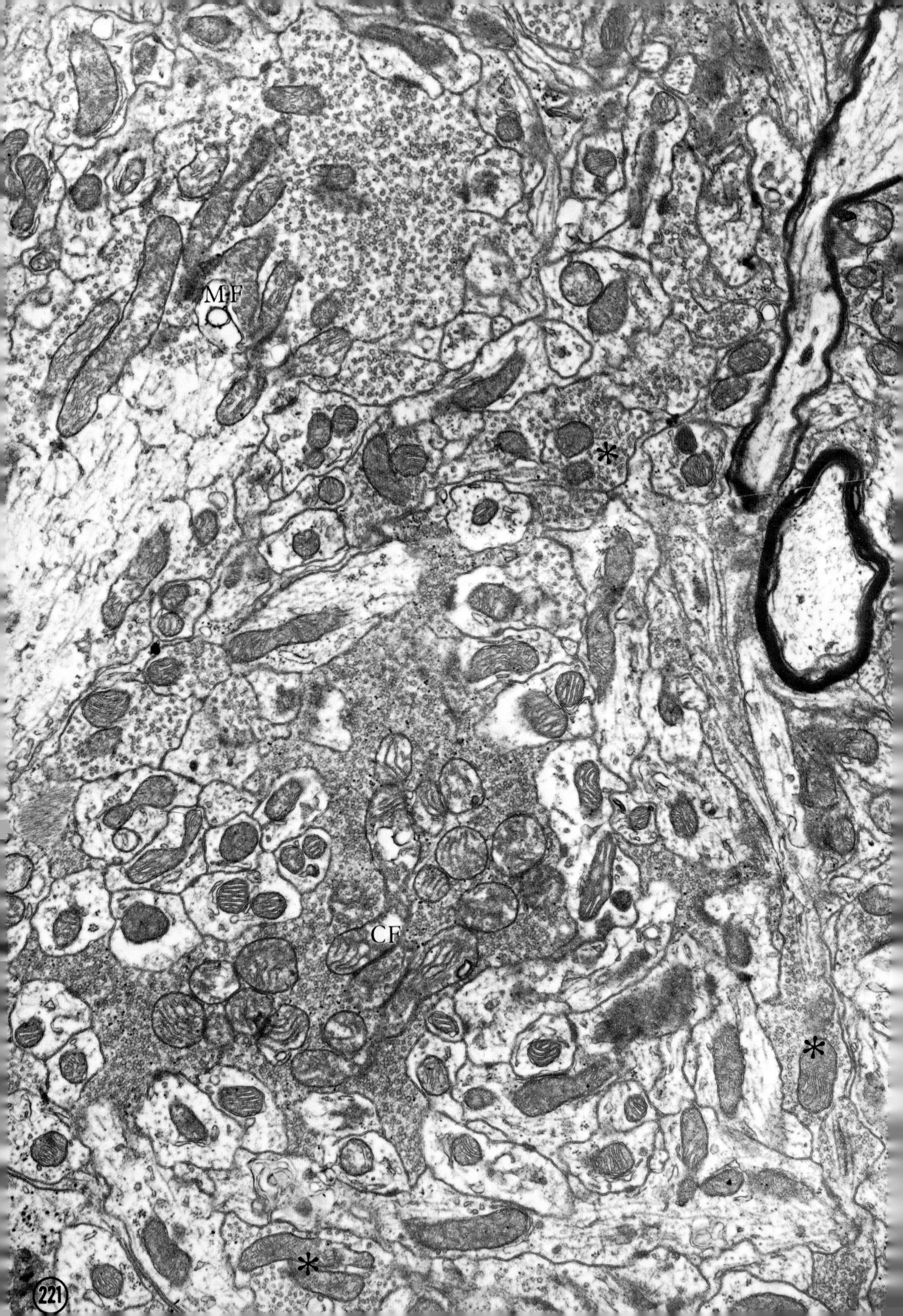

M F
*
CF
*
*
221

about the climbing fiber terminal (Fig. 223). The dendritic profiles are commonly attached to one another by puncta adhaerentia.

The dendrites are lightly impressed into the surface of the axon, which is recessed reciprocally to receive them. As may be seen in Figs. 225a, b, some of the interfaces between the climbing fiber terminal and the surrounding dendrites are marked by a flat-topped or slightly convex protuberance from the axon. In these spots the interstitial cleft is slightly widened from the usual 125 to 240 Å. At the center of the zone of apposition is a small patch of dense fibrillar material, attached to the cytoplasmic side of the dendritic plasmalemma (Fig. 225b). These dense patches have the same dimensions whether seen in longitudinally or transversely sectioned dendrites. Evidently they are roughly circular in outline. As only one dense patch always appears in any cross-sectional profile of a dendrite, it is probably not annular or very irregular in shape, but macular. It is to be noted that the dense patch occupies only a small part of the interface between terminal and dendrite. This characteristic explains why relatively few of the junctions in the glomerulus display the typical marks of synapses. These synaptic junctions conform to Gray's type 1, like those in mossy fiber glomeruli (GRAY, 1959, 1961).

The Golgi cell axonal plexus also participates in the climbing fiber glomerulus. In electron micrographs, small axonal profiles belonging to the Golgi cell are encountered twining round the terminal of the climbing fiber among the granule cell dendrites. These profiles contain small round, elliptical, and flat synaptic vesicles, loosely arranged in a light matrix with occasional mitochondria and a few tubules of the endoplasmic reticulum (Fig. 225). In the synaptic junctions that have occasionally been observed between the Golgi cell axons and granule cell dendrites, the synaptic cleft is widened and is occupied by dense fibrous material. Dense aggregates of filaments also adhere to the cytoplasmic surface of the postsynaptic membrane in some instances, but appear to be fragmentary in others. The form of this synapse is not consistent with either of Gray's types. Similar synaptic junctions are formed between Golgi axon terminals and granule cell dendrites in mossy fiber glomeruli.

Like many other glomeruli in the rat's cerebellar cortex, the climbing fiber glomerulus is not ensheathed by the processes of neuroglial cells. Complete neuroglial sheaths surrounding and investing glomeruli, like those that have been described in the cat (ECCLES, ITO, and SZENTÁGOTHAI, 1967), are exceedingly rare. Slips of neuroglial cytoplasm, however, can be found within most climbing fiber glomeruli, insinuated between the Golgi axons and the dendrites, but rarely extending to the central axonal terminal. These are the velamentous processes of proto-

◄

Fig. 221. A mossy fiber glomerulus and a climbing fiber glomerulus. Two glomeruli, one belonging to a climbing fiber (*CF*) and the other to a mossy fiber (*MF*) lie adjacent to each other in the granular layer. The climbing fiber efflorescence can be recognized by the presence of the numerous densely packed synaptic vesicles and the central cluster of mitochondria. The mossy fiber rosette shows a core of neurofilaments and a population of more loosely dispersed synaptic vesicles. Swirls of granule cell dendrites and axons of Golgi cells (*asterisks*) enwrap these central afferent axon terminals. Lobule V. ×14000

►

Fig. 222. Climbing fiber glomeruli in the granular layer. This montage of several electron micrographs shows a glomerular collateral of a climbing fiber (*arrowheads*) coursing through the granular layer to make two complex glomeruli in the neuropil. The fiber enters the field in the upper section and runs down the length of the figure. A large glomerulus (*CF* gl_1) is formed first; the stalk of the climbing fiber then twists out of the section and returns to form a second glomerulus (*CF* gl_2) with a synapse *en marron* on a Golgi cell (*Go*). Both of these glomeruli are repeated at higher magnifications in Figs. 223 and 224 respectively. Granule cells (*gr*) and other glomeruli occupy the remainder of the field. Lobule V. ×8000

Fig. 223. The climbing fiber glomerulus. This montage is made from a set of electron micrographs that pass through the long axis of a climbing fiber glomerulus (*CF gl*) and is an enlargement of the glomerulus (*CF* gl_1) shown in Fig. 222. The stem of the glomerular collateral of the climbing fiber enters the figure at the bottom right hand corner (*arrow*). It contains many longitudinally oriented microtubules and terminates abruptly in a large efflorescence that is the central structure of the glomerulus. Profiles of this core can be recognized by the many round synaptic vesicles that are packed together in a dense matrix with many slender mitochondria. The dendrites of granule cells (*gr d*) encircle this central core, participating in the formation of the glomerulus. Profiles of boutons of Golgi axons (*Go ax*) recognized by the presence of round and elliptical vesicles loosely arranged in a light matrix, intertwine with granule cell dendrites around the climbing fiber termination. Synapses occur between the climbing fiber core and the dendrites of granule cells (*asterisks*). Membrane specializations are also found at some of the junctions between Golgi axons and granule cell dendrites (Δ). This figure provides the electron microscopic correlation for a similar climbing fiber glomerulus shown in the light micrograph of Fig. 220. Lobule V. ×16000

Fig. 224. The synapse *en marron* and the climbing fiber glomerulus. The stem of the climbing fiber enters from the upper left hand corner as a thick stem with numerous longitudinally oriented microtubules (*CF*). This fiber twists behind a myelinated axon and emerges to spray into a large efflorescence, the climbing fiber glomerulus (*CF gl*). The terminal synapses upon a Golgi cell body (*Go*) in a large synapse *en marron*. The surface of the Golgi cell is thrown into a series of ridges and furrows, and the climbing fiber terminal is pressed against it. The long synaptic zones (*white arrows*) lie in the shallow furrows interrupted only by low ridges of Golgi cell cytoplasm that is bare of synapses (*black arrow*). The postsynaptic fibrillar zone (*pfz*) extends for about 0.75 μm into the Golgi cell. Dendrites of granule cells (*gr d*) are entwined around the surface of the climbing fiber terminal not in contact with the Golgi cell to form a glomerulus. Lobule V. ×20000

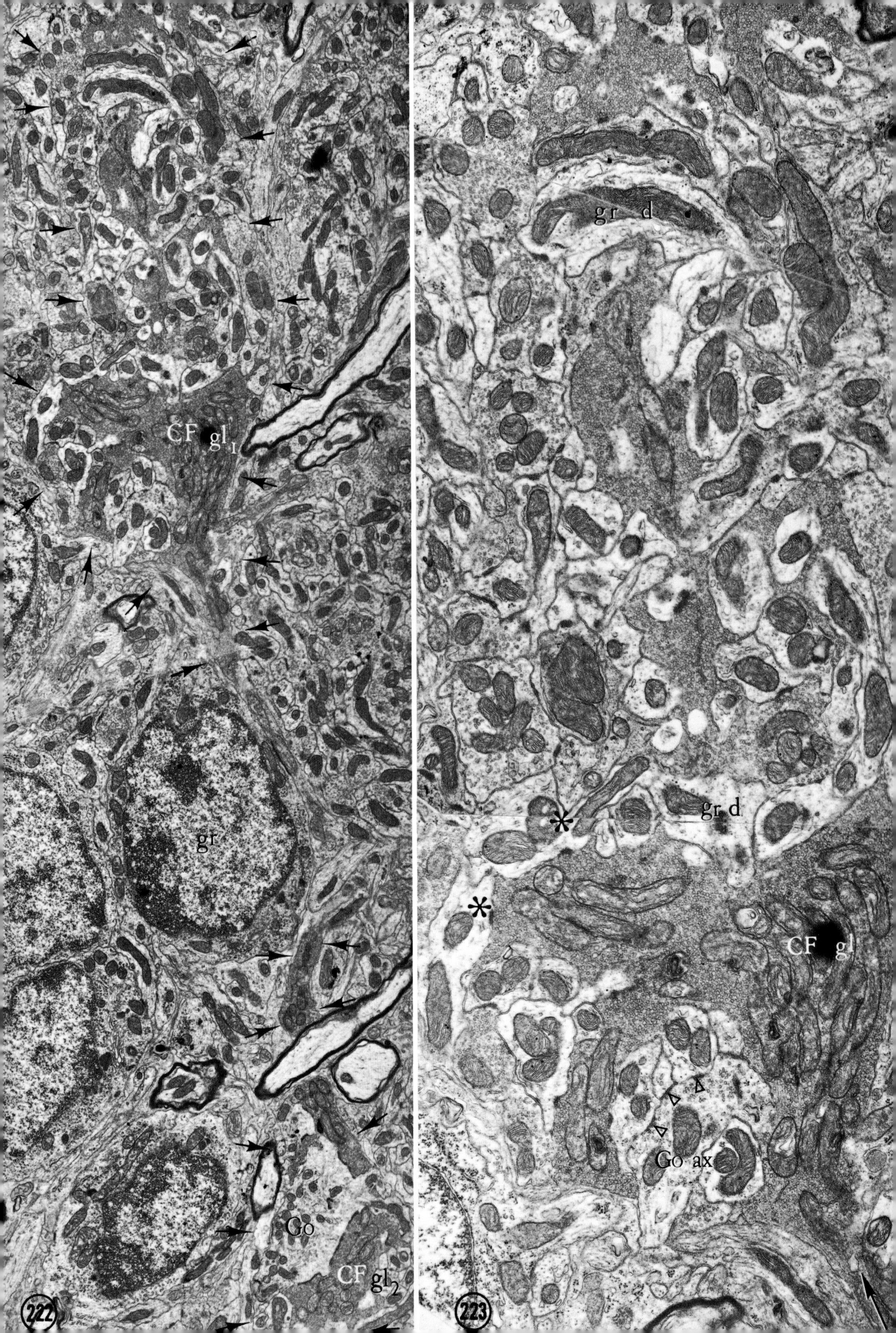

CF gl_1
gr
Go
CF gl_2
222
gr d
gr d
*
*
CF gl
Go ax
223

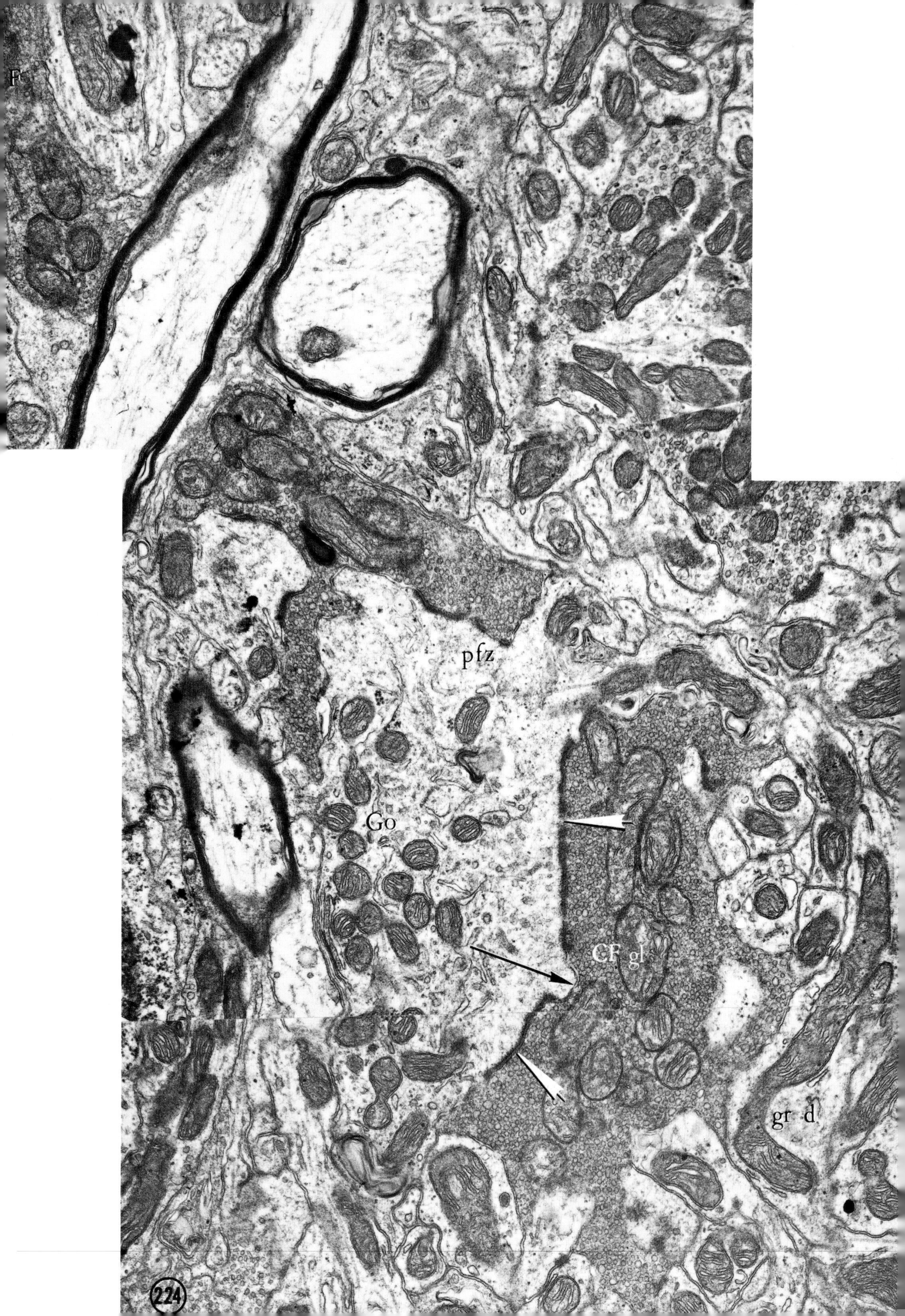
F
pfz
Go
CF gl
gr d
224

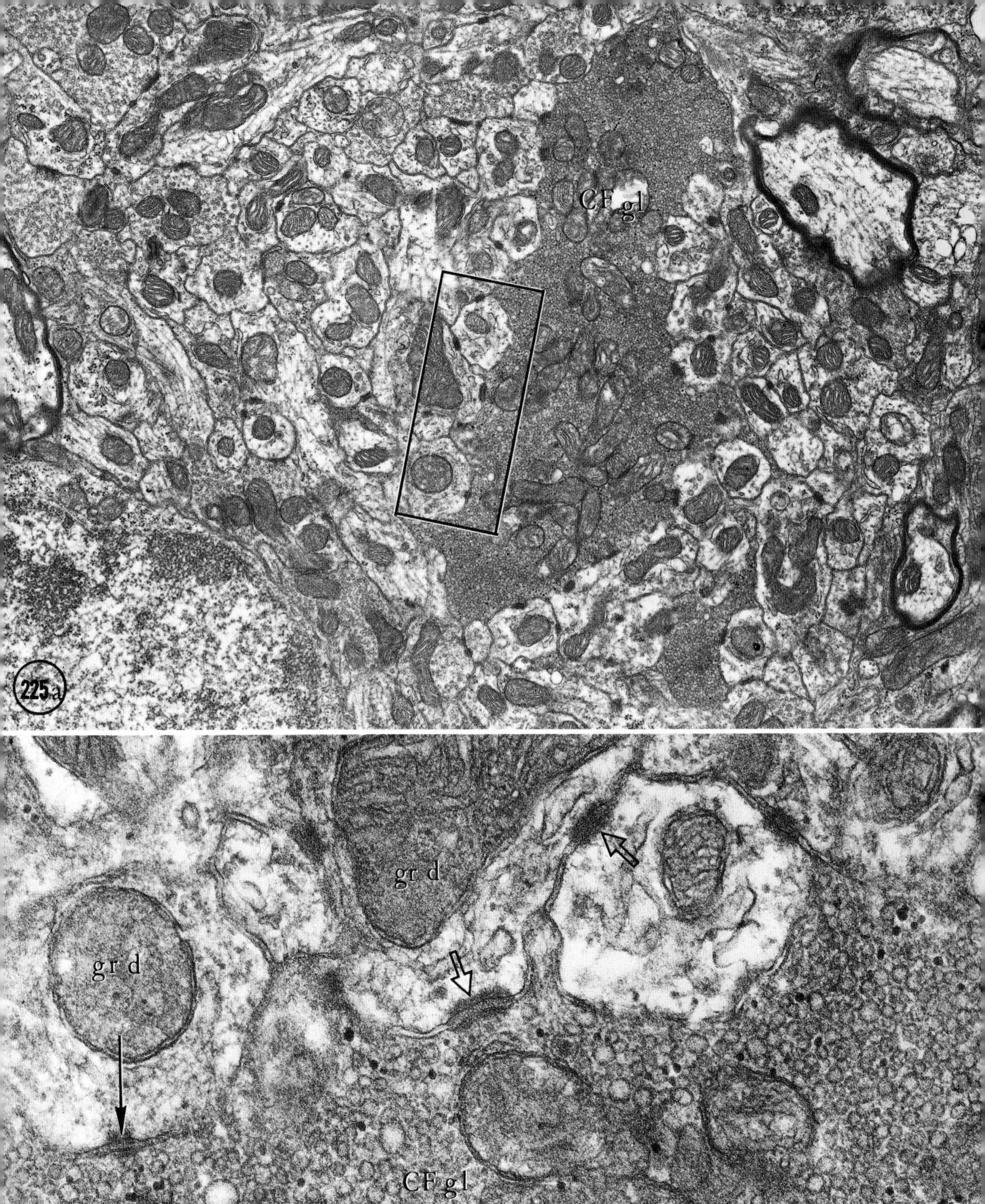

CF gl
225a
gr d
gr d
CF gl
225b

plasmic astrocytes, which partially separate glomeruli sharing the same protoplasmic islet, and compartmentalize the granular layer (Chap. XI).

The Climbing Fiber Synapse en Marron. Not infrequently the glomerular formation includes the soma of a Golgi cell as well as the usual axons and dendrites. A broad articulation, the synapse *en marron*, is developed between the Golgi cell and the efflorescent terminal of the climbing fiber (Figs. 222 and 224). The surface of the cell is thrown up into ridges and blunt projections, while the axonal terminal is reciprocally wrinkled and puckered. Extensive ribbons of synaptic junctions occur in the depths and sides of the broad, shallow furrows in the Golgi cell. Electron micrographs of such areas, in which the section runs parallel with the furrow, show long, straight synaptic interfaces interrupted by the slopes or tops of ridges, which protrude from the cell and are bare of synaptic junctions. In sections passing perpendicular to the furrows a series of short synaptic interfaces appear in their depths. These ribbon-like synaptic interfaces have the characteristics of fairly typical type 1 junctions (GRAY, 1959, 1961) with widened synaptic clefts and dense accumulations of filamentous material adherent to the cytoplasmic side of the postsynaptic membrane. Deeper within the Golgi cell there is a fringe, from 0.3 to 1.0 μm deep, beneath the junctional complex that is devoid of the usual cytoplasmic organelles but is filled with a dense fibrillar matrix. The component filaments of this matrix are predominantly oriented in a radial direction spreading toward the wrinkled surface (Fig. 224). Still deeper in the cell the normal cytoplasmic organelles begin to appear: first the tubules of the agranular endoplasmic reticulum, some of which radiate into the postsynaptic fibrillar zone, then successively the rough endoplasmic reticulum, the mitochondria, and finally all the other organelles as the perinuclear zone of the Golgi cell is approached. Often mitochondria accumulate at the margin of the fibrillar postsynaptic zone. In all particulars, then, except for the character of the axonal terminal, these climbing fiber marron synapses resemble the marron synapses formed by mossy fibers (CHAN-PALAY and PALAY, 1971b). As in the latter, the surfaces of the climbing fiber not in contact with the Golgi cell are entwined with granule cell dendrites and Golgi cell axons in the usual glomerular formation. This design allows a single climbing fiber efflorescence to make extensive contact with a Golgi cell perikaryon as well as multiple contacts with granule cells.

An occasionally encountered variant of the marron synapse is shown in the extraordinarily complicated assemblage of profiles in Fig. 226. In this formation the somata of two Golgi cells lie close enough together to interact in the same climbing fiber glomerulus. Their thick dendrites penetrate right through it. The highly irregular profiles of a climbing fiber efflorescence, interlaced with granule cell dendrites, clamber over the Golgi cell dendrites. An interesting feature of this articulation is that the Golgi cell dendrites send out numerous polypoidal excrescences, which protrude into the climbing fiber terminal. Since these dendritic appendages are somewhat like thorns, and contain only a thin filamentous matrix with an occasional mitochondrion or tubule, they appear as pale, ovoidal profiles within the darker, densely vesiculated climbing fiber. Punctate or narrow, elongated synaptic complexes occur between these excrescences and the axonal terminal. Such appendages were first noted by HÁMORI and SZENTÁGOTHAI (1966b), who described them in the cat. Profiles of the polypoid excrescences might be confused with the profiles of granule cell dendrites, which also synapse with the climbing fiber in this formation, but these dendrites are thicker than the appendages on the Golgi cell dendrites and contain longitudinally oriented microtubules and mitochondria, in addition to clusters of ribosomes. These dendrites also synapse with the axonal plexus of the Golgi cell, which pervades this kind of glomerulus.

Essentially these complicated formations are synapses *en marron* in which the junctional interface has been expanded to include the initial portions of the large dendrites of the Golgi cells. The climbing fiber and the Golgi cell dendrite running along together form the axial structure in a fusiform glomerulus, with granule cell dendrites swirling gently round them and Golgi cell axons interweaving with these dendrites. The whole structure is bounded by the dendrites approaching obliquely or more or less longitudinally from the surrounding granule cells. Slips of neuroglial processes, the velamentous processes of velate astrocytes, invade the formation from without, but are surprisingly sparse compared with those of other glomeruli.

◀

Fig. 225. a The climbing fiber glomerulus. This section passes through the center of a climbing fiber glomerulus (*CF gl*). The axon is surrounded by a swirl of granule cell dendrites. A small part of this field (*box*) is repeated in the next figure at higher magnification. Lobule III. × 14000. **b** Synapses in the climbing fiber glomerulus. The climbing fiber terminal (*CF gl*) synapses upon the dendrites of granule cells (*gr d, black arrow*). Puncta adhaerentia (*open arrows*) also occur between granule cell dendrites and the climbing fiber terminal. The extremely dense packing of synaptic vesicles in the climbing fiber terminal is seen clearly in this illustration. Lobule III. × 72000

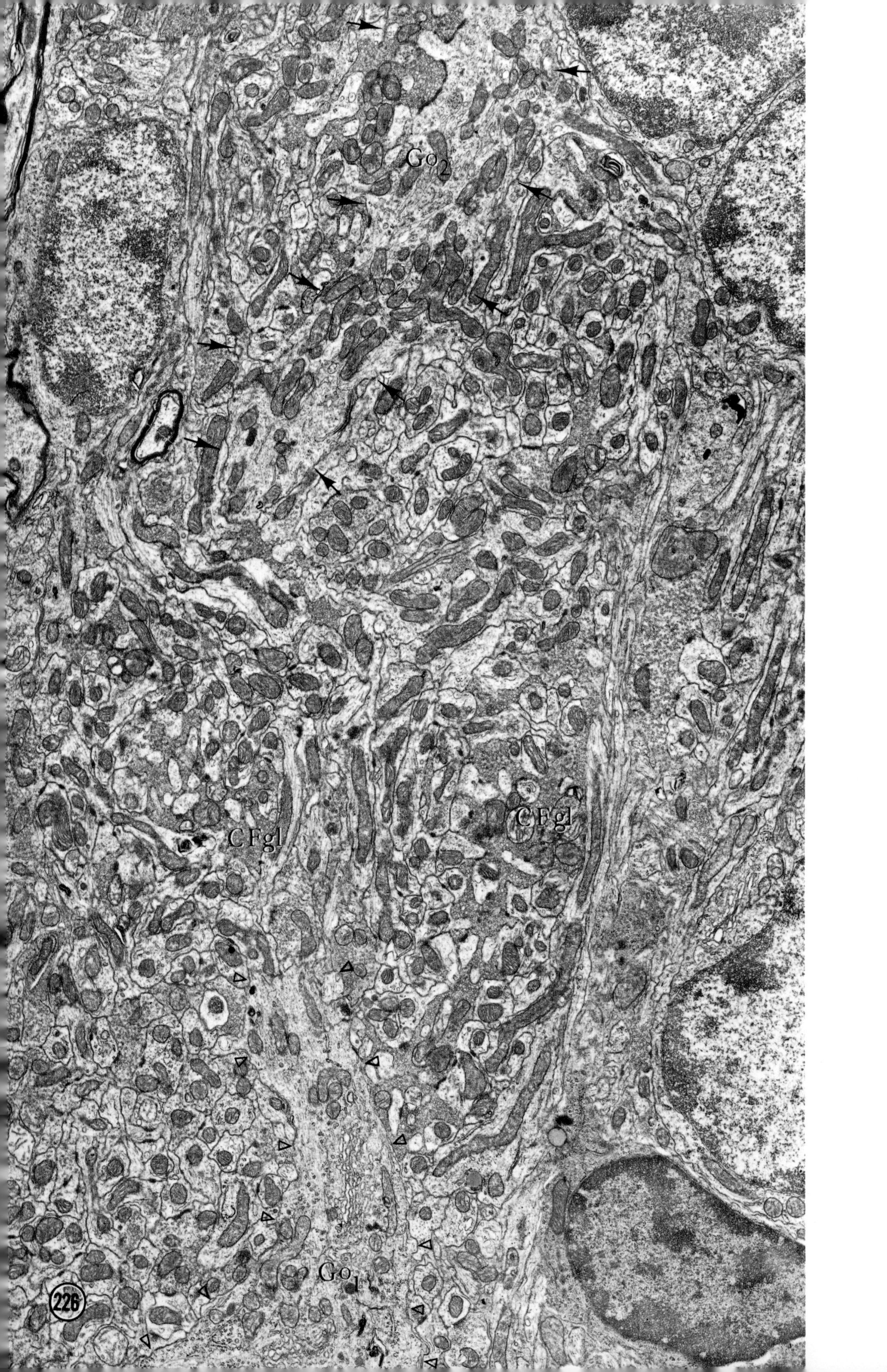
Go2
CFgl
CFgl
Go1
226

The Tendril Collaterals in the Granular Layer. The tendril collaterals of the climbing fiber in the granular layer resemble their counterparts in the molecular layer. In electron micrographs they appear as trains of round or elliptical profiles, about 2 to 3 μm in diameter and packed with round synaptic vesicles in a dense filamentous matrix. Occasionally mitochondria and several microtubules are encountered. These profiles derive from the chain of varicosities in the tendril collateral. The connecting thread between them is very fine and contorted, and is seldom found in continuity with the varicosities in electron micrographs. It contains a few microtubules running longitudinally and sometimes a long mitochondrion. Examination of the granular layer in electron micrographs shows that these tendril collaterals synapse with the dendrites of granule cells, with the somata of Golgi cells and Lugaro cells, and with the somatic thorns of Purkinje cells. All of these junctions display the characteristics of Gray's type 1.

Small profiles of tendril varicosities can be found in the upper granular layer beneath the level of the Purkinje cell bodies and around the pinceaux formed by the basket cell axons. These profiles are easily distinguished from the profiles of the basket cell axons, which usually contain numerous neurofilaments and round or ellipsoidal synaptic vesicles loosely arranged in a light matrix. They can also be distinguished from the boutons of the Purkinje cell axons, the recurrent collaterals of which are common in the infraganglionic neuropil. These terminals contain rectangular, flattened, and elliptical vesicles loosely arranged in a dense filamentous matrix.

The tendril varicosity of the climbing fiber is frequently encountered in apposition to dendrites of granule cells, but typical synaptic specializations at this interface are difficult to find. Perhaps the reason for this is that the synaptic complex occupies only a small part of the junction. When found, this junction resembles the articulations of the climbing fiber and granule cell dendrites in the glomerulus, a small spot, probably round, with dense filamentous aggregates on the postsynaptic side. The synaptic cleft is slightly widened, completing the list of features for a type 1 junction. Collaterals of climbing fibers have never been observed to synapse on the perikarya of granule cells, even though the two may actually be in apposition.

Similar varicosities also occur on the surface of Golgi cells. The varicosity, about 2 μm long and packed with round synaptic vesicles, presents a somewhat flattened face to the soma of the Golgi cell. Sometimes the terminal is more deeply set into the cell. Several small synaptic complexes occur along the junctional interface. These are short stretches in which the interspace is widened and filled with dense filamentous material. Filamentous material also accumulates on the cytoplasmic surfaces of the apposed plasmalemmas, with slightly more on the postsynaptic surface. Such images suggest that the synaptic complexes are rather small, macular or ribbon-like spots dotted over the interface. They contrast with the extensive and complex synapse *en marron* engaged in by the glomerular collateral of the climbing fiber and the Golgi cell perikaryon. There are profiles, however, which suggest that tendril varicosities can also participate in synapses *en marron*. One of these appears in Fig. 227, in which a small varicosity filled with the typical round vesicles seems to be molded against the deeply corrugated surface of a Golgi cell soma or its principal dendrite. Except in extent, this synapse *en marron* resembles the others already described above.

In the rat, varicosities of the climbing fiber stalk or tendrils are rarely encountered in synapse with thorns emitted from the somata of Purkinje cells. Such somatic thorns are fairly uncommon in the adult. They resemble those found on the so-called smooth parts of the major dendrites of Purkinje cells that synapse with the climbing fibers in the molecular layer. As Fig. 228 shows, the varicosity in this deeper position also synapses with two or more thorns.

c) The Fine Structure of Climbing Fiber Terminals and Their Synaptic Junctions

The most remarkable characteristic of a climbing fiber terminal in any position is the high density of its round synaptic vesicles in a dark filamentous matrix. This feature occurs in the glomerular collaterals as well as in the tendrils of both the molecular and the granular layers. Their appearance is so distinctive that the experienced eye can readily identify terminals of the climbing fiber on the electron microscope screen. In all three kinds of terminal the vesicles appear to be alike.

◄

Fig. 226. Two Golgi cells and a climbing fiber glomerulus. This immensely complicated assemblage of profiles is formed by two Golgi neurons (*Go*$_1$ and *Go*$_2$) interacting in one climbing fiber glomerulus (*CF gl*). The first Golgi cell (*Go*$_1$) sends out a large dendrite with polypoidal excrescences (*arrowheads*) which protrude into the climbing fiber glomerulus. A second Golgi cell (*Go*$_2$) produces a similar dendrite; in addition, it participates in a synapse *en marron* with the climbing fiber glomerulus (*arrows*). The entire assemblage is further enwrapped with dendrites of granule cells and axons of Golgi cells. Lobule IX. ×11000

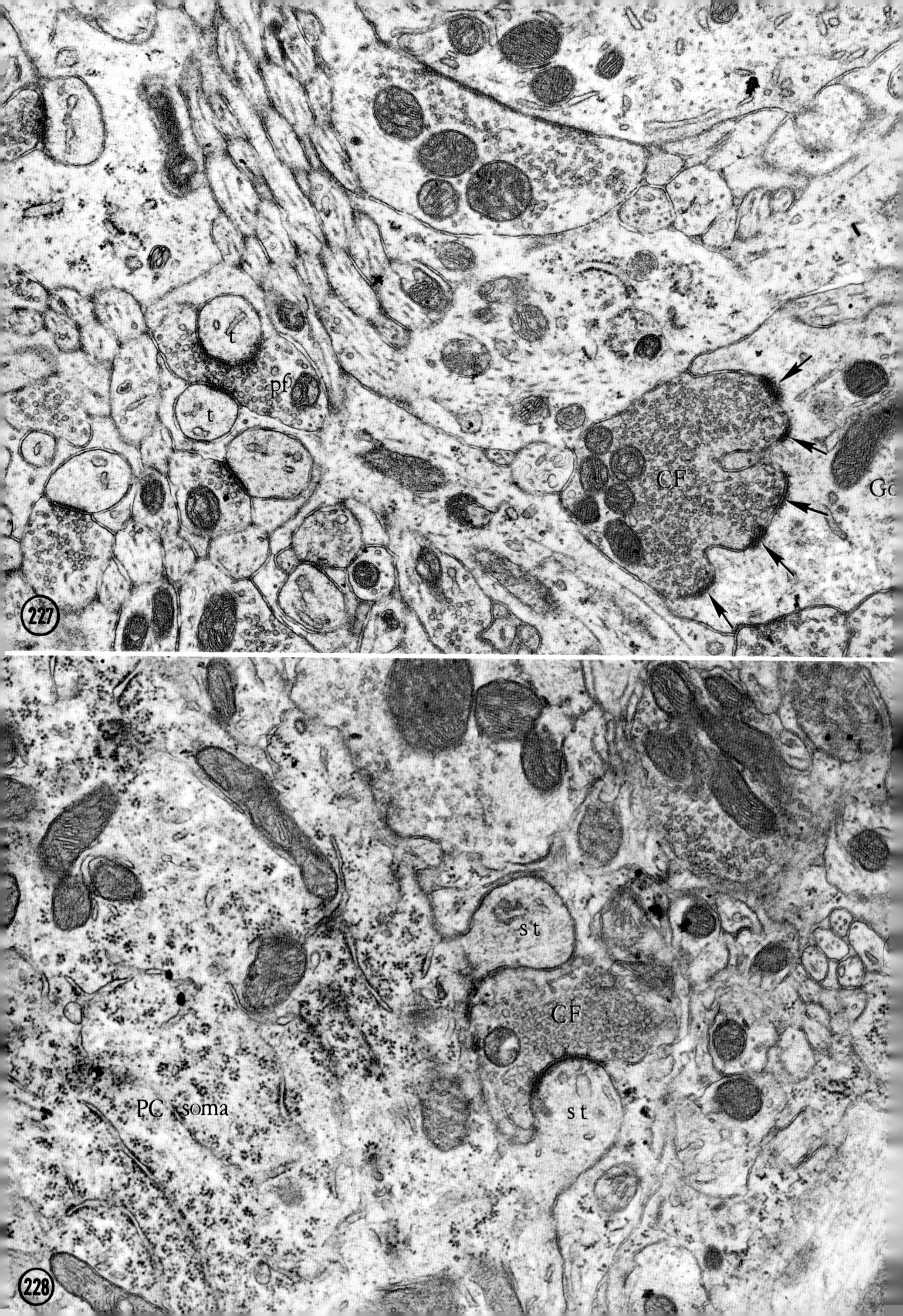
t
t
pf
CF
Go
c
227
st
CF
st
PC soma
228

Fig. 229. Sizes of synaptic vesicle profiles in climbing fiber terminals in the granular and molecular layers. Approximately 200 synaptic vesicles in each of the three climbing fiber terminal formations—varicosities in tendrils of the molecular layer (white bars), the granular layer (black bars), and in glomeruli (stippled bars)—were measured. The electron micrographs were obtained from twelve specimens of four different lobules of the cerebellum in several animals. These synaptic vesicle profiles were all round regardless of whether they occurred in varicosities of the molecular or granular layer or in glomeruli. The sizes of almost all the profiles of vesicles fall into a narrow zone from 420 Å to 620 Å, and the average diameter of the vesicles is estimated to be in excess of 520 Å

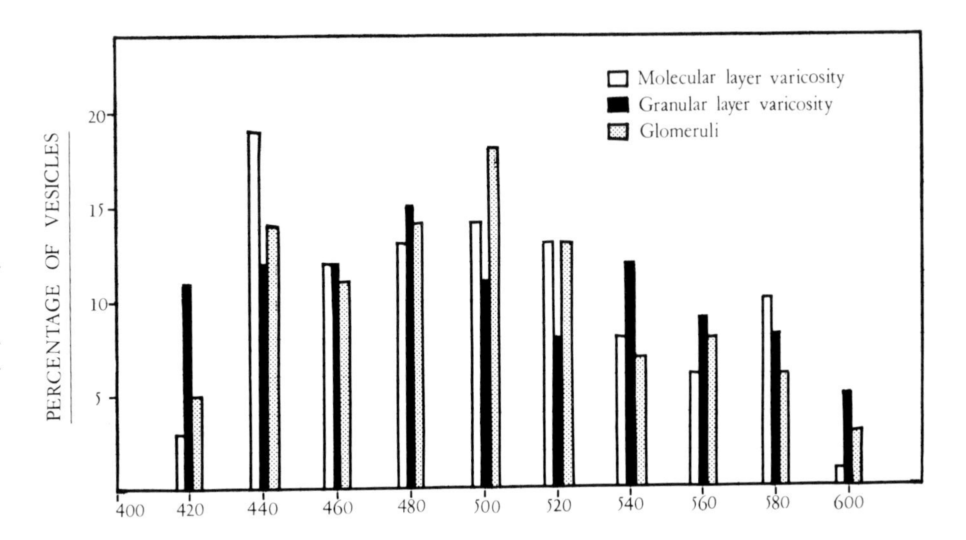

Chan-Palay and Palay (1971a) reported measurements on about 700 profiles of vesicles in climbing fibers of all kinds in the molecular and granular layers. All of the vesicles were round regardless of their location, and the diameters of their profiles fell between 420 and 600 Å. The average diameter of the vesicles must be in excess of 520 Å (see Fig. 229). When the synaptic vesicles in three terminals of climbing fibers in the granular and molecular layers were counted (Chan-Palay, unpublished results), it was found that the population density was 342, 366, and 390 vesicles per μm^2 respectively (Fig. 139). If circular profiles 500 Å in diameter were packed so that they touched one another, the maximum number that can be accommodated in a square µm is 400. Thus, the vesicles in climbing fiber terminals are crowded together almost as much as possible. These more recent measurements should be contrasted with those made on mossy fiber terminals in the same specimens. As the mossy fiber rosette is the only other kind of terminal with spherical synaptic vesicles in the granular layer, it would be useful to compare it with the climbing fiber with respect to the distinctive characteristics of the latter. In mossy fibers, the profiles of synaptic vesicles range in diameter from 320 to 500 Å (Fig. 136), and the diameter of the vesicles should be about 450 Å. The synaptic vesicles are therefore somewhat smaller in mossy fibers than in climbing fibers. Nevertheless, they are much less densely packed into the presynaptic zones of the terminals (see Fig. 138). It will be remembered that the synaptic vesicles in mossy fiber rosettes can be either dispersed or clustered. In three samples of the dispersed variety, counts showed that there were from 106 to 122 vesicles per μm^2, whereas in two samples of the more tightly clustered variety there were 205 and 240 vesicles per μm^2 respectively (Fig. 139). These counts substantiate the impression given by simple inspection. Climbing fibers have nearly twice the density of synaptic vesicles in tightly packed mossy fibers, and thrice the density in loosely packed mossy fibers.

Although all climbing fiber terminals have a common pattern of internal fine structure—most notably, their round synaptic vesicles packed into a dense matrix and their numerous microtubules—the form of their synaptic interfaces varies according to the site of the synapse. The existence of these differences suggests that *the style of synaptic junction formed by a collateral is determined by the postsynaptic cell, although the final morphological arrangement must necessarily reflect the interaction between the pre- and postsynaptic elements.* In the molecular layer, for example, climbing fiber tendrils synapse with thorns on the major Purkinje cell dendrites, whereas basket axons synapse on the shafts of the same dendrites. This remarkable selectivity for certain parts of the dendritic surface clearly attests to the interaction of both pre- and postsynaptic elements. The pattern of climbing fiber synapses in the granular layer also supports this idea. For example, tendril varicosities contact the Golgi

◀

Fig. 227. The climbing fiber tendril in the infraganglionic region. A tendril varicosity of a climbing fiber (*CF*) synapses *en marron* with the perikaryon of a Golgi cell (*Go*) through a set of five synaptic complexes (*arrows*). At the left a parallel fiber (*pf*) synapses with two Purkinje cell dendritic thorns (*t*). Lobule V. ×26000

Fig. 228. Synapses of a climbing fiber varicosity on somatic thorns of a Purkinje cell. The Purkinje cell soma (*PC soma*) gives rise to two somatic thorns (*st*) that clasp and synapse with a varicosity of a climbing fiber (*CF*). Somatic thorns are common in Purkinje cells of the developing cerebellar cortex, but persisting ones such as these are rare in the adult animal. Lobule X. ×33000

cell perikaryon with a series of short macular synapses, while the glomerular collateral terminates in an extensive synapse *en marron* with a peculiarly wrinkled sector of the perikaryal surface. *The surface of the Golgi cell, like that of the Purkinje cell, appears to be a mosaic with specific sites matched to each type of synaptic afferent.*

There is, however, another side to this argument. It must be remembered that the same collateral that forms a synapse *en marron* with the Golgi cell also furnishes the central axonal element in a glomerulus with the dendrites of granule cells entwined around it. Thus, the surface of an axonal terminal must be a mosaic of specific sites that match those in the postsynaptic element. However, mossy fibers also engage in exactly the same kinds of synapses with Golgi cells and granule cell dendrites as climbing fibers. Therefore, it appears that the postsynaptic element may be the critical determinant of the kind of synapse that is permitted on any part of the cell. It may be assumed that both chemical and morphogenetic factors must come into play in order to establish and maintain these specific configurations. Embryological and experimental investigations are needed to explore the complex factors that are only adumbrated by the morphological observations given here.

MUGNAINI (1970) has expressed similar thoughts in his discussion of the role of the "recipient neuron in the determination of synaptic configuration." Indeed, this determination is but an expression of a general architectural principle in the nervous system, that of neuronal specificity (see discussion by PALAY, 1967). It is interesting that so many examples of the morphological specificity of the neuronal surface can be assembled from an investigation of the cerebellar cortex, which is generally regarded as relatively simple and uniform.

4. The Connections of the Climbing Fiber

The correlated study of the climbing fiber with optical and electron microscopy has greatly expanded the list of known connections that a climbing fiber might make. Its major articulation with the Purkinje cell, of course, remains an extraordinary device for concentrating the full power of a single afferent fiber upon a single postsynaptic neuron. Its secondary collaterals, however, are equally extraordinary in diffusing the influence of the climbing fiber to every other type of nerve cell in the cerebellar cortex. These cells, both excitatory and inhibitory, participate themselves in a variety of feed-back and feed-forward circuits involving the Purkinje cell.

It is important to note that the terminal field of the climbing fiber in the granular layer has certain features that differ sharply from its pattern in the molecular layer. In the first place, it is a widely divergent field. Instead of focussing almost entirely on the dendritic tree of a single cell, as in the molecular layer, it distributes to a large number of cells, interneurons, of two different kinds. Each of these cell types distributes its axonal branches over a large area, one in the granular layer, the other in the molecular layer. Therefore the influence that a climbing fiber can exert transsynaptically is very considerably broadened by its synaptic contacts with these interneurons. Second, the terminal field of the climbing fiber collaterals in the granular layer is not confined to the thin parasagittal plane of the Purkinje cell dendritic tree as it is in the molecular layer. Instead, it is a three-dimensional field (Fig. 204d) centered on the main axis of the climbing fiber. Finally, the cells with which the collaterals synapse are interneurons with opposite functional roles, the granule cells being in the excitatory pathway to the Purkinje cells and being themselves inhibited by the Golgi cells, which are excited by granule cells, mossy fibers, and climbing fibers. This last feature of the terminal field in the granular layer is comparable in effect to the relations of the climbing fiber in the molecular layer, where the climbing fiber is itself directly excitatory to the Purkinje cell and also synapses on the basket and stellate cells, which inhibit Purkinje cells.

It will be recognized that these synaptic connections are achieved by means of four different styles of synaptic formations, each characteristic for its location and for the postsynaptic structure concerned. In the molecular layer the climbing fiber generates a complicated plexus of fine, varicose terminals, which articulate mainly with specific thorns on Purkinje cell dendrites (style 1) and to a small extent on the somata and dendritic shafts of basket and stellate cells (style 2). In the granular layer the stem of the climbing fiber gives rise again to varicose tendrils and also to glomerular efflorescences, both of which synapse with the same cell types. Tendril varicosities synapse with the somata and dendrites of Golgi cells as well as with granule cell dendrites (style 2), while the glomerular collaterals synapse *en marron* with the somata and dendrites of Golgi cells (style 3) and with granule cell dendrites in the glomerulus (style 4). The strange redundancy of the connections in the granular layer may be more apparent than real. Our observations do not permit us to make any statement about the exclusiveness of these relations: whether, for example, a branch of a particular climbing fiber synapses in these different ways with the same cell, or whether a single climbing fiber has all of these possible styles of connection. But it does seem clear that certain physiological consequences must follow from the different styles of synaptic formation. A signal coming over a single glomerular collateral will be subjected to a much wider divergence than one coming over a varicose tendril, because the former is usually in contact both with a Golgi cell body and with large numbers of granule cell dendrites. The form of these contacts suggests that they are devices for synchronizing the activation of a Golgi cell with that of numerous granule cells in the surround, but since these cells send only one or two dendrites each to any particular glomerulus, the activating effect on them is likely to be weak. The form of the varicose tendrils in the granular layer suggests that they might reinforce the activating effect of the glomerular collateral by selectively influencing other sites on the same granule cells or by dispersing the signal to still other cells.

5. Some Functional Correlations

Because the climbing fiber is in a position to activate through its varied connections in the granular and molecular layers both excitatory and inhibitory pathways leading to the Purkinje cell, as well as the Purkinje cell itself, it becomes a difficult and complicated undertaking to correlate the morphological findings with the interpretation of physiological records. This would be especially true of attempts to interpret in terms of possible neural circuits the results of experiments in which natural stimuli are used (THACH, 1970a, b, 1972). Nevertheless, the necessity for making these correlations in order to understand the operation of the cerebellar cortex emboldens us to make the following observations about the effects of climbing fiber activation.

Starting with the excitatory pathways alone, we can develop a hypothetical train of events leading from the stimulation of collaterals in the granular layer and ultimately affecting the Purkinje cell supplied by any particular climbing fiber. For example, when a climbing fiber is stimulated it excites not only a single Purkinje cell but also a considerable number of granule cells, which in turn send volleys of excitatory impulses along their axons, the parallel fibers in the molecular layer. These excite numerous basket and stellate cells, as well as the same and other Purkinje cells, in a broad beam along the longitudinal axis of the folium. Because of the three-dimensional spread of climbing fiber collaterals in the granular layer, the focus of excited granule cells will be large and the excitation carried by their parallel fibers will influence Purkinje cells for considerable distances around the particular Purkinje cell directly excited by this climbing fiber. The actual efficiency of activation of the granule cells may depend on summation of the effects of climbing fiber excitation with the effects of concurrent discharges from mossy fibers synapsing on their other dendrites in other glomeruli. Furthermore, the inhibitory influence of Golgi axons on these granule cells must be taken into account. The resultant of this modulated parallel fiber output may be the priming of neighboring Purkinje cells for sequential action by their specific climbing fibers in a temporal pattern that reflects muscle contraction and joint movement.

When the connections of the climbing fiber with inhibitory cells are considered (Fig. 230), this rather simple sequence has to be elaborately revised. Activation of a climbing fiber by stimulating the inferior olive usually results in a burst of afferent impulses rather than a single action potential (CRILL, 1970). This incoming volley spreads rapidly over the terminal arborizations in the granular and molecular layers, arriving slightly later at the Purkinje cell than at the Golgi and granule cells. The Purkinje cell reacts with a complex unitary response—the climbing fiber response (ECCLES, LLINÁS, and SASAKI, 1966d). This response is characterized by a strong initial spike followed by a prolonged depolarization upon which several smaller spikes are superimposed. During this period of depolarization the regular spike-generating mechanism of the soma and initial segment of the Purkinje cell is suppressed (GRANIT and PHILLIPS, 1956; ECCLES, LLINÁS, and SASAKI, 1966d; MARTINEZ, CRILL, and KENNEDY, 1971). The long-lasting depolarization is evidently due to a "sustained and intense active current source in the (major) dendrites. The wavelets superimposed on the climbing fiber response represent dendritic spikes recorded electrotonically in the soma" (MARTINEZ *et al.*, 1971, p. 355; see also ECCLES, ITO, and SZENTÁGOTHAI, 1967, p. 165). Evidence that the Purkinje cell dendrites are capable of firing spikes has been presented by FUJITA (1968) for the rabbit, and by LLINÁS and NICHOLSON (1969, 1971; NICHOLSON and LLINÁS, 1971) for the alligator. It will be remembered that the varicosities of the climbing fiber plexus synapse almost exclusively with the thorns projecting from the major dendrites. The summed inputs from these thorn synapses evoke in the Purkinje cell a large EPSP, which can be as much as 25 mv (MARTINEZ, CRILL, and KENNEDY, 1971), and result in the generation of the initial large spike and

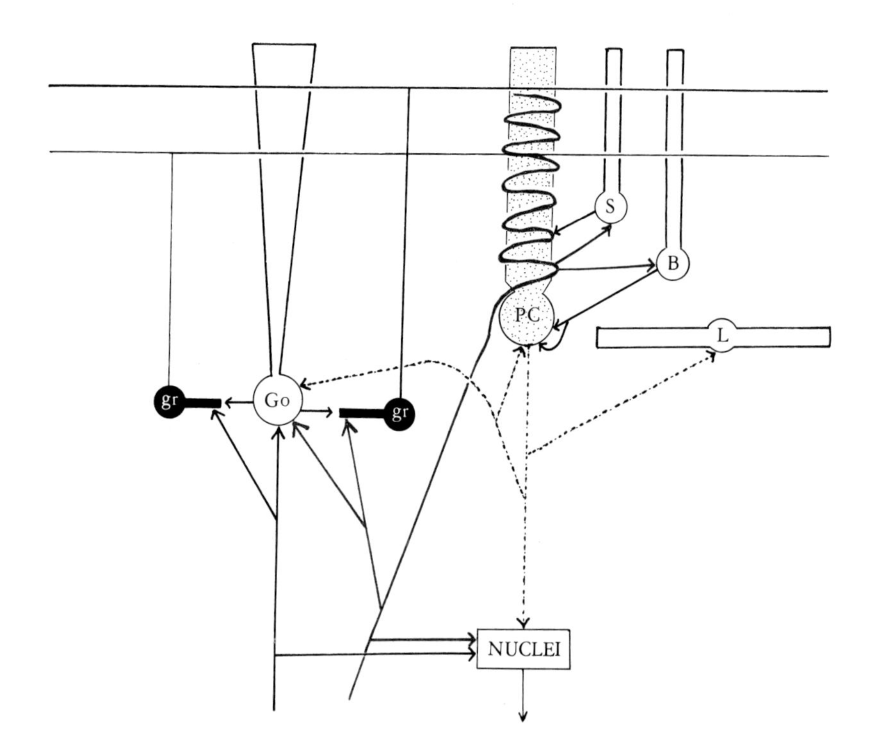

Fig. 230. Circuit diagram of the climbing fiber, Purkinje cell, and mossy fiber connections and interneuronal relays. The connections shown in this diagram are based on synaptic contacts that have been confirmed by electron miscroscopy. This diagram repeats the connections portrayed in Fig. 202 and adds the connections of the climbing fiber. The climbing fiber (red) contacts Golgi cells and granule cells in the granular layer. In the molecular layer it has multiple contacts with the Purkinje cell dendritic tree and in addition synapses with basket cells and stellate cells in the surround

the succeeding sustained depolarization during which a burst of dendritic spikes is discharged. A single climbing fiber impulse evokes a similar but briefer climbing fiber response. It will also be remembered, however, that both climbing fibers and basket cell axon collaterals entwine together round the bifurcation points in the major dendrites and cover them with festoons of fibers, each synapsing with the underlying dendrite in its own fashion. The strong convergence of excitatory and inhibitory fibers on these critical sites might be correlated with the high effectiveness of basket cell stimulation in suppressing the secondary dendritic spikes of the climbing fiber response (ECCLES, LLINÁS, SASAKI, and VOORHOEVE, 1966; ECCLES, ITO, and SZENTÁGOTHAI, 1967).

The same climbing fiber volley that elicits this climbing fiber response in the Purkinje cell also activates Golgi and granule cells in the granular layer, as indicated above (Fig. 230). The granule cells in a broad zone are stimulated to send impulses up into the molecular layer where they contribute to the excitatory background of the Purkinje cell. But at the same time, and perhaps more important, these parallel fiber impulses also activate stellate and basket cells. These in turn train upon the Purkinje cell a more slowly developing inhibitory barrage, which with appropriate timing may suppress the secondary dendritic spikes of the climbing fiber responses and through the periaxonal pinceau may be capable of quenching further spike discharges by the Purkinje cell. Pauses in the firing of Purkinje cells have been described by a number of investigators as a regular occurrence following either spontaneous or experimentally evoked climbing fiber responses (GRANIT and PHILLIPS, 1956; BELL and GRIMM, 1969; MURPHY and SABAH, 1970; BLOEDEL and ROBERTS, 1971; MARTINEZ, CRILL, and KENNEDY, 1971). BLOEDEL and ROBERTS (1971) found suppression of impulse activity lasting at least 45 msec following the application of a stimulus to the inferior olive. MURPHY and SABAH (1970) observed variable pauses in firing by spontaneous or evoked climbing fiber responses lasting up to 570 msec. As climbing fiber responses are often followed by hyperpolarization of varying size and lasting 50–200 msec (MARTINEZ *et al.*, 1971; MURPHY and SABAH, 1970), these pauses in firing appear to be due at least partly to synaptic inhibition of the Purkinje cell. As we have seen, the climbing fiber collaterals offer two possible pathways ending in synaptic inhibition of the Purkinje cell. The first is by way of granule cells activated in the granular layer and sending their impulses through parallel fibers to stimulate stellate and basket cells in the molecular layer. The second path is by way of direct collaterals of the climbing fiber in the molecular layer stimulating stellate and basket cells. The latter would be especially important in quenching gentler discharges of the Purkinje cell because of its peridendritic, perisomatic, and periaxonal collaterals.

The initial climbing fiber volley also activates Golgi

cells[25] in the granular layer, and their output inhibits practically all of the granule cells in the vicinity that are not strongly excited simultaneously by mossy or climbing fibers (ECCLES, LLINÁS, and SASAKI, 1966c; ECCLES, ITO, and SZENTÁGOTHAI, 1967). The inactivation of the granule cells removes not only the excitation of the Purkinje cell dendritic tree, but also the excitation of the basket and stellate cells, and the consequent relaxation of their inhibitory influence permits the restoration of the Purkinje cell to its previous state. The silence of large numbers of granule cells could also contribute to the prolongation of the pauses in the discharges of the Purkinje cell following the climbing fiber response. The very short latency of the Golgi cell enables its inhibitory discharges to take effect on the granule cells before the more slowly developing inhibition of Purkinje cells by basket and stellate cells, which is one synapse removed. Thus, the inhibitory effect of these interneurons on the Purkinje cells would be relatively short-lived because of the quick suppression of the granule cells by the Golgi cells. Some support for these suggestions is given by an experiment reported by ECCLES, LLINÁS, and SASAKI (1966c). They recorded the inhibitory postsynaptic activity ("synaptic noise") elicited in a Purkinje cell by synaptic bombardment from basket cells. Stimulation of climbing fibers through the inferior olive produced a brief suppression of this background inhibitory discharge.

The effective operation of this hypothetical scheme would require an exacting chronology for the activation of the synapses in sequence and for the maintenance of their postsynaptic effects. It would be adapted to follow rapidly changing events, such as might be expected to accompany joint and muscular movement. A physiological study of these integrations would necessitate recordings from multiple electrodes in a set of related neurons so that the sequences of their responses to a single climbing fiber volley might be monitored. Although such experiments are difficult to achieve in vertebrates, they are necessary for an understanding of the integrated operation of multicellular clusters. A beginning has already been made in the study of responses of several Purkinje cells to peripheral nerve stimulation (ECCLES, FABER, MURPHY, SABAH, and TÁBOŘÍKOVÁ, 1971a and b).

25 In experiments on the cerebellar cortex of cats, ECCLES, LLINÁS, and SASAKI (1966a) were able to elicit only a weak and inconstant excitation of large cells in the deeper cortical layers by stimulating the inferior olive. They interpreted their records as representing the activity of Golgi cells excited by climbing fibers in the granular layer. In view of the large synaptic interface between glomerular collaterals of climbing fibers and Golgi cells synapsing *en marron*, it would appear that the experimental design was not favorable for the study of this synapse. A much more potent excitatory effect on the Golgi cell would be expected.

In this connection, BLOEDEL and ROBERTS (1971) have made the astute suggestion that the pauses in Purkinje cell activity following on the climbing fiber response may be more important to overall function of the cerebellum than the large spikes evoked by climbing fiber activation. They write (p. 29):

"In the light of the convergence of Purkinje cells on neurons in the cerebellar nuclei and since most Purkinje cells are spontaneously active in unanesthetized preparations, one would expect that the initiation of a single spike by the climbing fiber input to Purkinje cells would be of less consequence to the activity of neurons in the cerebellar nuclei than the prolonged period of suppressed activity which usually follows the climbing fiber response. Since the Purkinje cells exert an inhibitory action on the activity of neurons in the cerebellar nuclei, a transient increase in the output of the latter could result from a synchronous climbing fiber input to the cortex."

They also proposed that the climbing fiber response and the subsequent quiescent period would erase the polarization built up in the Purkinje cell as a result of the thousands of excitatory and inhibitory inputs immediately preceding. Such a sequence would "prevent the output of a given Purkinje cell from reflecting transient instabilities arising in the neuronal circuitry. When the Purkinje cell was again capable of responding to its dendritic and somatic inputs, the firing frequency would be determined by those inputs arriving somewhat after the time of occurrence of the climbing fiber response" (BLOEDEL and ROBERTS, 1971, p. 29). Such a resetting mechanism would ensure that the Purkinje cell responds to the immediate balance of afferent impulses without residual influences from previous events. A similar idea is presented by HARMON *et al.* (in press) in a mathematical model of the cerebellar cortex. Such proposals can only be tested by further experimental work with refined recording methods in intact waking animals, work such as THACH has reported in primates trained to simple, repetitive tasks.

6. Summary of Intracortical Synaptic Connections of Climbing Fibers

climbing fiber ——————————— Purkinje cell
climbing fiber ——— granule cell ——— Purkinje cell
climbing fiber – Golgi cell — granule cell – Purkinje cell
climbing fiber ——— basket cell ——— Purkinje cell
climbing fiber ——— stellate cell ——— Purkinje cell
climbing fiber – granule cell – basket cell – Purkinje cell
climbing fiber – granule cell – stellate cell – Purkinje cell

Chapter XI

The Neuroglial Cells of the Cerebellar Cortex

The three-layered neuronal structure of the cerebellar cortex is reflected in the architectonics of the neuroglia. Each layer has a different population of neuroglial cell types, and one type, the Golgi epithelial cell, is peculiar to the cerebellar cortex, although it resembles the Müller cells of the retina and certain ependymal cells in the spinal cord and in the floor of the third ventricle.

Both astrocytes and oligodendrogliocytes are represented in this cortex. The astrocytes are principally of the protoplasmic variety, with which the cerebellar cortex is richly endowed. These can be divided into three subclasses: the Golgi epithelial cell, the lamellar or velate astrocyte, and the smooth astrocyte (CHAN-PALAY and PALAY, 1972c). The Golgi epithelial cell is characteristic of the Purkinje and molecular layers, and does not occur elsewhere. The lamellar astrocyte inhabits chiefly the granular layer, while the smooth astrocyte, the least common of the three, occurs throughout the cortex but most often in the granular layer. Oligodendrogliocytes are also seen throughout the cortex, but since they are intimately concerned with myelination they have the same distribution as the myelinated fibers, that is, mostly in the granular layer and the depths of the molecular layer.

All of these cells engage in intricate relations with the nerve cells and fibers of the cerebellar cortex. As in the interneuronal connections, the focal cell in these neuroglial relations is the Purkinje cell. Its cell body and all of its dendrites and its initial segment are ensheathed in lamellar processes of protoplasmic astrocytes. Its axon and recurrent collaterals are, of course, myelinated and, therefore, in the care of oligodendrogliocytes. None of the other nerve cells have so complete an investment as the Purkinje cell. The insulating role of such neuroglial processes was suggested by RAMÓN Y CAJAL (1909), and has been recently elaborated by PETERS and PALAY (1965) and PALAY (1966) (see also PETERS, PALAY, and WEBSTER, 1970). The several cell types will now be taken up sequentially in the following sections.

1. The Golgi Epithelial Cells

a) A Little History

GOLGI (1885) noted that although the molecular layer of the cerebellar cortex is permeated by large numbers of radially directed neuroglial fibers, this layer contains relatively few neuroglial cell bodies. Most of the small perikarya located there belong to nerve cells. The cerebellar radial fibers were already well known when GOLGI applied his chromate and silver impregnation method to the study of the neuroglia and discovered its completely cellular constitution (GOLGI, 1870, 1871, 1885). BERGMANN (1857) had discovered these fibers in the cerebellar cortex of cat and dog and in a human case of cerebellar atrophy, but he did not know their source. According to BERGMANN'S somewhat ambiguous description, the radial striae in the molecular layer were fine fibers passing vertically through the cortex and forming a limiting membrane at the surface by the fusion of the conical swellings at their tips. During their trajectory through the cortex, these fibers gave off lateral processes that joined together into a network. GOLGI (1871 and 1885) recognized that these so-called Bergmann fibers were the vertically directed processes of neuroglial cells located in three positions: (a) horizontal, flattened cells lying under the pia mater at the surface of the cerebellar cortex; (b) peculiar cells lying at the upper border of the granular layer, and (c) stellate cells lying deep in the granular layer and even

Fig. 231. Golgi epithelial cells in a folium. This illustration shows fourteen Golgi epithelial (*Go ep*) cells that have been impregnated by the rapid Golgi method. Their cell bodies have a ragged outline and are arranged in a layer at about the level of the Purkinje cell bodies. Each cell, like a candelabrum, has processes with warty excrescences that rise, like corroded wires, to the pial surface (*pia*). The limits of the molecular layer (*mol*), Purkinje cell layer (*PC*), granular layer (*gr*), and white matter (*wm*) are indicated. Rapid Golgi, 90 μm, sagittal, adult cerebellar cortex. Camera lucida drawing

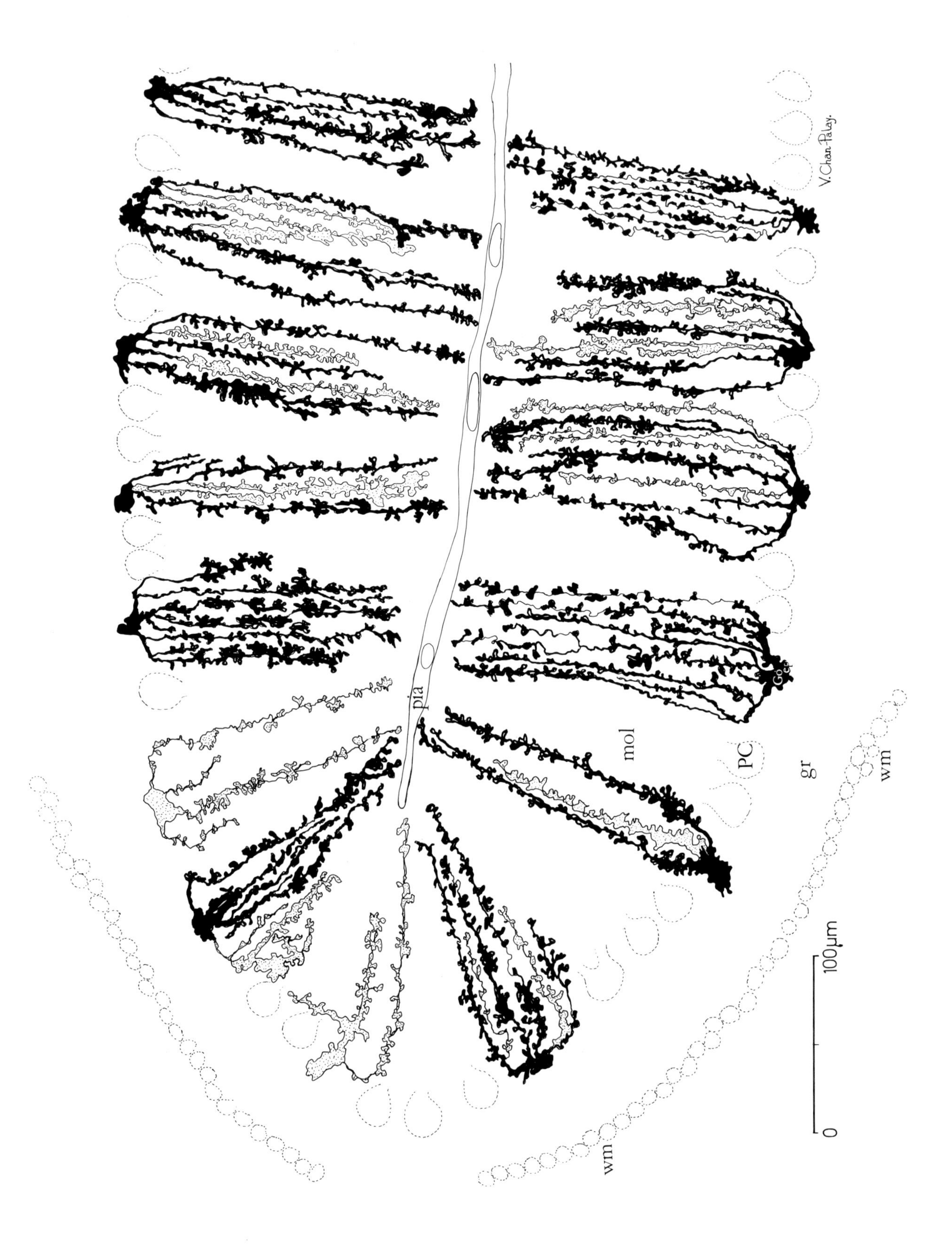

V. Chan-Palay.
pia
mol
PC
gr
wm
wm
Go
100 µm
0

in the subjacent white matter. Golgi's superficial cells, also described by GIERKE (1886), probably correspond to the flattened stellate cells later described by TERRAZAS (1897) as occurring rather rarely just under the limiting membrane. They do not contribute descending vertical processes. The deep stellate cells in the granular layer do send smooth processes up into the molecular layer, but according to TERRAZAS (1897) and RAMÓN Y CAJAL (1911) these processes fail to reach the limiting membrane. Instead, they recurve downwards from beneath the surface and end freely in the molecular layer. Consequently, it must be concluded that the overwhelming majority of the Bergmann fibers are actually the externally directed processes of the peculiar cells lying at the upper border of the granular layer, the Golgi epithelial cells.

GOLGI (1885) gave these cells only a sketchy description, saying that their cell bodies occur in several rows among the Purkinje cells and that their ascending processes penetrate through the molecular layer to the pial surface. These processes are characterized by their robust form, a certain rigid appearance, and their bifurcations. His figure shows spheroidal cell bodies arranged in the Purkinje cell layer and giving rise to one or several smooth ascending processes. Two kinds of cells are indicated, one with a small number of vertical processes that bifurcate once or twice and ascend to the limiting membrane, and the other with a multitude of unbranched wavy processes that ascend only as far as the middle third of the molecular layer.

RAMÓN Y CAJAL (1888 a, 1889 a, and 1890 a) gave a more complete description of these cells. Besides affirming the origin of the Bergmann fibers from the neuroglial cells in the Purkinje cell layer, he studied their development in foetal mammals and chick embryos. In these young animals the cells possess not only ascending processes but also extensive descending processes, sometimes penetrating into the white matter. After birth (or hatching) the descending processes atrophy, and in the adult animal they are absent or at most rudimentary. Besides, the perikarya of these cells apparently move from a deep location at various levels in the developing internal granular layer to a more regular array at the margin between the granular and molecular layers in the mature cerebellar cortex. These observations led RAMÓN Y CAJAL to infer that these neuroglial cells are displaced ependymal cells, still retaining the characteristics of their epithelial origin. In later publications (e.g., 1911) he referred to them as epithelial cells (the same name that he used for the components of the ependyma), and this name has become generally accepted in the literature on the neuroglia, despite its confusing connotations. Further ambiguities are added by the common designation "Golgi epithelial cells" in honor of their discoverer.

RAMÓN Y CAJAL (1909 and 1911) also gave names to the two varieties of epithelial cells shown in Golgi's drawing. Those with one to three bifurcating processes he called "forked cells" (*cellules fourchues*), and those with numerous ascending processes he called "horsetail" or "broom" cells (*cellules en queue de cheval, cellules en balai*). He did not, however, consider the difference of any importance. The processes of both varieties, in contrast to the indications of Golgi's drawing, reach the surface of the cerebellum, where they terminate in small conical enlargements with their bases turned toward the pia, forming the limiting membrane. In fact, as both RAMÓN Y CAJAL (1911) and RETZIUS (1894) show, the broom cells are the more important contributors to this membrane.

Perhaps more significant than these taxonomic explanations was RAMÓN Y CAJAL'S discovery that the ascending processes of the epithelial cells are, as he put it at first (1890 a), varicose, Previous authors, from Golgi onward, had all pictured these processes as smooth fibers of uniform caliber except at their tips (for example, VAN GEHUCHTEN, 1891). RAMÓN Y CAJAL, it seems, had a better technique than his contemporaries that brought out or preserved the appendages not only on dendrites, but also on neuroglial cell processes. His early figures (1890 a), however, show these only as irregularities in the caliber of the neuroglial processes. RETZIUS (1892 a) gave a hardly more detailed picture and mentioned that "most of the processes bear small lateral prongs and prickles".[26] But two years later, in an extensive review of his studies on the neuroglia, RETZIUS (1894) not only confirmed RAMÓN Y CAJAL'S embryological conclusions with a much more thorough examination of the neuroglia in development, but he also portrayed the "mossy" form of the processes in numerous drawings. In the younger animal foetuses he drew the cell bodies and ascending processes as shaggy, with fine granular and hair-like appendages that decreased in number and size as the processes entered the external granular layer. With advancing maturity, the cell body became smooth as the descending processes shrank, and the mossy character became restricted to the ascending branches. In the human foetus, however, RETZIUS depicted the ascending processes as varicose at first and later fitted with a few short bristles, which have almost completely vanished in his adult specimen. As a result of further

26 „... die meisten tragen aber kleine seitliche Zacken und Stacheln" (RETZIUS, 1892, p. 24).

studies by RAMÓN Y CAJAL (1896) and TERRAZAS (1897), it became evident that the ascending processes are almost completely "studded with a multitude of lamellae and granular, moderately branching appendages. These appendages and lamellae, after a short, sinuous course, come into contact with one another. In this way they form ... a sponge-like fabric in which the neural elements are lodged".[27] RAMÓN Y CAJAL was at pains to point out that these collateral processes were not artifacts of the Golgi method but were also easily shown in material stained with methylene blue.

In 1916 FAÑANAS, using Cajal's new gold-sublimate method, carried out a detailed study of the neuroglia in the cerebellar cortex of man, cat, and rabbit. With this method he was able to bring out the finest processes of the neuroglial cells in the molecular layer. He described a variety of cell types, with one, two, or more vertical ascending processes—all fitted, at least at their beginning, with fine, grumose appendages that gave them a feathery appearance. The deeper examples of these cells, with somata lying in the Purkinje cell layer and with long processes ascending to the pial surface, were obviously the same as the familiar Golgi epithelial cells. Others of a similar appearance, but with cell bodies strewn through the inner half of the molecular layer and with ascending processes that did not quite reach the surface, constituted a new class, which has since become known as Fañanas cells (JAKOB, 1928; SCHROEDER, 1929; PENFIELD, 1932). They are differentiated from the Golgi epithelial cells only by their location and by their shorter vertical processes. The close relation of all of these cell types with one another and with the Bergmann fibers was recognized by FAÑANAS. In view of the similarities among them, the necessity for adding yet another name seems somewhat forced. These results were later confirmed by SCHROEDER (1929) in a comprehensive re-examination of the cerebellar neuroglia, and the Fañanas cell took its place beside the Golgi epithelial cell as one of the macroglial elements in this cortex (JAKOB, 1928; PENFIELD, 1932; JANSEN and BRODAL, 1958).

The Fañanas cells are said to appear only in preparations made with the gold-sublimate method (JAKOB, 1928; SCHROEDER, 1929). The form of their processes is of particular interest since it is the same as that of the Golgi epithelial cells in the same preparations. The beginning of the process is beset with fine, round sprouts or gemmules, whereas the more distal stretches are completely smooth. FAÑANAS also described and figured in the molecular layer small stellate cells with a few radiating processes that were likewise studded with grumose appendages. This appearance is quite at variance with that of the neuroglial cells in Golgi preparations. There is good reason to think that the Golgi impregnation procedures give a more accurate representation of the detailed form of the epithelial cell and its processes than the gold-sublimate method.

27 «Il ressort de nos recherches que les expansions radiées des cellules épithéliales sont hérissées d'une multitude de lamelles et d'appendices granuleux, modérément ramifiés. Appendices et lamelles, après un court trajet flexueux, viennent au contact les uns des autres; ils forment ainsi dans l'épaisseur de la première couche une sorte de tissu spongieux dont les cavités hébergent les éléments nerveux» (RAMÓN Y CAJAL, 1911, p. 69).

b) The Golgi Epithelial Cell in the Optical Microscope

Fig. 231 shows a row of epithelial cells in a cerebellar folium impregnated according to the rapid Golgi method. Each cell gives rise to two or more processes, which bifurcate and ascend in almost parallel lines to the pial surface. The cells resemble miniature candelabra with their branches either clustered or spread out in the parasagittal plane (Fig. 232). RAMÓN Y CAJAL (1911) pointed out that the neuroglial processes are intercalated between the dendritic trees of successive Purkinje cells. Such an alternation of arborizations does indeed appear when one considers the distribution of only the ascending stems or Bergmann fibers, but, as will be seen from electron microscopy, the finer ramifications and appendages of the neuronal and neuroglial arborizations completely interdigitate and overlap.

In the rat as in other mammals, the cell bodies of Purkinje cells are surrounded by bouquets of Golgi epithelial cells and Fañanas cells. Since the differences between these two related neuroglial cells are not evident in the thin sections examined with the electron microscope, we shall not attempt to distinguish between them. These neuroglial cells are true satellites of the Purkinje cells. As will be seen in the following description and from an examination of Figs. 231, 237, and 241–245, they lie directly against the surface of the Purkinje cells and form a nearly complete sheath around them (Figs. 174, 178, and 180). In the rat, Purkinje cells are occasionally seen that are in immediate apposition with one another, but usually they are spaced far enough apart to permit the intrusion between them of at least some neuroglial processes, if not whole cell bodies of the satellite cells (Figs. 8, 9, 233, and 234).

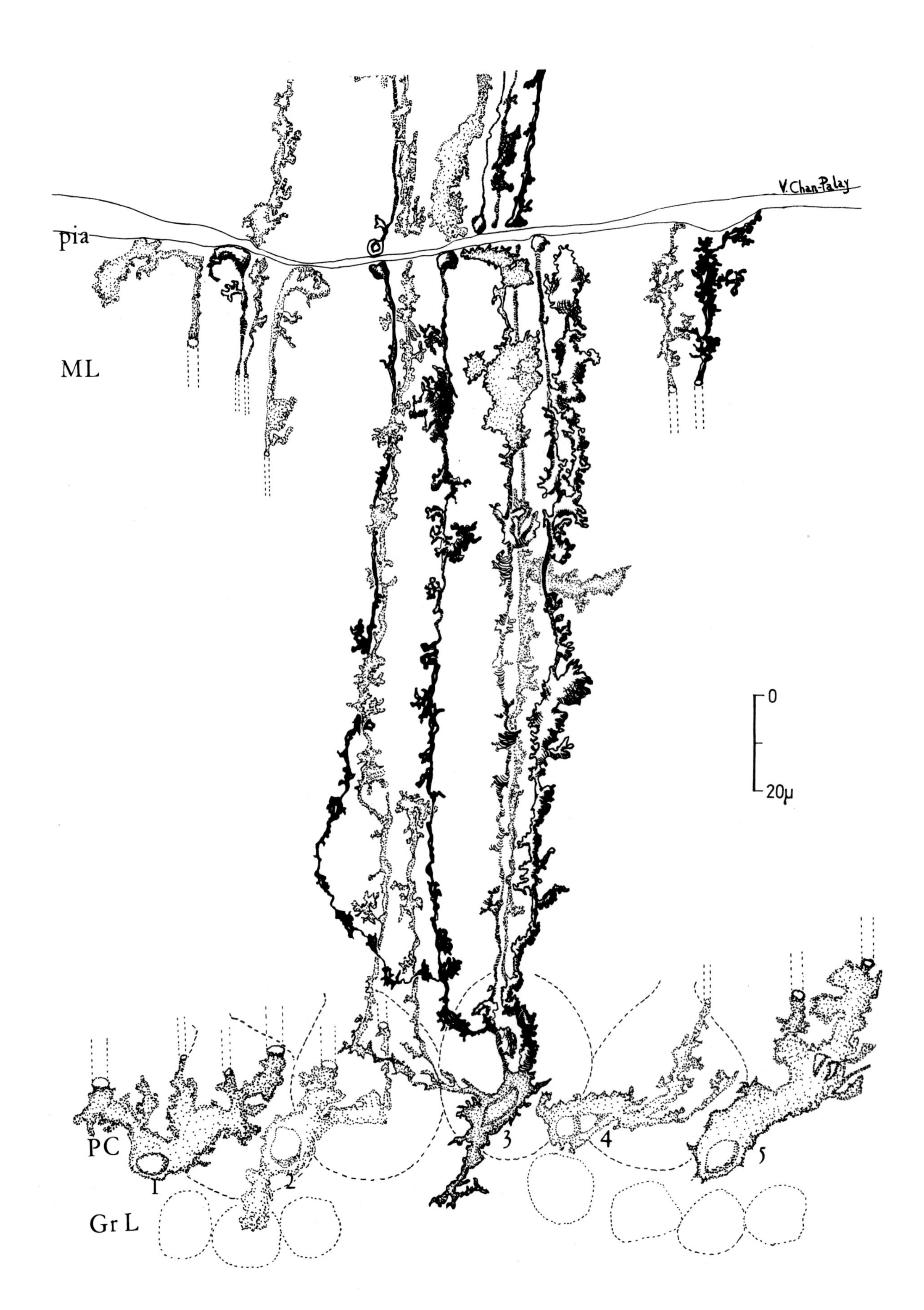
V. Chan-Palay
pia
ML
0
20μ
PC
1
2
3
4
5
Gr L

c) The Golgi Epithelial Cell in the Electron Microscope

The Golgi epithelial cell has a slightly elongated or bean-shaped nucleus, usually oriented with its long axis in the vertical, or radial, direction (Figs. 8 and 9). Although indented, the nucleus is almost never creased or folded. The nucleolus, consisting of the usual pars fibrosa and pars granularis, is located either centrally or peripherally at the nuclear envelope and because of its compactness contrasts sharply with the rather pale chromatin, which is homogeneously dispersed throughout the nucleus. Fine chromatin threads adhere to the inner surface of the nuclear envelope, accentuating the membrane in low power electron micrographs. These threads extend into the karyoplasm, where they become entwined with their neighbors. As they shun the region subjacent to the nuclear pores, they leave little cylindrical clearings in the karyoplasm at the margin of the nucleus, such as have been described in other cell types.

While the inner membrane of the nuclear envelope is smooth and regular, the outer membrane is ruffled and undulating. It sends out short streamers that join the granular endoplasmic reticulum. The outer membrane itself is studded with short, irregularly distributed groups of ribosomes.

The cytoplasm of the Golgi epithelial cell is typically pale and homogeneous. During the preparative procedures it is particularly susceptible to swelling, which causes the intracellular organelles to be widely spread apart. In the light microscope such swelling results in the appearance of clear halos around the Purkinje cells. In electron micrographs a moderate swelling of the neuroglia is sometimes helpful in segregating the elements of the neuropil, but it should not be considered representative of well-preserved tissue.

The cytoplasm is characterized by the presence of a loose granular endoplasmic reticulum (Fig. 235), numerous glycogen particles (Fig. 236), and a few bundles of fine filaments. The endoplasmic reticulum appears in the form of widely dispersed, randomly arranged tubules and small cisternae. Many of these membranous organelles are studded with rows, volutes, and clumps of ribosomes, but many are completely agranular. Ribosomes also occur singly and in small clusters, rosettes, and helices, unattached to membranes. They are so widely distributed over the cytoplasm that in low power electron micrographs the cell matrix appears to be peppered with fine, dark granules haphazardly scattered. This impression is enhanced by the presence of β glycogen particles, dense spherical granules, about 250 Å in diameter, stippled in pairs, clumps, or singly all over the cytoplasm.

Fine filaments, about 80 Å thick course in all directions through the cytoplasm (Figs. 246, 247, and 252), often in the company of microtubules. In favorable planes of section thin fascicles of these filaments can be found, oriented parallel to the long axis of the cell. These fascicles continue into the ascending processes, where they constitute the cores of the Bergmann fibers. The mitochondria of the neuroglial cell bodies are generally small rodlets 0.2–0.5 µm in diameter and 1–2 µm long. They are numerous in some perikarya and rather sparse in others, apparently varying according to the amount of cytoplasm pooled around the nucleus. Lysosomes are also highly variable in number and form. A commonly encountered form is the lipofuscin granule, a lumpy mass, like pudding stone, consisting of fine, dense particles and including pale, spherical or ovoid droplets (Figs. 234 and 238). The Golgi apparatus usually lies at the upper pole of the nucleus, where it contributes to the formation of a centro-

◀

Fig. 232. The Golgi epithelial cell of the cerebellar cortex. The limit of the cortex is marked by the pial surface (*pia*); the molecular layer (*ML*), the Purkinje cell layer (*PC*), and the granular layer (*Gr L*) are indicated. Five Golgi epithelial cells (1–5, stippled) are intercalated among the cell bodies of Purkinje cells (in dotted lines). The somata of these cells bear ragged, irregular, expanded appendages. The processes of cell No. 3 are shown completely, extending from the cell body upwards to the pial surface as in a miniature candelabrum. These processes, the Bergmann fibers, bear highly irregular leaf-like appendages which project horizontally. At the pial surface these processes end in bulbous tips or irregular terminations. Camera lucida drawing, rapid Golgi, rat, 90 µm

▶

Fig. 233. Golgi epithelial cells and an oligodendrocyte in the Purkinje cell layer. Two Golgi epithelial cells (*Go ep*) are recognizable by their thin homogeneous nucleoplasm. The cytoplasm of these cells ramifies into the neuropil (*arrowheads*). The oligodendrogliocyte shows a characteristically dark nucleoplasm and cytoplasm (*olig*). These two types of glial cells are to be compared to the neighboring granule cell (*gr*) with its clumped chromatin dispersed in cartwheel fashion. Two Purkinje cells (*PC*) and a basket cell flank the other sides of these glial cells. The basket cell (*Ba*) displays two nucleoli and several synapses with axons of other basket cells and parallel fibers upon its surface (*arrows*). Lobule III. ×11000

Fig. 234. The Golgi epithelial cell. This Golgi epithelial cell (*Go ep*) lies against a small contingent of parallel fibers (*pf*) passing between it and a Purkinje cell (*PC*). The chromatin in the nucleus is homogeneously dispersed with a thin rim condensed beneath the nuclear envelope. The cytoplasm of the neuroglial cell is thrown up (*arrows*) into pseudopods that insinuate themselves between the nervous elements of the neuropil. A large Golgi apparatus (*Go*) as well as several lysosomes (*Ly*) are conspicuous. To the left of this cell, a vertical column of neuroglial cytoplasm (Bergmann fiber) passes through the molecular layer. Lobule IX. ×14000

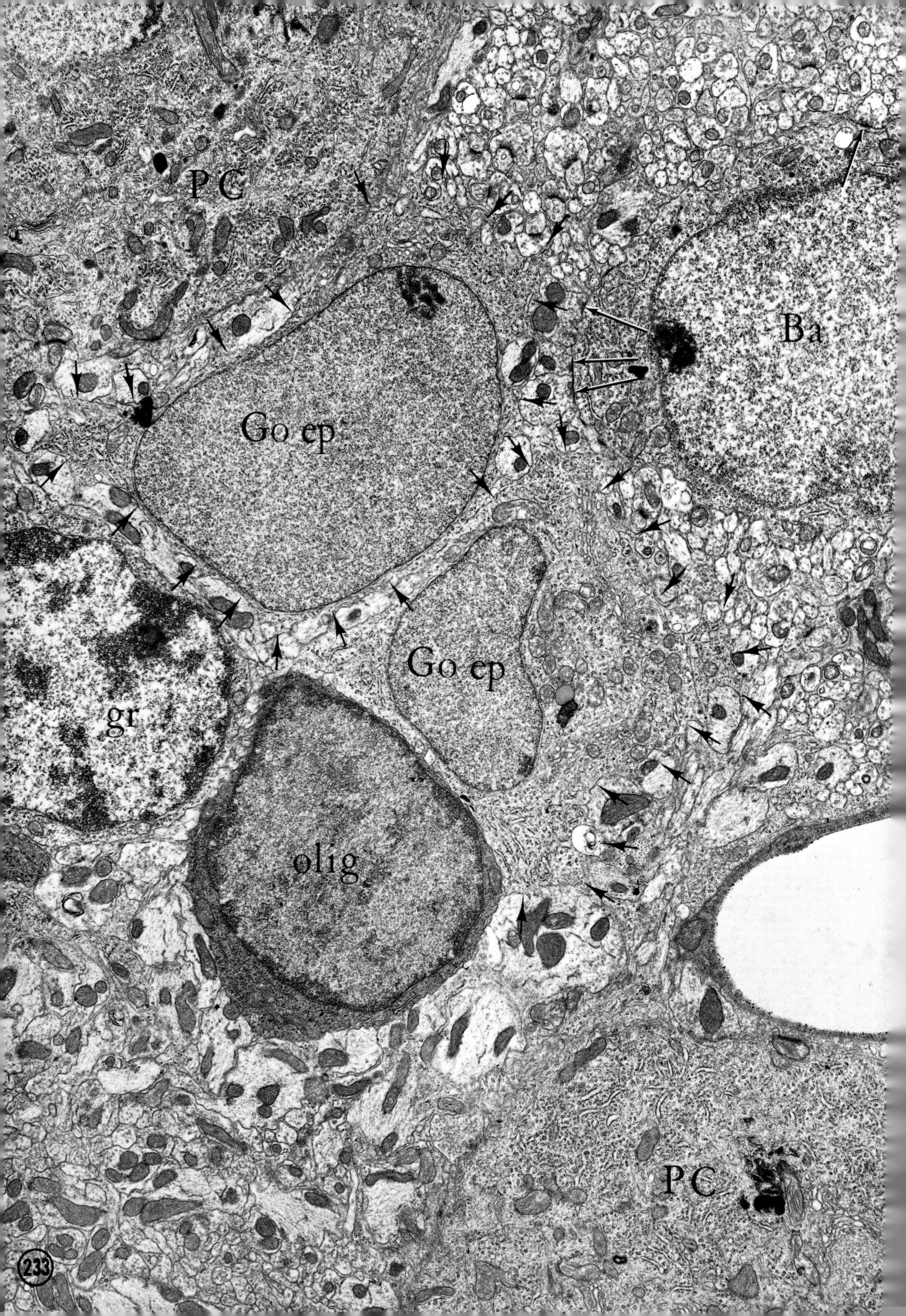
PC
Ba
Go ep
Go ep
gr
olig
PC
233

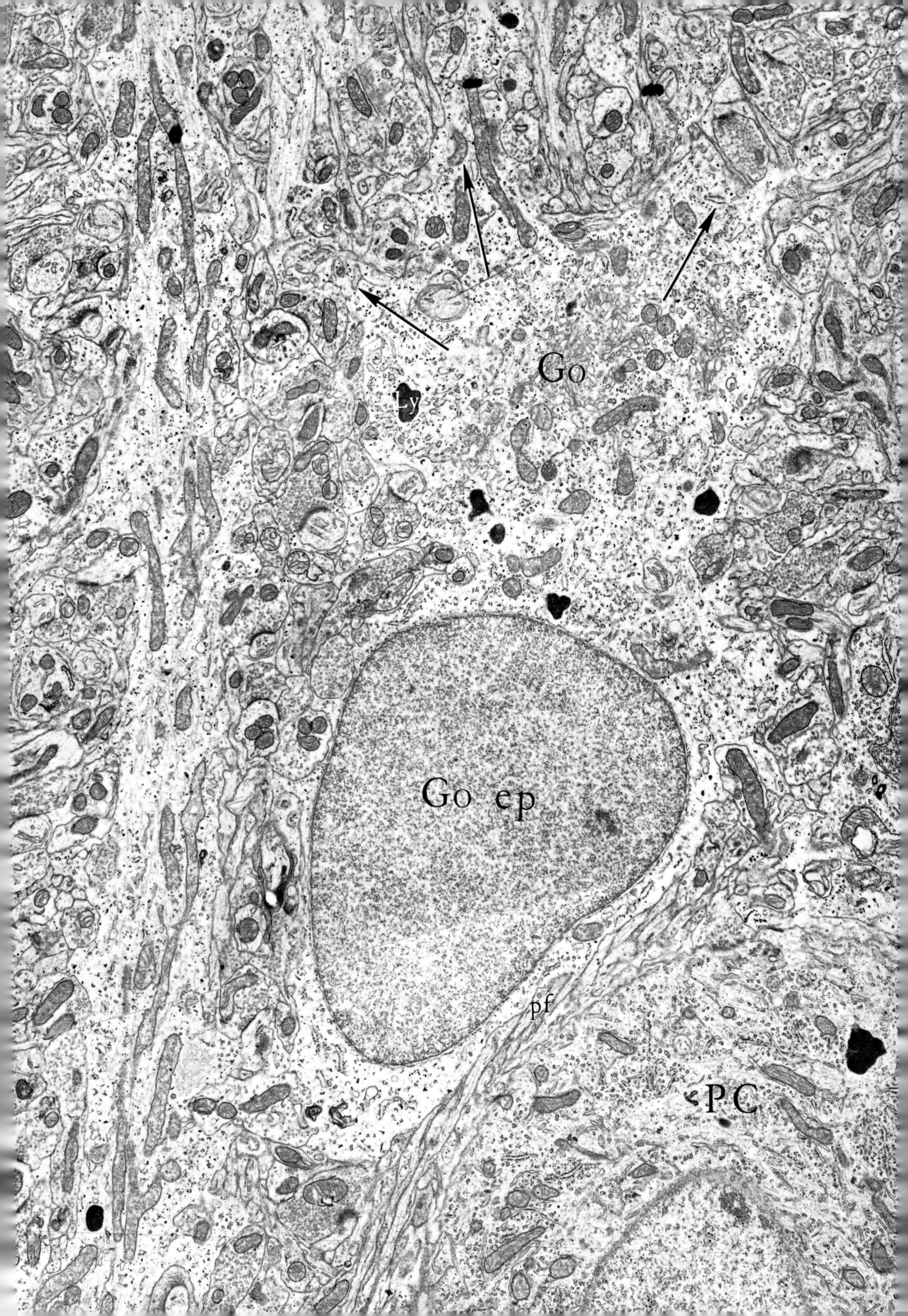
Go
Ly
Go ep
pf
PC

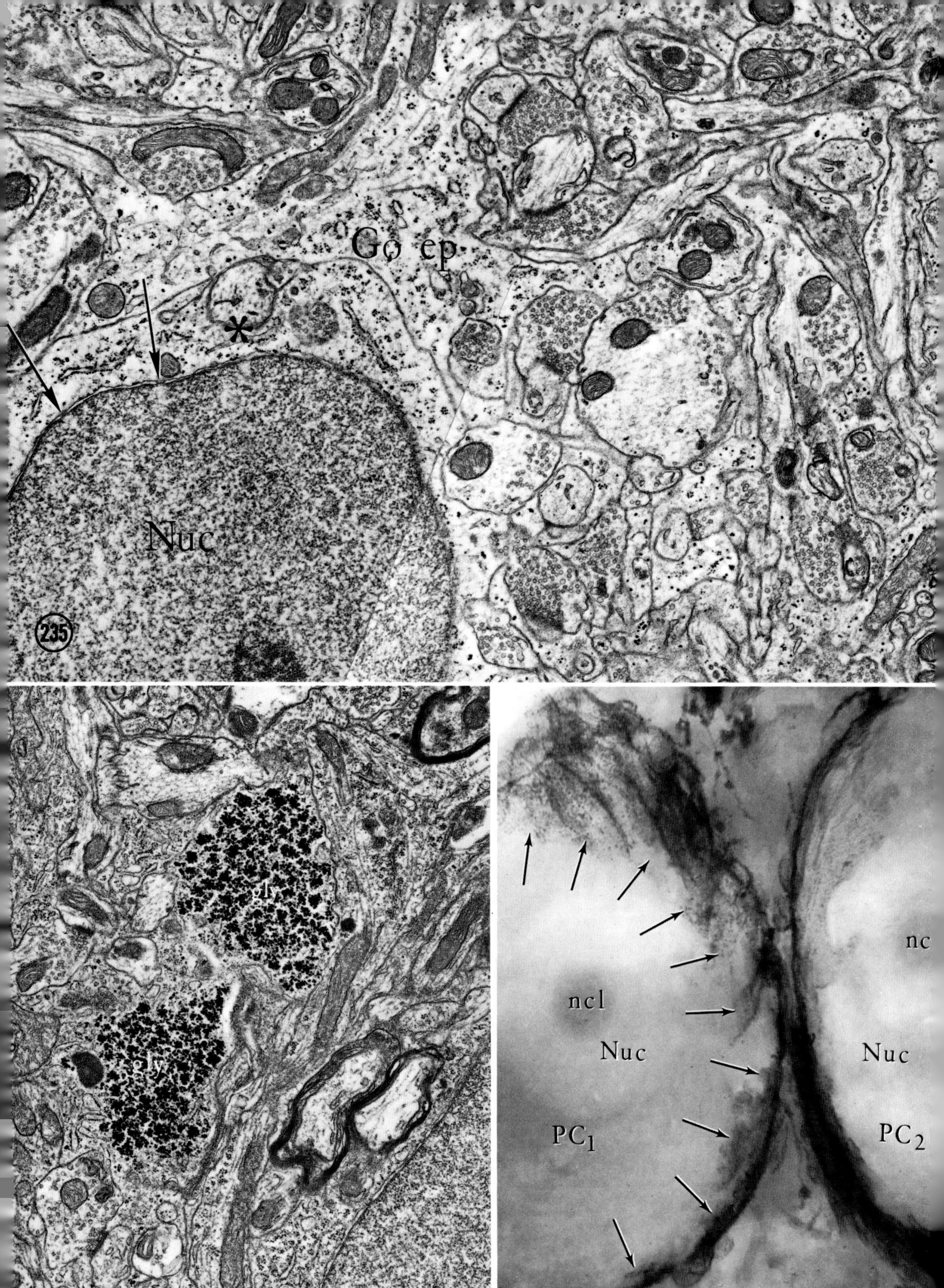

Go ep
*
Nuc
235
gly
gly
236
ncl
Nuc
PC1
nc
Nuc
PC2
237

sphere with its associated vesicles, lysosomes, multivesicular bodies, and centrioles.

The Perikaryal Processes. According to the old accounts and drawings (RETZIUS, 1894; VAN GEHUCHTEN, 1895 and 1897; RAMÓN Y CAJAL, 1911; JAKOB, 1928; SCHROEDER, 1929; PENFIELD, 1932), the cell body of the Golgi epithelial cell is a rather smooth globular structure with few processes except for the stems of the Bergmann fibers. With the gold-sublimate method (FAÑANAS, 1916; SCHROEDER, 1929), these perikarya appear to be studded with tiny granular sprouts. In our own Golgi material, the cell bodies display highly irregular, ragged contours caused by numerous folial expansions, which obscure the cell body of the neuroglial cell in their folds (Fig. 232). Electron micrographs of thin sections show that the cell body gives off numerous lamellar processes that penetrate into the surrounding neuropil. These thin sheets flow over the surface of the Purkinje cell body, overlapping with similar lamellae from other satellite cells to form a continuous sheath, one to several layers thick, completely enshrouding that perikaryon (Figs. 174, 178, 180, and 237). In order to articulate with the Purkinje cell surface, basket fibers descending around the cell body must slip between the neuroglial lamellae, establish their synapses *en passant* by means of slight enlargements, and then slip out again on their way to the pinceau. The ramifying terminals of Purkinje cell recurrent collaterals must execute a similar maneuver at the base of the perikaryon in order to synapse there, and climbing fibers and other neural elements that usually do not synapse on the cell body in the adult are kept at a significant distance by the neuroglial sheath.

This delicate lamellar sheath can be seen in Golgi preparations at the light microscope level, in which the impregnated structures are too readily interpreted as a nonspecific metallic deposit. In such preparations the cell body of the Purkinje cell appears to be encrusted with a thin red coat. Electron microscopy of thin sections of the Golgi preparation shows that the deposit is within the neuroglial sheath (BLACKSTAD, 1970).[28] The sheath is displayed most dramatically in stereograms of Golgi preparations taken with the high voltage electron microscope (CHAN-PALAY and PALAY, 1972c). As can be seen in Fig. 237, the Purkinje cell body is enveloped, like an egg in its shell, in a closely applied membranous sheath that is impregnated with the Golgi deposit. The high voltage electron micrographs clearly show the folds in the sheath, the presence of several layers, and hoop-like ribs that thicken the lamellae here and there. Since the complete sheath is formed by the imbrication of lamellae from several satellite cells, the Golgi method, which impregnates only selected cells, tends to give a partial rendition of the sheath around any particular Purkinje cell.

A more detailed picture is given by electron microscopy of thin sections of material prepared in the conventional way. The lamellae arise as tongues of cytoplasm given off from the different aspects of the cell body. They insinuate themselves among the neural processes in the neuropil, and, arriving at the surface of a Purkinje cell perikaryon, they expand over its surface, following its contours and sending rib-like extensions down along the initial segment of the axon (Figs. 40 and 184). Each lamella can vary from a few hundred angstroms in thickness, barely more than the reduplicated plasmalemma of the neuroglial cell, to a thick layer of cytoplasm replete with ribosomes and glycogen granules. In some lamellae, a fine fascicle of filaments is encountered, usually lodged in a local thickening or ribbon of cytoplasm. These, apparently, are the basis for the ribs seen in the high voltage electron micrographs. Adjacent lamellae are tied together from place to place by short gap junctions. At these sites the apposing plasmalemmas appear rigidly parallel and close together. The junctions are narrow ribbons, zonulae, and are associated with little cytoplasmic differentiation. It is not known whether these junctions occur between lamellae of different cells or different lamellae of the same cell, or both.

The Bergmann Fibers. The vertical, ascending processes, the Bergmann fibers, given off from the apical poles of

◀

Fig. 235. Golgi epithelial cell body. The nucleus (*Nuc*) of this Golgi epithelial cell (*Go ep*) shows several nuclear pores (*arrows*). Portions of the outer leaflet of the nuclear envelope bear short strings of ribosomes (*asterisk*). Small amounts of granular endoplasmic reticulum and polysomes are strewn about in the thin cytoplasm. Processes of the neuroglial cell pervade the surrounding neuropil. Lobule X. ×17000

Fig. 236. Glycogen in neuroglia. Glycogen accumulations are common in the processes of neuroglial cells and are flags for their identification. These two deposits of glycogen (*gly*) are unusually large for processes of Golgi epithelial cells. Lobule X. ×13000

Fig. 237. The neuroglial sheaths around the somata of two Purkinje cells. Although the bodies of two Purkinje cells (PC_1 and PC_2) are unstained, the shadowy images of the cytoplasm, nuclei (*Nuc*) and nucleoli (*ncl*) can be seen. The thin, veil-like neuroglial sheaths form a wrinkled partial covering over the surface of these two cells (*arrows*). These sheaths represent intracellular impregnation of the ragged appendages belonging to Golgi epithelial cell somata (Fig. 232) by the rapid Golgi method. High voltage electron micrograph, rhesus monkey, 5 µm. Lobule X. ×2500

28 BLACKSTAD, however, interpreted his figure as showing an *extracellular* deposit on the surface of the Purkinje cell body.

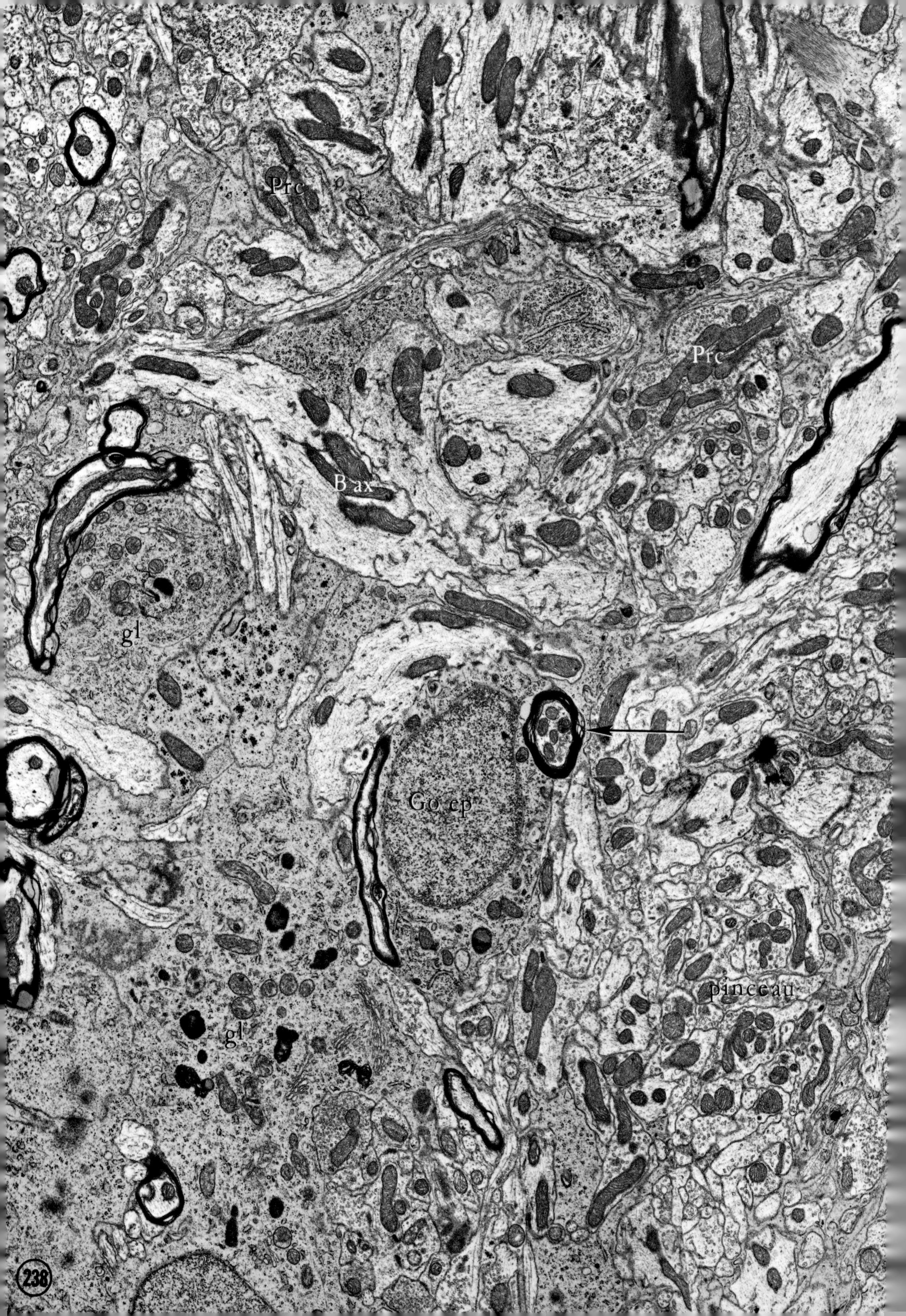

Prc
Prc
B ax
gl
Go ep
pinceau
gl
238

the Golgi epithelial cells, run nearly straight through the neuropil of the molecular layer toward the pial surface (Figs. 231 and 232). The cytoplasm of these processes contains few organelles besides an axial fascicle of fine filaments and single, very long, slender mitochondria. Even in thin sections such a mitochondrion can be followed for 5 or 10 microns. In addition to these structures, axially oriented microtubules and tubular elements of the endoplasmic reticulum run alongside the fascicle of filaments. Finally, a generous sprinkling of glycogen particles fills in otherwise empty areas of the cytoplasmic matrix.

Although in Golgi preparations the Bergmann fibers usually are as discrete as the candles in a candelabrum, electron micrographs of thin sections show them to be closely associated with one another (Fig. 239). They are usually encountered in pairs or small bundles, an observation that suggests that the ascending processes of neighboring cells are all interdigitated to form a dense forest rather than the alternating or intercalated palisades described by RAMÓN Y CAJAL (1911). Adjacent processes share a long, straight interface (Figs. 239 and 240), which is marked by densities adherent to the cytoplasmic surfaces of the apposed plasmalemmas and filling up the interstice between them (Fig. 240). This junction resembles the zonula adhaerens of other epithelia (FARQUHAR and PALADE, 1963) and the ependyma (BRIGHTMAN and PALAY, 1963; BRAWER, 1972). Gap junctions have also been seen at this level of the ascending processes. One of these junctions is shown in Fig. 240A, a freeze-fracture preparation of a bundle of ascending Bergmann fibers. An extensive gap junction is exposed in the lower half of the figure, with particles arranged in hexagonal array on the A face corresponding to an array of pits in the coapted B face of the adjacent fiber. This is the typical substructure of the membranes in gap junctions between epithelial or cardiac muscle cells (GOODENOUGH and REVEL, 1970; MCNUTT and WEINSTEIN, 1970) and differs from these only in the extent of the surface involved in the junction. Similar gap junctions occur between neuroglial cells in the granular layer. This figure also shows small groups of 4, 6 or 9 particles arrayed in a more rectilinear array, dispersed around the edges of the main gap junction and over the A faces of neighboring neuroglial processes. Such minute rectilinear arrays have been described by STAEHELIN (1972) in association with gap junctions in the intestinal epithelium of the rat. We have also seen them on neuroglial surfaces in the granular layer. Their significance is unknown.

It is not possible at the present time to tell whether the ascending processes involved in the zonulae adhaerentes and the gap junctions originate from the same cell, neighboring cells, or both. The junctions may be vestiges of the early ependymal stage of these cells, and may be considered a confirmation of RAMÓN Y CAJAL's (1889a and 1890a) surmise concerning their origin. The fascicles of Bergmann fibers are usually accompanied by various ascending neural processes: bundles of granule cell axons (Figs. 73a and b), climbing fibers, ascending collaterals of basket cell axons, spiny branchlets of Purkinje cell dendrites, Golgi cell dendrites, and basket cell dendrites. Frequently synaptic articulations are effected between pairs of these neural processes within the shelter of the vertical neuroglial fibers. For example, the varicosities in the course of the ascending granule cell axon synapse *en passant* with thorns of Purkinje cell dendrites that project through their neuroglial capsule. It is important to rec-

Fig. 238. Neuroglia in the infraganglionic region. This illustration provides a view of the infraganglionic region—certainly the most complex neuropil in the cerebellar cortex. Here the neuroglial profiles have been colored yellow in order to render them more conspicuous. The cell bodies of two Golgi epithelial cells (*Go ep*) are present. Their processes as well as those of other neuroglial cells invade the surround. Many profiles of basket axons (*B ax*) provide a tangle in the neuropil among unidentified myelinated axons, a myelinated Purkinje axon (*arrow*), two terminals of recurrent collaterals of Purkinje axons (*Prc*), and the fine terminals of basket cell axons in a pinceau. Lobule V. ×12000

Fig. 239. Ascending bundles of neuroglial processes in the molecular layer. The ascending processes of Golgi epithelial cells, or the Bergmann fibers, ascend in bundles from the Purkinje cell layer to the pial surface. Here, one such bundle is shown (*arrowheads*). Adjacent neuroglial processes in a bundle are bound together by specialized intercellular junctions (*between arrows*). See next figure. Lobule V. ×15000

Fig. 240. Intercellular junctions between ascending neuroglial processes. Adjacent columns of neuroglial processes on the Bergmann fibers have junctional specializations (*between arrows*). The interstitial space between apposed cell membranes is filled with a dense, filamentous material. The cell membranes are held rigidly parallel with each other, and filamentous densities adhere to the cytoplasmic surfaces symmetrically on both sides of the junction. Osmium tetroxide fixation. ×49000

Fig. 240A. Gap junction between Bergmann fibers. Two Bergmann fibers ascend through the center of the field. Their broken ends are visible at the bottom edge of the field. In the lower half of the picture an extensive gap junction occupies most of the exposed fracture face of one of these fibers. This junction is characterized by hexagonal arrays of close-packed A face particles upon which fragments of the B face of the coapted adjacent cell are superimposed. A complementary hexagonal array of fine pits occurs in this B face. The arrows point to small rectilinear clusters of A face particles found in association with interneuroglial gap junctions. The large arrow indicates the direction of shadowing. Replica, freeze fracture. ×90000

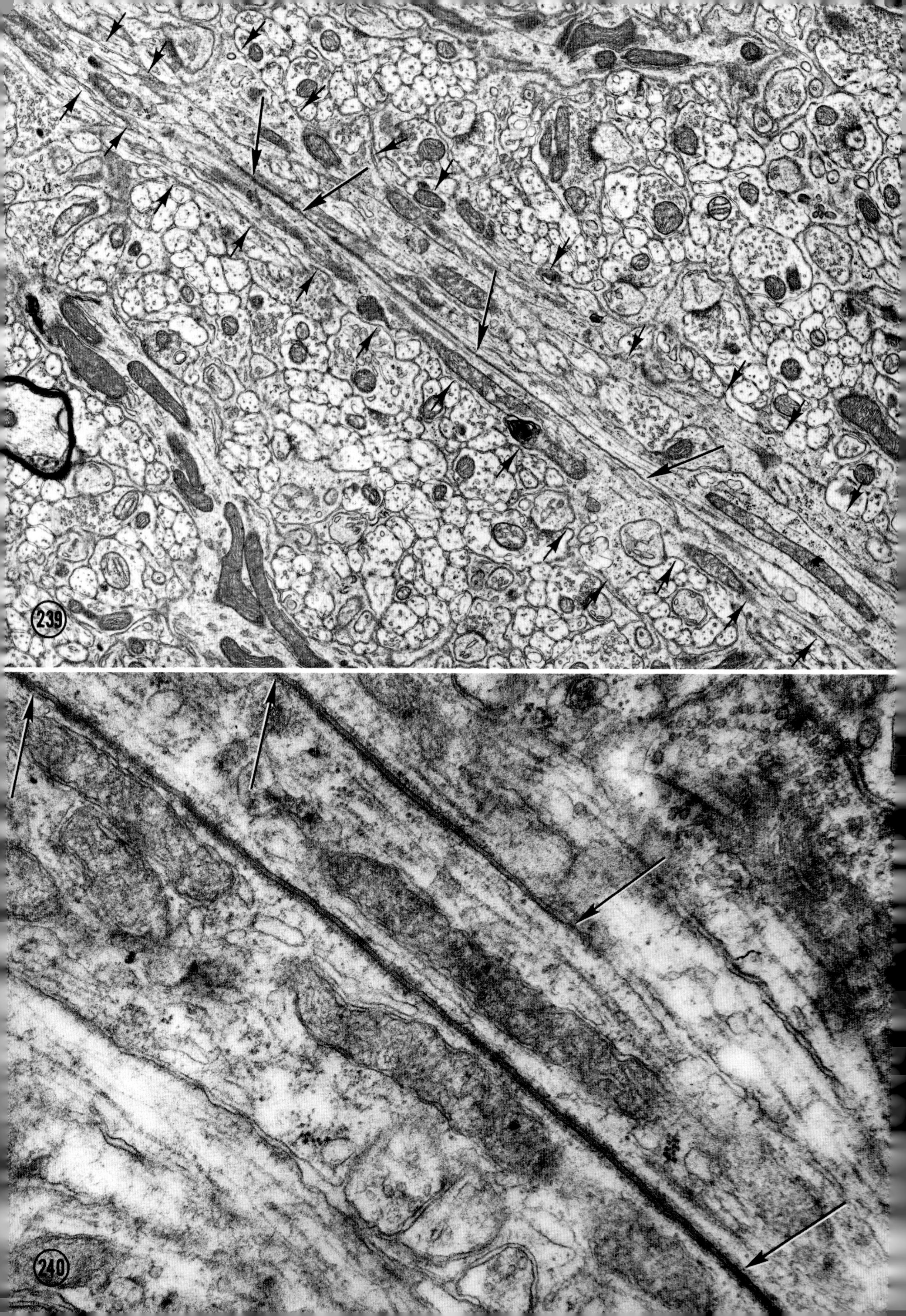

239
240

A
B
240 A

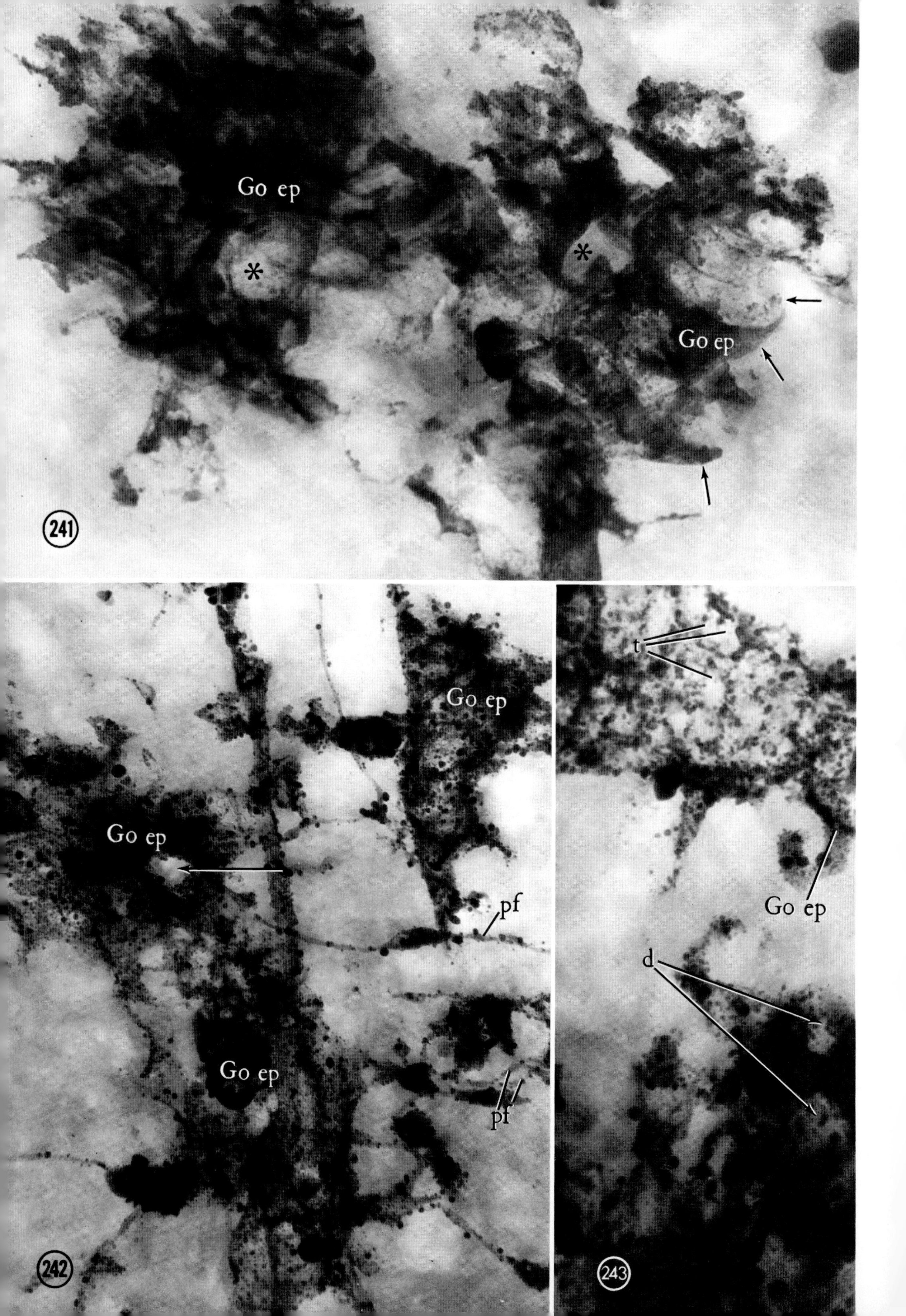
Go ep
*
*
Go ep
241
Go ep
Go ep
pf
Go ep
pf
242
t
Go ep
d
243

ognize that the ascending neural structures are not actually ensheathed by the Bergmann fibers; they are merely accompanied by them. Often the neural structures run in grooves in the surface of the Bergmann fibers. This arrangement is consistent with observations made on the immature cerebellar cortex suggesting that the Bergmann fibers guide the migrating granule cells (and presumably other neural processes) in their descent from the embryonic external granular layer through the molecular layer (RAMÓN Y CAJAL, 1911; MUGNAINI and FORSTRØNEN, 1966, 1967; RAKIC, 1971).

Each Bergmann fiber, as originally noted by RAMÓN Y CAJAL, is ornamented with small lateral offshoots of irregular shape and dimensions. When Golgi preparations are examined at low magnifications in the light microscope, the fibers appear to be encrusted with scale-like and crystalline deposits, like corroded wires. It would seem reasonable at first sight to suspect that these deposits are simply artifacts of overimpregnation. At higher magnifications, however, these incrustations prove to be leaf-like appendages similar to those given off by the cell bodies. They are highly polymorphic, some very small with a short, thin petiole and a rounded blade, others broad and sessile, and still others thick and fleshy. When examined in the high voltage electron microscope (Figs. 241–243), these foliar processes appear as extremely thin, involuted lamellae, filled with the fine-textured meshwork that is characteristic of the red Golgi impregnation (CHAN-PALAY and PALAY, 1972b).

Fig. 241. Laminar processes of the Golgi epithelial cell. The two dark areas in this micrograph represent scrolls of the leaf-like processes of a Golgi epithelial cell (*Go ep*). These veil-like sheets curl around (*arrows*), ensheathing the dendrites of Purkinje cells and, to a much smaller extent, the other cells of the molecular layer. The two channels marked (*asterisks*) in this illustration are probably occupied by the larger unstained dendrites of a Purkinje cell. Rapid Golgi preparation, high voltage electron micrograph, rhesus monkey, 5 μm. Lobule X. ×14000

Fig. 242. Processes of the Golgi epithelial cell. This high voltage electron micrograph shows three sets of laminar processes belonging to the Golgi epithelial cell (*Go ep*). They are thin and veil-like, often with a perforation through which a Purkinje cell dendrite may pass (*arrow*). Several parallel fibers (*pf*) course through the field. Rapid Golgi preparation, rhesus monkey, 1 μm. Lobule X. ×13000

Fig. 243. Neuroglial sheaths for Purkinje cell dendrites and thorns. Two sets of laminar processes of Golgi epithelial cells (*Go ep*) are indicated. In one, a honeycomb of small circular perforations is visible throughout the scroll of neuroglial cytoplasm. These are probably the sites occupied by Purkinje cell dendritic thorns (*t*) protruding through the neuroglial sheath. In the second set of processes, the larger perforations probably represent sites occupied by Purkinje cell dendrites (*d*). Lobule X. ×20000

Paired stereoscopic electron micrographs reveal that the curved or scrolled appendages enclose tubular channels of various sizes (Figs. 241–243) and are themselves formed of a fine lacework through which the thorns of Purkinje cell dendrites protrude (Fig. 243). RAMÓN Y CAJAL (1911) explains that the spaces in between the appendages of the Bergmann fibers are either small and circular, carrying, "like the insulators on telegraph poles," the nonsynaptic portions of the parallel fibers, or large and irregular, stuffed with the somata and dendrites of stellate and basket cells.[29] He inferred that the neuroglial processes prevent contacts between axons or between dendrites, while permitting axons and dendrites to articulate with each other. Electron microscopy of thin sections shows that although he was partially correct in principle, RAMÓN Y CAJAL had misidentified the structures that lie in the spaces enclosed by the neuroglial appendages. It is the Purkinje cell dendritic tree that is ensheathed with the neuroglial processes and not the parallel fibers, basket, and stellate cells, as he imagined.

The lateral appendages resemble those given off by the cell bodies. They appear in electron micrographs as extraordinarily thin sheets of cytoplasmic matrix penetrating between the neural processes to reach the surface of the Purkinje cell dendrites, which they clothe in a tightly fitting veil. Particles of glycogen, thin fascicles of neuroglial filaments, and occasionally mitochondria occur in the thicker ribs of these processes. Most of the appendages consist of little more than the reduplicated plasmalemma. This sheath is essentially confined to the Purkinje cell and its processes. The other cells of the cerebellar cortex are not ensheathed at all, or at least not so completely. As was mentioned earlier, the initial segments of stellate and basket cell axons are often enclosed in sheets of neuroglial cytoplasm, but since the rest of the cell is naked, this sheathing may be only a coincidence due to the passage of the initial segment close to the Purkinje cell dendrites and their neuroglial envelope. The sheath about the Purkinje cell dendritic tree is usually only one

29 «Les espaces réservés par les appendices latéraux des cellules épithéliales sont de deux sortes: les uns sont étroits et circulaires et soutiennent, comme les isolateurs des poteaux télégraphiques, les fibres parallèles dans les points où celles-ci ne s'articulent pas avec des expansions protoplasmiques; les autres sont grands, irréguliers et très nombreux dans le tiers inférieur de la couche moléculaire, où ils logent les cellules étoilées, leurs arborisations protoplasmiques, ainsi que celles des neurones à cylindre-axe court … En résumé, les choses sont disposées ici comme dans les autres centres gris, de façon qu'il ne puisse s'établir aucun contact entre les fibrilles axiles ou entre les expansions dendritiques. Il existe, au contraire, toutes sortes de facilités pour que ces deux espèces différentes de conducteurs entrent en connexion l'une avec l'autre» (RAMÓN Y CAJAL, 1911, p. 70).

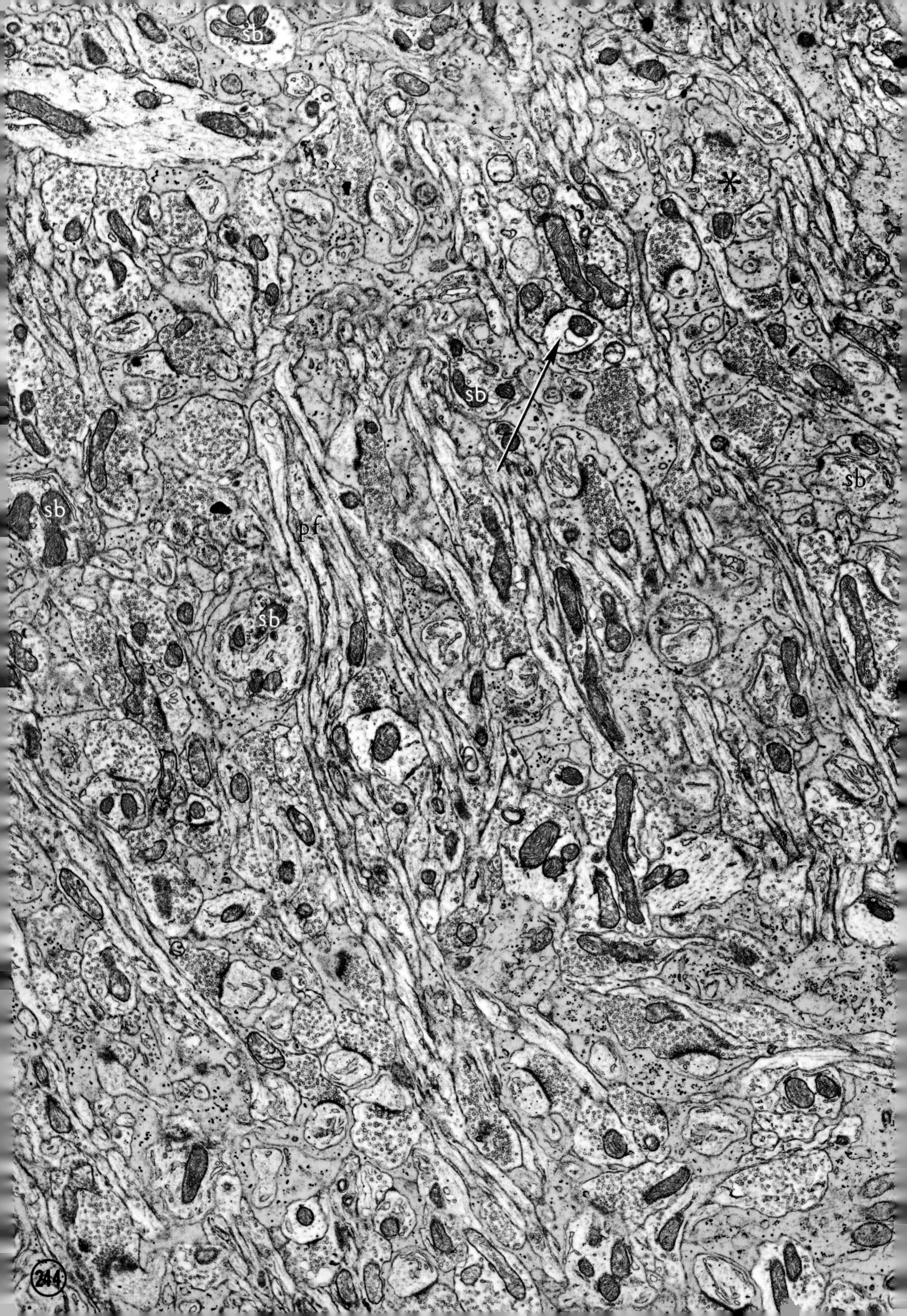

sb
*
sb
sb
sb
pf
sb
244

or two layers thick, and in some rare places it is absent altogether. It follows all the projections of the dendritic system including the thorns (Figs. 30 and 243–245). Synapses must be effected by penetrating this sheath. In the case of the parallel fiber synapses, the dendritic thorns project through the sheath, poking their heads just far enough out of it to articulate with the passing parallel fiber (Figs. 244 and 245). In the case of the climbing fiber synapse, which involves a larger presynaptic varicosity, the thorns may be more exposed, often protruding into recesses in the climbing fiber (Fig. 211). Stellate and basket axons, in contrast, are obliged to penetrate the sheath in order to articulate with the shaft of the dendrite (Fig. 169). The neuroglial investment is, therefore, much more perforate than might appear at first sight. The perforations, however, are nearly all related to the passage of elements participating in synaptic articulations.

The Subpial Terminals. The Bergmann fibers terminate just beneath the pia mater with rounded expansions, which RAMÓN Y CAJAL (1911) described as conical swellings. The flat bases of these swellings are turned toward the pia, and together they constitute the external glial limiting membrane of the cerebellar cortex. In our Golgi preparations of the rat's cerebellum two sorts of terminal enlargements can be distinguished (Figs. 232 and 248): first, a shapeless, lumpy terminal segment that spreads out at the surface of the brain, and second, a smooth, bulbous or conical tip that abuts against the pia. Both of these types can be found topping the ascending processes of any particular cell.

The first kind of terminal is different from the ones traditionally described. Its irregular contours make it hardly less shaggy than the rest of the process bearing it. In some instances numerous leaf-like or bud-like appendages cluster so thickly on its surface that they obscure the underlying rounded swelling. In electron micrographs of thin sections this type of terminal appears as a broad foot process of irregular outline, with its smoother face thrust against the basal lamina at the surface of the brain. A variety of profiles is encountered, from a simple leaf-like fold of the plasmalemma to a robust ridge or hillock of cytoplasm containing any of several organelles. Fascicles of neuroglial filaments and loose cisternae and tubules of the endoplasmic reticulum can be present, amidst a scattering of ribosomes and glycogen particles. The thin leaf-like processes, like those forming the perineuronal sheaths around Purkinje cells, often associate into imbricated laminae. The glial limiting membrane formed in this way has a wide range of thicknesses and can, in fact, be absent altogether. Neural processes can therefore approach exceedingly close to the surface of the brain and even lie directly against the basal lamina (Fig. 82). The flattened foot processes of the Golgi epithelial cells also contribute to the glial limiting membrane of the perivascular spaces within their territories (Figs. 249 and 252).

The second type of terminal is a flattened, spheroidal, cuboidal, or prismatic button corresponding precisely to the conical enlargement described by RAMÓN Y CAJAL (1889a and 1890a) and other early authors. This terminal usually contains a more or less complicated whorl of membranes derived from the endoplasmic reticulum. One or two small mitochondria (Fig. 251) or lysosomes (Figs. 246, 247, and 250) lie in the center of the whorl, which consists of a collapsed cisterna loosely wound around them or of a set of concentric, anastomosing cisternae enclosed one within the other (Figs. 246, 247, 250, and 251). Sometimes both configurations exist in the same whorl. The membranes are usually free of ribosomes, but pale granules of similar size are suspended in the matrix intervening between cisternae. Neuroglial terminals of this second type are often arrayed in rows. They can also be interspersed with the foot processes of the first type

◄

Fig. 244. Neuroglia in the molecular layer. This horizontal section through the molecular layer shows streams of parallel fibers (*pf*) coursing through and curving around thorns and dendrites in their paths. The neuroglial profiles in this figure have been identified and painted in yellow. Spiny branchlets of Purkinje cell dendrites (*sb*) and thorns occur regularly throughout the field. Most of the axonal profiles belong to parallel fibers. A synaptic varicosity (*asterisk*) of one parallel fiber synapses with two Purkinje cell dendritic thorns. A Golgi dendrite (*arrow*) synapses with two parallel fibers and a stellate cell axon. Lobule IX. ×14000

►

Fig. 245. The neuropil of the molecular layer. This illustration shows the complex interrelations of the molecular layer. The profiles of neuroglial cytoplasm have been identified and painted yellow. Neuroglia covers the surface of Purkinje cell dendrites (*Pc d*), their spiny branchlets (*sb*) and thorns. In order to effect synapses, axons have to penetrate this sheath. Several profiles of climbing fibers (*c*) illustrate this. Two bundles of ascending axons of granule cells (*gr ax*) are shown, as well as packets of parallel fibers in cross section (*pf*). Lobule V. ×11500

Figs. 246 and 247. Neuroglial end feet at the pial surface. The molecular layers of two folia meet at a common pial surface (*pia*). Two types of terminal processes of the Golgi epithelial cells are present at this surface. One of these is a flat, leaf-like end foot (*foot*), the other is a conical knob containing whorls of collapsed cisternae derived from endoplasmic reticulum (*knob*). Both of these terminal processes contain packets of filaments (*f*). The processes are pressed against the basal lamina of the pia forming the sub-pial membrane or the glial limiting membrane. Fig. 246, Lobule I. ×17000; Fig. 247, Lobule I. ×13000

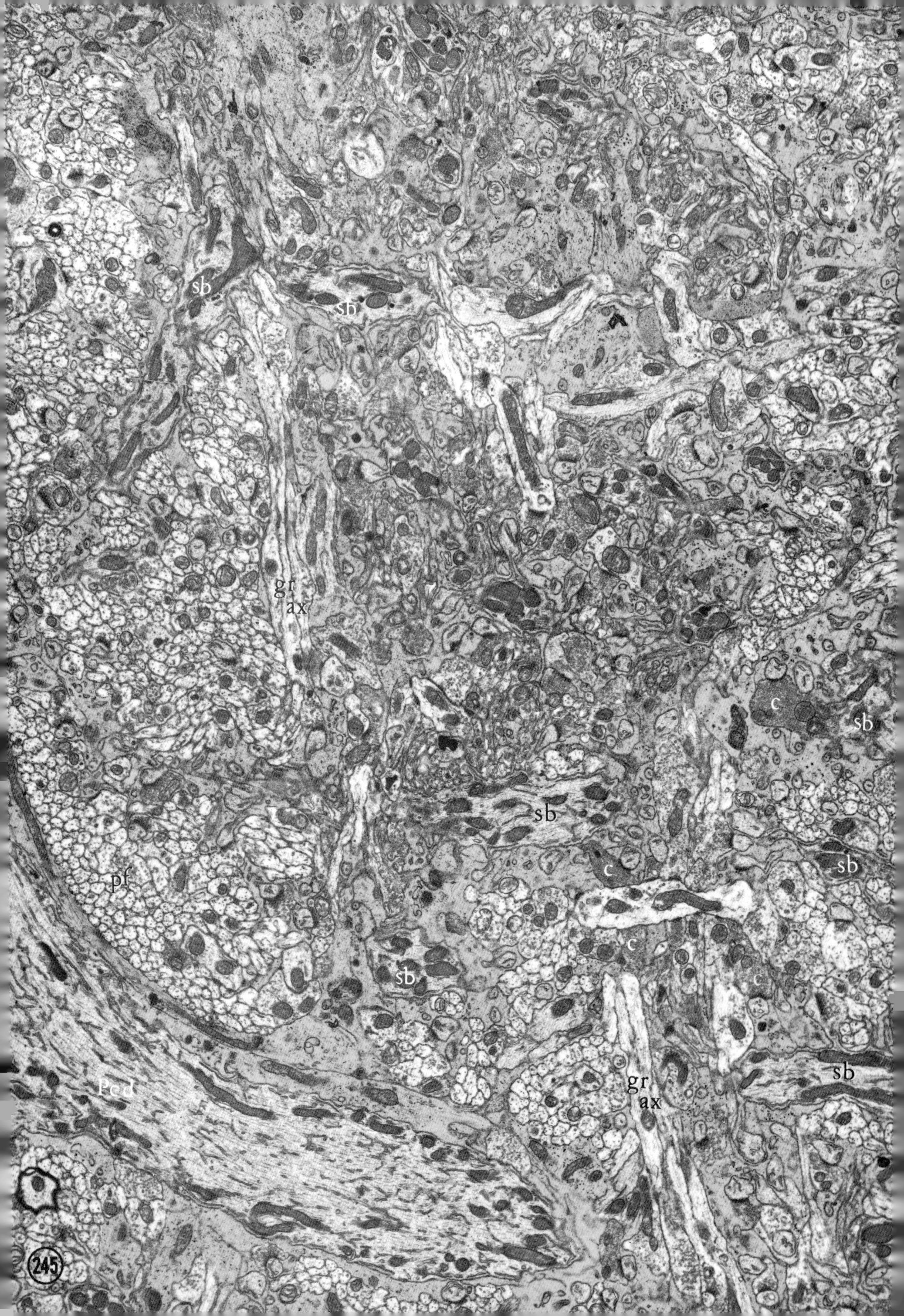
sb
sb
gr
ax
c
sb
sb
sb
c
c
pf
sb
c
sb
gr
ax
Ped
245

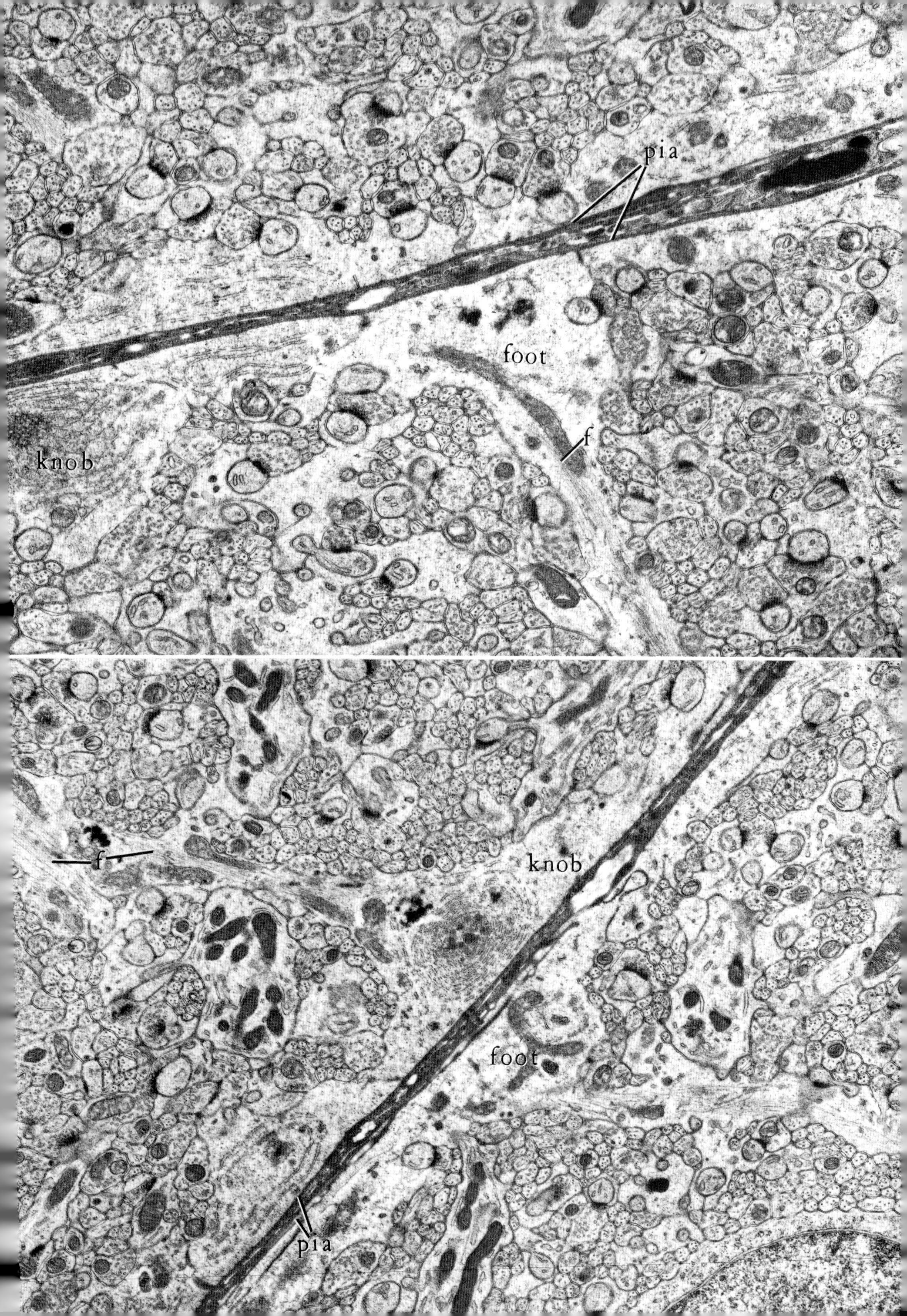
pia
foot
f
knob
f
knob
foot
pia

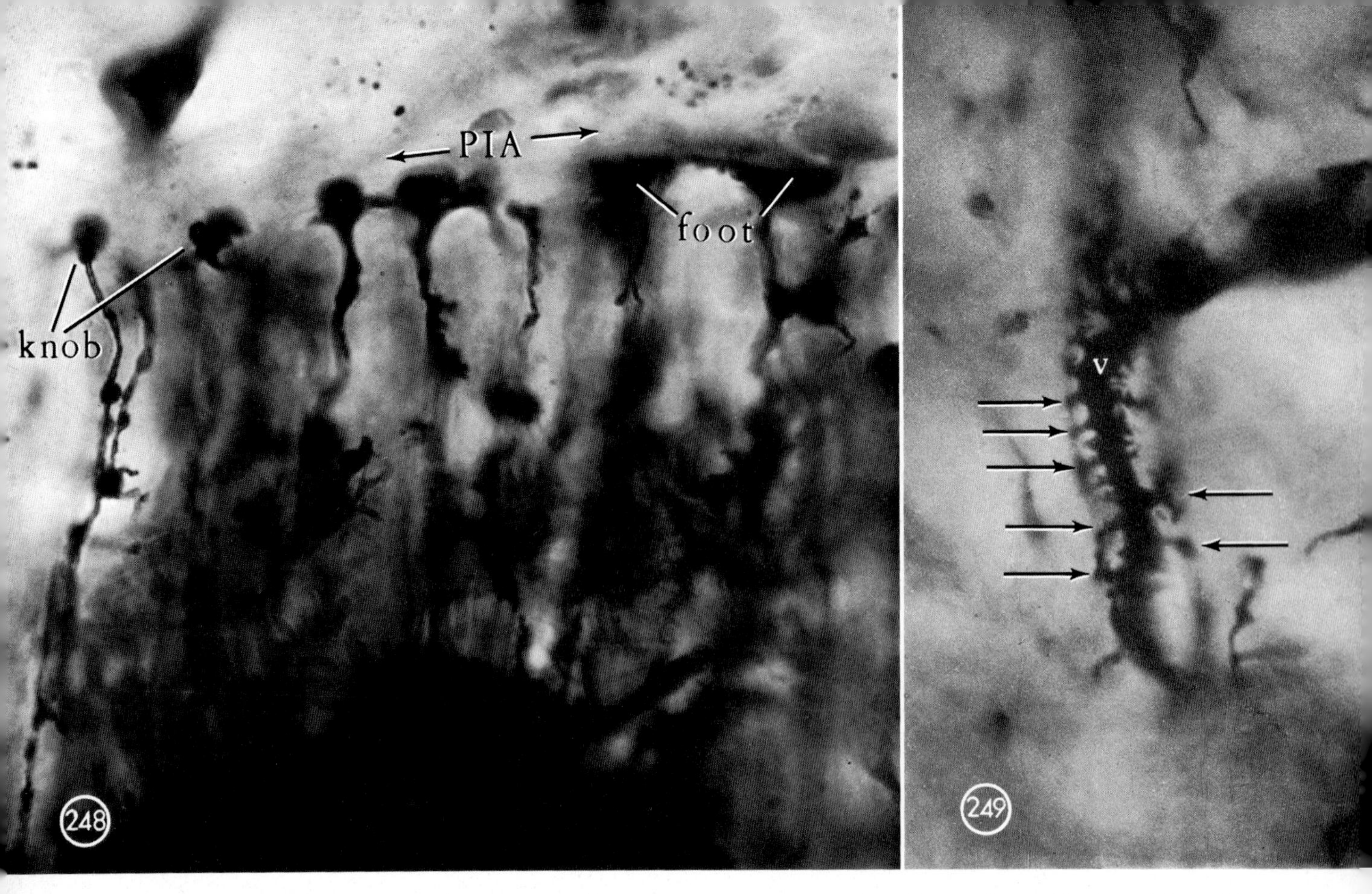

Fig. 248. Terminal processes of Golgi epithelial cells at the pial surface. The pial surface (*PIA*) is lined with the terminal processes of Golgi epithelial cells. Two kinds are present, the end knob (*knob*) and the end foot (*foot*). See Figs. 232, 246, and 247. Rapid Golgi, 90 μm, sagittal, adult rat, Lobule V. ×1725

Fig. 249. Terminals of neuroglia on a blood vessel. Larger blood vessels within the cerebellar cortex are covered by endfeet and knob terminals of neuroglial cells. This particular vessel (*v*) in the granular layer shows both of these terminations. Rapid Golgi, 90 μm, sagittal, adult rat. ×1900

(Fig. 232). These terminals appear not only at the outer surface of the cerebellar cortex (Figs. 232 and 246–248), but also in the glial limiting membrane surrounding the larger blood vessels after they have penetrated into the brain substance (Figs. 250 and 251). However, they are not seen in the pericapillary neuroglial investment (Figs. 252). In fact, on closer examination, the distinction between these two types of terminal formation becomes less sharp. Certain foliated bulbous terminals that appear in the Golgi preparations (Fig. 232) seem to be intermediate forms. Membranous whorls are encountered in either type, although they are less frequent and less complicated in the foot processes (Fig. 247) than in the bulbous terminals. Consequently the two kinds of terminals that have been distinguished in Golgi preparations should be considered as extremes in a continuous spectrum of forms. This pleomorphism is consistent with the fluid configuration of the entire appendicular apparatus of the Golgi epithelial cell. It suggests that these cells, like neuroglial cells explanted into culture chambers, are constantly changing their form and their relation with neighboring cells. They are thus not to be conceived of as confined to the rigid profiles captured in our preparations. Similar considerations suggest that the role of the membranous whorls in the terminals is intimately related to their shape and their surface area.

As Figs. 246 and 247 show, the expanded tips of the Bergmann fibers abut one against the other forming a limiting membrane, a kind of tessellated pavement at the surface of the brain. Each enlargement is bound to its immediate neighbors by gap junctions, which describe

Fig. 250. Terminal knob on a pial blood vessel. This end knob of a Golgi epithelial cell sits upon the basal lamina (*B*) of a small artery at the surface of the molecular layer. The knob contains whorls of collapsed cisternae wrapped around several small lysosomes (*Ly*). Lobule V. ×16000

Fig. 251. Terminal knob. This end knob of a Golgi epithelial cell also sits upon the basal lamina (*B*) of a blood vessel in the molecular layer. It contains a whorl of collapsed cisternae (*c*) surrounding two mitochondria. Lobule III. ×18000

Fig. 252. The glia limitans around a capillary in the granular layer. The astrocyte (*As*) is flanked on either side by two granule cells (*gr*). The astrocytic processes flow around the nerve fibers and separate them from the basal lamina (*B*). Three packets of filaments are found in the neuroglial cytoplasm. Lobule VIII. ×19500

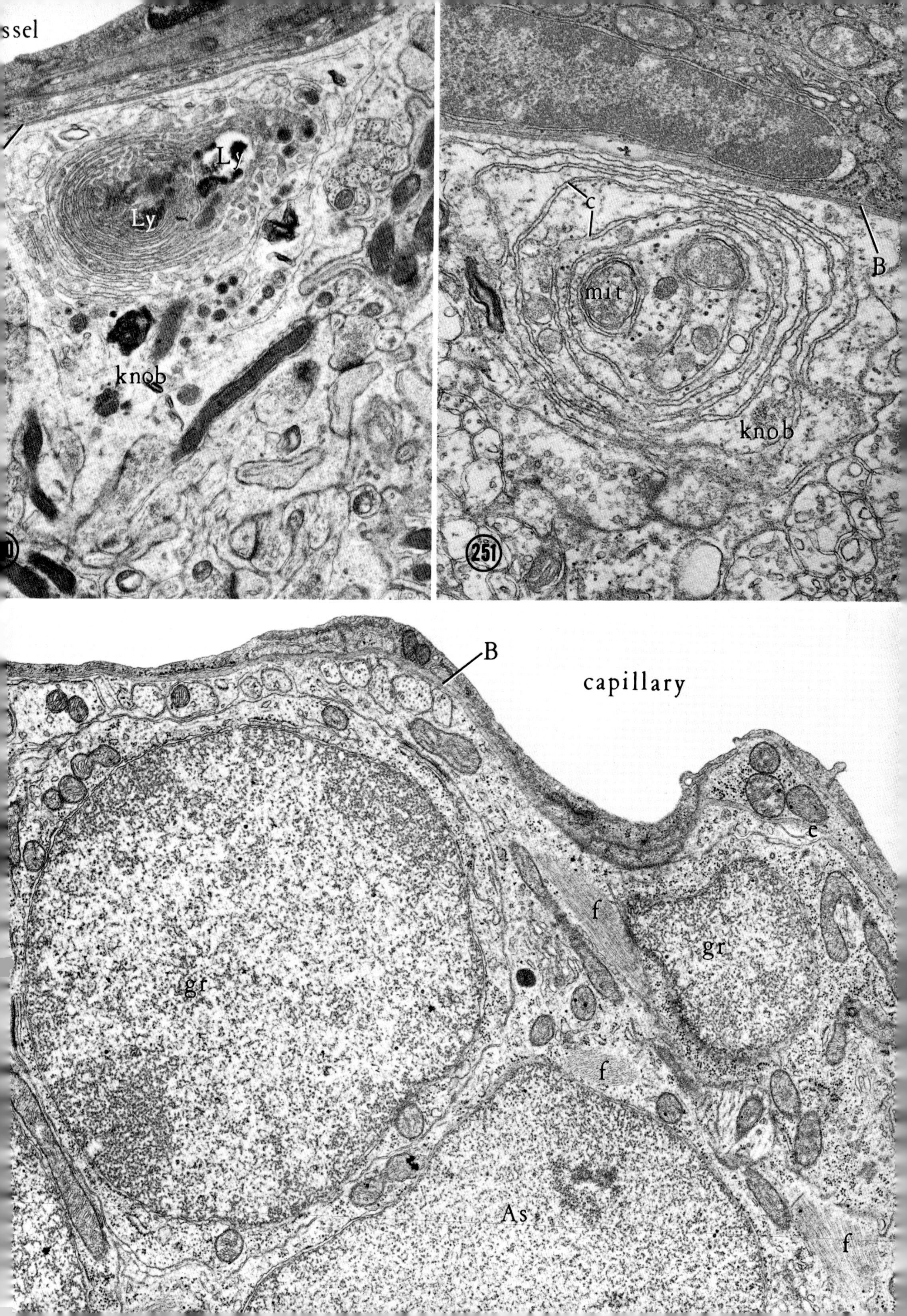

ssel
Ly
Ly
knob
c
mit
B
knob
251
B
capillary
e
f
gr
gr
f
As
f

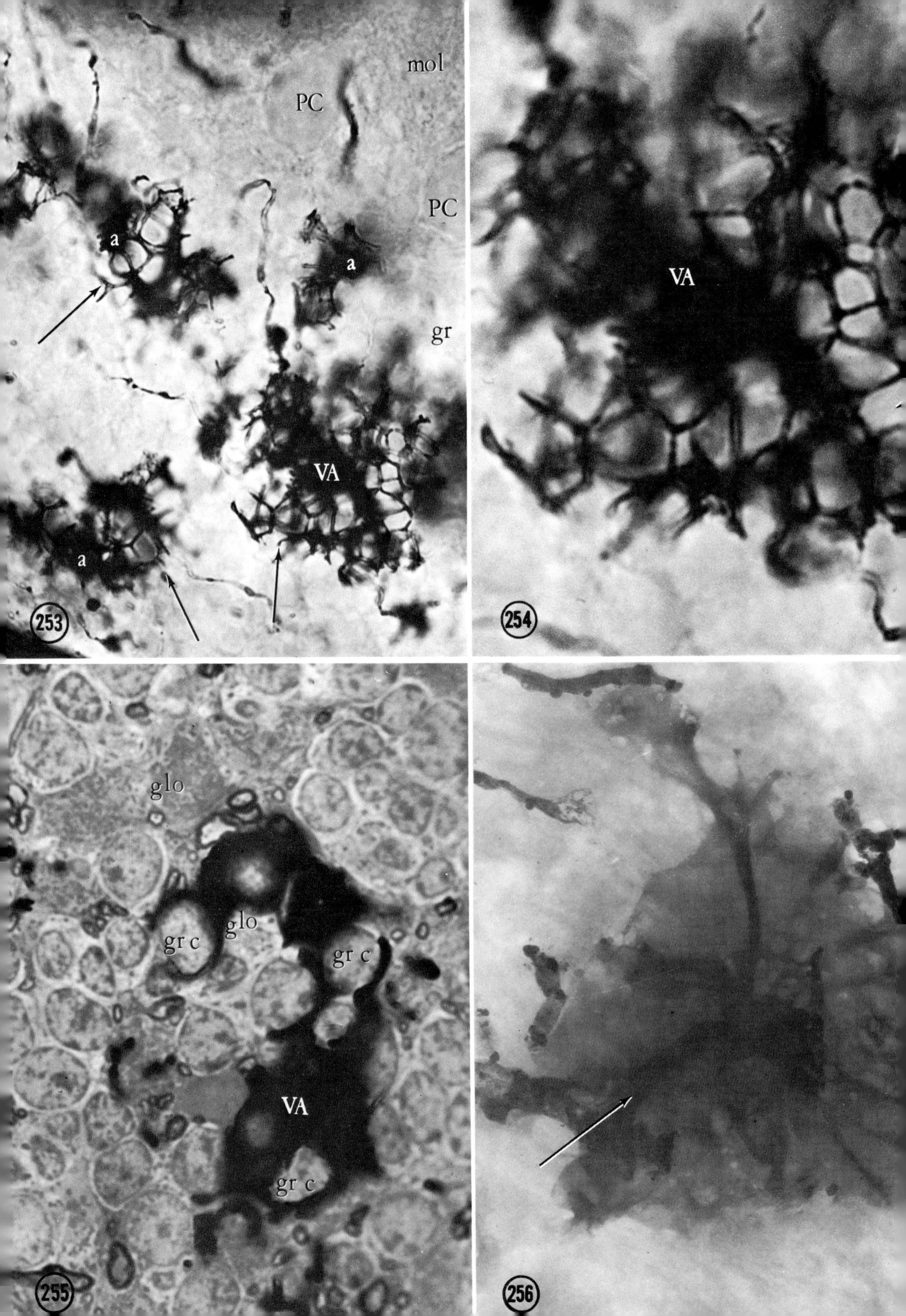
mol
PC
PC
a
a
gr
VA
a
253
VA
254
glo
glo
gr c
gr c
VA
gr c
255
256

discontinuous arcs around its circumference. Since these junctions form an imperfect seal between the processes (REVEL and KARNOVSKY, 1967; BRIGHTMAN and REESE, 1969), the external limiting membrane is not impervious to substances placed in the subarachnoid space. The free surface of this membrane is covered by a thin basal lamina continuous with that investing the entire nervous system. This extracellular layer may be considered as the proper delimiting membrane of the brain.

2. The Velate Protoplasmic Astrocyte

The more conventional type of protoplasmic astrocyte, the cell with a stellate rather than with a forked or umbellate profile, is only occasionally encountered in the molecular layer. It is, however, commonly seen in the granular layer, where its cell body must be distinguished from that of the granule cell. It is of about the same size as the granule cell or a little larger. The nucleus is usually bean-shaped and is characterized by a diffuse, homogeneous chromatin pattern that is quite different from the clockface or checkerboard pattern of the granule cell nucleus. This contrasting feature is evident in both light and electron microscopic preparations. One or two nucleoli are usually seen. The cytoplasm of the protoplasmic astrocyte is so similar to that of the Golgi epithelial cell that the description need not be repeated. Perhaps the only difference in internal structure between the two cell types is that the fascicles of neuroglial filaments are often thicker and more numerous in the astrocytes than in the more delicate Golgi epithelial cells. In fact, a detailed comparison of these two cell types—both of their internal structure and their external form—indicates that aside from the characteristic branching patterns of their major processes there is no important distinction between them. Their close similarities underscore their genetic affinities and explain why they should be considered simply variations of a single cell type (CHAN-PALAY and PALAY, 1972c).

The point of greatest interest about these cells is the form and distribution of their processes. GOLGI (1885), KÖLLIKER (1896), VAN GEHUCHTEN (1895), and other early authors on this subject described and figured their processes as smooth, unbranching rays that extend in all directions from the cell body. RAMÓN Y CAJAL (1890a, 1911), RETZIUS (1894), and TERRAZAS (1897) distinguished two kinds of processes, ones that are long and smooth and others that are short, thick, and ruffled or frilled. JAKOB (1928) and SCHROEDER (1929) describe the astrocytes as giving off radiating processes that "initially thick and robust, rapidly become thin and divide into innumerable branches" (SCHROEDER, p. 241, 1929). According to these authors, the processes form a rich network throughout the granular layer, in which it is difficult to ascertain how they end. In the meshes of this network lie granule cells and the protoplasmic islets. The latter are permeated by fine astroglial processes. Many of the processes end on blood vessels, either abutting simply with an expanded foot against the vessel wall or coiling round it.

Sections of cerebellum prepared with the Weigert neuroglial method disclose very few astrocytic fibers crossing the cortical layers, a result that WEIGERT (1895) recorded with evident disappointment. The reason for the paucity of stained neuroglial fibers is the delicacy of the filament bundles in these protoplasmic astrocytes. The rapid Golgi method, however, frequently displays these cells in considerable detail, often interfering with the clear delineation of nerve cells and their processes. A common appearance is seen in Figs. 253 and 254, which show apparently reticulated deposits in the granular layer. A careful study of these spots in the light microscope usually discloses the silhouette of a nucleated cell body

◄

Fig. 253. Velate protoplasmic astrocytes in the granular layer. An entire velate astrocyte (*VA*) and portions of the processes of several other velate astrocytes (*a*) are shown in this field. The processes of these cells form septa that wall off compartments in the granular layer (*arrows*). Note that each velate astrocyte has a distinct territory of the granular layer into which its processes expand. Two Purkinje cells (*PC*) and the granular and molecular layers are indicated. Rapid Golgi, 90 μm, sagittal, adult rat. Lobule VIII. ×1500

Fig. 254. The velate protoplasmic astrocyte. This higher magnification shows the cell body of the velate astrocyte (*VA*) and from it the various processes of the cell that ramify into the granular layer. The compartments (*arrow*) made by these tenuous sheets of neuroglial cytoplasm have the appearance of a honeycomb. Rapid Golgi, 90 μm, sagittal, adult rat, Lobule VIII. ×4025

Fig. 255. Laminar processes of velate protoplasmic astrocytes in the granular layer. This clump of processes (*VA*) probably belongs to one protoplasmic astrocyte. The sheets of neuroglial cytoplasm enwrap or partially surround granule cells (*gr c*) and individual glomeruli in the neuropil. This intracellular deposit within neuroglial cells appears as thin red coats that are often mistaken for nonspecific extracellular precipitate. Compare this with Figs. 253 and 254. Rapid Golgi, rhesus monkey, counterstained with toluidine blue, 1 μm, Lobule X. ×2400

Fig. 256. A velate astrocytic process. This process of the velate astrocyte flows from a rib-like stalk (*arrow*) to form an umbrella-like canopy or septum in the granular layer. This is a part of the septa that form the neuroglial compartments. High voltage electron micrograph, rapid Golgi preparation, monkey, 5 μm, Lobule X. ×10000

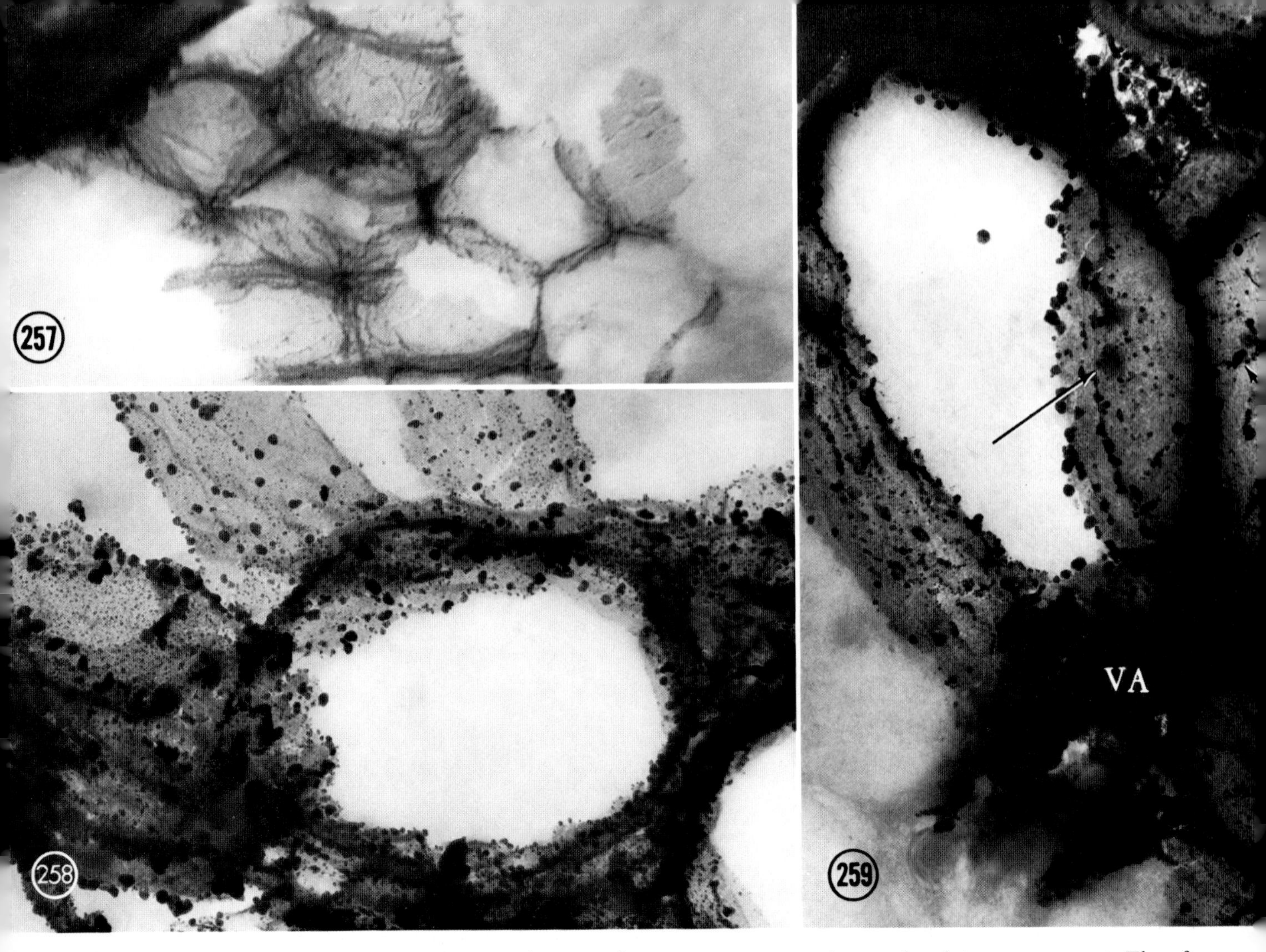

Fig. 257. Alveolate compartments formed by the laminar expansions of astrocytic processes in the granular layer. The veil-like processes in this clump belong to a single protoplasmic astrocyte. These neuroglial expansions form floors and walls that divide up this portion of the neuropil into compartments. Glomeruli and some granule cells occupy these compartments, and are partially segregated from neighboring cells and glomeruli by these neuroglial partitions. Rhesus monkey, high voltage electron micrograph, 5 μm, Lobule X. × 3500

Fig. 258. The laminar processes of protoplasmic astrocytes in the granular layer. The several contiguous compartments have thin, velamentous walls composed of neuroglial processes. The intracellular rapid Golgi precipitate shows up as a background grey fibrillar material upon which very dense particles are deposited. These particles often accentuate wrinkles or creases in the impregnated neuroglial processes. High voltage electron micrograph, rhesus monkey, 5 μm, Lobule X. × 10000

Fig. 259. A septum formed by a velate astrocytic process. The velate astrocytic process starts as a rib (*VA*) that expands in several planes into veil-like sheets. These sheets have been impregnated by the rapid Golgi method and are shown here in a high voltage electron micrograph. Two adjacent veils of cytoplasm (*arrows*) meet to form a septum between compartments. Rapid Golgi, rhesus monkey, 5 μm, Lobule X. × 13500

in the midst of the deposit (Fig. 254). Focussing up and down through the preparation reveals that the reticulations are really tenuous, velamentous walls bounding compartments in an alveolate arrangement. Therefore, these strange deposits can be recognized as velate astrocytes with their extensive lamellar processes displayed because of a nearly complete impregnation. When one-micron sections of such material are counterstained with toluidine blue, the contents of the compartments can also be readily recognized. They are single granule cells or portions of protoplasmic islets. Fig. 255 presents a clump of granule cells counterstained with toluidine blue and enmeshed in the lamellar processes of a velate protoplasmic astrocyte that has been impregnated by the rapid Golgi method. Similar figures are found in Golgi-Kopsch preparations (Fig. 261), in which, however, the impregnation is more erratic.

This kind of image could be readily dismissed as an uninterpretable splotch, presumably an area of precipitation or overimpregnation. But when these splotches are examined in 1–5 μm sections by means of high voltage electron microscopy (Figs. 256–260), they prove to be intersecting diaphanous veils compartmentalizing the granular layer, not random precipitates. Stereoscopic pairs of micrographs as well as light microscopy indicate that individual granule cells are lodged in many of these compartments.

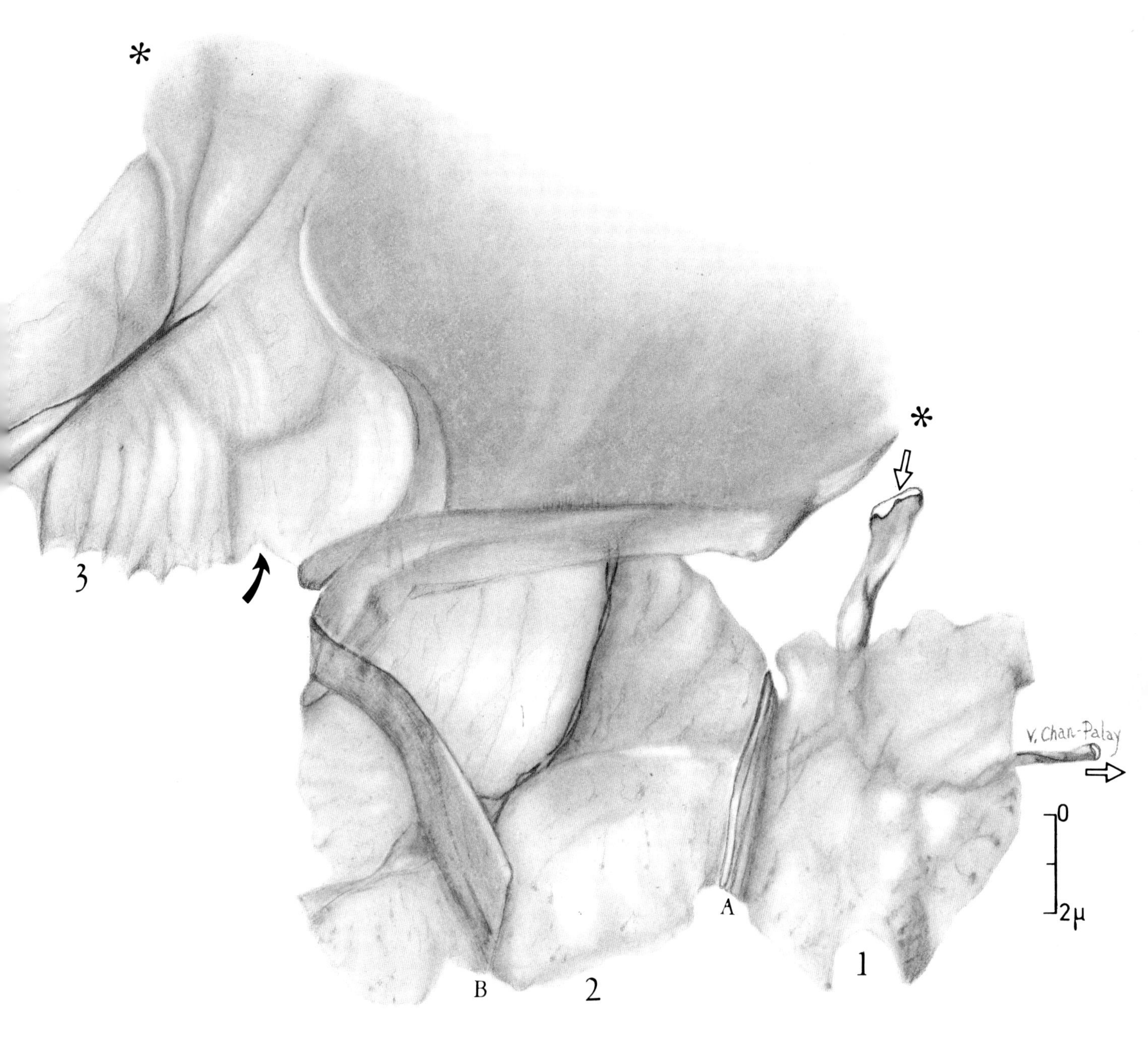

Fig. 260. A diagram of the compartments formed by processes of two velate protoplasmic astroglial cells in the granular layer. This drawing has been made from the actual images in a pair of high voltage electron stereograms from rhesus monkey cerebellar cortex in a section 5 μm thick. Compartment 1 is lined by a thin veil that expands from a cord-like process (*arrows*) of one astrocyte. This process helps to make septum A between compartments 1 and 2. Compartments 2 and 3 are formed from processes of a second astrocyte, between the asterisks. Compartment 2 is an open three-sided room that is separated from an adjoining space by septum B. The third compartment (*3*) is an open prismatic space bounded by a creased astrocytic veil (*curved arrow*) as the floor and a tent-like arrangement of two neuroglial lamellae as the roof. Compare this with Figs. 253–258

This surprising observation—surprising because granule cells are generally so closely packed together that the possibility of intervening neuroglial septa has been overlooked—can be easily confirmed by turning to the conventional electron micrographs of thin sections. Numerous lamellar neuroglial processes can be seen slipping between adjacent granule cells, interweaving with dendrites and axons, and invading the protoplasmic islets. The thin profiles of the lamellae have been filled in with light yellow in both Figs. 262 and 263. The interrupted processes delineated in this way correspond to the velamentous partitions seen in the high voltage electron micrographs of thicker sections (Figs. 256–260). Several features should be emphasized. First, the number and extent of these appendages are much greater than

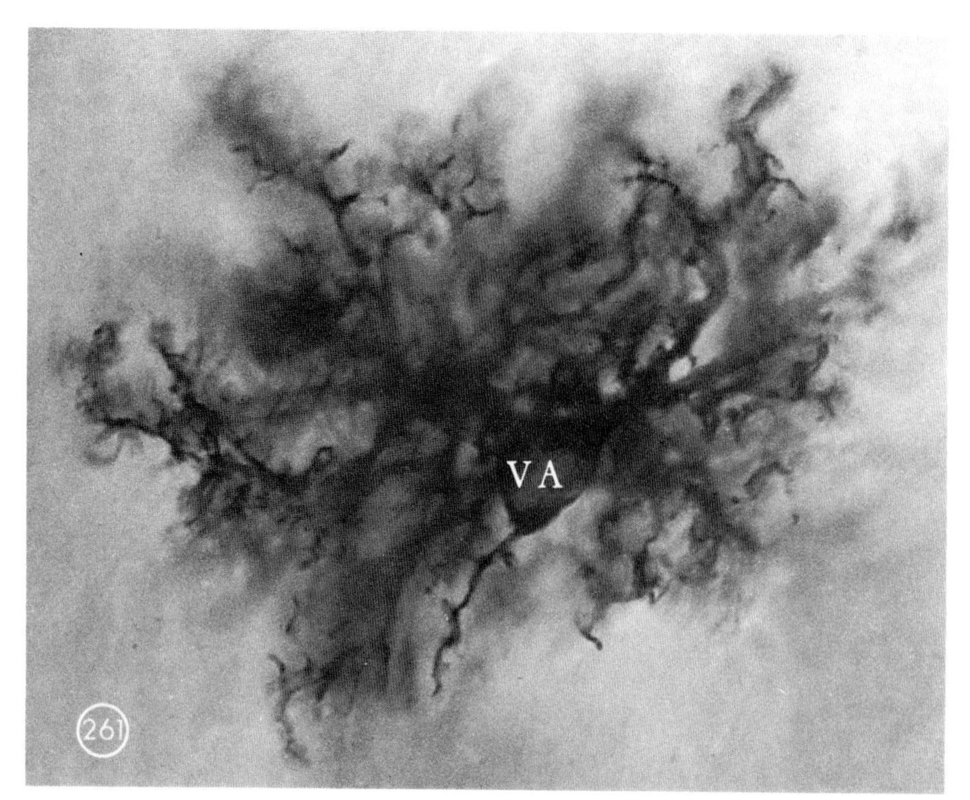

Fig. 261. The velate astrocyte. This cell has been impregnated by a variation of the Rio-Hortega method for neuroglial cells (see Methods, pp. 331–335). The processes of the cell radiate profusely into the granular layer, forming veil-like laminar expansions. This astrocyte is similar to the velate astrocyte shown in Figs. 253 and 254. In the latter cells, treated with the rapid Golgi method, all of the astrocytic processes have been impregnated, including the veil-like septa. In this cell stained by the Golgi-Kopsch method, only the rib-like portions of the processes are visible. Golgi-Kopsch, 90 µm, monkey. × 2000

one would have expected from a consideration of the number of astrocytes in the granular layer. Second, the lamellar processes are interposed between neighboring or adjacent granule cells and between groups of granule cells, forming open and confluent compartments.[30] In most instances any particular granule cell body has at least one of its faces free to confront another cell body or components of the neuropil. Third, the protoplasmic islets are similarly partitioned. Lamellar processes separate one glomerulus from another in the islet, and interweave with the dendrites and Golgi cell axons at the periphery of the glomerulus. Neuroglial processes seldom penetrate deep into the glomerulus or approach the mossy fiber or climbing fiber terminal at its center. The distribution of these appendages is therefore similar to that shown by Jakob (1928) and Schroeder (1929) with the gold sublimate method. Their procedure, however, must have demonstrated only the thick cytoplasmic ribs in the laminar processes and possibly only the locations and trajectories of the bundles of neuroglial filaments. Apparently the rapid Golgi method is capable of displaying the laminar processes in their entirety, as has been indicated by high voltage electron microscopy. Both this method and reconstructions from conventional electron microscopy vindicate the usefulness of the rapid Golgi method as a cytological tool. Similar velamentous processes can also be demonstrated with the Kopsch variant of the Golgi method, but this procedure appears to be less successful, as the processes are often incompletely impregnated (compare Figs. 261 and 264 with Figs. 253 to 260).

It may be relevant to point out here, as has already been suggested, that these appendages need not be con-

30 Ramón y Cajal (1911, p. 71) wrote that the lamellar appendages «sont destinées à séparer les groupes de grains entre eux; ...».

Fig. 262. Portions of two glomeruli and three granule cells (*gr c*) in the granular layer. The profiles of neuroglial processes have been colored yellow. In thin sections the neuroglial sheath appears irregular and incomplete. It forms a perforate capsule around two glomeruli. The neuroglial processes do not penetrate into the deeper parts of these glomeruli (MF_1 and MF_2) where the mossy fiber terminals are. Instead, the neuroglia surround the periphery, through which dendrites of granule cells, Golgi axons, and the afferent fibers enter. Rat, Crus I, VII. × 10000

Fig. 263. A protoplasmic astrocyte in the granular layer. The profile of this protoplasmic astrocyte (*VA*) as well as that of other astrocytic processes has been filled in with yellow. A granule cell ($gr\ c_1$) is practically ensheathed with these neuroglial processes. Two other granule cells in the field ($gr\ c_2$ and $gr\ c_3$) are partially encapsulated. A large dendrite of a Golgi cell (*Go d*) courses through the field and is also well enwrapped with neuroglia. The cytoplasm of the astrocyte PA is characteristic, with its packets of fine filaments (*f*), many ribosomes and other cellular organelles. Lobule IV. × 10000

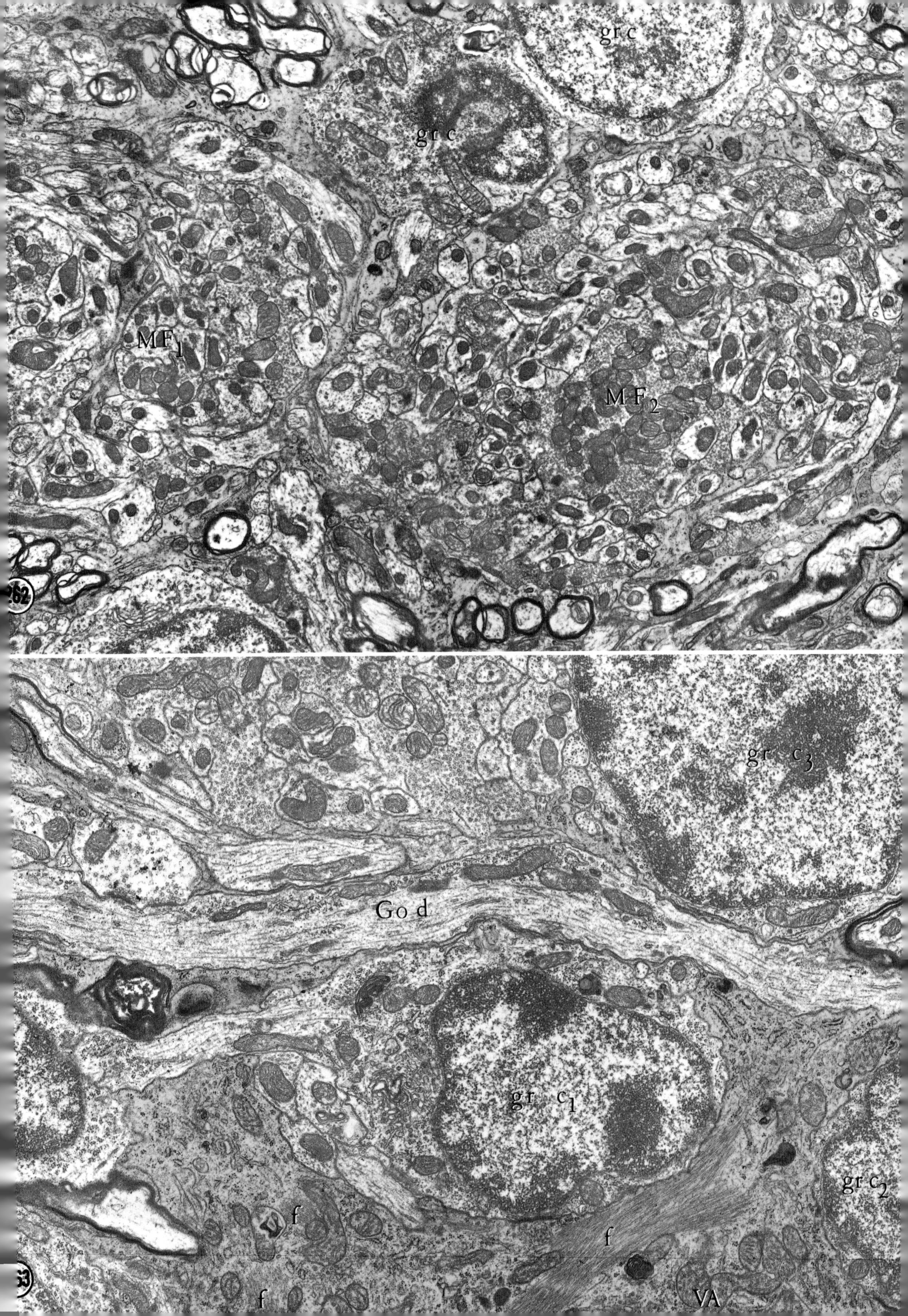
gr c
gr c
MF1
MF2
262
gr c3
Go d
gr c1
gr c2
f
f
f
VA
263

sidered permanent fixtures either in time or in place. In the tissue culture chamber neuroglial processes have been observed to change form, direction, and location (POMERAT, 1952). Similarly, in the intact organism they can be extended and withdrawn in a continuously changing pattern according to slight alterations in the microenvironment and in response to intrinsic and extrinsic stimuli of as yet unknown nature.

In concluding this section on the velate astrocytes, it is interesting to recall the pericellular network that was discovered by RAMÓN Y CAJAL (1890b) in cerebellar cortex stained with a modified Golgi method. The same structure was later demonstrated by GOLGI himself (1898a), who used a different heavy metal impregnation procedure. GOLGI called it the external reticular apparatus, distinguishing it from his now better known internal reticular apparatus. This pericellular network is an apparently extracellular envelope that covers the perikaryon and dendritic tree of the Purkinje cell and other large neurons. RAMÓN Y CAJAL (1911, Fig. 3) shows a similar material in the granular layer, where it encloses granule cells but leaves the glomeruli free. He describes it as a granular substance, an intercellular cement, furrowed by linear depressions that probably represent nerve fibers passing through it. From the observations presented in this chapter, it seems evident that this pericellular network corresponds not to some extracellular substance, but to the velamentous appendages of the Golgi epithelial cells in the molecular layer and the velate astrocytes in the granular layer.

3. The Smooth Protoplasmic Astrocyte

Not all of the astrocytes in the cerebellar cortex have lamellar appendages like those described above. In our Golgi preparations we have encountered another, far less common variety of astrocyte with smooth, long, radiating processes (Fig. 264). They are found in both the granular and the molecular layers. These cells resemble the stellate neuroglial cells figured by the early authors, but appear in much smaller number than their drawings indicate. The smooth astrocytes seem to be allied to the fibrous astrocytes lying among the myelinated fibers in the white matter. They have kinky, contorted processes that branch repeatedly and taper towards their ends. It is difficult to identify with certainty in electron micrographs the profiles that correspond to these cells in the Golgi preparations. Typical fibrous astrocytes having thick processes crowded with neuroglial filaments have not been found in the cerebellar cortex of the rat. Rarely, however, one comes across neuroglial perikarya or processes that contain fairly stout fascicles of filaments and arch about glomeruli or between granule cells. These may be the processes of the smooth protoplasmic astrocyte. More of these profiles are evidently encountered in the cat's cerebellum (HÁMORI and SZENTÁGOTHAI, 1966b).

4. The Oligodendrogliocyte

Another neuroglial cell that generates lamellar processes is the oligodendrogliocyte. In the rat this cell type is relatively rare in the outer two layers of the cerebellar cortex, being chiefly encountered in the deeper part of the molecular layer and the Purkinje cell layer where it occurs in the vicinity of myelinated axons and the Purkinje cell soma. It is more common in the granular layer, where there are more myelinated fibers. SCHROEDER (1929) found a similar distribution in man. Although SCHROEDER (1929) considered oligodendrogliocytes to occur frequently as satellites of the Purkinje cell bodies, this relation is rarely encountered in the rat at the electron microscope level. It is the Golgi epithelial cells that are the usual satellites, as was explained earlier. On the rare occasions, however, when it is seen in this position, the oligodendrogliocyte lies in direct apposition to the surface of the Purkinje cell body or dendritic trunk, displacing other neuroglial and neural elements. The more usual situation is that the oligodendrogliocyte lies at a small remove, with astrocytic processes filling the interval. We have also seen an example of an oligodendrogliocyte in the satellite position in relation to the soma of a basket cell.

In the granular layer the oligodendrogliocytes must be differentiated from both granule cells and astrocytes because of their similar size and shape. They exhibit no preferred relationship with any of the neurons of the granular layer, but are most frequently seen in association with myelinated nerve fibers. No unexpectedly, therefore, they are more common near the white matter in the depths of the granular layer than elsewhere.

Fig. 265 is a drawing of two oligodendrogliocytes discovered in a Golgi-Kopsch preparation at the deep margin of the granular layer. The lower cell is entirely within the lamella of white matter in the center of a folium, and the upper cell is entirely within the granular layer. Although the form of these cells is spectacular, it is closely allied to that of the Golgi epithelial cells and

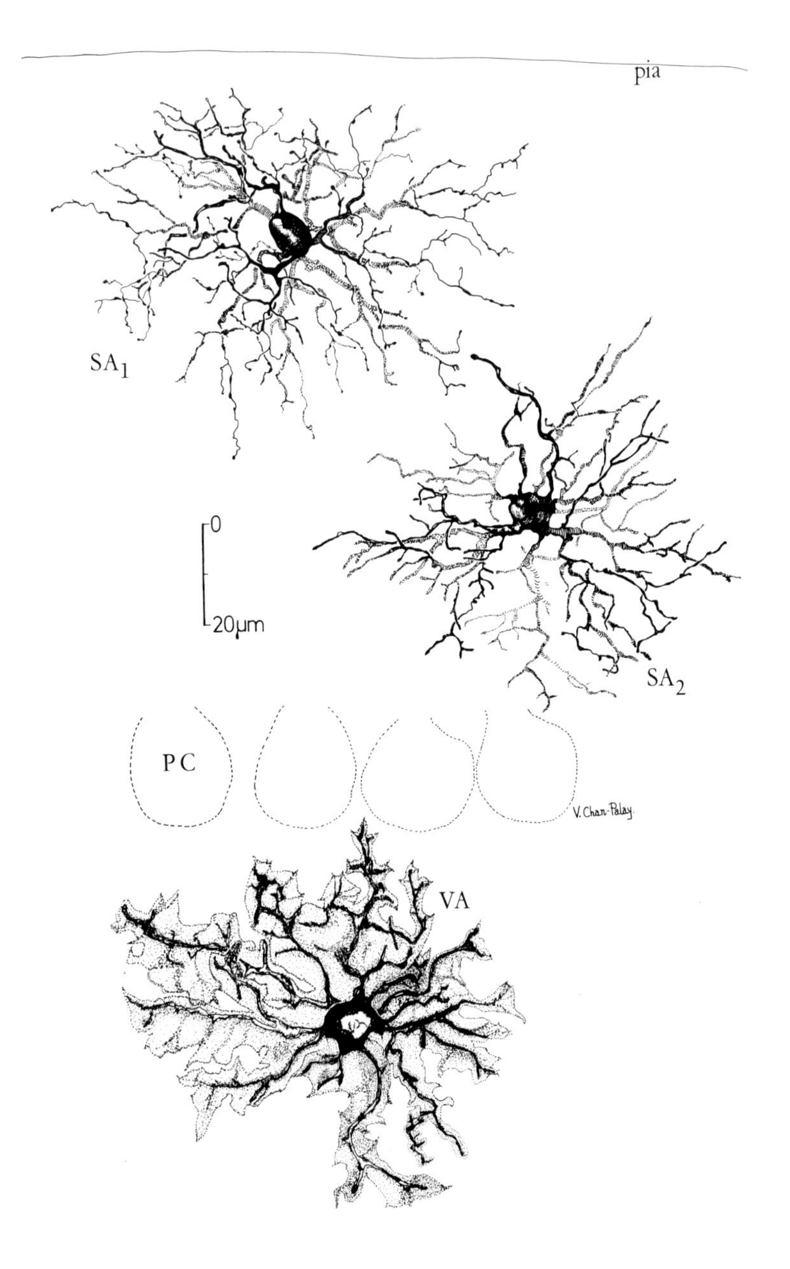

Fig. 264. Smooth protoplasmic astrocytes and the velate protoplasmic astrocyte. Two protoplasmic astrocytes are shown in the molecular layer (SA_1 and SA_2). Both cells have radiating processes that are slender, branched, and contorted. These astrocytes are not common, but have been seen in both the molecular and the granular layer. The smooth astrocyte is to be contrasted with the velate protoplasmic astrocyte (*VA*) of the granular layer with its laminar, veil-like processes extending in umbrella-like fashion from the cell body. The Purkinje cell layer (*PC*) is indicated. Golgi-Kopsch (modified Rio-Hortega method; see Chap. XII, Methods, pp. 331–335). 90 µm, adult rat cerebellar cortex. Camera lucida drawing

the velate astrocytes. All of these cells generate scrolled velamentous processes. Both of the cells in Fig. 265 display the long, coiled processes which were beautifully shown by DEL RÍO HORTEGA (1928) and which, we now know, represent the continuous cytoplasmic ribbons in the tongue processes of the myelin sheaths around nerve fibers (see PETERS, PALAY, and WEBSTER, 1970). The lower cell produces the sheaths of at least twelve different nerve fibers, while the upper one, which is more complex, services ten different fibers within the field drawn. Myelinated fibers in the white matter are plaited in regular, crisscrossing groups, so that they appear in clusters of alternating transverse and longitudinal sections. The lower oligodendrogliocyte in the drawing services fibers that extend in both directions. In the upper cell, the processes curve mostly in a longitudinal direction. Actually they are less parallel with one another than the drawing could indicate, a reflection of the more diverse trajectories of myelinated fibers in the granular layer than in the white matter. Most of these fibers are probably mossy fibers, for they tend to run up to the top of the folium along the central lamella and give off their collaterals on the way.

The oligodendrogliocyte is a small rounded cell with a subspherical or reniform nucleus. The cytoplasm is not voluminous. The cell body is about as large as the nucleus of a granule cell, i.e., 5–6 µm in diameter. The outline of the perikaryon in sections is usually smooth and regular, a consequence of the relatively small number of processes. Myelinated fibers can be found abutting directly against

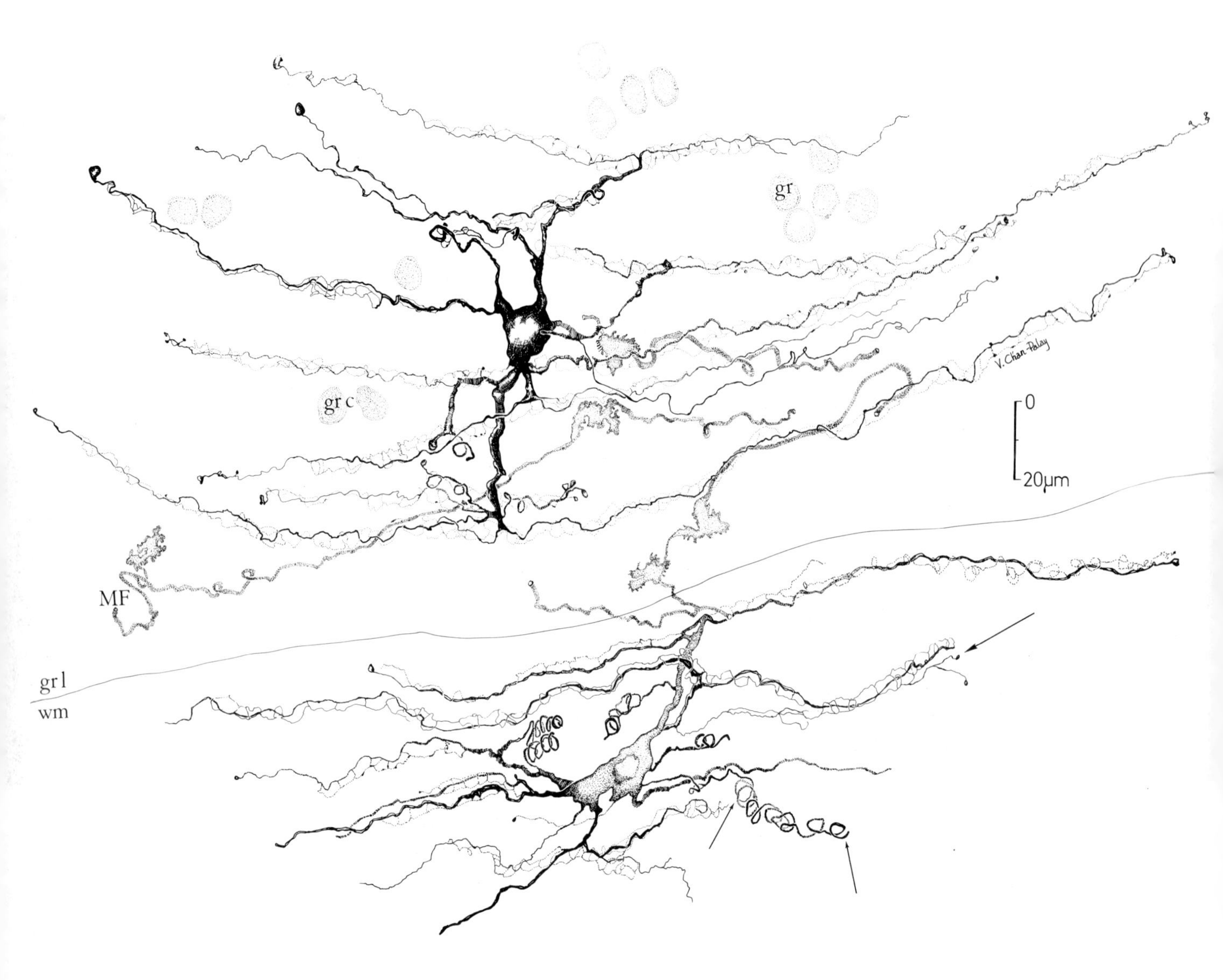

Fig. 265. Oligodendrogliocytes in the white matter and granular layer. The cell in the lower half of this illustration lies entirely within the white matter. Here the myelinated fibers are plaited in regular groups that criss-cross at right angles to one another. The long, coiled processes extending from the oligodendrogliocyte are the continuous cytoplasmic ribbons in the myelin sheaths. These processes serve axons that are running in the longitudinal (*large arrow*) and transverse directions (*fine arrows*). The cell in the upper half of the illustration has processes that run approximately longitudinally, although some run transversely. These processes follow the trajectory of myelinated mossy fibers and Purkinje cell axons in the granular layer. The limits of the granular layer (*gr l*) and white matter (*wm*) are indicated. Several mossy fibers (*MF*) with rosettes are present in the field among a few granule cells (*gr c*). Golgi-Kopsch (modified Rio-Hortega, see Methods, pp. 331–335), 90 µm, adult rat, sagittal

the surface of the perikaryon or invaginated into it, thus proclaiming the formation of myelin by this cell. In toluidine blue preparations, in the light microscope, both the nucleus and the cytoplasm are conspicuous because of their basophilia. In the electron microscope this characteristic is represented by their noticeable density compared to that of the other cells in the layer (Figs. 233 and 266). Because of the general density of the nucleus, the nucleolus may not be prominent. The chromatin, however, is distinctly marginated and the nuclear pores are quite conspicuous because of the abrupt clearings in the marginated chromatin masses opposite them. The outer membrane of the nuclear envelope is studded with ribosomes in discontinuous arrays.

The cytoplasm is characterized by hordes of fine granules, which give the cell matrix a dark tone (PETERS, PALAY, and WEBSTER, 1970). Suspended in this matrix are small numbers of short, slender mitochondria and a few cisternae of the granular endoplasmic reticulum, either isolated or in short stacks. Free ribosomes are also clustered in the spaces intervening between cisternae. The Golgi apparatus is diminutive, a compact clump of agranular cisternae and vesicles. Occasionally lysosomes

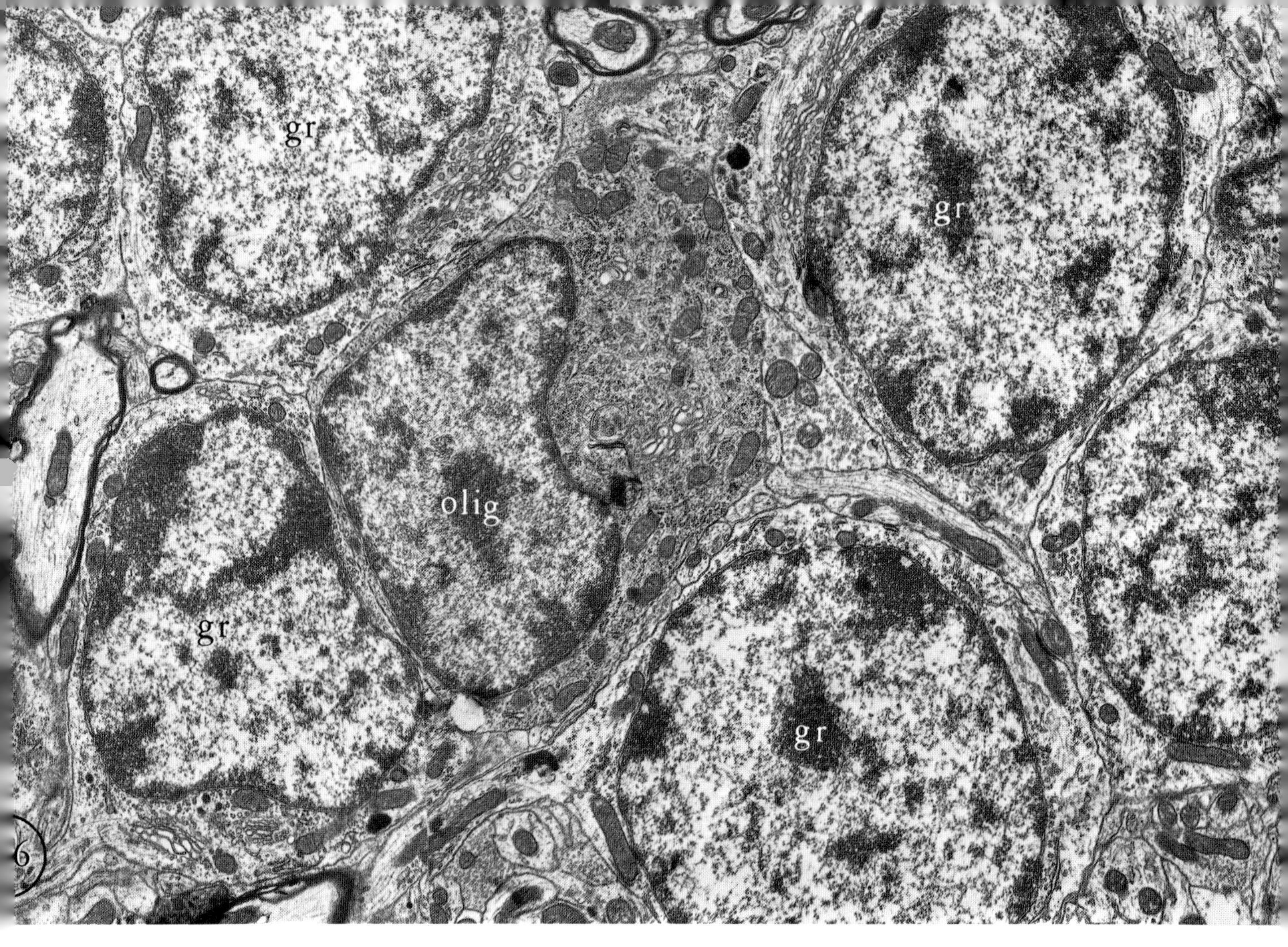

Fig. 266. The oligodendrogliocyte. This oligodendrogliocyte (*olig*) is readily distinguished from a cluster of granule cells (*gr*) by its dark cytoplasmic matrix. The nucleus of this cell has dense heterochromatin dispersed within the rim of the nuclear envelope, resembling to some extent the granule cells of the surround. The cytoplasmic differences, however, are unmistakable. Lobule V. $\times$ 10000

with a dense, granular content appear. A frequent inclusion is a round or elongated body composed of a whorl of thin myelin-like lamellae, usually somewhat disrupted by swelling. Such inclusions have been described by BUNGE, BUNGE, and RIS (1960) as typical of oligodendrogliocytes.

Another characteristic cytoplasmic component is the microtubules, which course in all directions through the perikaryon and collect into regularly spaced fascicles in the initial segments of the processes. The processes originate from the perikaryon as conical extensions that rapidly taper to a uniformly thin caliber and project stiffly from the cell body. They are filled with microtubules running parallel to their long axes. Occasionally a mitochondrion and tubules of the smooth endoplasmic reticulum are encountered, but few other organelles. Transverse sections of these processes are easily identified. They are circular profiles 0.1 to 0.2 μm in diameter and contain 4 to 6 microtubules in cross section. Closer to the cell body they can have a mitochondrion and elements of the granular endoplasmic reticulum. The concentration of microtubules is, however, the characteristic that distinguishes them from axons of the same size range. We have not made a special study of the relationship between these processes and the myelin sheath in the cerebellar cortex, but there is no reason to suspect that it differs from that in other parts of the central nervous system (BUNGE, 1968; PETERS, PALAY, and WEBSTER, 1970). SCHROEDER (1929) reports that like the astrocytes, the oligodendrogliocytes sometimes attach their processes to the walls of blood vessels. We have seen nothing in our electron micrographs indicating that these cells form any part of the perivascular boundary.

5. The Microglia

Microglial cells are not common inhabitants of the normal adult cerebellar cortex. Occasionally cells that resemble resting macrophages appear in the perivascular tissue between the basal lamina of the capillary endothelium and the basal lamina investing the neural tissue. Such cells are by definition outside of the brain substance. They contain elongated nuclei and a dense cytoplasm characterized by scanty endoplasmic reticulum and numerous variegated lysosomal inclusions. In some sections one or two such cells are encountered within the brain substance, and in the brains of animals that have been subjected to remote lesions, such as spinal cord section (see Figs. 150–157), the numbers of such cells in the granular layer is greatly increased twelve or more hours after the lesion. Their origin is presumably the perivascular mononuclear cells that are found resting in all normal specimens.

6. Functional Correlations

The role of the oligodendroglia and the microglia in the economy of the central nervous system is clearly signaled by their relation to the myelin sheath and their response to injury respectively. Although these activities may not be the only functions of these two types of cells, they are at least incontestable. The role of the astrocytes, especially the protoplasmic astrocytes is, however, still a subject of much speculation and controversy based on little solid evidence.

During the past two decades, after an initial period of confusion, electron microscopy has demonstrated that the gray matter of the central nervous system is pervaded by protoplasmic astrocytes, which insinuate their tenuous processes seemingly into every cranny of the tissue (PETERS and PALAY, 1965; WOLFF, 1965, 1968; PALAY, 1966; CHAN-PALAY and PALAY, 1972c). With improved preservation, the delicate veil-like expansions of these cells have become progressively more conspicuous, until microscopists have been obliged to acknowledge that the protoplasmic astrocytes in the intact tissue hardly differ in form from the expanded neuroglial cells growing freely in culture chambers. These observations add still another bit of evidence to substantiate the claim that tissue explants in culture can be useful models of the conditions *in situ*. As we have seen in this chapter, however, the velamentous expansions and "delicate cytoplasmic ruffles" of the cell in culture (POMERAT, 1962) were visible all along in the preparations impregnated with heavy metals and examined under the light microscope (CHAN-PALAY and PALAY, 1972c). Unfortunately, they had been misinterpreted and rejected out of hand. Electron microscopy shows clearly that the Golgi method is capable of rendering with great fidelity the form and disposition of the most delicate neuroglial appendages. In order to emphasize the profusion of filmy expansions generated by these astrocytes, we have given them the name *velate protoplasmic astrocytes*. When considered with this important characteristic in mind, the Golgi epithelial cells, Fañanas cells, and protoplasmic astrocytes of the cerebellar cortex all become variants of a single cell type. This is not just an idle exercise in taxonomy. Any cell that generates such enormous amounts of surface membrane must be understood in terms of the function of that membrane. (The myelin-producing Schwann cell and oligodendrogliocytes are examples of this rule.) Unfortunately, there is little evidence to guide us in thinking about the functions of this membrane in neural tissue. But that evidence, coming from both anatomy and physiology, implicates the neuroglial velamentous processes in the maintenance of equilibrium in the interstitial spaces of the central nervous system (KUFFLER and NICHOLLS, 1966; ORKAND *et al.*, 1966; DENNIS and GERSCHENFELD, 1969; ORKAND, 1969; TRACHTENBERG and POLLEN, 1970; HENN *et al.*, 1972). The broad surfaces of neuroglial plasmalemma presented to the interstitial fluid by the expanses of velamentous lamellae would be appropriate instruments for this function.

The insulating function of the neuroglia was originally suggested by RAMÓN Y CAJAL (1909, pp. 248–249). More recently, on the basis of electron microscopic studies, it has been proposed that the velamentous processes of astrocytes isolate the postsynaptic surfaces of cell bodies and dendrites from nonspecific synaptic influences, such as diffusion of transmitter chemicals and ions from other synaptic sites (PETERS and PALAY, 1965; PALAY, 1966; PETERS, PALAY, and WEBSTER, 1970). According to this hypothesis, the delicate, filmy expansions of velate astrocytes are concerned with confining the effect of any one synaptic action to a particular synaptic site, thus promoting neural specificity. This concept was deduced from the clearly nonrandom distribution of the neuroglial processes in the neuropil. The Golgi epithelial cell in the molecular layer of the cerebellar cortex exhibits this property especially well. This cell produces extensive lamellae which scroll around the cell bodies and dendrites of the Purkinje cell, almost completely isolating it. Other neurons of the molecular layer are only fragmentarily invested with neuroglia or not at all. Similarly, velamentous processes of the Bergmann fibers accompany the ascending axonal stems of the granule cells, but their branches, the parallel fibers, are completely uncovered and course next to one another without any intervening neuroglia.

In the granular layer, the velamentous processes of protoplasmic astrocytes partially encircle glomeruli and slip between granule cells, subdividing the gray matter into compartments. This habit illustrates the organizing function of the neuroglia at a somewhat grosser level. The separation of glomeruli in a protoplasmic islet by thin neuroglial septa suggests that afferent mossy fibers of adjoining glomeruli may be segregated according to the different kinds of information they bear. The clustering of granule cells into groups partially enclosed in a common sheath or enfolded into the loculi in a common velate astrocyte could indicate that these cells share common functional relations and operate as a set or unit. Such compartments need not be regarded as static, fixed assemblies, but could be undergoing constant alteration according to the flow of information through them and the shifting of their astrocytic envelope.

Chapter XII

Methods

1. Electron Microscopy

During the past decade procedures for the preparation of nervous tissue for electron microscopy have undergone considerable change. They have become more complex and more refined than they were in the previous decade. Since the descriptions of methods in most journal articles are usually not detailed enough to enable the reader to reproduce many of the procedures, the present chapter is intended as a guide to the technical procedures we have followed in preparing material for the electron micrographs or photomicrographs in this book. The procedures that are described and the solutions used in them are exactly those currently used in our laboratory. These methods are presented as a set of detailed instructions. Whenever appropriate, the equipment needed is listed, and the source of any special materials is given. No detail of a maneuver that is central to the success of any procedure has been considered too trivial to be included.

Bibliographic citations have been omitted from this chapter. Nearly everyone who has worked in this field has contributed some small or important step in the procedures we use. This chapter is not intended to document the origins of each detail, but to present the methods now in use in our laboratory.

a) Equipment for Perfusion of Adult Rats

Place on a moveable trolley for convenience (see Fig. 267a).

Glassware and Instruments

Fixative perfusion flask, 1 liter capacity (or an aspirator flask with a single outlet at the bottom)
Transfer flask or filter funnel, small
Beaker, 200 ml
Finger bowl for clamps
Battery jar for brains in the skulls
1 glass cannula for perfusion, $4\frac{1}{2}$ cm long with bulbous tip 3 mm wide
3 disposable tuberculin 1 ml syringes, 3 hypodermic needles, 27 gauge
perfusion (surgical rubber) tubing, 6 mm inside diameter; about 225 cm connected at one end to a glass Y tube, and at the other to the fixative perfusion flask. On the remaining arm of the Y another piece of rubber tubing about 10 cm long, and on the end of the Y a piece of rubber tubing about 45 cm long. This end is to be connected to the glass cannula.

Fig. 267. a Equipment for the perfusion: the perfusion flask and tubing (1), surgical instruments (2), paper towels (3), dissecting board, animal gloves, notepad and pencil (4), hair clippers (5), disposable gloves (6), and animal weighing pan (7) are some of the equipment necessary for the perfusion procedure. These are placed neatly upon a movable trolley for ease of transporting from laboratory to perfusion facilities and back. **b and c** The dissection. The skin over the head of the perfused animal is slit (b) and the skin and soft tissue (c) are removed from the skull. The bones are then dissected away from the brain to expose it as shown in Figs. 2a–c (photos by R. DE CASTRO). **d** A simple evacuation system for rapid dehydration. The slabs of tissue for electron microscopy can be rapidly processed during dehydration by the use of such a system. An evacuation flask (1) is connected to a rapid flow of water from a faucet. The flask is attached to a Y-shaped tube and more rubber tubing, and then to a Pasteur pipette (2). Fluids may be suctioned off through the pipette from small disposable vials (3), each holding a slab of tissue. **e** A block containing a slab of cerebellum (vermis). This block of tissue has been embedded in a hard mixture of epoxy resin. The block face contains the midsagittal vermis of the cerebellum. The excess Epon has been trimmed away to leave only a thin rim surrounding the block. From this, survey 1 µm sections are cut with a glass knife; ×7. **f** The 1 µm survey section. This section of the entire midsagittal vermis of the cerebellum has been obtained from the block shown in the previous figure. It is stained in toluidine blue (see text). An area of interest in the section is picked for study in the electron microscope (boxed); ×7. **g** The pyramid for thin sectioning. The area selected for electron microscopy (e.g., as chosen in the previous figure, 267f) is made into a pyramid-shaped plateau for thin sectioning. The material around the selected area is cut away and trimmed (see text); ×7. **h and i** Montages of electron micrographs. Enormous montages, some up to 20 feet long, are assembled showing the detailed topography of the cerebellar cortex stretching from white matter to the pial surface. Fig. 267h shows several of these montages lying against the wall and on the floor (office of SLP). Fig. 267i shows a montage of the pial surface, lying against a wall of books on shelves

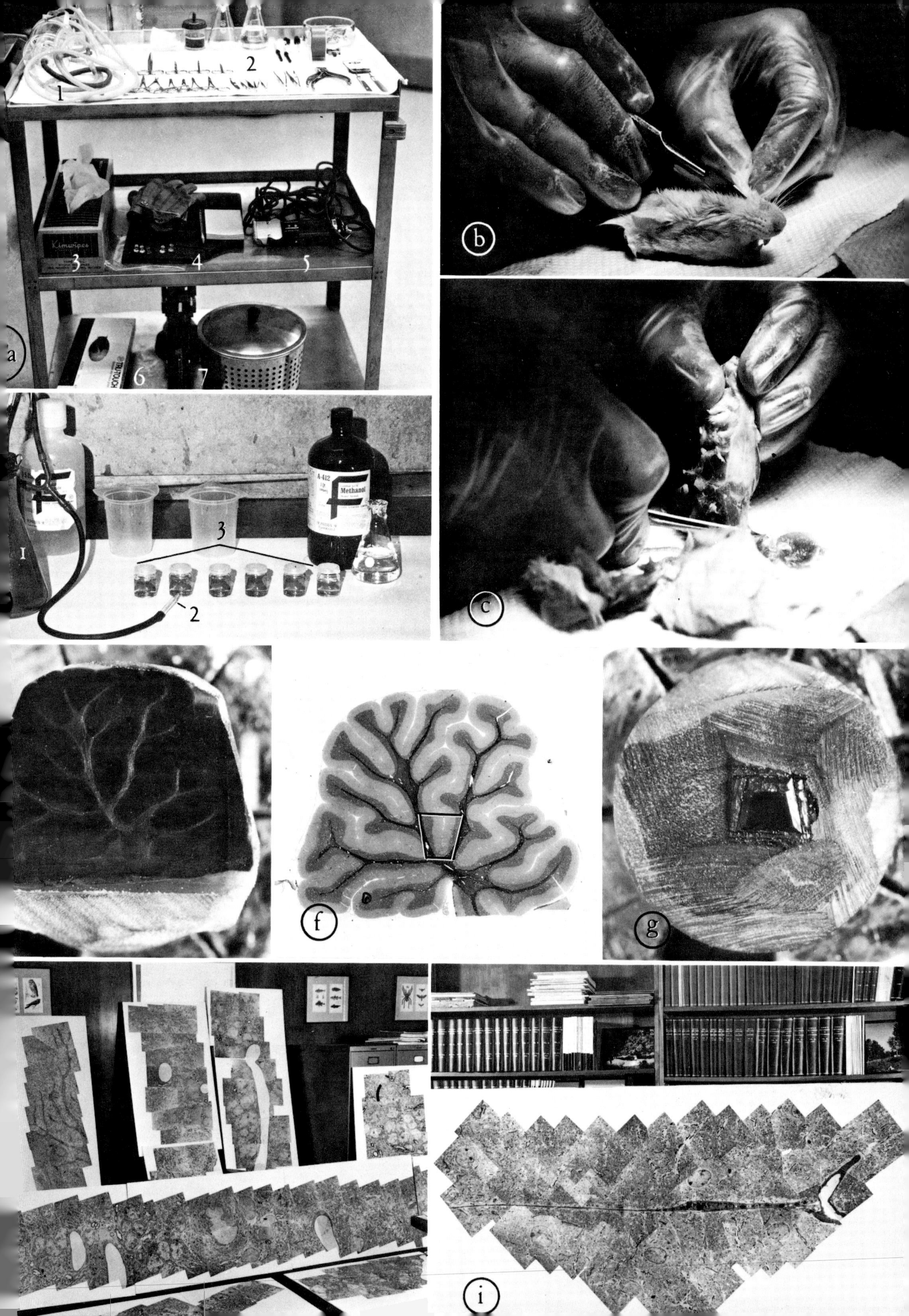

1
2
Kimwipes
3
4
5
a
6
7
b
A-412
Methanol
3
I
2
c
f
g
i

Instruments

Surgical scissors, large, blunt-tipped Mayo
Surgical scissors, large, sharp-tipped Mayo
Iridectomy scissors, small, sharp-tipped
Forceps, 2 large straight, 1 with tooth, 1 without
Forceps, 2 small curved, 1 with tooth, 1 without
1 small, straight forceps, toothed
7 haemostats: 1 large curved, 1 large straight, 5 small
Bone-cutting forceps
2 scalpel handles, Bard-Parker size No. 3
2 scalpel blades, Bard-Parker size No. 10
Clamps for tubing, 2 kinds: pinch clamps and pressure clamps, 2 each
Animal hair clippers
Large paper towels
Linen towels
Surgical thread, silk No. 2
Plastic bags
Cotton-tipped applicators
Small gauze pledgets
Animal gloves, leather
Disposable plastic gloves
Weighing cage
Cork dissection board
Disposable surgical masks
Labelling tape
Paper scissors
Felt pen for labelling
Notepad and pencil
Stereoscopic microscope on stand for dissection
Adjustable light weight tackle for suspending perfusion flask, about 200 cm high
Overhead surgical light

Equipment for Respiration

Tank of 95% oxygen, 5% carbon dioxide with control valves. Rubber tubing 60 cm long is attached to the outlet and connected to one arm of a glass Y tube. The other arm is left free. Another piece of rubber tubing 10 cm long is attached to the stem of the Y. The free end of this tubing is fitted with a metal No. 18 hypodermic needle 1 cm long, the tip of which has been filed square and smooth.

Wrench to release valves on tank.

Solutions at Room Temperature

Fixative (1% glutaraldehyde, 1% formaldehyde in 0.12 M phosphate buffer with 0.02 mM $CaCl_2$; see Solutions section C, p. 327).

1% sodium nitrite (1 g sodium nitrite in 100 ml of distilled water), freshly made up before perfusion.

35% chloral hydrate (35 g chloral hydrate in 100 ml of distilled water), dosage 0.1 ml/100 g body weight.

Heparin—check expiration date.

b) The Perfusion Procedure

1. Animals should be fasted for 24 hours prior to perfusion time. Water should be provided. An extra animal should be available for the perfusion in any experiment, in case of failure during the procedure.

2. An air exhaust system should be provided for pulling away the toxic fumes of the fixative. Ideally, this should be an inverted hood beneath the dissection board.

3. Weigh the animal in order to estimate the dose of anesthetic. Anesthetize the animal (100–200 g adult rats) by an intraperitoneal injection of 35% chloral hydrate. Dosage is 0.1 ml/100 g of body weight. The animals are insensitive to painful stimuli within 5 minutes.

4. Shave the rat from mid-abdomen to snout on the ventral surface and from mid-thorax to snout on the dorsal surface. Secure the rat to the dissection board with 4 pins or clamps, one on each footpad.

5. The perfusion apparatus consists of a glass Dewar flask that is closed except for a narrow funnel inlet and air outlet at the top and an outlet at the bottom, to which rubber tubing is attached. (An ordinary aspirator flask with an outlet at the bottom will do as well.) The fixative is poured into the perfusion flask through a small funnel. The rubber outlet tube is interrupted after 225 cm by a glass Y tube, the side arm of which leads to a second piece of rubber tubing 10 cm long. The stem of the Y tube leads to a third piece of tubing 45 cm long, to which a glass cannula is attached. To fill the apparatus, pour the fixative into the perfusion flask while the outlet tube is clamped close to its point of attachment and again just below the Y tube. Removal of the first clamp allows the fixative to flow down the tube and up the side arm of the Y. Any entrapped air bubbles can be seen through the tubing and may be removed by compressing the rubber tubing intermittently or by allowing the solution to flow back and forth several times. Air bubbles must be removed or the perfusion will fail. The tubing above the side arm is clamped. When the perfusion is started, the clamp above the cannula is removed. Perfusion is carried out under a head of pressure equivalent to 152 cm of water. Care must be taken to avoid inhalation of fixative vapors and to avoid contact with skin and eyes.

6. Fill one syringe with 1 ml of 1% sodium nitrite (use a 27 gauge needle). Fill a second syringe with 0.1 ml Heparin (use a 27 gauge needle). Hold in readiness on the dissection table.

7. Check respiration apparatus. A tank of 95% oxygen and 5% carbon dioxide is used. A rubber tube 60 cm long leads from the tank to a glass Y tube. The other side arm of the Y is left free. An 18 gauge needle filed down to a 1 cm length is affixed to a 10 cm length of rubber tubing, which is then attached to the Y tube. Artificial respiration is administered at a flow of 2 to 3 liters per minute at a rate of about one puff per second. This is achieved by fluttering the forefinger over the tip of the free limb of the tube. The respiration is conducted by an assistant at the perfusion.

8. Expose the trachea through a midline incision in the neck and separate the muscles overlying the trachea with a scalpel blade. Hold the muscles away with hemostats, one on each side. Care must be taken not to incise the thyroid gland nor the thyroid arteries, or profuse bleeding will occur. Make a T-shaped hole in the trachea by cutting into one or two cartilaginous rings below the level of the thyroid gland. Insert the tracheal cannula. Begin the respiration, regulating carefully to simulate the depth and rate of normal respiration.

9. Continue the midventral incision down over the abdomen, and reflect the skin on either side. Nick the abdominal wall in the midline, and place a hemostat across the midline, one above and one below the level of incision. Cut the abdominal wall to the midaxillary line on either side and along the costal attachment of the diaphragm with blunt-tipped Mayo scissors.

10. Avoid injuring the lungs by deflating them momentarily before placing the blades of the scissors in position for each cut. Insert the blunt-tipped blade of the scissors into the thorax along the midaxillary line and cut the ribs, first on one side, then the other. Elevate the rib cage, hinging at the first costosternal joint, and place a hemostat across it in order to clamp the internal mammary arteries. Remove the rib cage by cutting along the hemostat with scissors. Care is to be exercised during this maneuver to avoid obstructing the trachea.

11. This procedure exposes the heart and the great vessels. The pericardium is then slit with iridectomy scissors. A fine curved forceps is inserted under the arch of the aorta, and a length of No. 2 suture thread is drawn through. Loop the thread about the root of the aorta as a loose knot in preparation for tying the perfusion cannula in place. At this point allow a few moments for recovery of respiration and cardiac action. Then grasp the apex of the heart with a fine toothed forceps, and inject 1 ml of 1.0% sodium nitrite slowly over a period of a minute into the left ventricular cavity from a syringe and No. 27 needle. Following this, inject 0.1 ml of Heparin from a syringe using a No. 27 needle.

12. Next place a hemostat upon the descending aorta at about the level of the diaphragm. Amputate the tip of the heart, and discontinue artificial respiration. Check the glass cannula which forms the end of the perfusion apparatus for air bubbles. Insert it quickly into the left ventricle, through the aortic valve, and tie it into place by means of the previously placed ligature. This is to be done by the assistant previously administering artificial respiration. The right atrium is nicked to assist rapid outflow of fixative.

13. 250 ml of fixative (1% glutaraldehyde, 1% formaldehyde in 0.12 M phosphate buffer with 0.02 mM $CaCl_2$; see Solutions, section C) are allowed to flow at maximum pressure over the first 5 to 10 minutes, and another 250 ml of fixative are allowed to drip more slowly. This is done either by the introduction of a pressure flow clamp or by reducing the height of the perfusion flask. The procedure from the opening of the chest to the beginning of perfusion takes 2 to 3 minutes. Less than a minute should elapse from the cessation of artificial respiration to the first signs of fixation (intense muscular spasm, stiffening in head and upper limbs). If multiple perfusions are done in one day, the flask and tubing should not be allowed to empty, as this avoids problems with air bubbles in the succeeding perfusions.

14. After the perfusion decapitate the head and leave in fixative overnight for up to 18 hours. The brain is to be dissected out the following morning.

c) Equipment for the Dissection of Rat Brains for Electron Microscopy

Instruments

Disposable vials with snap caps, each with 5 ml or 2.5 ml capacity
Absorbent paper towels
Centimeter rule
Bone clippers, 2 small, and 1 large
Forceps, small straight, and small toothed, curved
Large forceps, toothed
Scissors, large sharp end, Mayo
Scissors, fine iridectomy
Double-edged razor blades, cleaned with detergent, alcohol, split in two
Linen towel
Plastic bag

Disposable plastic gloves
Dissecting microscope on stand
Beeswax plates
Labelling tape
Labelling pen
Notepad
Pencil

Solutions

Rinse solution (0.12 M phosphate buffer with 0.02 mM calcium chloride and 8% dextrose; see Solutions, section D, p. 327).
2% osmium tetroxide in 0.12 M phosphate buffer and 7% dextrose (see Solutions, section F, p. 328).

d) The Dissection Procedure

1. This is done the morning following the perfusion. Do the dissection under adequate lighting, a dissecting microscope, and with the head from the perfused animal set on absorbent paper toweling. Make a midline incision on the dorsal surface from snout to spine (Fig. 267b), and peel away skin from underlying muscle (Fig. 267c).

2. Remove all skin and soft tissue to expose the skull. The snout and spinal column are left attached in order to provide a handle during the dissection.

3. With large bone clippers remove the mandible on either side. Remove the eyes. The separation of the brain from the skull of an animal without pronounced cervical and pontine flexures is best done from the ventral surface.

4. The exposed brain is always to be kept wet with fixative.

5. Remove the hard palate and break into the tympanic cavity. Beginning at the spinal column, with a small pair of bone clippers sever the vertebrae at their transverse processes along the entire length on either side. Separate the row of vertebral arches from the centrum. Lift up the dorsal half of the vertebral column to expose the dorsal surface of the spinal cord. Remove the atlas and axis.

6. Chip around the bone encasing the flocculi and paraflocculi. Do not make any sudden or rough breaks, as these structures are attached to the main mass of the cerebellum only by very thin peduncles.

7. With the flocculi and paraflocculi exposed, lift off the occipital bone that caps the dorsum of the cerebellum. The rest of the brain may then be dissected out by carefully removing the remaining ventral skull bones (see Figs. 2a, b, c).

8. Note that the object of the dissection is the careful removal of small pieces of bones of the skull from an intact brain, not vice versa. The finer parts of the dissection are best done under a dissecting microscope.

e) The Postfixation of Tissue Slabs

1. The cerebellum of an adult rat can be divided into 6 slabs, approximately 2 to 3 mm wide each, in any one of the sagittal, frontal, or horizontal planes.

2. Each slab of cerebellum is then processed individually in disposable glass vials with snap caps that have been labeled both on the bottle and on the cap with indelible ink or by means of a stone on a dental drill.

3. Each bottle containing a piece of cerebellum is rinsed in 3 rapid changes of rinse solution containing 0.12 M phosphate buffer with 0.02 mM $CaCl_2$ and 8% dextrose (see Solutions, section D, p. 327). These rinses take place over 15 minutes.

4. Immerse each piece of tissue in the postfix solution, 2% OsO_4 in 0.12 M phosphate buffer and 7% dextrose (see Solutions, section F, p. 328). The tissue requires approximately 2 ml of this fluid. The bottles are then placed on a mechanical rotator for slow, continuous agitation for at least 2 to 3 hours.

f) In-Block Staining, Dehydration, and Embedding

1. After 2 to 3 hours in the osmicating fluid, decant the solutions of osmium tetroxide from the vials. Each block is then rinsed in 0.1 M sodium acetate in preparation for in-block staining with aqueous uranyl acetate. The sodium acetate used must be a cold solution, and the next three steps are to be done with the bottles on ice or in the refrigerator. There are 2 sodium acetate rinses, each for 2 minutes.

2. The blocks are then immersed in 0.5% aqueous uranyl acetate for 30 minutes on ice. Cooling is necessary to prevent precipitates of uranyl acetate throughout the tissue.

3. Rinse each block rapidly in cold 0.1 M sodium acetate for 2 changes at 2 minutes each.

The remaining steps in the dehydration with methanol are also done with cold solutions.

4. 50% cold methanol, 15 minutes.
5. 70% cold methanol, 15 minutes.
6. 80% cold methanol, 15 minutes.
7. 95% cold methanol, 15 minutes.
8. 100% cold methanol, 15 minutes, and 2 changes.
9. The blocks are then further dehydrated in fresh propylene oxide for 2 changes of 15 minutes each.

Note that steps 1 to 9 are usually carried out in this laboratory with the use of a vacuum evacuation flask connected to a water pressure suction system. This makes the process of changing solutions and pouring solutions out of small vials a neater and much more rapid process; the system is particularly useful where large slabs are used. A photograph of this set-up can be seen in Fig. 267d. It is important to ensure that the blocks are never left even momentarily without fluid.

10. The Epon mixture (see Solutions, section H) is prepared at this point. It must be made fresh before use.

11. Propylene oxide: Epon, a 1:1 mixture. Tissue blocks are left in this mixture for 3 hours. The fluid is decanted and a fresh mixture of 1:1 propylene oxide: Epon is poured in. The tissue blocks are left in this solution overnight on the rotator. Agitation is important for good infiltration.

12. The following morning another batch of fresh Epon mixture (see Solutions, section H, p. 328) is made.

13. The tissue blocks then are transferred to the Epon embedding mixture and left for 6 hours on the rotator. Agitation is necessary for good penetration.

14. Each tissue block is embedded individually, either in gelatin capsules or in flat disposable plastic dishes. They are appropriately labeled with typed identification numbers, which have been rinsed in absolute alcohol and air dried. A label is included in each capsule.

15. The blocks are left overnight in a 37° oven for slow curing.

16. They are then removed to a 45° oven for 8 hours.

17. The blocks are then moved to a 60° oven and left for 2 days. After this time, 1 µm sections or semi-thin sections may be cut from these blocks with glass knives. The blocks are then returned to the oven for another 2 days or more at 60° before attempting to thin-section with either glass or diamond knives.

g) Solutions and Other Formulas

A. Purification of Glutaraldehyde for Use in Fixatives for Electron and Light Microscopy. 50% biological grade glutaraldehyde (Fisher Scientific Co., 1-pint bottles) should be diluted upon opening with equal quantities of distilled water and stored in the refrigerator. This will make a stock solution of 25% glutaraldehyde that is good for 6 months.

1. Mix 20 g of Norit (neutral charcoal) with 200 ml of 25% glutaraldehyde. Vacuum filter through a Büchner funnel using Schleicher and Schuell filter paper No. 576.

2. To the filtrate add 20 g of Norit A (alkaline decolorizing carbon) and filter again.

3. Repeat step 2 and store in the refrigerator. Concentration should be 25%.

4. The true concentration of glutaraldehyde can be checked with a UV spectrophotometer using the following information:

The molar extinction coefficient of glutaraldehyde at 27° C, 280 µm, is 6.00 per mole per cm.

Measure the glutaraldehyde, using a 25× dilution at 27° C and 280 µm in a cell with a path length of 1 cm.

$\%$ glutaraldehyde = optical density $\times \frac{250}{6}$.

Note that the optical density of glutaraldehyde changes 3.2% for each degree between 18° C and 28° C.

5. 25% glutaraldehyde (purified) is used as the stock solution from which fixatives utilizing glutaraldehyde in lower concentrations are made.

B. 0.4 M Standard Phosphate Buffer

$NaH_2PO_4 \cdot H_2O$	5.3 g
K_2HPO_4	28.0 g

add H_2O to make 500 ml

C. Fixative: 1% Glutaraldehyde and 1% Formaldehyde in 0.12 M Phosphate Buffer with 0.02 mM $CaCl_2$

paraformaldehyde	5 g
distilled water	200 ml
1 N NaOH	4–6 drops
25% glutaraldehyde (purified, see above)	20 ml
0.4 M standard phosphate buffer	150 ml
0.5% $CaCl_2$	2 ml

add water to make 500 ml

To dissolve paraformaldehyde, heat 200 ml of water until it steams slightly, but doesn't boil. Add the paraformaldehyde to the hot water, stir constantly, and add 4–6 drops of 1 N NaOH until solution is just clear but not crystal clear. Allow the solution to cool and add 150 ml of buffer. Then add the glutaraldehyde and make up to almost the final concentration. Slowly add the 2 ml of 0.5% $CaCl_2$ while stirring the solution. Complete dilution, and filter.

D. Rinse Solution for Electron Microscopy: 0.12 M phosphate buffer with 0.02 mM $CaCl_2$ and 8% dextrose

dextrose	8 g
0.4 M standard phosphate buffer	30 ml
0.5% $CaCl_2$	0.4 ml

water to make 100 ml

Postfixation wash: add dextrose and buffer and make up to almost final concentration with water. Slowly add the calcium and complete the dilution.

E. Double-Strength Buffer for Osmium Tetroxide

dextrose	7 g
0.4 M standard phosphate buffer	30 ml
add water to make 50 ml of double-strength buffer	

F. 2% Osmium Tetroxide in 0.12 M Phosphate Buffer and 7% Dextrose

double-strength buffer	15 ml
4% osmium tetroxide in water	15 ml
0.5% $CaCl_2$	0.15 ml

Tissue blocks are postfixed in this solution. One gram of osmium tetroxide is dissolved in 25 ml of distilled water in a *chemically clean* stock bottle. The OsO_4 is stirred with a magnetic flea or sonicated until dissolved completely. Each slab of tissue from the cerebellum requires 2 ml of this 2% OsO_4 solution. Osmium tetroxide crystals can be removed from vials by the following procedure. Wash vial in detergent and rinse thoroughly in triple-distilled water. Avoid touching vial with bare fingers. Fingers can be protected with plastic gloves or by wrapping vials in lens tissue. Place vials upright in a small container with fragments of solid carbon dioxide. When vials are thoroughly chilled, the crystals can be shaken free from the walls of the vials by gentle tapping. Score vials and break open. Tap the crystals into the bottle of distilled water prepared in advance. The procedure should be done in a chemical hood with glass door lowered. No glass particles should be allowed to fall into the stock bottle.

G. 0.5% Uranyl Acetate for in Block Staining

uranyl acetate	0.5 g
distilled water	100 ml

Dissolve by placing in a sonicator.

H. Epoxy Resin (Epon 812), Hard, for Glass and Diamond knife sectioning

Epon 812	47 g
Dodecenyl succinic anhydride (DDSA)	21 g
Nadic methyl anhydride (NMA)	32 g
Total	100 g
Add DMP-30	1.4 ml

The three constituents of this epoxy mixture are weighed out successively in one disposable beaker in the order set out above. Each one of the ingredients is to be added slowly in order not to pour too much into the beaker by accident. (Each drop weighs approximately 0.1 g.) The DMP-30 is delivered to the mixture through a glass syringe that has been set aside and is used only for this purpose. These 4 ingredients are then put under a rotary mixer and mixed for 15 minutes. The Epon turns first yellow, then to a deep golden color. Care must be taken that the epoxy is not beaten. As little air as possible should be included in the epoxy mixture for best hardening properties. It is important to be certain that the container, syringe, and rotary mixer used for mixing epon by completely dry and free of moisture. Since these compounds are allergenic, it is essential to carry out this procedure and the changing of solutions with caution and neatness. The skin may be protected by wearing disposable gloves, but there is no substitute for careful handling. The components of this epoxy resin may be purchased in small quantities from Biodynamics Research Corporation, 6010 Executive Boulevard, Rockville, Maryland.

h) The Cutting of 1 μm Semithin Sections of Epon-Embedded Cerebellum

1. The entire face of a block of flat-embedded cerebellum measuring almost 9 mm in diameter can be cut with glass knives on any rotary, manual or automated microtome. One-inch glass knives are made either by breaking by hand or on an LKB knife-making machine. A scoring angle of 45° is used. Each knife is checked carefully for a good cutting edge under a good light source and a dissecting microscope. Each block is trimmed with a dental drill fitted with a diamond cutting wheel (Diaflex 86X, both sides, Hopf Ringleb and Co., Berlin), or a micro hand saw. The blocks are trimmed so that the extra resin surrounding each slab of tissue is cut away, leaving only the tissue itself and reducing the total surface area of the block to be sectioned (see Fig. 267e). The block is mounted in the chuck of the microtome with its largest dimension parallel with the edge of the knife. The block is then trimmed by sectioning until the entire face of the tissue is exposed. It is important that the sections should be free of knife scratches or any irregularities due to compression or other defects of sectioning.

2. The glass knife has to be changed often, sometimes after every tenth or fifteenth section. The section is picked up, dry, from the edge of the glass knife with a pair of fine-tipped forceps, and carefully placed on a drop of distilled water on a clean glass slide. The slide is then placed on an electric hot plate set at a low temperature. It should be only warm to the touch. A hot plate that is too warm will cause bubbling of the drop of water and

wrinkling or folding of the section. The section is allowed to flatten slowly on the drop of water and then to dry out thoroughly for about half an hour. If the section has not completely adhered to the glass slide, it will come off during the staining process. Sections at this point can even be allowed to stay overnight on the hot plate.

3. The sections on the slides are stained in 0.1% toluidine blue in 1% sodium borate solution (0.1 g of toluidine blue and 1 g of sodium borate in 100 ml of water). A drop of the stain is placed on the section, and the slide, section, and drop of stain are warmed on the hot plate for a few seconds. Wait until the edges of the drop of stain on the glass slide begin to recede slowly. At this time the drop of stain begins to steam very gently.

4. Remove the slide from the hot plate. Carefully wash with distilled water from a wash bottle. Squeeze the water gently over the section and wash the excess stain from the slide. Wipe the back of the slide dry completely before returning the slide to the hot plate. Allow the slide to dry thoroughly. Ideally, the section should be flat, unscratched, should include the entire face of the block, and be stained intensely and evenly throughout.

5. The section can then be mounted in a drop of epoxy resin (see Solutions, section H) under a coverglass. This Epon can be made up, then put in 5 ml disposable capped syringes to be frozen for future use. In this case, thaw before using. Deliver directly from syringe. In order to facilitate rapid drying of the epoxy resin, the slide can be placed in a 60° oven for a few hours or overnight. Any minor imperfections on the surface of the section will be unnoticeable after mounting with epoxy resin. Sections for photomicrography should be stained strongly with toluidine blue, and are often best when thicker than 1 μm, for example, 2–3 μm thick.

i) Thin Sectioning

Survey sections 1 μm thick of the entire face of each slab are cut and stained with toluidine blue. From these sections areas can be selected for electron microscopy (see Fig. 267f). The area chosen is identified on the face of the original block. This area is then trimmed into a flat-topped pyramid for thin sections (see Fig. 267g).

Shaping the Pyramid. In order to form a pyramid like the one shown in Fig. 9, it is necessary to plane away a thin layer of Epon-embedded tissue from the portion of the block not to be included in the pyramid. This is accomplished by using either a jeweler's handsaw or the flat surface of a diamond cutting wheel mounted on a dental drill. Only a layer about 0.5 mm deep is removed from the surface of the block, being careful to stay away from the selected area. This leaves a low plateau of tissue out of which the pyramid is carved. Using half of a double-edged razor blade that has been cleaned by washing in detergent, water, and acetone, carefully cut off the rough edges of the tissue to shape the pyramid. This must be done in the following manner: 1) under a dissecting microscope, 2) with the block held firmly in a chuck that is stable under the illumination, 3) with both hands holding onto the razor blade (one hand on each end) and elbows well placed on the countertop. Shaping the pyramid requires considerable skill, which comes with practice.

This pyramid *must* have absolutely parallel edges on its top and bottom. This ensures that straight ribbons will form during thin sectioning. The face of the pyramid should be approximately 1 mm across the bottom edge for sectioning with diamond knives. Slightly larger blocks have been used successfully. Ideally, it should be possible to place three perfectly oriented thin sections from such a pyramid face onto a 3 mm grid. Very small pyramids are less useful because of the limited tissue present, but are necessary when using glass knives for thin sectioning. In selecting the area for the pyramid, care should be taken to avoid including blank Epon. Also, heavily myelinated areas should be kept to the leading edge of the block. This reduces the possibility of sectioning problems. In order to re-use the same block for another area, simply resume semi-thin sectioning, select another area, and make another pyramid. Or if two or three areas are needed at the same level or depth in the block, then the block can be cut into separate parts with a diamond bit or handsaw and remounted on a blank block with Epon. A pyramid can then be made on each part and thin sections may be obtained. This method also allows a block to be turned to various planes.

Orientation of Sections. Thin serial sections are best cut with a diamond knife. Proper thin sectioning necessitates scrupulous cleanliness and strict concentration and attention to details. Casual conversation and slovenliness are incompatible with successful sectioning. All instruments and the knife trough in particular must be clean. Serial sections are mounted directly on uncoated copper grids. The grids are cleaned in formic acid followed by three rinses of distilled water and one of absolute alcohol. The grids are then treated by floating them overnight on a solution of 10^{-5} M of benzalkonium chloride (Zephiran chloride, 17%, Winthrop Laboratories). A more concentrated solution (0.0135%; 0.1 ml in 250 ml of deionized water) may also be used to save time.

Grids are floated for at least an hour prior to use. This treatment facilitates the secure attachment of serial sections to the grids. It also makes the grid hydrophilic and thus makes the orienting of the serial sections directly on the grids much simpler. No loops or other devices need be used. Just before picking up a ribbon of sections, the treated grid should be dried by slipping a piece of filter paper between the tines of the forceps and withdrawing the drop of solution.

The sections are oriented so that a view of the entire thickness of the cortex can be obtained without interference from crossbars. Two types of grids have proved very useful in our laboratory: 1) the 4829 A-7 75/300 mesh (3 mm) grids, otherwise known as the slit grids, made by LKB; and 2) the 3 mm R300A Minimesh EM copper grids made by Mason & Morton, Ltd., England. In this laboratory, sections are routinely oriented so that a strip of the entire cortex from white matter to pial surface lies uninterrupted across the windows in the grid. Thus, the combination of 1 μm sections of the entire block face and precise orientation of the sections on slit grids permits unequivocal recognition of the folium and the cortical layer to be studied in the electron microscope.

j) The Staining of Thin Sections on Grids

Equipment

2 graduated centrifuge tubes	ice
1 50 ml Erlenmeyer flask	2 staining dishes
1 125 ml Erlenmeyer flask	6 10 ml beakers
1 100 ml volumetric flask	1 large beaker
test tube rack	1 wash bottle
	timer

1. Thin sections are stained soon after cutting, preferably in the first 24 hours.

2. All glassware is to be cleaned and rinsed throughly at least twice with deionized water.

Solutions

3% uranyl acetate in 50% ethanol

0.3 g of uranyl acetate is weighed out and placed into the 50 ml Erlenmeyer flask. 5 ml of water and 5 ml of ethanol are then pipetted into the flask. The solution is then sonicated until dissolved. This usually takes 5–10 minutes. Care should be taken to ensure that only chemically clean glassware is used.

0.075% lead citrate in 0.1 N NaOH

0.075 g of lead citrate is weighed out and placed into a 100 ml volumetric flask. About 50 ml of deionized water is then added to the flask. Then a 1 ml solution of 10 N NaOH is pipetted into the solution. Make up to 100 ml. The solution is then poured into the 125 ml Erlenmeyer flask and sonicated for a few minutes.

3. Pour 10 ml of each solution into the graduated centrifuge tubes.

4. Spin in a clinical centrifuge at one half speed for 15–20 minutes. Watch that the tubes do not get hot.

5. The tubes are then placed on ice. The entire staining procedure, from the making of solutions of stain through the actual staining, should take no more than 30 minutes. Uranyl acetate in ethanol is rather unstable and tends to precipitate after that time. This, of course, causes precipitates of stain on the grids.

Staining Procedure

6. The grids are stained in uranyl acetate for 2–5 minutes. Each grid is then washed 3 times in 50% ethanol, washed once in distilled water from a wash bottle, and dried on filter paper before being stained with lead citrate. The grid is then floated on the surface of a drop of the lead citrate staining solution. Sections are stained in lead citrate for 30 seconds to 2 minutes. Each grid is washed twice in water and then once in running water from the wash bottle.

Note: it is important to stress that each grid must be stained individually in order to maintain the right amount of staining time for each one. The grids also have to be washed carefully and to be handled carefully throughout the staining procedure. The copper grids should not be mangled or bent at the edges. They should be picked up lightly by the rim. During washing they should be held firmly so that the grids do not drop to the bottom of the washing container. Most important, the grids must not be agitated in the wash solution, as the resultant pressures tend to break the delicate thin sections between grid bars. After the uranyl acetate staining and washing procedure, a piece of filter paper, Schleicher & Schuell No. 576, is slipped between the tines of the forceps in order to withdraw the water lying on the grid surface.

When washing grids with a wash bottle, the flow of water from the wash bottle must not be too fast. Instead, a slow, steady, and continuous flow should be maintained for about 1 minute. The grid should be held with its edge towards the flow of the water in order not to subject the thin sections to undue turbulence. This prevents the tearing of thin sections. A filter paper is then again inserted between the tines of the forceps to dry the grid.

7. The grids are carefully lined up on a piece of filter paper in a Petri dish, labelled, and left to dry in an oven at 60° overnight. This tends to stabilize sections that are not supported by Formvar films so that even under intense electron bombardment they do not tear.

k) Electron Microscopy

In order to take the most advantage of the precise orientation of the thin sections, the following procedure was devised for microscopy. Low power overlapping electron micrographs ($\sim 3000\times$) are taken in order to survey the area. These are assembled into large montages, sometimes stretching from the white matter to the pial surface. Such montages provide a wealth of details about cell-to-cell connections and the general topography of each cortical area (see Figs. 267h and i). Then the area is restudied on the screen of the electron microscope. When a previously selected area or a small field of especial interest is found, high magnification micrographs are made. These may be single plates or part of a smaller montage. A systematic approach of this type is much preferable to the recording of random micrographs in unoriented sections. These detailed topographic surveys of the nervous system necessitate large fields of well-preserved tissue in well-stained perfect thin sections that are exactly oriented. It cannot be stressed too strongly that attention to every detail is necessary.

The following list contains a few suggestions for dealing with some common problems that might occur during the procedures leading to microscopy.

1. If a perfusion goes poorly, e.g., the animal dies before fixation or the brain is still somewhat soft after perfusion, then abandon the tissue and begin again. More time is saved in doing this than is gained by trying to salvage poorly fixed tissue. Successful perfusion is a result of skill acquired through practice and attention to small details. Novices are inclined to blame the fixing solution and to waste much time ringing the changes on the pH, buffer, concentrations, and every possible parameter. By the time they have returned to the recommended conditions, their results improve. Meanwhile they have learned how to carry out the perfusion quickly and smoothly.

2. When the weather is very humid, as it often is in the summer in Boston, it is usually not suitable for embedding tissue. The Epon hardens slowly and unevenly, to the detriment of the tissue. Every September is clean-out time in our laboratory. Open bottles of epoxy resin, and other perishable items should be discarded.

3. If the tissue under the microscope is full of large holes and susceptible to stretching and tearing under the beam, then either the Epon is inadequate and thus giving way too easily or the tissue has been poorly dehydrated and infiltrated.

4. If the sections are coated with stain precipitate or dirt, most often the source is a dirty trough, dirty forceps, instruments, or glassware. Do not use a grid containing sections that has been dropped on the floor or on an uncovered counter. During the handling of grids with sections, it is advisable to cover benchtops in use for that procedure either with lens paper or with lintless laboratory mats (BENCHKOTE, WHATMAN). Loading and unloading of specimen holders are always done on such a surface.

2. The Golgi Methods

a) Introduction

It is impossible to predict or to recommend which one, if any, of the Golgi procedures will be successful on a particular area of the brain in any animal. It is thus necessary to try some of these methods out. Usually, each method produces its own peculiar impregnation of a certain population of cells. Thus, trying out various procedures will rapidly result in a collection of slides with different populations of cell types, neurons or neuroglia, in the area of the brain studied.

In general, we have found perfusions with aldehydes the most simple and practical approach to solving the problem of good fixation for Golgi perparations. They are simple to perform, and the background of these finished preparations is generally light gold, even in adult tissue. Most important, perfusions with aldehyde mixtures provide potential material for both Golgi preparations and electron microscopy. In dealing with brains of animals that are not readily available, e.g., primates and animals from difficult experiments, we have found that perfusions with aldehydes, particularly 1% glutaraldehyde/1% formaldehyde, are invaluable. This allows us to use alternate slabs of well-preserved tissue from one particular area of the brain first for Golgi preparations (both light and high voltage electron microscopy) and then for electron microscopy of thin sections. Material from immersion fixation and dichromate/osmium tetroxide perfusion is less versatile. In both cases, the initial fixative is not the fixative of choice for best results from thin section electron microscopy, as the preservation is generally inadequate for ultrastructure.

The choice of the Golgi impregnation procedure to be used after fixation is also empirical. Both the multiple impregnation Golgi method and the zinc chromate/formic acid procedure have been most successful in our hands for adult as well as young brains. The potassium dichromate/chloral hydrate method has provided neuroglial populations and many short-axon nerve cells in the cerebellar cortex. The length of time during which the tissue slabs are immersed in solutions of chromate or silver salts depends a great deal on the tissue, area, age of the animal, and other more intangible circumstances. Thus, the times indicated in the procedures given below are only suggestions for a beginning—the actual times have to be worked out by the investigator by trial and error.

This process is not so hopeless as it might seem from reading these comments. In fact, a great deal of time and effort can be saved by interposing a test system into all of the Golgi procedures. The slabs of tissue are tested for their state of impregnation after each cycle of chromation and silvering. This is done by taking a thin (about 100 µm) slice or two from one of the slabs of tissue by means of a very sharp single-edged razor blade. If the tissue appears well populated with reddish black cells and fibers, then it is dehydrated and embedded. If the tissue is poorly impregnated, the slab is returned and again put through another cycle of chromation and silvering.

Lastly, the embedding of material in nitrocellulose aids routine serial sectioning and allows sections to be examined with oil immersion lenses under the light microscope. Embedding of Golgi material in plastics such as araldite (Durcupan, see below) enhances the final preservation of tissue even more, and allows both thin section electron microscopy and high voltage electron microscopy to be carried out on the same material.

Golgi preparations of adult rat cerebellum have been made with all of the following methods. Slabs of the entire brain cut in any one of the three cardinal planes can be used for the impregnating steps. The brains may be taken from animals that have been perfused with any one of the three following solutions. Brains have also been used without perfusion. In this case the brain was removed directly after a lethal dose of anaesthetic and processed beginning with the first chromation solution of the Golgi procedures. Inexplicably, these particular specimens tended to give excellent examples of the Purkinje axon collateral plexus.

b) Perfusion Solutions—Freshly Prepared

(a) 1% glutaraldehyde and 1% formaldehyde in 0.12 M phosphate buffer with 0.02 mM $CaCl_2$ (see instructions in electron microscopy section). The brain slices may be processed 24 hours after perfusion.

(b) 4% formaldehyde solution

paraformaldehyde	40 g
distilled water	500 ml

(i) Heat water but do not boil. Cover flask while heating.

(ii) Add paraformaldehyde, stir. Cool

$NaH_2PO_4 \cdot H_2O$ (monosodium acid phosphate anhydride)	4 g
Na_2HPO_4 (disodium acid phosphate)	6 g
distilled water	500 ml

(iii) Make up buffer. Add to paraformaldehyde solution. Stir. Adjust pH to 7.3 with 0.1 N sodium hydroxide. Filter before using. After perfusion, brain slices may be left in this solution for up to a week.

(c) Osmium tetroxide/potassium dichromate perfusion solutions

(i) *Wash out—0.1 M phosphate buffer, pH 7.3+0.5% PVP-40*

$NaH_2PO_4 \cdot H_2O$	2.65 g
K_2HPO_4	14.0 g
distilled water to 1000 ml	
	pH 7.3
PVP-40	1 g
0.1 M phosphate buffer	200 ml

Dissolve and filter (No. 3 Whatman paper)

(ii) *Perfusion fixative 0.25% OsO_4+3% $K_2Cr_2O_7$, pH 6.8, +0.5% PVP-40*

4% $K_2Cr_2O_7$ pH 6.8	
$K_2Cr_2O_7$	72 g
distilled water	1800 ml
pH is now 4.0	
add ~26 ml of 40% NaOH to pH 6.8	

Fixative

PVP-40	1 g
4% $K_2Cr_2O_7$ pH 6.8	150 ml
1% OsO_4	50 ml

Dissolve PVP-40 well in the dichromate solution, then add osmium tetroxide. Filter using Whatman No. 3 paper.

After perfusion the brain slices are processed immediately. Following the perfusion with either of the above solutions, the brains are dissected from the skull and cut into 3–5 mm thick slabs in any of the three planes. These slabs of tissue are then processed according to any one of the following impregnation procedures.

c) Procedures for the Golgi Methods

A. Rapid Golgi Method—Multiple Impregnation Process

Used after perfusion with aldehyde or osmium tetroxide/potassium dichromate solutions or directly without perfusion.

2.33% potassium dichromate and 0.19% osmium tetroxide

potassium dichromate	12 g
osmium tetroxide	1 g
distilled water	500 ml

This solution is good for several weeks if stored in the refrigerator.

1. Each piece of tissue is put into a small jar with a tight cap containing at least 20 ml of this solution per piece of tissue. (Glass jars used for Golgi preparations may be obtained from A.H. Thomas Co. as ointment jars, No. 6286-A, with screw caps, 2 oz size.) At the bottom of each jar there should be a layer of glass beads, either 3 mm for small pieces of tissue or 6 mm for larger ones, to ensure adequate penetration and fluid space around the tissue. Make sure that the beads do not macerate the tissue. Place jars on rotator for 7 days at room temperature. Save solution in refrigerator for re-use in step 5.
2. With a section lifter, remove the tissue to a gauze pad, pat briefly, and rinse in a small volume of 0.75% aqueous silver nitrate. Rinse out bottle and beads with distilled water and place tissue in freshly prepared 0.75% aqueous silver nitrate.
3. Store tissue in 0.75% silver nitrate for 24 hours, on the rotator.
4. Test the tissue slab for state of impregnation by cutting thin slices from the surface with a single edged razor. Put a section or two on a glass slide in 70% alcohol under a coverslip for observation. If impregnation is adequate as judged by the presence of reddish brown cells and fibers, then proceed with dehydration and embed in nitrocellulose (see below).
5. If impregnation is not adequate (i.e., too few fibers and cells visible), then continue thus. Blot, and without rinsing transfer to the same osmium tetroxide/potassium dichromate solution used in step No. 1. Store for 6 days, at room temperature, on rotator.
6. Store in fresh 0.75% silver nitrate for 48 hours, repeating steps 2 and 3. After this point tissues may be tested again for their state of impregnation. If further chromating and silvering are needed, then proceed with step 7. If not, dehydrate and embed, step 9.
7. Blot and transfer to new osmium tetroxide/potassium dichromate solution (same volume) for 5 days.
8. Store in fresh 0.75% silver nitrate for 3 days.
9. Tissue is now ready for dehydration and embedding (see below).

B. Zinc Chromate-Formic Acid Golgi Method

Usually used after aldehyde perfusion.

Chromating Solution

zinc chromate ($ZnCrO_4$)	60 mg
formic acid	32 ml
distilled water	make up to 1000 ml

adjust pH to 3.1 with formic acid

Silvering Solution (Stock)

10% silver nitrate in distilled water

Take 75 ml stock solution. Add distilled water to make up to 1000 ml. Shake well. The solution immediately becomes opaque. Allow to settle. Disregard any precipitate formed.

1. Remove the slices of tissue from the perfusion solution with a section lifter and place them gently on gauze. Place tissue slices in the chromating solution (20 ml). Close jars tightly, making sure that the glass beads are *not* macerating the tissue, and place jars on the rotator for 3 days.
2. Take tissues out of the jars in the chromating step and pat them gently on gauze. Set up bottles, each with one layer of glass beads, plus one slice of tissue and the silvering solution (20 ml). Place them on the rotator for one day.
3. Rinse out bottle and beads with distilled water and add more silvering solution. Keep on rotator for another day.
4. Tissues are now ready to be tested for their state of impregnation using a single-edged razor. If the impregnation is adequate and many cells and fibers are colored reddish black, then proceed with the dehydration and embedding in nitrocellulose.
5. If the impregnation is not adequate, then the entire procedure may be repeated again. That is, chromation and silvering are done once more for an equivalent number of days, or for shorter, or longer, period.
6. Dehydrate and embed tissue blocks.

C. Potassium Dichromate/Chloral Hydrate Method for Neuroglial Cells and Some Short Axon Neurons

Chromating Solution

potassium dichromate	5 g
chloral hydrate	3 g
formalin (40% formaldehyde)	10 ml
distilled water	90 ml

1. Prepare afresh before use. Dissolve the potassium dichromate in water, add formalin, then dissolve the chloral hydrate.
2. After perfusion, place 3–4 mm thick slices of brain in chromating solution in jars with glass beads and tight-fitting lids. After each 24 hour period, fresh solution should be used.
3. Two solutions of silver nitrate are used for silvering: 1% silver nitrate; 1.5% silver nitrate. Blocks are transferred from the chromating solution to the 1% silver nitrate for a quick rinse, then left in the 1.5% silver nitrate for 18–24 hours or more.
4. The blocks are then sampled for impregnation of cells and fibers by cutting several sections with a single edged razor blade and examining them under the microscope. If impregnation is acceptable, the tissue may be dehydrated and then embedded in nitrocellulose.

d) Dehydration and Infiltration of Slabs of Golgi-Impregnated Tissue for Embedding in Nitrocellulose

(a) Solutions

0.5% silver nitrate in 50% ethanol

Dissolve 5 g of silver nitrate in water, make up to 450 ml with distilled water in a graduated liter flask. Add 500 ml of 95% ethanol, to make 950 ml of solution.

(b) Nitrocellulose

Nitrocellulose (type RS 1/2 second) may be purchased in 5-gallon drums from Randolph Products Co., Carlstadt, N.J. 07072.

5% nitrocellulose	*10% nitrocellulose*
40 g nitrocellulose	80 g nitrocellulose
254 ml ethanol (absolute)	240 ml ethanol (absolute)
254 ml diethyl ether	240 ml diethyl ether
254 ml diethyl ether	240 ml diethyl ether

20% nitrocellulose	*30% nitrocellulose*
160 g nitrocellulose	240 g nitrocellulose
214 ml ethanol (absolute)	186 ml ethanol (absolute)
214 ml diethyl ether	186 ml diethyl ether
214 ml diethyl ether	186 ml diethyl ether

1. Add equal volumes of ether and ethanol slowly to 1-liter brown bottles with tightly fitting caps. MUST BE DONE IN CHEMICAL HOOD. AVOID FLAMES AND ELECTRIC SPARKS.
2. Weigh out nitrocellulose and add to ethanol-ether in bottles. Nitrocellulose is added in small quantities at a time and carefully stirred with a mixing rod until all lumps are wetted.
3. Add remaining volume of ether. Stir. Agitate gently over the next 24 hours, until dissolved.
4. All bottles of nitrocellulose must be taped up tightly to prevent solvent evaporation. Ordinary hospital adhesive tape is useful for this purpose.

Procedure for Dehydration and Infiltration

1. The glass beads used in the chromating and silvering procedures are now removed. The tissue blocks are rinsed in 0.5% silver nitrate in 50% ethanol and processed according to the following schedule. The jars are sealed throughout the various steps of this procedure with adhesive tape. Dehydration and infiltration with nitrocellulose are carried out in wide-mouthed sealed jars left on a commercial rotator for gentle continuous agitation.

2.

0.5% $AgNO_3$ in 50% ethanol	— ~1 hour
70% ethanol	— ~2 hours
90% ethanol	— ~1 hour
absolute ethanol	— ~1 hour
diethyl ether/ethanol (2:1)	— ~3 hours
diethyl ether/ethanol (2:1)	— overnight
5% nitrocellulose	— minimum of 8 hours
10% nitrocellulose	— minimum of 24 hours
20% nitrocellulose	— minimum of 2 days
30% nitrocellulose	— minimum of 2 days

3. The tissue blocks are ready to be embedded at this stage. However, should this be an inconvenient time for embedding, the blocks may be left in fresh 30% nitrocellulose, in sealed jars in a freezer. The blocks may remain under such conditions for about 30 days. Before embedding, the frozen blocks should be thawed out for several hours in their sealed jars.

Nitrocellulose Embedding Procedure

1. The tissue blocks are to be carefully oriented and embedded in a paper boat in 30% nitrocellulose. The blocks are then hardened in a closed chamber (e.g., a desiccator) saturated with chloroform vapors. In order to prevent holes caused by air bubbles in the final block, the following procedure is recommended. Place the block in nitrocellulose in paper boat and put entire assembly into a covered Stender dish for 30 minutes before hardening in chloroform.

2. In order to hasten the hardening process, the blocks may be set upon a rack of fine wire mesh over a deep dish of chloroform, and dipped occasionally into the fluid.
3. The hardened blocks are removed from the paper boats, trimmed to a 60° angle pyramid leaving a rim of nitrocellulose around the tissue.
4. The trimmed block is then attached to a sliding microtome block holder in the following manner. Dip upper surface of the holder into 30% nitrocellulose, dip the bottom of the block in 30% nitrocellulose. Dip both nitrocellulose wetted surfaces in an ether-alcohol (1:1) mixture and press firmly together. This promotes adherence of block to holder. Leave in chloroform vapors for an hour to harden. Place blocks in a closed jar of 70% alcohol or in chloroform overnight.
5. Cut sections next day.

Sectioning and Mounting of Nitrocellulose-Embedded Golgi Material

1. Sections 90–120 µm in thickness are cut on a sliding microtome. As they come off the knife edge, the sections are placed individually in serial order on discs of thin lens paper cut slightly smaller than the Stender dish in which the sections are to be stored. The block is kept wet with 85% ethanol during the sectioning process, and the sections are held in the same solution until dehydration begins.
2. Immediately after sectioning, the sections (now in serial order and interleaved with lens paper discs) are dehydrated and cleared according to the following schedule. Make sure that there is good exchange of the various fluids between the layers of lens paper, by inverting the dishes several times.
3. 85% ethanol — 3 changes for 15 minutes each
95% ethanol — 3 changes for 15 minutes each
butanol — 3 changes for 15 minutes each
cedarwood oil — overnight

Fresh cedarwood oil (purchased from A. H. Thomas) must be used.
4. Sections must be thoroughly dehydrated. They may be held in fresh cedarwood oil for about two weeks, or they may be mounted immediately.
5. Mounting. Sections are moved individually through 2 changes of xylene, arranged in serial order, and mounted on slides in Harleco Coverbond mounting medium (VWR Scientific Co.) and covered with No. 0 coverslips.
6. Slides are then allowed to dry and are labelled. Slides examined when still wet with mounting medium invariably foul the microscope. It is not advisable to examine wet or fresh slides. Patience and forbearance are important ingredients of the Golgi method!

3. High Voltage Electron Microscopy

a) Embedding and Sectioning of Golgi Material

1. The brains of animals after 1% glutaraldehyde/1% formaldehyde perfusion may be used for high voltage electron microscopy of Golgi preparations as well as for conventional electron microscopy of thin sections and light microscopy of Golgi preparations. Alternate slabs of one particular area of the brain of such well-preserved tissue may be processed for Golgi preparations (light and high voltage microscopy) and the next for electron microscopy of thin sections, and so forth.

2. Slabs, 3 mm^3, of brain tissue may be processed in one of the above Golgi methods. The multiple impregnation rapid Golgi method has been our method of choice. The blocks are then dehydrated and embedded in Durcupan-ACM (Fluka, from Roboz Surgical Instrument Co., 810 18th St., N.W., Washington, D.C. 20006) according to the following schedule:

30% acetone — 15 minutes
50% acetone — 30 minutes
70% acetone — 30 minutes
90% acetone — 30 minutes
dry acetone — 30 minutes
dry acetone — 30 minutes

Durcupan Mixture No. 1

23 g A/M
20 g B
0.3 ml D

Place in disposable beaker, stir for 10 minutes with a rotary mixer.

Durcupan Mixture No. 2

23 g A/M
20 g B
0.7 ml C accelerator
0.3 ml D dibutyl phthalate

Place in disposable beaker. Stir for 10 minutes with a rotary mixer.

3:1 acetone/Durcupan mixture No. 1 — 1 hour
2:2 acetone/Durcupan mixture No. 1 — 1 hour
1:3 acetone/Durcupan mixture No. 1 — overnight
Durcupan mixture No. 1 — 2 hours
Durcupan mixture No. 1 — 2 hours
Durcupan mixture No. 2 — 2 hours

Throughout this procedure the blocks in their respective solutions should be gently agitated on a commercial rotator. This ensures complete infiltration. Embed in flat dishes or flat bottom capsules in Durcupan mixture No. 2 in a 60° oven overnight.

3. Glass knives, freshly broken and inspected for a perfect cutting edge, are used for sectioning. Sections are cut on a dry knife edge and picked up directly thereafter by means of fine-tipped forceps.

4. Each section is transferred to the surface of a drop of warm distilled water on a clean glass slide set on a hot plate. Avoid spreading the section too long on the water in order to prevent dissolution of the impregnation.

5. Serial sections are cut and placed sequentially on clean glass slides for examination in the light microscope. When an interesting well-impregnated area within the block is found, the next section is prepared for high voltage microscopy in the following way.

6. Sections 0.25–5 μm thick may be used. Contrast is best in the 5 μm sections, but 1 μm sections have better resolution. In sections less than 1 μm thick, axons and dendrites are too fragmentary.

7. A section is cut and floated onto a drop of warm distilled water. An uncoated clean copper grid (150 mesh or LKB slit grids) prepared in benzalkonium chloride (see Sectioning procedure for electron microscopy, p. 329) is carefully dropped onto the upper face of the floating section. Orient the longitudinal elements in the tissue along the lengths of the slits for best visibility and least interference from crossbars in the microscope.

8. Pick up the grid, blot dry, place grid section side up on the same drop of warm distilled water. Allow the drop to dry down on the hot plate. Leave the grid on the dry slide to heat for another half hour in order to fix the section to the grid. Sections may be examined with or without counterstaining.

b) Counterstaining

3% uranyl acetate in 50% ethanol, 0.075% lead citrate in 0.1 N sodium hydroxide for 4–8 minutes is suitable for 1 μm sections. Stronger solutions (e.g., 20% uranyl acetate, 5% lead citrate) may be used to give greater electron density. Long staining times may cause impregnation to dissolve from sections into staining solutions. Place grids in 60° C oven overnight.

4. Electron Microscopy of Freeze-Fractured Material

1. The animals are perfused with 1% glutaraldehyde/1% formaldehyde in 0.12 M phosphate buffer with 0.02 mM $CaCl_2$; see Solutions section C p. 327. The procedure used is described in detail in section 1 b. The only modification is that the fixation of the tissue should not be carried on for more than 20 minutes beginning with the perfusion and ending with washing of the tissue blocks in 20% buffered glycerol. A short fixation time is necessary to ensure good cleaving properties of the tissue during the fracturing procedure.

2. The perfusion is continued for 5 minutes; during this time about 200 ml of fixative should be washed through the animal.

3. The brain is quickly dissected from the skull and carefully sliced into slabs. Each slab is about 2 mm thick and is carefully trimmed during the dissection in order to maintain a predetermined orientation of the cortex during the rest of the procedure.

4. The blocks are rinsed in 3 changes of 0.12 M phosphate buffer over a few minutes. A maximum of only 20 minutes should elapse between the start of the perfusion and the end of the buffer rinse.

5. Then the blocks are infiltrated for 2 hours with 20% glycerol in 0.12 M phosphate buffer at room temperature.

6. The tissue blocks are then frozen in liquid Freon 22 (chlorodifluoromethane) for storage in liquid nitrogen, and for use in the fracturing procedure.

7. A BALZER's apparatus is used for freeze fracturing and platinum-carbon shadowing. The replicas are then cleaned of organic material in bleach (chlorox) and mounted on 350 mesh grids.

8. Micrographs of the replica taken in a transmission electron microscope at a primary magnification of about 20000 diameters are the most useful for both detail and orientation.

9. There may be a need to give the recording emulsion less exposure in order to broaden the range of gray tones in the image. Satisfactory contrast is less difficult to achieve here than in electron micrographs of thin sections.

10. A correct impression of the relief is obtained by looking at the positive prints of the electron micrographs along the shadow direction in the replica.

References

AKERT, K., PFENNINGER, K., SANDRI, C.: The fine structure of synapses in the subfornical organ of the cat. Z. Zellforsch. **81**, 537–556 (1967).

AKERT, K., PFENNINGER, K., SANDRI, C., MOOR, M.: Freeze etching and cytochemistry of vesicles and membrane complexes in synapses of the central nervous system. In: Structure and function of synapses (G.D. PAPPAS and D.P. PURPURA, eds.), p. 67–86. New York: Raven Press 1972.

AKERT, K., SANDRI, C.: Identification of the active synaptic region by means of histochemical and freeze-etching techniques. In: Excitatory synaptic mechanisms (P. ANDERSEN and J.K.S. JANSEN, eds.), p. 27–41. Oslo: Universitetsforlaget 1970.

ALTMAN, J.: Coated vesicles and synaptogenesis. A developmental study in the cerebellar cortex of the rat. Brain Res. **30**, 311–322 (1971).

ARMSTRONG, D.M., SCHILD, R.F.: A quantitative study of the Purkinje cells in the cerebellum of the albino rat. J. comp. Neurol. **139**, 449–456 (1970).

ANDERSEN, P., ECCLES, J.C., LØYNING, Y.: Recurrent inhibition in the hippocampus with identification of the inhibitory cell and its synapses. Nature (Lond.) **198**, 540–542 (1963).

ANDERSEN, P., ECCLES, J.C., VOORHOEVE, P.E.: Postsynaptic inhibition of cerebellar Purkinje cells. J. Neurophysiol. **27**, 1138–1153 (1964).

ANDRES, K.H.: Untersuchungen über den Feinbau von Spinalganglien. Z. Zellforsch. **55**, 1–48 (1961).

ANDRES, K.H.: Über die Feinstruktur besonderer Einrichtungen in markhaltigen Nervenfasern des Kleinhirns der Ratte. Z. Zellforsch. **65**, 701–712 (1965).

BAFFONI, G.M.: Contributo alla conoscenza della morfogenesi e dell'istogenesi cerebellare. Arch. Zool. ital. Napoli **41**, 1–112 (1956).

BARR, M.L., BERTRAM, L.F., LINDSAY, H.A.: The morphology of the nerve cell nucleus, according to sex. Anat. Rec. **107**, 283–297 (1950).

BAUD, C.A.: Untersuchungen der Körnerschicht der Kleinhirnrinde mit dem Elektronenmikroskop. Verh. Anat. Ges. (Jena) 201–205 (1959).

BECHTEREW, V.M.: Die Lehre von den Neuronen und die Entladungstheorie. (Untersuchungsresultate des Nervensystems nach der Golgischen Methode.) Neurol. Zbl. **15**, 50–57, 103–111 (1896).

BELL, C.C., DOW, R.S.: Cerebellar circuitry. Neurosci. Res. Progr. Bull. **5**, 121–221 (1967).

BELL, C.C., GRIMM, R.J.: Discharge properties of Purkinje cells recorded on single and double microelectrodes. J. Neurophysiol. **32**, 1044–1055 (1969).

BERGMANN, C.: Notiz über einige Structurverhältnisse des Cerebellum und Rückenmarks. Z. rationelle Med. (HENLE und PFEUFER), N.F. **8**, 360–363 (1857).

BETHE, A.: Über die Neurofibrillen in den Ganglienzellen von Wirbeltieren und ihre Beziehungen zu den Golginetzen. Arch. mikr. Anat. **55**, 513–558 (1900).

BIELSCHOWSKY, M., WOLFF, M.: Zur Histologie der Kleinhirnrinde. J. Psychol. Neurol. (Lpz.) **4**, 1–23 (1904).

BILLINGS, S.M., SWARTZ, F.J.: DNA content of Mauthner cell nuclei in *Xenopus laevis*: A spectrophotometric study. Z. Anat. Entwickl.-Gesch. **129**, 14–23 (1969).

BIRCH-ANDERSEN, A., DAHL, V., OLSEN, S.: Elektronenmikroskopische Untersuchungen über die Struktur der Kleinhirnrinde des Menschen. In: Proceedings, IVth Internat. Congr. of Neuropathology (H. JACOB, ed.), p. 71–77. Stuttgart: Thieme 1962.

BLACKSTAD, T.W.: Electron microscopy of Golgi preparations for the study of neuronal relations. In: Contemporary research methods in neuroanatomy (W.J.H. NAUTA and S.O.E. EBBESON, eds.), p. 186–216. Berlin-Heidelberg-New York: Springer 1970.

BLOEDEL, J.R., ROBERTS, W.J.: Action of climbing fibers in the cerebellar cortex of the cat. J. Neurophysiol. **34**, 17–31 (1971).

BLOOM, F.E., HOFFER, B.J., SIGGINS, G.R.: Studies on norepinephrine containing afferents to Purkinje cells of rat cerebellum. I. Localization of the fibers and their synapses. Brain Res. **25**, 501–521 (1971).

BODIAN, D.: Development of fine structure of spinal cord in monkey fetuses. I. The motoneuron neuropil at the time of onset of reflex activity. Bull. Johns Hopk. Hosp. **119**, 129–149 (1966).

BRAITENBERG, V., ATWOOD, R.P.: Morphological observations on the cerebellar cortex. J. comp. Neurol. **109**, 1–33 (1958).

BRANTON, D.: Fracture faces of frozen membranes. Proc. nat. Acad. Sci. (Wash.) **55**, 1048–1056 (1966).

BRAWER, J.R.: The fine structure of the ependymal tanycytes at the level of the arcuate nucleus. J. comp. Neurol. **145**, 25–42 (1972).

BRIGHTMAN, M.W., PALAY, S.L.: The fine structure of ependyma in the brain of the rat. J. Cell Biol. **19**, 415–439 (1963).

BRIGHTMAN, M.W., REESE, T.S.: Junctions between intimately apposed cell membranes in the vertebrate brain. J. Cell Biol. **40**, 648–677 (1969).

BRODAL, A.: Neurological anatomy in relation to clinical medicine, second ed. New York: Oxford University Press 1969.

BRODAL, A., DRABLØS, P.A.: Two types of mossy fiber terminals in the cerebellum and their regional distribution. J. comp. Neurol. **121**, 173–187 (1963).

BRODAL, A., GRANT, G.: Morphology and temporal course of degeneration in cerebellar mossy fibers following transection of spinocerebellar tracts in the cat. Exp. Neurol. **5**, 67–87 (1962).

BROWN, W.J., PALAY, S.L.: Acetylcholinesterase activity in certain glomeruli and Golgi cells of the granular layer of the rat cerebellar cortex. Z. Anat. Entwickl.-Gesch. **137**, 317–334 (1972).

BUNGE, R.P.: Glial cells and the central myelin sheath. Physiol. Rev. **48**, 197–251 (1968).

BUNGE, R.P., BUNGE, M.B., RIS, H.: Electron microscopic study of demyelination in an experimentally induced lesion in adult cat spinal cord. J. biophys. biochem. Cytol. **7**, 685–696 (1960).

CECCARELLI, B., HURLBUT, W.P., MAURO, A.: Depletion of vesicles from frog neuromuscular junctions by prolonged tetanic stimulation. J. Cell Biol. **54**, 30–38 (1972).

CHANDLER, R.L., WILLIS, R.: An intranuclear fibrillar lattice in neurons. J. Cell Sci. **1**, 283–286 (1966).

CHAN-PALAY, V.: The recurrent collaterals of Purkinje cell axons: A correlated study of the rat's cerebellar cortex with electron microscopy and the Golgi method. Z. Anat. Entwickl.-Gesch. **134**, 200–234 (1971).

CHAN-PALAY, V.: The tripartite structure of the undercoat in initial segments of Purkinje cell axons. Z. Anat. Entwickl.-Gesch. **139**, 1–10 (1972a).

CHAN-PALAY, V.: Arrested granule cells and their synapses with mossy fibers in the molecular layer of the cerebellar cortex. Z. Anat. Entwickl.-Gesch. **139**, 11–20 (1972b).

CHAN-PALAY, V.: Central nuclei of cerebellum. In preparation, 1973a.

CHAN-PALAY, V.: Early stages in the degeneration of mossy fibers after hemi-section of the spinal cord. In preparation, 1973b.

CHAN-PALAY, V., PALAY, S.L.: Interrelations of basket cell axons and climbing fibers in the cerebellar cortex of the rat. Z. Anat. Entwickl.-Gesch. **132**, 191–227 (1970).

CHAN-PALAY, V., PALAY, S.L.: Tendril and glomerular collaterals of climbing fibers in the granular layer of the rat's cerebellar cortex. Z. Anat. Entwickl.-Gesch. **133**, 247–273 (1971a).

CHAN-PALAY, V., PALAY, S.L.: The synapse *en marron* between Golgi II neurons and mossy fiber in the rat's cerebellar cortex. Z. Anat. Entwickl.-Gesch. **133**, 274–289 (1971b).

CHAN-PALAY, V., PALAY, S.L.: The stellate cells of the rat's cerebellar cortex. Z. Anat. Entwickl.-Gesch. **136**, 224–248 (1972a).

CHAN-PALAY, V., PALAY, S.L.: High voltage electron microscopy of rapid Golgi preparations. Neurons and their processes in the cerebellar cortex of monkey and rat. Z. Anat. Entwickl.-Gesch. **137**, 125–152 (1972b).

CHAN-PALAY, V., PALAY, S.L.: The form of velate astrocytes in the cerebellar cortex of monkey and rat: High voltage electron microscopy of rapid Golgi preparations. Z. Anat. Entwickl.-Gesch. **138**, 1–19 (1972c).

COLONNIER, M.: On the nature of intranuclear rods. J. Cell Biol. **25**, 646–653 (1965).

COLONNIER, M.: Synaptic patterns on different cell types in the different laminae of the cat visual cortex. An electron microscope study. Brain Res. **9**, 268–287 (1968).

CONRADI, S.: Ultrastructural specialization of the initial axon segment of cat lumbar motoneurons. Preliminary observations. Acta Soc. Med. upsalien. **71**, 281–284 (1966).

CONRADI, S.: Observations on the ultrastructure of the axon hillock and initial axon segment of lumbosacral motoneurons in the cat. Acta physiol. scand., Suppl. **332**, 65–84 (1969).

COUTEAUX, R.: Principaux critères morphologiques et cytochimiques utilisables aujourd'hui pour définir les divers types de synapses. In: Actualités neurophysiologiques, troisième série, p. 145–173 (A.M. MONNIER, ed.). Paris: Masson et Cie 1961.

CRAIGIE, E.H.: Notes on the morphology of the mossy fibers in some birds and mammals. Trab. Lab. Invest. biol. (Madrid) **24**, 319–331 (1926).

CRILL, W.E.: Unitary, multiple spiked responses in cat inferior olive nucleus. J. Neurophysiol. **33**, 199–209 (1970).

CSILLIK, B., JOÓ, F., KÁSA, P.: Cholinesterase activity of archicerebellar mossy fiber apparatuses. J. Histochem. Cytochem. **11**, 113–114 (1963).

DAHL, V., OLSEN, S., BIRCH-ANDERSEN, A.: The fine structure of the granular layer in the human cerebellar cortex. Acta neurol. scand. **38**, 81–97 (1962).

DENISSENKO, G.: Zur Frage über der Kleinhirnrinde bei verschiedenen Klassen von Wirbelthieren. Arch. mikr. Anat. **14**, 203–242 (1877).

DENNIS, M.J., GERSCHENFELD, H.M.: Some physiological properties of identified mammalian neuroglial cells. J. Physiol. (Lond.) **203**, 211–222 (1969).

DIAMOND, J., GRAY, E.G., YASARGIL, G.M.: The function of dendritic spines: an hypothesis. J. Physiol. (Lond.) **202**, 116 (1969).

DIAMOND, J., GRAY, E.G., YASARGIL, G.M.: The function of the dendritic spine: an hypothesis. In: Excitatory synaptic mechanisms (P. ANDERSEN and J.K.S. JANSEN, eds.), p. 213–222. Oslo: Universitetsforlaget 1970.

DOGIEL, A.S.: Die Nervenelemente im Kleinhirne der Vögel und Säugethiere. Arch. mikr. Anat. **47**, 707–719 (1896).

DOW, R.S.: Action potentials of cerebellar cortex in response to local electrical stimulation. J. Neurophysiol. **12**, 245–256 (1949).

DOW, R.S., MORUZZI, G.: The physiology and pathology of the cerebellum. Minneapolis: University of Minnesota Press 1958.

DUNCAN, D., WILLIAMS, V.: On the occurrence of a precise order in axoplasm. Tex. Rep. Biol. Med. **20**, 503–505 (1962).

ECCLES, J.C., FABER, D.S., MURPHY, J.T., SABAH, N.H., TÁBOŘÍKOVÁ, H.: Afferent volleys in limb nerves influencing impulse discharges in cerebellar cortex. II. In Purkyně cells. Exp. Brain Res. **13**, 36–53 (1971a).

ECCLES, J.C., FABER, D.S., MURPHY, J.T., SABAH, N.H., TÁBOŘÍKOVÁ, H.: Investigations on integration of mossy fiber inputs to Purkyně cells in the anterior lobe. Exp. Brain Res. **13**, 54–77 (1971b).

ECCLES, J., ITO, M., SZENTÁGOTHAI, J.: The cerebellum as a neuronal machine. Berlin-Heidelberg-New York: Springer 1967.

ECCLES, J.C., LLINÁS, R., SASAKI, K.: The inhibitory interneurones within the cerebellar cortex. Exp. Brain Res. **1**, 1–16 (1966a).

ECCLES, J.C., LLINÁS, R., SASAKI, K.: Parallel fibre stimulation and the responses induced thereby in the Purkinje cells of the cerebellum. Exp. Brain Res. **1**, 17–39 (1966b).

ECCLES, J.C., LLINÁS, R., SASAKI, K.: The mossy fibre-granule cell relay of the cerebellum and its inhibitory control by Golgi cells. Exp. Brain Res. **1**, 82–101 (1966c).

ECCLES, J.C., LLINÁS, R., SASAKI, K.: The excitatory synaptic action of climbing fibres on the Purkinje cells of the cerebellum. J. Physiol. (Lond.) **182**, 268–296 (1966d).

ECCLES, J., LLINÁS, R., SASAKI, K.: The action of antidromic impulses on the cerebellar Purkinje cells. J. Physiol. (Lond.) **182**, 316–345 (1966e).

ECCLES, J.C., LLINÁS, R., SASAKI, K., VOORHOEVE, P.E.: Interaction experiments on the responses evoked in Purkinje cells by climbing fibers. J. Physiol. (Lond.) **182**, 297–315 (1966).

ECCLES, J.C., SASAKI, K., STRATA, P.: The potential fields generated in the cerebellar cortex by a mossy fiber volley. Exp. Brain Res. **3**, 58–80 (1967a).

ECCLES, J.C., SASAKI, K., STRATA, P.: A comparison of the inhibitory actions of Golgi cells and of basket cells. Exp. Brain Res. **3**, 81–94 (1967b).

ECCLES, J.C., TÁBOŘÍKOVÁ, H., TSUKAHARA, N.: Responses of the Purkinje cells of a selachian cerebellum (Mustelus canis). Brain Res. **17**, 57–86 (1970).

EIDE, E., FEDINA, L., JANSEN, J., LUNDBERG, A., VYKLICKÝ, L.: Unitary excitatory postsynaptic potentials in Clarke's column neurones. Nature (Lond.) **215**, 1176–1177 (1967).

ESTABLE, C.: Notes sur la structure comparative de l'écorce cérébelleuse, et dérivées physiologiques possibles. Trab. Lab. Invest. biol. (Madrid) **21**, 169–256 (1923).

FAÑANÁS, J. RAMÓN Y: Contribución al estudio de la neuroglia del cerebelo. Trab. Lab. Invest. biol. (Madrid) **14**, 163–179 (1916).

FARQUHAR, M.G., PALADE, G.E.: Junctional complexes in various epithelia. J. Cell Biol. **17**, 375–412 (1963).

FAWCETT, D.W.: On the occurrence of a fibrous lamina on the inner aspect of the nuclear envelope in certain cells of vertebrates. Amer. J. Anat. **119**, 129–146 (1966a).

FAWCETT, D.W.: The cell: Its organelles and inclusions. Philadelphia: W.B. Saunders Co. 1966b.

FONNUM, F., STORM-MATHISEN, J., WALBERG, F.: Glutamate decarboxylase in inhibitory neurons. A study of the enzyme in Purkinje cell axons and boutons in the cat. Brain Res. **20**, 259–275 (1970).

FOX, C.A.: The intermediate cells of Lugaro in the cerebellar cortex of monkey. J. comp. Neurol. **112**, 39–51 (1959).

FOX, C.A., ANDRADE, A., SCHWYN, R.C.: Climbing fiber branching in the granular layer. In: Neurobiology of cerebellar evolution and development (R. LLINÁS, ed.), p. 603–611. Chicago: AMA-ERF Institute for Biomedical Research 1969.

FOX, C.A., BARNARD, J.W.: A quantitative study of the Purkinje cell dendritic branchlets and their relationship to afferent fibres. J. Anat. (Lond.) **91**, 299–313 (1957).

FOX, C.A., BERTRAM, E.G.: Connections of the Golgi cells and the intermediate cells of Lugaro in the cerebellar cortex of the monkey (Abstract). Anat. Rec. **118**, 423–424 (1954).

FOX, C.A., DUTTA, C.R., HILLMAN, D.E., SIEGESMUND, K.A.: The granular layer of the cerebellar cortex (Abstract). Anat. Rec. **151**, 487 (1965).

FOX, C.A., HILLMAN, D.E., SIEGESMUND, K.A., DUTTA, C.R.: The primate cerebellar cortex: A Golgi and electron microscopic study. In: The cerebellum (C.A. FOX and R.S. SNIDER, eds.). Progr. Brain Res. **25**, 174–225 (1967).

FOX, C.A., SIEGESMUND, K.A., DUTTA, C.R.: The Purkinje cell dendritic branchlets and their relation with the parallel fibers: light and electron microscopic observations. In: Morphological and biochemical correlates of neural activity (M.M. COHEN and R.S. SNIDER, eds.), p. 112–141. New York: Hoeber-Harper and Row 1964.
FOX, C.A., UBEDA-PURKISS, M., MASSOPUST, L.C.: Structure of the cerebellar cortex in the adult monkey (*Macaca mulatta*): a Golgi study. In: Abstracts of communications, Fifth Internat. Anatomical Congr., p. 69–70. Oxford: Oxford University Press 1950.
FRIEDE, R.: Quantitative Verschiebungen der Schichten innerhalb des Windungsverlaufes der Kleinhirnrinde und ihre biologische Bedeutung. Acta anat. (Basel) **25**, 65–72 (1955).
FRIEND, D.S., FARQUHAR, M.G.: Functions of coated vesicles during protein absorption in the rat vas deferens. J. Cell Biol. **35**, 357–376 (1967).
FRIEND, D.S., GILULA, N.B.: A distinctive cell contact in the rat adrenal cortex. J. Cell Biol. **53**, 148–163 (1972).
FUJITA, Y.: Activity of dendrites of single Purkinje cells and its relationship to so-called inactivation response in rabbit cerebellum. J. Neurophysiol. **31**, 131–141 (1968).
FULTON, J.F.: Functional localization in the frontal lobes and cerebellum. London: Oxford University Press 1949.
FURUKAWA, T., FURSHPAN, E.J.: Two inhibitory mechanisms in the Mauthner neurons of goldfish. J. Neurophysiol. **26**, 140–176 (1963).
FUSARI, R.: Sull'origine delle fibre nervose nello strato molecolare delle circonvoluzione cerebellari dell'uomo. Atti Reale Accad. Sci. (Torino) **19**, 47–51 (1883).
GEHUCHTEN, A. VAN: La structure des centres nerveux. La moelle épinière et le cervelet. Cellule **7**, 83–122 (1891).
GEHUCHTEN, A. VAN: La neuroglie dans le cervelet de l'homme. Bibl. anat. (Basel) **2**, 146–152 (1895).
GEHUCHTEN, A. VAN: Anatomie du système nerveux de l'homme. Louvain: Uystpruyst (Dieudonné) 1897.
GIERKE, H.: Die Stützsubstanz des Centralnervensystems. II. Theil. Arch. mikr. Anat. **26**, 129–228 (1886).
GILULA, N.B., BRANTON, D., SATIR, P.: The septate junction: a structural basis for intercellular coupling. Proc. nat. Acad. Sci. (Wash.) **67**, 213–220 (1970).
GOBEL, S.: Electron microscopical studies of the cerebellar molecular layer. J. Ultrastruct. Res. **21**, 430–458 (1967).
GOBEL, S.: Axo-axonic septate junctions in the basket formations of the cat cerebellar cortex. J. Cell Biol. **51**, 328–333 (1971).
GOLGI, C.: Sulla sostanza connettiva del cervello. Rend. Ist. Lomb. Sci. Lett. (Milano) **3**, Ser. II, April 1870. Reprinted as Ch. I in: Opera Omnia, vol. I: Istologia normale, 1870–1883, p. 1–4. Milan: Ulrico Hoepli 1903. (1870).
GOLGI, C.: Contribuzione alla fina anatomia degli organi centrali del sistema nervoso. Riv. clin. (Bologna), Ser. II, **1**, No 11 (Nov.), p. 338–350 and No 12 (Dec.), p. 371–380. Reprinted as Ch. II in: Opera Omnia, vol. I: Istologia normale, 1870–1883, p. 5–70. Milan: Ulrico Hoepli 1903. (1871).
GOLGI, C.: Sulla fina anatomia del cervelletto umano. Lecture, Istituto Lombardo di Sci. e Lett. 8 Jan. 1874. Ch. V in: Opera Omnia, vol. I: Istologia normale, 1870–1883, p. 99–111. Milan: Ulrico Hoepli 1903. (1874).
Same as: Untersuchungen über den feineren Bau des centralen und peripherischen Nervensystems, translated from Italian by R. TEUSCHER, Ch. III, p. 39–45. Jena: Gustav Fischer 1894.
GOLGI, C.: Sulla fina anatomia degli organi centrali del sistema nervoso. I. Note preliminari sulla struttura, morfologia e vicendevoli rapporti delle cellule gangliari. Riv. sper. Freniat. **8**, 165–195 (1882).
Same as: Arch. ital. Biol. **3**, 285–299 (1882).
Same as: Untersuchungen über den feineren Bau des centralen und peripherischen Nervensystems, translated from Italian by R. TEUSCHER, ch. VIII, p. 81–96. Jena: Gustav Fischer 1894.
Same as: Opera Omnia, vol. I: Istologia normale, 1870–1883, ch. XVI, p. 295–324. Milan: Ulrico Hoepli 1903.
GOLGI, C.: Sulla fina anatomia degli organi centrali del sistema nervoso. IV. Sulla fina anatomia delle circonvoluzioni cerebellari. Riv. sper. Freniat. **9**, 1–17 (1883).
Same as: Arch. ital. Biol. **4**, 92–123 (1883).
Same as: Untersuchungen über den feineren Bau des centralen und peripherischen Nervensystems, translated from Italian by R. TEUSCHER, ch. VIII, p. 111–119. Jena: Gustav Fischer 1894.
Same as: Opera Omnia, vol. I: Istologia normale, 1870–1883, ch. XVI, p. 353–372. Milan: Ulrico Hoepli 1903.
GOLGI, C.: Sulla fina anatomia degli organi centrali del sistema nervoso. VIII. Tessuto interstiziale degli organi nervosi centrali (Nevroglia). Riv. sper. Freniat. **11**, 72–123 (1885).
Same as: Untersuchungen über den feineren Bau des centralen und peripherischen Nervensystems, translated from Italian by R. TEUSCHER, ch. VIII, p. 144–169. Jena: Gustav Fischer 1894.
Same as: Opera Omnia, vol. II: Istologia normale, 1883–1902, ch. XVI, p. 435–485. Milan: Ulrico Hoepli 1903.
GOLGI, C.: Sur la structure des cellules nerveuses. Arch. ital. Biol. **30**, 60–71 (1898a).
GOLGI, C.: Sur la structure des cellules nerveuses des ganglions spinaux. Arch. ital. Biol. **30**, 278–286 (1898b).
GOODENOUGH, D.A., REVEL, J.-P.: A fine structural analysis of intercellular junctions in the mouse liver. J. Cell Biol. **45**, 272–290 (1970).
GRANIT, R., PHILLIPS, C.G.: Excitatory and inhibitory processes acting upon individual Purkinje cells of the cerebellum of cats. J. Physiol. (Lond.) **133**, 520–547 (1956).
GRAY, E.G.: Axo-somatic and axo-dendritic synapses of the cerebral cortex: An electron microscope study. J. Anat. (Lond.) **93**, 420–433 (1959).
GRAY, E.G.: The granule cells, mossy synapses and Purkinje spine synapses of the cerebellum: light and electron microscope observations. J. Anat. (Lond.) **95**, 345–356 (1961).
GRAY, E.G.: Electron microscopy of presynaptic organelles of the spinal cord. J. Anat. (Lond.) **97**, 101–106 (1963).
GRAY, E.G.: Electron microscopy of excitatory and inhibitory synapses: a brief review. In: Mechanisms of synaptic transmission (K. AKERT and P.G. WASER, eds.). Progr. Brain Res. **31**, 141–155 (1969).
GRAY, E.G., GUILLERY, R.W.: A note on the dendritic spine apparatus. J. Anat. (Lond.) **97**, 389–392 (1963).
GRAY, E.G., WILLIS, R.A.: On synaptic vesicles, complex vesicles, and dense projections. Brain Res. **24**, 149–168 (1970).
HÁMORI, J.: Identification in the cerebellar isles of Golgi II axon endings by aid of experimental degeneration. In: Electron microscopy 1964. Proceedings of the Third European Regional Conference held in Prague (M. TITLBACH, ed.), vol. B, p. 291–292. Prague: Czechoslovak Academy of Sciences 1964.
HÁMORI, J., SZENTÁGOTHAI, J.: The Purkinje cell baskets: Ultrastructure of an inhibitory synapse. Acta biol. Acad. Sci. hung. **15**, 465–476 (1965).
HÁMORI, J., SZENTÁGOTHAI, J.: Identification under the electron microscope of climbing fibers and their synaptic contacts. Exp. Brain Res. **1**, 65–81 (1966a).
HÁMORI, J., SZENTÁGOTHAI, J.: Participation of Golgi neuron processes in the cerebellar glomeruli: An electron microscopic study. Exp. Brain Res. **2**, 35–48 (1966b).
HÁMORI, J., SZENTÁGOTHAI, J.: Identification of synapses formed in the cerebellar cortex by Purkinje axon collaterals: an electron microscopic study. Exp. Brain Res. **15**, 118–128 (1968).
HAND, A.R., GOBEL, S.: The structural organization of the septate and gap junctions of Hydra. J. Cell Biol. **52**, 397–408 (1972).
HARMON, L.D., KADO, R.T., LEWIS, E.R.: Cerebellar modelling problems. Unpublished ms. 1972.
HAY, E.D.: Structure and function of the nucleolus in developing cells. In: Ultrastructure in biological systems, vol. 3: The nucleus (A.J. DALTON and F. HAGUENAU, eds.), p. 1–79. New York: Academic Press 1968.

HEBB, C. O.: Chemical agents of the nervous system. Int. Rev. Neurobiol. **1**, 165–193 (1959).
HELD, H.: Beiträge zur Struktur der Nervenzellen und ihrer Fortsätze III. Arch. Anat. Physiol., Anat. Abt., Suppl. 273–312 (1897).
HENN, F. A., HALJAMAE, H., HAMBERGER, A.: Glial cell function: active control of extracellular K^+ concentration. Brain Res. **43**, 437–443 (1972).
HERNDON, R. M.: The fine structure of the Purkinje cell. J. Cell Biol. **18**, 167–180 (1963).
HERNDON, R. M.: The fine structure of the rat cerebellum. II. The stellate neurons, granule cells, and glia. J. Cell Biol. **23**, 277–293 (1964).
HEUSER, J., REESE, T. S.: Stimulation induced uptake and release of peroxidase from synaptic vesicles in frog neuromuscular junctions (Abstract). Anat. Rec. **172**, 329–330 (1972).
HILLMAN, D. E.: Neuronal organization of the cerebellar cortex in amphibia and reptilia. In: Neurobiology of cerebellar evolution and development (R. LLINÁS, ed.), p. 279–324. Chicago: AMA-ERF Institute for Biomedical Research 1969.
HÖKFELT, T.: The possible ultrastructural identification of tubero-infundibular dopamine-containing nerve endings in the median eminence of the rat. Brain Res. **5**, 121–123 (1967a).
HÖKFELT, T.: On the ultrastructural localization of noradrenaline in the central nervous system. Z. Zellforsch. **79**, 110–117 (1967b).
HÖKFELT, T., FUXE, K.: Cerebellar monoamine nerve terminals, a new type of afferent fibers to the cortex cerebelli. Exp. Brain Res. **9**, 63–72 (1969).
HÖKFELT, T., LJUNGDAHL, Å.: Cellular localization of labeled gamma-aminobutyric acid (^{3}H-GABA) in rat cerebellar cortex: an autoradiographic study. Brain Res. **22**, 391–396 (1970).
HÖKFELT, T., LJUNGDAHL, Å.: Autoradiographic identification of cerebral and cerebellar cortical neurons accumulating labeled Gamma-aminobutyric acid (^{3}H-GABA). Exp. Brain Res. **14**, 354–362 (1972).
HOFFER, B. J., SIGGINS, G. R., BLOOM, F. E.: Studies on norepinephrine-containing afferents to Purkinje cells of rat cerebellum. II. Sensitivity of Purkinje cells to norepinephrine and related substances administered by microiontophoresis. Brain Res. **25**, 523–534 (1971).
HOLTZMAN, E.: Cytochemical studies of protein transport in the nervous system. Phil. Trans. B **261**, 407–421 (1971).
HOLTZMAN, E., NOVIKOFF, A. B., VILLAVERDE, H.: Lysosomes and GERL in normal and chromatolytic neurons of rat ganglion nodosum. J. Cell Biol. **33**, 419–435 (1967).
HORTEGA, P. DEL RÍO: Estudios sobre el centrosoma de las células nerviosas y neuróglicas de los vertebrados, en sus formas normal y anormales. Trab. Lab. Invest. biol. (Madrid) **14**, 117–153 (1916).
HORTEGA, P. DEL RÍO: Tercera aportación al conocimiento morfológico e interpretación funcional de la oligodendroglía. Mem. R. Soc. Exp. Hist. Nat. (Madrid) **14**, 5–122 (1928).
ITO, M., YOSHIDA, M.: The cerebellar-evoked monosynaptic inhibition of Deiters' neurones. Experientia (Basel) **20**, 515–516 (1964).
ITO, M., YOSHIDA, M.: The origin of cerebellar-induced inhibition of Deiters' neurones. I. Monosynaptic initiation of the inhibitory post-synaptic potentials. Exp. Brain Res. **2**, 330–349 (1966).
ITO, M., YOSHIDA, M., OBATA, K.: Monosynaptic inhibition of the intracerebellar nuclei induced from the cerebellar cortex. Experientia (Basel) **20**, 575–576 (1964).
JAKOB, A.: Das Kleinhirn. In: Handbuch der mikroskopischen Anatomie des Menschen, Bd. IV/1 (W. v. MÖLLENDORFF, ed.), p. 674–916. Berlin: Springer 1928.
JAMIESON, J. D., PALADE, G. E.: Intracellular transport of secretory proteins in the pancreatic exocrine cell. I. Role of the peripheral elements of the Golgi complex. J. Cell Biol. **34**, 577–596 (1967a).
JAMIESON, J. D., PALADE, G. E.: Intracellular transport of secretory proteins in the pancreatic exocrine cell. II. Transport to condensing vacuoles and zymogen granules. J. Cell Biol. **34**, 597–615 (1967b).
JAMIESON, J. D., PALADE, G. E.: Condensing vacuole conversion and zymogen granule discharge in pancreatic exocrine cells: metabolic studies. J. Cell Biol. **48**, 503–522 (1971a).
JAMIESON, J. D., PALADE, G. E.: Synthesis, intracellular transport, and discharge of secretory proteins in stimulated pancreatic exocrine cells. J. Cell Biol. **50**, 135–158 (1971b).
JANSEN, J., BRODAL, A.: Das Kleinhirn. In: Handbuch der mikroskopischen Anatomie des Menschen, Bd. IV/8 (Ergänzung zu Bd. IV/1) (W. BARGMANN, ed.), p. 101–103. Berlin-Göttingen-Heidelberg: Springer 1958.
JANSEN, J., BRODAL, A.: Experimental studies on the intrinsic fibers of the cerebellum. The cortico-nuclear projection in the rabbit and the monkey (*Macacus rhesus*). Skr. norske Vidensk.-Akad., I. mat.-nat. Kl., **11**, No 3, 1–50 (1942).
JONES, E. G., POWELL, T. P. S.: Morphological variations in the dendritic spines of the neocortex. J. Cell Sci. **5**, 509–529 (1969).
KAISERMAN-ABRAMOF, I. R., PALAY, S. L.: Fine structural studies of the cerebellar cortex in a mormyrid fish. In: Neurobiology of cerebellar evolution and development (R. LLINÁS, ed.), p. 171–204. Chicago: AMA-ERF Institute for Biomedical Research 1969.
KANASEKI, T., KADOTA, K.: The "vesicle in a basket". A morphological study of the coated vesicle isolated from the nerve endings of the guinea pig brain with special reference to the mechanism of membrane movements. J. Cell Biol. **42**, 202–220 (1969).
KIRSCHE, W., DAVID, H., WINKELMANN, E., MARX, I.: Elektronenmikroskopische Untersuchungen an synaptischen Formationen im Cortex cerebelli von Rattus rattus norvegicus, Berkenhoot. Z. mikr.-anat. Forsch. **72**, 49–80 (1964).
KITAI, S. T., SHIMONO, T., KENNEDY, D. T.: Inhibition in the cerebellar cortex of the lizard, Lacerta viridis. In: Neurobiology of cerebellar evolution and development (R. LLINÁS, ed.), p. 481–492. Chicago: AMA-ERF Institute for Biomedical Research 1969.
KÖLLIKER, A.: Zur feineren Anatomie des zentralen Nervensystems. I. Das Kleinhirn. Z. wiss. Zool. **49**, 663–689 (1890).
KÖLLIKER, A.: Handbuch der Gewebelehre des Menschen, Bd. II: Nervensystem, p. 349–350. Leipzig: Wilhelm Engelmann 1896.
KOHNO, K.: Neurotubules contained within the dendrite and axon of Purkinje cell of frog. Bull. Tokyo med. dent. Univ. **11**, 411–442 (1964).
KORNGUTH, S. E., ANDERSON, J. W., SCOTT, G.: The development of synaptic contacts in the cerebellum of Macaca mulatta. J. comp. Neurol. **132**, 531–545 (1968).
KRUGER, L., MAXWELL, D. S.: Cytoplasmic laminar bodies in the striate cortex. J. Ultrastruct. Res. **26**, 387–390 (1969).
KUFFLER, S. W., NICHOLLS, J. G.: The physiology of neuroglial cells. Ergebn. Physiol. **57**, 1–90 (1966).
KUNO, M., MIYAHARA, J. T.: Factors responsible for multiple discharge of neurons in Clarke's column. J. Neurophysiol. **31**, 624–638 (1968).
KUNO, M., MIYAHARA, J. T.: Analysis of synaptic efficacy in spinal motoneurones from 'quantum' aspects. J. Physiol. (Lond.) **201**, 479–493 (1969).
KURIYAMA, K., HABER, B., SISKEN, B., ROBERTS, E.: The γ-aminobutyric acid system in rabbit cerebellum. Proc. nat. Acad. Sci. (Wash.) **55**, 846–852 (1966).
LANDAU, E.: Über cytoarchitektonische Bauunterschiede in der Körnerschicht des Kleinhirns. Z. Anat. Entwickl.-Gesch. **87**, 551–557 (1928).
LANDAU, E.: La cellule synarmotique. Bull. Histol. appl. **9**, 159–168 (1932).
LANDAU, E.: La cellule synarmotique dans le cervelet humain. Arch. Anat. **17**, 273–285 (1933).
LAPHAM, L. W.: Tetraploid DNA content of Purkinje neurons of human cerebellar cortex. Science **159**, 310–312 (1968).
LARRAMENDI, L. M. H.: Purkinje axo-somatic synapses at seven and fourteen post-natal days in the mouse: an electronmicroscopic study (Abstract). Anat. Rec. **151**, 460 (1965).
LARRAMENDI, L. M. H.: Morphological characteristics of extrinsic and intrinsic nerve terminals and their synapses in the cerebellar cortex of the mouse. In: The cerebellum in health and disease (W. S. FIELDS and W. D. WILLIS, Jr., eds.), p. 63–110. St. Louis: W. H. M. Green Inc. 1969a.
LARRAMENDI, L. M. H.: Analysis of synaptogenesis in the cerebellum of the mouse. In: Neurobiology of cerebellar evolution and develop-

ment (R. Llinás, ed.), p. 803–843. Chicago: AMA-ERF Institute for Biomedical Research 1969b.

Larramendi, L. M. H., Lemkey-Johnston, N. J.: The distribution of recurrent Purkinje collateral synapses in the mouse cerebellar cortex: An electron microscopic study. J. comp. Neurol. **138**, 451–482 (1970).

Larramendi, L. M. H., Victor, T.: Synapses on the Purkinje cell spines in the mouse. An electronmicroscopic study. Brain Res. **5**, 15–30 (1967).

Larsell, O.: The morphogenesis and adult pattern of the lobules and fissures of the cerebellum of the white rat. J. comp. Neurol. **97**, 281–356 (1952).

Larsell, O., Jansen, J.: The comparative anatomy and histology of the cerebellum: The human cerebellum, cerebellar connections, and cerebellar cortex. Minneapolis: University of Minnesota Press 1972.

Lemkey-Johnston, N., Larramendi, L. M. H.: Morphological characteristics of mouse stellate and basket cells and their neuroglial envelope: An electron microscopic study. J. comp. Neurol. **134**, 39–72 (1968a).

Lemkey-Johnston, N., Larramendi, L. M. H.: Types and distribution of synapses upon basket and stellate cells of the mouse cerebellum. J. comp. Neurol. **134**, 73–112 (1968b).

Lenhossék, M. von: Centrosom und Sphäre in den Spinalganglienzellen des Frosches. Arch. mikr. Anat. **46**, 345–369 (1895).

Lentz, R. D., Lapham, L. W.: Postnatal development of tetraploid DNA content in rat Purkinje cells: A quantitative cytochemical study. J. Neuropath. exp. Neurol. **29**, 43–56 (1970).

Llinás, R., Ayala, J. S.: Purkinje axon collateral action upon interneurons in the cerebellar cortex. In: Symposium on Neurophysiological Basis of Normal and Abnormal Motor Activities, Parkinson's Disease Information and Research Center (M. D. Yahr and D. P. Purpura, eds.). Hewlett, N. Y.: Raven Press 1967.

Llinás, R., Hillman, D. E.: Physiological and morphological organization of the cerebellar circuits in various vertebrates. In: Neurobiology of cerebellar evolution and development (R. Llinás, ed.), p. 43–73. Chicago: AMA-ERF Institute for Biomedical Research 1969.

Llinás, R., Nicholson, C.: Electrophysiological analysis of alligator cerebellar cortex: A study on dendritic spikes. In: Neurobiology of cerebellar evolution and development (R. Llinás, ed.), p. 431–464. Chicago: AMA-ERF Institute for Biomedical Research 1969.

Llinás, R., Nicholson, C.: Electrophysiological properties of dendrites and somata in alligator Purkinje cells. J. Neurophysiol. **34**, 532–551 (1971).

Llinás, R., Nicholson, C., Freeman, J. A., Hillman, D. E.: Dendritic spikes and their inhibition in alligator Purkinje cells. Science **160**, 1132–1135 (1968).

Llinás, R., Nicholson, C., Precht, W.: Preferred centripetal conduction of dendritic spikes in alligator Purkinje cells. Science **163**, 184–187 (1969).

Llinás, R., Precht, W.: Recurrent facilitation by disinhibition in Purkinje cells of the cat cerebellum. In: Neurobiology of cerebellar evolution and development (R. Llinás, ed.), p. 619–627. Chicago: AMA-ERF Institute for Biomedical Research 1969.

Lugaro, E.: Sulle connessioni tra gli elementi nervosi della corteccia cerebellare con considerazioni generali sul significato fisiologico dei rapporti tra gli elementi nervosi. Riv. sper. Freniat. **20**, 297–331 (1894).

Marinesco, G.: La cellule nerveuse, vols. I and II. Paris: O. Doin et Fils 1909.

Marr, D.: A theory of cerebellar cortex. J. Physiol. (Lond.) **202**, 437–470 (1969).

Martinez, F. E., Crill, W. E., Kennedy, T. T.: Electrogenesis of cerebellar Purkinje cell responses in cats. J. Neurophysiol. **34**, 348–356 (1971).

Masurovsky, E. B., Benitez, H. H., Kim, S. U., Murray, M. R.: Origin, development, and nature of intranuclear rodlets and associated bodies in chicken sympathetic neurons. J. Cell Biol. **44**, 172–191 (1970).

Maul, G. G., Price, J. W., Lieberman, M. W.: Formation and distribution of nuclear pore complexes in interphase. J. Cell Biol. **51**, 405–418 (1971).

McNutt, N. S., Weinstein, R. S.: The ultrastructure of the nexus. A correlated thin-section and freeze-cleave study. J. Cell Biol. **47**, 666–688 (1970).

Meldolesi, J., Jamieson, J. D., Palade, G. E.: Composition of cellular membranes in the pancreas of the guinea pig. III. Enzymatic activities. J. Cell Biol. **49**, 150–158 (1971).

Milhaud, M., Pappas, G. D.: Postsynaptic bodies in the habenula and interpeduncular nuclei of the cat. J. Cell Biol. **30**, 437–441 (1966).

Mlonyeni, M.: The number of Purkinje cells and inferior olivary neurones in the cat. J. comp. Neurol. **147**, 1–10 (1973).

Morales, R., Duncan, D.: Multilaminated bodies and other unusual configurations of endoplasmic reticulum in the cerebellum of the cat. An electron microscopic study. J. Ultrastruct. Res. **15**, 480–489 (1966).

Morales, R., Duncan, D., Rehmet, R.: A distinctive laminated cytoplasmic body in the lateral geniculate body neurons of the cat. J. Ultrastruct. Res. **10**, 116–123 (1964).

Mugnaini, E.: Ultrastructural aspects of cerebellar morphology in the chick embryo (Abstract). Anat. Rec. **154**, 391 (1966).

Mugnaini, E.: Ultrastructural studies on the cerebellar histogenesis. II. Maturation of nerve cell populations and establishment of synaptic connections in the cerebellar cortex of the chick. In: Neurobiology of cerebellar evolution and development (R. Llinás, ed.), p. 749–782. Chicago: AMA-ERF Institute for Biomedical Research 1969.

Mugnaini, E.: Neurones as synaptic targets. In: Excitatory synaptic mechanisms (P. Andersen and J. K. S. Jansen, Jr., eds.), p. 149–169. Oslo: Universitets Forlaget 1970.

Mugnaini, E.: The histology and cytology of the cerebellar cortex. In: The comparative anatomy and histology of the cerebellum: The human cerebellum, cerebellar connections and cerebellar cortex (O. Larsell and J. Jansen, eds.), p. 201–265. Minneapolis: University of Minnesota Press 1972.

Mugnaini, E., Forstrønen, P.: Ultrastructural observations on the astroglia in the cerebellar folia of the chick embryo. J. Ultrastruct. Res. **14**, 415–416 (1966).

Mugnaini, E., Forstrønen, P.: Ultrastructural studies on the cerebellar histogenesis. I. Differentiation of granule cells and development of glomeruli in the chick embryo. Z. Zellforsch. **77**, 115–143 (1967).

Müller, J.: Jahresbericht über die Fortschritte der anatomisch-physiologischen Wissenschaften im Jahre 1836. Müller's Arch. i-cxxxxii (1837).

Müller, J.: Jahresbericht über die Fortschritte der anatomisch-physiologischen Wissenschaften im Jahre 1837. Müller's Arch. xci-cxviii (1838).

Murphy, J. T., Sabah, N. H.: The inhibitory effect of climbing fiber activation on cerebellar Purkinje cells. Brain Res. **19**, 486–490 (1970).

Nicholson, C., Llinás, R.: Field potentials in the alligator cerebellum and theory of their relationship to Purkinje cell dendritic spikes. J. Neurophysiol. **34**, 509–531 (1971).

Nicholson, C., Llinás, R., Precht, W.: Neural elements of the cerebellum in elasmobranch fishes: Structural and functional characteristics. In: Neurobiology of cerebellar evolution and development (R. Llinás, ed.), p. 215–243. Chicago: AMA-ERF Institute for Biomedical Research 1969.

Nieuwenhuys, R., Nicholson, C.: Aspects of the histology of the cerebellum of mormyrid fishes. In: Neurobiology of cerebellar evolution and development (R. Llinás, ed.), p. 135–169. Chicago: AMA-ERF Institute for Biomedical Research 1969.

Nosal, G., Radouco-Thomas, C.: Ultrastructural study on the differentiation and development of the nerve cell; the "nucleus-ribosome" system. In: Advances in cytopharmacology, vol. 1, p. 434–456 (1971). First Internat. Symposium on Cell Biology and Cytopharmacology, Venice (F. Clementi and B. Ceccarelli, eds.). New York: Raven Press 1969.

NOVIKOFF, A.B.: Enzyme localization and ultrastructure of neurons. In: The neuron (H. HYDÉN, ed.), p. 255–318. Amsterdam: Elsevier 1967a.

NOVIKOFF, A.B.: Lysosomes in nerve cells. In: The neuron (H. HYDÉN, ed.), p. 319–377. Amsterdam: Elsevier 1967b.

OBATA, K., ITO, M., OCHI, R., SATO, N.: Pharmacological properties of the postsynaptic inhibition by Purkinje cell axons and the action of γ-aminobutyric acid on Deiters neurones. Exp. Brain Res. **4**, 43–57 (1967).

OBATA, K., OTSUKA, M., TANAKA, Y.: Determination of gamma-aminobutyric acid in single nerve cells of cat central nervous system. J. Neurochem. **17**, 697–698 (1970).

O'LEARY, J.L., INUKAI, J., SMITH, J.M.: Histogenesis of the cerebellar climbing fiber in the rat. J. comp. Neurol. **142**, 377–392 (1971).

O'LEARY, J.L., PETTY, J., SMITH, J.M., O'LEARY, M., INUKAI, J.: Cerebellar cortex of rat and other animals. A structural and ultrastructural study. J. comp. Neurol. **134**, 401–432 (1968).

OLSON, L., FUXE, K.: On the projections from the locus coeruleus noradrenaline neurons: the cerebellar innervation. Brain Res. **28**, 165–171 (1971).

ORKAND, R.K.: Neuroglial-neuronal interactions. In: Basic mechanisms of the epilepsies (H.H. JASPER, A.A. WARD, and A. POPE, eds.), p. 737–746. Boston: Little, Brown and Co. 1969.

ORKAND, R.K., NICHOLLS, J.G., KUFFLER, S.W.: Effect of nerve impulses on the membrane potential of glial cells in the central nervous system of amphibia. J. Neurophysiol. **29**, 788–806 (1966).

PALAY, S.L.: Synapses in the central nervous system. J. biophys. biochem. Cytol. **2** Suppl., 193–202 (1956).

PALAY, S.L.: The morphology of synapses in the central nervous system. Exp. Cell Res., Suppl. **5**, 275–293 (1958).

PALAY, S.L.: The electron microscopy of the glomeruli cerebellosi. In: Cytology of nervous tissue. Proceedings of the Anatomical Society of Great Britain and Ireland, p. 82–84. London: Taylor and Francis 1961.

PALAY, S.L.: Alveolate vesicles in Purkinje cells of the rat's cerebellum. J. Cell Biol. **19**, 89A–90A (1963).

PALAY, S.L.: Fine structure of cerebellar cortex of the rat (Abstract). Anat. Rec. **148**, 419 (1964a).

PALAY, S.L.: The structural basis for neural action. In: Brain function, vol. II: RNA and brain function; memory and learning (M.A.B. BRAZIER, ed.). UCLA Forum Med. Sci. No. 2, p. 69–108. Los Angeles: University of California Press 1964b.

PALAY, S.L.: The role of neuroglia in the organization of the central nervous system. In: Nerve as a tissue (K. RODAHL, and B. ISSEKUTZ, Jr., eds.), p. 3–10. New York: Hoeber-Harper 1966.

PALAY, S.L.: Principles of cellular organization in the nervous system. In: The neurosciences (G.C. QUARTON, T. MELNECHUK, and F.O. SCHMITT, eds.), p. 24–31. New York: Rockefeller University Press 1967.

PALAY, S.L., PALADE, G.E.: The fine structure of neurons. J. biophys. biochem. Cytol. **1**, 69–88 (1955).

PALAY, S.L., RAVIOLA, E., CHAN-PALAY, V.: Studies on synapses in the cerebellar cortex by the freeze-fracturing method. In preparation (1973).

PALAY, S.L., SOTELO, C., PETERS, A., ORKAND, P.M.: The axon hillock and the initial segment. J. Cell Biol. **38**, 193–201 (1968).

PALKOVITS, M., MAGYAR, P., SZENTÁGOTHAI, J.: Quantitative histological analysis of the cerebellar cortex in the cat. I. Number and arrangement in space of the Purkinje cells. Brain Res. **32**, 1–13 (1971a).

PALKOVITS, M., MAGYAR, P., SZENTÁGOTHAI, J.: Quantitative histological analysis of the cerebellar cortex in the cat. II. Cell numbers and densities in the granular layer. Brain Res. **32**, 15–30 (1971b).

PALKOVITS, M., MAGYAR, P., SZENTÁGOTHAI, J.: Quantitative histological analysis of the cerebellar cortex in the cat. III. Structural organization of the molecular layer. Brain Res. **34**, 1–18 (1971c).

PALKOVITS, M., MAGYAR, P., SZENTÁGOTHAI, J.: Quantitative histological analysis of the cerebellar cortex in the cat. IV. Mossy fiber-Purkinje cell numerical transfers. Brain Res. **45**, 15–29 (1972).

PENFIELD, W.: Neuroglia: normal and pathological. In: Cytology and cellular pathology of the nervous system (W. PENFIELD, ed.), vol. 2, p. 421–479. New York: Paul B. Hoeber 1932.

PENSA, A.: Osservazioni e considerazioni sulla struttura della corteccia cerebellare dei mammiferi. Reale Accad. Naz. Lincei, Ser. VI, **5**, 1–26 (1931).

PETERS, A., KAISERMAN-ABRAMOF, I.R.: The small pyramidal neuron of the rat cerebral cortex. The perikaryon, dendrites and spines. Amer. J. Anat. **127**, 321–355 (1970).

PETERS, A., PALAY, S.L.: An electron microscope study of the distribution and patterns of astroglial processes in the central nervous system (Abstract). J. Anat. (Lond.) **99**, 419 (1965).

PETERS, A., PALAY, S.L., WEBSTER, H. DE F.: The fine structure of the nervous system. The cells and their processes. New York: Hoeber-Harper and Row 1970.

PETERS, A., PROSKAUER, C.C., KAISERMAN-ABRAMOF, I.R.: The small pyramidal neuron of the rat cerebral cortex. The axon hillock and initial segment. J. Cell Biol. **39**, 604–619 (1968).

PFENNINGER, K., AKERT, K., MOOR, H., SANDRI, C.: The fine structure of freeze-fractured presynaptic membranes. J. Neurocytology **1**, 129–149 (1972).

PFENNINGER, K., SANDRI, C., AKERT, K., ENGSTER, C.H.: Contribution to the problem of structural organization of the presynaptic area. Brain Res. **12**, 10–18 (1969).

PINTO DA SILVA, P., BRANTON, D.: Membrane splitting in freeze etching. Covalently bound ferritin as a membrane marker. J. Cell Biol. **45**, 598–605 (1970).

POMERAT, C.M.: Dynamic neurogliology. Tex. Rep. Biol. Med. **10**, 885–913 (1952).

POMERAT, C.M., HENDELMAN, W.J., RAIBORN, C.W., MASSEY, J.F.: Dynamic activities of nervous tissue *in vitro*. In: The neuron (H. HYDÉN, ed.), p. 119–178. Amsterdam: Elsevier 1967.

PONTI, U.: Sulla corteccia cerebellare della Cavia. Monit. zool. ital. **8**, 36–40 (1897).

PURKINJE, J.E.: Neueste Untersuchungen aus der Nerven- und Hirn-Anatomie. In: Bericht über die Versammlung deutscher Naturforscher und Aerzte in Prag im Sept. 1837 (K. STERNBERG and J.V. VON KROMBHOLTZ, eds.), p. 177–180, 1837.

RAKIC, P.: Neuron-glia relationship during granule cell migration in developing cerebellar cortex. A Golgi and electronmicroscopic study in Macacus rhesus. J. comp. Neurol. **141**, 283–312 (1971).

RAKIC, P.: Extrinsic cytological determinants of basket and stellate cell dendritic pattern in the cerebellar molecular layer. J. comp. Neurol. **146**, 335–354 (1972).

RAMÓN Y CAJAL, S.: Estructura de los centros nerviosos de las aves. Rev. trimestr. Histol. No 1 (May), 1–10 (1888a).

RAMÓN Y CAJAL, S.: Sobre las fibras nerviosas de la capa molecular del cerebelo. Rev. trimestr. Histol. No 2 (August), 33–41 (1888b).

RAMÓN Y CAJAL, S.: Sobre las fibras nerviosas de la capa granulosa del cerebelo. Rev. trimestr. Histol. No 4 (March), 107–118 (1889a).

RAMÓN Y CAJAL, S.: Sur l'origine et la direction des prolongations nerveuses de la couche moléculaire du cervelet. Int. Mschr. Anat. Physiol. **6**, 158–174 (1889b).

RAMÓN Y CAJAL, S.: Sur les fibres nerveuses de la couche granuleuse du cervelet et sur l'évolution des éléments cérébelleux. Int. Mschr. Anat. Physiol. **7**, 12–30 (1890a).

RAMÓN Y CAJAL, S.: À propos de certains éléments bipolaires du cervelet avec quelques détails nouveaux sur l'évolution des fibres cérébelleuses. Int. Mschr. Anat. Physiol. **7**, 447–468 (1890b).

RAMÓN Y CAJAL, S.: Sur la structure de l'écorce cérébrale de quelques mammifères. Cellule **7**, 125–176 (1891).

RAMÓN Y CAJAL, S.: Neue Darstellung vom histologischen Bau des Centralnervensystems. Arch. Anat. Physiol., Anat. Abth., 319–428 (1893).

RAMÓN Y CAJAL, S.: Croonian Lecture. La fine structure des centres nerveux. Proc. roy. Soc. (Lond.) **55**, 444–468 (1894).

RAMÓN Y CAJAL, S.: El azul de metileno en los centros nerviosos. Rev. trimestr. micr. **1**, 151–204 (1896).

RAMÓN Y CAJAL, S.: Les cellules étoilées de la couche moléculaire du cervelet et quelques faits contraires à la notion de la fonction exclusivement conductrice des neurofibrilles. Trab. Lab. Invest. biol. (Madrid) **4**, 33–43 (1905).

RAMÓN Y CAJAL, S.: L'hypothèse de la continuité d'Apathy. Réponse aux objections de cet auteur contre la doctrine neuronale. Trab. Lab. Invest. biol. (Madrid) **6**, 21–89 (1908).

RAMÓN Y CAJAL, S.: Histologie du système nerveux de l'homme et des vertébrés (translated by L. AZOULAY), vols. I and II. Paris: Maloine. Reprinted 1952 and 1955. Madrid: Consejo Superior de Investigaciones Cientificas 1909–1911.

RAMÓN Y CAJAL, S.: Sobre ciertos plexos pericelulares de la capa de los granos del cerebelo. Trab. Lab. Invest. biol. (Madrid) **10**, 273–276 (1912).

RAMÓN Y CAJAL, S.: Sur les fibres mousseuses et quelques points douteux de la texture de l'écorce cérébelleuse. Trab. Lab. Invest. biol (Madrid) **24**, 215–251 (1926).

RAMÓN Y CAJAL, S.: Les preuves objectives de l'unité anatomique des cellules nerveuses. Trab. Lab. Invest. biol. (Madrid) **29**, 1–137 (1934). Translated from the Spanish version into English by M. U. PURKISS and C. A. FOX, as Neuron theory or reticular theory? Objective evidence of the anatomical unity of nerve cells. Madrid: Consejo Superior de Investigaciones Cientificas 1954.

RAMÓN Y CAJAL, S., ILLERA, R.: Quelques nouveaux détails sur la structure de l'écorce cérébelleuse. Trab. Lab. Invest. biol. (Madrid) **5**, 1–22 (1907).

RETZIUS, G.: Die nervösen Elemente der Kleinhirnrinde. In: Biologische Untersuchungen. Neue Folge, Bd. III, p. 17–24. Stockholm: Samson & Wallin 1892a.

RETZIUS, G.: Kleinere Mittheilungen von dem Gebiete der Nervenhistologie. I. Ueber die Golgi'schen Zellen und die Kletterfasern Ramón y Cajal's in der Kleinhirnrinde. Biologische Untersuchungen. Neue Folge, Bd. IV, p. 57–59. Stockholm: Samson & Wallin 1892b.

RETZIUS, G.: Die Neuroglia des Gehirns beim Menschen und bei Säugethieren: Die Neuroglia des Kleinhirns. Biologische Untersuchungen. Neue Folge, Bd. VI, p. 16–20. Jena: Gustav Fischer 1894.

REVEL, J.-P., KARNOVSKY, M. J.: Hexagonal array of subunits in intercellular junctions of the mouse heart and liver. J. Cell Biol. **33**, C 7–C 12 (1967).

RICHARDSON, K. C., JARETT, L., FINKE, E. H.: Embedding in epoxy resins for ultrathin sectioning in electron microscopy. Stain Technol. **35**, 313–323 (1960).

ROBERTSON, J. D., BODENHEIMER, T. S., STAGE, D. E.: The ultrastructure of Mauthner cell synapses and nodes in goldfish brains. J. Cell Biol. **19**, 159–199 (1963).

RONCORONI, L.: Su un nuovo reperto nel nucleo delle cellule nervose. Arch. Psichiat. **16**, 447–450 (1895).

ROSENBLUTH, J.: Subsurface cisterns and their relationship to the neuronal plasma membrane. J. Cell Biol. **13**, 405–421 (1962).

ROSENBLUTH, J.: Redundant myelin sheaths and other ultrastructural features of the toad cerebellum. J. Cell Biol. **28**, 73–93 (1966).

RUSHMER, D. S., WOODWARD, D. J.: Inhibition of Purkinje cells in the frog cerebellum. I. Evidence for a stellate cell inhibitory pathway. Brain Res. **33**, 83–90 (1971).

SALA Y PONS, C.: La neuroglia de los vertebrados. Barcelona 1894.

SAMORAJSKI, T., ORDY, J. M., KEEFE, J. R.: The fine structure of lipofuscin age pigment in the nervous system of aged mice. J. Cell Biol. **26**, 779–795 (1965).

SÁNCHEZ, M.: Recherches sur le réseau endocellulaire de Golgi dans les cellules de l'écorce du cervelet. Trab. Lab. Invest. biol. (Madrid) **14**, 87–99 (1916).

SANDRI, C., AKERT, K., LIVINGSTON, R. B., MOOR, H.: Particle aggregations at specialized sites in freeze-etched postsynaptic membranes. Brain Res. **41**, 1–16 (1972).

SANDRITTER, W., NOVÁKOVÁ, V., PILNY, J., KIEFER, G.: Cytophotometrische Messungen des Nukleinsäure- und Proteingehaltes von Ganglienzellen der Ratte während der postnatalen Entwicklung und im Alter. Z. Zellforsch. **80**, 145–152 (1967).

SASAKI, K., STRATA, P.: Responses evoked in the cerebellar cortex by stimulating mossy fiber pathways to the cerebellum. Exp. Brain Res. **3**, 95–110 (1967).

SCHAFFER, K.: Zum normalen und pathologischen Fibrillenbau der Kleinhirnrinde. Z. ges. Neurol. Psychiat. **21**, 1–48 (1913).

SCHEIBEL, M. E., SCHEIBEL, A. B.: Observations on the intracortical relations of the climbing fibers of the cerebellum. J. comp. Neurol. **101**, 733–763 (1954).

SCHEIBEL, M. E., SCHEIBEL, A. B.: On the nature of dendritic spines. Communications in Behavioral Biology, part A, **1**, 231–265 (1968).

SCHROEDER, A. H.: Die Gliaarchitektonik des menschlichen Kleinhirns. J. Psychol. Neurol. **38**, 234–257 (1929).

SEITE, R.: Étude ultrastructurale de divers types d'inclusions nucléaires dans les neurones sympathiques du Chat. J. Ultrastruct. Res. **30**, 152–165 (1970).

SEITE, R., ESCAIG, J., COUINEAU, S.: Microfilaments et microtubules nucléaires et organisation ultrastructurale des bâtonnets intranucléaires des neurones sympathiques. J. Ultrastruct. Res. **37**, 449–478 (1971).

SEITE, R., MEI, N., COUINEAU, S.: Modification quantitative des bâtonnets intranucléaires des neurones sympathiques sous l'influence de la stimulation électrique. Brain Res. **34**, 277–290 (1971).

SHANKLIN, W. M., ISSIDORIDES, M., NASSAR, T. K.: Neurosecretion in the human cerebellum. J. comp. Neurol. **107**, 315–337 (1957).

SHIMIZU, N., ISHII, S.: Electron microscopic histochemistry of acetylcholinesterase of rat brain by Karnovsky's method. Histochemie **6**, 24–33 (1966).

SHUTE, C. C., LEWIS, P. R.: Cholinesterase-containing pathways of the hindbrain; Afferent cerebellar and centrifugal cochlear fibres. Nature (Lond.) **205**, 242–246 (1965).

SIEGESMUND, K. A., DUTTA, C. R., FOX, C. A.: The ultrastructure of the intranuclear rodlet in certain nerve cells. J. Anat. (Lond.) **98**, 93–97 (1964).

SILVER, A.: Cholinesterases of the central nervous system with special reference to the cerebellum. Int. Rev. Neurobiol. **10**, 57–109 (1967).

SMIRNOW, A. E.: Über eine besondere Art von Nervenzellen der Molecularschicht des Kleinhirns bei erwachsenen Säugetieren und beim Menschen. Anat. Anz. **13**, 636–642 (1897).

SMOLYANINOV, V. V.: Some special features of organization of the cerebellar cortex. In: Models of the structural-functional organization of certain biological systems (I. M. GELFAND, V. S. GURFINKEL, S. V. FOMIN, and M. L. TSETLIN, eds.), p. 250–423. Cambridge: MIT Press 1971. Originally published as Modeli Strukturno-Funktionalnoy Organisatsii Nyekotoyh Biologichyeskikh Sistem, Moscow, 1966.

SOTELO, C.: Ultrastructural aspects of the cerebellar cortex of the frog. In: Neurobiology of cerebellar evolution and development (R. LLINÁS, ed.), p. 327–367. Chicago: AMA-ERF Institute for Biomedical Research 1969.

SOTELO, C.: Stellate cells and their synapses on Purkinje cells in the cerebellum of the frog. Brain Res. **17**, 510–514 (1970).

SOTELO, C., LLINÁS, R.: Specialized membrane junctions between neurons in the vertebrate cerebellar cortex. J. Cell Biol. **53**, 271–289 (1972).

SOTELO, C., PALAY, S. L.: The fine structure of the lateral vestibular nucleus in the rat. I. Neurons and neuroglial cells. J. Cell Biol. **36**, 151–179 (1968).

SOTELO, C., PALAY, S. L.: Altered axons and axon terminals in the lateral vestibular nucleus of the rat: possible example of axonal remodeling. Lab. Invest. **25**, 653–671 (1971).

STAEHELIN, L. A.: Three types of gab junctions interconnecting intestinal epithelial cells visualized by freeze-etching. Proc. nat. Acad. Sci. (Wash.) **69**, 1318–1321 (1972).

STIEDA, L.: Zur vergleichenden Anatomie und Histologie des Cerebellum. Arch. Anat. Physiol. (Lpz.) 407–433 (1864).

STREIT, P., AKERT, K., SANDRI, C., LIVINGSTON, R. B., MOOR, H.: Dynamic ultrastructure of presynaptic membranes at nerve terminals in the spinal cord of rats. Anesthetized and unanesthetized preparations compared. Brain Res. **48**, 11–26 (1972).

SZENTÁGOTHAI, J.: The use of degeneration methods in the investigation of short neuronal connexions. In: Degeneration patterns in the nervous system (M. SINGER and J.P. SCHADÉ, eds.). Progress in Brain Research **14**, 1–32 (1965).

SZENTÁGOTHAI, J., RAJKOVITS, U.: Über den Ursprung der Kletterfasern des Kleinhirns. Z. Anat. Entwickl.-Gesch. **121**, 130–141 (1959).

TERRAZAS, R.: Notas sobre la neuroglia del cerebelo y el crecimiento de los elementos nerviosos. Rev. trimestr. micr. **2**, 49–65 (1897).

THACH, W.T.: Discharge of cerebellar neurons related to two maintained postures and two prompt movements. I. Nuclear output. J. Neurophysiol. **33**, 527–536 (1970a).

THACH, W.T.: Discharge of cerebellar neurons related to two maintained postures and two prompt movements. II. Purkinje cell output and input. J. Neurophysiol. **33**, 537–547 (1970b).

THACH, W.T.: Cerebellar output: properties, synthesis and uses. Brain Res. **40**, 89–97 (1972).

TOWER, D.B.: Structural and functional organization of mammalian cerebral cortex: the correlation of neurone density with brain size. J. comp. Neurol. **101**, 19–51 (1954).

TRACHTENBERG, M.C., POLLEN, D.A.: Neuroglia: biophysical properties and physiological function. Science **167**, 1248–1252 (1970).

UCHIZONO, K.: Characteristics of excitatory and inhibitory synapses in the central nervous system of the cat. Nature (Lond.) **207**, 642–643 (1965).

UCHIZONO, K.: Inhibitory synapses on the stretch receptor neurone of the crayfish. Nature (Lond.) **214**, 833–834 (1967a).

UCHIZONO, K.: Synaptic organization of the Purkinje cells in the cerebellum of the cat. Exp. Brain Res. **4**, 97–113 (1967b).

UCHIZONO, K.: Synaptic organization of the mammalian cerebellum. In: Neurobiology of cerebellar evolution and development (R. LLINÁS, ed.), p. 549–581. Chicago: AMA-ERF Institute for Biomedical Research 1969.

VALDIVIA, O.: Methods of fixation and the morphology of synaptic vesicles. J. comp. Neurol. **142**, 257–277 (1971).

VOOGD, J.: The cerebellum of the cat. Assen, Netherlands: Van Gorcum and Co. 1964.

VOOGD, J.: The importance of fiber connections in the comparative anatomy of the mammalian cerebellum. In: Neurobiology of cerebellar evolution and development (R. LLINÁS, ed.), p. 493–514. Chicago: AMA-ERF Institute for Biomedical Research 1969.

WEIGERT, C.: Beiträge zur Kenntnis der normalen menschlichen Neuroglia. Abhandl. senckenberg. naturforsch. Ges. (Frankfurt a.M.) **19**, H. II (1895).

WEISS, P., HISCOE, H.B.: Experiments on the mechanism of nerve growth. J. exp. Zool. **107**, 315–396 (1948).

WESTRUM, L.E.: Observations on initial segments of axons in the prepyriform cortex of the rat. J. comp. Neurol. **139**, 337–355 (1970).

WILSON, E.B.: The cell in development and heredity, third ed. New York: Macmillan 1927.

WINKLER, C.: Anatomie du système nerveux, vol. 3: Le cervelet, p. 108–367. Haarlem: E.F. Bohn 1927.

WOLFF, J.: Elektronenmikroskopische Untersuchungen über Struktur und Gestalt von Astrozytenfortsätzen. Z. Zellforsch. **66**, 11–28 (1965).

WOLFF, J.: Die Astroglia im Gewebsverband des Gehirns. Acta neuropath., Suppl. **4**, 33–39 (1968).

WOOD, R.L.: Intercellular attachments in the epithelium of *Hydra* as revealed by electron microscopy. J. biophys. biochem. Cytol. **6**, 343–352 (1959).

WOODWARD, D.J., HOFFER, B.J., SIGGINS, G.R., BLOOM, F.E.: The ontogenetic development of synaptic junctions, synaptic activation and responsiveness to neurotransmitter substances in rat cerebellar Purkinje cells. Brain Res. **34**, 73–97 (1971).

WOODWARD, D.J., HOFFER, B.J., SIGGINS, G.R., OLIVER, A.P.: Inhibition of Purkinje cells in the frog cerebellum. II. Evidence for GABA as the inhibitory transmitter. Brain Res. **33**, 91–100 (1971).

WUERKER, R.B., PALAY, S.L.: Neurofilaments and microtubules in anterior horn cells of the rat. Tissue & Cell **1**, 387–402 (1969).

Subject Index

Figure numbers are given in *italics*